THE GEOMETRY
PROBLEM SOLVER®

REGISTERED TRADEMARK

PLANE · SOLID · ANALYTIC

THE GEOMETRY
PROBLEM SOLVER®

REGISTERED TRADEMARK

A Complete Solution Guide to Any Textbook

Staff of Research and Education Association
Dr. M. Fogiel, Director

special chapter reviews by
Ernest Woodward, Ed.D.
Professor of Mathematics
Austin Peay State University
Clarkesville, Tennessee

 Research and Education Association
61 Ethel Road West
Piscataway, New Jersey 08854

THE GEOMETRY
PROBLEM SOLVER®

Printed in the United States of America

Library of Congress Control Number 00-112213

International Standard Book Number 0-87891-510-9

WHAT THIS BOOK IS FOR

For as long as geometry has been taught in schools, students have found this subject difficult to understand and learn. Despite the publication of hundreds of textbooks in this field, each one intended to provide an improvement over previous textbooks, students continue to remain perplexed, and the subject is often taken in class only to meet school/departmental requirements for a selected course of study.

In a study of the problem, REA found the following basic reasons underlying students' difficulties with geometry taught in schools:

(a) No systematic rules of analysis have been developed which students may follow in a step-by-step manner to solve the usual problems encountered. This results from the fact that the numerous different conditions and principles which may be involved in a problem, lead to many possible different methods of solution. To prescribe a set of rules to be followed for each of the possible variations, would involve an enormous number of rules and steps to be searched through by students, and this task would perhaps be more burdensome than solving the problem directly with some accompanying trial and error to find the correct solution route.

(b) Textbooks currently available will usually explain a given principle in a few pages written by a professional who has an insight in the subject matter that is not shared by students. The explanations are often written in an abstract manner which leaves the students confused as to the application of the principle. The explanations given are not sufficiently detailed and extensive to make the student aware of the wide range of applications and different aspects of the principle being studied. The numerous possible variations of principles and their applications are usually not discussed, and it is left for the students to discover these for themselves while doing the exercises. Accordingly, the average student is expected to rediscover that which has been long known and practiced, but not published or explained extensively.

(c) The examples usually following the explanation of a topic are too few in number and too simple to enable the student to obtain a thorough grasp of the principles involved. The explanations do not provide sufficient basis to enable a student to solve problems that may be subsequently assigned for homework or given on examinations.

The examples are presented in abbreviated form which leaves out much material between steps, and requires that students derive the omitted material themselves. As a result, students find the examples difficult to understand--contrary to the purpose of the examples.

Examples are, furthermore, often worded in a confusing manner. They do not state the problem and then present the solution. Instead, they pass through a general discussion, never revealing what is to be solved for.

Examples, also, do not always include diagrams/graphs, wherever appropriate, and students do not obtain the training to draw diagrams or graphs to simplify and organize their thinking.

(d) Students can learn the subject only by doing the exercises themselves and reviewing them in class, to obtain experience in applying the principles with their different ramifications.

In doing the exercises by themselves, students find that they are required to devote considerably more time to geometry than to other subjects of comparable credits, because they are uncertain with regard to the selection and application of the theorems and principles involved. It is also often necessary for students to discover those "tricks" not revealed in their texts (or review books), that make it possible to solve problems easily. Students must usually resort to methods of trial-and-error to discover these "tricks", and as a result they find that they may sometimes spend several hours to solve a single problem.

(e) When reviewing the exercises in classrooms, instructors usually request students to take turns in writing solutions on the boards and explaining them to the class. Students often find it difficult to explain in a manner that holds the interest of the class, and does not enable the remaining students to follow the material written on the boards. The remaining students seated in the class are, furthermore, too occupied with copying the material from the boards, to listen to the oral explanations and concentrate on the methods of solution.

This book is intended to aid students in geometry to overcome the difficulties described, by supplying detailed illustrations of the solution methods which are usually not apparent to students. The solution methods are illustrated by problems selected from those that are most often assigned for class work and given on examinations. The problems are arranged in order of complexity to enable students to learn and understand a particular topic by reviewing the problems in sequence. The problems are illustrated with detailed step-by-step explanations, to save the students the large amount of time that is often needed to fill in the gaps that are usually found between steps of illustrations in textbooks or review/outline books.

The staff of REA considers geometry a subject that is best learned by allowing students to view the methods of analysis and solution techniques themselves. This approach to learning the subject matter is similar to that practiced in various scientific laboratories, particularly in the medical fields.

In using this book, students may review and study the illustrated problems at their own pace; they are not limited to the time allowed for explaining problems on the board in class.

When students want to look up a particular type of problem and solution, they can readily locate it in the book by referring to the index which has been extensively prepared. It is also possible to locate a particular type of problem by glancing at just the material within the boxed portions. To facilitate rapid scanning of the problems, each problem has a heavy border around it. Furthermore, each problem is identified with a number immediately above the problem at the right -hand margin.

To obtain maximum benefit from the book, students should familiarize themselves with the section, "How To Use This Book," located in the front pages.

To meet the objectives of this book, staff members of REA have selected problems usually encountered in assignments and examinations, and have solved each problem meticulously to illustrate the steps which are usually difficult for students to comprehend. Gratitude for their patient work in this area is due to Barbara Dunkle, Sheldon Lipski, Arpard Fazakas, Robert Long, Anthony Longhitano, Stephen Zuckerman and the contributors who devoted short periods of time to this work. Barbara Dunkle and Sheldon Lipski deserve special praise for their outstanding efforts.

Gratitude is also expressed to the many persons involved in the difficult task of typing the manuscript with its endless changes, and to the REA art staff who prepared the numerous detailed illustrations together with the layout and physical features of the book.

Finally, special thanks are due to Helen Kaufmann for her unique talents to render those difficult border-line decisions and constructive suggestions related to the design and organization of the book.

Max Fogiel, Ph. D.
Program Director

HOW TO USE THIS BOOK

This book can be an invaluable aid to students in geometry as a supplement to their textbooks. The book is divided into 52 chapters, each dealing with a separate topic. The subject matter is developed beginning with lines and angles and extending through analytic (coordinate) and solid geometry. Sections on constructions, coordinate conversions, polygons, surface areas, and volumes have also been included.

Each chapter in the book starts with a section titled "Basic Attacks and Strategies for Solving Problems in this Chapter." This section explains the principles that are applicable to the topics in the chapter. By reviewing these principles, students can acquire a good grasp of the underlying techniques and strategies through which problems related to the chapter may be solved.

HOW TO LEARN AND UNDERSTAND
A TOPIC THOROUGHLY

1. Refer to your class text and read the section pertaining to the topic. You should become acquainted with the principles discussed there. These principles, however, may not be clear to you at the time.

2. Then locate the topic you are looking for by referring to the Table of Contents in the front of this book. After turning to the beginning of the appropriate chapter, read the section titled "Basic Attacks and Strategies for Solving Problems in this Chapter." This section is a review of the important principles related to the chapter, and it will help you to understand further how and why problems in the chapter are solved in the manner shown.

3. Turn to the page where the topic begins and review the problems under each topic, in the order given. For each topic, the problems are arranged in order of complexity, from the simplest to the more difficult. Some problems may appear similar to others, but each problem has been selected to illustrate a different point or solution method.

To learn and understand a topic thoroughly and retain its contents, it will generally be necessary for students to review the problems several times. Repeated review is essential in order to gain experience in recognizing the principles that should be applied and to select the best solution technique.

HOW TO FIND A PARTICULAR PROBLEM

To locate one or more problems related to particular subject matter, refer to the index. In using the index, be certain to note that the numbers given there refer to problem numbers, not to page numbers. This arrangement of the index is intended to facilitate finding a problem more rapidly, since two or more problems may appear on a page.

If a particular type of problem cannot be found readily, it is recommended that the student refer to the Table of Contents and then turn to the chapter which is applicable to the problem being sought. By scanning or glancing at the material that is boxed, it will generally be possible to find problems related to the one being sought, without consuming considerable time. After the problems have been located, the solutions can be reviewed and studied in detail.

For the purpose of locating problems rapidly, students should acquaint themselves with the organization of the book as found in the Table of Contents.

In preparing for an exam, it is useful to find the topics to be covered in the exam from the Table of Contents, and then review the problems under those topics several times. This should equip the student with what might be needed for the exam.

Glossary of Symbols and Abbreviations

$\overline{AB}$	line segment AB
$\overrightarrow{AB}$	ray AB
$m(\overline{AB})$	measure of line segment AB
AB	line segment AB or the measure of line segment AB
$\angle 1$	angle 1 or measure of angle 1
$m(\angle 1)$	measure of angle 1
$=$	equal
$\neq$	not equal
$\perp$	perpendicular
bis	bisector
supp	supplementary
comp	complementary
$\cong$	congruent
$\triangle$	triangle
cpcte	corresponding parts of congruent triangles are equal
adj	adjacent
$\parallel$	parallel
$\nparallel$	not parallel
$\square$	parallelogram
$A(\ \ )$	the area of
π	pi (3.14159 . . .)
$\sqrt{\ }$	the square root of
$\overset{\frown}{AB}$	arc AB
$<$	less than
$>$	greater than
$\rightarrow$	if . . . then . . .
$\therefore$	therefore
$\sim$	not or similar
m	slope of a nonvertical line
$\circ$	degree(s)

CONTENTS

> *Every Chapter Begins with a Section of Basic Attacks and Strategies for Solving Problems Pertaining to that Chapter*

Chapter No.   **Page No.**

1 METHODS OF PROOF   1
   Deductive Reasoning   1
   Logic   5
   Direct Proof   11
   Indirect Proof   15
   Inductive Reasoning   16

2 LINES AND ANGLES   19
   Lines   19
   Angles   22

3 PERPENDICULARITY   26

4 TRIANGLES   32
   Scalene Triangles   32
   Isosceles Triangles   34
   Right Triangles   36
   Equilateral Triangles   38
   Angle Bisectors   40
   Exterior Angles   43

CONGRUENCE

5 CONGRUENT ANGLES AND LINE SEGMENTS   48
   Congruent Angles   48
   Congruent Line Segments   58

6 SIDE-ANGLE-SIDE POSTULATE   62

7 ANGLE-SIDE-ANGLE POSTULATE 85

8 SIDE-SIDE-SIDE POSTULATE 95

9 HYPOTENUSE–LEG POSTULATE 107

10 PARALLELISM 114
Proving Lines Parallel 114
Properties of Parallel Lines 122

11 INEQUALITIES 142
Lines and Angles 142
Exterior Angle Theorem 152
Triangle Inequality Theorem 155

12 QUADRILATERALS 160
Parallelograms 160
Rhombi 172
Squares 181
Rectangles 190
Trapezoids 201

13 GEOMETRIC PROPORTIONS AND SIMILARITY 204
Ratios and Proportions 204
Proving Triangles Similar 206

14 COMPUTATIONS INVOLVING SIMILAR TRIANGLES 224
Lengths 224
Proportions Involving Angle Bisectors 229
Proportions in Right Triangles 232
Proportions Involving Medians 241

CIRCLES

15 CENTRAL ANGLES AND ARCS 246

16 INSCRIBED ANGLES AND ARCS 258

17 CHORDS 272

18 TANGENTS AND INTERSECTING CIRCLES 293
      Tangents 293
      Circles that are Tangent to Other Circles 300
      Intersecting Circles 305

19 ANGLES FORMED BY TANGENTS, SECANTS, AND CHORDS 313
      Tangents 313
      Secants 321

20 AREAS 337

21 PYTHAGOREAN THEOREM AND APPLICATIONS 350

22 TRIGONOMETRIC RATIOS 371

23 AREAS OF QUADRILATERALS AND TRIANGLES 386
      Squares and Rectangles 386
      Triangles 390
      Similar Triangles 400
      Parallelograms 407

POLYGONS

24 INTERIOR AND EXTERIOR ANGLES 420
      Exterior Angles 420
      Interior Angles 421

25 CYCLIC QUADRILATERALS 432

26 CIRCLES INSCRIBED IN POLYGONS 442

27 CIRCLES CIRCUMSCRIBING POLYGONS 451

28 PERIMETER 472

29 AREAS 483
        Solving for Area Using the Pythagorean Theorem 483
        Solving for Area Using Trigonometric Functions 500

CONSTRUCTIONS

30 LINES AND ANGLES 506

31 TRIANGLES 523

32 CIRCLES 530

33 POLYGONS 536

34 COMPLEX CONSTRUCTIONS 542

COORDINATE GEOMETRY

35 DISTANCE 558
        Plotting Points 558
        Determining Distance 560
        Midpoints 564

36 SLOPE 571

37 LINEAR EQUATIONS 585
        The Equation of a Line 585
        Graphing Equations 601

38 AREA 606

39 LOCUS 615

40 COORDINATE PROOFS 628

41 POLAR/ANALYTIC COORDINATES 640
Graphing Polar Coordinates 640
Conversion of Polar to Cartesian Coordinates 659
Translation and Rotation 665

42 CONIC SECTIONS: CIRCLES AND ELLIPSES 671
Circles 671
Ellipses 688

43 CONIC SECTIONS: PARABOLAS AND HYPERBOLAS 705
Parabolas 705
Hyperbolas 718

SOLID GEOMETRY

44 LOCUS 732

45 INTERSECTIONS OF LINES AND PLANES 752
Lines and Planes 752
Polyhedral Angles 768

46 RECTANGULAR SOLIDS, PRISMS, AND PYRAMIDS 774
Rectangular Solids and Prisms 774
Pyramids 779

47 SPHERES 792

48 SURFACE AREAS 806
Rectangular Solids and Cubes 806
Tetrahedrons, Pyramids and Prisms 809
Spheres 817
Cylinders and Cones 822

49 VOLUMES I 829
Rectangular Solids and Cubes 829
Prisms 832
Pyramids and Tetrahedrons 836

50 VOLUMES II 852
       Spheres 852
       Cylinders 858
       Cones 859

51 THREE-DIMENSIONAL COORDINATE/ANALYTIC GEOMETRY 877
       Lines and Planes 877
       Solids 884

52 GEOMETRY OF THE EARTH 889

SUMMARY OF ESSENTIAL GEOMETRIC THEOREMS & PROPERTIES 899

INDEX 913

CHAPTER 1

METHODS OF PROOF

Basic Attacks and Strategies for Solving Problems
in this Chapter. See pages 1 to 18 for step-by-step
solutions to problems.

The writing of proofs is the primary task of a mathematician. Since most
proofs build on other proofs, a math student must be able to use the tools of
logic to correctly draw conclusions from things that are already known.
Methods of proof are different ways of organizing sequences of logically
connected statements which show the truth of what we want to prove.

The two simplest methods of proof are **direct proofs** and **indirect proofs**.
Often, statements to be proved are expressed in terms of an if-then sentence.
Suppose that p, q, and r are mathematical statements and the "theorem" to be
proved is

"If p and q, then r."

A direct two-column proof of this theorem would have the following
basic structure:

STATEMENTS	REASONS
1. p	1. Given
2. q	2. Given
.	
.	
.	
n. r	n. ???

This means that the statements p and q (called the **hypotheses**) are taken as
true. Then follows a sequence of statements which lead from p and q to the
conclusion r. Each statement in this sequence must be justified by a reason,
and the process is often complicated. The person constructing the proof
should ask the following two questions:

(1) What follows from statements p and q?

(2) How can r be established as true?

Someone once said that in situations like this a mathematician should have one eye in his/her forehead and one eye below the mouth. With the top eye you look at where you are and with the bottom eye you look at where you want to be.

An indirect proof, or proof by contradiction of the same theorem, would have the following structure, where s represents a statement:

STATEMENTS		REASONS	
1.	p	1.	Given
2.	q	2.	Given
3.	not r	3.	Assumed
	.		
	.		
	.		
m.	s	m.	??
	.		
	.		
	.		
n–1.	not s	n–1.	??
n.	r	n.	by contradiction?

In an indirect proof, the negation of the conclusion (not r) is assumed, and as a result of that assumption two statements, s and "not s," are both derived. Since both of these statements cannot be true, a contradiction has resulted. This contradiction is a direct result of assuming that the statement "not r" is true. Thus, it must be concluded that "not r" is false, so r is true.

These two kinds of proofs are illustrated in this chapter and throughout the entire book. Another, more sophisticated type of proof is called **proof by mathematical induction,** whereby a statement is shown to be true in a specific instance then generalized for all cases. This technique is illustrated in Problems 18 and 19 at the end of this chapter.

Step-by-Step Solutions to Problems in this Chapter, "Methods of Proof"

DEDUCTIVE REASONING

● PROBLEM 1

How may the following three statements be arranged so that the first two will make it possible to deduce the third? (a) An eagle has feathers. (b) All birds have feathers. (c) An eagle is a bird.

<u>Solution</u>: An arrangement of statements that would allow you to deduce the third one from the preceding two is called a syllogism. A syllogism has three parts.

The first part is a general statement concerning a whole group. This is called the major premise.

The second part is a specific statement which indicates that a certain individual is a member of that group. This is called the minor premise.

The third and final part of a syllogism is a statement to the effect that the general statement concerning the group also applies to the individual. This statement is called a deduction.

The technique of employing a syllogism to arrive at a conclusion is called deductive reasoning. This technique will be used in this problem.

Statement (b) "All birds have feathers" is the major premise since it assigns a general characteristic to the whole group, in this case birds.

Statement (c) "An eagle is a bird" links the eagle to the larger group mentioned in the major premise and, therefore, (c) is the minor premise.

Statement (a) "An eagle has feathers" assigns the

general trait, feathers, to the individual, eagles, and, thus, qualifies as the deduction.

Therefore, the order of the statements that allows you to deduce the third from the first two is (b), (c), (a).

● **PROBLEM** 2

All residents of this state who are registered voters are 18 years of age or older. If John is a resident of this state, by valid reasoning which of the following can we conclude: (a) If John is 18 or over, he is a registered voter; (b) If John is a registered voter, he is 18 or over. (c) If John is not a registered voter, he is not 18 or over.

<u>Solution</u>: To find the logically correct statement, we find which of the choices forms a valid syllogism with the given statement. The given statement "All residents of this state who are registered voters are 18 years of age or older" is the major premise, because it attributes a characteristic ("being 18 or older") to a general group ("residents of this state who are registered voters"). A minor premise links a particular example, John, to the general group. Note that there are two conditions to being a member of the group: a member must (1) be a resident of the state; and (2) be a registered voter. Therefore, any valid minor premise must include both conditions.

One condition of the minor premise is common to all the choices and therefore written in the front - "If John is a resident of this state". The second condition of the minor premise is different in each choice.

For choice (a), the minor premise is "John is a resident of this state and John is 18 or over". This is not enough to make John a resident of the state who is a registered voter. Thus, John is not linked to the group and choice (a) is not valid reasoning. Thinking about this in another way, it is definitely possible that apathetic John is over 18, and a resident of the state; and yet John may not have bothered to register.

For choice (b), the minor premise is "John is a resident of this state and John is a registered voter." This is enough to qualify John as a member of the group. Thus, the quality of the group can be linked with John in the conclusion - John is 18 or over. This is, indeed, the conclusion of (b). Thus choice (b) leads to valid reasoning.

For choice (c), the minor premise is "John is a resident of this state and John is not a registered voter." This is not enough to make John a member of the general group. The reasoning is invalid. It is possible for an apathetic 50 year old citizen not to register to vote.

2

(1) Azuela, Browning, Conway, and Degas met at the 40th reunion of their kindergarten class. (2) Browning and Conway knew each other because they had both attended Oxford University but neither of them remembered the artist who went to Cambridge. (3) Azuela was pleased to find out that the geometer and the artist had heard of his work. (4) Browning told the writer that he and the geometer had once written a song in honor of congruent triangles. (5) With the given information, find the name of the writer, the poet, the geometer, and the artist.

	writer	poet	geometer	artist
Azuela				
Browning				
Conway				
Degas				

Figure 1

Solution: The best method for solving such problems is to list all the possibilities and cross off all possibilities that cannot be true. After eliminating all impossible possibilities, there should only be four possibilities - matching one person with one profession. In this problem, we shall not list each possibility in words - Azuela-writer and Azuela-poet, for example. Instead (see figure 1), we draw a graph. Each box represents a possibility. The upper left box represents the possibility Azuela-writer, for example. The two methods - writing out each possibility, and using boxes - are equivalent. Using boxes, though, involves less writing and is clearer.

From (1), we know the names. From (5), we have their professions. The names are listed in columns, and the professions in the top row. In this way, there is a box for each person matched to each profession. From statement (2), we have that Browning and Conway went to Oxford but the artist went to Cambridge. Therefore, neither Browning nor Conway is the artist. We cross out the possibilities Browning-artist and Conway-artist.

From statement (3), Azuela is pleased that the geometer and the artist had heard of his work. Therefore, Azuela-geometer, Azuela-artist cannot be true possibilities.

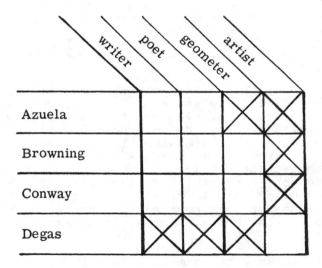

Note that Azuela, Browning, and Conway cannot be
artists. Thus only Degas can be the artist. Since Degas
is the artist (we assume he has only one job), Degas cannot
be a writer, poet, and geometer. We cross out these three
possibilities.

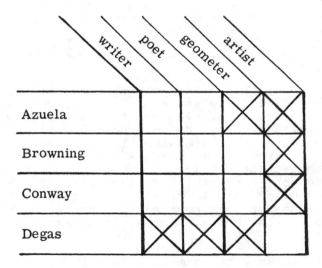

From statement (4), Browning talks to the writer about
the geometer; therefore Browning cannot be the writer or
the geometer. We eliminate these two possibilities.

In the row labelled Browning, only Browning-poet remains. Therefore Browning is the poet, and we can eliminate Azuela-poet, and Conway-poet.

In the row Azuela, only Azuela-writer remains. There-

fore, we eliminate Conway-writer, leaving only Conway-geo-
meter.

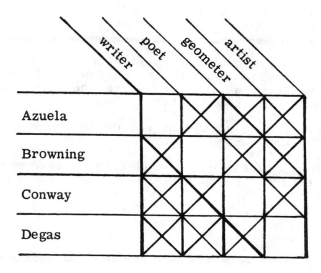

So Azuela is the writer, Browning is the poet; Conway
is the geometer; and Degas is the artist.

LOGIC

● **PROBLEM** 4

Identify the hypothesis and conclusion of the following
statement: Two non-congruent angles are not vertical angles.

<u>Solution</u>: In any reasoning process, there is a set of known
facts and assumptions from which another fact can be deduced.
The set of known facts are the axioms, definitions, and previously established theorems. The sets of assumptions are
the hypothesis, and the fact that is deduced is the conclusion.

Let us rewrite this statement in If-then form: If two
angles are not congruent, then they are not vertical angles.
This type of rewrite can be most helpful in identifying the
hypothesis and conclusion of a statement.

The hypothesis is "two angles are not congruent" and
the conclusion that logically follows is "they are not
vertical angles."

Identify the hypothesis and conclusion in the statement, "If Mary has a temperature, then she is probably sick."

Solution: In a statement of the "If-then" type, called a conditional statement, the phrase following the word "if", upon which the conclusion depends, is the hypothesis. It is sometimes called the dependent clause.

The part of the statement following the word "then" is called the conclusion.

Therefore, in the given conditional statement, "Mary has a temperature" is the hypothesis, whereas "she is probably sick" is the conclusion.

Identify the hypothesis and conclusion in the statement, "The median to the base of an isosceles triangle is perpendicular to the base."

Solution: The statement dealt with in this problem is a "simple" sentence, as opposed to the "if-then" type of statement. In general, if a statement is written as a simple sentence, then the subject of the sentence is the hypothesis, and the predicate is the conclusion.

In this example, the phrase "The median to the base of an isosceles triangle" is the subject of the sentence and, consequently, the hypothesis is "a line segment is the median to the base of an isosceles triangle." The phrase "is perpendicular to the base" is the predicate and, therefore, the conclusion is "the line is perpendicular to the base."

State whether the condition in the hypothesis is: both necessary and sufficient, necessary but not sufficient, not necessary but sufficient, neither necessary nor sufficient. (a) If a man is sick, he is a patient in a hospital. (b) If a man is a senator, he is a member of Congress. (c) If a figure is a square, then the figure is a quadrilateral.

Solution: A necessary condition is a fact or set of facts which must be given in the hypothesis of a conditional statement to make it possible to deduce the conclusion.

A sufficient condition is a fact or set of facts which, when given in the hypothesis, supplies enough information to make it possible to deduce the conclusion.

If a condition is necessary, it is essential that the fact be given in the hypothesis to have any chance of deducing the conclusion. A necessary condition will not always be enough to allow one to arrive at a conclusion by formal logic. On the other hand, a sufficient condition, when given, will always lead to a logical deduction of a certain conclusion.

(a) In this statement the hypothesis is "a man is sick," the conclusion is "he is a patient in a hospital." Certainly, if a man is a patient in a hospital, he must be sick. Therefore, it is at least necessary that he be sick before he can be in the hospital.

However, it is not sufficient that he be merely sick as the sole prerequisite for being in the hospital because one has many ways of being sick and not ending up in the hospital.

We conclude then that the hypothesis is necessary but not sufficient to deduce the stated conclusion.

In this part we showed the converse of that statement to be true which is a standard way to decide if the hypothesis is necessary. To prove sufficiency we must prove that the statement is true.

(b) The hypothesis is "the man is a senator," while the conclusion is "he is a member of Congress."

The converse is "If he is a member of Congress, he is a senator." This is false, because members of the House of Representatives are members of Congress but certainly not senators. Therefore, in the original statement, the hypothesis is not necessary because the conclusion can be reached under much different hypothetical conditions, i.e. if the hypothesis were "he is a member of the House of Representatives."

The statement is true and this lets us conclude that the hypothesis is sufficient for the conclusion. Once a person is a senator he is automatically a member of Congress.

Therefore, the hypothesis is not necessary but sufficient for the stated conclusion.

(c) In this statement, the hypothesis is "a figure is a square" and the conclusion is "the figure is a quadrilateral."

The statement is true since every square is a four sided polygon and every four sided polygon is a quadrilateral. Thus, the hypothetical condition is sufficient.

The converse is "if the figure is a quadrilateral, then the figure is a square." This is false since the figure could also be a rectangle or trapezoid. Since the converse is false the hypothesis of the original statement is not necessary.

The hypothesis is sufficient but not necessary.

● **PROBLEM** 8

Write the converse of each of the following statements. Consider whether each converse is true or false. (a) If a man drives a Pontiac, he drives an American car. (b) If two triangles are congruent, they have three pairs of congruent angles.

Solution: The converse of a given statement is another statement which is formed by interchanging the hypothesis and the conclusion in the given statement. If the stament is true, the converse may be true but does not necessarily have to be.

(a) In this statement, "a man drives a Pontiac," is the hypothesis and, "he drives an American car," is the conclusion. Therefore, the converse is the statement, "If a man drives an American car, he drives a Pontiac."

The converse is false in this case, while the original statement was true, because a man can certainly drive an American car and be driving one of many makes other than a Pontiac.

(b) In this statement, "two triangles are congruent," is the hypothesis and, "they have three pairs of congruent angles," is the conclusion. The converse of the original statement is, "If two triangles have three pairs of congruent angles, they are congruent."

The only methods for proving congruence between triangles are $sss \cong sss$, $a \cdot s \cdot a \cong a \cdot s \cdot a$, $a \cdot a \cdot s \cong a \cdot a \cdot s$ and $s \cdot a \cdot s \cong s \cdot a \cdot s$. Having three pairs of congruent angles is a proof for triangle similarity, not congruence. Therefore the converse is false.

● **PROBLEM** 9

Write the negation for each of the following statements. (a) I am Chinese. (b) He is not good. (c) John is unfriendly. (d) She is a thief and a liar. (e) Tom and Jerry pitch for the New York Mets. (f) You are very lucky or very smart.

Solution: The negation of a statement is a statement that is true whenever the original statement is false and false whenever the original statement is true. Since a statement usually connects a person or thing with a group or quality, the negation is usually of the form: person or thing does not belong to the group; or person or thing does not have that quality.

(a) The negation of "I am Chinese" is "I am not Chinese."

(b) The statement "He is not good" connects "he" with the quality "not good." The negation would be "He is not 'not good'." Since double negatives are not good grammar, the negation can be changed to "He is good."

(c) The double negative is encountered in a different form in this case. The negation of "John is unfriendly" could be "John is not unfriendly." However, grammatically speaking, the correct answer is "John is friendly."

(d) There are three ways for this statement to be false. The statement "She is a thief and a liar" is true only if she is both a thief and a liar. Thus the statement is false if she is not a thief whether or not she is a liar. The statement is also false if she is not a liar - whether or not she is a thief.

Since the negation must be true every time the statement is false, the negation is "She is not a thief or she is not a liar." In general given two conditions a and b (for example, a = "she is a thief"; b = "she is a liar"), the negation of the statement "a and b" is "not a or not b."

(e) The statement "Tom and Jerry pitch for the New York Mets" can be rewritten as "Tom pitches for the New York Mets and Jerry pitches for the New York Mets." From part (d), we know that the negation of "statement a and statement b" is "not statement a or not statement b." Thus, the negation is "Either Tom does not pitch for the New York Mets or Jerry does not pitch for the New York Mets." This can be rewritten more compactly as "Either Tom or Jerry does not pitch for the New York Mets."

(f) The statement "You are very lucky or very smart" is true even if you are very lucky but not very smart, or if you are very smart but not very lucky. The only way this statement can be false is if you are not very lucky and you are not very smart. Thus, the negation is "You are not very lucky and you are not very smart," which can be shortened to "You are neither very lucky nor very smart."

In general, given that a and b are two conditions, the negation of the statement "a or b" is "not a and not b."

Summing up our knowledge, we have: Let a and b be simple statements. Then,

STATEMENT	NEGATION
a	not a
a and b	not a or not b
a or b	not a and not b

● **PROBLEM** 10

Write the inverse for each of the following statements.
Determine whether the inverse is true or false. (a) If a
person is stealing, he is breaking the law. (b) If a line
is perpendicular to a segment at its midpoint, it is the
perpendicular bisector of the segment. (c) Dead men tell
no tales.

Solution: The inverse of a given conditional statement is
formed by negating both the hypothesis and conclusion of
the conditional statement.

(a) The hypothesis of this statement is "a person is
stealing"; the conclusion is "he is breaking the law."
The negation of the hypothesis is "a person is not stealing." The negation of the conclusion is "he is not breaking
the law." The inverse is "if a person is not stealing, he
is not breaking the law."

The inverse is false, since there are more ways to
break the law than by stealing. Clearly, a murdered may
not be stealing but he is surely breaking the law.

(b) In this statement, the hypothesis contains two
conditions: (1) the line is perpendicular to the segment;
and (2) the line intersects the segment at the midpoint.
The negation of (statement a and statement b) is (not
statement a or not statement b). Thus, the negation of the
hypothesis is "The line is not perpendicular to the segment
or it doesn't intersect the segment at the midpoint." The
negation of the conclusion is "the line is not the perpendicular bisector of a segment."

The inverse is "if a line is not perpendicular to the
segment or does not intersect the segment at the midpoint,
then the line is not the perpendicular bisector of the
segment."

In this case, the inverse is true. If either of the
conditions hold (The line is not perpendicular; the line
does not intersect at the midpoint.), then the line cannot
be a perpendicular bisector.

(c) This statement is not written in if-then form
which makes its hypothesis and conclusion more difficult
to see. The hypothesis is implied to be "the man is dead;"
the conclusion is implied to be "the man tells no tales."

The inverse if, therefore, "If a man is not dead, then he will tell tales."

The inverse is false. Many witnesses to crimes are still alive but they have never told their stories to the police, either out of fear or because they didn't want to get involved.

In each of the following examples of reasoning, the truth of an inverse is assumed. Comment whether or not this leads to an incorrect conclusion. (a) Stars shine at night. Since the stars are not shining, it is not night. (b) Since a polygon with more than four sides can be a pentagon, a polygon with more than four sides is a pentagon.

Solution: (a) In this deductive reasoning the inverse, "If stars don't shine, then it is not night," is assumed to be true. If this were indeed true, not seeing a star would make one conclude that it is not night.

However, on cloudy nights stars have a difficult time shining and being seen and,therefore, when they don't shine we cannot definitely determine on that information alone that it is not night. Accordingly, the inverse is false and assuming it true led to an erroneous conclusion.

(b) The true statement whose inverse is pivotal to this reasoning is, "If a figure is a pentagon, then it is a polygon with more than four sides."

This original statement describes a basic characteristic of pentagons, i.e., that all pentagons have more than four sides. The inverse is false, though, since not all polygons that have more than four sides are pentagons. A hexagon, for example, is not a pentagon, and yet it has more than four sides.

Since an inverse was assumed true that was actually false, the conclusion reached is false.

DIRECT PROOF

If $\angle x$ has a measure of 40° and $\angle y$ has a measure of 40°, prove m$\angle x$ = m$\angle y$ using the standard two column proof method.

Solution: To write a two column proof, begin by stating

the given facts. In this case, the facts are:

(1)   m∡x = 40°

(2)   m∡y = 40°.

Then, state what is to be proven:

m∡x = m∡y

The proof is a series of logical steps that all have valid reasons for their statement. The logical steps are listed in the first column; the reasons in the second column. To clearly identify each step, number the steps. Place the number in front of each step and in front of the corresponding reasons.

STATEMENT	REASON
1. m∡x = 40°	1. Given
2. m∡y = 40°	2. Given
3. m∡x = m∡y	3. The Transitive Property of Equality: If quantities are equal to the same quantity, then they are equal to each other.

● **PROBLEM** 13

In the figure shown, the measure of ∡ DAC equals the measure of ∡ ECA and the measure of ∡1 equals the measure of ∡2. Show that the measure of ∡3 equals the measure of ∡4.

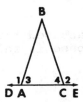

Solution: This proof will require the subtraction postulate, which states that if equal quantities are subtracted from equal quantities, the differences are equal.

GIVEN: ∡ DAC ≅ ∡ ECA, ∡1 ≅ ∡2

PROVE: ∡3 ≅ ∡4

STATEMENTS	REASONS
1. m∢DAC = m∢ECA m∢1 = m∢2	1. Given.
2. m∢DAC - m∢1 = m∢ECA - m∢2	2. Subtraction Postulate.
3. m∢3 = m∢4	3. Substitution Postulate.

● **PROBLEM 14**

In the diagram AB = CD, RS = 2AB, and LM = 2CD. Prove that RS = LM.

A●——●B R●————●S

C●——●D L●————●M

Solution: This proof will involve an application of the Multiplication Postulate, which states that if equal quantities are multiplied by equal quantities, the products are equal.

STATEMENT	REASON
1. AB = CD	1. Given.
2. 2AB = 2CD	2. Multiplication Postulate.
3. RS = 2 · AB LM = 2 · CD	3. Given.
4. RS = LM	4. Doubles of equal quantities are equal.

● **PROBLEM 15**

If line segment AD is divided by points B and C, as shown in the diagram, such that $\overline{AB} \cong \overline{CD}$, prove that $\overline{AC} \cong \overline{BD}$.

A●——●——●——●D
   B   C

Solution: We show that AB + BC = DC + CB. Since AC = AB + BC and BD = CD + BC, it then follows that $\overline{AC} \cong \overline{BD}$. To do this, we use the Reflexive Property of Equality and the Addition Postulate. The Reflexive Property of Line Segments states that a segment is congruent to itself. The Addition

Postulate states that equal quantities added to equal quantities yield equal quantities.

GIVEN: Line segment $\overline{ABCD}$; $\overline{AB} \cong \overline{CD}$

PROVE: $\overline{AC} \cong \overline{BD}$

STATEMENTS	REASONS
1. $\overline{ABCD}$; $\overline{AB} \cong \overline{CD}$	1. Given.
2. $\overline{BC} \cong \overline{BC}$	2. A line segment is congruent to itself.
3. AB = CD BC = BC	3. Congruent segments have equal lengths.
4. AB + BC = BC + CD or AC = BD	4. Addition Postulate. Also Substitution Postulate.
5. $\overline{AC} \cong \overline{BD}$	5. Segments of equal length are congruent.

● **PROBLEM** 16

Every line in a proof requires justification. In the proof below the reasons have been omitted. Give the reasons.

GIVEN: $\angle 1 \cong \angle 2$; $\angle 3 \cong \angle 4$

PROVE: $\angle H \cong \angle F$

STATEMENTS	REASONS
1. $\angle 1 \cong \angle 2$; $\angle 3 \cong \angle 4$	1.
2. m$\angle 1$ + m$\angle 3$ = m$\angle 2$ + m$\angle 4$	2.
3. m$\angle 1$ + m$\angle 3$ + m$\angle F$ = 180	3.
and m$\angle 2$ + m$\angle 4$ + m$\angle H$ = 180	
4. m$\angle 1$ + m$\angle 3$ + m$\angle F$ = m$\angle 2$ + m$\angle 4$ + m$\angle H$	4.
5. m$\angle F$ = m$\angle H$	5.

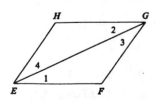

Solution: For any statement, several reasons may be possible. In steps 2, 3 and 4, below, alternative reasons are given. Any one will do. Even for a given reason, there may be a choice between citing or writing out the theorem. Generally, the shorter the reason, the better. Postulates and famous theorems can be named; however, theorems that are less well known should be written out.

The reasons for the proof in this question are:

(1) Given.

(2) The Angle Addition Postulate or the Addition Property of Equality; if a = b and c = d, then a + c = b + d.

(3) The measures of the interior angles of a triangle sum to 180° degrees or the interior angles of a triangle form a linear pair.

(4) Transitivity Postulate or Substitition Postulate or if a = b and b = c, then a = c.

(5) Subtraction Property of Equality.

INDIRECT PROOF

● **PROBLEM** 17

Prove, by indirect method, that if two angles are not congruent, then they are not both right angles.

Solution: Indirect proofs involve considering two possible outcomes, the result we would like to prove and its negative, and then showing, under the given hypothesis, that a contradiction of prior known theorems, postulates, or definitions is reached when the negative is assumed.

In this case the outcomes can be that the two angles are not right angles or that the two angles are right angles. Assume the negative of what we want to prove - that the two angles are right angles.

The given hypothesis in this problem is that the two angles are not congruent. A previous theorem states that all right angles are congruent. Therefore, the conclusion we have assumed true leads to a logical contradiction. As such, the alternative conclusion must be true. Therefore, if two angles are not congruent, then they are not both right angles.

INDUCTIVE REASONING

● **PROBLEM** 18

Prove by mathematical induction

$$1^2 + 2^2 + 3^2 + \ldots + n^2 = \frac{1}{6}n(n+1)(2n+1).$$

<u>Solution:</u> Mathematical induction is a method of proof. The steps are:
(1) The verification of the proposed formula or theorem for the smallest value of n. It is desirable, but not necessary, to verify it for several values of n.
(2) The proof that if the proposed formula or theorem is true for n = k, some positive integer, it is true also for n = k+1. That is, if the proposition is true for any particular value of n, it must be true for the next larger value of n.
(3) A conclusion that the proposed formula holds true for all values of n.

<u>Proof:</u> Step 1. Verify:

For n = 1: $1^2 = \frac{1}{6}(1)(1+1)[2(1)+1] = \frac{1}{6}(1)(2)(3) = \frac{1}{6}(6) = 1$

$$1 = 1$$

For n = 2: $1^2 + 2^2 = \frac{1}{6}(2)(2+1)[2(2)+1] = \frac{1}{6}(2)(3)(5) = \frac{1}{6}(6)(5)$

$$1 + 4 = (1)(5)$$

$$5 = 5$$

For n = 3: $1^2 + 2^2 + 3^2 = \frac{1}{6}(3)(3+1)[2(3)+1]$

$$1 + 4 + 9 = \frac{1}{6}(3)(4)(7) = \frac{1}{6}(12)(7) = 14$$

$$14 = 14$$

Step 2. Let k represent any particular value of n. For n = k, the formula becomes

$$1^2 + 2^2 + 3^2 + \ldots + k^2 = \frac{1}{6}k(k+1)(2k+1). \qquad \text{(A)}$$

For n = k+1, the formula is

$$1^2 + 2^2 + 3^2 + \ldots + k^2 + (k+1)^2 = \frac{1}{6}(k+1)[(k+1) + 1][2(k+1) + 1]$$

$$= \frac{1}{6}(k+1)(k+2)(2k+3). \qquad \text{(B)}$$

We must show that if the formula is true for n = k, then it must be true for n = k+1. In other words, we must show that (B) follows from (A). The left side of (A) can be converted into the left side of (B) by merely adding $(k+1)^2$. All that remains to be demonstrated is that when $(k+1)^2$ is added to the right side of (A), the result is the right side of (B).

$$1^2 + 2^2 + \ldots + k^2 + (k+1)^2 = \frac{1}{6}k(k+1)(2k+1) + (k+1)^2$$

Factor out (k+1):

$$1^2 + 2^2 + 3^2 + \ldots + k^2 + (k+1)^2 = (k+1)\left[\frac{1}{6}k(2k+1) + (k+1)\right]$$

$$= (k+1)\left[\frac{k(2k+1)}{6} + \frac{(k+1)6}{6}\right]$$

$$= (k+1)\frac{2k^2 + k + 6k + 6}{6}$$

$$= \frac{(k+1)(2k^2 + 7k + 6)}{6}$$

$$= \frac{1}{6}(k+1)(k+2)(2k+3),$$

since $\qquad 2k^2 + 7k + 6 = (k+2)(2k+3)$.

Thus, we have shown that if we add $(k+1)^2$ to both sides of the equation for $n = k$, then we obtain the equation **or** formula for $n = k+1$. We have thus established that if (A) is true, then (B) must be true; that is, if the formula is true for $n = k$, then it must be true for $n = k+1$. In other words, we have proved that if the proposition is true for a certain positive integer k, then it is also true for the next greater integer $k+1$.

Step 3. The proposition is true for $n = 1,2,3$ (Step 1). Since it is true for $n = 3$, it is true for $n = 4$ (Step 2, where $k = 3$ and $k+1 = 4$). Since it is true for $n = 4$, it is true for $n = 5$, and so on, for all positive integers n.

● **PROBLEM** 19

Prove by mathematical induction that

$$1 + 7 + 13 + \ldots + (6n - 5) = n(3n - 2).$$

Solution: (1) The proposed formula is true for $n = 1$, since $1 = 1(3 - 2)$.

(2) Assume the formula to be true for $n = k$, a positive integer; that is, assume

(A) $\qquad 1 + 7 + 13 + \ldots + (6k - 5) = k(3k - 2)$.

Under this assumption we wish to show that

(B) $\qquad 1 + 7 + 13 + \ldots + (6k - 5) + (6k + 1) = (k+1)(3k+1)$.

When $(6k+1)$ is added to both members of (A), we have on the right

$$k(3k-2) + (6k+1) = 3k^2 + 4k + 1 = (k+1)(3k+1);$$

hence, if the formula is true for $n = k$ it is true for $n = k + 1$.

(3) Since the formula is true for $n = k = 1$ (Step 1), it is true for $n = k + 1 = 2$; being true for $n = k = 2$ it is true for $n = k + 1 = 3$; and so on, for every positive integral value of n.

Prove by mathematical induction that the number of straight lines determined by $n > 1$ points, no 3 on the same straight line, is $\frac{1}{2}n(n-1)$.

Solution:

(1) The theorem is true when $n = 2$, since $\frac{1}{2} \cdot 2(2-1) = 1$ and two points determine one line.

(2) Let us assume that k points, no 3 on the same straight line, determine $\frac{1}{2}k(k-1)$ lines.

When an additional point is added (not on any of the lines already determined) and is joined to each of the original k points, k new lines are determined. Thus, altogether we have $\frac{1}{2}k(k-1) + k = \frac{1}{2}k(k-1+2) = \frac{1}{2}k(k+1)$ lines and this agrees with the theorem when $n = k + 1$.

Hence, if the theorem is true for $n = k$, a positive integer greater than 1, it is true for the next one $n = k + 1$.

(3) Since the theorem is true for $n = k = 2$ (Step (1)), it is true for $n = k + 1 = 3$; being true for $n = k = 3$, it is true for $n = k+1 = 4$; and so on, for every possible integral value > 1 of n.

CHAPTER 2

LINES AND ANGLES

Basic Attacks and Strategies for Solving Problems in this Chapter. See pages 19 to 25 for step-by-step solutions to problems.

In geometry the term **line** is left undefined. To say that a line is the shortest distance between two points is incorrect, because a line is not a distance, it is a geometrical object. When we speak of a line, we mean a straight line, as shown below:

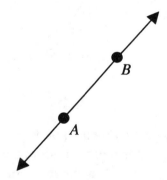

The arrows indicate that the line does not stop, but continues in both directions indefinitely. This line is denoted $\overleftrightarrow{AB}$ and also $\overleftrightarrow{BA}$. An important axiom concerning lines is that *two points determine exactly one line.*

Two other important concepts in this chapter are (line) **segment** and **ray**. A segment and a ray are pictured on the next page:

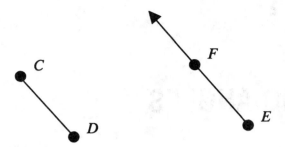

The segment is denoted $\overline{CD}$ and also $\overline{DC}$. The ray is denoted $\overrightarrow{EF}$. The arrow in both the notation and the picture is an indication that the ray continues in one direction indefinitely.

An **angle** is defined as the union of two rays with a common endpoint (called the **vertex**), as shown below:

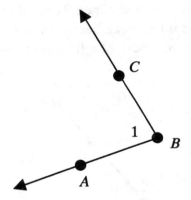

The angle pictured above can be denoted equivalently $\angle ABC$, $\angle CBA$, $\angle B$ or $\angle 1$. (Note that in three-letter representation, the vertex must be in the middle. For example, the angle pictured above is not $\angle ACB$.)

Two angles are said to be **complementary** if the sum of their measures is 90°, and two angles are called **supplementary** if the sum of their measures is 180°.

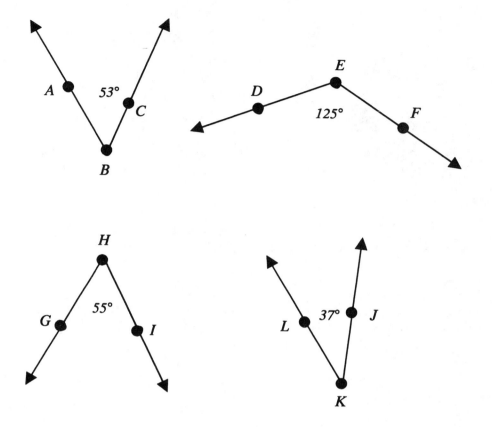

Note that the numerical value is not the angle itself. For example, 53° is the **measure** of ∡ B.

Of the angles pictured above ∡ B and ∡ K are complementary, and ∡ E and ∡ H are supplementary.

An angle is called **acute** if its measure is less than 90°, and **obtuse** if its measure is greater than 90°. ∡ABC, ∡GHI, and ∡JKL are all acute; ∡DEF is obtuse.

LINES

How many lines can be found that contain (a) one given point (b) two given points (c) three given points?

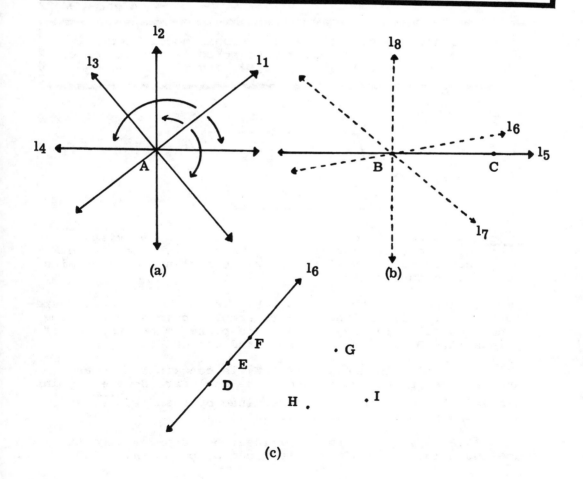

(a)

(b)

(c)

Solution: (a) Given one point A, there are an infinite number of distinct lines that contain the given point. To see this, consider line l_1 passing through point A. By rotating l_1 around A like the hands of a clock, we obtain different lines l_2, l_3, etc. Since we can rotate l_1 in infinitely many ways, there are infinitely many lines containing A.

(b) Given two distinct points B and C, there is one and only one distinct line. To see this, consider all the lines containing point B; l_5, l_6, l_7 and l_8. Only l_5 contains both points B and C. Thus, there is only one line containing both points B and C. Since there is always at least one line containing two distinct points and never more than one, the line passing through the two points is said to be determined by the two points.

(c) Given three distinct points, there may be one line or none. If a line exists that contains the three points, such as D, E, and F, then the points are said to be collinear. If no such line exists - as in the case of points G, H, and I, then the points are said to be noncollinear.

● **PROBLEM** 22

Using the accompanying three-dimensional model, determine whether each of the following sets is collinear and/or coplanar: {A,F,D}, {P,A,F}, {A,B,C,E}, and {A,F,D,E}.

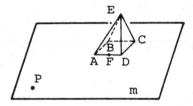

Solution: A set of points is collinear if there exists a straight line that contains every point in the set. A set of points is coplanar if there exists a plane that contains every point in the set.

Each point in the set {A,F,D} lie on the line $\overleftrightarrow{AD}$. Therefore, the set is collinear. Furthermore, points on the same line automatically lie on the same plane. Thus, the set of points {A,F,D} is both collinear and coplanar.

Each point in the set {P,A,F} is contained in plane m and, as such, points P, A, and F are coplanar. However, point P does not lie on the line determined by F and P, $\overleftrightarrow{FP}$; and, hence, the points are not collinear.

{A,B,C,E} are neither collinear nor coplanar. Point E does not lie on the plane determined by points A, B, and C (plane m) - thus the set cannot be coplanar. Points E and C do not lie on the line determined by points A and B - thus the set cannot be collinear.

The set of points {A,F,D,E} is not collinear. Although F lies on the line determined by points A and D, point E does not. Therefore, the set is noncollinear. To determine whether or not the set is coplanar, consider the plane that contains points A, E, and D. Therefore, the set is coplanar if point F lies on this plane.

Since points A and D are contained in this new plane, the line determined by these points, $\overleftrightarrow{AD}$, and every point on the line lies on the new plane. Since F is a point on $\overleftrightarrow{AD}$, point F lies on the new plane. Therefore, points A,F, D, and E are coplanar.

● **PROBLEM 23**

Find point C between A and B in the figure below such that $\overline{AC} \cong \overline{CB}$.

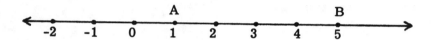

Solution: We must determine point C in such a way that $\overline{AC} \cong \overline{CB}$, or AC = CB. We are first given that C is between A and B. Therefore, since the measure of the whole is equal to the sum of the measure of its parts:

(I)   AC + CB = AB

Using these two facts, we can find the length of AC. From that we can find C.

First, since AC = CB, we substitute AC for CB in equation (I)

(II) AC + AC = AB

(III) 2(AC) = AB

Dividing by 2 we have

(IV)   AC = ½ AB

To find AC, we must know AB. We can find AB from the coordinates of A and B. They are 1 and 5, respectively. Accordingly,

(V)   AB = |5 - 1|

(VI)   AB = 4

We substitute 4 for AB in equation (IV)

(VII)   AC = ½(4)

(VIII)   AC = 2.

Therefore, C is 2 units from A. Since C is between A and B, the coordinate of C must be 3.

● **PROBLEM** 24

The measure of the complement of a given angle is four times the measure of the angle. Find the measure of the given angle.

Solution: By the definition of complementary angles, the sum of the measures of the two complements must equal 90°.

Accordingly,

(1)   Let   x = the measure of the angle
(2)   Then 4x = the measure of the complement of this angle.
Therefore, from the discussion above,

$$x + 4x = 90°$$

$$5x = 90°$$

$$x = 18°$$

Therefore, the measure of the given angle is 18°.

ANGLES

● **PROBLEM** 25

In the following quadrilateral, find the angles from vertices A and B that are subtended by side $\overline{CD}$.

Solution: To do this problem, we first must define "subtend." Consider a point X and a line segment $\overline{YZ}$. Now, draw rays from the point through the endpoints of the segment (see figure.) There are two ways of defining the relation between the angle ∢YXZ and segment $\overline{YZ}$. The angle ∢YXZ "intercepts" segment $\overline{YZ}$. From another point of view, the segment $\overline{YZ}$ "subtends" ∢YXZ.

In this problem, we are given that $\overline{DC}$ is the segment. We are asked first to consider A as the outside point, and then B. If A is the outside point, and $\overline{CD}$ is the segment, then $\overline{CD}$ subtends ⊰DAC. If B is the outside point, and $\overline{CD}$ is the segment, then $\overline{CD}$ subtends ⊰DBC.

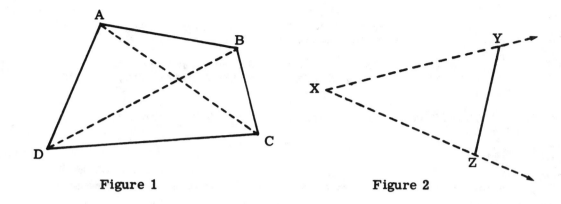

Figure 1   Figure 2

● PROBLEM 26

What is the measure of a given angle whose measure is half the measure of its complement?

Solution: When two angles are said to be complementary we know that their measures must sum, by definition, to 90°.

If we let x = the measure of the given angle,
then   2x = the measure of its complement.

To determine the measure of the given angle, set the sum of the two angle measures equal to 90 and solve for x. Accordingly,   x + 2x = 90

$$3x = 90$$

$$x = 30$$

Therefore the measure of the given angle is 30° and its complement is 60°.

● PROBLEM 27

In the figure, we are given $\overset{\leftrightarrow}{AB}$ and triangle ABC. We are told that the measure of ⊀1 is five times the measure of ⊀2. Determine the measures of ⊀1 and ⊀2.

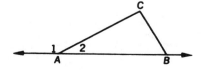

23

Solution: Since ∡1 and ∡2 are adjacent angles whose non-common sides lie on a straight line, they are, by definition, supplementary. As supplements, their measures must sum to 180°.

If we let x = the measure of ∡2
then,   5x = the measure of ∡1.

To determine the respective angle measures, set x + 5x = 180 and solve for x. 6x = 180. Therefore, x = 30 and 5x = 150.

Therefore, the measure of ∡1 = 150 and the measure of ∡2 = 30.

● **PROBLEM 28**

Find the measure of the angle whose measure is 40° more than the measure of its supplement.

Solution: By the definition of supplementary angles, the sum of the measure of two supplements must equal 180°. Accordingly,

Let   x = the measure of the supplement of the angle.
Then x + 40° = the measure of the angle.

Therefore, x + (x + 40°) = 180°

$$2x + 40° = 180°$$

$$2x = 140°$$

$$x = 70°   \text{ and } x + 40° = 110°.$$

Therefore the measure of the angle is 110°.

● **PROBLEM 29**

Given that straight lines $\overleftrightarrow{AB}$ and $\overleftrightarrow{CD}$ intersect at point E, that ∡BEC has measure 20° greater than 5 times a fixed quantity, and that ∡AED has measure 60° greater than 3 times this same quantity; Find a) the unknown fixed quantity, b) the measure of ∡BEC, and c) the measure of∡ CEA. (For the actual angle placement refer to the accompanying diagram.)

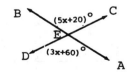

24

Solution: a) Since $\overleftrightarrow{AB}$ and $\overleftrightarrow{CD}$ are straight lines intersecting at point E, ∢BEC and ∢AED are, by definition, vertical angles. As such, they are congruent and their measures are equal. Therefore, if we let x represent the fixed quantity, we can set up the following equality, and solve for the unknown quantity.

$$5x + 20° = 3x + 60°$$

$$5x - 3x = 60° - 20°$$

$$2x = 40°$$

$$x = 20°$$

Therefore, the value of the unknown quantity is 20°.

b) From the information given about ∢BEC, we know that m∢BEC = 5x + 20°. By substitution we have

$$m∢BEC = 5(20°) + 20° = 100° + 20° = 120°.$$

Therefore, the measure of ∢BEC is 120°.

c) We know that $\overleftrightarrow{AB}$ is a straight line; therefore, ∢CEA is the supplement of ∢BEC. Since the sum of the measure of two supplements is 180°, the following calculation can be made:

$$m∢CEA + m∢BEC = 180°$$

$$m∢CEA = 180° - m∢BEC,$$

Substituting in our value for m∢BEC, we obtain:

$$m∢CEA = 180° - 120° = 60°$$

Therefore, the measure of ∢CEA is 60°.

25

CHAPTER 3

PERPENDICULARITY

Basic Attacks and Strategies for Solving Problems
in this Chapter. See pages 26 to 31 for step-by-step
solutions to problems.

When two lines intersect to form **right angles** (angles with measures of 90°), the lines are said to be perpendicular. For example, in the figure below ∡ *AED*, ∡ *DEB*, ∡ *BEC*, and ∡*CEA* are all right angles.

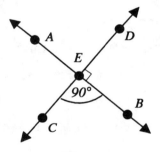

Thus $\overleftrightarrow{AB}$ is perpendicular to $\overleftrightarrow{CD}$, and this is written as $\overleftrightarrow{AB} \perp \overleftrightarrow{CD}$. The intersection of any pair of perpendicular lines forms four right angles. When two segments intersect to form at least one right angle, the segments are perpendicular.

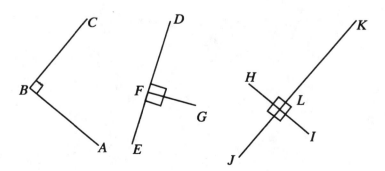

In the figures on the previous page, $\overline{AB} \perp \overline{BC}$, $\overline{FG} \perp \overline{DE}$, and $\overline{HI} \perp \overline{JK}$. If the lengths of the segments $\overline{HL}$ and $\overline{LI}$ are equal, then L is the midpoint of $\overline{HI}$, so we call $\overline{JK}$ (or JK) a **perpendicular bisector** of $\overline{HI}$. Note also that, since $\overline{JL}$ is not the same length as $\overline{KL}$, $\overline{HI}$ is not a bisector of $\overline{JK}$.

Step-by-Step Solutions to
Problems in this Chapter,
"Perpendicularity"

● **PROBLEM** 30

In the figure, AE = n + p, ED = m + p (measured in inches) and m ✪ AED = m + n (measured in degrees). Find lengths AB and CD given that $\overline{AB}$ and $\overline{CD}$ are perpendicular bisectors of each other, $\frac{AE}{ED} = \frac{4}{3}$, and AE · ED = 3888.

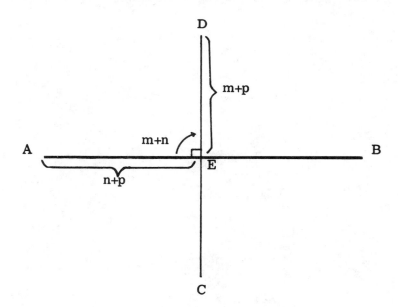

Solution: Because $\overline{AB}$ and $\overline{CD}$ are perpendicular bisectors of each other, E is the midpoint of both $\overline{AB}$ and $\overline{CD}$. Thus, AB = 2·AE = 2(n + p) and CD = 2 · ED = 2(m + p). To find AB and CD, we must find the values of m, n and p.

To solve for m, n, and p, we collect all available data on the three unknowns. First, since $\overline{AB}$ is the perpendicular bisector of $\overline{CD}$, ✪AED is a right angle. Since m ✪

AED = m + n, we have

(I)   m + n = 90°

Furthermore, we are given that $\frac{AE}{ED} = \frac{4}{3}$. Remember AE = n + p and ED = m + p. Thus,

(II) $\frac{n + p}{m + p} = \frac{4}{3}$

Finally, we also know that AE · ED = 3888. Thus,

(III)   (n + p) · (m + p) = 3888.

We now have three equations and three unknowns. We solve (I) for m in terms of n, m = 90 - n, and substitute this result in (II) and (III) to obtain two equations in two unknowns, n and p.

(IV) $\frac{n + p}{90 - n + p} = \frac{4}{3}$

(V)   (n + p)(90 - n + p) = 3888.

Crossmultiplying the proportion in (IV), we obtain:

(VI)   3n + 3p = 360 - 4n + 4p

Solving for p in terms of n, we obtain

(VII)   7n = 360 + p

(VIII)   p = 7n - 360

Substituting this result in (V), we obtain a single equation in the unknown n:

(IX)   (n + 7n - 360)(90 - n + 7n - 360) = 3888

(X)   (8n - 360)(6n - 270) = 3888.

Before blindly multiplying through, note that (8n - 360) = 8(n - 45) and that (6n - 270) = 6(n - 45). Thus, (8n - 360)(6n - 270) = 8 · 6 · (n - 45)(n - 45) = 48(n - 45)^2.

(XI)   48(n - 45)^2 = 3888

(XII)   (n - 45)^2 = $\frac{3888}{48}$ = 81

Before plunging ahead and using the quadratic formula, let us note that (XII) is true if and only if $\pm\sqrt{(n - 45)^2} = \sqrt{81}$. Thus, +(n - 45) = 9 or -(n - 45) = 9.

Consider the first case. n - 45 = 9 implies n = 54. To find p, we use eq. (VIII).

(XIII)   p = 7(54) - 360,   or p = 18.

The value of m can be found by another substitution. m = 90 - n = 90 - 54 = 36. Then, AE = n + p = 54 + 18 = 72; ED = m + p = 36 + 18 = 54, and m $\sphericalangle$ AED = 36 + 54 = 90°. As a check note, $\frac{AE}{ED} = \frac{72}{54} = \frac{4}{3}$, which agrees with the given.

If we consider the second case - (n - 45°) = 9, then n = 36. Since m + n = 90, then m = 90 - n = 54, and p = 7n - 360 = - 108. This leads to a negative value for AE (AE = n + p = 36 + (- 108) = - 72). Since no segment can have negative length, the second case leads to no valid answer and, hence, must not be considered.

Thus, n = 54; m = 36; and p = 18. Substituting these values in our expression for AB and CD, we obtain the required lengths AB = 2(n + p) = 2(54 + 18) = 144 in. and CD = 2(m + p) = 2(36 + 18) = 108 in.

● **PROBLEM** 31

We are given straight lines $\overleftrightarrow{AB}$ and $\overleftrightarrow{CD}$ intersecting at point P. $\overrightarrow{PR} \perp \overleftrightarrow{AB}$ and the measure of $\sphericalangle$APD is 170. Find the measures of $\sphericalangle$1, $\sphericalangle$2, $\sphericalangle$3, and $\sphericalangle$4.

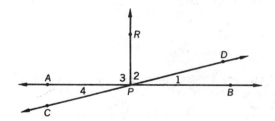

<u>Solution</u>: This problem will involve making use of several of the properties of supplementary and vertical angles, as well as perpendicular lines.

$\sphericalangle$APD and $\sphericalangle$1 are adjacent angles whose non-common sides lie on a straight line, $\overleftrightarrow{AB}$. Therefore, they are supplements and their measures sum to 180°.

m$\sphericalangle$APD + m$\sphericalangle$1 = 180.

We know m$\sphericalangle$APD = 170. Therefore, by substitution, 170 + m$\sphericalangle$1 = 180. This implies m$\sphericalangle$1 = 10.

$\sphericalangle$1 and $\sphericalangle$4 are vertical angles because they are formed by the intersection of two straight lines, $\overleftrightarrow{CD}$ and $\overleftrightarrow{AB}$, and their sides form two pairs of opposite rays. As vertical angles, they are, by theorem, of equal measure. Since m$\sphericalangle$1 = 10, then m$\sphericalangle$4 = 10.

Since $\overleftrightarrow{PR} \perp \overleftrightarrow{AB}$, at their intersection the angles formed must be right angles. Therefore, ∢3 is a right angle and its measure is 90. m∢3 = 90.

The figure shows us that ∢APD is composed of ∢3 and ∢2. Since the measure of the whole must be equal to the sum of the measures of its parts, m∢APD = m∢3 + m∢2. We know the m∢APD = 170 and m∢3 = 90, therefore, by substitution, we can solve for m∢2, our last unknown.

$$170 = 90 + m∢2$$

$$80 = m∢2$$

Therefore, m∢1 = 10,   m∢2 = 80

m∢3 = 90,   m∢4 = 10.

● **PROBLEM** 32

In the accompanying figure $\overline{SM}$ is the perpendicular bisector of $\overline{QR}$, and $\overline{SN}$ is the perpendicular bisector of $\overline{QP}$. Prove that SR = SP.

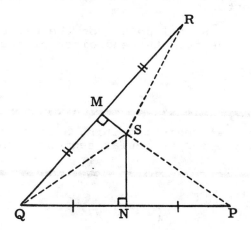

<u>Solution</u>: Every point on the perpendicular bisector of a segment is equidistant from the endpoints of the segment.

Since point S is on the perpendicular bisector of $\overline{QR}$,
(I)   SR = SQ

Also, since point S is on the perpendicular bisector of $\overline{QP}$,
(II)   SQ = SP

By the transitive property (quantities equal to the same quantity are equal), we have:

(III)   SR = SP.

If, in the figure shown, AD = AE and B is the midpoint of $\overline{DE}$, prove DC = EC.

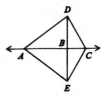

Solution: Any point that is equidistant from the endpoints of a line segment will always lie on the perpendicular bisector of that segment.

Since B is the midpoint of $\overline{DE}$, DB = BE. We are given that AD = AE. Hence, points B and A are equidistant from endpoints E and D and are, therefore, on the perpendicular bisector of $\overline{DE}$.

Two points determine a line. Therefore, $\overleftrightarrow{AB}$ is the perpendicular bisector.

C is a point on $\overleftrightarrow{AB}$. All points on the perpendicular bisector are equidistant from the endpoints of the segment. Hence, DC = EC.

Show that the shortest segment joining a line with an external point is the perpendicular segment from the point to the line.

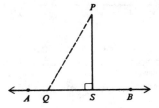

Solution: To prove the statement, choose any other line segment joining the point and the line; and show that it must be of greater length than the perpendicular. Note that the new segment, the perpendicular, and the given line form a right triangle with the right angle opposite the new segment. Since the right angle is the largest angle of the triangle, the side opposite it, the new segment, must be greater than either of the other sides. In particular, the new segment is longer than the perpendicular. Since the new segment was any possible segment, the perpendicular is thus shown to be shorter than any other segment.

GIVEN: $\overleftrightarrow{AB}$ with the external point P; $\overleftrightarrow{PS} \perp \overleftrightarrow{AB}$; $\overline{PQ}$ is any segment from P to $\overleftrightarrow{AB}$ other than the perpendicular from P to $\overleftrightarrow{AB}$.

PROVE: PS < PQ

STATEMENT	REASONS
1. $\overleftrightarrow{AB}$ with the external point P; $\overleftrightarrow{PS} \perp \overleftrightarrow{AB}$	1. Given
2. ⊁ PSA is a right angle	2. Perpendicular lines form right angles.
3. Select any point Q on $\overleftrightarrow{AB}$	3. Point Uniqueness Postulate.
4. ⊁ PQS is acute	4. If a triangle has one right angle, then its other two angles must be acute.
5. m ⊁ PQS < m ⊁ PSA	5. Definition of an acute angle.
6. PS < PQ	6. If two angles of a triangle are not congruent, then the sides opposite those angles are not congruent, the longer side being opposite the angle with the greater measure.

CHAPTER 4

TRIANGLES

> **Basic Attacks and Strategies for Solving Problems in this Chapter. See pages 32 to 47 for step-by-step solutions to problems.**

Triangles are classified both in terms of the sizes of their angles and the relative lengths of their sides. The words **right, acute, obtuse, scalene, isosceles**, and **equilateral** are all names given to triangles with special properties. Here are the definitions of those kinds of triangles:

Right triangle – a triangle with one right angle

Acute triangle – a triangle with three acute angles

Obtuse triangle – a triangle with one obtuse angle

Scalene triangle – a triangle with no pair of sides of equal measure

Isosceles triangle – a triangle that has at least two sides of equal measure

Equilateral triangle – a triangle with all three sides of equal measure

Since the sum of the measures of the angles of a triangle is 180°, a triangle can have at most one right angle or one obtuse angle.

A triangle can fall into more than one of these categories. For example, if a triangle is equilateral, it is also acute (see Problems 46 and 47). Also, since an equilateral triangle has three congruent sides, it has at least two congruent sides. Thus, every equilateral triangle is isosceles. (However, not every isosceles triangle is equilateral.)

A tree diagram which illustrates the various kinds of triangles follows:

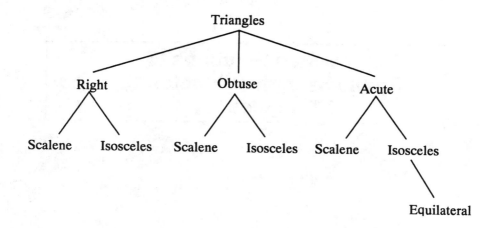

This figure may help with the understanding of the solutions given to problems in this chapter.

Since a triangle is formed by three line segments, it is also interesting to consider the angles formed by the lines containing these segments, which are not inside the triangle.

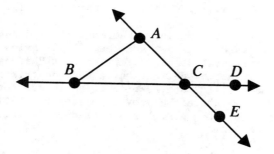

In the picture above, $\angle ACD$ and $\angle BCE$ are **exterior angles** of the triangle ABC (notated: $\triangle ABC$).

SCALENE TRIANGLES

● **PROBLEM** 35

Prove that a scalene triangle has no 2 angles congruent.

<u>Solution</u>: We shall prove the given theorem by the method of "proof by contradiction". This method entails proving that the contrapositive of the given theorem is true. If this is so, then the original theorem is true.

The contrapositive of the given theorem is: If 2 angles of a triangle are congruent, then it is not scalene. If 2 angles of a triangle are congruent, it is isosceles. This implies that 2 sides of the triangle are congruent. But, by definition, a scalene triangle has no sides congruent. Hence, if 2 angles of a triangle are congruent, it is not scalene. We have therefore proven the contrapositive of the given theorem to be true; hence, the given theorem is true.

● **PROBLEM** 36

Prove that if a triangle has no 2 angles congruent, then it is scalene.

<u>Solution</u>: We prove this theorem by the method of contradiction. That is, we prove that the contrapositive of the given theorem is true. It then follows that the theorem itself is true.

The contrapositive of the given theorem is: If a triangle is non-scalene, then at least 2 of its angles are congruent.

By definition, a scalene triangle has no 2 sides congruent. Hence, a non-scalene triangle has at least 2 sides congruent. This means that a non-scalene triangle is either an isosceles triangle or an equilateral triangle. Each of the latter triangles has at least 2 angles congruent. Hence, we have shown that a non-scalene triangle has at least 2 angles congruent. Since the contrapositive of the given theorem is true, the given theorem is true.

In △ ABC, m ⊁ C = 125° and m ⊁ B = 35°. As drawn in the figure, which is the shortest side of the triangle?

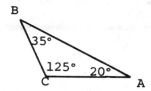

Solution: When the angles of a triangle are of unequal measure, the sides are, correspondingly, of unequal length - that is, the shortest side will be opposite the smallest angle.

Since the sum of the measures of the angles of a triangle is 180°, and since m ⊁ B = 35° and m ⊁ C = 125°, the m ⊁ A = 180° - (35°+125°) = 180° - 160° = 20°. Therefore, the side opposite < A will be the short-est. $\overline{BC}$ is this side.

Therefore, $\overline{BC}$ is the shortest side of △ ABC.

Prove that a triangle can have, at most, one obtuse angle.

Solution: We prove this result indirectly. We assume that a triangle can have more than one obtuse angle. Then we show this leads to a contra-diction.

Assume that there exists a △ ABC with more than one, say two, obtuse angles. Let ⊁ A and ⊁ B be the obtuse angles. Since the measure of an obtuse angle must be greater than 90° but less than 180°, m ⊁ A > 90 and m ⊁ B > 90. Thus, by the Addition Property of Inequality, we have

(i)   m ⊁ A + m ⊁ B > 180.

Since the measure of ⊁ C must be greater than zero, that is, m ⊁ C > 0, it follows that

(ii)   m ⊁ A + m ⊁ B + m ⊁ C > 180.

This contradicts the theorem that states that the sum of the measures of the angles of a triangle must equal 180°. Therefore, our assumption is false, and a triangle has, at most, one obtuse angle.

ISOSCELES TRIANGLES

The measure of the vertex angle of an isosceles triangle exceeds the measure of each base angle by 30°. Find the value of each angle of the triangle.

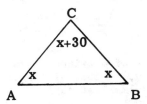

Solution: We know that the sum of the values of the angles of a triangle is 180°. In an isosceles triangle, the angles opposite the congruent sides (the base angles) are, themselves, congruent and of equal value.

Therefore,
(1) Let x = the measure of each base angle.
(2) Then x + 30 = the measure of the vertex angle.
We can solve for x algebraically by keeping in mind the sum of all the measures will be 180°.

$$x + x + (x+30) = 180$$
$$3x + 30 = 180$$
$$3x = 150$$
$$x = 50$$

Therefore, the base angles each measure 50°, and the vertex angle measures 80°.

Given: AC = BC .   AD = BD.
Prove: m ∡ CAD = m ∡ CBD .

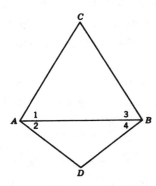

Solution: The required results can be derived by proving that the angles that make up ∡ CAD and ∡ CBD are equal to each other.

Statements	Reasons
1. AC = BC AD = BD	1. Given
2. m ∡ 2 = m ∡ 4 m ∡ 1 = m ∡ 3	2. If two sides of a triangle are of equal length, then the angles opposite those sides have equal measure.
3. m ∡ 1 + m ∡ 2 = m ∡ 3 + m ∡ 4	3. If equal quantities are added to equal quantities, the sums are equal quantities.
4. m ∡ CAD = m ∡ 1 + m ∡ 2 m ∡ CBD = m ∡ 3 + m ∡ 4	4. The measure of the whole is equal to the sum of the measures of its parts.
5. m ∡ CAD = m ∡ CBD	5. A quantity may be substituted for its equal.

● **PROBLEM** 41

Given: m ∡ A = m ∡ B. AD = BE .
Prove: m ∡ CDE = m ∡ CED.

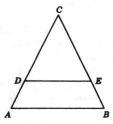

Solution: This proof will revolve around the theorem stating that two angles of a triangle are of equal measure if and only if the sides opposite them are of equal length.

Statements	Reasons
1. m ∡ A = m ∡ B	1. Given
2. CA = CB	2. If a triangle has two angles of equal measure, then the sides opposite those angles are equal in length.
3. CB = CE + BE CA = CD + AD	3. The measure of the whole is equal to the sum of the measures of its parts.
4. CD + AD = CE + BE	4. A quantity may be substituted for its equal.
5. AD = BE	5. Given
6. CD = CE	6. If equal quantities are subtracted from equal quantities, the differences are equal quantities.
7. m ∡ CDE = m ∡ CED	7. If two sides of a triangle are equal, then the angles opposite those sides are equal.

RIGHT TRIANGLES

As seen in the accompanying diagram, △ABC is constructed in such a way that the measure of ⦜ A equals 9x, m ⦜ B equals 3x - 6, and m ⦜ C equals 11x + 2, x being some unknown. Show that △ABC is a right triangle.

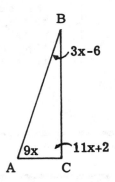

Solution: A triangle is a right triangle if one of its angles is a right angle. The best way to determine the "rightness" of this triangle would be to sum the measures of all its angles, set this sum equal to 180°, and solve for the unknown x. If the measure of one angle turns out to be 90°, then it is a right angle and the triangle is a right triangle. The algebra is as follows:

$$m ⦜ A + m ⦜ B + m ⦜ C = 180$$
$$9x + 3x - 6 + 11x + 2 = 180$$
$$23x - 4 = 180$$
$$23x = 184$$
$$x = 8$$

Therefore,

⦜ A measures (9)(8) or 72
⦜ B measures (3)(8) - 6 or 18

and

⦜ C measures (11)(8) + 2 or 90

Therefore, since ⦜ C measures 90°, △ABC is a right triangle.

a) Let ABC be a right triangle with m ⦜ BCA = 90° and m ⦜ CAB = 30°. What is m ⦜ ABC ?

b) Prove that in a right triangle the sum of the measures of the angles adjacent to the hypotenuse is 90°.

Solution: a) △ ABC is illustrated in figure (a). By the angle sum theorem,

(1)   m ⦜ BCA + m ⦜ CAB + m ⦜ ABC = 180°.

But, from the statement of the problem, m $\sphericalangle$ CAB = 30° and m $\sphericalangle$ BCA = 90°. Substituting these results into equation (1),

$$m \sphericalangle ABC + 30° + 90° = 180°$$

or $\quad$ m $\sphericalangle$ ABC = 60°.

b) An arbitrary right triangle (Δ A'B'C') is shown in figure (b). By the angle sum theorem,

(2) $\qquad$ m $\sphericalangle$ A'B'C' + m $\sphericalangle$ B'C'A' + m $\sphericalangle$ B'A'C' = 180° .

Since Δ A'B'C' is a right triangle, m $\sphericalangle$ B'C'A' = 90°. Hence, equation (2) becomes

$$m \sphericalangle A'B'C' + m \sphericalangle B'A'C' + 90° = 180°$$

or $\quad$ m $\sphericalangle$ A'B'C' + m $\sphericalangle$ B'A'C' = 90° .

Since $\sphericalangle$ A'B'C' and $\sphericalangle$ B'A'C' are the angles adjacent to the hypotenuse of Δ A'B'C', we have obtained the required result.

Figure (a)

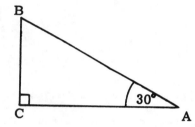

Figure (b)

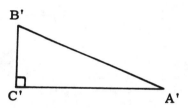

● **PROBLEM 44**

Prove that the base angles of an isosceles right triangle have measure 45° .

<u>Solution</u>: As drawn in the figure, Δ ABC is an isosceles right triangle with base angles BAC and BCA. The sum of the measures of the angles of any triangle is 180°. For Δ ABC, this means

(1) $\quad$ m $\sphericalangle$ BAC + m $\sphericalangle$ BCA + m $\sphericalangle$ ABC = 180°.

But m $\sphericalangle$ ABC = 90° because ABC is a right triangle. Furthermore, m $\sphericalangle$ BCA = m $\sphericalangle$ BAC, since the base angles of an isosceles triangle are congruent. Using these facts in equation (1)

$$m \angle BAC + m \angle BCA + 90° = 180°$$

or
$$2m \angle BAC = 2m \angle BCA = 90°$$

or
$$m \angle BAC = m \angle BCA = 45°.$$

Therefore, the base angles of an isosceles right triangle have measure 45°.

● **PROBLEM** 45

The length of the median drawn to the hypotenuse of a right triangle is 12 inches. Find the length of the hypotenuse. (See figure).

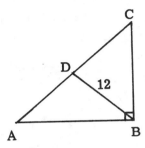

Solution: A theorem tells us that the length of the median to the hypotenuse of a right triangle is equal to one-half the length of the hypotenuse. We must identify the median, the hypotenuse, their respective lengths, and substitute them according to the rule cited, and to solve for any unknowns.

$\overline{AC}$ is the hypotenuse, $\overline{BD}$ the median of length 12", and the length of $\overline{AC}$ is unknown. By applying the above theorem we know,

(1). BD = ½AC and, by, substitution

(2). 12" = ½AC which implies that AC = 24".

Therefore, the length of hypotenuse $\overline{AC}$ is 24 in.

EQUILATERAL TRIANGLES

● **PROBLEM** 46

Prove that an equilateral triangle has three equal angles.

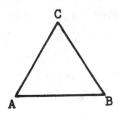

Solution: Draw equilateral △ABC.

Hence, the problem can be restated as:

Given: equilateral △ABC

Prove: m∡A = m∡B = m∡C.

STATEMENTS	REASONS
1. equilateral △ABC	1. Given.
2. AC ≅ BC ≅ AB	2. Definition of an equilateral triangle.

3. ∡A ≅ ∡B
   ∡A ≅ ∡C
   ∡B ≅ ∡C
	3. If two sides of a triangle are congruent, then the angles opposite those sides are congruent.

4. ∡A ≅ ∡B ≅ ∡C	4. Transitive Property of Congruence.

● **PROBLEM 47**

Prove that each angle of an equilateral triangle has measure 60°.

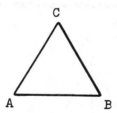

Solution: We will apply the angle sum theorem to an equilateral triangle to prove that the measure of each angle is 60°.

Let the equilateral triangle be △ ABC. By the angle sum theorem,

(1) $m\angle A + m\angle B + m\angle C = 180°$.

But, the angles of an equilateral triangle are congruent. That is,

$$\angle A \cong \angle B \cong \angle C .$$

Thus,

(2) $m\angle A = m\angle B = m\angle C.$

Substituting (2) into (1),

$$3m\angle A = 180°$$

or $m\angle A = 60°$.

By equation (2), all the angles of an equilateral triangle have measure 60°.

Is every equilateral triangle isosceles? Is every isosceles triangle equilateral? Is every nonscalene triangle equilateral?

<u>Solution</u>: An equilateral triangle has all sides congruent. Since an isosceles triangle has at least 2 sides congruent, every equilateral triangle is isosceles.

An isosceles triangle isn't necessarily equilateral, because the former need have only 2 sides congruent, whereas the latter must have 3 sides congruent.

A scalene triangle has no 2 sides congruent. Hence, a non-scalene triangle has at least 2 sides congruent.

A non-scalene triangle is, therefore, definitely isosceles, but not necessarily equilateral.

ANGLE BISECTORS

● **PROBLEM** 49

For the following statement, draw a figure, label it, and state, in terms of the letters of the figure, the hypothesis and the conclusion.

If the bisector of the vertex angle of an isosceles triangle is drawn, then the bisector is perpendicular to the base of the triangle.

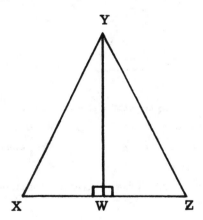

<u>Solution</u>: The hypothesis gives the specific guidelines for the diagram to be drawn with one exception. This exception is filled in by the conclusion of the statement.

The hypothesis is "the bisector of the vertex angle of an isosceles triangle is drawn." Draw an isosceles triangle and then the bisector of its vertex angle. The bisector will intersect the base of the triangle, but in what specific manner is not yet known.

The conclusion tells us that the bisector will be perpendicular to the base of the triangle.

In terms of the diagram, the hypothesis can be written as: "If $\overline{YW}$ is the bisector of ∢Y in ΔXYZ."

Conclusion: "then $\overline{YW} \perp \overline{XZ}$."

● **PROBLEM** 50

Given: DA bisects ∢ CAB. DB bisects ∢ CBA.
m ∢ 1 = m ∢ 2 .
Prove: CA = CB.

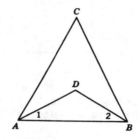

Solution: This proof will involve using the definition of angle bisector and the substitution and addition postulates to prove m ∢ CAB = m ∢ CBA. The desired results will be obtained recalling that the sides of a triangle opposite angles of equal measure are of equal length.

Statements	Reasons
1. DA bisects ∢ CAB DB bisects ∢ CBA	1. Given
2. m ∢ CAD = m ∢ 1 m ∢ CBD = m ∢ 2	2. An angle bisector divides an angle into two angles of equal measure.
3. m ∢ = 1 = m ∢ 2	3. Given
4. m ∢ CBD = m ∢ 1	4. A quantity may be substituted for its equal.
5. m ∢ CAD = m ∢ CBD	5. If two quantities are equal to the same quantity, or equal quantities, then they are equal to each other.
6. m ∢ CAD + m ∢ 1 = m ∢ CBD + m ∢ 2	6. If equal quantities are added to equal quantities, the sums are equal quantities.
7. m ∢ CAB = m ∢ CAD + m ∢ 1 m∢ CBA = m∢ CBD + m∢2	7. The measure of the whole is equal to the sum of the measures of its parts.
8. m ∢ CAB = m ∢ CBA	8. If two quantities are equal to the same quantity, or equal quantities, then they're equal to each other.
9. CA = CB	9. If a triangle has two angles of equal measure then the sides opposite those angles are equal in length.

Show that the angle bisectors of a triangle are concurrent at a point equidistant from the sides of the triangle.

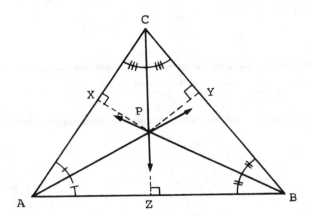

Solution: The angle bisector of any angle is the set of points in the interior of the angle that are equidistant from the sides of the angle.

In △ABC, the intersection of the bisector of ∢ A and the bisector of ∢ B is a point P. Because P is on the bisector of ∢ A, the distance from P to $\overline{AC}$ equals the distance from P to $\overline{AB}$. Because P is on the bisector of ∢ B, the distance from P to $\overline{BC}$ equals the distance from P to $\overline{AB}$. By transitivity, the distance from P to $\overline{AC}$ equals the distance from P to $\overline{BC}$, i.e. P is equidistant from the sides of ∢ C. Therefore, point P, is also on the bisector of ∢ C, and the three bisectors are concurrent at a point equidistant from the sides of the triangle.

Given: The angle bisectors of △ABC.
Prove: The bisectors are concurrent at a point equidistant from the sides.

Statements	Reasons
1. The angle bisectors of △ABC	1. Given
2. Let P be the point of intersection of the bisectors of ∢A and ∢B.	2. In a plane, two non-parallel, non-coincident lines intersect in a unique point.
3. P is in the interior of ∢ A. P is in the interior of ∢ B.	3. All points (except for the vertex) of the angle bisector lie in the interior of the angle.
4. P is in the interior of △ABC	4. If a point is in the interior of two angles of a triangle, then it is in the interior of the triangle.
5. Let X,Y, and Z be the points on sides $\overline{AC}$, $\overline{CB}$, and $\overline{AB}$ such that $\overline{PX} \perp \overline{AC}$, $\overline{PY} \perp \overline{CB}$, and $\overline{PZ} \perp \overline{AB}$.	5. From a given external point, only one line can be drawn perpendicular to a given line.

6. PX, PY, and PZ are the distances from P to the sides	6. The distance from an external point to a line is the length of the perpendicular segment from the point to the line.
7. PX = PZ PY = PZ	7. All points on the angle bisector are equidistant from the sides of the angle.
8. PX = PY	8. Transitivity Postulate.
9. P is in the interior of ∡ C.	9. All points in the interior of a triangle are in the interior of each of the angles.
10. P is on the angle bisector of ∡ C.	10. All points in the interior of an angle that are equidistant from the sides of the angle are on the angle bisector.
11. P is equidistant from the sides of ΔABC.	11. Follows from Steps 7 and 8.
12. The angle bisectors are concurrent at a point equidistant from the sides.	12. Lines are concurrent if their intersection consists of at least one point. Also, Step 11.

EXTERIOR ANGLES

• **PROBLEM** 52

In the figure, what can be said about ∡3 and ∡1? About ∡2 and ∡4? If m∡4 > m∡3, prove that m∡2 > m∡1.(See figure.)

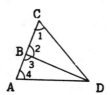

Solution: We use the Exterior Angle Theorem to find relationships between the given angles. This theorem states that an exterior angle of a triangle has greater measure than either of its remote interior angles. (An exterior angle of a triangle is any angle outside of the triangle which forms a linear pair with an angle of the triangle. A remote interior angle of a triangle is an angle of a triangle which does not form a linear pair with the given exterior angle in question.)

In the figure, ∡3 is an exterior angle of ΔBCD. ∡1 is a remote interior angle of the triangle, relative to angle 3. Hence, m∡3 > m∡1. Furthermore, ∡2 is an exterior angle of ΔBAD, and ∡4 is a remote interior angle of ΔBAD relative to ∡2. Hence, m∡2 > m∡4.

Given that m∡4 > m∡3, and using the conclusions drawn

above, we may write the following string of inequalities:

m∡2 > m∡4
m∡4 > m∡3   (1)
m∡3 > m∡1

Coupling the first two of these inequalities yields

m∡2 > m∡3   and   m∡3 > m∡1.

Combining these 2 inequalities gives

m∡2 > m∡1.

● PROBLEM 53

In △ABC, $\overline{AC} \cong \overline{BC}$. The measure of an exterior angle of vertex C is represented by 5x + 10°. If ∡A measures 30°, find the value of x.

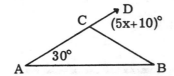

Solution: To solve this problem, we relate m∡DCB to ∡A and ∡B. First, since $\overline{AC} \cong \overline{BC}$, we know that △ACB is isosceles. Because the base angles of an isosceles triangle are congruent, m∡A = m∡B = 30°.

Secondly, ∡DCB forms a linear pair with one of the interior angles of △ACB. Thus ∡DCB is an exterior angle and must be equal in measure to the sum of the remote interior angles, ∡A and ∡B.

(I)   m∡DCB = m∡A + m∡B

(II)   5x + 10 = 30 + 30

(III)   5x = 50°

(IV)   x = 10°.

● PROBLEM 54

Find the numerical measure of each of the angles of △ABC if m∡A = 19x − 15, m∡C = 9x + 25, and m∡ABD = 26x + 20.

Solution: To find m∡A, m∡B and m∡C, we first solve for
> x can be solved for by using the Exterior Angle Theorem: the measure of an exterior angle of the triangle equals the sum of the remote interior angles. Therefore,

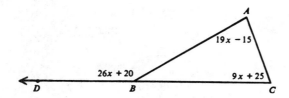

m∡ABD = m∡A + m∡C

26x + 20 = (19x - 15) + (9x + 25) = 28x + 10

To isolate the variable x, we subtract 26x and 10 from both sides.

2x = 10   or   x = 5.

Then m∡A = 19x - 15 = 19(5) - 15 = 80°

m∡C = 9x + 25 = 9(5) + 25 = 70°

Since, ∡A, ∡B, and ∡C are angles of the same triangle,

m∡A + m∡B + m∡C = 180°

or,   m∡B = 180 - m∡A - m∡C = 180 - 80 - 70 = 30°.

CONGRUENCE

(Chapters 5 to 9)

CHAPTER 5

CONGRUENT ANGLES AND LINE SEGMENTS

Basic Attacks and Strategies for Solving Problems
in this Chapter. See pages 48 to 61 for step-by-step
solutions to problems.

The two angles, $\angle ABC$ and $\angle DEF$, pictured below are **congruent**. This is noted, in the diagram, by the single arcs across each angle, and in text by $\angle ABC \cong \angle DEF$. This means that $m \angle ABC = m \angle DEF$, which is necessary for two angles to be congruent. Specifically,

$\angle ABC \cong \angle DEF$ if and only if $m \angle ABC = m \angle DEF$

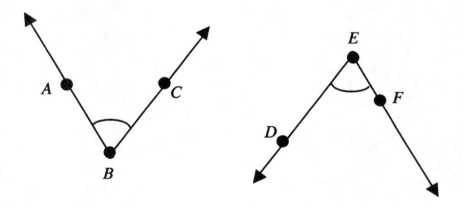

Note that $\angle ABC \neq m \angle DEF$. In order for the angles to be equal, they must be composed of the same pair of rays and that is not the case.

In the picture below $m \angle GHJ \cong m \angle JHI$, which means that $\overrightarrow{HJ}$ is the angle bisector of $\angle GHI$:

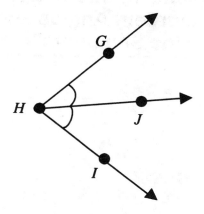

In the picture below, $\angle 1$ and $\angle 2$ are called **vertical angles,** and vertical angles are always congruent:

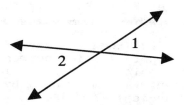

The definition of congruence as applied to line segments is similar to the definition of congruence applied to angles. Two segments are congruent if and only if the two segments have the same measure (length). Since AB is used to denote the length of $\overline{AB}$, and CD is used to denote the length of $\overline{CD}$, then $\overline{AB} \cong \overline{CD}$ if and only if $AB = CD$. If $\overline{AB}$ and $\overline{CD}$ are different segments, then $\overline{AB} \ne \overline{CD}$.

CONGRUENT ANGLES

● **PROBLEM** 55

Which of the following are true? a) If $\angle A = \angle B$, then $\angle A \cong \angle B$. b) If $\angle A \cong \angle B$, then $\angle A = \angle B$. c) If $AB = \overline{CD}$, then $AB \cong \overline{CD}$.

Solution: a) If $\angle A = \angle B$, then the set of points composing $\angle A$ is the same set of points which composes $\angle B$. (That is, each point of $\angle A$ is a point of $\angle B$, and vice versa.) If this is the case, then $m\angle A = m\angle B$, and, by definition of the congruence of 2 angles, $\angle A \cong \angle B$.

b) If $\angle A \cong \angle B$, then $m\angle A = m\angle B$ by definition of the congruence of 2 angles. However, each point of $\angle A$ is not necessarily a point of $\angle B$, and each point of $\angle B$ is not necessarily a point of $\angle A$. Hence, it is not absolutely necessary for $\angle A = \angle B$.

c) The hypothesis, $AB = \overline{CD}$, is meaningless because it states that the set consisting of the length of AB is equal to the set consisting of the points of $\overline{CD}$. These 2 sets can't be equal, because the former has one element, and the latter has an infinite number of elements. A correct statement is "If $AB = CD$, then $\overline{AB} \cong \overline{CD}$." This follows from the definition of the congruence of 2 line segments.

● **PROBLEM** 56

Prove that if $\angle A \cong \angle B$ and $\angle B \cong \angle C$, then $\angle A \cong \angle C$.

Solution: We use the transitive property of real numbers (if $a = b$ and $b = c$, then $a = c$) and the definition of congruence of angles to obtain the required result.

If $\angle A \cong \angle B$ and $\angle B \cong \angle C$, then $m\angle A = m\angle B$ and $m\angle B = m\angle C$, by definition of the congruence of angles. By the transitive property of real numbers, $m\angle A = m\angle C$. Since angles of equal measure are congruent, it follows that $\angle A \cong \angle C$, as was to be shown.

● **PROBLEM** 57

State the RST properties for congruence of angles.

Solution: The RST properties for congruence of angles are:

1. Reflexive Property: $\angle A \cong \angle A$

2. Symmetric Property; $\angle A \cong \angle B \Rightarrow \angle B \cong \angle A$

3. Transitive Property: $\angle A \cong \angle B$ and $\angle B \cong \angle C \Rightarrow$

$$\angle A \cong \angle C$$

● **PROBLEM** 58

Prove that if an angle is congruent to one of two complementary angles, then it is complementary to the other angle.

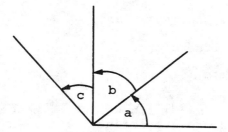

Solution: Let $\angle a$ and $\angle b$ be the two complementary angles and suppose $\angle c \cong \angle a$. We must show $\angle c$ and $\angle b$ are complementary.

Because $\angle a$ and $\angle b$ are complementary, it follows that:

(I) $m\angle a + m\angle b = 90°$

Because $\angle c \cong \angle a$, it follows that

(II) $m\angle c \cong m\angle a$

Substituting (II) into (I), we obtain:

(III) $m\angle c + m\angle b = 90°$

Because the measures of ∢c and ∢b sum to 90°, ∢c and ∢b are complementary. Thus, an angle congruent to one of two complementary angles is complementary to the other.

In the figure, ∢b and ∢a form a linear pair. a) Prove that if ∢a ≅ ∢c then ∢b and ∢c are supplementary; b) prove that if ∢b and ∢c are supplementary, then ∢a ≅ ∢c.

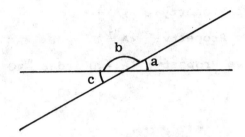

Solution: a) Two angles are supplementary if and only if the sum of their measures is 180°. Since ∢a and ∢b form a linear pair (that is, they share one common ray and the second rays of each angle are oppositely directed), they are automatically supplementary. That is,

$$m∢a + m∢b = 180°. \tag{1}$$

But ∢c ≅ ∢a, and, therefore,

$$m∢a = m∢c \tag{2}$$

Using (2) in (1)   m∢c + m∢b = 180°,   hence, by definition, ∢b and ∢c are supplementary.

b) We are now told that ∢b and ∢c are supplementary. By definition, this means that

$$m∢b + m∢c = 180°.$$

Since ∢a and ∢b are a linear pair

$$m∢a + m∢b = 180°$$

Equating the last 2 equations

$$m∢a + m∢b = m∢b + m∢c \qquad or \qquad ∢a ≅ ∢c.$$

50

∢ABE is intersected by rays BD and BC in such a way that ∢ABC ≅ ∢DBE. Prove that ∢ABD ≅ ∢CBE.

Solution: This proof will first employ the reflexive property of angles (An angle is congruent to itself) and then the Subtraction Postulate (Equal quantities subtracted from equal quantities yield equal quantities).

GIVEN: ∢ABC ≅ ∢DBE

PROVE: ∢ABD ≅ ∢CBE

STATEMENTS	REASONS
1. ∢ABC ≅ ∢DBE	1. Given
2. ∢DBC ≅ ∢DBC	2. An angle is congruent to itself.
2. m∢ABC = m∢DBE m∢DBC = m∢DBC	2. Congruent angles have equal measures.
3. m∢ABC−m∢DBC=m∢DBE−m∢DBC	3. Subtraction Postulate.
4. m∢ABD = m∢CBE	4. Substitution Posulate.
5. ∢ABD ≅ ∢CBE	5. Angles of equal measure are congruent.

In parallelogram ABCD one of the diagonals intersects both ∢BAD and ∢BCD. If ∢r ≅ ∢s, m∢BAD = 2m∢r, and m∢BCD = 2m∢s, prove ∢BAD ≅ ∢BCD.

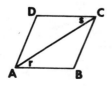

<u>Solution</u>: The Multiplication Postulate states that equal quantities multiplied by equal quantities yield equal quantities. Thus, if m⊀r = m⊀s, then 2m⊀r = 2m⊀s. By substitution it follows that ⊀BAD ≅ ⊀BCD.

GIVEN: ⊀r ≅ ⊀s; m⊀BAD = 2m⊀r; m⊀BCD = 2m⊀s

PROVE: ⊀BAD ≅ ⊀BCD

STATEMENT	REASON
1. ⊀r≅⊀s; m⊀BAD=2m⊀r; m⊀BCD = 2m⊀s	1. Given.
2. m⊀r = m⊀s	2. Congruent angles have equal measure.
3. 2m⊀r = 2m⊀s	3. Multiplication Postulate.
4. m⊀BAD = m⊀BCD	4. Transitive Property of Equality.
5. ⊀BAD ≅ ⊀BCD	5. Angles of equal measure are congruent.

● **PROBLEM** 62

Present a formal proof of the following conditional statement: If ⊀ACB is a right angle and ⊀DAC is complementary to ⊀ACD, then ⊀BCD ≅ ⊀DAC.

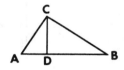

<u>Solution</u>: In this proof, the fact that a right angle measures 90° will be needed. We also need to know that if the measures of two angles sum to 90°, then they are complements. The 2 fundamental rules: (1) the whole is equal to the sum of its parts, and (2) equal quantities may be substituted without changing any equalities, are also needed. The congruence of the two angles will be arrived at by showing that they are both complements of the same angle.

STATEMENT	REASON
1. ⊀ACB is a right angle	1. Given.
2. m⊀ACB = 90°	2. The measure of a right angle is 90.

3. m∡ACB = m∡ACD + m∡BCD	3. The measure of a whole quantity is equal to the sum of the measures of all its parts, (see figure).
4. m∡ACD + m∡BCD = 90°	4. Substitution Postulate.
5. ∡BCD is complementary to ∡ACD	5. If the measure of two angles is 90°, the angles are complementary.
6. ∡DAC is complementary to ∡ACD	6. Given.
7. ∡BCD ≅ ∡DAC	7. If two angles are complements of the same angle, then they are congruent.

● **PROBLEM** 63

Present a formal proof of the following conditional statement: If $\overleftrightarrow{CE}$ bisects ∡ADB, and if $\overleftrightarrow{FDB}$ and $\overleftrightarrow{CDE}$ are straight lines, then ∡a ≅ ∡x. (Refer to the accompanying figure).

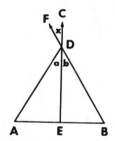

Solution: In this problem, it will be necessary to recognize vertical angles and be knowledgeable of their key properties. Furthermore, we will need the definition of the bisector of an angle.

Vertical angles are two angles which have a common vertex, and whose sides are two pairs of opposite rays. Vertical angles are always congruent.

Lastly, the bisector of any angle divides the angle into two congruent angles.

STATEMENT	REASON
1. $\overleftrightarrow{CE}$ bisects ∡ADB	1. Given.

53

2. ∢a ≅ ∢b	2. A bisector of an angle divides the angle into two congruent angles.
3. $\overleftrightarrow{FDB}$ and $\overleftrightarrow{CDE}$ are straight lines	3. Given.
4. ∢x and ∢b are vertical angles	4. Definition of vertical angles.
5. ∢b ≅ ∢x	5. Vertical angles are congruent.
6. ∢a ≅ ∢x	6. Transitivity property of congruence of angles.

Note that step 3 is essential because without $\overleftrightarrow{FDB}$ and $\overleftrightarrow{CDE}$ being straight lines the definition of vertical angles would not be applicable to ∢x and ∢b.

● **PROBLEM 64**

In the figure, we are given that $\overline{CB} \cong \overline{DB} \cong \overline{DA} \cong \overline{CA}$. Prove ∢CAD ≅ ∢CBD. (See figure.)

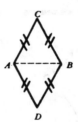

<u>Solution</u>: Our strategy is to note that △ACB and △ADB are isosceles. Thus ∢CAB ≅ ∢CBA, and ∢DAB ≅ ∢DBA. Hence, ∢CAB + ∢DAB ≅ ∢CBA + ∢DBA, or ∢CAD ≅ ∢CBD.

Since we are given that $\overline{CB} \cong \overline{CA}$, and $\overline{DB} \cong \overline{DA}$, we know that △CAB and △DAB are isosceles. Because the base angles of an isosceles triangle are congruent,

    ∢CAB ≅ ∢CBA

and   ∢DAB ≅ ∢DBA.

Using the definition of the congruence of 2 angles, we can rewrite the last 2 equations as

    m∢CAB = m∢CBA

    m∢DAB = m∢DBA.

Adding

$$m\angle CAB + m\angle DAB = m\angle CBA + m\angle DBA \tag{1}$$

Looking at the figure, and using the Angle Addition Postulate, we obtain:

$$m\angle CAD = m\angle CAB + m\angle DAB$$

$$m\angle CBD = m\angle CBA + m\angle DBA \tag{2}$$

Hence, using (2) in (1) yields

$$m\angle CAD = m\angle CBD$$

or $\angle CAD \cong \angle CBD$ as was to be shown.

Given: Collinear points A, B, C, and D. $m\angle 2 = m\angle A$.
$\overline{CE} \cong \overline{AE}$. Prove: $\triangle BCF$ is isosceles.

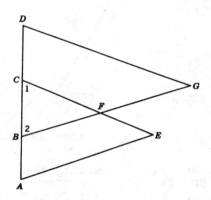

<u>Solution:</u> To prove $\triangle BCF$ is isosceles, it is sufficient to prove $CF = BF$. We can do this by proving $\angle 1 \cong \angle 2$ and then applying the theorem that states that if two angles of a triangle have the same measure, then the sides opposite them are equal in length.

STATEMENTS	REASONS
1. $\overline{CE} \cong \overline{AE}$	1. Given.
2. $\angle 1 \cong \angle A$ or $m\angle 1 = m\angle A$	2. If two sides of a triangle are congruent, then the angles opposite those sides are congruent.
3. $m\angle 2 = m\angle A$	3. Given.

55

4. m∡1 = m∡2	4. If two quantities are equal to the same quantity, then they are equal to each other.
5. CF = BF	5. If a triangle has two angles of equal measure, then the sides opposite those angles are equal in length.
6. ΔBCF is isosceles	6. If two sides of a triangle have equal measure, then the triangle is isosceles.

● **PROBLEM** 66

Given: Isosceles triangles:ΔAFC with base $\overline{AC}$. ΔBGD with base $\overline{BD}$. ΔBHC with base $\overline{BC}$. Prove: ΔAED is isosceles.

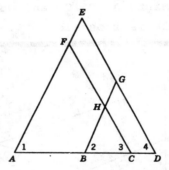

<u>Solution</u>: Isosceles triangles have two congruent sides. Therefore, to prove ΔAED is isosceles, it is sufficient to show $\overline{EA} \cong \overline{ED}$. We will do this by proving the angles opposite $\overline{EA}$ and $\overline{ED}$ are congruent. This latter result will follow directly from the given information about Δ's AFC, BGD and BHC as related by transitivity.

STATEMENTS	REASONS

1. ΔAFC isosceles
   ΔBGD isosceles
   ΔBHC isosceles
	1. Given.

2. $\overline{FA} \cong \overline{FC}$
   $\overline{GB} \cong \overline{GD}$
   $\overline{HB} \cong \overline{HC}$
	2. Definition of an isosceles triangle.

3. ∡1 $\cong$ ∡3
   ∡2 $\cong$ ∡4
   ∡2 $\cong$ ∡3
	3. If two sides of a triangle are congruent, then the angles opposite those sides are congruent.

4. ∡4 ≅ ∡3	4. If two quantities are congruent to the same quantity then they are congruent to each other.
5. ∡1 ≅ ∡4	5. Same as reason 4.
6. $\overline{EA}$ ≅ $\overline{ED}$	6. If a triangle has two congruent angles, then the sides opposite those angles are congruent.
7. △AED is isosceles	7. A triangle with two congruent sides is isosceles.

● **PROBLEM** 67

In the figure shown, △ABC is an isosceles triangle, such that $\overline{BA}$ ≅ $\overline{BC}$. Line segment $\overline{AD}$ bisects ∡BAC and $\overline{CD}$ bisects ∡BCA. Prove that △ADC is an isosceles triangle.

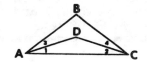

Solution: In order to prove △ADC is isosceles, we must prove that 2 of its sides, $\overline{AD}$ and $\overline{CD}$, are congruent. To prove $\overline{AD}$ ≅ $\overline{CD}$ in △ADC, we have to prove that the angles opposite $\overline{AD}$ and $\overline{CD}$, ∡1 and ∡2, are congruent.

STATEMENT	REASON
1. $\overline{BA}$ ≅ $\overline{BC}$	1. Given.
2. ∡BAC ≅ ∡BCA or m∡BAC = m∡BCA	2. If two sides of a triangle are congruent, then the angles opposite them are congruent.
3. $\overline{AD}$ bisects ∡BAC $\overline{CD}$ bisects ∡BCA	3. Given.
4. m∡1 = ½m∡BAC m∡2 = ½m∡BCA	4. The bisector of an angle divides the angle into two angles whose measures are equal.
5. m∡1 = m∡2	5. Halves of equal quantities are equal.

6. $\angle 1 \cong \angle 2$	6. If the measure of two angles are equal, then the angles are congruent.
7. $\overline{CD} \cong \overline{AD}$	7. If two angles of a triangle are congruent, then the sides opposite these angles are congruent.
8 $\triangle ADC$ is an isosceles triangle	8. If a triangle has two congruent sides, then it is an isosceles triangle.

CONGRUENT LINE SEGMENTS

● **PROBLEM** 68

In the accompanying figure, point B is between points A and C, and point E is between points D and F. Given that $\overline{AB} \cong \overline{DE}$ and $\overline{BC} \cong \overline{EF}$. Prove that $\overline{AC} \cong \overline{DF}$.

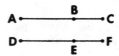

Solution: Two important postulates will be employed in this proof. The Point Betweenness Postulate states that if point Y is between point X and Z, then XY + YZ = XZ. Furthermore, the Postulate states that the converse is also true - that is, if XY + YZ = XZ, then point Y is between point X and Z.

The Addition Postulate states that equal quantities added to equal quantities yield equal quantities. Thus, if a = b and c = d, then a + c = b + d.

GIVEN: Point B is between A and C; point E is between points D and F; $\overline{AB} \cong \overline{DE}$; $\overline{BC} \cong \overline{EF}$

PROVE: $\overline{AC} \cong \overline{DF}$.

STATEMENTS	REASONS
1. (For the given, see above).	1. Given.
2. AB = DE BC = EF	2. Congruent segments have equal lengths.

3. AB + BC = DE + EF	3. Addition Postulate.
4. AC = DF	4. Point Between Postulate.
5. $\overline{AC} \cong \overline{DF}$	5. Segments of equal length are congruent.

● **PROBLEM** 69

In the figure shown, the length of line segment $\overline{AB}$ is equal to two times the length of line segment $\overline{AD}$. Prove that, if segment $\overline{AD}$ is congruent to segment $\overline{DB}$, then the length of $\overline{AB}$ equals twice the length of $\overline{DB}$.

Solution: This problem makes use of the Substitution Postulate, which states that a quantity may be substituted for its equal in any expression.

STATEMENT	REASON
1. AB = 2AD	1. Given.
2. $\overline{AD} \cong \overline{DB}$	2. Given.
3. AD = DB	3. Congruent segments have equal lengths.
4. AB = 2DB	4. The Substitution Postulate (as stated above).

● **PROBLEM** 70

Given isosceles triangle ABC with sides $\overline{AB} \cong \overline{AC}$, and the fact that $\overline{DB} \cong \overline{EC}$, prove that $\overline{AD} \cong \overline{AE}$.

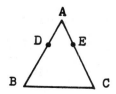

Solution: In this problem, we make use of the Subtraction Postulate as applied to congruency. The Subtraction Postulate states that when equal quantities are subtracted from equal quantities, the differences are equal.

STATEMENT	REASON
1. $\overline{AB} \cong \overline{AC}$ $\overline{DB} \cong \overline{EC}$	1. Given.
2. $AB = AC$ $DB = EC$	2. Congruent segments are equal in length.
3. $AB - DB = AC - EC$	3. Subtraction Postulate.
4. $AD = AE$	4. Substitution Postulate.
5. $\overline{AD} \cong \overline{AE}$	5. Segments of equal lengths are congruent.

● **PROBLEM** 71

Lines have been drawn in rectangle ABCD from point A bisecting $\overline{CD}$, and from point C, bisecting $\overline{AB}$ (as in the figure). In a rectangle, opposite sides are of equal length. Therefore, $\overline{AB} \cong \overline{DC}$. Prove $\overline{AF} \cong \overline{EC}$.

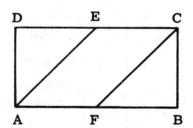

Solution: The Division Postulate states that congruent quantities divided by equal nonzero quantities result in quotients that are equal. This postulate will be employed in the following proof.

GIVEN: $\overline{AB} \cong \overline{DC}$; $\overline{AE}$ bisects $\overline{DC}$; $\overline{CF}$ bisects $\overline{AB}$

PROVE: $\overline{AF} \cong \overline{EC}$

STATEMENT	REASON
1. $\overline{AB} \cong \overline{DC}$; $\overline{AE}$ bisects $\overline{DC}$; $\overline{CF}$ bisects $\overline{AB}$	1. Given.
2. $EC = \frac{1}{2} DC$ $AF = \frac{1}{2} AB$	2. The bisector of a segment divides it into two congru-

	ent segments, each of whose measure is one half the measure of the original segment.
3. EC = AF	3. The Division Postulate.
4. $\overline{AF} \cong \overline{EC}$	4. Line segments of equal length are congruent.

● **PROBLEM** 72

Given isosceles triangle RST in the figure shown. (By definition of isosceles triangles $\overline{RT} \cong \overline{ST}$). Points A and B lie at the midpoint of $\overline{RT}$ and $\overline{ST}$, respectively. Prove that $\overline{RA} \cong \overline{SB}$.

Solution: This solution is best presented as a formal proof.

STATEMENT	REASON
1. $\overline{RT} \cong \overline{ST}$ or RT = ST	1. Given.
2. A is the midpoint of $\overline{RT}$	2. Given.
3. RA = ½ RT	3. The midpoint of a line segment divides the line segment into two equal halves.
4. B is the midpoint of $\overline{ST}$	4. Given.
5. SB = ½ TS	5. The midpoint of a line segment divides the line segment into two equal halves.
6. RA = SB	6. Division Postulate: Halves of equal quantities are equal. Statements 3 and 5.
7. $\overline{RA} \cong \overline{SB}$	7. If two line segments are of equal length, then they are congruent.

CHAPTER 6

SIDE-ANGLE-SIDE POSTULATE

> **Basic Attacks and Strategies for Solving Problems in this Chapter. See pages 62 to 84 for step-by-step solutions to problems.**

Congruence is the most important of all geometric relations. The definition of congruence as it applies to two triangles:

$$\triangle ABC \cong \triangle DEF$$

if and only if

$$\angle A \cong \angle D, \ \angle B \cong \angle E, \ \angle C \cong \angle F, \ \overline{AB} \cong \overline{DE}, \ \overline{BC} \cong \overline{EF}, \text{ and}$$

$$\overline{AC} \cong \overline{DF}$$

It would be difficult to prove that $\triangle ABC \cong \triangle DEF$ if all three of the angle congruence statements and all three of the side congruence statements had to be established. In the following four chapters, we'll see that in some cases, three congruences may be sufficient.

In this chapter we consider the **Side-Angle-Side (SAS)** postulate:

If, in two triangles, two sides and the included angle of one triangle are congruent to two sides and the included angle of the other triangle, then the triangles are congruent.

All the problems in this chapter involve application of this postulate. One convenient way to see that the postulate applies is to mark sides and angles of two triangles to see how the sides and angles are related. Suppose, for instance, in $\triangle ABC$ and $\triangle DEF$, $\overline{AB} \cong \overline{DE}$, $\angle B \cong \angle E$, and $\overline{BC} \cong \overline{EF}$. Then, as shown on the next page, one mark can be made through $\overline{AB}$ and $\overline{DE}$ to show that they are congruent, two marks can be made through $\overline{BC}$ and $\overline{EF}$ to show that they are congruent and small arcs can be made across angles B and E to show that they are congruent:

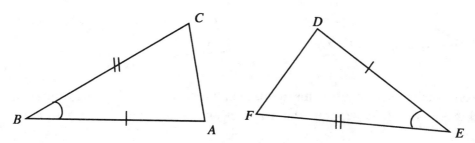

Pictorially, this shows that two sides and the included angle in one triangle are indeed congruent to two sides and the included angle of the other triangle. This marking procedure is highly recommended.

Usually you will only be given explicitly some of the congruences necessary to apply the SAS postulate. In this case, you should examine other information given about the triangles in an effort to establish the remaining congruences.

For example, if the following information is given (see figure below), you will need to establish that $\angle A \cong \angle D$ in order to apply the SAS postulate.

Or, if the following is given (see figure below), you will need to establish that $\overline{AC} \cong \overline{DF}$ to apply the SAS postulate. Note that showing that $\overline{AB} \cong \overline{DE}$ is not sufficient to meet the criteria of the SAS postulate; the angle must be included between the line segments.

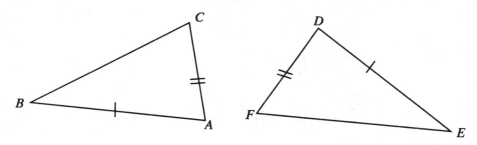

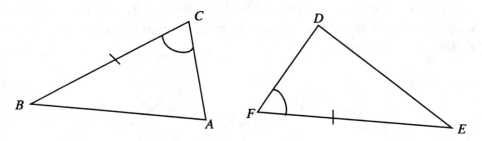

In some instances, the problems in this chapter involve overlapping triangles, and it is desirable to separate the triangles as is illustrated on the bottom of page 63.

Once the congruence of the two triangles has been proven (by the SAS postulate, or postulates discussed in later chapters), it follows that the remaining corresponding sides and angles of the triangles are congruent. The reason commonly given in proofs for this is "CPCTC," or Corresponding Parts of Congruent Triangles are Congruent.

● **PROBLEM** 73

In △ABC, (see figure) assume that $\overline{AC} \cong \overline{BC}$ and $\overline{CD}$ bisects ⊀ACB. Present a formal proof to show that △ACD $\cong$ △BCD.

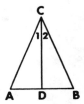

Solution: The SAS postulate states that two triangles are congruent if there exists a congruence between two sides of the triangles and the included angle between these two sides. This postulate will be employed in this problem. In addition, we will use the fact that an angle bisector divides the angle into two congruent angles.

GIVEN: $\overline{AC} \cong \overline{BC}$; $\overline{CD}$ bisects ⊀ACB

PROVE: △ACD $\cong$ △BCD

STATEMENT	REASON
1. $\overline{AC} \cong \overline{BC}$	1. Given.
2. $\overline{CD}$ bisects ⊀ACB	2. Given.
3. ⊀ACD $\cong$ ⊀BCD	3. An angle bisector of an angle divides that angle into two congruent angles.
4. $\overline{CD} \cong \overline{CD}$	4. Reflexivity of congruence.
5. △ACD $\cong$ △BCD	5. S.A.S. $\cong$ S.A.S.

a) In the figure, let AB = AC and BD = CD. Prove that ΔABD ≅ ΔACD. What does this tell one about ⦠1 and ⦠2? About ⦠3 and ⦠4? b) Let AB = AC and ⦠1 ≅ ⦠2. Prove that ΔABD ≅ ΔACD. What can be said about BD and CD? About ⦠3 and ⦠4?

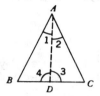

<u>Solution</u>:   a) We prove that ΔABD ≅ ΔACD by the S.A.S. (side angle side) Postulate. We are given that AB = AC. By definition of the congruence of 2 segments, $\overline{AB} ≅ \overline{AC}$. This also means that ΔABC is isosceles. Since the base angles of an isosceles triangle are congruent, ⦠ABD ≅ ⦠ACD. Lastly, the problem gives us BD = CD, which implies $\overline{BD} ≅ \overline{DC}$. Hence, by the S.A.S. Postulate, ΔABD ≅ ΔACD. Since corresponding parts of congruent triangles are congruent, this means that (1) ⦠BAD ≅ ⦠CAD, or (using the notation of the figure), ⦠1 ≅ ⦠2, and (2) ⦠BDA ≅ ⦠CDA, or ⦠4 ≅ ⦠3.

b) We again use the S.A.S. Postulate to show that ΔABD ≅ ΔACD. We are given that AB = AC, which means that $\overline{AB} ≅ \overline{AC}$. Furthermore, ⦠1 ≅ ⦠2. Lastly, by the reflexive property, $\overline{AD} ≅ \overline{AD}$. Hence, by the S.A.S. Postulate, ΔABD ≅ ΔACD. Therefore, $\overline{BD} ≅ \overline{CD}$ (or BD = CD), and ⦠ADB ≅ ⦠ADC (or, using the figure's notation, ⦠3 ≅ ⦠4).

In isosceles triangle ABC, $\overline{CD}$ and $\overline{BE}$ are the medians to $\overline{AB}$ and $\overline{AC}$. Prove that ΔABE ≅ ΔACD.

Separate the Triangles

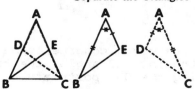

<u>Solution</u>: The given information conveys specifications about one angle and one side of each triangle (the angle is ∢A and the side of each triangle is one of the equal legs of the original isosceles triangle). Rules concerning medians will allow us to conclude something about a second side.

We use the congruence theorem which states that if two sides and the included angle of a triangle are congruent, respectively, to two sides and the included angle of another triangle, then the triangles are congruent (SAS Postulate).

STATEMENT	REASON
1. $\overline{AB} \cong \overline{AC}$ or AB = AC	1. Given by definition of an isosceles triangle.
2. $∢A \cong ∢A$	2. Reflexive property of congruence.
3. $\overline{CD}$ and $\overline{BE}$ are medians	3. Given.
4. AD = ½ AB AE = ½ AC	4. A median divides the side to which it is drawn into to congruent parts, each half the length of the side.
5. AD = AE	5. Halves of equal quantities are equal.
6. $\overline{AD} \cong \overline{AE}$	6. Definition of congruent segments.
7. △ABE $\cong$ △ACD	7. S.A.S. $\cong$ S.A.S.

● **PROBLEM** 76

Given: ∢A $\cong$ ∢B. AF = BE. AC= BC. Prove △AFC $\cong$ △BEC.

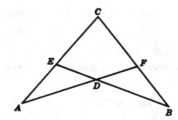

<u>Solution</u>: Since we are given congruence between two corresponding sides and the inclued angle, the triangles can be proved congruent using the SAS postulate.

STATEMENTS	REASONS
1. ∢A ≅ ∢B	1. Given.
2. AF = BE	2. Given.
3. AC = BC	3. Given.
4. ΔAFC ≅ ΔBEC	4. SAS = SAS.

● **PROBLEM** 77

Given: $\overline{AD} \cong \overline{AC}$. $\overline{AB} \cong \overline{AE}$. Prove: ΔADB ≅ ΔACE.

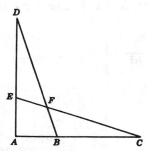

Solution: The fact that the relevant triangles overlap will aid us in completing this proof. We are given congruence between two pairs of corresponding sides. Furthermore the angle included between the sides is shared by the two triangles. By the SAS Postulate, the two triangles are congruent.

STATEMENTS	REASONS
1. $\overline{AD} \cong \overline{AC}$	1. Given.
2. $\overline{AB} \cong \overline{AE}$	2. Given.
3. ∢A ≅ ∢A	3. Reflexive Property of Congruence; a quantity is congruent to itself.
4. ΔADB ≅ ΔACE	4. SAS ≅ SAS; if two sides and the included angle of one triangle are congruent, respectively, to two sides and the included angle of a second triangle, then the triangles are congruent.

● **PROBLEM** 78

Given: $\overline{AB} \cong \overline{AC}$ and $\overline{AQ} \cong \overline{AP}$. Prove ∢BQA ≅ ∢CPA.

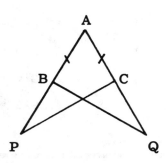

Solution: ⦦BQA and ⦦CPA are corresponding angles of triangles ΔBQA and ΔCPA. We prove ΔBQA ≅ ΔCPA by the SAS Postulate, and ⦦BQA ≅ ⦦CPA will follow by corresponding parts.

STATEMENTS	REASONS
1. $\overline{AB} \cong \overline{AC}$ and $\overline{AQ} \cong \overline{AP}$	1. Given.
2. ⦦A ≅ ⦦A	2. Every angle is congruent to itself.
3. ΔABQ ≅ ΔACP	3. If two sides and the included angle of one triangle are congruent to two sides and included angle of a second triangle, then the two triangles are congruent (SAS Postulate).
4. ⦦BQA ≅ ⦦CPA	4. Corresponding angles of congruent triangles are congruent.

● **PROBLEM** 79

Given: ΔABE, $\overline{AE} \cong \overline{BE}$, $\overline{AC} \cong \overline{BD}$. Prove: ⦦3 ≅ ⦦4.

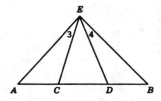

Solution: Angles 3 and 4 are corresponding parts of Δ's EAC
and EBD. We can prove the required congruence, therefore, by
showing ΔEAC ≅ ΔEBD.

From the given information, we have congruence between
two pairs of corresponding sides. We can use the fact that

the angles opposite the congruent sides of a triangle are congruent to show congruence of the corresponding included angles. Hence, by the SAS posulate, we will have proved ΔEAC ≅ ΔEBD.

STATEMENTS	REASONS
1. $\overline{AE} \cong \overline{BE}$	1. Given.
2. ⦦A ≅ ⦦B	2. If two sides of a triangle are congruent, then the angles opposite those sides are congruent.
3. $\overline{AC} \cong \overline{BD}$	3. Given.
4. ΔEAC ≅ ΔEBD	4. SAS ≅ SAS
5. ⦦3 ≅ ⦦4	5. Corresponding parts of congruent triangles are congruent.

● **PROBLEM** 80

In the figure, $\overline{SR}$ and $\overline{ST}$ are straight line segments. $\overline{SX} \cong \overline{SY}$ and $\overline{XR} \cong \overline{YT}$. Prove that ΔRSY ≅ ΔTSX.

<u>Solution</u>: The triangles are overlapping in this problem because they share a common angle.

This proof will employ the S.A.S. Postulate for showing congruence between two triangles.

STATEMENT	REASON
1. $\overline{SX} \cong \overline{SY}$	1. Given.
2. $\overline{XR} \cong \overline{YT}$	2. Given.
3. SX = SY; XR = YT	3. Congruent segments are of equal length.
4. SX + XR = SY + YT or SR = ST	4. Addition Postulate
5. $\overline{SR} \cong \overline{ST}$	5. Segments of equal length are congruent.
6. ⦦S ≅ ⦦S	6. Reflexive Property of Congruence.

7. $\triangle RSY \cong \triangle TSX$　　　|　7. S.A.S. $\cong$ S.A.S.

● **PROBLEM** 81

Given: $\triangle ADE$, $\overline{AE} \cong \overline{DE}$, $\overline{AB} \cong \overline{DC}$. Prove $\angle 1 \cong \angle 2$.

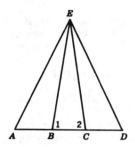

Solution: Angles 1 and 2 are angles of $\triangle BCE$. Since the angles opposite congruent sides of a triangle are congruent, it will be sufficient to prove $\overline{BE} \cong \overline{CE}$ in order to derive the required results.

$\overline{BE}$ and $\overline{CE}$ are corresponding parts of $\triangle$'s ABE and DCE. By proving $\triangle ABE \cong \triangle DCE$, we can derive our desired results.

STATEMENTS	REASONS
1. $\overline{AE} \cong \overline{DE}$	1. Given.
2. m$\angle$D = m$\angle$A	2. If two sides of a triangle are congruent, then the angles opposite those sides are congruent.
3. $\overline{AB} \cong \overline{DC}$	3. Given.
4. $\triangle BAE \cong \triangle CDE$	4. SAS = SAS
5. $\overline{BE} \cong \overline{CE}$	5. Corresponding parts of congruent triangles are equal.
6. $\angle 1 \cong \angle 2$	6. If two sides of a triangle are congruent, then the angles opposite those sides are congruent.

● **PROBLEM** 82

$\overline{AD}$ is a straight line segment. Triangles AEB and DFC are drawn in such a way that $\angle A \cong \angle D$ and $\overline{AE} \cong \overline{DF}$. Additionally, it is given that $\overline{AC} \cong \overline{BD}$. Prove that $\triangle AEB \cong \triangle DFC$.

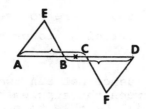

Solution: Since we are given some information about two sides and the included angle, it seems logical to use the postulate which states that if two sides of a triangle and the included angle are congruent to two sides and the included angle of another triangle, then the triangles are congruent.

GIVEN: ∢A ≅ ∢D; $\overline{AE}$ ≅ $\overline{DF}$; $\overline{AC}$ ≅ $\overline{BD}$

Prove: ΔAEB ≅ ΔDFC

STATEMENT	REASON
1. $\overline{AE}$ ≅ $\overline{DF}$	1. Given.
2. ∢A ≅ ∢D	2. Given.
3. $\overline{AD}$ is a straight line segment	3. Given.
4. $\overline{AC}$ ≅ $\overline{DB}$	4. Given.
5. $\overline{BC}$ ≅ $\overline{BC}$	5. Reflexive property of congruence
6. $\overline{AC}$ − $\overline{BC}$ ≅ $\overline{DB}$ − $\overline{BC}$ or $\overline{AB}$ ≅ $\overline{DC}$	6. If congruent segments are subtracted from congruent segments the differences are congruent segments.
7. ΔAEB ≅ ΔDFC	7. S.A.S. ≅ S.A.S.

● **PROBLEM 83**

Given: ΔADE, ∢1 ≅ ∢2, $\overline{AB}$ ≅ $\overline{DC}$, $\overline{BE}$ ≅ $\overline{CE}$. Prove: $\overline{AE}$ ≅ $\overline{DE}$.

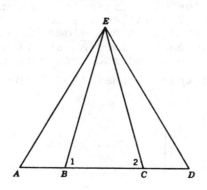

<u>Solution:</u> Segments $\overline{AE}$ and $\overline{DE}$ are corresponding parts of
△'s ABE and DCE. It is sufficient to prove △ABE ≅ △DCE, to
conclude AE ≅ DE.

We are given congruence between two pairs of corresponding sides and between the supplements of the included
angle. Therefore, we will be able to prove the triangles
congruent using the SAS postulate.

<u>STATEMENTS</u>	<u>REASONS</u>
1. ∡1 ≅ ∡2	1. Given.
2. ∡1 and ∡ABE are supplements, ∡2 and ∡DCE are supplements	2. Adjacent angles whose non-common sides form a straight line are supplements.
3. ∡ABE ≅ ∡DCE	3. Supplements of congruent angles are congruent.
4. $\overline{AB}$ ≅ $\overline{DC}$	4. Given.
5. $\overline{BE}$ ≅ $\overline{CE}$	5. Given.
6. △ABE ≅ △DCE	6. SAS ≅ SAS
7. $\overline{AE}$ ≅ $\overline{DE}$	7. Corresponding parts of congruent triangles are congruent.

● **PROBLEM** 84

Given: ∡1 ≅ ∡2; ∡3 ≅ ∡4.   Prove △ADC ≅ △BEC.

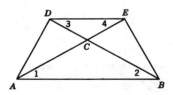

<u>Solution:</u> If two angles of a triangle are congruent, then
the sides opposite the two angles are congruent. Using this
theorem, we can show that in △DEC, sides $\overline{DC}$ and $\overline{EC}$ are
congruent; and that in △ABC, sides $\overline{AC}$ and $\overline{BC}$ are congruent.
Since opposite angles ∡DCA and ∡ECB are congruent, it
follows by the SAS Postulate, that △ADC ≅ △BEC.

<u>STATEMENTS</u>	<u>REASONS</u>
1. ∡1 ≅ ∡2 ∡3 ≅ ∡4	1. Given.

2. $\overline{CB} \cong \overline{CA}$
   $\overline{CE} \cong \overline{CD}$

2. If a triangle has two congruent angles, then the sides opposite those angles are congruent.

3. ∢DCA ≅ ∢ECB

3. Vertical angles are congruent.

4. △ADC ≅ △BEC

4. SAS ≅ SAS

● **PROBLEM** 85

Given: △ABE, ∢1 ≅ ∢2, $\overline{AC} = \overline{BD}$. Prove: ∢A ≅ ∢B.

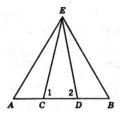

Solution: ∢A and ∢B are corresponding angles of △ACE and △BDE, respectively. Hence, it will suffice to show △ACE ≅ △BDE. We will do this by using the SAS posulate.

We are given congruence of one pair of corresponding sides, $\overline{AC} \cong \overline{BD}$. Since ∢1 ≅ ∢2, their respective supplements will provide a pair of corresponding angles, ∢ACE ≅ ∢BDE. Finally, $\overline{EC} \cong \overline{ED}$, because they are opposite a pair of congruent angles included within one triangle.

STATEMENTS	REASONS
1. ∢1 ≅ ∢2	1. Given.
2. $\overline{EC} \cong \overline{ED}$	2. If a triangle has two congruent angles, then the sides opposite those angles are congruent.
3. ∢1 and ∢ACE are supplements ∢2 and ∢BDE are supplements	3. Angles with a common side whose non-adjacent sides form a straight line are supplements.
4. ∢ACE ≅ ∢BDE	4. Supplements of equal angles are congruent.
5. $\overline{AC} \cong \overline{BD}$	5. Given.
6. △ACE ≅ △BDE	6. SAS = SAS
7. ∢A ≅ ∢B	7. Corresponding parts of congruent triangles are congruent.

We are given quadrilateral LMNP with $\overline{LS}$ and $\overline{MR}$ drawn as shown in the figure. If $\overline{LP} \perp \overline{PN}$, $\overline{MN} \perp \overline{PN}$, $\overline{LP} \cong \overline{MN}$ and $\overline{PR} \cong \overline{NS}$, then prove that $\triangle LPS \cong \triangle MNR$.

Solution: Triangles LPS and MNR overlap in the sense that $\overline{RS}$ is shared by both side PS of $\triangle LPS$ and side RN of $\triangle MNR$. This feature will aid us in proving the congruence required by the problem.

STATEMENT	REASON
1. $\overline{LP} \cong \overline{MN}$	1. Given.
2. $\overline{LP} \perp \overline{PN}$ and $\overline{MN} \perp \overline{PN}$	2. Given.
3. ⟩LPS and ⟩MNR are right angles	3. Perpendicular lines inter-sect and form right angles.
4. ⟩LPS $\cong$ ⟩MNR	4. All right angles are congruent.
5. $\overline{RS} \cong \overline{SR}$	5. Reflexive Property of Congruence.
6. $\overline{PR} \cong \overline{NS}$	6. Given.
7. RS = SR; PR = NS	7. Congruent segments are of equal length.
8. RS + PR = SR + NS or SP = NR	8. Addition Postulate.
9. $\overline{SP} \cong \overline{NR}$	9. Segments of equal length are congruent.
10. $\triangle LPS \cong \triangle MNR$	10. SAS Postulate.

Given: ⟩A $\cong$ ⟩D, $\overline{AE} \cong \overline{DE}$, $\overline{AC} \cong \overline{DB}$. Prove: $\triangle ABE \cong \triangle DCE$.

Solution: Since we are given certain information concerning two corresponding sides and the included angle of $\triangle ABE$ and $\triangle DCE$, the required congruence can be shown using the SAS postulate.

In ΔABE and ΔDCE, we are given $\overline{AE} \cong \overline{DE}$ and $\angle A \cong \angle D$. To find the third necessary congruence, we will apply the reflexive property and the subtraction postulate to the given statement, $\overline{AC} \cong \overline{DB}$.

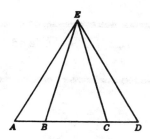

STATEMENT	REASON
1. $\overline{AC} \cong \overline{DB}$	1. Given.
2. $\overline{BC} \cong \overline{BC}$	2. A quantity is congruent to itself (reflexive property).
3. AC = DB BC = BC	3. Congruent segments have equal lengths.
4. AC - BC = DB - BC	4. Subtraction Postulate.
5. AB = DC	5. Substitution Postulate.
6. $\overline{AB} \cong \overline{DC}$	6. Segments of equal length are congruent.
7. $\angle A \cong \angle D$ $\overline{AE} \cong \overline{DE}$	7. Given.
8. ΔABE ≅ ΔDCE	8. The SAS Postulate.

● **PROBLEM** 88

In the figure, D is the midpoint of $\overline{CB}$ and D is the midpoint of $\overline{AE}$. Prove that ΔADC ≅ ΔEDB.

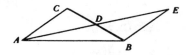

<u>Solution</u>: We shall prove that ΔADC ≅ ΔEDB by using the S.A.S. Postulate.

D is the midpoint of $\overline{CB}$, hence $\overline{CD} \cong \overline{BD}$. Since D is the midpoint of $\overline{AE}$, $\overline{AD} \cong \overline{ED}$. ⊀CDA $\cong$ ⊀BDE, because they are vertical angles. By the S.A.S. Postulate, $\triangle ADC \cong \triangle EDB$.

● **PROBLEM** 89

In the figure, given that $\overline{AB} \cong \overline{AC}$ and $\overline{AD} \cong \overline{AE}$, prove that $\overline{BE} \cong \overline{CD}$.

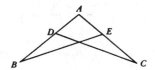

Solution: We will use the SAS Postulate to prove that $\triangle ABE \cong \triangle ACD$. This will then insure that $\overline{BE} \cong \overline{CD}$.

By the given facts, $\overline{AB} \cong \overline{AC}$ and $\overline{AD} \cong \overline{AE}$. Furthermore, ⊀BAE $\cong$ ⊀CAD, since ⊀A is common to both triangles. Using the SAS Postulate, $\triangle ABE \cong \triangle ACD$. By the definition of the congruence of 2 triangles, corresponding parts are congruent, hence $\overline{BE} \cong \overline{CD}$.

● **PROBLEM** 90

In the diagram shown, we are given straight line segment $\overline{AB}$, $\overline{CE} \cong \overline{DF}$, ⊀1 $\cong$ ⊀2 and $\overline{AE} \cong \overline{BF}$. Prove that $\triangle AFD \cong \triangle BEC$.

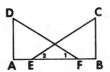

Solution: Triangles $\triangle AFD$ and $\triangle BEC$ are overlapping, and extra care must be taken in analyzing the figure. Other than this, the proof is a straightforward triangle congruence proof. We will use the SAS Postulate.

STATEMENT	REASON
1. $\overline{CE} \cong \overline{DF}$	1. Given.
2. ⊀1 $\cong$ ⊀2	2. Given.
3. $\overline{AE} \cong \overline{BF}$	3. Given.

4. $\overline{EF} \cong \overline{FE}$	4. Reflexive Property of Congruence.
5. AE = BF EF = FE	5. Congruent segments have equal lengths.
6. AE + EF = BF + FE or AF = BE	6. Addition Postulate.
7. $\overline{AF} \cong \overline{BE}$	7. Segments of equal length are congruent.
8. $\triangle AFD \cong \triangle BEC$	8. S.A.S. $\cong$ S.A.S.

● **PROBLEM** 91

Given: $\overline{DC} \cong \overline{AB}$. $\angle CDA \cong \angle BAD$. Prove: $\overline{AC} \cong \overline{BD}$.

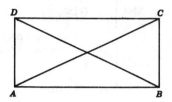

Solution: Noticing that $\overline{AC}$ and $\overline{BD}$ are corresponding parts of overlapping triangles CDA and BAD will assist us in completing this proof. We are given congruence between two corresponding parts and will use the fact that the triangles overlap to gain the third congruence.

STATEMENTS	REASONS
1. $\overline{DC} \cong \overline{AB}$	1. Given.
2. $\angle CDA \cong \angle BAD$	2. Given.
3. $\overline{DA} \cong \overline{DA}$	3. Reflexive Property of Congruence.
4. $\triangle CDA \cong \triangle BAD$	4. SAS $\cong$ SAS.
5. $\overline{AC} \cong \overline{BD}$	5. Corresponding parts of congruent triangles are congruent.

● **PROBLEM** 92

Given isosceles $\triangle ABC$ with $\overline{CA} \cong \overline{CB}$. M is the midpoint of $\overline{AB}$ and points D and E are placed on $\overline{CA}$ and $\overline{CB}$, respectively, so as to make $\overline{AD} \cong \overline{BE}$. Prove that $\overline{MD} \cong \overline{ME}$.

Solution: In this problem, we have to prove $\triangle ADM \cong \triangle BEM$ in order to prove $MD \cong ME$. The triangles can be proven congruent by the S.A.S. $\cong$ S.A.S. method.

The theorem stating that if two sides of a triangle are congruent then the angles opposite them are congruent will be essential to the proof.

STATEMENT	REASON
1. $\overline{AD} \cong \overline{BE}$	1. Given.
2. M is the midpoint of $\overline{AB}$	2. Given.
3. $\overline{AM} \cong \overline{BM}$	3. A midpoint divides a line segment into two congruent parts.
4. $\overline{CA} \cong \overline{CB}$	4. Given.
5. $\angle A \cong \angle B$	5. Base angles of an isosceles triangle are congruent.
6. $\triangle AMD \cong \triangle BEM$	6. S.A.S. $\cong$ S.A.S.
7. $\overline{MD} \cong \overline{ME}$	7. Corresponding parts of congruent triangles are congruent.

● **PROBLEM** 93

In the figure shown, $\overline{CB} \cong \overline{DA}$, $\angle 1 \cong \angle 2$ and $\overleftrightarrow{ABM}$ is a straight line. Prove $\triangle ABC \cong \triangle BAD$.

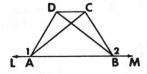

Solution: The congruence proof will be assisted by the fact that the triangles in question overlap and, as such, share a common side.

The S.A.S. $\cong$ S.A.S. Postulate can be used in this example.

STATEMENT	REASON
1. $\overline{AB} \cong \overline{BA}$	1. Reflexive Property of Congruence.
2. $\overleftrightarrow{LM}$ is a straight line	2. Given.
3. $\angle LAB \cong \angle MBA$	3. All straight angles are congruent.
4. $\angle 1 \cong \angle 2$	4. Given.
5. $m\angle LAB = m\angle MBA$ $m\angle 1 = m\angle 2$	5. Congruent angles are of equal measure.
6. $m\angle LAB - m\angle 1 = m\angle MBA - m\angle 2$	6. Equal quantities subtracted from equal quantities yield equal quantities.
7. $m\angle DAB = m\angle CBA$ or $\angle DAB \cong \angle CBA$	7. Substitution Postulate.
8. $\overline{DA} \cong \overline{CB}$	8. Given.
9. $\triangle ABC \cong \triangle BAD$	9. SAS Postulate.

● **PROBLEM** 94

In the figure, $\overline{AB} \cong \overline{AC}$ and $\overline{A'B} \cong \overline{A'C}$. Using the SAS Postulate prove that $\triangle AA'B \cong \triangle AA'C$.

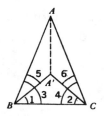

Solution: We shall use the fact that the base angles of an isosceles triangle are congruent to prove that $\angle 5 \cong \angle 6$ (see figure). Using this and the given facts we can then prove that $\triangle AA'B \cong \triangle AA'C$ by the S.A.S. Postulate.

Because $\overline{AB} \cong \overline{AC}$ and $\overline{A'B} \cong \overline{A'C}$, $\triangle ABC$ and $\triangle A'BC$ are both isosceles. Since the base angles of an isosceles triangle are congruent, we obtain (see figure)

$$\angle 1 \cong \angle 2 \tag{1}$$

$$\angle 3 \cong \angle 4 \tag{2}$$

Subtracting (1) from (2) yields

$(\not{4}3 - \not{4}1) \stackrel{\sim}{=} (\not{4}4 - \not{4}2)$ or $\not{4}5 \stackrel{\sim}{=} \not{4}6$.

From the given facts, $\overline{AB} \stackrel{\sim}{=} \overline{AC}$ and $\overline{A'B} \stackrel{\sim}{=} \overline{A'C}$. Hence, by the S.A.S. Postulate, $\triangle AA'B \stackrel{\sim}{=} \triangle AA'C$.

● **PROBLEM** 95

In the accompanying figure, D, E, and F are the midpoints of the sides of $\triangle ABC$. If AB = BC, $\not{4}A \stackrel{\sim}{=} \not{4}C$, prove that DF = EF.

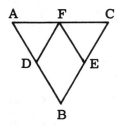

Solution: $\overline{DF}$ and $\overline{EF}$ are corresponding parts of $\triangle ADF$ and $\triangle CEF$. To show DF = EF, we first show $\triangle ADF \stackrel{\sim}{=} \triangle CEF$ by using the SAS Postulate. The three corresponding congruences are (1) $\overline{AF} \stackrel{\sim}{=} \overline{FC}$ (since F is the midpoint of $\overline{AC}$); (2) $\not{4}A \stackrel{\sim}{=} \not{4}C$ (given); and (3) $\overline{AD} \stackrel{\sim}{=} \overline{EC}$ (Since D and E are midpoints, AD = ½ AB and CE = ½ CB. Because $\triangle ABC$ is isosceles, AB = BC. Then it follows that AD = EC.)

GIVEN: D, E, and F are midpoints of the sides of $\triangle ABC$;
AB = BC; $\not{4}A \stackrel{\sim}{=} \not{4}C$

Prove: DF = EF

STATEMENTS	REASONS
1. D, E, and F are midpoints of sides of $\triangle ABC$; AB = BC; $\not{4}A \stackrel{\sim}{=} \not{4}C$	1. Given.
2. AD = ½ AB CE = ½ CB	2. The midpoint of a segment divides the segment into two congruent segments each one-half the length of the original.
3. ½AB = ½CB	3. If a = b and c ≠ 0, then $\frac{a}{c} = \frac{b}{c}$ (Division Property of Equality).

4. AD = CE	4. Transitivity Postulate.
5. AF = FC	5. The midpoint of a segment divides the segment into two congruent segments.
6. △ADF ≅ △CEF	6. The SAS Postulate.
7. $\overline{DF} \cong \overline{FE}$	7. Corresponding parts of congruent triangles are congruent.
8. DF = FE	8. Congruent segments have equal lengths.

● **PROBLEM 96**

Given: $\overline{AC}$ is the perpendicular bisector of $\overline{BD}$ at point O.
Prove: $\overline{DC} \cong \overline{BC}$.

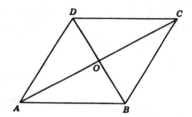

Solution: Segments $\overline{DC}$ and $\overline{BC}$ are corresponding parts of triangles DOC and BOC. We will use the definition of a perpendicular bisector and the reflexive property of congruence to prove △DOC ≅ △BOC, by the SAS postulate.

STATEMENTS	REASONS
1. $\overline{AC}$ bisects $\overline{DB}$ at O	1. Given.
2. $\overline{DO} \cong \overline{BO}$	2. Definition of a bisector.
3. $\overline{AC}$ is perpendicular to $\overline{BD}$	3. Given.
4. ⟩DOC ≅ ⟩BOC	4. If two lines are perpendicular, they meet and form congruent adjacent angles.
5. $\overline{OC} \cong \overline{OC}$	5. Reflexive Property of Congruence
6. △DOC ≅ △BOC	6. SAS ≅ SAS
7. $\overline{DC} \cong \overline{BC}$	7. Corresponding parts of congruent triangles are congruent.

Prove that any point on the perpendicular bisector of a line segment is equidistant from the endpoints of the segment.

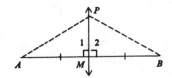

Solution: Let A and B be the endpoints and M the midpoint of the line segment. Let P be any point on the bisector. The distances from P to the endpoints are AP and BP.

AP and BP are corresponding sides of triangles ΔAPM and ΔBPM. We show ΔAPM ≅ ΔBPM by the SAS Postulate and thus AP must equal BP. Since P is an arbitrary point on the bisector, then all points of the perpendicular bisector must be equidistant from the endpoints of the bisected segment.

GIVEN: P is any point on the perpendicular bisector of $\overline{AB}$; the bisector and $\overline{AB}$ intersect at point M.

PROVE: PA = PB

STATEMENTS	REASONS
1. P is any point on the perpendicular bisector of $\overline{AB}$. The bisector and $\overline{AB}$ intersect at M	1. Given.
2. $\overleftrightarrow{PM} \perp \overline{AB}$	2. The perpendicular bisector of a segment is the line perpendicular to the segment at its midpoint.
3. ∡ 1 and ∡ 2 are right angles	3. Perpendicular lines intersect at right angles.
4. ∡ 1 ≅ ∡ 2	4. All right angles are congruent.
5. M is the midpoint of $\overline{AB}$	5. Definition of perpendicular bisector.
6. AM = MB	6. The midpoint of a segment divides the segment into two equal segments.
7. Draw the auxiliary segments $\overline{PA}$ and $\overline{PB}$.	7. Any two distinct points determine a line (Line Postulate).

8. $\overline{PM} \cong \overline{PM}$	8. Any segment is congruent to itself.
9. $\triangle PAM \cong \triangle PBM$	9. The SAS Postulate.
10. PA = PB	10. Corresponding sides of congruent triangles are congruent.

● **PROBLEM** 98

Given: Isosceles $\triangle ABC$, with $\overline{AC} \cong \overline{BC}$; $\overline{CE}$ bisects $\overline{AB}$. Prove: $\triangle ABD$ is isosceles.

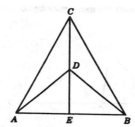

Solution: To prove that $\triangle ABD$ is isosceles, we need to show that two of its sides are congruent, say $\overline{AD} \cong \overline{BD}$. Since these segments are corresponding parts of $\triangle$'s ACD and BCD, by proving $\triangle ACD \cong \triangle BCD$ we will be able to conclude the desired congruence.

We will use the fact that $\overline{CE}$ is the perpendicular bisector of the base of isosceles triangle ABC to derive the necessary congruence conditions. The SAS postulate will be applied.

STATEMENTS	REASONS
1. $\overline{CE}$ bisects $\overline{AB}$	1. Given.
2. $\overline{CE}$ is the perpendicular bisector of $\overline{AB}$	2. The perpendicular bisector of the base of an isosceles triangle passes through the vertex and the midpoint of the base.
3. $\angle ACD \cong \angle BCD$	3. The perpendicular bisector of the base of an isosceles triangle is the bisector of the vertex angle.
4. $\overline{AC} \cong \overline{BC}$	4. Given.
5. $\overline{CD} \cong \overline{CD}$	5. Reflexive property of congruence.

6. ΔACD $\overset{\sim}{=}$ ΔBCD	6. SAS = SAS
7. $\overline{AD} \overset{\sim}{=} \overline{BD}$	7. Corresponding parts of congruent triangles are congruent.
8. ΔABD is isosceles	8. A triangle with two congruent sides is an isosceles triangle.

● **PROBLEM** 99

Prove by an indirect method that the bisector of an angle of a scalene triangle is not perpendicular to the opposite side.

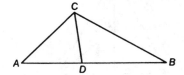

Solution: The procedure of indirect proof consists of assuming that the negative of the desired conclusion is true and then showing that a logical contradiction of some known theorems, definitions, axioms, or hypotheses arises. This then shows that the negative of the desired conclusion is false, or that the desired conclusion is true.

In the figure, scalene triangle ABC is shown with ∤C bisected by $\overline{CD}$.

We want to show $\overline{CD}$ not ⊥ $\overline{AB}$. Therefore, assume $\overline{CD}$ ⊥ $\overline{AB}$. Then, ∤ADC and ∤BDC are right angles, because when perpendicular lines intersect they form right angles. As such, ∤ADC $\overset{\sim}{=}$ ∤BDC. Also, since $\overline{CD}$ bisects angle ACB, ∤ACD $\overset{\sim}{=}$ ∤BCD. By reflexivity, CD $\overset{\sim}{=}$ CD. Since we have two angles and the included side of one triangle congruent to the corresponding parts of another triangle, by s.a.s $\overset{\sim}{=}$ s.a.s., the triangles must be congruent, i.e. ΔACD $\overset{\sim}{=}$ ΔBDC. By corresponding parts of congruent triangles being congruent, we have $\overline{CA} \overset{\sim}{=} \overline{CB}$. This contradicts the hypothesis that ΔABC is scalene, i.e. no two sides congruent. Therefore, we can conclude that our assumption is false and, therefore, $\overline{CD}$ is not ⊥ AB.

Thus, the angle bisector of a scalene triangle is not perpendicular to the opposite side.

When looking at the reflection of an object in the mirror, it seems as if the object is actually behind the mirror. The place where the object seems to be is called the virtual image. In the accompanying figure, light from the object at point O reflects off the mirror $\overline{MN}$ at point B and strikes the eye at point D. Given (by the laws of physics) that the angle of incidence equals the angle of reflection ($\angle$OBM $\stackrel{\sim}{=}$ $\angle$DBN) and that the total distance from the eyes to the virtual image, I, must equal the total distance traveled by the light from the eyes to the object, show that the virtual image is diametrically opposite the object with respect to the mirror.

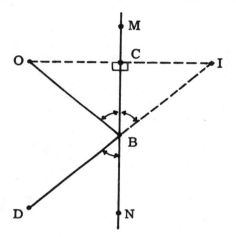

Solution: For two points to be diametrically opposite, they must be on different sides of the mirror. The perpendicular distances from the mirror must be the same, and the perpendiculars from each point to the mirror must intersect the mirror at the same point.

We construct the perpendicular from point O to $\overline{MN}$, $\overline{OC}$. We must show $\overline{IC} \perp \overline{MN}$ and IC = OC. To do this, we show $\triangle OCB \stackrel{\sim}{=} \triangle ICB$.

We are given that light reflects off the mirror at the same angle it hits. Thus, $\angle$OBM $\stackrel{\sim}{=}$ $\angle$DBN. Secondly since the total distance to the virtual image must equal the total distance traveled by the light, it must be true that

(I)   DI = DB + BO.

Since DI = DB + BI, we have

(II)   DB + BI = DB + BO

(III)   BI = BO.

Furthermore, $\overline{BC} \cong \overline{BC}$. By the SAS Postulate, we have $\triangle OCB \cong \triangle ICB$. By corresponding parts, $\angle OCB \cong \angle ICB$ and $OC = CI$. Because $\angle OCB$ is formed by perpendicular $\overline{OC}$, $\angle OCB$ is a right angle and $\angle ICB$ is also a right angle. Thus $\overline{IC}$ is the perpendicular distance from point I to the mirror. Also by corresponding parts, $IC = OC$. Thus the virtual image, I, is the same distance behind the mirror as the real object, O, is in front.

CHAPTER 7

ANGLE-SIDE-ANGLE POSTULATE

> **Basic Attacks and Strategies for Solving Problems in this Chapter. See pages 85 to 94 for step-by-step solutions to problems.**

As we stated in the beginning of Chapter 6, proving that $\triangle ABC \cong \triangle DEF$ by showing that all three pairs of corresponding angles are congruent and all three pairs of corresponding sides are congruent would be quite difficult. The **Angle-Side-Angle (ASA)** postulate which follows makes this job easier:

If, in two triangles, two angles and the included side of one triangle are congruent to two angles and the included side of the other triangle, then the triangles are congruent.

All the problems in this chapter involve application of the ASA postulate. A convenient way to see that this postulate applies is to mark the sides and angles of two triangles to see how the sides and angles are related concerning given congruence statements. Suppose, for example, that in $\triangle ABC$ and $\triangle DEF$, $\angle B \cong \angle E$, $\angle C \cong \angle F$, and $\overline{BC} \cong \overline{EF}$. Then as shown on the following page, one mark can be made through $\overline{BC}$ and one mark through $\overline{EF}$ to show they are congruent. Also, arcs can be drawn across angles B, C, E, and F, one mark can be made through the arcs with centers at B and E to show $\angle B \cong \angle E$, and two marks can be made through arcs with centers at C and F to show that $\angle C \cong \angle F$.

From the diagrams on the next page it is easy to see that two angles and the included side of $\triangle ABC$ are congruent to two angles and the included side of $\triangle DEF$.

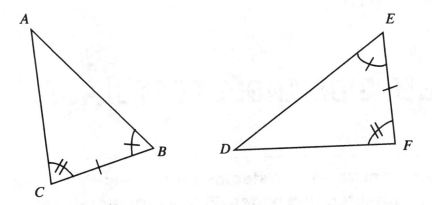

Suppose that the only information given directly in the problem is ∡ B ≅ ∡ E and ∡ C ≅ ∡ F. Then the picture would be as follows:

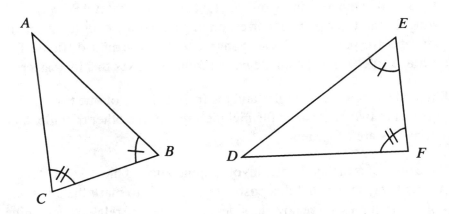

From the picture, it is obvious that $\overline{BC}$ ≅ $\overline{EF}$ must be established before the ASA postulate can be applied. It would thus be necessary to investigate other information given in the problem to determine whether these two sides are in fact congruent.

Next, assume that the only information given directly in the problem is ∡ C ≅ ∡ F and $\overline{BC}$ ≅ $\overline{EF}$. The picture for this situation follows. It illustrates that before the ASA postulate can be applied, ∡ B ≅ ∡ E must be established.

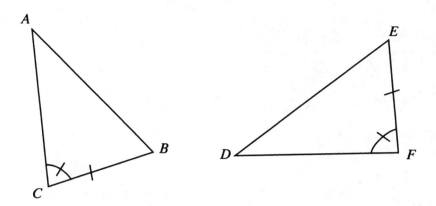

Note that showing ∠ A ≅ ∠ D would not meet the criteria of the ASA postulate; the angles must share the edge as a common side.

Step-by-Step Solutions to
Problems in this Chapter,
"Angle-Side-Angle Postulate"

● **PROBLEM** 101

We are given, as shown in the figure, △ACB and △ADB and are told that straight line $\overleftrightarrow{AE}$ bisects ⫝̸CAD. Also given is ⫝̸CBE ≅ ⫝̸DBE. Prove △ACB ≅ △ADB.

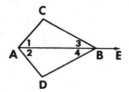

<u>Solution</u>: The most direct way to prove congruence in this case is to make use of the ASA Postulate. This postulate tells us that if two angles and the included side of one triangle are congruent, respectively, to two angles and the included side of another triangle, then the two triangles are congruent.

STATEMENT	REASON
1. $\overleftrightarrow{AE}$ bisects ⫝̸CAD	1. Given.
2. ⫝̸1 ≅ ⫝̸2	2. An angle bisector divides the angle into two congruent angles.
3. $\overline{AB}$ ≅ $\overline{AB}$	3. Reflexivity of congruence.
4. $\overleftrightarrow{AE}$ is a straight line	4. Given.
5. ⫝̸CBE ≅ ⫝̸DBE	5. Given.
6. ⫝̸3 is supplementary to ⫝̸CBE and ⫝̸4 is supplementary to ⫝̸DBE	6. If the non-common sides of two adjacent angles lie on a straight line, then the angles are supplementary.

| 7. $\angle 3 \stackrel{\sim}{=} \angle 4$ | 7. Supplements of congruent angles are congruent. |
| 8. $\triangle ACB \stackrel{\sim}{=} \triangle ADB$ | 8. A.S.A. Postulate. |

● **PROBLEM** 102

Let $\triangle ABC$ be a triangle such that $\angle B \stackrel{\sim}{=} \angle C$. Use the ASA Theorem to prove that AB = AC. (Don't use the Isosceles Triangle Theorem.)

Solution: By proving that $\triangle ABC \stackrel{\sim}{=} \triangle ACB$, we will be able to show that $\overline{AB} \stackrel{\sim}{=} \overline{AC} \Rightarrow AB = AC$.

First, $\angle B \stackrel{\sim}{=} \angle C$ by the given facts. Similarly, $\angle C \stackrel{\sim}{=} \angle B$. Lastly, $\overline{BC} \stackrel{\sim}{=} \overline{CB}$, since this segment is common to both triangles. By the ASA Postulate, $\triangle ABC \stackrel{\sim}{=} \triangle ACB$, hence $\overline{AB} \stackrel{\sim}{=} \overline{AC}$ and AB = AC.

● **PROBLEM** 103

Given: $\overline{AE}$ and $\overline{BD}$ are straight lines intersecting at C. $\overline{BC} \stackrel{\sim}{=} \overline{DC}$; $\angle B \stackrel{\sim}{=} \angle D$. Prove: $\triangle ABC \stackrel{\sim}{=} \triangle EDC$.

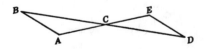

Solution: Since we are given congruence between one pair of corresponding sides and one pair of angles adjacent to these sides, we can use the ASA or SAS postulate to complete the proof. However, because the other pair of adjacent angles are vertical angles and, hence, congruent we will apply the ASA postulate to arrive $\triangle ABC \stackrel{\sim}{=} \triangle EDC$.

STATEMENTS	REASONS
1. $\overline{BC} \stackrel{\sim}{=} \overline{DC}$	1. Given.
2. $\angle B \stackrel{\sim}{=} \angle D$	2. Given.

3. $\overline{BD}$ and $\overline{AE}$ are straight lines	3. Given.
4. ⊀BCA and ⊀DCE are vertical angles	4. Definition of vertical angles.
5. ⊀BCA ≅ ⊀ECD	5. Vertical angles are congruent.
6. △ABC ≅ △EDC	6. ASA ≅ ASA.

Given: △ABC. DC = EC. ⊀1 = ⊀2. Prove: $\overline{AC}$ ≅ $\overline{BC}$.

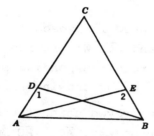

Solution: The segments in question, $\overline{AC}$ and $\overline{BC}$, are corresponding parts of overlapping △'s CDB and CEA. The proof is set up using the ASA postulate to show △CDB ≅△CEA.

STATEMENTS	REASONS
1. ⊀1 = ⊀2	1. Given.
2. ⊀1 and ⊀CDB are supplements ⊀2 and ⊀CEA are supplements	2. Adjacent angles whose non-common sides form a straight line are supplements.
3. ⊀CDB = ⊀CEA	3. Supplements of equal angles are equal.
4. ⊀C = ⊀C	4. Reflexive law.
5. $\overline{DC}$ ≅ $\overline{EC}$	5. Given.
6. △CDB ≅ △CEA	6. ASA ≅ ASA.
7. $\overline{AC}$ ≅ $\overline{BC}$	7. Corresponding parts of congruent triangles are congruent.

87

Show that the angle bisectors of the base angles of an isosceles triangle are congruent.

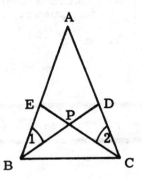

Solution: In the figure, ΔABC is isosceles with base angles ≮B and ≮C bisected by $\overline{BD}$ and $\overline{CE}$. We are required to show $\overline{BD} \cong \overline{EC}$. Since $\overline{BD}$ and $\overline{EC}$ are corresponding parts of ΔABD and ΔACE, we can prove $\overline{BD} \cong \overline{EC}$ by showing ΔABD $\cong$ ΔACE.

This condition can be proved using the ASA Postulate: ≮A $\cong$ ≮A by the Reflexive Property. $\overline{AB} \cong \overline{AC}$ because the sides of an isosceles triangle opposite the base angles are congruent.

As the final condition for congruence, note that because base angles are congruent, ≮ABC $\cong$ ≮ACB. We are given that $\overline{BD}$ and $\overline{EC}$ are angle bisectors. Hence, m≮ABD = ½ m≮ABC and m≮ACE = ½ m≮ACB. Since ≮ABC $\cong$ ≮ACB, then ½ m≮ABC = ½ m≮ACB, and thus m≮ABD = m≮ACE or ≮ABD $\cong$ ≮ACE. Thus, we have an angle (≮A), side ($\overline{AB}$ and $\overline{AC}$), angle (≮ABD and ≮ACE) correspondence between two triangles, and, as such, ΔABD $\cong$ ΔACE. Because corresponding parts of congruent triangles are congruent, $\overline{BD} \cong \overline{EC}$.

Thus, the angle bisectors of the base angles of an isosceles triangle are congruent.

If ≮3 $\cong$ ≮4 and $\overline{QM}$ bisects ≮PQR, prove that M is the midpoint of $\overline{PR}$.

Solution: To show that M is the midpoint of $\overline{PR}$, we prove that M divides $\overline{PR}$ into two congruent segments - that is, $\overline{PM} \cong \overline{MR}$. If we prove ΔPQM $\cong$ ΔRQM, $\overline{PM} \cong \overline{MR}$ follows by corresponding parts.

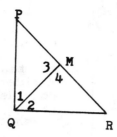

We prove $\triangle PQM \cong \triangle RQM$ by the ASA Postulate: (1) $\angle 1 \cong \angle 2$ because $\overline{QM}$ bisects $\angle PQR$; (2) $\overline{QM} \cong \overline{QM}$; and (3) $\angle 3 \cong \angle 4$ is given. We present our results in the formal, two column proof format..

GIVEN: $\angle 3 \cong \angle 4$; $\overline{QM}$ bisects $\angle PQR$.

PROVE: M is the midpoint $\overline{PR}$.

STATEMENTS	REASONS
1. $\angle 3 \cong \angle 4$; $\overline{QM}$ bisects $\angle PQR$	1. Given.
2. $\angle 1 \cong \angle 2$	2. The angle bisector of an angle divides the angle into two congruent angles.
3. $\overline{QM} \cong \overline{QM}$	3. Every segment is congruent to itself.
4. $\triangle PQM \cong \triangle RQM$	4. ASA Postulate.
5. $\overline{PM} \cong \overline{RM}$	5. Corresponding sides of congruent triangles are congruent.
6. M is the midpoint of PR	6. Definition of a midpoint.

● **PROBLEM** 107

Prove that in $\triangle ABC$, as shown in the figure, if $\overline{BD}$ bisects $\angle ABC$ and $\overline{BD} \perp \overline{AC}$, then $\overline{BD}$ bisects $\overline{AC}$.

<u>Solution</u>: To prove $\overline{BD}$ bisects $\overline{AC}$, it is sufficient to prove $\overline{AD} \cong \overline{CD}$. This can be done by proving $\triangle ABD \cong \triangle CDB$. Since information is given about two angles of each triangle, it seems logical to employ the A.S.A. $\cong$ A.S.A. method of proving congruence of triangles.

STATEMENT	REASON
1. $\overline{BD}$ bisects ⊀ABC	1. Given.
2. <3 ≅ <4	2. The bisector of an angle divides that angle into two congruent angles.
3. $\overline{BD} \perp \overline{AC}$	3. Given.
4. ⊀1 and ⊀2 are right angles	4. Perpendicular lines intersect to form right angles.
5. ⊀1 ≅ ⊀2	5. If two angles are right angles, they are congruent.
6. $\overline{BD} ≅ \overline{BD}$	6. Reflexive property of congruence.
7. ΔABD ≅ ΔCBD	7. A.S.A. ≅ A.S.A.
8. $\overline{AD} ≅ \overline{CD}$	8. Corresponding parts of congruent triangles are congruent.

● **PROBLEM** 108

In the accompanying figure, line segments $\overline{CD}$ and $\overline{AB}$ intersect at point E in such a way that $\overline{BA}$ bisects $\overline{CD}$. The perpendiculars from points A and B to $\overline{CD}$ have been drawn.

Prove, formally, that ΔACE ≅ ΔBDE.

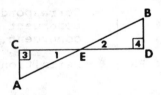

Solution: This proof will involve making use of the ASA Postulate. This postulate tells us that if two angles and the included side of one triangle are congruent, respectively, to two angles and the included side of another triangle, then the two triangles are congruent. In short notation,

this is written as A.S.A. ≅ A.S.A. or the ASA Postulate.

In this problem, to show congruence between the respective parts, we will employ rules concerning congruence between halves of bisected line segments, opposite vertical angles, and right angles.

STATEMENT	REASON
1. $\overline{CD}$ and $\overline{AB}$ are straight line segments intersecting at E	1. Given.
2. ⊀1 and ⊀2 are vertical angles	2. Definition of vertical angles.

3. $\angle 1 \cong \angle 2$	3. Vertical angles are congruent.
4. $\overline{BA}$ bisects $\overline{CD}$	4. Given
5. $\overline{CE} \cong \overline{ED}$	5. A bisector divides a line segment into two congruent parts.
6. $\overline{AC} \perp \overline{CD}$ and $\overline{BD} \perp \overline{CD}$	6. Given.
7. $\angle 3$ and $\angle 4$ are right angles	7. Perpendicular lines intersect to form right angles.
8. $\angle 3 \cong \angle 4$	8. Right angles are congruent.
9. $\triangle ACE \cong \triangle BDE$	9. ASA Postulate.

● **PROBLEM** 109

Given: $\angle 1 \cong \angle 2$; $\angle 3 \cong \angle 4$. Prove: RM = RN.

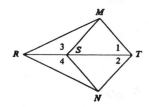

Solution: In most proofs, the necessary congruences are not always given. Frequently, to show two triangles congruent, we have to prove the congruence of corresponding parts by proving another set of triangles congruent. Here, we must show RM = RN or $\overline{RM} \cong \overline{RN}$. $\overline{RM}$ and $\overline{RN}$ are corresponding sides of $\triangle RMS$ and $\triangle RNS$. In these triangles, we have that $\angle 3 \cong \angle 4$ and $\overline{RS} \cong \overline{RS}$. To prove congruence, we must also show $\overline{MS} \cong \overline{NS}$. We can show $\overline{MS} \cong \overline{NS}$ by proving $\triangle MST \cong \triangle NST$. Thus we (1) prove $\triangle MST \cong \triangle NST$ by the ASA Postulate; (2) show $\overline{MS} \cong \overline{NS}$ by corresponding parts; and (3) show $\triangle RMS \cong \triangle RNS$ by the SAS Postulate.

STATEMENTS	REASONS
1. $\angle 1 \cong \angle 2$; $\angle 3 \cong \angle 4$	1. Given.
2. $\angle MST$ and $\angle 3$ are supplementary $\angle NST$ and $\angle 4$ are supplementary	2. Two angles that form a linear pair are supplementary.
3. $\angle MST \cong \angle NST$	3. Angles supplementary to congruent angles are congruent.

4. $\overline{ST} \cong \overline{ST}$	4. A segment is congruent to itself.
5. $\triangle SMT \cong \triangle SNT$	5. The ASA Postulate.
6. $\overline{MS} \cong \overline{NS}$	6. Corresponding sides of congruent triangles are congruent.
7. $\overline{RS} \cong \overline{RS}$	7. A segment is congruent to itself.
8. $\triangle RMS \cong \triangle RNS$	8. The SAS Postulate.
9. $\overline{RM} \cong \overline{RN}$	9. Corresponding sides of congruent triangles are congruent.
10. RM = RN	10. Congruent segments are of the same length.

● **PROBLEM** 110

Given: $\overline{AB} \cong \overline{AC}$; $\overline{BE} \cong \overline{CE}$. Prove: $\overline{DE} \cong \overline{FE}$.

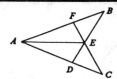

Solution: In this proof of congruence, we locate the congruent triangles of which $\overline{FE}$ and $\overline{DE}$ are corresponding parts, △BEF and △CED. It is already known that $\overline{BE} \cong \overline{CE}$, but no other information is given about these two triangles. ∢ABD can be proven congruent to ∢ACE by proving △AEC and △AEB congruent. This can be done by using the SSS Postulate. ∢BEF and ∢CED are congruent because they are vertical angles. △BEF and △CED are, thus, congruent by the ASA Postulate.

STATEMENTS	REASONS
1. $\overline{AB} \cong \overline{AC}$; $\overline{BE} \cong \overline{CE}$	1. Given.
2. $\overline{AE} \cong \overline{AE}$	2. A segment is congruent to itself.
3. $\triangle ABE \cong \triangle ACE$	3. The SSS Postulate.
4. ∢ABE $\cong$ ∢ACE	4. Corresponding parts of congruent triangles are congruent.
5. ∢CED $\cong$ ∢BEF	5. Vertical angles are congruent.
6. $\triangle BEF \cong \triangle CED$	6. The ASA Postulate.
7. $\overline{DE} \cong \overline{FE}$	7. Corresponding parts of congruent triangles are congruent.

Given: ∆ABC, $\overline{AC} \stackrel{\sim}{=} \overline{BC}$, ⦨1 $\stackrel{\sim}{=}$ ⦨2. Prove: ⦨3 $\stackrel{\sim}{=}$ ⦨4.

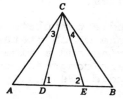

Solution: Angles 3 and 4, which are to be proved congruent, are corresponding parts of ∆'s ADC and BEC.

We prove ∆ADC $\stackrel{\sim}{=}$ ∆BEC by the AAS Postulate.

STATEMENTS	REASONS
1. ⦨1 $\stackrel{\sim}{=}$ ⦨2	1. Given.
2. ⦨ADC is supplementary to ⦨1; ⦨BEC supplementary to ⦨2	2. Supplementary angles are angles whose non-adjacent sides form a straight line.
3. ⦨ADC $\stackrel{\sim}{=}$ ⦨BEC	3. Two angles supplementary to congruent angles are congruent.
4. $\overline{AC} \stackrel{\sim}{=} \overline{BC}$	4. Given.
5. ⦨A $\stackrel{\sim}{=}$ ⦨B	5. If two sides of a triangle are congruent, the angles opposite those sides are congruent.
6. ∆ADC $\stackrel{\sim}{=}$ ∆BEC	6. AAS $\stackrel{\sim}{=}$ AAS.
7. ⦨3 $\stackrel{\sim}{=}$ ⦨4	7. Corresponding parts of congruent triangles are congruent.

● **PROBLEM** 112

Prove that the altitudes drawn to the legs of an isosceles triangle are congruent.

Solution: The accompanying figure shows an isosceles triangle ABC with $\overline{BA} \stackrel{\sim}{=} \overline{BC}$, and altitudes $\overline{CD}$ and $\overline{AE}$. By the definition of altitudes, $\overline{CD} \perp \overline{AB}$ and $\overline{AE} \perp \overline{BC}$.

We must prove that $\overline{CD} \cong \overline{AE}$. This can be done by proving $\triangle AEC \cong \triangle CDA$ and employing the corresponding parts rule of congruent triangles.

The congruent triangle postulate that can be best used in this problem is the one which states that two triangles are congruent if two angles and a side opposite one of the angles in one triangle are congruent to the corresponding parts of the other triangle. We shall refer to this rule as A.A.S. Postulate.

STATEMENT	REASON
1. In $\triangle ABC$, $\overline{BA} \cong \overline{BC}$	1. Given.
2. $\angle BAC \cong \angle BCA$	2. If two sides of a triangle are congruent, the angles opposite these sides are congruent.
3. $\overline{AE} \perp \overline{BC}$, $\overline{CD} \perp \overline{BA}$	3. Given.
4. $\angle CDA$ and $\angle AEC$ are right angles	4. When two perpendicular lines intersect, they form right angles.
5. $\angle CDA \cong \angle AEC$	5. All right angles are congruent.
6. $\overline{AC} \cong \overline{AC}$	6. Reflexive property of congruence.
7. $\triangle ADC \cong \triangle CEA$	7. A.A.S. $\cong$ A.A.S.
8. $\overline{CD} \cong \overline{AE}$	8. Corresponding parts of congruent triangles are congruent.

CHAPTER 8

SIDE-SIDE-SIDE POSTULATE

> **Basic Attacks and Strategies for Solving Problems in this Chapter. See pages 95 to 106 for step-by-step solutions to problems.**

Another tool for proving the congruence of triangles is the **Side-Side-Side** (SSS) postulate:

> If, in two triangles, the three sides of one triangle are congruent to the corresponding three sides of the other triangle, then the triangles are congruent.

The problems in this chapter involve the use of this important postulate.

A convenient way to see that this postulate applies is to mark the sides of the two triangles. Suppose for example, $\overline{AB} \cong \overline{DE}$, $\overline{BC} \cong \overline{EF}$ and $\overline{AC} \cong \overline{DF}$. Then, as shown below, one mark can be made on $\overline{AB}$ and $\overline{DE}$, two marks on $\overline{BC}$ and $\overline{EF}$, and three marks on $\overline{AC}$ and $\overline{DF}$.

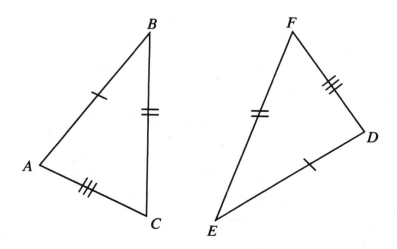

This marking indicates that the SSS postulate can be applied to $\triangle ABC$ and $\triangle DEF$.

Occasionally the information directly given in a problem involves the congruence of only two sides, such as $\overline{BC} \cong \overline{EF}$ and $\overline{AC} \cong \overline{DF}$. Then the triangles would be marked as shown below.

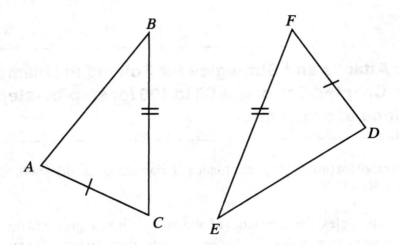

It is obvious that in order to apply SSS, $\overline{AB} \cong \overline{DE}$ must be established. It will be necessary to examine other information given in the problem in an effort to prove that these segments are actually congruent.

If a median is drawn to the base of an isosceles triangle, prove that the median divides the triangle into two congruent triangles.

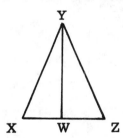

Solution: Congruence of two triangles can be proved if it can be shown that the three sides of one triangle are congruent, respectively, to three sides of another triangle. Notationally, this is written as S.S.S. $\cong$ S.S.S.

It is necessary to know that a median drawn in a triangle divides the side to which it is drawn into two congruent parts.

STATEMENT	REASON
1. In isosceles $\triangle$XYZ, $\overline{YX} \cong \overline{YZ}$	1. Given in definition of isosceles triangle.
2. $\overline{YW}$ is the median to base $\overline{XZ}$	2. Given.
3. $\overline{XW} \cong \overline{ZW}$	3. A median in a triangle divides the side to which it is drawn into two congruent parts.
4. $\overline{YW} \cong \overline{YW}$	4. Reflexivity of congruence.
5. $\triangle$XYW $\cong$ $\triangle$ZYW	5. S.S.S. $\cong$ S.S.S.

Prove: the median drawn to the base of an isosceles triangle bisects the vertex angle.

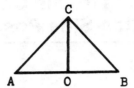

Solution: Draw isosceles △ABC with median $\overline{CO}$, as in the figure shown.

We are asked to prove ⦨ACO ≅ ⦨BCO. By proving △ACO ≅ △BCO, we can derive the desired result since the angles are corresponding angles of the two triangles.

Given: isosceles △ABC; median $\overline{CO}$

Prove: $\overline{CO}$ bisects ⦨C

STATEMENTS	REASONS
1. $\overline{CO}$ is a median	1. Given.
2. $\overline{AO} \cong \overline{BO}$	2. Definition of median.
3. △ABC is isosceles	3. Given.
4. $\overline{AC} \cong \overline{BC}$	4. Definition of an isosceles triangle.
5. $\overline{CO} \cong \overline{CO}$	5. Reflexive property.
6. △AOC ≅ △BOC	6. SSS ≅ SSS
7. ⦨ACO ≅ ⦨BCO	7. Corresponding parts of congruent triangles are congruent.
8. $\overline{CO}$ bisects ⦨C	8. A segment that divides an angle into two angles that have equal measure bisects that angle.

Prove the median to the base of an isosceles triangle is perpendicular to the base. (In the figure, △ ABC is isosceles, such that $\overline{CA} \cong \overline{CB}$, and $\overline{CM}$ is the median to base $\overline{AB}$. Prove $\overline{CM} \perp \overline{AB}$.)

Solution: Two methods can be used to solve this problem. The first method will utilize the theorem which states that two lines intersecting to form two congruent adjacent angles are perpendicular. The second method will employ the fact that two points each equidistant from the endpoints of a line segment determine the perpendicular bisector of the line segment.

Method 1: This method involves proving ∆CMA ≃ ∆CMB, and by coresponding parts, showing that ⊀1 ≃ ⊀2. Since two adjacent angles are congruent, CM ⊥ AB. To prove ∆CMA ≃ ∆CMB, use the S.S.S. ≃ S.S.S. method.

STATEMENT	REASON
1. $\overline{CA} \cong \overline{CB}$	1. Given.
2. $\overline{CM}$ is the median to base $\overline{AB}$	2. Given.
3. $\overline{AM} \cong \overline{BM}$	3. A median in a triangle divides the side to which it is drawn into two congruent parts.
4. $\overline{CM} \cong \overline{CM}$	4. Reflexive property of congruence.
5. ∆CMA ≃ ∆CMB	5. S.S.S. ≃ S.S.S.
6. ⊀CMA ≃ ⊀CMB	6. Corresponding parts of congruent triangles are congruent.
7. ⊀CMA and ⊀CMB are adjacent angles	7. Two angles are adjacent if they have a common vertex and a common side but do not have common interior points.
8. $\overline{CM} \perp \overline{AB}$	8. Two line segments are perpendicular if they intersect and form congruent adjacent angles.

Method 2: This method involves proving that point C and point M are both equidistant from the endpoints, A and B, of $\overline{AB}$. By the second theorem mentioned above, this will prove that $\overline{CM}$ is perpendicular to $\overline{AB}$.

STATEMENT	REASON
1. $\overline{CA} \cong \overline{CB}$ or C is equidistant from A and B	1. Given.
2. $\overline{CM}$ is a median	2. Given

3. $\overline{MA} \stackrel{\sim}{=} \overline{MB}$ or M is equidistant from A and B

4. $\overline{CM} \perp \overline{AB}$

3. A median in a triangle
divides the side to which it
is drawn into two congruent
parts.

4. Two points, each equidistant
from the endpoints of a line
segment, determine the perpendicular bisector of the line
segment.

● **PROBLEM** 116

Let △ ABC be an equilateral triangle and let D be the midpoint of $\overline{AB}$. In △ DCB, what are the measures of ⊀ BDC, ⊀ DCB, and ⊀ DBC? If BC = 2, what does DB equal?

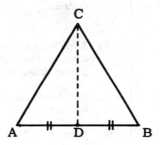

Solution: In order to find the measures of the given angles, we first prove that △ ADC $\stackrel{\sim}{=}$ △ BDC by the S.S.S. (side side side) Postulate. Being that △ ABC is equilateral, we will then have enough information to calculate m ⊀ BDC, m ⊀ DCB, and m ⊀ DBC.

$\overline{AD} \cong \overline{DB}$ due to the fact that D is the midpoint of $\overline{AB}$. $\overline{DC}$ is common to △ ADC and △ BDC, i.e. $\overline{DC} \stackrel{\sim}{=} \overline{DC}$. $\overline{AC} \cong \overline{BC}$ because △ ABC is equilateral (all sides congruent). By the S.S.S. Postulate, △ ADC $\stackrel{\sim}{=}$ △ BDC. From this, we conclude, by corresponding parts, that

$$⊀ ACD \stackrel{\sim}{=} ⊀ BCD$$
$$⊀ ADC \stackrel{\sim}{=} ⊀ BDC$$

or

$$m ⊀ ACD = m ⊀ BCD$$
$$m ⊀ ADC = m ⊀ BDC. \qquad (1)$$

But,

$$m ⊀ ACD + m ⊀ BCD = 60°$$
$$m ⊀ ADC + m ⊀ BDC = 180° \qquad (2)$$

The first equation in (2) is a result of the fact that m ⊀ ACB = 60°

(because △ ABC is equilateral). The second equation in (2)   comes from the fact that ⊁ ADC and ⊁ BDC form a linear pair, and are therefore supplementary angles.

Substituting equations (1) in equations (2),

$$2m \angle BCD = 60°$$
$$2m \angle BDC = 180°$$

or

$$m \angle BCD = 30°$$
$$m \angle BDC = 90° .$$

The m ⊁ DBC = 60° since △ ABC is equilateral.

If BC = 2, AB = 2 because △ ABC is equilateral. But, if AB = 2, DB = 1 because D is the midpoint of $\overline{AB}$, which implies that DB = ½ AB.

● **PROBLEM** 117

Let ABC be an equilateral triangle. Let D be the midpoint of $\overline{AB}$. Prove that △DCB $\overset{\sim}{=}$ △DCA. What kind of triangle is △DCB? What can be said about ⊁ACD and ⊁BCD? (See figure.)

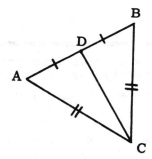

Solution: We use the S.S.S. Postulate to prove that △DCB $\overset{\sim}{=}$ △DCA. We know that D is the midpoint of $\overline{AB}$. Hence, $\overline{DA} \overset{\sim}{=} \overline{DB}$. Also, ABC is an equilateral triangle, thus $\overline{AC} \overset{\sim}{=} \overline{BC}$. Lastly, $\overline{DC} \overset{\sim}{=} \overline{DC}$ by the reflexive property. By the S.S.S. Postulate △DCB $\overset{\sim}{=}$ △DCA.

Since △DCB $\overset{\sim}{=}$ △DCA,

then   m⊁ADC $\overset{\sim}{=}$ m⊁BDC.   (1)

But, from the figure

m⊁ADC + m⊁BDC = 180°.   (2)

Using (1) in (2) yields

m∢ADC = m∢BDC = 90°.

Hence, ΔDCB is a right triangle.

Again, because ΔDCB ≅ ΔDCA,

∢ACD ≅ ∢BCD   (3)

As shown above,

m∢ADC = 90°.   (4)

Since ΔABC is equilateral,

m∢DAC = 60°.   (5)

By the angle sum theorem,

m∢DAC + m∢ACD + m∢ADC = 180°.   (6)

Inserting (4) and (5) in (6) gives

m∢ACD = 30°.   (7)

Hence, using (7) in (3)

m∢ACD = m∢BCD = 30°.

● **PROBLEM** 118

In triangle ABC, lines are drawn from points B and A to sides $\overline{AC}$ and $\overline{BC}$, respectively. As shown in the figure, two smaller triangles are formed. It is given that $\overline{CA} ≅ \overline{CB}$, $\overline{CE} ≅ \overline{CD}$, and $\overline{BE} ≅ \overline{AD}$.

    (a) Prove that ∢EAB ≅ ∢DBA; (b) Find the measure of ∢EAB, if it is represented by 5x - 8 and the measure of ∢DBA equals 3x + 12.

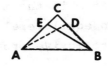

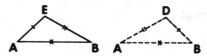

Separate the Triangles

Solution: (a) By proving that overlapping triangles EAB and DBA are congruent, and employing the theorem which

100

states that corresponding parts of congruent triangles are congruent, we can show $\angle EAB \cong \angle DBA$. Since most of the given information concerns sides of the triangles, the logical plan of attack is to prove congruence through the S.S.S. $\cong$ S.S.S. method.

STATEMENT	REASON
1. $\overline{BE} \cong \overline{AD}$	1. Given.
2. $\overline{CA} \cong \overline{CB}$; $\overline{CE} \cong \overline{CD}$	2. Given.
3. CA= CB; CE = CD	3. Congruent segments are of equal measure.
4. CA- CE = CB - CD	4. Subtraction of equal quantities from equal quantities yields equals.
5. EA = DB	5. Substitution Postulate. Point Betweenness Postulate.
6. $\overline{EA} \cong \overline{DB}$	6. Segments of equal length are congruent.
7. $\overline{AB} \cong \overline{AB}$	7. Reflexive Property of Congruence.
8. $\triangle EAB \cong \triangle DBA$	8. S.S.S. $\cong$ S.S.S.
9. $\angle EAB \cong \angle DBA$	9. Corresponding parts of congruent triangles are congruent.

(b) We proved in part (a) that $\angle EAB \cong \angle DBA$. Thus, the measures of the two angles are equal, m$\angle$EAB = m$\angle$DBA. We have an algebraic representation for each measure and, by setting them equal to each other, we can solve for the unknown and substitute to determine the measure of $\angle$EAB.

m$\angle$EAB = m$\angle$DBA

$5x - 8 = 3x + 12$

$2x = 20$

$x = 10$

m$\angle$EAB = 5(10) - 8 = 50 - 8 = 42.

Therefore, the measure of $\angle$EAB is 42°.

● **PROBLEM 119**

Given: $\overline{DC} \cong \overline{BA}$, $\overline{AD} \cong \overline{CB}$. Prove: $\triangle ABD \cong \triangle CDB$.

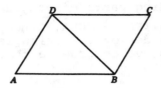

Solution: The required congruence can be proved by the SSS Postulate. We are given congruence between two pairs of corresponding sides. The third congruence $\overline{BD} \cong \overline{DB}$ follows from the reflexive property.

STATEMENTS	REASONS
1. $\overline{DC} \cong \overline{BA}$	1. Given.
2. $\overline{AD} \cong \overline{CB}$	2. Given.
3. $\overline{DB} \cong \overline{BD}$	3. A quantity is congruent to itself.
4. $\triangle ABD \cong \triangle CDB$	4. S.S.S. $\cong$ S.S.S.; if three sides of one triangle are congruent, respectively, to three sides of a second triangle, the triangles are congruent.

● **PROBLEM** 120

LET ABCD be a quadrilateral in which AD = BC and AB = CD. Let its diagonals, $\overline{AC}$ and $\overline{BD}$, intersect at point E. a) Prove that $\triangle ABC \cong \triangle CDA$; b) Prove that $\angle DAC \cong \angle BCA$; c) Prove that $\triangle ABD \cong \triangle CDB$; d) Prove that $\angle ADB \cong \angle CBD$.

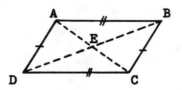

Solution: The situation is shown in the accompanying figure.
a) To prove that $\triangle ABC \cong \triangle CDA$, we use the S.S.S. Postulate. We are given that AD = BC and AB = CD. Hence, by definition of the congruence of 2 line segments, $\overline{AD} \cong \overline{BC}$ and $\overline{AB} \cong \overline{CD}$. Furthermore, by the reflexive property, $\overline{AC} \cong \overline{CA}$, hence $\triangle ABC \cong \triangle CDA$ by the S.S.S. Postulate.

b) Since $\triangle ABC \cong \triangle CDA$, $\angle DAC \cong \angle BCA$ by definition of the congruence of 2 triangles.

c) We again use the S.S.S. Postulate to show that
$\triangle ABD \cong \triangle CDB$. From the given facts, AD = BC and AB = DC which
implies $\overline{AD} \cong \overline{BC}$ and $\overline{AB} \cong \overline{CD}$. Furthermore, by the reflexive
property, $\overline{DB} \cong \overline{BD}$. By the S.S.S. Postulate, $\triangle ABD \cong \triangle CDB$.

d) Because $\triangle ABD \cong \triangle CDB$, ⦨ADB $\cong$ ⦨CBD, by the definition
of the congruence of 2 triangles.

● **PROBLEM** 121

Given: $\overline{CA} \cong \overline{DB}$. $\overline{CB} \cong \overline{DA}$. Prove $\triangle ABC \cong \triangle BAD$.

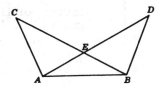

<u>Solution</u>: Since we are given congruence between two pairs
of corresponding sides, and because the triangles overlap,
the required congruence can be shown using the SSS Postulate.

STATEMENTS	REASONS
1. $\overline{CA} \cong \overline{DB}$	1. Given.
2. $\overline{CB} \cong \overline{DA}$	2. Given.
3. $\overline{AB} \cong \overline{AB}$	3. A quantity is congruent to itself.
4. $\triangle ABC \cong \triangle BAD$	4. SSS $\cong$ SSS; if three sides of one triangle are congruent, respectively, to three sides of a second triangle, the triangles are congruent.

● **PROBLEM** 122

Given: $\overline{AB} \cong \overline{DC}$; $\overline{BD} \cong \overline{CA}$. Prove: ⦨ABC $\cong$ ⦨DCB.

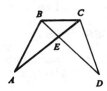

Solution: First, show that △ABC ≅ △DĊB by the SSS posulate. Congruence of ⊀ABC and ⊀DCB follows by corresponding parts.

STATEMENTS	REASONS
1. $\overline{AB}$ ≅ $\overline{DC}$ and $\overline{BD}$ ≅ $\overline{CA}$	1. Given.
2. $\overline{BC}$ ≅ $\overline{CB}$	2. Every segment is congruent to itself.
3. △ABC ≅ △DCB	3. If three sides of one triangle are congruent to three sides of a second triangle, then the triangles are congruent. (SSS Postulate).
4. ⊀ABC ≅ ⊀DCB	4. Corresponding angles of congruent triangles are congruent.

● **PROBLEM** 123

Given: $\overline{DB}$ ≅ $\overline{EA}$, $\overline{AD}$ ≅ $\overline{BE}$. Prove: ⊀DAB ≅ ⊀EBA.

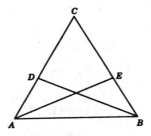

Solution: ⊀DAB and ⊀EBA are corresponding angles of △DAB
and △EBA. First, we prove △DAB ≅ △EBA by the SSS Postulate.
⊀DAB ≅ ⊀EBA follows by corresponding parts.

STATEMENT	REASON
1. $\overline{DB}$ ≅ $\overline{EA}$	1. Given.
2. $\overline{AD}$ ≅ $\overline{BE}$	2. Given.
3. $\overline{AB}$ ≅ $\overline{BA}$	3. Reflexive Property of congruence.
4. △DAB ≅ △EBA	4. SSS ≅ SSS
5. ⊀DAB = ⊀EBA	5. Corresponding parts of congruent triangles are congruent.

104

In the diagram shown, $\overline{AEB}$ is a straight line segment. We are given that $\overline{AC} \cong \overline{AD}$ and $\overline{BC} \cong \overline{BD}$. Prove that $\overline{CE} \cong \overline{DE}$.

Solution: If we can prove $\triangle ACE \cong \triangle ADE$, then by corresponding parts we can show $\overline{CE} \cong \overline{DE}$. We know that $\overline{AC} \cong \overline{AD}$, by the given facts, and $\overline{AE} \cong \overline{AE}$, by reflexivity. However, before we can prove congruence we need to show that the included angles are congruent. The best way to do this is to prove $\triangle ACB \cong \triangle ADB$ and, by corresponding parts, show $\angle EAC \cong \angle EAD$ (the included angles needed for the first part). Triangles ACB and ADB can be proved congruent by the SSS Postulate.

STATEMENT	REASON
1. $\overline{AC} \cong \overline{AD}$	1. Given.
2. $\overline{BC} \cong \overline{BD}$	2. Given.
3. $\overline{AB} \cong \overline{AB}$	3. Reflexive property of congruence.
4. $\triangle ACB \cong \triangle ADB$	4. S.S.S. Postulate.
5. $\angle CAE \cong \angle DAE$	5. Corresponding parts of congruent triangles are congruent.
6. $\overline{AC} \cong \overline{AD}$	6. Given.
7. $\overline{AE} \cong \overline{AE}$	7. Reflexive property of congruence.
8. $\triangle ACE \cong \triangle ADE$	8. S.A.S. Postulate.
9. $\overline{CE} \cong \overline{DE}$	9. Corresponding parts of congruent triangles are congruent.

Given: $\overline{QS}$ intersects $\overline{PR}$ at T such that RQ = RS and QT = ST.
Prove: $\overline{TP}$ bisects $\angle SPQ$.

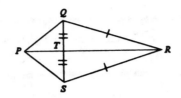

Solution: $\overline{TP}$ bisects ⦨SPQ only if ⦨SPT $\cong$ ⦨QPT. Showing
$\triangle PTQ \cong \triangle PTS$ or $\triangle PQR \cong \triangle PSR$ would be sufficient to show
⦨SPT $\cong$ ⦨QPT. Consider $\triangle PQR$ and $\triangle PSR$. $\overline{PR} \cong \overline{PR}$. $\overline{QR} \cong \overline{SR}$.
To show $\triangle PQR \cong \triangle PSR$, we need one more congruence - either
a pair of sides or else ⦨QRT $\cong$ ⦨SRT.

We can show ⦨QRT $\cong$ ⦨SRT by proving $\triangle QTR \cong \triangle STR$ by the
SSS Postulate. Thus, $\triangle PQR \cong \triangle PSR$ by the SAS Postulate.
⦨QPT $\cong$ ⦨SPT by corresponding angles, and $\overline{TP}$ bisects ⦨QPS.

STATEMENTS	REASONS
1. $\overline{QS}$ intersects $\overline{PR}$ at T such that RQ=RS;QT=ST	1. Given
2. $\overline{QT} \cong \overline{ST}$ $\overline{RQ} \cong \overline{RS}$	2. Two segments of equal lengths are congruent.
3. $\overline{TR} \cong \overline{TR}$	3. A segment is congruent to itself.
4. $\triangle QTR \cong \triangle STR$	4. The SSS Postulate.
5. ⦨QRT $\cong$ ⦨SRT	5. Corresponding parts of congruent triangles are congruent.
6. $\overline{PR} \cong \overline{PR}$	6. A segment is congruent to itself.
7. $\triangle PQR \cong \triangle PSR$	7. The SAS Postulate.
8. ⦨QPR $\cong$ ⦨SPR	8. Corresponding parts of congruent triangles are congruent.
9. $\overline{TP}$ bisects ⦨SPQ	9. If a segment divides an angle into two congruent angles, then the segment bisects the angle.

CHAPTER 9

HYPOTENUSE-LEG POSTULATE

> **Basic Attacks and Strategies for Solving Problems in this Chapter. See pages 107 to 113 for step-by-step solutions to problems.**

In the three previous chapters the SAS, ASA, and SSS postulates are described and applied. These postulates apply to any pairs of triangles. When the triangles happen to be right triangles, a special postulate, called the **Hypotenuse-Leg** (HL) postulate, can be applied.

If, in two right triangles, the hypotenuse and a leg in one triangle are congruent to the corresponding parts of the other triangle, the two right triangles are congruent.

For the triangles pictured below, the segments with one mark on them are congruent and the segments with two marks are congruent:

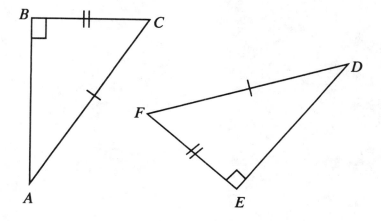

Since $\overline{AC}$ and $\overline{DF}$ are both opposite the right angles, each is the hypotenuse of the triangle in which it appears. Thus the HL postulate applies, and $\triangle ABC$ $\cong \triangle DEF$.

Notice that in right triangles *GHI* and *JKL* there are two congruent sides, but the triangles are not congruent because neither $\overline{KL}$ nor $\overline{LJ}$ is the hypotenuse of $\triangle JKL$:

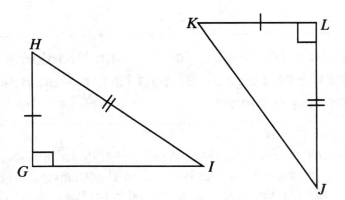

● **PROBLEM** 126

Quadrilateral ABCD is, as shown in the figure, cut by the line $\overline{AC}$. If $\overline{CB} \perp \overline{AB}$, $\overline{CD} \perp \overline{AD}$, and $\overline{BC} \cong \overline{DC}$. Prove that $\overline{AC}$ bisects ⟩BAD.

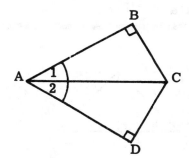

Solution: To prove $\overline{AC}$ is the bisector of ⟩BAD, it is necessary to prove ⟩1 $\cong$ ⟩2. This can be done by proving △ABC $\cong$ △ADC and then using the rule which assigns congruence to corresponding parts of congruent triangles in order to prove ⟩1 $\cong$ ⟩2.

The hypotenuse leg theorem is used in this proof. It states that two right triangles are congruent if the hypotenuse and the leg of one triangle are congruent to the corresponding parts of the other triangle. Notationally, it can be written hy. leg. $\cong$ hy. leg.

STATEMENT		REASON
1. $\overline{CB} \perp \overline{AB}$, $\overline{CD} \perp \overline{AD}$		1. Given
2. ⟩ABC and ⟩ADC are right triangles		2. When perpendicular lines intersect, they form right angles.

3. Δ's ABC and ADC are right triangles	3. A triangle which contains a right angle is a right triangle.
4. $\overline{BC} \cong \overline{CD}$	4. Given.
5. $\overline{AC} \cong \overline{AC}$	5. Reflexive property of congruence.
6. ΔABC $\cong$ ΔADC	6. Hy. leg $\cong$ Hy. leg.
7. ⊀1 $\cong$ ⊀2	7. Corresponding parts of congruent triangles are congruent.
8. AC bisects ⊀BAD	8. A segment which divides an angle into two congruent angles bisects the angle.

● **PROBLEM** 127

In the accompanying figure, $\overleftrightarrow{PM} \perp \overline{RS}$ and $\overline{PR} \cong \overline{PS}$. Prove that $\overleftrightarrow{PM}$ is the perpendicular bisector of $\overline{RS}$.

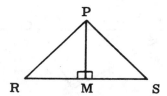

Solution: To show that $\overleftrightarrow{PM}$ is the perpendicular bisector of $\overline{RS}$, we must show (1) $\overleftrightarrow{PM} \perp \overline{RS}$; and (2) $\overleftrightarrow{PM}$ intersects $\overline{RS}$ at the midpoint of $\overline{RS}$. We are given $\overleftrightarrow{PM} \perp \overline{RS}$. To show $\overleftrightarrow{PM}$ intersects $\overline{RS}$ at the midpoint, we prove ΔPRM $\cong$ ΔPSM.

Since $\overleftrightarrow{PM} \perp \overline{RS}$, ⊰PMR and ⊰PMS are right angles. Thus, ΔPRM and ΔPSM are right triangles. Furthermore, ⊰PMR $\cong$ ⊰PMS. We are given that $\overline{PR} \cong \overline{PS}$. By the reflexive property, $\overline{PM} \cong \overline{PM}$. Thus, one leg and the hypotenuse of right ΔPRM are congruent to one leg and the hypotenuse of right ΔPSM. By the Hypotenuse-Leg Theorem, ΔPRM $\cong$ ΔPSM. By corresponding parts, $\overline{RM} \cong \overline{MS}$. Since M divides $\overline{RS}$ into two congruent segments, M is the midpoint. Thus $\overleftrightarrow{PM}$ intersects $\overline{RS}$ at the midpoint.

Therefore, $\overleftrightarrow{PM}$ is the perpendicular bisector of $\overline{RS}$.

On diameter $\overline{AB}$ of semicircle O, two points C and D are located on opposite sides of the center O so that $\overline{AC} \cong \overline{BD}$. At C and D, perpendiculars are constructed from $\overline{AB}$ and extended to meet arc $\overarc{AB}$ at points E and F, respectively. Using the fact that all radii of a semicircle are congruent, prove $\overline{CE} \cong \overline{DF}$.

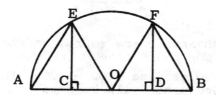

Solution: If we can prove $\triangle ECO \cong \triangle FDO$, then we will be able to conclude that $\overline{CE} \cong \overline{DF}$. Since the triangles include a right angle, they are right triangles and, as such, we can use the hypotenuse-leg theorem to prove congruence between them. Hypotenuse-leg is similar to the SSS Postulate because, when you know the hypotenuse and leg of a right triangle the third leg is uniquely determined by Pythagoras' Theorem.

STATEMENT	REASON
1. $\overline{EC} \perp \overline{AO}$ and $\overline{FD} \perp \overline{BO}$	1. Given
2. ∢ECO and ∢FDO are right angles	2. Perpendicular lines intersect to form right angles.
3. $\triangle ECO$ and $\triangle FDO$ are right triangles	3. A triangle which contains a right angle is a right triangle.
4. $\overline{OA} \cong \overline{OB}$	4. Radii of the same circle are congruent.
5. $\overline{AC} \cong \overline{BD}$	5. Given.
6. OA - AC = OB - BD or OC = OD or $\overline{OC} \cong \overline{OD}$	6. Subtraction Property of Equality
7. $\overline{OE} \cong \overline{OF}$	7. Same as reason 4.
8. $\triangle ECO \cong \triangle FDO$	8. Two right triangles are congruent if the hypotenuse and leg of one triangle are congruent to the hypotenuse and leg of the second.
9. $\overline{CE} \cong \overline{DF}$	9. Corresponding sides of congruent triangles are congruent.

Given: DB and AC intersect at O. m∡1 = m∡2 = 90°. $\overline{CD} \cong \overline{CB}$.
Prove: $\overline{OD} \cong \overline{OB}$.

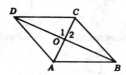

Solution: Since $\overline{OD}$ and $\overline{OB}$ are corresponding parts of right triangles COD and COB respectively, we can derive the required congruence by proving $\triangle COD \cong \triangle COB$. The hypotenuse leg theorem for congruence will be used.

STATEMENTS	REASONS
1. m∡1 = m∡2 = 90°	1. Given.
2. ∡1 and ∡2 are right angles	2. A right angle has a measure of 90°.
3. △COD and △COB are right triangles	3. A right triangle is a triangle with one right angle.
4. $\overline{CD} \cong \overline{CB}$	4. Given.
5. $\overline{CO} \cong \overline{CO}$	5. Reflexive property of congruence.
6. △COD $\cong$ △COB	6. hy.leg $\cong$ hy. leg
7. $\overline{OD} \cong \overline{OB}$	7. Corresponding parts of congruent triangles are congruent.

Given: $\overline{AC} \cong \overline{BD}$; m∡CDA = m∡BAD = 90°. Prove $\overline{AB} \cong \overline{DC}$.

Solution: By establishing that triangles CDA and BAD are right triangles, we will be able to derive the desired congruence between their corresponding sides $\overline{AB}$ and $\overline{DC}$ using the hypotenuse leg theorem for congruence of right triangles.

STATEMENTS	REASONS
1. m⊀CDA = m⊀BAD = 90°	1. Given.
2. ⊀CDA and ⊀BAD are right angles	2. A right angle has a measure of 90°.
3. △CDA and △BAD are right triangles	3. A right triangle is a triangle with one right angle.
4. $\overline{AC} \cong \overline{BD}$	4. Given.
5. $\overline{DA} \cong \overline{DA}$	5. Reflexive property of congruence.
6. △CDA $\cong$ △BAD	6. hy.leg = hy. leg.
7. $\overline{AB} \cong \overline{DC}$	7. Corresponding parts of congruent triangles are congruent.

● **PROBLEM** 131

Given: △ABC is isosceles with base $\overline{AB}$. $\overline{BD} \perp \overline{AC}$. $\overline{AE} \perp \overline{BC}$. ⊀1 $\cong$ ⊀2. Prove ⊀3 $\cong$ ⊀4.

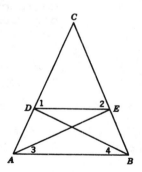

Solution: In this problem, we must first notice that ⊀3 and ⊀4 are corresponding parts of rt.△'s AEB and BDA, respectively. Hence, by proving △BDA $\cong$ △AEB, we can deduce that ⊀3 $\cong$ ⊀4. The hypotenuse leg theorem for congruence will be employed.

STATEMENTS	REASONS
1. $\overline{BD} \perp \overline{AC}$; $\overline{AE} \perp \overline{BC}$	1. Given.

2. ∡BDA and ∡AEB are right angles	2. Perpendicular lines meet to form right angles.
3. ΔBDA and ΔAEB are right triangles	3. A right triangle is a triangle with one right angle.
4. ΔABC is isosceles	4. Given.
5. $\overline{AC} \cong \overline{BC}$	5. Definition of an isosceles triangle.
6. AD + DC = AC BE + EC = BC	6. The whole is congruent to the sum of its parts.
7. AD + DC = BE + EC	7. Addition Postulate.
8. ∡1 $\cong$ ∡2	8. Given.
9. $\overline{DC} \cong \overline{EC}$ or DC = EC	9. If a triangle has two congruent angles, then the sides opposite those angles are congruent.
10. AD = BE	10. If congruent quantities are subtracted from congruent quantities, then the differences are congruent quantities. (Steps 9 and 7).
11. AB = AB	11. Reflexive law.
12. ΔBDA $\cong$ ΔAEB	12. hy. leg $\cong$ hy. leg.
13. ∡3 = ∡4	13. Corresponding parts of congruent triangles are equal.

CHAPTER 10

PARALLELISM

Basic Attacks and Strategies for Solving Problems in this Chapter. See pages 114 to 141 for step-by-step solutions to problems.

When one line intersects two other lines, eight angles are formed:

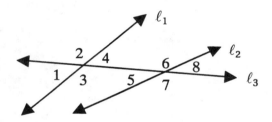

In the picture above, line ℓ_3 intersects line ℓ_1 and line ℓ_2, and the angles formed are $\angle 1$, $\angle 2$, $\angle 3$, $\angle 4$, $\angle 5$, $\angle 6$, $\angle 7$, and $\angle 8$. ℓ_3 is called a **transversal**. Since $\angle 3$, $\angle 4$, $\angle 5$, and $\angle 6$ are "between" ℓ_1 and ℓ_2, they are called **interior angles**, and $\angle 1$, $\angle 2$, $\angle 7$, and $\angle 8$ are called **exterior angles**. Angles on opposite sides of the transversal ℓ_3 are called **alternate angles**, so:

$\angle 1$ and $\angle 6$ are alternate angles,

$\angle 3$ and $\angle 6$ are alternate interior angles, and

$\angle 1$ and $\angle 8$ are alternate exterior angles.

Angles which are in a "corresponding" position relative to the transversal are called **corresponding angles**. Thus, the pairs of corresponding angles are:

$\angle 1$ and $\angle 5$

$\angle 2$ and $\angle 6$

∡ 3 and ∡ 7

∡ 4 and ∡ 8.

  Some interesting things happen when a transversal intersects two parallel lines.

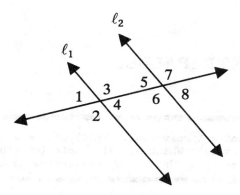

In this case, all the angles which appear to be congruent actually are congruent. This means that every angle in the set {∡ 1, ∡ 4, ∡ 5, ∡ 8} is congruent to every other angle in that set. Likewise, the angles {∡ 2, ∡ 3, ∡ 6, ∡ 7} are all congruent to each other. Moreover, every angle in the first set is supplementary to every angle in the second set.

  Conversely, if those same relationships hold between the angles formed, the lines are parallel. More specifically, to prove that two lines are parallel it is only necessary to show that one pair of corresponding angles is congruent, and that generalization is used several times in the problems in this chapter.

Step-by-Step Solutions to Problems in this Chapter, "Parallelism"

PROVING LINES PARALLEL

● **PROBLEM** 133

Draw two lines and a transversal. Which of the angles in the drawing are corresponding angles? Which are alternate interior angles? Which are alternate exterior angles? Which are consecutive interior angles?

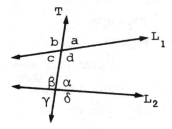

<u>Solution</u>: The situation is shown in the figure. Line T is said to be a transversal of lines L_1 and L_2 provided T, L_1, and L_2 are coplanar and not concurrent.

There are four pairs of corresponding angles: a and α, b and β, c and γ; and d and δ. a and α are a pair of corresponding angles because the position of ∢ a with respect to T and L_1 (top left) is the same as the position of ∢ α with respect to T and L_2 (top left).

Similar logic applies for other pairs of corresponding angles.

There are two pairs of alternate interior angles: β and d; and c and α. The name results from the fact that the pair of angles, say b and δ, are on "alternate" sides of the transversal and lie in the "interior" of L_1 and L_2.

Angles a and γ and b and δ are alternate exterior angles. The name results from the fact that each pair of alternate exterior angles lies on "alternate" sides of the transversal and exterior to L_1 and L_2.

Angles b and α, and c and δ are consecutive interior angles. The word "consecutive" indicates that the angles lie on the same side of the transversal, and "interior" indicates the location of the angles relative to L_1 and L_2.

Given: ∡ 2 is supplementary to ∡ 3 .
Prove: $\ell_1 \parallel \ell_2$.

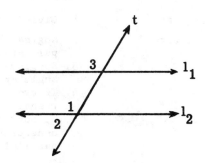

Solution: Given two lines intercepted by a transversal, if a pair of corresponding angles are congruent, then the two lines are parallel. In this problem, we will show that since ∡ 1 and ∡ 2 are supplementary and ∡ 2 and ∡ 3 are supplementary, ∡ 1 and ∡ 3 are congruent. Since corresponding angles ∡ 1 and ∡ 3 are congruent, it follows $\ell_1 \parallel \ell_2$.

Statement	Reason
1. ∡ 2 is supplementary to ∡ 3	1. Given
2. ∡ 1 is supplementary to ∡ 2	2. Two angles that form a linear pair are supplementary.
3. ∡ 1 ≅ ∡ 3	3. Angles supplementary to the same angle are congruent.
4. $\ell_1 \parallel \ell_2$	4. Given two lines intercepted by a transversal, if a pair of corresponding angles are congruent, then the two lines are parallel.

Given: ∡ 1 and ∡ 2 are supplementary.   Prove: $\ell_1 \parallel \ell_2$.

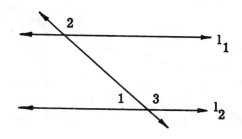

<u>Solution</u>: Given two lines cut by a transversal, if a pair of cor-
responding angles are congruent, then the lines are parallel. Here,
we are given that ∢ 1 and ∢ 2 are supplementary. From this, we
will prove that ∢ 2 ≅ ∢ 3. Since congruent angles ∢ 2 and ∢ 3
are corresponding angles, the lines must be parallel.

<u>Statements</u>	<u>Reasons</u>
1. ∢ 1 and ∢ 2 supplementary	1. Given
2. ∢1 and ∢ 3 supplementary	2. Angles that form a linear pair are supplementary.
3. ∢ 2 ≅ ∢ 3	3. Supplements of the same angle are congruent.
4. $\ell_1 \parallel \ell_2$	4. If two lines are cut by a transversal so that a pair of corresponding angles are congruent, the lines are parallel.

● **PROBLEM** 136

If $\ell_1 \parallel \ell_2$, prove ∢ 1 is supplementary to ∢ 2.

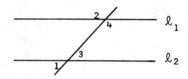

<u>Solution</u>: From an earlier theorem, we know that interior angles on the
same side of the transversal are supplementary. Thus, ∢ 3 and ∢ 4 are
supplementary. Using the fact that opposite angles are congruent, we
will show that ∢ 3 ≅ ∢ 1 and ∢ 4 ≅ ∢ 2. Since ∢ 3 and ∢ 4 are
supplementary, ∢ 1 and ∢ 2 must also be supplementary.

    Given: $\ell_1 \parallel \ell_2$.

    Prove: ∢ 1 is supplementary to ∢ 2.

<u>Statements</u>	<u>Reasons</u>
1. $\ell_1 \parallel \ell_2$	1. Given
2. ∢ 3 supplementary to ∢ 4	2. If two parallel lines are cut by a transversal, interior angles on the same side of the transversal are supplementary.
3. m ∢ 3 + m ∢4 = 180°	3. Definition of supplementary angles.
4. m ∢ 3 = m ∢ 1 and m ∢ 4 = m ∢ 2	4. Vertical angles are equal in measure.
5. m ∢ 1 + m ∢ 2 = 180°	5. A quantity may be sub-

stituted for its equal in
any mathematical expres-
sion. (Substitution post-
ulate).

6. ∡ 1 is supplementary to ∡ 2.

6. If the measures of two
angles sum to 180°, then
the angles are supplementary.

● **PROBLEM** 137

If $\ell_1 \parallel \ell_2$ and m ∡ 1 = 30°, how many degrees are there in ∡ 2 ?
In ∡ 3 ?

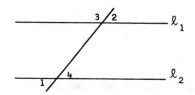

Solution: Since ∡ 1 and ∡ 4 are vertical angles, m ∡ 1 = m ∡ 4.
Since ∡ 2 and ∡ 4 are corresponding angles, m ∡ 2 = m ∡ 4. By the
transitive property, m ∡ 1 = m ∡ 2. (If two quantities are equal to
the same quantity, they are equal to each other.) We are given that
m ∡ 1 = 30°. Thus m ∡ 2 = 30°. ∡ 2 and ∡ 3 are supplementary;
thus, m ∡ 2 + m ∡ 3 = 180°. Substituting 30° for m ∡ 2, we obtain:

$$30° + m ∡ 3 = 180°$$
$$m ∡ 3 = 150°$$

● **PROBLEM** 138

If line $\overleftrightarrow{AB}$ is parallel to line $\overleftrightarrow{CD}$ and line $\overleftrightarrow{EF}$ is parallel to
line $\overleftrightarrow{GH}$, prove that m ∡ 1 = m ∡ 2.

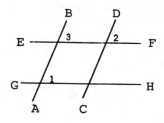

Solution: To show ∢ 1 ≅ ∢ 2, we relate both to ∢ 3. Because $\overline{EF} \parallel \overline{GH}$, corresponding angles 1 and 3 are congruent. Since $\overline{AB} \parallel \overline{CD}$, corresponding angles 3 and 2 are congruent. Because both ∢ 1 and ∢ 2 are congruent to the same angle, it follows that ∢ 1 ≅ ∢ 2 .

Statements	Reasons
1. $\overleftrightarrow{EF} \parallel \overleftrightarrow{GH}$	1. Given
2. m ∢ 1 = m ∢ 3	2. If two parallel lines are cut by a transversal, corresponding angles are of equal measure.
3. $\overleftrightarrow{AB} \parallel \overleftrightarrow{CD}$	3. Given
4. m ∢ 2 = m ∢ 3	4. If two parallel lines are cut by a transversal, corresponding angles are equal in measure.
5. m ∢ 1 = m ∢ 2	5. If two quantities are equal to the same quantity, they are equal to each other.

● **PROBLEM** 139

The transversal of parallel lines $\overleftrightarrow{AB}$ and $\overleftrightarrow{CD}$ intersects the lines at points F and E, respectively. Thus, m∢LED = m∢LFB. $\overleftrightarrow{EG}$ and $\overleftrightarrow{FH}$ bisect the angles they pass through. Prove m∢1 = m∢2. (See figure.)

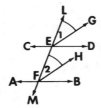

Solution: To complete this problem , we will need the Division Postulate and the definition of an angle bisector. The Division Postulate states that if equal quantities are divided by equal nonzero quantities, the quotients are equal. The angle bisector of an angle is the ray in the interior of the angle that divides the angle into two congruent angles each half the measure of the original.

GIVEN: $\overleftrightarrow{AB} \parallel \overleftrightarrow{CD}$; m∢LED = m∢LFB; $\overleftrightarrow{EG}$ bisects ∢LED:
$\overleftrightarrow{FH}$ bisects ∢LFB.

PROVE: m∢1 = m∢2.

STATEMENT	REASON

1. m∢LED = m∢LFB;
   $\overleftrightarrow{EG}$ bisects ∢LED:
   $\overleftrightarrow{FH}$ bisects ∢LFB
	1. Given.

2. m∢1 = ½m∢LED m∢2 = ½m∢LFB	2. Definition of angle bisector.
3. m∢1 = m∢2	3. Division Postulate – equal quantities divided by equal nonzero quantities yield equal quantities.

● **PROBLEM** 140

As shown in the accompanying figure, if $\overline{BD}$ bisects ∢ ABC and $\overline{BC} \cong \overline{CD}$, prove formally that $\overleftrightarrow{CD} \parallel \overleftrightarrow{BA}$.

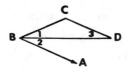

Solution: If two lines are cut by a transversal which form a pair of alternate interior angles that are congruent, then the two lines are parallel. The transversal of lines $\overleftrightarrow{CD}$ and $\overleftrightarrow{BA}$ is $\overleftrightarrow{BD}$, and ∢ 2 and ∢ 3 are alternate interior angles of this transversal. Our task is to prove ∢ 2 $\cong$ ∢ 3. From the above theorem, it then follows that $\overleftrightarrow{BA} \parallel \overleftrightarrow{CD}$.

Statement	Reason
1. $\overline{BC} \cong \overline{CD}$	1. Given
2. ∢ 3 $\cong$ ∢ 1	2. If two sides of a triangle are congruent, the angles opposite these sides are congruent.
3. ∢ 1 $\cong$ ∢ 2	3. An angle bisector divides the angle into two congruent angles.
4. ∢ 3 $\cong$ ∢ 2	4. Transitive property of congruence.
5. $\overleftrightarrow{CD} \parallel \overleftrightarrow{BA}$	5. If two lines are cut by a transversal which forms a pair of congruent alternate interior angles, then the two lines are parallel.

Given that $\overleftrightarrow{AB} \| \overleftrightarrow{CD}$, $\overleftrightarrow{ACE}$ is a line, ∡ A measures 60°, and ∡ B measures 80° (refer to figure), determine a) the measure of ∡ x and b) the measure of ∡ y.

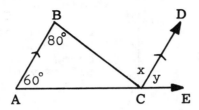

Solution: a) A theorem tells us that when two parallel lines are cut by a transversal, the alternate interior angles are congruent. We are given that $\overleftrightarrow{AB}$ and $\overleftrightarrow{DC}$ are parallel. Line $\overleftrightarrow{BC}$ fulfills the definition of a transversal. Therefore, ∡ B and ∡ x, being alternate interior angles, must be congruent and of equal measure. Since the measure of ∡ B is 80°, the measure of ∡ x is also 80°.

b) There is a theorem that states that when parallel lines are cut by a transversal, any two corresponding angles are congruent. Transversal $\overleftrightarrow{AC}$ forms the corresponding angles, ∡ y and ∡ A, with parallel lines $\overleftrightarrow{AB}$ and $\overleftrightarrow{CD}$. According to the theorem, then, ∡ A and ∡ y are congruent and of equal measure. ∡ A measures 60° and, therefore, ∡ y measures 60°.

If the measure of ∡ A is 100 + 3x and the measure of ∡ B is 80 - 3x, explain why we can conclude $\overline{AD} \| \overline{BC}$. (Refer to the figure for positioning of angles and lines.)

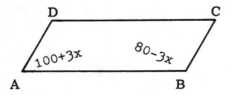

Solution: If two lines are cut by a transversal such that a pair of interior angles on the same side of the transversal are supplementary, then the two lines are parallel. $\overline{AB}$ is a transversal of $\overline{AD}$ and $\overline{BC}$.

Angles A and B are interior angles on the same side of the transversal and will be supplementary if their measures sum to 180°. By
algebra, it is seen that m ∡ A + m ∡ B = 100 + 3x + 80 - 3x = 180.
Therefore, ∡ A and ∡ B are supplementary and, by the above theorem,
$\overline{AD} \parallel \overline{BC}$

● PROBLEM 143

In the accompanying figure, given that $\overline{AB} \parallel \overline{DC}$ and $\overline{AD} \parallel \overline{BC}$, what
are the measures of ∡ B, ∡ 1, ∡ D, and ∡ 2 ?

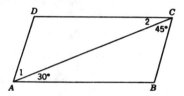

Solution: From the accompanying figure we see that ∡ 2 and ∡ BAC
are alternate interior angles, as are ∡ 1 and ∡ ACB. Since $\overline{AB} \parallel \overline{DC}$,
the alternate interior angles are equal. Therefore, m ∡ 2 = m ∡ BAC
and m ∡ 1 = m ∡ ACB. Since m ∡ BAC = 30° and m ∡ ACB = 45°,
then m ∡ 2 = 30° and m ∡ 1 = 45°. Now, Δ ACB consists of a 30°
angle, a 45° angle, and ∡ B. Since the sum of the interior angles of
a triangle is 180°, m ∡ B = 105°.

Δ DAC also consists of ∡ 1, ∡ 2, and ∡ D with m ∡ 1 = 45° and
m ∡ 2 = 30°. Therefore, m ∡ D = 105°.

● PROBLEM 144

Given: Δ ABC is isosceles with base $\overline{AB}$.
          ∡ A ≅ ∡ 1
Prove: AB ∥ ED .

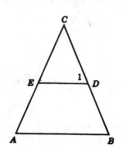

<u>Solution</u>: To show two lines parallel, it is sufficient to show that a pair of corresponding angles are congruent. Here, we use the fact that △ ACB is isosceles to show ∡ A ≅ ∡ B. Combining this with the given, ∡ 1 ≅ ∡ A, we obtain ∡ B≅∡1. Since these are corresponding angles of $\overline{ED}$ and $\overline{AB}$, it follows that $\overline{ED} \parallel \overline{AB}$.

Statements	Reasons
1. △ ABC is isosceles with base $\overline{AB}$	1. Given
2. $\overline{AC} \cong \overline{BC}$	2. Definition of an isosceles triangle.
3. ∡ A ≅ ∡ B	3. If two sides of a triangle are congruent, then the angles opposite those sides are congruent.
4. ∡ A ≅ ∡ 1	4. Given
5. ∡ B ≅ ∡ 1	5. If two quantities are congruent to the same quantity, they are congruent to each other.
6. $\overline{AB} \parallel \overline{ED}$	6. If two lines are cut by a transversal so that a pair of corresponding angles are congruent, the lines are parallel.

PROPERTIES OF PARALLEL LINES

● **PROBLEM** 145

Referring to the figure, answer the following questions:

(a) If ∡ 1 ≅ ∡ 2, can ED intersect line AB?

(b) If ∡ 1 ≅ ∡ 3, can ED intersect line AB?

(c) If ∡ 1 ≅ ∡ 3, must ED intersect line AB?

(d) If △ ABC is isosceles and ∡ 1 ≅ ∡ 4, can ED intersect line AB?

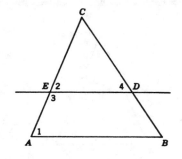

Solution: (a) ∡ 1 and ∡ 2 are corresponding angles of parallel lines $\overline{ED}$ and $\overline{AB}$, with $\overline{CA}$ the transversal. Recall that if two lines are cut by a transversal so that corresponding angles are congruent, the lines are parallel. If ∡ 1 ≅ ∡ 2, then a pair of corresponding angles are congruent. Hence, $\overline{ED} \parallel \overline{AB}$ and $\overline{ED}$ cannot intersect line $\overline{AB}$.

(b) We are given that ∡ 1 ≅ ∡ 3. We cannot conclude that the lines are parallel, since there is no theorem relating congruent alternate interior angles to parallelism. Therefore, $\overline{ED}$ _can_ intersect line $\overline{AB}$.

(c) Knowing only that ∡ 1 = ∡ 3, we cannot conclude that the lines are parallel. However, neither can we conclude that $\overline{ED}$ must intersect $\overline{AB}$. In the special case, m ∡ 1 = m ∡ 3 = 90°, the two lines would be parallel.

(d) No, because $\overleftrightarrow{AB} \parallel \overleftrightarrow{ED}$. To see this, note that since △ ABC is isosceles, AC = BC. If two sides of a triangle are congruent, then the angles opposite these sides are congruent. Thus, ∡ 1 ≅ ∡ B. Furthermore, we are given ∡ 1 ≅ ∡ 4. Combining the two, we obtain ∡ B ≅ ∡ 4. Given two lines, $\overleftrightarrow{AB}$ and $\overleftrightarrow{ED}$, cut by a transversal $\overleftrightarrow{CB}$, if a pair of corresponding angles, ∡ 1 and ∡ B, are congruent, then the two lines are parallel. Thus, $\overleftrightarrow{AB} \parallel \overleftrightarrow{ED}$, and lines $\overleftrightarrow{AB}$ and $\overleftrightarrow{ED}$ cannot intersect.

● **PROBLEM** 146

A collapsible ironing board is constructed so that the supports bisect each other. Show why the board will always be parallel to the floor.

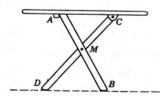

Solution: Assume that there is a plane determined by the four non-collinear points A, B, C, and D. Then the edge of the ironing board, $\overline{AC}$, will be parallel to the line segment determined by the endpoints of the supports, $\overline{DB}$, if either pair of alternate interior angles formed by the supports, ∢ MCA and ∢ MDB for example, is congruent.

Since M is given to be the midpoint of both supports, $\overline{AM} \cong \overline{MB}$ and

123

$\overline{CM} \cong \overline{MD}$. Hence, two pairs of corresponding sides are congruent.

$\overline{AB}$ and $\overline{CD}$ intersect to form vertical angles $\not\lessgtr$ AMC and $\not\lessgtr$ BMD. Since vertical angles are congruent, $\not\lessgtr$ AMC $\cong$ $\not\lessgtr$ BMD. Therefore, $\triangle$ AMC $\cong \triangle$BMD by the S.A.S. = S.A.S. postulate.

By corresponding parts, $\not\lessgtr$ MCA $\cong$ $\not\lessgtr$ MDB.

Consequently, $\overline{AC} \parallel \overline{DB}$ because the alternate interior angles formed by transversal $\overline{CD}$ are congruent.

● **PROBLEM** 147

Let L_1, L_2, and L_3 be lines in plane M. Prove that if $L_1 \parallel L_2$ and $L_3 \perp L_2$, then $L_3 \perp L_1$.

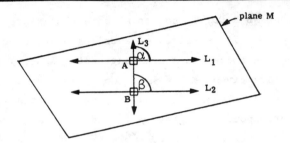

Solution: There are 2 cases to consider:

I) L_3 intersects L_1 ;

II) L_3 does not intersect L_1 .

Each case is treated separately.

Case I: The figure shows the situation for this case. We shall use the fact that the consecutive interior angles of 2 parallel lines cut by a transversal are supplementary to prove that m $\not\lessgtr$ α = 90°. (i.e., $L_3 \perp L_1$).

First note that L_3 acts as a transversal for parallel lines L_1 and L_2. Angles α and BAC are consecutive interior angles for L_1 and L_2, and are therefore supplementary. Hence, their measures must sum to 180°, and we may write:

$$m \not\lessgtr BAC + m \not\lessgtr \alpha = 180^\circ \qquad (1)$$

But, as the problem indicates, $\not\lessgtr$ BAC is a right angle. Therefore,

$$m \not\lessgtr BAC = 90^\circ . \qquad (2)$$

Substituting (2) in (1) yields

$$m \not\lessgtr \alpha + 90^\circ = 180^\circ$$

or

$$m \not\lessgtr \alpha = 90^\circ .$$

This implies that $L_3 \perp L_1$, as was to be shown.

Case II: In this situation, L_3 doesn't intersect L_1. But, if 2 lines don't intersect, they are parallel, i.e., $L_3 \parallel L_1$. However, we know (from the statement of the problem), that $L_1 \parallel L_2$. Coupling these 2 facts leads to the conclusion that $L_2 \parallel L_3$, which is contrary to the facts as given by the statement of the problem. Hence, this case need not be considered.

● **PROBLEM** 148

Prove that if L_1, L_2, and L_3 are lines in a plane, M, such that $L_1 \perp L_3$ and $L_2 \perp L_3$, then $L_1 \parallel L_2$ or $L_1 = L_2$. (See figure).

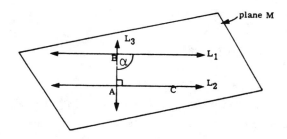

Solution: We will use the fact that if the corresponding angles of 2 lines cut by a transversal are congruent, then the 2 lines are parallel. From this, we can show that $L_1 \parallel L_2$ (or $L_1 = L_2$).

Let point A be the intersection of L_1 and L_3 and point B be the intersection of L_3 and L_2. Two cases must be considered:

(1) A not coincident with B (A $\neq$ B)
(2) A coincides with B (A = B)

(1) We first assume A $\neq$ B.

Note that angles α and β are corresponding angles for lines L_1 and L_2 cut by transversal, L_3.

From the statement of the problem, $L_1 \perp L_3$ and $L_2 \perp L_3$ and thus both α and β are right angles. Hence, $\angle \alpha \cong \angle \beta$. This implies that $L_1 \parallel L_2$.

(2) However, suppose points A and B coincide in the figure so that A equal B. Then, if the given facts still hold, both L_1 and L_2 must be perpendicular to L_3 at the same point. Since there is only 1 line perpendicular to a given line at a given point, L_1 and L_2 must coincide (i.e., $L_1 = L_2$).

Prove the angle sum theorem: the sum of the measures of the angles of a triangle is $180°$.

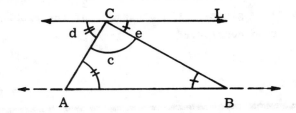

Solution: In the figure, L is constructed parallel to $\overline{AB}$ and it passes through C. Since $\overline{LC} \parallel \overline{AB}$, $\angle d \cong \angle CAB$ because they are alternate interior angles. Similarly, $\angle e \cong \angle CBA$.

The sum of the angles of the triangle is

$$S = m \angle CAB + m \angle ACB + m \angle CBA .$$

Using the results found above,

$$S = m \angle d + m \angle ACB + m \angle e. \tag{1}$$

But

$$m \angle d + m \angle e + m \angle c = 180°$$

since they sum to a straight angle (segment $\overline{CL}$). Hence,

$$m \angle d + m \angle e = 180° - m \angle c . \tag{2}$$

Substituting (2) into (1),

$$S = 180° - m \angle c + m \angle ACB.$$

But, $\angle ACB = \angle c$. Thus,

$$S = 180°.$$

Prove that if two adjacent angles of a quadrilateral are right angles, the bisectors of the other two angles are perpendicular to each other.

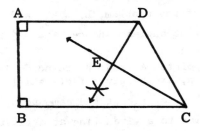

Solution: In the accompanying figure, quadrilateral ABCD has two adjacent right angles, ∡ A and ∡ B. Point E is the intersection of the angle bisectors of ∡ D and ∡ C. We must show that m ∡ DEC = 90°.

Note, that the only connection of ∡ E with the angles of the quadrilateral is that ∡ E is an angle of ΔDEC and, therefore, m ∡ DEC + m ∡ EDC + m ∡ ECD = 180, or m ∡ DEC = 180 - (m ∡ EDC + m ∡ ECD). We know that because $\overline{ED}$ and $\overline{EC}$ are angle bisectors, m ∡ EDC = ½m ∡ ADC and m ∡ ECD = ½m ∡ BCD. Thus, m ∡ DEC = 180 - (½m ∡ ADC + ½m ∡ BCD) or m ∡ DEC = 180 - ½(m ∡ ADC + m ∡ BCD). We do not know either m ∡ ADC or m ∡ BCD, but we can determine their sum.

Since $\overline{AD} \perp \overline{AB}$ and $\overline{BC} \perp \overline{AB}$, $\overline{AD} \parallel \overline{BC}$ and thus ∡ ADC and ∡ BCD are interior angles on the same side as the transversal. Thus, ∡ ADC and ∡ BCD are supplementary, or m ∡ ADC + m ∡ BCD = 180°.

Substituting this result in our equation for m ∡ DEC, we obtain: m ∡ DEC = 180 - ½(180) = 90°. Thus, ∡ DEC is a right angle and the angle bisectors of ∡ ADC and ∡ BCD are perpendicular.

● **PROBLEM** 151

If segment $\overline{AB}$ is parallel to segment $\overline{CD}$ and BC = DC, prove that $\overline{BD}$ bisects ∡ CBA.

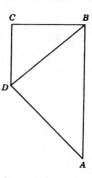

Solution: To show $\overline{BD}$ bisects ∡ CBA, we will show that ∡ CBD ≅ ∡ ABD. Our strategy is to relate both ∡ CBD and ∡ ABD to ∡ CDB.

Statements	Reasons
1. $\overline{AB} \parallel \overline{CD}$	1. Given
2. m ∡ CDB = m ∡ ABD	2. If two parallel lines are cut by a transversal, alternate interior angles have equal measure.
3. BC = DC	3. Given
4. m ∡ CDB = m ∡ CBD	4. If two sides of a triangle are equal in length, then the angles

opposite those sides are equal
in measure.

5. m ⊀ ABD = m ⊀ CBD

5. If two quantities are equal to
    the same quantity, they are equal
    to each other.

6. BD bisects ⊀ CBA

6. A segment bisects an angle if it
    divides the angle into two angles
    that have equal measures.

● **PROBLEM** 152

Given: △ ABE

$\overline{AB} \parallel \overline{CD}$

$\overline{CE} \cong \overline{DE}$

Prove: AE = BE .

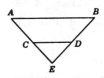

Solution: If two angles of a triangle are congruent, then the two sides
opposite them are congruent. To show AE = BE in △ABE, we show ⊀ A ≅
⊀ B. We prove this by the following steps: (1) Since CE = ED, ⊀ DCE
≅ ⊀ CDE; (2) Since $\overline{AB} \parallel \overline{CD}$, the corresponding angles are congruent.
Thus, ⊀ A ≅ ⊀ ECD and ⊀ B ≅ ⊀ CDE; (3) Since ⊀ A and ⊀ B are
congruent to congruent angles, ⊀ A ≅ ⊀ B.

Statements	Reasons
1. $\overline{CE} \cong \overline{DE}$	1. Given
2. m ⊀ ECD = m ⊀ CDE	2. If two sides of a triangle have equal length, then the angles opposite those sides are of equal measure.
3. $\overline{AB} \parallel \overline{CD}$	3. Given
4. m ⊀ ECD = m ⊀ A	4. If two parallel lines are cut by a transversal, the corresponding angles are of equal measure.
5. m ⊀ CDE = m ⊀ A	5. If two quantities are equal to the same quantity, they are equal to each other.
6. m ⊀ CDE = m ⊀ B	6. If two parallel lines are cut by a transversal, the corresponding angles are of equal measure.

7. m ∡ A = m ∡ B

7. If two quantities are equal to the same quantity, they are equal to each other.

8. AE = BE

8. If a triangle has two angles of equal measure, then the sides opposite those angles are of equal lengths.

● **PROBLEM** 153

Given: $\overline{AB} \parallel \overline{DC}$ and $\overline{AB} \cong \overline{CD}$.
Prove: ∡ A ≅ ∡·C.

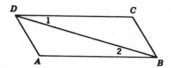

Solution: ∡ A and ∡ C are corresponding angles of △ BDC and △ DBA. Thus, to prove ∡ A ≅ ∡ C, we first prove △ BDC ≅ △ DBA by the SAS Postulate.

Statements	Reasons
1. $\overline{AB} \parallel \overline{DC}$	1. Given.
2. ∡ 1 ≅ ∡ 2	2. If two parallel lines are cut by a transversal, alternate interior angles are congruent.
3. $\overline{AB} \cong \overline{CD}$	3. Given.
4. $\overline{DB} \cong \overline{DB}$	4. Reflexive property.
5. △ ADB ≅ △ CBD	5. SAS ≅ SAS
6. ∡ A ≅ ∡ C	6. Corresponding parts of congruent triangles are congruent.

● **PROBLEM** 154

Given: $\overline{AC}$ and $\overline{EB}$ bisect each other at D.
Prove: $\overline{AE} \parallel \overline{BC}$.

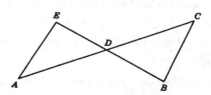

Solution: To show two lines are parallel, it is sufficient to show that a pair of alternate interior angles, such as ∡ A and ∡ C, are congruent.

We first prove $\triangle AED \cong \triangle CBD$ by the SAS postulate. Because corresponding parts of congruent triangles are congruent, $\angle A \cong \angle C$, and thus the lines are parallel.

Statements	Reasons
1. $\overline{AC}$ and $\overline{EB}$ bisect each other at D.	1. Given
2. $\overline{ED} \cong \overline{BD}$ $\overline{AD} \cong \overline{CD}$	2. Definition of bisector.
3. $\angle EDA \cong \angle BDC$	3. Vertical angles are congruent.
4. $\triangle EDA \cong \triangle BDC$	4. SAS $\cong$ SAS
5. $\angle A \cong \angle C$	5. Corresponding parts of congruent triangles are congruent.
6. $\overline{AE} \parallel \overline{BC}$	6. If two lines are cut by a transversal so that alternate interior angles are congruent, the lines are parallel.

● **PROBLEM** 155

Prove that if both pairs of opposite sides of a quadrilateral are congruent, then they are also parallel.

Given: Quadrilateral ABCD; $\overline{AB} \cong \overline{CD}$; $\overline{AD} \cong \overline{BC}$

Prove: $\overline{AD} \parallel \overline{BC}$; $\overline{AB} \parallel \overline{CD}$

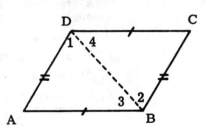

Solution: In the accompanying figure, the opposite sides of quadrilateral ABCD are congruent. Thus, $\overline{AB} \cong \overline{CD}$ and $\overline{AD} \cong \overline{BC}$. We must show $\overline{DC} \parallel \overline{AB}$ and $\overline{AD} \parallel \overline{BC}$.

To do this, we draw diagonal $\overline{DB}$. Remember that if alternate interior angles are congruent, the two lines are parallel. Thus, to show AB $\parallel$ CD, we prove $\angle 3 \cong \angle 4$. To show $\angle 3 \cong \angle 4$, we prove $\triangle ADB \cong \triangle CBD$ by the SSS Postulate. Thus, by corresponding parts, $\angle 3 \cong \angle 4$ and $\overline{AB} \parallel \overline{DC}$.

By corresponding parts, we can also say $\angle 1 \cong \angle 2$. Since $\angle 1$ and $\angle 2$ we alternate interior angles of $\overleftrightarrow{AD}$ and $\overleftrightarrow{BC}$, it follows that $\overline{AD} \parallel \overline{BC}$.

Statements	Reasons
1. Quadrilateral ABCD; $\overline{AB} \cong \overline{CD}$; $\overline{AD} \cong \overline{CD}$	1. Given.
2. $\overline{DB} \cong \overline{DB}$	2. A segment is congruent to itself.
3. $\triangle$ ADB $\cong \triangle$ CBD	3. The SSS Postulate.
4. $\angle 1 \cong \angle 2$ $\angle 3 \cong \angle 4$	4. Corresponding angles of congruent triangles are congruent.
5. $\overline{AD} \parallel \overline{BC}$, $\overline{AB} \parallel \overline{CD}$	5. If two coplanar lines are cut by a transversal such that the alternate interior angles are congruent, then the lines are parallel.

● **PROBLEM** 156

In the accompanying figure, $\overline{AD} = \overline{DC}$ and $\overline{BD} = \overline{DE}$. Prove that $\overline{AB} \parallel \overline{EC}$.

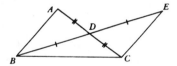

Solution: Our strategy is to prove that $\triangle$ ADB $\cong \triangle$ CDE, thereby proving that $\angle$ BAD $\cong \angle$ ECD. Since the latter 2 are alternate interior angles of segments $\overline{AB}$ and $\overline{EC}$, $\overline{AB} \parallel \overline{EC}$.

From the given fact, $\overline{AD} \cong \overline{DC}$ and $\overline{BD} \cong \overline{DE}$. $\angle$ ADB $\cong \angle$ CDE, since they are vertical angles. By the S.A.S. (Side Angle Side) Postulate, $\triangle$ ADB $\cong \triangle$ CDE. This implies that $\angle$ BAD $\cong \angle$ ECD. But these are alternate interior angles of the segments $\overline{AB}$ and $\overline{EC}$ which are cut by transversal $\overline{AC}$. Hence, $\overline{AB} \parallel \overline{EC}$.

● **PROBLEM** 157

Given: $\overline{AD} \cong \overline{BC}$; $\overline{AEB}$ and $\overline{DFC}$; AE = FC ; EB = DF ; $\angle$ A $\cong \angle$ C. Prove: $\overline{DE} \parallel \overline{BF}$.

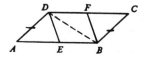

Solution: To show these two lines are parallel, we begin by drawing transversal $\overline{DB}$. Then, $\overline{DE} \parallel \overline{BF}$ if alternate interior angles $\angle$ EDB and ∡ FBD are congruent. To prove ∡ EDB ≅ ∡ FBD, note ∡ EDB and ∡ FBD are also corresponding angles of Δ DEB and Δ BFD.

We show Δ DEB ≅ Δ BFD by the SSS Postulate.

Statements	Reasons
1. $\overline{AD}$ ≅ $\overline{BC}$; $\overline{AEB}$; $\overline{DFC}$; AE = FC; EB = DF; ∡ A ≅ ∡ C.	1. Given
2. Δ ADE ≅ Δ CBF	2. The SAS Postulate
3. $\overline{DE}$ ≅ $\overline{BF}$	3. Corresponding parts of congruent triangles are congruent.
4. $\overline{DB}$ ≅ $\overline{DB}$	4. A segment is congruent to itself.
5. $\overline{DF}$ ≅ $\overline{BE}$	5. Segments that are equal in length are congruent.
6. Δ DEB ≅ Δ BFD	6. The SSS Postulate.
7. ∡ EDB ≅ ∡ FBD	7. Corresponding parts of congruent triangles are congruent.
8. $\overline{DE}$ ∥ $\overline{BF}$	8. Given two lines intercepted by the transversal; if the alternate interior angles are congruent, then the two lines are parallel.

● **PROBLEM** 158

Given: $\overline{AD}$ and $\overline{BC}$ intersect at E.
$\overline{AB}$ ∥ $\overline{CD}$.
CE = DE.
Prove: AE = BE.

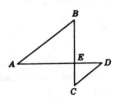

Solution: If two angles of a triangle are congruent, then the sides opposite them are congruent. Therefore, in Δ ABE, we prove AE = BE by showing ∡ A ≅ ∡ B.

We show the congruence of ∡ A and ∡ B by relating them to the congruent base angles of isosceles Δ ECD.

Statements	Reasons
1. $\overline{AB}$ ∥ $\overline{CD}$	1. Given

2. m ⟡ D = m ⟡ A	2. If two parallel lines are cut by a transversal, alternate interior angles are of equal measure.
3. CE = DE	3. Given
4. m ⟡ D = m ⟡ C	4. If two sides of a triangle are of equal length, then the angles opposite those sides are of equal measure.
5. m ⟡ A = m ⟡ C	5. If two quantities are equal to the same quantity, they are equal to each other.
6. m ⟡ B = m ⟡ C	6. If two parallel lines are cut by a transversal, alternate interior angles are of equal measure.
7. m ⟡ A = m ⟡ B	7. If two quantities are equal to the same quantity, they are equal to each other.
8. AE = BE	8. If a triangle has two angles of equal measure, then the sides opposite those angles are of equal length.

● **PROBLEM** 159

We are given two triangles, ABC and EHD, as shown in the diagram, and the straight line segment AE . If AD ≅ EC, BC ≅ HD and BC ∥ HD, prove AB ∥ EH .

Solution: To prove $\overline{AB} \parallel \overline{EH}$, we first show that the alternate interior angles, ⟡ BAC and ⟡ HED, formed by transversal $\overline{AE}$ and segments $\overline{AB}$ and $\overline{EH}$, are congruent. Since a pair of congruent alternate interior angles are sufficient to prove two lines parallel, we may conclude that the $\overline{AB} \parallel \overline{EH}$.

To prove ⟡ BAC and ⟡ HED are congruent, we must first prove the triangles containing them, Δ's ABC and EHD, are congruent by the SAS ≅ SAS method.

Statement	Reason
1. $\overline{BC} \cong \overline{HD}$	1. Given
2. $\overline{BC} \parallel \overline{HD}$	2. Given
3. ⟡ BCA ≅ ⟡ HDE	3. If two parallel lines are cut by a transversal, then alternate interior angles are congruent.
4. $\overline{AE}$ is a line segment.	4. Given

5. $\overline{AD} \cong \overline{EC}$	5. Given
6. $\overline{DC} \cong \overline{DC}$	6. Reflexive property of congruence.
7. AD = EC; DC = DC	7. Congruent segments are of equal lengths.
8. AD + DC = EC + CD	8. Addition Postulate
9. AC = ED	9. Point Betweenness Postulate.
10. $\overline{AC} \cong \overline{ED}$	10. Segments of equal length are congruent.
11. $\triangle ABC \cong \triangle EHD$	11. SAS $\cong$ SAS
12. $\not{\times}$ BAC $\cong \not{\times}$ HED	12. Corresponding parts of congruent triangles are congruent.
13. $\overline{AB} \parallel \overline{EH}$	13. If two lines are cut by a transversal making a pair of alternate interior angles congruent, then the lines are parallel.

● **PROBLEM** 160

Show that $\triangle AEB \cong \triangle BFC \cong \triangle CGD$; given that AE = BF = CG and $L_1 \parallel L_2$, $L_2 \parallel L_3$, $L_3 \parallel L_4$, (see figure) and $\overline{AE} \perp L_2$, $\overline{BF} \perp L_3$, $\overline{CG} \perp L_4$.

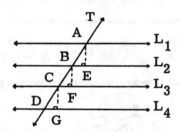

Solution: To show $\triangle AEB \cong \triangle BFC \cong \triangle CGD$, we note the available congruences. We are given AE = BF = CG. Since $\overline{AE} \perp L_2$, $\overline{BF} \perp L_3$ and $\overline{CG} \perp L_4$, the angles $\not{\times}$ AEB, $\not{\times}$ BFC, and $\not{\times}$ CGD are right angles and thus all congruent. $\not{\times}$ ABE, $\not{\times}$ BCF, and $\not{\times}$ CDG are corresponding angles of parallel lines L_1, L_2, L_3, and L_4. Therefore, $\not{\times}$ ABE $\cong \not{\times}$ BCF $\cong \not{\times}$ CDG. We thus have an A.A.S. correspondence. By the A.A.S. Postulate, $\triangle AEB \cong \triangle BFC \cong \triangle CGD$.

● **PROBLEM** 161

Given: $\overline{AD} \parallel \overline{BE}$, $\overline{BD} \parallel \overline{CE}$, and B is midpoint of $\overline{AC}$.
Prove: BE = AD.

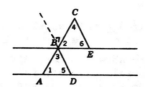

Solution: Our strategy is to prove △ BCE ≅ △ ABD. From corresponding parts, it then follows that BE ≅ AD. To prove △ BCE ≅ △ ABD, we use the ASA Postulate.

Statements	Reasons
1. AD ∥ BE	1. Given
2. ∡ 1 ≅ ∡ 2	2. If two parallel lines are cut by a transversal, corresponding angles are congruent.
3. B is the midpoint of AC	3. Given
4. AB ≅ CB	4. Definition of midpoint.
5. BD ∥ CE	5. Given
6. ∡ 3 ≅ ∡ 4	6. If two parallel lines are cut by a transversal, the alternate interior angles are congruent.
7. △ ABD ≅ △ BCE	7. The ASA Postulate
8. BE ≅ AD	8. Corresponding parts of congruent triangles are congruent.
9. BE = AD	9. Congruent segments are of equal length.

● **PROBLEM 162**

In Figure 1, ∡ B measures 30°, $\overleftrightarrow{BQ} \parallel \overleftrightarrow{AP}$ and $\overleftrightarrow{BP} \parallel \overleftrightarrow{AC}$. Determine m ∡ 1, m ∡ 2, m ∡ 3, m ∡ 4, and m ∡ 5.

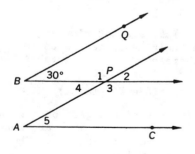

Figure 1

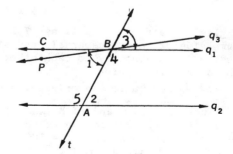

Figure 2

Solution: Prior to attacking the question at hand, we will establish several important facts about angles formed by parallel lines cut by a transversal. We will employ two characteristics about parallel lines:
(1) Through a given external point, only one line can be drawn parallel

135

to a given line; and (2) If <u>one</u> pair of alternate interior angles are congruent, then the two lines must be parallel. (See Figure 2).

The first fact we wish to show is that, given two parallel lines q_1 and q_2 cut by transversal t, every pair of alternate interior angles are congruent. We do this by indirect proof. Suppose $q_1 \| q_2$ but alternate interior angles ∡ 1 and ∡ 2 are not congruent. Draw q_3 through point B such that it forms ∡ ABP so that ∡ ABP ≅ ∡ 2. This leads to two contradictory lines of reasoning:

(A) Because ∡ ABP ≅ ∡ 2 and ∡ 1 is not ≅ ∡ 2, it follows that ∡ ABP is not ≅ ∡ 1. Since q_3 and q_1 make different angles with transversal t, lines q_1 and q_3 must be different lines.

(B) Because alternate interior angles ∡ ABP and ∡ 2 are congruent, (by assumption (2)) it follows that $q_3 \| q_2$. Since both q_3 and q_1 are lines parallel to q_2 and passing through external point B, (by assumption (1)) it follows that q_1 and q_3 are the same line.

Because the two lines of reasoning lead to different conclusions, our original assumption - ∡ 1 is not ≅ ∡ 2 - must be false. We can thus conclude that if two parallel lines are cut by a transversal, <u>any</u> pair of alternate interior angles are congruent.

Next we shall establish that if two parallel lines are cut by a transversal, each pair of corresponding angles are congruent. From the above theorem, we know ∡ 1 ≅ ∡ 2. Since ∡ 1 and ∡ 3 are, by definition, vertical angles, then ∡ 1 ≅ ∡ 3. Therefore, by transitivity, ∡ 2 ≅ ∡ 3 and we have the desired results.

Finally, we will prove that interior angles on the same side of the transversal are supplementary.

Again, since alternate interior angles of parallel lines cut by a transversal are congruent, we have ∡ 1 ≅ ∡ 2. Adjacent angles, whose exterior sides are contained in a straight line, are supplementary. Thus, their measures sum to 180°. Therefore,

$$m \angle 1 + m \angle 4 = 180$$

$$m \angle 2 + m \angle 5 = 180 .$$

Since $m \angle 1 = m \angle 2$, the quantities can be interchanged to obtain the following:

$$m \angle 2 + m \angle 4 = 180$$

$$m \angle 1 + m \angle 5 = 180 .$$

Angles 2 and 4, as well as ∡ 1 and ∡ 5, are interior angles on the same side of the transversal, and since their measures sum to 180°, they must be supplementary to each other. Therefore, we have established our third and final point.

Now we return to the original question. (See Figure 1.)

Since $\overleftrightarrow{BQ} \| \overleftrightarrow{AP}$ and ∡ B and ∡ 1 are interior angles on the same side of the transversal, by a theorem above, ∡ B and ∡ 1 are supplements, i.e. $m \angle B + m \angle 1 = 180$. We are given $m \angle B = 30$. Therefore, by substitution, $30 + m \angle 1 = 180$, and $m \angle 1 = 150$.

Angles B and 2 are corresponding angles formed by a transversal cutting parallel lines $\overleftrightarrow{BQ}$ and $\overleftrightarrow{AP}$. As such, $m \angle B = m \angle 2$. Therefore, $m \angle 2 = 30$.

For the parallel lines $\overleftrightarrow{BP}$ and $\overleftrightarrow{AC}$ cut by transversal $\overleftrightarrow{AP}$, ∡ 5 and ∡ 2 are corresponding angles and, accordingly, are congruent. Therefore, $m \angle 5 = 30$.

On the same set of parallel lines, ∡ 5 and ∡ 4 are alternate interior angles and, thus are congruent. Therefore, m ∡ 4 = 30.

The last angle we need to determine is ∡ 3. Angle 3 is on the same side of the transversal as ∡ 5 and on the interior of $\overline{BP} \| \overline{AC}$. Therefore, ∡ 3 is supplementary to ∡ 5. As such, m ∡ 3 + m ∡ 5 = 180. By substitution, m ∡ 3 + 30 = 180. Thus, m ∡ 3 = 150.

Collecting our results, we have: m ∡ 1 = 150, m ∡ 2 = 30, m ∡ 3 = 150, m ∡ 4 = 30, and m ∡ 5 = 30.

● **PROBLEM** 163

Show that any two medians of a triangle intersect at a point in the interior of the triangle.

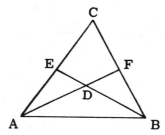

Solution: In the accompanying figure, △ABC has medians $\overline{AF}$ and $\overline{EB}$ and D is the point of intersection. To show that lines intersect in a point in the interior of a triangle, we must show (1) the lines intersect in a point; and (2) the lines intersect in the interior of the triangle.

(1) To show the lines intersect in a point: Two lines can be related in four ways; they can be skew, parallel, coincident, or intersecting in a point. We disprove the first three: Since only the last case remains, then it must be true that the lines intersect in a point.

A triangle is a planar figure. Therefore all points, including the vertices and the midpoints of the sides, must be coplanar. The medians connect a midpoint of a side to the opposite vertex. Since these two points of each median are on the plane of the triangle, all medians of the triangle are coplanar, and the medians cannot be skew.

Suppose median $\overline{AF} \|$ median $\overline{EB}$. Then ∡ EBA and ∡ FAB are interior angles on the same side of the transversal, $\overline{AB}$ and must be supplementary, or m ∡ EBA + m ∡ FAB = 180.

The median $\overline{AF}$ must be in the interior of ∡ A. To see this, remember the requirements for being in the interior of the angle. If point P is in the interior of the angle, then there can be found a point on each of the lines forming the angle such that a segment determined by these points contain P. Point F is on $\overline{BC}$. B and C are points on the lines forming angle A. Therefore F is in the interior of angle A. Furthermore, the ray with the vertex as an endpoint which contains an interior point of the angle lies entirely in the interior of a triangle. Then, median $\overline{AF}$ must be in the interior of ∡ A. Therefore, m ∡ FAB < m ∡ A. Similarly m ∡ EBA < m ∡ B. Thus m ∡ A + m ∡ B > 180. But no two angles of a triangle (in this case, ΔABC) can sum to greater than 180°; therefore, our original supposition $\overline{AF} \parallel \overline{EB}$ is incorrect. The medians cannot be __parallel__.

Suppose medians $\overline{AF}$ and $\overline{BE}$ are coincident. Then points A,B,E, and F all lie on the same line. By definition, the midpoint E of side $\overline{AC}$ must be collinear with A and C or, vice versa, C must be collinear with A and E. Since A and E are collinear with B and F, A,B, and C are collinear. But the vertices of a triangle cannot be collinear. By definition, a triangle is determined by three __non__-colinear points. Therefore, our supposition about the medians is incorrect. The medians cannot be coincident.

By elimination we have shown that the medians must intersect at a point.
(2) To show the point of intersection is in the interior of the triangle: Here there are three cases to be considered: the point is exterior to the triangle, on a side of the triangle, or in the interior of the triangle.

The median is a segment that joins two points, the vertex and the midpoint of the opposite side, on the boundary of a triangle. All triangles are convex - they contain no angles greater than 180° - and thus share the special property of convex figures that any line segment drawn between two points on the boundary or in the interior of the convex figure lies entirely in the interior of the figure. The median is such a line segment. Therefore all points on the median lie in the interior of the triangle, except for the vertex and the midpoint which lie on the boundary. Consequently the intersection of medians can only include points on the boundary or in the interior of the triangle.

Suppose the medians intersect at a point on the triangle. There are three cases.

Case I: $\overline{AF}$ intersects $\overline{BE}$ at point D such that $\overline{ADC}$ is a line segment
Case II: $\overline{AF}$ intersects $\overline{BE}$ at D such that $\overline{BDC}$ is a line segment.
Case III: $\overline{AF}$ intersects $\overline{BE}$ at D such that $\overline{ADB}$ is a line segment. Without loss of generality it is sufficient to consider just Case I for it is exactly typical of the others and the others pose no additional problems. Since points A and D of median $\overline{AF}$ are collinear with $\overline{AC}$, then $\overline{AF}$ must be collinear with $\overline{AC}$. Using an argument similar to the one used to show that the medians are not coincident, we note that, since F is the midpoint of $\overline{CB}$, B must be collinear with $\overline{CF}$. Since $\overline{ACF}$ is a line, it follows that $\overline{ABC}$ is a line which contradicts our assumption that the vertices of ΔABC are non-collinear. Therefore, the point of intersection is not on the boundary of the triangle.

By elimination, the intersection of the medians must be in the interior of the triangle.

● **PROBLEM** 164

Show that the perpendicular bisectors of the sides of a triangle are concurrent at a point equidistant from the vertices of the triangle.

__Solution__: First, we have to show that the perpendicular bisectors must intersect. Second, we will show that any point that is the intersection of two of the perpendicular bisectors must be on the third perpendiculer bisector. Third, the point common to the bisectors is equidistant from

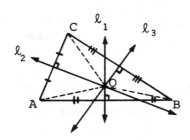

any pair of vertices and therefore is equidistant from all three.

The method of proof to be followed here is one of contradiction. We will assume two perpendicular bisectors are not concurrent and show that this leads to a logical contradiction of a known theorem, definition or hypothesis.

Given: Lines ℓ_1, ℓ_2, ℓ_3 are the perpendicular bisectors of the sides of $\triangle ABC$: AB, AC, BC, respectively.

Prove: ℓ_1, ℓ_2, ℓ_3 have point Q in common such that AQ = BQ = CQ.

Statements	Reasons
1. Lines ℓ_1, ℓ_2, ℓ_3 are the perpendicular bisectors of $\overline{AB}$, $\overline{AC}$, and $\overline{BC}$, sides of $\triangle ABC$.	1. Given.
2. Either (1) ℓ_1 does not intersect ℓ_2; or (2) ℓ_1 intersects ℓ_2 .	2. Either a statement or its negation must be true.
3. Case I: (1) ℓ_1, ℓ_2 do not intersect. $\ell_1 \parallel \ell_2$	3. Lines in the same plane that do not intersect are parallel.
4. $\overline{AC} \perp \ell_2$, $\overline{AB} \perp \ell_1$.	4. Definition of perpendicular bisector.
5. $\overline{AC} \parallel \overline{AB}$.	5. In a plane, lines perpendicular to parallel lines are parallel.
6. $\overline{AC} \nparallel \overline{AB}$.	6. Adjacent sides of a triangle cannot be parallel.
7. "(1) ℓ_1, ℓ_2 do not intersect" is not a true statement.	7. If a statement leads to, by logically valid reasoning, to a contradiction, then that statement must be false.
8. ℓ_1 intersects ℓ_2 at point Q.	8. Follows from Step 2, and: the intersection of two non-parallel distinct lines is a point.
9. AQ = BQ AQ = CQ	9. Since Q is in both perpendicular bisectors, it is equidistant from the endpoints of both segments.
10. AQ = BQ = CQ	10. Transitivity Property.
11. AQ = CQ	11. From Step 10.
12. Q is on ℓ_3	12. The perpendicular bisector of a segment is the locus of all points that are equidistant from the endpoints of the segment.

13. ℓ_1, ℓ_2, and ℓ_3 are concurrent at point Q such that AQ = BQ = CQ

13. Follows from Steps 12, 10, and 8.

● **PROBLEM** 165

Show that the lines containing the three altitudes are concurrent.

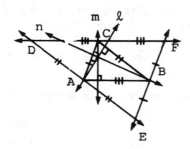

<u>Solution:</u> We have shown that the perpendicular bisectors are concurrent. Here we are asked to show that another set of lines are concurrent. If we show that the lines that contain the altitudes of a given triangle also contain the perpendicular bisectors of a second triangle, then the concurrency follows immediately.

Given: The altitudes of △ABC, ℓ, m, and n.
Prove: ℓ, m, n are concurrent.

Statements	Reasons
1. The altitudes of △ABC, ℓ, m, and n	1. Given
2. Construct $\overleftrightarrow{DF} \parallel AB$ through C, $\overleftrightarrow{BE} \parallel \overline{BC}$ through A, and $\overleftrightarrow{EF} \parallel \overleftrightarrow{AC}$ through B	2. Through a given external point only one line can be drawn parallel to a given line.
3. Quadrilateral ABFC is a parallelogram.	3. If opposite sides of a quadrilateral ($\overline{AC}$ and $\overline{BF}$, $\overline{CF}$ and $\overline{AB}$) are parallel, then the quadrilateral is a parallelogram.
4. Quadrilateral ACBE is a parallelogram. Quadrilateral ADCB is a parallelogram.	4. Same reason as Step 3.
5. (a) $\overline{DC} \cong \overline{AB}$, $\overline{CF} \cong \overline{AB}$ (b) $\overline{BE} \cong \overline{CA}$, $\overline{FB} \cong \overline{CA}$ (c) $\overline{DA} \cong \overline{CB}$, $\overline{AE} \cong \overline{CB}$	5. Opposite sides of a parallelogram are congruent.
6. (a) $\overline{DC} \cong \overline{CF}$ (b) $\overline{BE} \cong \overline{FB}$ (c) $\overline{DA} \cong \overline{AE}$	6. Segments congruent to the same segments are congruent.

7. (a) C is the midpoint of $\overline{DF}$

   (b) B is the midpoint of $\overline{FE}$

   (c) A is the midpoint of $\overline{DE}$

8. (a) m ⊥ AB
   (b) n ⊥ AC
   (c) ℓ ⊥ CB .

9. (a) m ⊥ $\overline{DF}$
   (b) n ⊥ $\overline{FE}$
   (c) ℓ ⊥ $\overline{DE}$

10. (a) m is the perpendicular bisector of $\overline{DF}$

   (b) n is the perpendicular bisector of $\overline{FE}$

   (c) ℓ is the perpendicular bisector of $\overline{DE}$

11. $\overline{DF}$, $\overline{FE}$, $\overline{DE}$ are sides of $\triangle DFE$

12. n,m, and ℓ are concurrent.

7. Definition of midpoint.

8. Definition of the altitude of a triangle.

9. In a plane, if a line is perpendicular to one of two parallel lines, then it is perpendicular to the other parallel line.

10. A line perpendicular to a segment at its midpoint is the perpendicular bisector of the segment.

11. In a plane, three non-colinear points determine a triangle

12. The perpendicular bisectors of the sides of a triangle are concurrent.

CHAPTER 11

INEQUALITIES

Basic Attacks and Strategies for Solving Problems in this Chapter. See pages 142 to 159 for step-by-step solutions to problems.

Most of the problems in this chapter involve establishing that the measure of one angle is greater than the measure of another angle, or one segment is longer than another segment. Since the lengths of segments and the measures of angles are represented by positive real numbers, the following properties involving inequalities apply to these problems. For all positive real numbers a, b c, and d:

if $a < b$ then $b > a$

if $a < b$ then $a + c < b + c$

if $a < b$ then $a - c < b - c$

if $a < b$ then $ac < bc$

if $a < b$ and $c \neq 0$ then $\dfrac{a}{c} < \dfrac{b}{c}$

if $a < b$ and $b < c$ then $a < c$

if $a < b$ and $c < d$ then $a + c < b + d$

if $a < b$ and $c < d$ then $ac < bd$

In this chapter, by substituting these variable names for angle measures or segment lengths, we can use these algebraic inequalities to represent geometric truths.

142-A

Step-by-Step Solutions to
Problems in this Chapter,
"Inequalities"

LINES & ANGLES

● **PROBLEM** 166

Triangles ABC and DBC, as shown in the accompanying figure, are constructed in a way such that m ∡ DBC > m ∡ ABC and m ∡ ABC > m ∡ ACB. Prove that m ∡ DBC > m ∡ ACB.

<u>Solution</u>: This proof will involve applying the Transitive Property of Inequality, which states that if the first of three quantities is greater than the second and the second is greater than the third, then the first is greater than the third.

<u>Statement</u>	<u>Reason</u>
1. m ∡ DBC > m ∡ ABC	1. Given
2. m ∡ ABC > m ∡ ACB	2. Given
3. m ∡ DBC > m ∡ ACB	3. Transitive Property of Inequality.

● **PROBLEM** 167

In the accompanying figure, line segment $\overline{CD}$ is drawn equal in length to $\overline{CB}$. If CB < CA, prove that CD < CA.

<u>Solution</u>: This proof will involve knowledge of the Substitution
Postulate for Inequalities, which states that a quantity may be substituted for its equal in any inequality.

Statement	Reason
1. CB < CA	1. Given
2. CD = CB	2. Given
3. CD < CA	3. Substitution Postulate for Inequalities.

● **PROBLEM** 168

In the figure, we are given $\angle$ DEF and $\angle$ ABC. m $\angle$ DEF = m $\angle$ ABC.
If m $\angle$ DEH > m $\angle$ ABG, then prove that m $\angle$ HEF < m $\angle$ GBC.

<u>Solution</u>: A postulate tells us that when unequal quantities are sub-
tracted from equal quantities, the differences are unequal in the opposite order. For example, if 10 = 10 and 8 > 4, then 10 - 8 < 10 - 4
or 2 < 6. This postulate will be utilized here.

Statements	Reasons
1. m $\angle$ DEF = m $\angle$ ABC	1. Given
2. m $\angle$ DEH > m $\angle$ ABG	2. Given
3. m $\angle$ DEF - m $\angle$ DEH < m $\angle$ ABC - m $\angle$ ABG or m $\angle$ HEF < m $\angle$ GBC	3. If unequal quantities are sub-tracted from equal quantities, the differences are unequal in the opposite order.

● **PROBLEM** 169

We are given, as the figures show, $\angle$ ABC and $\angle$ DEF. m $\angle$ ABC >
m $\angle$ DEF. If m $\angle$ ABG = m $\angle$ DEH, prove m $\angle$ GBC > m $\angle$ HEF.

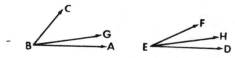

<u>Solution</u>: By realizing that when equal quantities are subtracted from
unequal quantities, the differences are unequal in the same order, we
can proceed directly to the proof.

Consult the diagram to clarify the steps of the proof.

Statements	Reasons
1. m ⦡ ABC > m ⦡ DEF	1. Given
2. m ⦡ ABG = m ⦡ DEH	2. Given
3. m ⦡ ABC - m ⦡ ABG > m ⦡ DEF - m ⦡ DEH or m ⦡ GBC > m ⦡ HEF	3. If equal quantities are subtracted from unequal quantities, the differences are unequal in the same order.

• **PROBLEM** 170

In the accompanying figure, unequal angles ABC and DEF are bisected by $\overleftrightarrow{BG}$ and $\overleftrightarrow{EH}$, respectively. If m ⦡ ABC > m ⦡ DEF, prove m ⦡ ABG > m ⦡ DEH.

Solution: The fact that unequal quantities remain unequal in the same order when they are divided by equal positive quantities will be employed in this proof. (Since we are dealing with an angle bisector, the specific form to be used will be that halves of unequal quantities are unequal in the same order.)

Statements	Reasons
1. m ⦡ ABC > m ⦡ DEF	1. Given
2. $\frac{m \angle ABC}{2} > \frac{m \angle DEF}{2}$ or ½m ⦡ ABC > ½m ⦡ DEF	2. Halves of unequal quantities are unequal in the same order.
3. $\overleftrightarrow{BG}$ bisects ⦡ ABC $\overleftrightarrow{EH}$ bisects ⦡ DEF	3. Given
4. m ⦡ ABG = ½m ⦡ ABC m ⦡ DEH = ½m ⦡ DEF	4. A bisector of an angle divides the angle into two congruent angles.
5. m ⦡ ABG > m ⦡ DEH	5. A quantity may be substituted for its equal in any inequality. (See steps (2) and (4)).

• **PROBLEM** 171

Prove that the supplement of an obtuse angle is an acute angle.

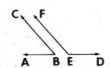

Solution: There are two subtraction postulates involving inequalities. The first, the Subtraction Postulate of Inequality, states that equal quantities subtracted from an inequality yield an inequality in the

144

same sense. The second postulate states that unequal quantities subtracted from an equality yield an inequality in the opposite sense-
that is, if a = b and c > d, then a - c < b - d. We will use the
second postulate in this proof.

Given: ∡ DEF is obtuse; ∡ ABC and ∡ DEF are supplements.

Prove: ∡ ABC is acute.

Statements	Reasons
1. ∡ DEF is obtuse; ∡ ABC and ∡ DEF are supplements	1. Given
2. m ∡ ABC + m ∡ DEF = 180	2. Supplementary angles sum to 180°.
3. m ∡ DEF > 90°	3. The measure of an obtuse angle is greater than 90° and less than 180°.
4. m ∡ ABC < 90°	4. If unequal quantities are subtracted from equal quantities, the differences are unequal in the opposite order (see steps (2) and (3)).
5. ∡ ABC is an acute angle	5. An angle whose measure is less than 90° and greater than 0° is acute.

● **PROBLEM** 172

Let ∡ AOB and ∡ COB be a linear pair and ∡ MPN and ∡ QPN be
another linear pair. Prove that m ∡ AOB > m ∡ MPN if and only if
m ∡ QPN > m ∡ COB.

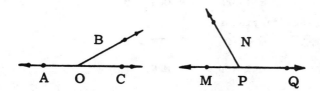

Solution: Two angles form a linear pair if they share a common ray,
and if their second rays are directly opposite to each other. In the
figure, ∡ AOB and ∡ COB are a linear pair. They share a common ray,
OB, and their second rays, OA and OC, are directly opposite. Similarly, ∡ MPN and ∡ QPN are a linear pair.

Since the given theorem is an "if and only if" theorem, we can
prove it by breaking it up into 2 parts and proving each part separately.

Part I: If m ∡ AOB > m ∡ MPN, then m ∡ QPN > m ∡ COB.

Part II: If m ∡ QPN > m ∡ COB, then m ∡ AOB > m ∡ MPN.

145

We proceed to prove Part I by noting that 2 angles which form a linear pair are supplementary. Since ∢ AOB and ∢ COB are a linear pair,

$$m \angle AOB + m \angle COB = 180° \tag{1}$$

Because ∢ MPN and ∢ QPN form a linear pair,

$$m \angle MPN + m \angle QPN = 180° \tag{2}$$

Equating (1) and (2) yields

$$m \angle MPN + m \angle QPN = m \angle AOB + m \angle COB$$

or

$$m \angle AOB - m \angle MPN = m \angle QPN - m \angle COB \tag{3}$$

But, the hypothesis of Part I states that

$$m \angle AOB > m \angle MPN$$

or

$$m \angle AOB - m \angle MPN > 0 \tag{4}$$

However, comparing (4) and (3) yields

$$m \angle QPN - m \angle COB > 0$$

or

$$m \angle QPN > m \angle COB$$

Therefore, Part I is proven. We next prove Part II.

Again, we use the fact that a linear pair of angles is supplementary. We can still use equations (1) and (2) above, because they are a statement of the fact that the angles shown in the figure are supplementary. Hence, equating (1) and (2) yields

$$m \angle MPN + m \angle QPN = m \angle AOB + m \angle COB$$

or

$$m \angle QPN - m \angle COB = m \angle AOB - m \angle MPN \tag{5}$$

From the hypothesis of Part II,

$$m \angle QPN > m \angle COB$$

or

$$m \angle QPN - m \angle COB > 0 \tag{6}$$

Comparing (6) and (5) yields

$$m \angle AOB - m \angle MPN > 0$$

or

$$m \angle AOB > m \angle MPN.$$

● **PROBLEM** 173

Point P is in the exterior of Δ ABC, in the opposite half plane of BC from A, such that BP = CP. m ∢ ABC > m ∢ ACB. Show m ∢ ABP > m ∢ ACP.

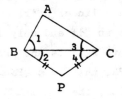

146

<u>Solution</u>: Adding equal quantities to both sides of an inequality does not change the sense of the inequality. We are given m ∢ 1 > m ∢ 3. We will show m ∢ 2 = m ∢ 4. Therefore, m ∢ 1 + m ∢ 2 > m ∢ 3 + m ∢ 4. Since m ∢ ABP = m ∢ 1 + m ∢ 2 and m ∢ ACP = m ∢ 3 + m ∢ 4, it follows that m ∢ ABP > m ∢ ACP.

Given: P is in the exterior of △ ABC, in the opposite half-plane of BC from A; m ∢ ABC > m ∢ ACB; BP = CP.

Prove: m ∢ ABP > m ∢ ACP.

Statement	Reason
1. P is in the exterior of △ ABC, in the opposite half-plane of BC from A; BP = CP; m ∢ ABC > m ∢ ACB .	1. Given
2. △ BPC is an isosceles triangle.	2. If two sides of a triangle are congruent, then the triangle is isosceles.
3. m ∢ PBC = m ∢ PCB	3. The base angles of an isosceles triangle are congruent.
4. m ∢ PBC + m ∢ ABC > m ∢ PCB + m ∢ ACB	4. For all real numbers m, n, a, and b, if m > n and a ≥ b then m + a > n + b. (Addition Postulate of Inequalities).
5. m ∢ ABP > m ∢ ACP	5. Angle Sum Postulate

● **PROBLEM** 174

Isosceles triangle ABC and triangle ADC share a side, $\overline{AC}$, as shown in the figure. If $\overline{AB} \cong \overline{BC}$, and m ∢ BAD > m ∢ BCD, prove m ∢ CAD > m ∢ ACD.

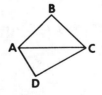

<u>Solution</u>: By the Isosceles Triangle Theorem, we conclude that the base angles of isosceles triangle ABC are equal in measure. We will then subtract these equal angles from the given unequal angles and conclude, by the Subtraction Postulate of Inequality that the differences are unequal in the same order.

Statements	Reasons
1. m ∢ BAD > m ∢ BCD	1. Given

147

2. $\overline{AB} \cong \overline{BC}$	2. Given
3. m ⦨ BAC = m ⦨ BCA	3. If two sides of a triangle are congruent, the angles opposite these sides are of equal measure.
4. m ⦨ BAD - m ⦨ BAC > m ⦨ BCD - m ⦨ BCA	4. If equal quantities are subtracted from unequal quantities, the differences are unequal in the same order. (See steps (1) and (3)).
5. m ⦨ CAD = m ⦨ BAD - m ⦨ BAC m ⦨ ACD = m ⦨ BCD - m ⦨ BCA	5. Angle Addition Postulate.
6. m ⦨ CAD > m ⦨ ACD	6. Substitution Postulate.

● **PROBLEM** 175

In triangle ABC, angles ABC and ACB are divided by $\overline{BD}$ and $\overline{CE}$, respectively. This results in m ⦨ BCE > m ⦨ DBC and m ⦨ ACE > m ⦨ ABD. Prove: m ⦨ ACB > m ⦨ ABC.

Solution: The Addition Postulate of Inequalities states that when un-equal quantities are added to unequal quantities of the same order, then the sums are unequal in the same order. (The three main orders are greater than/less than/equal to.) In the given problem angles ABC and ACB are each divided into two parts. Each part of ⦨ ACB is greater than a part of ⦨ ABC. By the above-mentioned Postulate, m ⦨ ACB > m ⦨ ABC.

Given: △ ABC; m ⦨ BCE > m ⦨ DBC; m ⦨ ACE > m ⦨ ABD.
Prove: m ⦨ ACB > m ⦨ ABC.

Statements	Reasons
1. m ⦨BCE > m ⦨DBC m ⦨ACE > m ⦨ABD	1. Given
2. m ⦨BCE + m ⦨ACE m ⦨DBC + m ⦨ABD	2. Addition Postulate of Inequality.
3. m ⦨ACB = m ⦨BCE + m ⦨ACE m ⦨ABC = m ⦨DBC + m ⦨ABD	3. Angle Addition Postulate.
4. m ⦨ACB > m ⦨ABC	4. Substitution Postulate.

148

In triangle ABC, which is shown in the figure, $\overline{AD}$ bisects ⊰ CAB and $\overline{BE}$ bisects ⊰ CBA. If m ⊰ DAB < m ⊰ EBA, prove that m ⊰ CAB < m ⊰ CBA.

Solution: When unequal quantities are multiplied by equal positive quantities, the products are unequal in the same order. (The special case of this, that doubles of unequal quantities are unequal in the same order, will be applied to this problem.) To reach our final conclusion, we recall that a quantity may be substituted for its equal in any inequality.

Statements	Reasons
1. m ⊰ DAB < m ⊰ EBA	1. Given
2. 2m ⊰ DAB < 2m ⊰ EBA	2. Doubles of unequal quantities are unequal in the same order.
3. $\overline{AD}$ bisects ⊰ CAB $\overline{BE}$ bisects ⊰ CBA	3. Given
4. m ⊰ CAB = 2m ⊰ DAB m ⊰ CBA = 2m ⊰ EBA	4. An angle bisector divides the angle into two congruent angles.
5. m ⊰ CAB < m ⊰ CBA	5. A quantity may be substituted for its equal in any inequality. (See step (2).)

Given: ⊰ 1 ≅ ⊰ 2; ⊰ 3 ≅ ⊰ 4; AB < AC; $\overline{AEB}$; $\overline{AFC}$.
Prove: EB < FC.

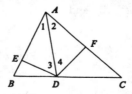

Solution: We can consider $\overline{EB}$ and $\overline{FC}$ as parts of sides $\overline{AB}$ and $\overline{AC}$ of △ ABC.

We are given AB < AC. Furthermore, by showing △ AED ≅ △ AFD, we

obtain AE = AF. By the Addition Postulate of Inequality, AB - AE <
AC - AF or EB < FC.

Statements	Reasons
1. ∡ 1 ≅ ∡ 2; ∡ 3 ≅ ∡ 4; AB < AC; $\overline{AEB}$; $\overline{AFC}$	1. Given
2. $\overline{AD} \cong \overline{AD}$	2. A segment is congruent to itself.
3. △ AED ≅ △ AFD	3. The ASA Postulate.
4. $\overline{AE} \cong \overline{AF}$	4. Corresponding parts of congruent triangles are congruent.
5. AE = AF	5. Definition of congruent segments.
6. AB - AE < AC - AF	6. Subtraction Property of Inequality.
7. EB = AB - AE FC = AC - AF	7. Definition of "betweenness" (if $\overline{XYZ}$, then XZ = XY + YZ.)
8. EB < FC	8. Substitution Postulate.

● **PROBLEM** 178

If D and E are the respective midpoints of sides $\overline{AB}$ and $\overline{AC}$ of
△ ABC, and AD < AE, prove that AB < AC.

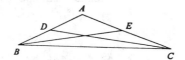

Solution: By the definition of midpoint, AB = 2(AD) and AC = 2(AE).
Since we are given AD < AE, then, by the multiplication postulate for
inequalities, AB < AC.

Given: △ ABC with D and E as the respective midpoints of $\overline{AB}$
and $\overline{AC}$; AD < AE.

Prove: AB < AC

Statements	Reasons
1. △ ABC, with D and E the respective midpoints of $\overline{AB}$ and $\overline{AC}$; AD < AE	1. Given
2. 2(AD) < 2(AE)	2. For all real numbers a, m, and n, if m > n and a > 0, then am > an. (Multiplication Postulate for Inequality).
3. AB = 2(AD); AC = 2(AE)	3. Definition of midpoint.
4. AB < AC	4. Substitution Postulate.

150

Given: Δ ABC; D is the midpoint of $\overline{AC}$ and $\overline{BP}$; P is in the interior of ⊀ ACR.

Prove: m ⊀ ACR > m ⊀ A.

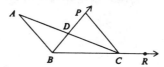

Solution: If we show ⊀ A is congruent to an angle smaller than ⊀ ACR, then m ⊀ ACR > m ⊀ A. By showing Δ ABD ≅ Δ CPD, we show ⊀ A ≅ ⊀ DCP. Since P is a point in the interior of ⊀ ACR, then m ⊀ DCP < m ⊀ ACR, and m ⊀ A < m ⊀ ACR.

Statements	Reasons
1. Δ ABC; D is the midpoint of $\overline{AC}$ and $\overline{BP}$; P is in the interior of ⊀ ACR.	1. Given
2. $\overline{BD} \cong \overline{DP}$; $\overline{AD} \cong \overline{DC}$	2. A midpoint of a segment divides the segment into two congruent segments.
3. ⊀ ADB ≅ ⊀ CDP	3. Opposite angles formed by intersecting lines are congruent. (Vertical Angle Theorem).
4. Δ ADB ≅ Δ CDP	4. The SAS Postulate.
5. ⊀ A ≅ ⊀ DCP	5. Corresponding parts of congruent triangles are congruent.
6. m ⊀ DCP < m ⊀ ACR	6. If P is in the interior of ⊀ ACR, then m ⊀ ACP < m ⊀ ACR.
7. m ⊀ A < m ⊀ ACR	7. If a = b and a < c, then b < c.

Given: ⊰ 1 ≅ ⊰ 2; ⊰ 3 ≅ ⊀ 4; $\overline{GD} \cong \overline{HD}$; EH = CG; $\overline{BHD}$; $\overline{EGHC}$; $\overline{APGD}$.
Prove: m ⊀ 5 < m ⊀ 6.

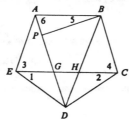

Solution: We can relate ⊀ 5 and ⊀ 6 as angles of Δ BAP or we can relate them to a third angle not in Δ BAP. Too many parts of Δ BAP are unknown for the first method to be useful to us. It is difficult to see how we can show PB > AP and, therefore, m ⊀ 6 > m ⊀ 5.

Using the second method, we note the symmetry of the figure:
∢ 1 ≅ ∢ 2; ∢ 3 ≅ ∢ 4. It would seem reasonable that Δ ADB is isosceles. If this were so, then base angles ∢ 6 and ∢ ABD would be
congruent. Since m ∢ 5 < m ∢ ABD, m ∢ 5 < m ∢ 6.

We show Δ ADB is isosceles by showing AD = DB. Since AD =
AG + GD, DB = DH + HB, and it is given that GD = HD, we need only show
AG = BH. We do this by showing Δ AEG ≅ Δ BHC by the ASA Postulate.

Statements	Reasons
1. ∢ 1 ≅ ∢ 2; ∢ 3 ≅ 4; $\overline{GD}$ ≅ $\overline{HD}$; EH = CG; $\overline{BHD}$; $\overline{EGHC}$; $\overline{APGD}$	1. Given. (We show Δ AEG ≅ Δ BHC by noting that ∢ 3 ≅ ∢ 4, $\overline{EG}$ ≅ $\overline{HC}$, and ∢ AGE ≅ ∢ BHC. In Step 2 to 4, we show ∢ AGE ≅ ∢ BHC.)
2. ∢ HGD ≅ ∢ GHD	2. The base angles of an isosceles triangle are congruent.
3. ∢ AGE ≅ ∢ HGD; ∢ BHC ≅ ∢ GHD	3. The opposite angles formed by intersecting lines are congruent. (Vertical Angle Theorem.)
4. ∢ AGE ≅ ∢ BHC	4. Angles congruent to congruent angles are congruent. (We now show $\overline{EG}$ ≅ $\overline{HC}$ in Steps 5 to 8.)
5. EH = EG + GH CG = CH + HG	5. Point Betweenness Postulate.
6. EG + GH = CH + HG	6. Substitution Postulate. (Steps 5, 1).
7. EG = HC	7. Subtraction Postulate.
8. $\overline{EG}$ ≅ $\overline{HC}$	8. Segments of equal length are congruent.
9. Δ AEG ≅ Δ BCH	9. The ASA Postulate.
10. $\overline{AG}$ ≅ $\overline{BH}$	10. Corresponding parts of congruent triangles are congruent.
11. AG = BH	11. Congruent segments are of equal length.
12. HD = GD	12. Same reason as Step 11.
13. AD = AG + GD BD = BH + HD	13. Definition of "betweenness."
14. AD = BD	14. Substitution Postulate.
15. ∢ 6 ≅ ∢ ABD	15. The base angles of an isosceles triangle are congruent.
16. m ∢ 5 < m ∢ ABD	16. If point P is in the interior of ∢ ABD, then m ∢ ABP < m ∢ ABD.
17. m ∢ 5 < m ∢ 6	17. If a = b and c < a, then c < b.

EXTERIOR ANGLE THEOREM

● **PROBLEM** 181

In isosceles triangle ABC, with $\overline{AC}$ ≅ $\overline{CB}$ base $\overline{AB}$ is extended
to D, and $\overline{CD}$ is drawn. Prove that CD > CA (see figure).

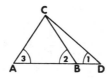

<u>Solution</u>: The cornerstone of this proof is a theorem
stating that the measure of an exterior angle of a triangle is greater than the measure of either remote interior angle. It can be derived from the fact that both
the measure of the exterior angle, and the sum of the
measures of the non-adjacent interior angles, when added
to the measure of the adjacent interior angle, will sum
to 180°. Since the non-adjacent angles must have positive
measure, the measure of the exterior angle must be greater
than either one alone.

The required conclusion, CD > CA, will be reached
by applying the theorem that if two angles of a triangle
are unequal, the sides opposite these angles are unequal,
and the greater side lies opposite the greater angle.

We will use the first theorem to show m⊀3 > m⊀1
and then, by the second theorem, conclude CD, opposite
⊀3, is greater than CA, opposite ⊀1.

STATEMENT	REASON
1. In △CBD, m⊀2 > m⊀1	1. The measure of an exterior angle of a triangle is greater than the measure of either non-adjacent interior angle.
2. $\overline{AC} \cong \overline{CB}$	2. Given.
3. ⊀3 $\cong$ ⊀2 or m⊀3 = m⊀2	3. Base angles of an isosceles triangle are congruent.
4. m⊀3 > m⊀1	4. A quantity may be substituted for its equal in any inequality.
5. CD > CA	5. If two angles of a triangle are unequal, then the sides opposite these angles are unequal, and the side of greater length lies opposite the angle of greater measure.

● **PROBLEM** 182

Given: $\overline{CD}$ bisects ⊀ACB. Prove: m⊀3 > m⊀2.

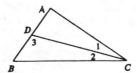

<u>Solution</u>: ⊀3 can be treated as either an angle in △BDC
or an exterior angle of △DAC. As an exterior angle,
m⊀3 > m⊀1. Since $\overline{CD}$ bisects ⊀ACB; ⊀1 $\cong$ ⊀2. Combining
these two facts, we conclude m⊀3 > m⊀2.

STATEMENTS	REASONS
1. $\overline{CD}$ bisects ⊁ACB	1. Given.
2. ⊁3 is an exterior angle of ⟨DAC	2. Any angle that forms a linear pair with an interior angle of the triangle is an exterior angle of the triangle.
3. m⊁3 > m⊁1	3. The measure of the exterior angle is greater than the measure of either of the remote interior angles.
4. m⊁1 = m⊁2	4. Definition of the angle bisector.
5. m⊁3 > m⊁2	5. If a = b and c > b, then c > a.

● **PROBLEM** 183

Given: Quadrilateral ABCD and straight rays $\overrightarrow{ADF}$ and $\overrightarrow{ABE}$.
Prove: m⊁EBC + m⊁FDC > ½(m⊁A + m⊁C). (Hint: Draw $\overline{AC}$)

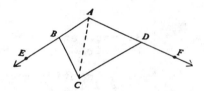

<u>Solution</u>: We have made use of the hint by drawing $\overline{AC}$ in the figure shown.

Notice, that m⊁A = m⊁BAC + m⊁CAD and m⊁C = m⊁BCA + m⊁ACD.

Since the measure of the whole is equal to the sum of the measures of its parts, we can substitute m⊁A = m⊁BAC + m⊁CAD and m⊁C = m⊁BCA + m⊁ACD in the equation to be proven. Hence, the equation becomes

m⊁EBC + m⊁FDC > ½(m⊁BAC + m⊁CAD + m⊁BCA + m⊁ACD).

We will be able to derive this form of the results using the Exterior Angle Theorem. Recall that the theorem tells us that the measure of an exterior angle of a triangle is greater than the measure of either remote interior angle.

Hence, in △ABC

m⊁EBC > m⊁BAC   and   m⊁EBC > m⊁BCA.

Similarly, in △DAC,

m⊁FDC > m⊁ACD   and   m⊁FDC > m⊁CAD.

We know that the sum of unequals are unequal in the same order. Accordingly,

$$2m\angle EBC + 2m\angle FDC > m\angle BAC + m\angle BCA + m\angle ACD + m\angle CAD.$$

Dividing by 2 and rearranging, we obtain

$$m\angle EBC + m\angle FDC > \tfrac{1}{2}(m\angle BAC + m\angle CAD + m\angle BCA + m\angle ACD).$$

These are exactly the results we set out to obtain.

TRIANGLE INEQUALITY THEOREM

● **PROBLEM** 184

Show that if two sides of a triangle are not congruent, then (1) the angles opposite these sides are not congruent, and (2) the angle with the greater measure is opposite the longer side.

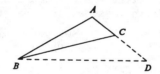

<u>Solution</u>: To compare two angles, we need a third angle whose relationship to the two angles is known. In the figure, in △ ABC, AB > AC. Consider the isosceles triangle △ ABD. ∠ ABD = ∠ ADB. We know m ∠ ABD > m ∠ ABC because the measure of the whole is greater than the measure of any part. Furthermore, ∠ ACB is an exterior angle of △ BCD; therefore m ∠ ACB > m ∠ ADB.

m ∠ ABC is less than the base angles of the isosceles triangle. m ∠ ACB is greater than the base angles. Therefore, m ∠ ACB > m ∠ ABC or m ∠ C > m ∠ B.

Given:　△ ABC; AB > AC
Prove:　m ∠ ACB > m ∠ ABC

Statements	Reasons
1. △ ABC; AB > AC	1. Given
2. Select D on $\overrightarrow{AC}$ so that $\overline{AD} \cong \overline{AB}$	2. Point Uniqueness Postulate.
3. AD > AC	3. Substitution Postulate.
4. $\overline{ACD}$	4. Point Betweenness Postulate.
5. m ∠ ABD = m ∠ ABC + m ∠ CBD	5. Angle Sum Postulate.
6. m ∠ ABD > m ∠ ABC	6. For any numbers k, m, and n, if n = m+k, and k > 0, then n > m.
7. ∠ ABD ≅ ∠ D	7. The base angles of an isosceles triangle are congruent.

8. m ∡ D > m ∡ ABC	8. Substitution Postulate
9. m ∡ ACB > m ∡ D	9. The measure of an exterior angle of a triangle is greater than the measure of either remote interior angle.
10. m ∡ ACB > m ∡ ABC	10. If a > b and b > c, then a > c.

● **PROBLEM** 185

Δ ABC is drawn in the accompanying figure with BC > BA. If $\overline{CD}$ bisects ∡ BCA and $\overline{AE}$ bisects ∡ BAC, prove that m ∡ EAC > m ∡ DCA.

Solution: By the definition of angle bisector, m ∡ EAC = ½m ∡ BAC and m ∡ DCA = ½m ∡ BCA. Therefore to show m ∡ EAC > m ∡ DCA, we show m ∡ BAC > m ∡ BCA. To accomplish this, we use the theorem that states: if two sides of a triangle are unequal in length then the opposite angles are unequal in measure, and the greater angle lies opposite the greater side.

Statements	Reasons
1. In Δ ABC, BC > BA	1. Given
2. m ∡ BAC > m ∡ BCA	2. If two sides of a triangle are unequal, the angles opposite these sides are unequal, and the greater angle lies opposite the greater side.
3. $\overline{AE}$ bisects ∡ BAC $\overline{CD}$ bisects ∡ BCA	3. Given
4. m ∡ EAC = ½m ∡ BAC m ∡ DCA = ½m ∡ BCA	4. A bisector of an angle divides the angle into two congruent angles.
5. m ∡ EAC > m ∡ DCA	5. Halves of unequal quantities are unequal in the same order. Also, Substitution Postulate.

● **PROBLEM** 186

Given: $\overline{AM}$ is the median of Δ ABC; m ∡ 2 > m ∡ 1 .

Prove: AC > AB

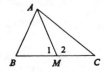

<u>Solution</u>: A theorem states that if two sides of a triangle are congruent respectively to two sides of a second triangle, and the measure of the included angle of the first triangle is greater than the measure of the included angle of the second triangle, then the measure of the third side of the first triangle is greater than the measure of the third side of the second triangle. Here, it can be shown that $\overline{BM} \cong \overline{MC}$ and $\overline{AM} \cong \overline{AM}$. Since m ∡ 2 > m ∡ 1, then AC > AB.

Statements	Reasons
1. $\overline{AM}$ is a median of △ ABC; m ∡ 2 > m ∡ 1.	1. Given
2. M is the midpoint of $\overline{BC}$	2. A median connects a vertex of a triangle to the midpoint of the opposite side.
3. $\overline{BM} \cong \overline{MC}$	3. Definition of a midpoint.
4. $\overline{AM} \cong \overline{AM}$	4. Every segment is congruent to itself.
5. AC > AB	5. If two sides of a triangle are congruent, respectively, to two sides of a second triangle, and the measure of the included angle of the first triangle is greater than the measure of the included angle of the second triangle, then the measure of the third side of the first triangle is greater than the measure of the third side of the second triangle.

● **PROBLEM** 187

Given: △ ABC; D is a point between A and C; BD > AB

Prove: BC > AB

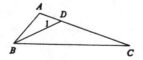

<u>Solution</u>: $\overline{BC}$ and $\overline{AB}$ are sides of △ ABC. BC > AB only if the angle opposite $\overline{BC}$ is greater in measure than the angle opposite $\overline{AB}$ – that is, m ∡ A > m ∡ C.

To show m ∡ A > m ∡ C, we relate both to m ∡ 1. Note in △ ABD, BD > AB. This implies m ∡ A > m ∡ 1. Furthermore in △ BDC, ∡ 1 is an exterior angle while ∡ C is a remote interior angle. Thus, m ∡ 1 > m ∡ C.

Combining m ∡ A > m ∡ 1 and m ∡ 1 > m ∡ C, we obtain m ∡ A > m ∡ C; and, therefore, BC > AB.

Statements	Reasons
1. △ ABC; D is a point between A and C; BD > AB	1. Given
2. m ∡ A > m ∡ 1	2. If two sides of a triangle are not congruent, the angle with the greater measure is opposite the longer side.
3. m ∡ 1 > m ∡ C	3. The measure of an exterior angle of

4. m ∢ A > m ∢ C
5. BC > AB

4. Transitive property of inequalities.
5. If two angles of a triangle are not congruent, the longer side is opposite the angle with the greater measure.

a triangle is greater than the measure of either remote interior angle.

Given: Point P is in the interior of △ ABC.
Prove: AP + PB + PC > ½(AB + AC + BC).

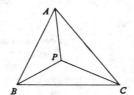

Solution: Whenever we deal with the inequality of the sides of a triangle, the Triangle Inequality Theorem is often useful. Here, we find triangles whose sides are AP, PB, PC, AB, AC, and BC. Consider △ APB. By the Triangle Inequality Theorem, AP + PB > AB. If we consider △ APC and △ BPC, then AP + PC > AC and PB + PC > BC. To obtain all sides of the triangle on one side of the inequality, we add our three equations. Thus, (2AP + 2PB + 2PC) > (AB + BC + AC). Dividing both sides of the equation by 2, we obtain our desired result.

Statements	Reasons
1. Point P is in the interior of △ ABC	1. Given

2. PA + PB > AB
   PA + PC > AC
   PB + PC > BC
	2. Triangle Inequality Theorem

3. 2PA + 2PB + 2PC > AB + AC + BC	3. Addition Property of Inequalities.
4. PA + PB + PC > ½(AB + AC + BC)	4. Division by a positive number does not affect the sense of the inequality.

If the lengths of two sides of a triangle are 10 and 14, the length of the third side may be which of the following: (a) 2 (b) 4 (c) 22 (d) 24?

Solution: By the Triangle Inequality Theorem, we know that the sum of the lengths of any two sides of a triangle must be greater than the length of the third.
    Therefore, in this example, we can discover which of the lengths are possible answers by determining if the sum of the lengths of the two shortest sides is greater than the length of the longest side.
(a) 2 cannot be the third side, because 2 + 10 is not > 14.
(b) 4 cannot be the third side, because 4 + 10 is not > 14.

(c) 22 can be the third, because 10 + 14 > 22.

(d) 24 cannot be the third side, because 10 + 14 is not > 24.

Therefore, the third side may be 22.

Given: Point P is an interior point of △ ABC.

Prove: AB + AC > BP + PC (Hint: extend $\overleftrightarrow{BP}$ so that it intersects $\overline{AC}$ at point N).

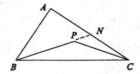

Solution: We are asked to prove an inequality involving the sides of triangles. Therefore, we apply the Triangle Inequality Theorem. This theorem states that the sum of the lengths of any two sides of a triangle is greater than the length of the third. We must now find triangles whose sides include AB, AC, BP, and PC. We use the hint and extend $\overleftrightarrow{BP}$ so that $\overleftrightarrow{BP}$ intersects $\overline{AC}$ at point N. Thus, we obtain △ ABN and △ PNC whose sides comprise all the lengths in question - plus an extra length PN. By the Triangle Inequality Theorem, we know in △ ABN that AB + AN > BN or (since BN = BP + PN) that AB + AN > BP + PN. In △ PNC, PN + NC > PC. Summing the two equations together, we obtain an inequality involving AB, AC, BP, and PC: AB + AN + NC + PN > BP + PC + PN. Combining AN + NC to obtain AC, and cancelling PN from both sides, we obtain the desired result: AB + AC > BP + PC.

Statements	Reasons
1. Point P is in the interior of △ ABC	1. Given
2. $\overleftrightarrow{BP}$ intersects $\overline{AC}$ at point N	2. Two noncoincident, nonparallel coplanar lines intersect at a point.
3. AB + AN > BN PN + NC > PC	3. Triangle Inequality Theorem.
4. BN = PN + PB AC = AN + NC	4. Point Betweenness Postulate.
5. AB + AN + PN + NC > BN + PC	5. Addition Postulate of Inequality.
6. AB + AC + PN > PN + PB + PC	6. Substitution Postulate.
7. AB + AC > PB + PC	7. If a > b, then a - c > b - c.

159

CHAPTER 12

QUADRILATERALS

> **Basic Attacks and Strategies for Solving Problems in this Chapter. See pages 160 to 203 for step-by-step solutions to problems.**

The problems in this chapter concern the properties of and the relationships between various kinds of quadrilaterals. The special kinds of quadrilaterals are **trapezoids, parallelograms, rectangles, rhombuses**, and **squares**. The definitions of these appear in the appendix but are repeated here:

Trapezoid – a quadrilateral with exactly two parallel sides

Parallelogram – a quadrilateral with two pairs of parallel sides

Rectangle – a parallelogram with (at least) one right angle

Rhombus – a parallelogram with (at least) two adjacent congruent sides

Square – a rectangle with (at least) two adjacent congruent sides

The problems in this chapter make use of these definitions. It will be shown (Problem 191) that adjacent angles of a parallelogram are supplementary. Thus, it is easy to see that if a parallelogram has one right angle, all four of the angles are right angles. It will also be shown (Problem 193) that opposite sides of a parallelogram are congruent. Thus, in the case of a rhombus or square, the two adjacent sides which are congruent, are also congruent to the oppposite two sides. Therefore, not just two, but all four sides are congruent.

In a broad context, it is desirable to first classify quadrilaterals in terms of the numbers of pairs of parallel sides. Here is a helpful way to do this.

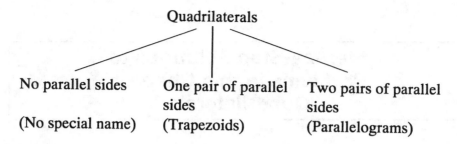

Quadrilaterals

No parallel sides   One pair of parallel   Two pairs of parallel
                           sides   sides

(No special name)   (Trapezoids)   (Parallelograms)

The most important type of quadrilateral is the parallelogram. One of the problems in this chapter involves proving that every rectangle is a parallelogram. From this information and the definitions given, the following is a Venn diagram illustrating how various parallelograms are related:

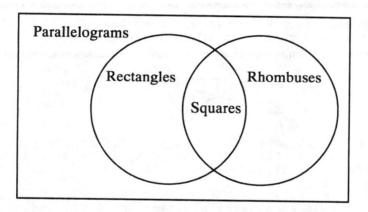

Step-by-Step Solutions to
Problems in this Chapter,
"Quadrilaterals"

PARALLELOGRAMS

● **PROBLEM** 191

Prove that all pairs of consecutive angles of a
parallelogram are supplementary. (See figure.)

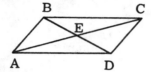

Solution: We must prove that the pairs of angles ⦨BAD
and ⦨ADC, ⦨ADC and ⦨DCB, ⦨DCB and ⦨CBA, and ⦨CBA and ⦨BAD
are supplementary. (This means that the sum of their
measures is 180°.)

Because ABCD is a parallelogram, $\overline{AB}$ || $\overline{CD}$. Angles
BAD and ADC are consecutive interior angles, as are
⦨CBA and ⦨DCB. Since the consecutive interior angles
formed by 2 parallel lines and a transversal are supplementary, ⦨BAD and ⦨ADC are supplementary, as are ⦨CBA
and ⦨DCB.

Similarly, $\overline{AD}$ || $\overline{BC}$. Angles ADC and DCB are con-
secutive interior angles, as are ⦨CBA and ⦨BAD. Since
the consecutive interior angles formed by 2 parallel
lines and a transversal are supplementary, ⦨CBA and ⦨BAD
are supplementary, as are ⦨ADC and ⦨DCB.

● **PROBLEM** 192

Prove that if any 2 consecutive angles of a quadrilateral
are not supplementary, then the quadrilateral is not a
parallelogram.

Solution: The figure shows an arbitrary quadrilateral,
ABCD, with angles a, b, c, and d. (Note that angles a and
b, b and c, c and d, and d and a are consecutive.) We will
show that if any 2 consecutive angles are not supplementary,

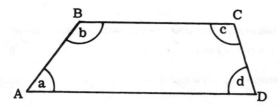

then there exist 2 sides of ABCD which are not parallel. Hence, we will have proven that ABCD is not a parallelogram.

Examine any pair of consecutive angles, say ∡a and ∡d. These angles are consecutive interior angles of line segments $\overline{AB}$ and $\overline{CD}$, cut by transversal $\overline{AD}$. If ∡a and ∡d are supplementary, then $\overline{AB} \parallel \overline{CD}$. However, if ∡a and ∡d are not supplementary, then $\overline{AB} \not\parallel \overline{CD}$. If this is the case, then ABCD cannot be a parallelogram, since opposite sides of a parallelogram must be parallel. This argument can be applied to any pair of consecutive angles of the quadrilateral. Hence, if any 2 consecutive angles of a quadrilateral are not supplementary, then the quadrilateral is not a paralellogram.

● **PROBLEM** 193

Prove that both pairs of opposite sides of a parallelogram are congruent.

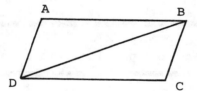

Solution: In the figure shown, let ABCD be the parallelogram and let $\overline{DB}$ be its diagonal. Using the definition of

a parallelogram, we will prove ΔDAB ≅ ΔBCD. It will then follow directly that $\overline{AD} \cong \overline{BC}$ and $\overline{AB} \cong \overline{DC}$.

By definition, the pairs of opposite sides of a parallelogram are parallel (i.e. $\overline{AB} \parallel \overline{DC}$, $\overline{AD} \parallel \overline{BC}$). This

means that ∡ADB ≅ ∡CBD, since these are alternate interior angles for parallel segments $\overline{AD}$ and $\overline{BC}$ cut by a transversal.

Also, $\overline{DB}$ is common to ΔDAB and ΔBCD, therefore, $\overline{DB} \cong \overline{DB}$. Lastly, since $\overline{AB} \parallel \overline{DC}$, then alternate interior angles ABD

and CDB are congruent. By the ASA Postulate, ΔDAB ≅ ΔBCD. It then follows, by corresponding parts, that $\overline{AD} \cong \overline{CB}$ and $\overline{AB} \cong \overline{CD}$.

Prove that the diagonals of a parallelogram bisect each other.

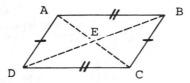

Solution: To prove that $\overline{AC}$ bisects $\overline{DB}$ (and vice versa) we must prove that $\overline{AE} \cong \overline{CE}$ and $\overline{DE} \cong \overline{BE}$. We shall do this by showing that $\triangle AED \cong \triangle CEB$ (see figure).

Because ABCD is a parallelogram, opposite sides $\overline{AD} \cong \overline{CB}$. Also, $\angle DAE \cong \angle BCE$ because they are alternate interior angles for the parallel segments $\overline{AD}$ and $\overline{CB}$ cut by transversal $\overline{AC}$. Lastly, $\angle AED \cong \angle CEB$, since they are vertical angles. By the AAS Postulate, $\triangle AED \cong \triangle CEB$. Hence, by corresponding parts, $\overline{AE} \cong \overline{CE}$ and $\overline{DE} \cong \overline{BE}$. From this, we conclude the diagonals of a paralellogram bisect each other.

Prove that a quadrilateral, in which one pair of opposite sides are both congruent and parallel, is a parallelogram.

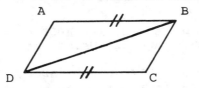

Solution: In quadrilateral ABCD (as shown in the figure) assume $\overline{AB} \cong \overline{CD}$ and $\overline{AB} \parallel \overline{CD}$. We shall prove that $\triangle DAB \cong \triangle BCD$ by the SAS Postulate. We will then know that $\overline{DA} \cong \overline{BC}$. Since we already know that $\overline{AB} \cong \overline{CD}$, we will have shown that both pairs of opposite sides of the quadrilateral are congruent, implying that ABCD is a parallelogram.

As given, $\overline{AB} \cong \overline{CD}$. Furthermore, since $\overline{AB} \parallel \overline{CD}$, then $\angle ABD \cong \angle CDB$. This follows because these angles are alternate interior angles of parallel segments out by a transversal. Lastly, $\overline{DB}$ is shared by both triangles, therefore $\overline{DB} \cong \overline{DB}$. By the SAS Postulate, $\triangle DAB \cong \triangle BCD$. This implies that $\overline{DA} \cong \overline{BC}$.

We have shown that both pairs of opposite sides of quadrilateral ABCD are congruent. This implies that ABCD is a parallelogram.

● **PROBLEM** 196

If ABCD is a quadrilateral such that $\overline{AB} \cong \overline{CD}$ and $\overline{AD} \cong \overline{BC}$, then prove that ABCD is a parallelogram. (See Figure.)

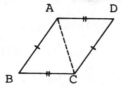

Solution: Our strategy is to show that $\triangle ADC \cong \triangle CBA$ by the SSS Postulate. This will tell us that $\angle BAC \cong \angle DCA$. Since these are alternate interior angles for segments $\overline{AB}$ and $\overline{CD}$, we will have shown that $\overline{AB} \parallel \overline{CD}$. A similar procedure will yield the fact that $\overline{AD} \parallel \overline{BC}$. By definition, this will mean that ABCD is a parallelogram.

By the given facts, $\overline{AB} \cong \overline{CD}$ and $\overline{AD} \cong \overline{BC}$; $\overline{AC}$ is shared by $\triangle ADC$ and $\triangle CBA$. By reflexivity, $\overline{AC} \cong \overline{AC}$. By the SSS Postulate, then, $\triangle ADC \cong \triangle CBA$. By corresponding parts,

$$\angle BAC \cong \angle DCA \qquad\qquad (1)$$

and
$$\angle BCA \cong \angle DAC. \qquad\qquad (2)$$

But the set of angles in (1) are alternate interior angles with respect to segments $\overline{BA}$ and $\overline{DC}$. Hence, $\overline{BA} \parallel \overline{DC}$. Similarly, the set of angles in (2) are alternate interior angles with respect to segments $\overline{AD}$ and $\overline{BC}$. Since the angles are congruent, $\overline{AD} \parallel \overline{BC}$.

ABCD has both pairs of opposite sides congruent and parallel and is, by definition, a parallelogram.

● **PROBLEM** 197

If both pairs of opposite angles of a quadrilateral ABCD are congruent, then prove that the quadrilateral is a parallelogram. (See Figure.)

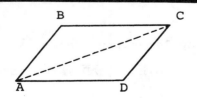

<u>Solution</u>: Given that ∠BAD ≅ ∠DCB, and ∠ABC ≅ ∠CDA, we are to prove that ABCD is a parallelogram. We shall do this by using the angle sum theorem, as applied to △ABC and △CDA, and the given facts about the angles of ABCD.

Applying the angle sum theorem separately to triangles ABC and CDA yields

m∠CAB + m∠ABC + m∠ACB = 180°.

m∠DAC + m∠ACD + m∠CDA = 180°.

Adding the previous 2 equations,

(m∠DAC + m∠CAB) + (m∠ACD + m∠ACB) + m∠CDA

$$+ \ m∠ABC = 360° \qquad (1)$$

But, from the figure,

m∠DAC + m∠CAB = m∠DAB

m∠ACD + m∠ACB = m∠DCB

Equation (1) becomes

$$m∠DAB + m∠DCB + m∠CDA + m∠ABC = 360° \qquad (2)$$

This shows that the sum of the angles of a quadrilateral is 360°.

However, the given fact is that

∠ABC ≅ ∠CDA

$$\text{or} \quad m∠ABC = m∠CDA \qquad (3)$$

and ∠DAB ≅ ∠DCB

or   m∠DAB = m∠DCB

Using (3) in (2)

2m∠DAB + 2m∠ABC = 360°

$$\text{or} \quad m∠DAB + m∠ABC = 180°. \qquad (4)$$

This means that ∠DAB and ∠ABC are supplementary and BC || AD. Similarly, substituting m∠ABC = m∠CDA in (4) yields

m∠DAB + m∠CDA = 180°

∠DAB and ∠CDA are supplementary. Hence, AB || CD.

Since AB || CD, and BC || AD, ABCD is a parallelogram, by definition.

164

Let ABCD be a parallelogram. Prove that △ABD ≅ △CDB (see figure).

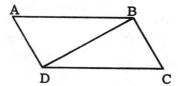

Solution: We shall use the definition of a parallelogram, and the properties of parallel lines, to show that △ABD ≅ △CDB by the ASA Postulate.

In a parallelogram, each pair of opposite sides is parallel (i.e. $\overline{AB}$ || $\overline{DC}$ and $\overline{AD}$ || $\overline{BC}$). Because $\overline{AB}$ || $\overline{DC}$, ∢ABD ≅ ∢CDB since they are alternate interior angles formed by transversal $\overline{BD}$. Similarly, $\overline{AD}$ || $\overline{BC}$, which implies that alternate interior angles ADB and CBD are congruent. $\overline{DB}$ is common to both △ABD and △CDB and, by reflextivity $\overline{DB}$ ≅ $\overline{DB}$. Hence, by the ASA Postulate, △ABD ≅ △CDB.

Prove that in a quadrilateral the lines which join the mid-points of the opposite sides and the line which joins the midpoints of the diagonals bisect one another at a common point.

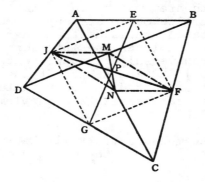

Solution: In the accompanying figure, the sides of quadrilateral ABCD have midpoints E, F, G, and J. Diagonals $\overline{AC}$ and $\overline{BD}$ have midpoints M and N. We are asked to show that $\overline{JF}$, $\overline{EG}$ and $\overline{MN}$ all intersect and bisect each other at a point P. The key to this problem is to be able to identify several parallelograms. In the problem, there is constant mention of midpoints and bisections. Recall, in a parallelogram, the diagonals bisect each other. Therefore, we will show that the lines joining midpoints are diagonals of some parallelogram and, thus bisect each other.

Since quadrilateral EFGJ is formed by joining the midpoints of a quadrilateral, EFGJ is a parallelogram and thus its diagonals $\overline{JF}$ and $\overline{GE}$ bisect each other. To see that EFGJ is a parallelogram, consider $\triangle ADB$ and $\triangle CDB$. In $\triangle ADB$, $\overline{JE}$ joins the midpoints of two sides, therefore, $\overline{JE} \parallel \overline{BD}$ and $\overline{JE} = \frac{1}{2} \overline{BD}$. In $\triangle CDB$, $\overline{GF}$ joins the midpoints of two sides, thus $\overline{GF} \parallel \overline{BD}$ and $\overline{GF} = \frac{1}{2} \overline{BD}$. Then $\overline{GF} \parallel \overline{JE}$ and $\overline{GF} \cong \overline{JE}$. Since a pair of opposite sides are congruent and parallel, then EFGJ is a parallelogram.

Furthermore, we show that JMFN is a parallelogram. Since $\overline{JM}$ joins the midpoints of two sides of $\triangle ADB$, then $\overline{JM} \parallel \overline{AB}$ and $\overline{JM} = \frac{1}{2} \overline{AB}$. Since $\overline{FN}$ joins the midpoints of two sides of $\triangle ABC$, then $\overline{FN} \parallel \overline{AB}$ and $\overline{FN} = \frac{1}{2} \overline{AB}$. Thus, $\overline{JM} \parallel \overline{FN}$, $\overline{JM} = \overline{FN}$, and JMFN is a parallelogram.

Thus, diagonals $\overline{JF}$ and $\overline{MN}$ bisect each other, and intersect at the midpoint of $\overline{JF}$, P. However, from the parallelogram EFGJ, we know that $\overline{JF}$ and $\overline{EG}$ bisect each other and intersect at the midpoint of $\overline{JF}$, P. Therefore, $\overline{JF}, \overline{GE},$ and $\overline{MN}$ are concurrent at their midpoint P.

● **PROBLEM** 200

In parallelogram ABCD, if the measure of ⊁B exceeds the measure of ⊁A by 50°, find the measure of ⊁B.

Solution: There is a theorem which states that the consecutive angles of a parallelogram are supplementary. This fact, more generally stated, tells us that when a transversal cuts across parallel lines the interior angles on the same side of the transversal are supplementary. In the parallelogram shown, $\overline{AD} \parallel \overline{BC}$ and $\overline{AB}$ is a transversal. Therefore, it follows that ⊁A and ⊁B, being interior angles on the same side of the transversal, are supplements and their measures sum to 180°.

Hence, to find the measure of ⊁B (1) Let x = the measure of ⊁A. (2) Then, x + 50 = the measure of ⊁B. It follows then that

$$x + (x + 50) = 180$$

$$2x + 50 = 180$$

$$2x = 130$$

$$x = 65 \quad \text{and } x + 50 = 115.$$

Therefore, the measure of ⊁B is 115°.

In the figure, ABCD is a parallelogram with diagonals $\overline{AC}$ and $\overline{BD}$. ⦣ABC is an obtuse angle. Prove that AC > BD.

Solution: The main theorem to be applied here is the one which states that if two sides of one triangle are congruent, respectively, to two sides of another triangle, and the included angles are not congruent, then the triangle which has the included angle of larger measure has the third side of greater length.

$\overline{AC}$ and $\overline{BD}$ are in different triangles. We will show AC > BD, under the above theorem, by showing two sides in △ABC, $\overline{BC}$ and $\overline{AB}$, are congruent to two sides in △BAD, $\overline{AD}$ and $\overline{BA}$ and that the measure of the included angle ABC is greater than the measure of the included angle BAD.

STATEMENT	REASON
1. ABCD is a parallelogram	1. Given.
2. $\overline{AD} \cong \overline{BC}$	2. Opposite sides of a parallelogram are congruent.
3. ⦣BAD is supplementary to ⦣ABC	3. Consecutive angles of a parallelogram are supplementary.
4. ⦣ABC is an obtuse angle	4. Given.
5. ⦣BAD is an acute angle	5. The supplement of an obtuse angle is an acute angle (see step (3)).
6. m⦣ABC > m⦣BAD	6. By definition, the measure of an obtuse angle is greater than the measure of an acute angle.
7. $\overline{AB} \cong \overline{BA}$	7. Reflexive Property of Congruence.
8. AC > BD	8. In two triangles, if two sides of one triangle are congruent, respectively, to two sides of the other, and the included angles are not congruent, then the triangle which has the included angle of larger measure has the greater third side.

Given: P, Q, R, and S are the respective midpoints of sides $\overline{AB}$, $\overline{BC}$, $\overline{CD}$, and $\overline{AD}$ of quadrilateral ABCD. Prove: **Quadrilateral PQRS is a parallelogram.**

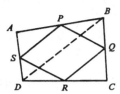

Solution: There are seven different ways to show that a quadrilateral is a parallelogram. We can prove that (1) both pairs of opposite sides are parallel, (2) both pairs of opposite sides are congruent, (3) one pair of opposite sides are both parallel and congruent, (4) the opposite angles are congruent, (5) the angles of one pair of opposite angles are congruent and the sides of a pair of opposite sides are parallel, (6) the angles of one pair of opposite angles and the sides of one pair of opposite sides are congruent, or (7) the diagonals bisect each other.

Here, it is not obvious at all how to proceed. There is nothing given about the quadrilateral with which we are concerned, neither parallelism nor congruence, that can help us.

By drawing diagonal $\overline{BD}$ of ABCD, we form triangles ABD and CBD. Since sides $\overline{SP}$ and $\overline{RQ}$ of PQRS join the midpoints of two sides of ∆ABD and ∆CBD respectively, they are the midlines of the triangles they lie in. According to several theorems on midlines, both $\overline{SP}$ and $\overline{RQ}$ are parallel to, and half as long as $\overline{BD}$, the side of the triangles not cross by them. By method (3), we will prove that quadrilateral PQRS is a parallelogram.

STATEMENTS	REASONS
1. P, Q, R, and S are the respective midpoints of sides $\overline{AB}$, $\overline{BC}$, $\overline{CD}$, and $\overline{AD}$ of quadrilateral ABCD.	1. Given.
2. $\overline{SP}$ is a midline of ∆ABD.	2. A midline of a triangle is the line segment joining the midpoints of two sides of the triangle.
3. $\overline{SP} \parallel \overline{DB}$	3. The midline of a triangle is parallel to the third side.
4. SP = ½ DB	4. The midline of a triangle is half as long as the third side of the triangle.

5. $\overline{QR}$ is a midline of $\triangle CDB$	5. Definition of midline.
6. $\overline{QR} \parallel \overline{DB}$	6. The midline of a triangle is parallel to the third side of the triangle.
7. $QR = \frac{1}{2} DB$	7. The midline of a triangle is half as long as the third side of the triangle.
8. $\overline{SP} \parallel \overline{QR}$	8. If each of two lines is parallel to a third line, then they are parallel to each other.
9. $SP = QR$	9. Transitive property.
10. Quadrilateral PQRS is a parallelogram.	10. A quadrilateral is a parallelogram if two of its sides are both congruent and parallel.

● **PROBLEM** 203

In triangle ABC (shown in the accompanying figure), D is the midpoint of $\overline{AC}$ and E is the midpoint of $\overline{CB}$. If $\overline{DE}$ is extended to F so that $\overline{DE} \cong \overline{EF}$ and $\overline{FB}$ is drawn, prove that ABFD is a parallelogram.

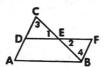

Solution: To prove that ABFD is a parallelogram it is necessary to show that one pair of its opposite sides are both congruent and parallel. The fact that alternate interior angles of parallel lines cut by a transversal are congruent will be used to prove two sides are parallel. To prove congruence of the same two sides, the corresponding parts rule for congruent triangles will be employed. To derive both of the properties, congruence between $\triangle$'s BEF and CED must be established.

STATEMENT	REASON
1. $\overline{CE} \cong \overline{BE}$	1. Given. (E is the midpoint of $\overline{CB}$.)
2. $\overline{DE} \cong \overline{FE}$	2. Given.
3. $\angle 1 = \angle 2$	3. If two angles are vertical angles, then they are congruent.
4. $\triangle CED \cong \triangle BEF$	4. S.A.S. $\cong$ S.A.S.
5. $\angle 3 \cong \angle 4$	5. Corresponding parts of congru-

ent triangles are congruent.

6. $\overline{FB} \parallel \overline{CA}$ or $\overline{FB} \parallel \overline{DA}$ | 6. If two lines are cut by a transversal making a pair of alternate interior angles congruent, then the lines are parallel.

7. $\overline{FB} \cong \overline{DC}$ | 7. Same as reason 5.

8. $\overline{DC} \cong \overline{DA}$ | 8. Given. (D midpoint of $\overline{CA}$.)

9. $\overline{FB} \cong \overline{DA}$ | 9. Transitivity of congruence.

10. ABFD is a parallelogram. | 10. A quadrilateral is a parallelogram if one pair of opposite sides are both congruent and parallel.

● **PROBLEM** 204

In the figure shown, ABCD is a parallelogram and CDFE is a parallogram. Prove (a) ABEF is a parallelogram and (b) $\overline{FA} \cong \overline{EB}$ and $\overline{FA} \parallel \overline{EB}$.

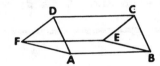

Solution: (a) We will employ the fact that if opposite sides of a quadrilateral are congruent and parallel, then the quadrilateral is a parallelogram, to prove that ABEF is a parallelogram.

STATEMENT	REASON
1. ABCD and CDFE are parallelograms.	1. Given.
2. $\overline{CD} \cong \overline{AB}$ and $\overline{CD} \cong \overline{EF}$	2. Opposite sides of a parallelogram are congruent.
3. $\overline{AB} \cong \overline{EF}$	3. If two segments are congruent to the same segment, they are congruent to each other.
4. $\overline{CD} \parallel \overline{AB}$ and $\overline{CD} \parallel \overline{EF}$	4. Opposite sides of a parallelogram are parallel.
5. $\overline{AB} \parallel \overline{EF}$	5. If two lines are parallel to the same line, they are parallel to each other.
6. ABEF is a parallelogram.	6. A quadrilateral which has one pair of opposite sides

congruent and parallel is
a parallelogram.

(b) $\overline{FA} \cong \overline{EB}$ and $\overline{FA} \parallel \overline{EB}$ because they are a pair of
opposite sides of a parallelogram and, by theorem, we know
that if a quadrilateral is a parallelogram, then opposite
sides are both congruent and parallel.

● **PROBLEM** 205

Use an indirect method of proof to prove: Two line segments
drawn inside a triangle from the endpoints of one side of
the triangle, and terminating in points located on the other
two sides, cannot bisect each other.

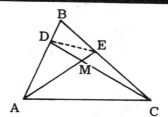

Solution: As in all indirect proofs, we will assume the
negative of the conclusion we wish to prove, and then show
that this leads to a logical contradiction of the given.

Assume that the line segments do, in fact, bisect
each other, i.e. $\overline{DM} \cong \overline{MC}$ and $\overline{AM} \cong \overline{ME}$.

These line segments are diagonals of ACED, and, if
diagonals of a quadrilateral bisect each other, then the
quadrilateral is a parallelogram. Therefore, ACED is a
parallelogram.

As opposite sides of ACED, $\overline{AD} \parallel \overline{EC}$. Side $\overline{AC}$ acts
as a transversal to these lines. Since ∢DAC and ∢ECA are
interior angles on the same side of the transversal, they
are supplementary. Therefore, m∢DAC + m∢ECA = 180°.

However, if these two angles have measures summing
to 180°, then the third angle, ∢ABC of ∆ABC, must measure
zero in order for all three angles of ∆ABC of sum to 180°.
But this leads to a contradiction, since each of the three
angles of any triangle must have a measure greater than
zero.

Upon arriving at this contradiction, we realize that
our original assumption is false. Therefore, the only
other alternative that does not contradict any part of the
hypothesis is that the line segments do not bisect each
other.

Therefore, we have proved the original statement
presented to us.

RHOMBUSES

> If the diagonals of a parallelogram meet at right angles,
> prove that the parallelogram is a rhombus.

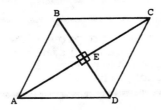

Solution: The accompanying figure shows parallelogram
ABCD, whose diagonals ($\overline{BD}$ and $\overline{AC}$) meet at right angles. By
definition, a rhombus is a quadrilateral, all of whose
sides are congruent. Our strategy will be to use the
properties of a parallelogram to show that $\overline{AB} \cong \overline{CD}$ and
$\overline{BC} \cong \overline{DA}$. We shall then prove that $\triangle AEB \cong \triangle AED$, which implies that $\overline{AB} \cong \overline{AD}$. Coupling these facts will show that
ABCD is a quadrilateral, all of whose sides are congru-
ent. (i.e. a rhombus).

Since ABCD is a parallelogram, it is a quadrilateral.
Furthermore, by the properties of paralellograms, opposite
sides are congruent. That is

$$\overline{AB} \cong \overline{CD}$$

and $\overline{BC} \cong \overline{DA}$.   (1)

Now, focus attention on $\triangle AEB$ and $\triangle AED$. $\overline{BE} \cong \overline{DE}$,
because the diagonals of a parallelogram bisect each other.
$\angle BEA \cong \angle DEA$, since both are right angles. Lastly, $\overline{AE} \cong \overline{AE}$,
since it is common to both triangles. By the SAS (side
angle side) Postulate, $\triangle AEB \cong \triangle AED$.

Therefore, by corresponding parts

$$\overline{AB} \cong \overline{AD}$$ (2)

Using (2) in (1) yields,

$$\overline{AB} \cong \overline{CD}$$

$$\overline{AB} \cong \overline{AD}$$

$$\overline{AD} \cong \overline{CB}$$

or $\overline{AB} \cong \overline{CD} \cong \overline{AD} \cong \overline{CB}$

which shows that ABCD is a quadrilateral, all of whose
sides are congruent. Hence, ABCD is a rhombus.

Use an indirect method of proof to prove: If a diagonal
of a parallelogram does not bisect the angles through whose
vertices the diagonal is drawn, the parallelogram is not
a rhombus.

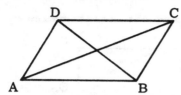

Solution: As the method of indirect proof suggests, we
have two possible results. Either the parallelogram is a
rhombus or it is not. Assume it is a rhombus and then show
that the hypothesis, as stated, leads to a logical contradiction, which will show that our assumption is wrong. By
the postulate of elimination, the parallelogram given in
the hypothesis will not be a rhombus.

STATEMENT	REASON
1. ABCD is a rhombus or it is not	1. There are only these two possibilities.
2. Assume ABCD is a rhombus	2. One of the two possibilities.
3. In $\triangle$'s ADC and ABC $\overline{AD} \cong \overline{BC}$	3. Opposite sides of a rhombus are congruent.
4. $\overline{AB} \cong \overline{CD}$	4. Same as reason 3.
5. $\angle ADC \cong \angle ABC$	5. Opposite angles in a rhombus are congruent.
6. $\triangle ADC \cong \triangle ABC$	6. S.A.S. $\cong$ S.A.S.
7. $\angle CAB \cong \angle CAD$	7. Corresponding angles in congruent triangles are equal.
8. This contradicts the hypothesis which states that $\overline{AC}$ is not an angle bisector	8. Only an angle bisector divides an angle into two congruent halves.
9. The assumption that ABCD is a rhombus is false.	9. Postulate of Contradiction: If a proposition contradicts a true proposition, then it is false.
10. ABCD is not a rhombus	10. Postulate of Elimination: If one of a given set of pro-

positions must be true, and all but one of those propositions have been proved to be false, then the one remaining proposition must be true.

● **PROBLEM** 208

Prove that the lines joining the midpoints of a rectangle form a rhombus.

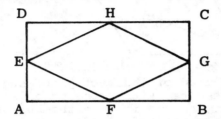

Solution: Since a rhombus is a quadrilateral in which all sides are congruent to each other, by showing $\overline{EF} \cong \overline{FG} \cong \overline{HG} \cong \overline{HE}$ we will be able to prove EFGH is a rhombus.

The sides in question are corresponding parts of Δ's DEH, AEF, BGF and CGH. By proving the four triangles congruent to each other, we can conclude the desired results.

The congruence proof follows:

$\overline{AB} \cong \overline{DC}$ and $\overline{AD} \cong \overline{CB}$ because ABCD is given to be a rectangle and opposite sides of a rectangle are congruent. Since E, F, G, and H are midpoints of these sides, they divide the sides into congruent segments. Hence, $\overline{AF} \cong \overline{FB} \cong \overline{HC} \cong \overline{HD}$ and $\overline{AE} \cong \overline{GB} \cong \overline{GC} \cong \overline{DE}$. This result gives us two corresponding congruent sides between the triangles.

The angles of a rectangle are right angles and, since all right angles are congruent, ⊬EAF $\cong$ ⊬FBG $\cong$ ⊬GCH $\cong$ ⊬HDE.

Therefore, ΔAEF $\cong$ ΔBGF $\cong$ ΔCGH $\cong$ ΔDEH by the SAS $\cong$ SAS postulate.

By corresponding parts, $\overline{EF} \cong \overline{FG} \cong \overline{HG} \cong$ HE.

Hence, quadrilateral EFHG is a rhombus since all its sides are congruent.

● **PROBLEM** 209

.In the figure shown, $\overrightarrow{BF}$ bisects angle CBA. $\overline{DE} \mid\mid \overrightarrow{BA}$ and $\overline{GE} \mid\mid \overrightarrow{BC}$. Prove GEDB is a rhombus.

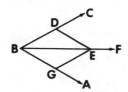

Solution: We want to show that GEDB is a parallelogram which has congruent adjacent sides, since this is, by definition, a rhombus.

We will prove that GEDB is a parallelogram, and then prove ΔBEG ≅ ΔBED in order to deduce that adjacent sides GE and DE are congruent.

STATEMENT	REASON
1. $\overrightarrow{DE}$ || $\overrightarrow{BA}$, $\overrightarrow{GE}$ || $\overrightarrow{BC}$	1. Given.
2. GEDB is a parallelogram	2. A quadrilateral in which all opposite sides are parallel is a parallelogram.
3. $\overrightarrow{BF}$ is the angle bisector of ⊀CBA	3. Given.
4. ⊀DBE ≅ ⊀GBE	4. An angle bisector divides the angle into two congruent angles.
5. $\overline{BG}$ ≅ $\overline{DE}$	5. Opposite sides of a parallelogram are congruent.
6. $\overline{BE}$ ≅ $\overline{BE}$	6. Reflexive Property of Congruence.
7. ΔBEG ≅ ΔBED	7. S.A.S. ≅ S.A.S.
8. $\overline{GE}$ ≅ $\overline{DE}$	8. Corresponding sides of congruent triangles are congruent.
9. GEDB is a rhombus	9. A rhombus is a parallelogram with adjacent sides congruent.

● **PROBLEM** 210

Let ABCD be a rhombus and let its diagonals, $\overline{AC}$ and $\overline{BD}$, intersect in point E. Prove that DE = BE.

Solution: $\overline{DE}$ and $\overline{BE}$ are corresponding parts of ΔCED and ΔAEB, respectively. We will prove that DE = BE by showing that ΔAEB ≅ ΔCED via the AAS Postulate.
Since ABCD is a rhombus, $\overline{AB}$ \|\| $\overline{CD}$. As alternate in-

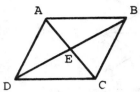

terior angles formed by transversal $\overline{AC}$ of $\overline{AB}$ and $\overline{CD}$, ⊀BAE ≅ ⊀DCE. Lastly, ⊀DEC ≅ ⊀BEA because they are vertical angles. By the AAS Postulate, $\triangle AEB \cong \triangle CED$. DE = BE, by corresponding parts.

Note: The result DE = BE could have been derived directly from the fact that diagonals of a rhombus bisect each other.

● **PROBLEM** 211

Let ABCD be a rhombus. Prove that diagonal $\overline{AC}$ bisects ⊀A.

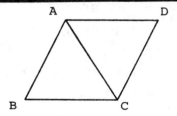

Solution: The figure shows rhombus ABCD with diagonal $\overline{AC}$. To show that ⊀A is bisected by $\overline{AC}$, we prove that ⊀BAC ≅ ⊀DAC. This can be accomplished by showing that $\triangle BAC \cong \triangle DAC$.

A rhombus is a quadrilateral, all of whose sides are congruent. Hence, $\overline{BA} \cong \overline{DA}$ and $\overline{BC} \cong \overline{DC}$. $\overline{AC} \cong \overline{AC}$, by the reflexive property. Therefore, by the SSS Postulate, $\triangle BAC \cong \triangle DAC$, which implies that ⊀BAC ≅ ⊀DAC. Hence, $\overline{AC}$ bisects ⊀A.

● **PROBLEM** 212

As drawn in the figure, ABCD is a parallelogram. The lengths of each side can be represented in the following algebraic manner:
$$AB = 2x + 1$$
$$DC = 3x - 11$$
$$AD = x + 13$$
Show that ABCD is a rhombus.

Solution: A rhombus is a quadrilateral in which all sides are congruent. Since ABCD is given to be a parallelogram we know it is a quadrilateral whose opposite sides are congruent. Therefore, to prove ABCD is a rhombus, it is sufficient to show that any two consecutive sides are congruent. (Any congruent sides are of equal length.)

In effect, we are given DC $\overset{\sim}{=}$ AB and want to show AD $\overset{\sim}{=}$ AB.

Set the length of $\overline{AB}$ equal to the length of $\overline{DC}$ and solve for x. Then, substitute back into the expressions for the length of $\overline{AB}$ and $\overline{AD}$ to determine whether they are equal.

$$DC = AB$$

$$3x - 11 = 2x + 1$$

$$3x - 2x = 1 + 11$$

$$x = 12.$$

Then, $$AB = 2x + 1 = 2(12) + 1 = 25$$

$$AD = x + 13 = 12 + 13 = 25.$$

Therefore, the length of $\overline{AB}$ = length of $\overline{AD}$ and $\overline{AB} \overset{\sim}{=} \overline{AD}$.

Therefore, ABCD is a rhombus because it is a quadrilateral (a parallelogram) in which two consecutive sides are congruent.

● **PROBLEM 213**

Given: $\overline{AN}$ is an angle bisector of $\triangle ABC$; $\overleftrightarrow{CNBH}$; $\overline{HA} \perp \overline{AN}$; $\overrightarrow{CAR}$; $\overrightarrow{HR} \parallel \overline{AB}$; $\overrightarrow{ABP}$; $\overrightarrow{HP} \parallel \overline{CA}$. Prove: quadrilateral APHR is a rhombus.

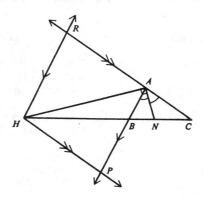

<u>Solution</u>: Aside from the various collinearities, we are given: (1) $\overrightarrow{HP}$ || $\overrightarrow{CA}$; $\overrightarrow{HR}$ || $\overline{AB}$; (2) $\overline{AN}$ is an angle bisector of $\triangle ABC$; and (3) $\overline{HA} \perp \overline{AN}$. From (1), we can show that quadrilateral APHR has both pairs of opposite sides parallel, and therefore APHR is a parallelogram.

Given that a figure is a parallelogram, we can show that it is a rhombus if we show that a diagonal of the parallelogram bisects an angle of the parallelogram. Most of the information we are given involves vertex A. Therefore, we prove the result by showing diagonal $\overline{AH}$ bisects ⦣RAP or m⦣RAH = m⦣HAP; we set out finding expressions for m⦣RAH and m⦣HAP; and try to show them equal.

STATEMENTS	REASONS
1. $\overline{AN}$ is an angle bisector of $\triangle ABC$; $\overleftrightarrow{CNBH}$; $\overline{HA} \perp \overline{AN}$; $\overleftrightarrow{CAR}$; $\overrightarrow{HR}$ || $\overline{AB}$; $\overleftrightarrow{ABP}$; $\overrightarrow{HP}$ || $\overline{CA}$	1. Given.
2. $\overline{RA}$ || $\overline{HP}$; $\overline{HR}$ || $\overline{AP}$	2. Two segments each on separate parallel lines are parallel.
3. $\square$ APHR is a parallelogram	3. If both pairs of opposite sides of a quadrilateral are parallel, then the quadrilateral is a parallelogram.
4. ⦣HAN is a right angle or m⦣HAN = 90°	4. Perpendicular lines intersect to form right angles.
5. ⦣HAB and ⦣BAN are complementary	5. Two angles that form a right angle are complementary.
6. m⦣RAC = m⦣RAH + m⦣HAN + m⦣NAC	6. Angle Sum Postulate.
7. m⦣RAC = 180°	7. The measure of a straight angle is 180°.
8. 180° = m⦣RAH + 90° + m⦣NAC	8. Substitution Postulate.
9. 90° = m⦣RAH + m⦣NAC	9. Subtraction Postulate.
10. ⦣RAH and ⦣NAC are complementary	10. Two angles whose measures sum to 90° are complementary angles.
11. ⦣NAC $\overset{\sim}{=}$ ⦣BAN	11. Definition of angle bisector.
12. ⦣RAH $\overset{\sim}{=}$ ⦣HAB	12. Angles complementary to congruent angles are congruent.

13. $\overline{HA}$ bisects ⟩RAP	13. Definition of angle bisector.
14. Quadrilateral APHR is a rhombus	14. If the diagonal of a parallelogram bisects an angle of the parallelogram, then the parallelogram is a rhombus.

● **PROBLEM** 214

We are given rhombus ABCD with its diagonals drawn. F is the midpoint of $\overline{DE}$, G is the midpoint of $\overline{BE}$ and H is a point on $\overline{AE}$. Prove that ΔFGH is an isosceles triangle.

Solution: To prove ΔFGH is isosceles, it is sufficient to prove $\overline{FH} \cong \overline{HG}$. We can do this by proving ΔFEH $\cong$ ΔGEH and applying corresponding parts to obtain the desired results.

We will make use of the fact that the diagonals of a rhombus intersect perpendicularly.

STATEMENT	REASON
1. ABCD is a rhombus	1. Given.
2. $\overline{ED} \cong \overline{EB}$	2. Diagonals of a parallelogram bisect each other.
3. $\overline{EG} \cong \overline{GB}$ and $\overline{EF} \cong \overline{FD}$	3. A midpoint divides a line segment into two congruent segments.
4. $\overline{EF} \cong \overline{EG}$	4. Halves of equal quantities are equal quantities (see steps (3) and (2)).
5. $\overline{AC} \perp \overline{BD}$	5. The diagonals of a rhombus are perpendicular to each other.
6. ⟩HEG and ⟩HEF are right angles	6. Perpendicular lines intersect at right angles.
7. ⟩HEG $\cong$ ⟩HEF	7. All right angles are congruent.
8. $\overline{EH} \cong \overline{EH}$	8. Reflexive Property of Congruence.

9. ΔFEH ≅ ΔGEH	9. SAS ≅ SAS.
10. $\overline{FH}$ ≅ $\overline{GH}$	10. Corresponding sides of congruent triangles are congruent.
11. ΔFHG is isosceles	11. A triangle that has two congruent sides is an isosceles triangle.

● **PROBLEM** 215

In rhombus WXYZ; A, B, and C are the midpoints of $\overline{WX}$, $\overline{XY}$, and $\overline{YZ}$, respectively. Prove: ΔABC is a right triangle.

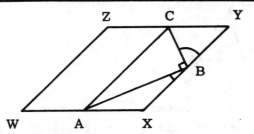

<u>Solution</u>: To show ΔABC is a right triangle, we show that any angle of ΔABC measures 90°. ⟩CBA seems to be a good candidate for this angle. Since they form a straight angle, m⟩CBY + m⟩CBA + m⟩ABX = 180°. As such, m⟩CBA = 180 - (m⟩CBY + m⟩ABX). Thus, if we show that ⟩CBY and ⟩ABX are complementary and sum to 90°, then ⟩CBA must be a right angle.

To show ⟩CBY and ⟩ABX are complementary, note that ⟩CBY and ⟩ABX are angles of triangles ΔCBY and ΔABX. Because WXYZ is a rhombus, the angles of ΔCBY and ΔABX can be related to each other. Note that (1) m⟩YCB + m⟩CYB + m⟩CBY = 180° and (2) m⟩BAX + m⟩AXB + m⟩ABX = 180°. ⟩CYB and ⟩AXB are adjacent angles of parallelogram WXYZ. Therefore, they are supplementary. Also, because the sides of a rhombus are congruent and A, B, and C are midpoints of the sides they lie on, $\overline{CY}$ ≅ $\overline{YB}$ ≅ $\overline{BX}$ ≅ $\overline{AX}$. Therefore, ΔCBY and ΔABX are isosceles. Thus, for the base angles, ⟩YCB ≅ ⟩YBC and ⟩XAB ≅ ⟩XBA. Substituting these results into (1) and (2), we obtain (3) 2 · m⟩XBA + m⟩AXB = 180, and (4) 2 · m⟩YCB + (180 - m⟩AXB) = 180. Adding the two equations we obtain (5) 2 · m⟩XBA + 2 · m⟩YCB = 180, or m⟩XBA + m⟩YCB = 90. Thus, ⟩XBA and ⟩YCB are complementary.

By substitution,

m⟩CBA = 180 - (m⟩CBY + m⟩ABX) = 180 - 90.

Therefore, m⟩CBA = 90° and ⟩CBA is a right angle. Thus, ΔCBA is a right triangle.

SQUARES

● **PROBLEM** 216

In the accompanying figure, ΔABC is given to be an isosceles right triangle with ⟩ABC a right angle and AB ≅ BC. Line segment $\overline{BD}$, which bisects $\overline{CA}$, is extended to E, so that $\overline{BD}$ ≅ $\overline{DE}$. Prove BAEC is a square.

Solution: A square is a rectangle in which two consecutive sides are congruent. This definition will provide the framework for the proof in this problem. We will prove that BAEC is a parallelogram that is specifically a rectangle with consecutive sides congruent, namely a square.

STATEMENT	REASON
1. $\overline{BD}$ ≅ $\overline{DE}$ and $\overline{AD}$ ≅ $\overline{DC}$	1. Given ($\overline{BD}$ bisect $\overline{CA}$)
2. BAEC is a parallelogram	2. If diagonals of a quadrilateral bisect each other, then the quadrilateral is a parallelogram.
3. ⟩ABC is a right angle	3. Given.
4. BAEC is a rectangle	4. A parallelogram, one of whose angles is a right angle, is a rectangle.
5. $\overline{AB}$ ≅ $\overline{BC}$	5. Given.
6. BAEC is a square	6. If a rectangle has two congruent consecutive sides, then the rectangle is a square.

● **PROBLEM** 217

Given: Square PQRS, with N on $\overline{RS}$ so that $\overline{TS}$ ≅ $\overline{SN}$. Prove: m⟩STN = 3(m⟩NTR).

Solution: Here, we can solve for the exact values of ⟩STN and ⟩NTR and then compare. We are given that ΔSTN is isosceles. Because PQRS is a rhombus (every square is a

rhombus) diagonal $\overline{SQ}$ bisects right angle ⊀PSR and m⊀TSR
= ½(90°) = 45°. ⊀STN is a base angle of isosceles ΔSTN,
and thus ⊀STN $\cong$ ⊀SNT. Since 180° = m⊀TSN + m⊀STN + m⊀SNT
and m⊀SNT = m⊀STN, by substitution, 2m⊀STN = 180° - 45°,
and thus m⊀STN = $\frac{135°}{2}$ = 67.5°.

Since the diagonals of a square are perpendicular to
each other, ⊀STN and ⊀NTR are complementary, and m⊀NTR =
90° - 67.5° = 22.5°. 67.5° = 3(22.5°), so m⊀STN = 3(m⊀NTR).

STATEMENTS	REASONS
1. Square PQRS, with N in $\overline{RS}$, so that $\overline{TS} \cong \overline{SN}$.	1. Given.
2. $\overline{QS}$ bisects ⊀PSR	2. The diagonals of a rhombus bisect the angles.
3. m⊀TSN = ½m⊀PSR	3. The bisector of an angle divides an angle into two equal parts.
4. m⊀PSR = 90°	4. The angles of a square. are right angles.
5. m⊀TSN = ½ 90° = 45°	5. Substitution Postulate.
6. m⊀TSN + m⊀STN + m⊀SNT = 180°	6. The angle sum of a tri- angle is 180°.
7. m⊀STN = m⊀SNT	7. The base angles of an isosceles triangle are congruent.
8. 2m⊀STN + 45° = 180°	8. Substititon Postulate.
9. m⊀STN = $\frac{180° - 45°}{2}$ = 67.5°.	9. Subtraction and Division Properties of Equality.
10. $\overline{QS} \perp \overline{PR}$	10. The diagonals of a rhombus are perpendicular to each other.
11. m⊀STN + m⊀NTR = 90°	11. Two angles that form a right angle are comple- mentary.
12. m⊀NTR = 90° - 67.5° = 22.5°	12. Substitution Postulate.
13. 67.5° = 3 · 22.5°	13. Factoring Postulate.

14. m∡STN = 3(m∡NTR) | 14. Substitution Postulate.

● **PROBLEM** 218

Given: Square ABCD; $\overline{AQC}$; $\overline{APB}$; $\overline{BC} \stackrel{\sim}{=} \overline{QC}$; $\overline{PQ} \perp \overline{AC}$. Prove:
AQ = PQ = PB. (Hint: Draw $\overline{PC}$.)

Solution: To show AQ = QP, we show ΔAQP is an isosceles
triangle. We will show ΔAQP is isosceles by showing base
angles ∡QAP $\stackrel{\sim}{=}$ ∡QPA. Secondly, $\overline{PQ}$ and $\overline{PB}$ are corresponding
parts of ΔQPC and ΔBPC. We show ΔQPC $\stackrel{\sim}{=}$ ΔBPC by the Hypotenuse-Leg Theorem and thus PQ = PB.

STATEMENTS	REASONS
1. Square ABCD; $\overline{AQC}$; $\overline{APB}$; $\overline{BC} \stackrel{\sim}{=} \overline{QC}$; $\overline{PQ} \perp \overline{AC}$	1. Given.
2. m∡QAP = m∡DAQ = ½m∡DAP	2. The diagonals of a square bisect the intersecting angles.
3. m∡DAP = 90°	3. The angles of a square are right angles.
4. m∡QAP = 45°	4. Substitution Postulate.
5. ΔAQP is a right triangle	5. If one of the angles of a triangle is a right angle, then the triangle is a right triangle.
6. m∡QAP + m∡QPA = 90°	6. The acute angles of a right triangle are complementary.
7. m∡QPA = 45°	7. Substitution Postulate and Subtraction Property of Equality.
8. ΔQPA is isosceles	8. If two angles of a triangle are congruent, then the triangles are isosceles.
9. $\overline{AQ} = \overline{QP}$ (or AQ = QP)	9. The legs of an isosceles triangle are congruent.
10. $\overline{PC} \stackrel{\sim}{=} \overline{PC}$	10. A segment is congruent to itself.

11. ΔPBC is a right triangle	11. Same as reason 5.
12. ΔQPC ≅ ΔBPC	12. If the hypotenuse and a leg of one right triangle are congruent to the hypotenuse and leg of a second right triangle, then the two triangles are congruent.
13. $\overline{QP} ≅ \overline{BP}$ (or QP = BP)	13. Corresponding parts of congruent triangles are congruent.
14. AQ = QP = BP	14. Transitivity Postulate.

● **PROBLEM** 219

Given: Square ABCD; P is any point of $\overline{AB}$, Q is any point on $\overline{AD}$, $\overline{CQ} \perp \overline{PD}$ at R. Prove $\overline{PD} ≅ \overline{QC}$.

Solution: $\overline{PD}$ and $\overline{QC}$ are corresponding parts of ΔAPD and ΔDQC. It is then sufficient to show ΔAPD and ΔDQC are congruent. We do this using the AAS Postulate.

STATEMENTS	REASONS
1. Square ABCD; P is any point of $\overline{AB}$ so that $\overline{APB}$; $\overline{CQ} \perp \overline{PD}$ at R; $\overline{AQD}$	1. Given.
2. ⊁DAP and ⊁ADC are right angles.	2. All angles of a square are right angles.
3. ⊁DAP ≅ ⊁ADC	3. All right angles are congruent.
4. $\overline{AD} ≅ \overline{DC}$	4. All sides of a square are congruent.
5. ⊁APD and ⊁ADP are complementary	5. The acute angles of a right triangle (ΔDAP) are complementary
6. ⊁DQR and ⊁ADP are complementary	6. The acute angles of a right triangle (ΔQDR) are complementary
7. ⊁APD ≅ ⊁DQR	7. Angles complementary to congruent angles are congruent.

| 8. $\triangle APD \cong \triangle DQC$ | 8. The AAS Postulate. |
| 9. $\overline{PD} \cong \overline{QC}$ | 9. Corresponding parts of congruent triangles are congruent. |

● **PROBLEM** 220

Given: Square PQRS; A is any point of $\overline{QMS}$; $\overline{QBC} \perp \overline{PCA}$, B and M are on $\overline{PR}$. Prove: $\overline{AS} \cong \overline{BP}$.

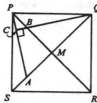

Solution: $\overline{AS}$ and $\overline{BP}$ are corresponding sides of triangles $\triangle PBQ$ and $\triangle SAP$. To prove that $\overline{AS} \cong \overline{BP}$, it is sufficient to show $\triangle SAP \cong \triangle PBQ$. This will be done using the ASA Postulate.

STATEMENTS	REASONS
1. Square PQRS; A is any point of $\overline{QMS}$; $\overline{QBC} \perp \overline{PCA}$; PBMR	1. Given.
2. $\overline{PS} \cong \overline{PQ}$	2. The sides of a square are congruent.
3. m⟩PSR = m⟩QPS = 90°	3. The angles of a square are right angles.
4. m⟩PSQ = ½m⟩PSR m⟩QPB = ½ m⟩QPS	4. The diagonals of a square bisect the angles of the square.
5. m⟩PSA = m⟩QPB	5. Transitivity Postulate.
6. $\triangle PCQ$ is a right triangle.	6. Definition of right triangle.
7. ⟩CPQ and ⟩PQC are complementary	7. The acute angles of a right triangle are congruent.
8. ⟩CPQ and ⟩SPA are complementary	8. Two angles of a right triangle are complementary
9. ⟩PQC $\cong$ ⟩SPA	9. Angles complementary to congruent triangles are congruent.
10. $\triangle PBQ \cong \triangle SAP$	10. The ASA Postulate.

| 11. $\overline{AS} \cong \overline{BP}$ | 11. Corresponding parts of congruent triangles are congruent. |

● **PROBLEM** 221

In the accompanying figure, WXYZ is a square. It has been placed in such a way inside quadrilateral ABCD that when its sides are extended to the vertices of ABCD, $\overline{AW} \cong \overline{BX} \cong \overline{CY} \cong \overline{DZ}$. All segments shown are straight line segments. Show that ABCD is a square.

Solution: In this problem we will set out to show Δ's AWB, BXC, CYD, and DZA are congruent. Then, by several applications of the fact that corresponding parts of congruent triangles are congruent, we will show that ABCD is a quadrilateral all of whose sides are congruent, and which contains a right angle (i.e., ABCD is a square).

STATEMENT	REASON
1. WXYZ is a square	1. Given.
2. $\overline{AW} \cong \overline{BX} \cong \overline{CY} \cong \overline{DZ}$	2. Given.
3. $\overline{WZ} \cong \overline{XW} \cong \overline{YX} \cong \overline{ZY}$	3. All four sides of a square are congruent.
4. $\overline{AW} + \overline{WZ} \cong \overline{BX} + \overline{XW}$ $\overline{CY} + \overline{YX} \cong \overline{DZ} + \overline{ZY}$ or $\overline{AZ} \cong \overline{BW} \cong \overline{CX} \cong \overline{DY}$	4. Congruent quantities added to congruent quantities result in congruent quantities.
5. $\overline{DZ} \cong \overline{AW} \cong \overline{BX} \cong \overline{CY}$	5. Given.
6. $\overleftrightarrow{BW}$, $\overleftrightarrow{CX}$, $\overleftrightarrow{DY}$ and $\overleftrightarrow{AZ}$ are straight lines	6. Given.
7. $\overline{BW} \perp \overline{AZ}$, $\overline{AZ} \perp \overline{DY}$, $\overline{DY} \perp \overline{CX}$ and $\overline{CX} \perp \overline{BW}$	7. The sides of a square meet to form right angles and are, therefore, perpendicular.
8. $\angle$AZD, $\angle$AWB, $\angle$CXB and $\angle$CYD are right angles	8. Perpendicular lines intersect to form right angles.
9. $\angle$AZD $\cong \angle$AWB $\cong \angle$CXB $\cong \angle$CYD	9. All right angles are congruent.

186

10. ΔAZD $\overset{\sim}{=}$ ΔBWA $\overset{\sim}{=}$ ΔCXB $\overset{\sim}{=}$ ΔDYC	10. SAS $\overset{\sim}{=}$ SAS.
11. $\overline{AD}$ $\overset{\sim}{=}$ $\overline{DC}$ $\overset{\sim}{=}$ $\overline{CB}$ $\overset{\sim}{=}$ $\overline{BA}$	11. Corresponding sides of congruent triangles are congruent.
12. ∢WAB is complementary to ∢WBA	12. The acute angles of right triangles are complementary.
13. ∢ZAD $\overset{\sim}{=}$ ∢WBA	13. Corresponding parts of congruent triangles are congruent.
14. ∢WAB is complementary to ∢ZAD	14. An angle congruent to a given angle's complement is complementary to that given angle.
15. m∢WAB + m∢ZAD = 90°	15. The sum of the measures of two complements is 90°.
16. m∢DAB = m∢WAB + m∢ZAD	16. The measure of the whole is equal to the sum of the measures of its parts.
17. m∢DAB = 90°	17. Substitution Postulate.
18. ∢DAB is a right angle	18. An angle that measures 90° is a right angle.
19. ABCD is a square	19. A quadrilateral that contains a right angle and all of whose sides are congruent is a square.

● **PROBLEM** 222

Given: Points E and F are exterior to square ABCD; $\overline{CF}$ $\overset{\sim}{=}$ $\overline{CE}$; ∢ACF $\overset{\sim}{=}$ ∢ACE. Prove: $\overline{EF}$ $||\overline{BD}$.

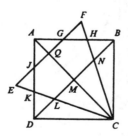

Solution: We are given that CF = CE. Therefore, ΔEFC is isosceles. Furthermore, ∢ACF $\overset{\sim}{=}$ ∢ACE, thus $\overline{AC}$ is the angle bisector of ∢FCE. In an isosceles triangle, the angle bisector of the vertex angle is the perpendicular bisector of the base. Thus, $\overline{EF}$ $\perp$ $\overline{AC}$. But $\overline{AC}$ is also a diagonal of the square, and thus $\overline{AC} \perp \overline{DB}$, $\overline{DB}$ being another diagonal of ABCD. Since $\overline{AC}$ is perpendicular to both $\overline{DB}$ and $\overline{EF}$, $\overline{EF}$ $||$ $\overline{BD}$.

STATEMENTS	REASONS
1. Points E and F are exterior to square ABCD; $\overline{CF} \cong \overline{CE}$; $\angle ACF \cong \angle ACE$	1. Given.
2. $\triangle ECF$ is isosceles	2. If two sides of a triangle are congruent, then the triangle is isosceles.
3. $\overrightarrow{CA}$ bisects $\angle ECF$	3. If a ray divides an angle into two angles such that the divided angles are congruent, then the ray bisects the angle.
4. $\overrightarrow{CA} \perp \overline{EF}$	4. The bisector of the vertex angle of an isosceles triangle is the perpendicular bisector of the base.
5. $\overline{CA} \perp \overline{BD}$	5. The diagonals of a square are perpendicular to each other.
6. $\overline{EF} \parallel \overline{BD}$	6. In a plane, two lines perpendicular to a given line are parallel.

● **PROBLEM** 223

Starting with any triangle ABC; construct the exterior squares BCDE, ACFG and BAHK; then construct parallelograms FCDQ and EBKP. Prove △PAQ is an isosceles right triangle. (Hint: Draw diagonals PB and CQ.)

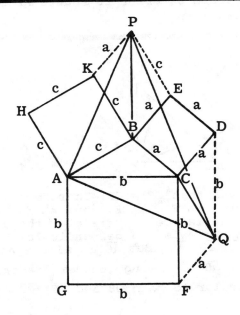

Solution: In the accompanying figure, ΔABC has sides of lengths a, b, and c. Using the fact that opposite sides of a parallelogram are congruent, we have labelled the remaining segments in the figure. The key to this proof is recognizing groups of congruent triangles: ΔPKB ≅ΔABC ≅ ΔFQC and ΔACQ ≅ΔPBA. Before proceeding further with the proof, we must prove these congruences. To show ΔPKB ≅ ΔABC, note that AB = KB = c and BC = KP = a. Thus $\overline{AB}$ ≅ $\overline{KB}$ and $\overline{BC}$ ≅ $\overline{KP}$. Furthermore, ∡ABC ≅ ∡BKP. (m∡ABC + m∡KBA + m∡KBE + m∡EBC = 360°. Since ∡KBA and ∡EBC are vertex angles of a square, they are right angles. Thus, m∡KBA + m∡EBC = 90° + 90° = 180°. Substituting in the first equation, we obtain m∡ABC + m∡KBE = 360° - 180° = 180°, or ∡ABC and ∡KBE are supplementary. ∡KBE and ∡BKP are adjacent angles of a parallelogram and are therefore, supplementary. Since both ∡ABC and ∡BKP are supplementary to the same angle, they are congruent.) By the SAS Postulate, ΔPKB ≅ ΔCBA.

In a similar method, we show $\overline{BC}$ ≅ $\overline{FQ}$, $\overline{AC}$ ≅ $\overline{CF}$, ∡CFQ ≅ ∡ACB, and thus ΔCBA ≅ ΔFQC. Combining the two results, we have ΔPKB ≅ ΔCBA ≅ ΔFQC.

To show ΔACQ ≅ ΔPBA, note that $\overline{BP}$ ≅ $\overline{AC}$ (corresponding parts of congruent triangles PKB and FQC) and $\overline{AB}$ ≅ $\overline{CQ}$ (corresponding parts of congruent triangles CBA and FQC).

To show that the included angles are congruent, note that m∡ABP=m∡ABK+m∡KBP and m∡ACQ=m∡ACF+m∡FCQ. ∡ABK and ∡ACF are vertex angles of a square and are therefore right angles. Thus, m∡ABK=m∡ACF. Furthermore, ∡KBP and ∡QFC are corresponding angles of congruent triangles PKB and FQC. Thus, m∡ABK+m∡KBP= m∡ACF+m∡FCQ, and ∡ABP ≅ ∡ACQ. By the SAS Postulate, ΔACQ = ΔPBA.

With these two proofs, we can now show that ΔPAQ is an isosceles right triangle.

To show that ΔPAQ is isosceles, we show that sides $\overline{PA}$ and $\overline{AQ}$ are congruent. Note that $\overline{PA}$ and $\overline{AQ}$ are corresponding sides of congruent triangles ACQ and PBA. Thus, $\overline{PA}$ ≅ $\overline{AQ}$ and ΔPAQ is isosceles.

To show that ΔPAQ is a right Δ, we show ∡PAQ is a right angle. Since m∡PAQ = m∡PAB + m∡BAC + m∡CAQ, we can show that m∡PAB + m∡BAC + m∡CAQ = 90°.

Consider these four facts:

(1) ∡BAC ≅ ∡PBK (corresponding parts of congruent ΔCBA and ΔPKB)

189

(2) ⊀CAQ ≅ ⊀BPA (corresponding parts of congruent ΔACQ and ΔPBA)

(3) m⊀ABP = m⊀ABK + m⊀PBK = 90° + m⊀PBK

(4) m⊀PAB + m⊀ABP + m⊀BPA = 180° (angle sum of ΔPBA must equal 180°)

Substituting (1) into (3), we have m⊀ABP = 90° + m⊀BAC. Substituting this result into (4) we have m⊀PAB + (90° + m⊀BAC) + m⊀CAQ = 180° or m⊀PAB + m⊀BAC + m⊀CAQ + 90° = 180°, or m⊀PAB + m⊀BAC + m⊀CAQ = 90°. Thus, m⊀PAQ = 90° and ΔPAQ is a right isosceles triangle.

RECTANGLES

● **PROBLEM** 224

Given: ABCD is a parallelogram. $\overline{DE}$ bisects ⊀D. $\overline{BF}$ bisects ⊀B. Prove: $\overline{DE}$ || $\overline{BF}$.

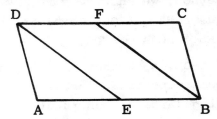

Solution: We will prove $\overline{DE}$ || $\overline{BF}$, by showing that when a transversal $\overline{AB}$ cuts these lines the corresponding angles formed (⊀DEA, ⊀EBF) are congruent.

Since opposite sides of a parallelogram are parallel, $\overline{AB}$ || $\overline{CD}$. $\overline{DE}$, then acts as a transversal forming congruent alternate interior angles: ⊀DEA ≅ ⊀EDF.

Since opposite angles of a parallelogram are congruent, ⊀ADC ≅ ⊀CBA. We are given that $\overline{DE}$ bisects ⊀ADC and $\overline{BF}$ bisects ⊀CBA. Therefore, ⊀ADE ≅ ⊀EDF and ⊀EBF ≅ ⊀CBF. Hence, since halves of congruent quantities are congruent, ⊀EDF ≅ ⊀EBF.

Since ⊀DEA ≅ ⊀EDF and ⊀EDF ≅ ⊀EBF, by transitivity, ⊀DEA ≅ ⊀EBF.

Because transversal $\overline{AB}$ forms congruent corresponding angles with segments $\overline{DE}$ and $\overline{BF}$, $\overline{DE}$ || $\overline{BF}$.

● **PROBLEM** 225

In parallelogram ABCD, as shown in the figure, $\overline{DE}$ ⊥ $\overline{AB}$ and $\overline{BF}$ ⊥ $\overline{DC}$. With this in mind, prove altitudes $\overline{DE}$ ≅ $\overline{BF}$.

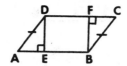

Solution: To prove that $\overline{DE} \cong \overline{BF}$, the best approach would
be to prove that the triangles, which have $\overline{DE}$ and $\overline{BF}$ as
corresponding parts, are congruent. Therefore, we want to
prove that $\triangle AED \cong \triangle CFB$. The method of proving congruence
that is best for this problem is the one in which two tri-
angles are shown to be congruent because two angles and
the side opposite one angle in one triangle are congruent
to the corresponding parts of another triangle. (A.A.S. $\cong$
A.A.S.)

Additionally, to prove the parts congruent it is
essential to remember that opposite sides and angles in
a parallelogram are congruent.

STATEMENT	REASON
1. ABCD is a parallelogram.	1. Given.
2. $\overline{AD} \cong \overline{CB}$	2. Opposite sides of a parallelogram are congruent.
3. ⅄A $\cong$ ⅄C	3. Opposite angles of a parallelogram are congruent.
4. $\overline{DE} \perp \overline{AB}$ and $\overline{BF} \perp \overline{DC}$	4. Given.
5. ⅄E and ⅄F are right angles.	5. When perpendicular lines intersect they form right angles.
6. ⅄E $\cong$ ⅄F	6. All right angles are congruent.
7. $\triangle AED \cong \triangle CFB$	7. a.a.s. $\cong$ a.a.s.
8. $\overline{DE} \cong \overline{BF}$	8. Corresponding parts of congruent triangles are congruent.

● **PROBLEM** 226

Prove that a rectangle is a parallelogram.

Solution: A rectangle is a quadrilateral, all of whose
angles measure 90° (see figure). We will show that ABCD
is a parallelogram by analyzing the pairs of angles ABC
and BCD, and DAB and ABC.

Since ⅄ABC, ⅄BCD, and ⅄DAB are all right angles,

191

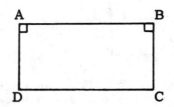

$$m \measuredangle ABC = 90° \tag{1}$$

$$m \measuredangle BCD = 90° \tag{2}$$

$$m \measuredangle DAB = 90° \tag{3}$$

Adding (1) and (2) shows that ⋋ABC and ⋋BCD are supplementary:

$$m \measuredangle ABC + m \measuredangle BCD = 180°. \tag{4}$$

Adding (1) and (3) shows that ⋋ABC and ⋋DAB are supplementary:

$$m \measuredangle ABC + m \measuredangle DAB = 180°. \tag{5}$$

Now, if the consecutive interior angles of 2 lines crossed by a transversal sum to 180°, then the 2 lines are parallel.

⋋ABC and ⋋BCD are consecutive interior angles of line segments $\overline{AB}$ and $\overline{DC}$. Also, ⋋ABC and ⋋DAB are consecutive interior angles of line segments $\overline{AD}$ and $\overline{BC}$. Using these facts, plus (4) and (5), we conclude that $\overline{AD} \parallel \overline{BC}$ and $\overline{AB} \parallel \overline{DC}$. By definition this means that ABCD, a rectangle, is a parallelogram.

● **PROBLEM** 227

Let ABCD be a rectangle such that $\overline{AD} \cong \overline{BC}$ and ⋋ADC ≅ ⋋BCD. Prove that its diagonals are congruent.

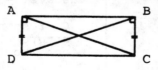

Solution: In the figure, we are given that $\overline{AD} \cong \overline{BC}$ and ⋋ADC ≅ ⋋BCD. We must show that $\overline{AC} \cong \overline{DB}$. The method we shall use to accomplish this is to show that △ADC ≅ △BCD.

First, we are given that $\overline{AD} \cong \overline{BC}$, and ⋋ADC ≅ ⋋BCD. Also, $\overline{DC} \cong \overline{CD}$, by the reflexive property. Hence, △ADC ≅ △BCD by the SAS Postulate. We may therefore conclude, by corresponding parts, that $\overline{AC} \cong \overline{BD}$, and the diagonals of ABCD are congruent.

In the figure shown, we are given rectangle ABCD.
(a) Prove $\overline{AC} \cong \overline{BD}$. (b) Prove △AEB is isosceles.

Solution: (a) In this part, we are asked to prove that
the diagonals $\overline{AC}$ and $\overline{BD}$ are congruent. Since the only
given fact is that ABCD is a rectangle, we are proving
that, in general, the diagonals of a rectangle are
congruent.

We will do this by proving △ACB $\cong$ △BDA, using the
SAS Postulate.

STATEMENT	REASON
1. ABCD is a rectangle	1. Given.
2. $\overline{AD} \cong \overline{BC}$	2. Opposite sides of a rectangle are congruent.
3. $\overline{AB} \cong \overline{BA}$	3. Reflexive Property of Congruence.
4. ⊀DAB and ⊀CBA are right angles	4. All angles of a rectangle are right angles.
5. ⊀DAB $\cong$ ⊀CBA	5. All right angles are congruent.
6. △ACB $\cong$ △BDA	6. SAS $\cong$ SAS
7. $\overline{AC} \cong \overline{BD}$	7. Corresponding sides of congruent triangles are congruent.

(b) To prove △AEB isosceles, we must prove that
$\overline{AE} \cong \overline{EB}$. We can do this by showing that the angles opposite these sides are congruent.

STATEMENT	REASON
1. △ACB $\cong$ △BDA	1. Results from part (a).
2. ⊀CAB $\cong$ ⊀DBA	2. Corresponding angles of congruent triangles are congruent.
3. In △AEB, $\overline{AE} \cong \overline{EB}$	3. In a triangle, if two angles are congruent, then

the sides opposite those
angles are congruent.

4. ΔAEB is isosceles

4. By definition, if a triangle has a pair of
congruent sides, then it
is isosceles.

● **PROBLEM** 229

Given: ΔABC with median BE; AE = BE. Prove: m⊁ABC = 90°.

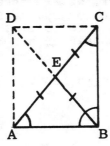

Figure 1

<u>Solution</u>: We are asked to show that an angle is a right
angle. We could solve for m⊁ABC algebraically. Since ΔAEB
and ΔECB are isosceles triangles, m⊁EAB = m⊁EBA and m⊁ECB
= m⊁EBC. Using the fact that m⊁EAB + m⊁ABC + m⊁ECB = 180°
m⊁ABC = m⊁EBA + m⊁EBC, by substitution, we could obtain
m⊁EBA + m⊁ABC + m⊁EBC = m⊁ABC + m⊁ABC = 2 m⊁ABC = 180° or
m⊁ABC = 90°.

A more elegant method would be to recall that the
angles of a rectangle are right angles and find a rectangle
such that ⊁B is an angle of the rectangle. By extending $\overline{EB}$
to point D, such that DE = EB, we would have determined a
quadrilateral ABCD in which (1) the diagonals bisect each
other (making ABCD a parallelogram) and (2) the diagonals
are congruent (ABCD is a rectangle). Thus ⊁ABC is a right
angle.

● **PROBLEM** 230

Prove that the bisectors of the angles of a rectangle
enclose a square.

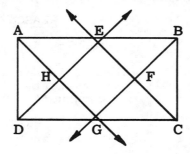

Solution: In the diagram, the angle bisectors of rectangle ABCD intersect at points E, F, G, and H. We must show that quadrilateral EFGH is a square. To do this we will show that (1) EFGH is a rectangle and (2) two adjacent sides of EFGH are congruent.

To show that EFGH is a rectangle, we show that it is a quadrilateral with four right angles. We will do this with several applications of the theorem stating that the angle sum of a triangle is 180°.

$\overline{AG}$ is an angle bisector of ∢DAB. Since quadrilateral ABCD is a rectangle, then ∢DAB is a right angle. Thus, ∢DAH = ½(90°) = 45°. Similarly, $\overline{DE}$ is an angle bisector of ∢ADC, and thus ∢HDA = ½(90°) = 45°. In ΔHDA, m∢AHD + m∢HDA + m∢DAH = 180°, or m∢AHD = 180°− m∢DAH − m∢HDA = 180° − 45° − 45° = 90°. Because ∢EHG and ∢AHD are opposite angles, m∢EHG = m∢AHD = 90°. By similar reasoning, we can show m∢EFG = 90°.

To show m∢DEC is 90°, note that m∢HDG = ½m∢ADG = ½(90°) = 45°. Also, m∢ECD = ½ m∢BCD = ½(90°) = 45°. In ΔDEC, m∢HDG + m∢ECD + m∢DEC = 180°. m∢DEC = 180° − m∢EDC − m∢ECD = 180° − 45° − 45° = 90°. By analogous reasoning we can show m∢AGB = 90°.

Since all four vertex angles are right angles, quadrilateral EFGH is a rectangle. The proof of the second part – that two adjacent sides of EFGH are congruent – follows:

(1) DE = EC: Note m∢EDC = m∢ECD = 45°. Thus, ΔDEC is isosceles and DE = EC.

(2) ΔAHD $\overset{\sim}{=}$ ΔBFC: m∢DAH = m∢CBF = 45°. m∢ADH = m∢BCF = 45°. Opposite sides of rectangle ABCD are congruent. Thus, $\overline{AD} \overset{\sim}{=} \overline{BC}$. By the ASA Postulate, ΔAHD $\overset{\sim}{=}$ ΔBFC.

(3) HD = FC: Corresponding parts of congruent triangles are congruent and equal in length.

(4) DE − HD = EC − FC :   By the Subtraction Property of Equality.

(5) HE = EF: Substitution Postulate.

Thus, in rectangle EFGH, two adjacent sides HE and EF are congruent. Therefore, EFGH is a square.

● **PROBLEM** 231

The accompanying figure shows, in parallelogram ABCD, $\overline{DE}$ drawn perpendicular to $\overline{AB}$ and, on the exterior of ABCD, $\overline{CF}$ dropped perpendicular to $\overline{AB}$. Prove that DEFC is a rectangle.

Solution: The fact that a rectangle is a parallelogram

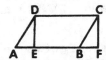

one of whose angles is a right angle will be the basis of
this proof. Specifically, we will show that DEFC is a
parallelogram with ∢E being a right angle. We will need
to use the theorem which states that if both pairs of
opposite sides of a quadrilateral are parallel, then the
quadrilateral is a parallelogram.

(The symbol ▱ will be used in lieu of the word
parallelogram.)

STATEMENT	REASON
1. ABCD is a ▱	1. Given.
2. $\overline{DC} \parallel \overline{AB}$	2. A pair of opposite sides of a ▱ are parallel.
3. $\overline{DE} \perp \overline{AB}$ and $\overline{CF} \perp \overline{AB}$	3. Given.
4. $\overline{DE} \parallel \overline{CF}$	4. Two lines perpendicular to the same line are parallel.
5. DEFC is a ▱	5. If both pairs of opposite sides of a quadrilateral are parallel, then the quadrilateral is a ▱.
6. ∢DEB is a right angle	6. When perpendicular lines intersect, they form a right angle.
7. DEFC is a rectangle	7. If one angle of a ▱ is a right angle, then the ▱ is a rectangle.

● **PROBLEM** 232

Given: Rectangle ABCD; $\overline{AE} \cong \overline{AC}$; $\overline{AC}$ meets $\overline{BD}$ at P; D is
the midpoint of $\overline{AE}$. Prove: $\overline{APC} \perp \overline{PE}$.

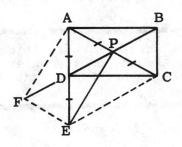

<u>Solution:</u> $\overline{APC}$ and $\overline{PE}$ intersect to form ⧖APE. It is sufficient to prove ⧖APE is a right angle to conclude $\overline{APC} \perp \overline{PE}$. To do this we find some rectangle of which ⧖APE is an angle.

ABCD is a rectangle. Therefore, its diagonals are congruent and bisect each other. AP = PC = DP = PB = ½AC.
Since D is a midpoint of $\overline{AE}$ and $\overline{AE} \overset{\sim}{=} \overline{AC}$, AP = DE = ½ AE = ½ AC, or AD = DE = PD. Locate point F such that $\overline{FDP}$ and FD = DP. Then in quadrilateral FAPE, the diagonals are congruent and bisect one another. Therefore, FAPE is a rectangle and m⧖APE = 90°. Thus, $\overline{APC} \perp \overline{PE}$.

STATEMENTS	REASONS
1. Rectangle ABCD; $\overline{AE} \overset{\sim}{=} \overline{AC}$; $\overline{AC}$ meets $\overline{BD}$ at P; D is the midpoint of $\overline{AE}$	1. Given.
2. AP = PC = PD = PB = ½AC	2. The diagonals of a rectangle are congruent and bisect each other.
3. AD = DE = ½AE	3. Definition of a midpoint.
4. AD = DE = PD	4. Transitivity Property.
5. Locate point F such that $\overline{FDP}$ and FD = PD	5. Point Uniqueness Postulate.
6. Quadrilateral FAPE is a rectangle	6. If the diagonals of a quadrilateral are congruent and bisect each other, then the quadrilateral is a rectangle.
7. ⧖APE is a right angle	7. All vertex angles of rectangles are right angles.
8. $\overline{APC} \perp \overline{PE}$	8. Lines that intersect to form right angles are perpendicular.

● **PROBLEM** 233

(a) In parallelogram ABCD, AE = 7x - 1 and EC = 5x + 5. Find AC. (b) If DB = 10x + 10, find DB. (c) What kind of parallelogram is ABCD? Why?

Solution: (a) The length of $\overline{AC}$ is equal to the sum of the lengths of the segments making it up, i.e. AC = AE + EC Therefore, AC = (7x - 1) + (5x + 5).

We know that the diagonals of a parallelogram bisect each other. As such, AE = EC. This provides us with a mechanism to solve for x and determine the total length AC.

AE = EC.

By substitution,   7x - 1 = 5x + 5

2x = 6

x = 3.

Returning to the equation for AC, and substituting, we obtain

AC = (7(3) - 1) + (5(3) + 5) = 20 + 20 = 40.

(b) We know x = 3. Therefore, if DB = 10x + 10, by substitution, DB = 10(3) + 10 = 40.

(c) Parallelogram ABCD is a rectangle since both diagonals are of the same length and, as such, congruent. A theorem tells that when a parallelogram has congruent diagonals, the parallelogram is a rectangle.

● **PROBLEM** 234

In an isosceles triangle, the sum of the lengths of the perpendiculars drawn to the legs from any point on the base is equal to the length of an altitude drawn to one of the legs. (Hint: Draw $\overline{PF} \perp \overline{AG}$, as shown in the figure.)

Solution: Let P represent an arbitrary point on the base of △ABC. Draw perpendiculars $\overline{PE}$ and $\overline{PD}$, as in the figure, and altitude $\overline{AG}$ to side $\overline{BC}$. We are asked to prove PE + PD = AG. We make use of the hint and draw $\overline{PF} \perp \overline{AG}$ to complete PEGF.

Since AG = AF + FG, we can arrive at the desired conclusion if we prove PE = FG and PD = AF. These results can be derived if we prove PEGF is a rectangle and △APF $\cong$ △PAD.

STATEMENT	REASON
1. $\overline{PE} \perp \overline{BC}$ and $\overline{AG} \perp \overline{BC}$	1. Given.

198

(or $\overline{PE} \perp \overline{GE}$ and $\overline{FG} \perp \overline{GE}$)

2. Draw $\overline{PF} \perp \overline{AG}$ or $\overline{PF} \perp \overline{FG}$	2. A perpendicular can be drawn to a line from a given point not on the line.
3. $\overline{PE} \parallel \overline{FG}$	3. Lines perpendicular to the same line are parallel.
4. $\overline{PF} \parallel \overline{GE}$	4. Same as reason 4.
5. ⦬FGE is a right angle	5. Perpendicular lines intersect to form a right angle.
6. PEGF is a rectangle	6. A quadrilateral in which all pairs of opposite sides are parallel, which also contains a right angle, is a rectangle.
7. $\overline{PE} \cong \overline{FG}$ or PE = FG	7. Opposite sides of a rectangle are congruent.
8. △ABC is isosceles	8. Given.
9. ⦬ECA $\cong$ ⦬DAC	9. Base angles of an isosceles triangle are congruent.
10. ⦬ECA $\cong$ ⦬FPA	10. If parallel lines ($\overline{CG}$ and $\overline{PF}$) are cut by a transversal ($\overline{CA}$), then corresponding angles are congruent.
11. In △APF and △PAD ⦬FPA $\cong$ ⦬DAP	11. Transitive Property of Congruence and steps (9) and (10).
12. $\overline{AP} \cong \overline{PA}$	12. Reflexive Property of Congruence.
13. ⦬PFA and ⦬ADP are right angles	13. Perpendicular lines intersect in right angles.
14. ⦬PFA $\cong$ ⦬ADP	14. All right angles are congruent.
15. △APF $\cong$ △PAD	15. AAS $\cong$ AAS.
16. $\overline{PD} \cong \overline{AF}$ or PD = AF	16. Corresponding sides of congruent triangles are congruent.
17. PE + PD = AF + FG or PE + PD = AG	17. Substitution Postulate and steps (16) and (7)

What should be the minimum length L of a wall mirror so that a person of height h can view herself from head to shoes?

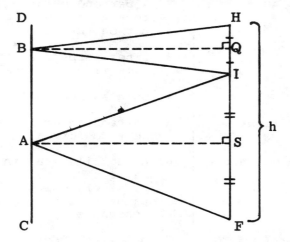

Solution: In the accompanying figure, the person $\overline{HF}$ has eyes at point I, and stands parallel to mirror AB. $\overline{BQ} \perp \overline{HF}$, $\overline{AS} \perp \overline{HF}$. To find the minimum length, the light from the head H must just touch the top of the mirror before reflecting into the eye. Similarly, light from the feet F must just touch the bottom of the mirror before reflecting into the eyes. We will show that the minimum mirror length AB is half the height h by

(1) showing QI = ½ HI and SI = ½ IF (2) showing AB = QS = QI + SI = ½HI + ½IF = ½HF = ½ h.

(1) To show QI = ½HI, we show that $\triangle BHQ \stackrel{\sim}{=} \triangle BIQ$. First, note that point Q is defined as the point such that $\overline{BQ}$ is perpendicular to HF. We are given $\overline{BA} \parallel \overline{QS}$. If a line is perpendicular to one of two parallel lines, it is perpendicular to the other. Thus, $\overline{BQ} \perp \overline{AB}$, and m⟩DBQ = m⟩ABQ = 90°.

From physics, we know that when light hits a reflecting surface the angle of incidence equals the angle of reflection. Thus, for the light travelling from point H to point B to point I, ⟩DBH = ⟩ABI. By the Angle Subtraction Postulate, m⟩DBQ - m⟩DBH = m⟩ABQ - m⟩ABI or ⟩HBQ $\stackrel{\sim}{=}$ ⟩IBQ.

By the reflexive property, $\overline{BQ} = \overline{BQ}$. Also, since $\overline{BQ} \perp \overline{HI}$, ⟩BQH and ⟩BQI are right angles. Since all right angles are congruent, ⟩BQH $\stackrel{\sim}{=}$ ⟩BQI. By the ASA Postulate, $\triangle BQH \stackrel{\sim}{=} BQI$. By corresponding parts, $\overline{HQ} \stackrel{\sim}{=} \overline{QI}$, and QI = HQ = ½ HI.

By similar argument, we show IS = SF = ½IF.

(2) To show the second part - that AB = ½ h, note
that since $\overline{BQ}$ and $\overline{AS}$ are both perpendicular to the same
line, $\overline{BQ} \parallel \overline{AS}$. We are given that $\overline{AB} \parallel \overline{QS}$. Since both
pairs of opposite sides of quadrilateral ABQS are paral-
lel, ABQS is a parallelogram, and opposite sides $\overline{AB}$ and
$\overline{QS}$ are congruent.

QS = QI + IS. From part (1), we know that QI = ½HI
and IS = ½IF. Thus, QS = ½HI + ½IF = ½(HI + IF). Substituting QS = AB and HI + IF = HF = h, we have the
final result: AB = ½h.

TRAPEZOIDS

● **PROBLEM** 236

We are given parallelogram ABCD and straight line segment
$\overline{AE}$. (See the figure.) ≮CBE $\overset{\sim}{=}$ ≮CEB. Present a formal proof
to show AECD is an isosceles trapezoid.

<u>Solution</u>: A trapezoid is a quadrilateral that has two and
only two sides parallel. It is isosceles if the two nonparallel sides are congruent.

In this problem, we will first prove AECD is a
trapezoid. Then, $\overline{DA}$ will be shown to be congruent to $\overline{CE}$
by first proving that both of the former are congruent to
the same segment and then applying the transitive property
of congruence.

STATEMENT	REASON
1. ABCD is a parallelogram	1. Given.
2. $\overline{DC} \parallel \overline{AB}$ (or $\overline{DC} \parallel \overline{AE}$)	2. Opposite sides of a paral-lelogram are parallel.
3. $\overline{DA} \parallel \overline{CB}$	3. Same as reason 2.
4. $\overline{DA} \not\parallel \overline{CE}$	4. Through a point not on a given line only one line can be drawn parallel to that given line.
5. AECD is a trapezpoid	5. A trapezoid is a quadri-lateral that has two and only two sides parallel.
6. $\overline{DA} \overset{\sim}{=} \overline{CB}$	6. A pair of opposite sides of a parallelogram are

	congruent, and statement 1.
7. ∢CBE ≅ ∢CEB	7. Given.
8. $\overline{CB}$ ≅ $\overline{CE}$	8. If two angles of a triangle are congruent, then the sides opposite those angles are congruent.
9. $\overline{DA}$ ≅ $\overline{CE}$	9. Transitive property of congruence.
10. AECD is an isosceles trapezoid	10. An isosceles trapezoid is a trapezoid whose non-parallel sides are congruent.

● **PROBLEM** 237

The lengths of the bases of an isosceles trapezoid are 8 and 14, and each of the base angles measures 45°. Find the length of the altitude of the trapezoid.

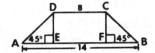

Solution: As can be seen in the figure, it is helpful to draw both altitudes $\overline{DE}$ and $\overline{CF}$. DCFE is a rectangle because $\overline{DC}$ || $\overline{AB}$ (given by the definition of a trapezoid), $\overline{DE}$ || $\overline{CF}$ (they are both ⊥ to the same line) and ∢E is a right angle ($\overline{DE}$ ⊥ $\overline{AB}$). Opposite sides of a rectangle are congruent. Therefore, DC = EF = 8.

ΔAED ≅ ΔBFC because ∢DEA ≅ ∢CFB (both are right angles).

∢DAE ≅ ∢CBF, and $\overline{DA}$ ≅ $\overline{BC}$.(The last two facts come from the definition of an isosceles trapezoid).Therefore, by corresponding parts, $\overline{AE}$ ≅ $\overline{BF}$. Then, AE = ½(AB - EF) and, substituting AE = ½(14 - 8) = ½(6) = 3.

In ΔAED, m∢E = 90 and m∢A = 45. By the angle sum postulate for **triangles,** m∢D = 45. Since ΔAED has two angles of equal measure, it must be an isosceles triangle. Therefore, AE = DE = 3.

Therefore, the length of the altitude of trapezoid ABCD is 3.

● **PROBLEM** 238

Prove that the line containing the median of a trapezoid bisects any altitude of the trapezoid.

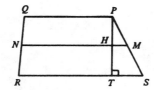

Solution: The median of a trapezoid is the line segment
joining the midpoints of the nonparallel sides. Since the
median is everywhere half way between the bases, it is
everywhere equidistant from the two bases and, as such,
is parallel to them. The nonparallel sides and any
altitude of a trapezoid are all transversals of the
parallel bases and median.

There is a theorem that states if three or more
parallel lines intercept congruent segments on one transversal, then they intercept congruent segments on any

other transversal. Since parallel lines $\overleftrightarrow{QP}$, $\overleftrightarrow{NM}$, and $\overleftrightarrow{RS}$ cut
equal segments on $\overline{QR}$ and $\overline{PS}$, they cut $\overline{PT}$ equally.

Given: Trapezoid PQRS with altitude $\overline{PT}$ intersecting
median $\overline{MN}$ at H.

Prove: $\overline{MN}$ bisects $\overline{PT}$.

STATEMENTS	REASONS
1. Trapezoid PQRS with altitude $\overline{PT}$ meeting median $\overline{MN}$ at H.	1. Given.
2. $\overline{MN} \parallel \overline{SR}$	2. The median of a trapezoid is parallel to the bases.
3. M is the midpoint of $\overline{PS}$	3. Definition of the median of a triangle.
4. $\overline{PM} \cong \overline{MS}$	4. Definition of the midpoint of a line segment.
5. $\overline{PH} \cong \overline{HT}$	5. If three or more parallel lines intercept congruent segments on a transversal, then they intercept congruent segments on any other transversal.
6. $\overline{MN}$ bisects $\overline{PT}$	6. Definition of bisection.

CHAPTER 13

GEOMETRIC PROPORTIONS AND SIMILARITY

Basic Attacks and Strategies for Solving Problems in this Chapter. See pages 204 to 223 for step-by-step solutions to problems.

After congruence, similarity is the next most important geometric relation. Similarity of geometric figures involves proportions. Therefore, proportions will be introduced first. A **proportion** is a statement which indicates the equality of two ratios. For example, $\frac{1}{2} = \frac{x}{6}$ is a proportion. Often it is necessary to solve for an unknown in a proportion. In the example above it is easy to see that $x = 3$. However, it will not usually be possible to solve for the unknown by inspection. When that is the case, the following generalization is often useful:

For any real numbers a, b, c, and d with $b \neq 0$ and $d \neq 0$,

$$\frac{a}{b} = \frac{c}{d} \text{ if and only if } ad = bc.$$

In the proportion $\frac{a}{b} = \frac{c}{d}$, a and d are called the **extremes** and b and c are called the **means**. The generalization above can be restated as "In a proportion, the product of the means equals the product of the extremes."

Below is the definition of similarity applied to triangles.

$$\triangle ABC \sim \triangle DEF \text{ if and only if}$$

(1) $\angle A \cong \angle D$

(2) $\angle B \cong \angle E$

(3) $\angle C \cong \angle F$

(4) $\dfrac{AB}{DE} = \dfrac{BC}{EF} = \dfrac{AC}{DF}$

This means that there is a correspondence between vertices of the triangles

$$A \leftrightarrow D$$
$$B \leftrightarrow E$$
$$C \leftrightarrow F$$

such that corresponding angles are congruent and the ratios of lengths of corresponding sides are equal.

It would be difficult to prove that a pair of triangles is similar if all these conditions had to be established. Below is a theorem which makes this job easier:

$\triangle ABC \sim \triangle DEF$ if and only if $\angle A \cong \angle D$, $\angle B \cong \angle E$, and $\angle C \cong \angle F$

This is called the **Angle-Angle-Angle (AAA) similarity theorem.** In Problem 243 you will see a proof where all that is required is the congruence of two pairs of angles.

It's worthwhile to note that any two congruent triangles are necessarily similar (since congruent triangles require congruent corresponding angles and congruent corresponding sides). Thus, the postulates used in Chapters 6 through 10 to establish congruence are also sufficient to show similarity between triangles.

Step-by-Step Solutions to
Problems in this Chapter,
"Geometric Proportions
and Similarity"

RATIOS AND PROPORTIONS

● **PROBLEM** 239

Is $\frac{12}{20} = \frac{36}{60}$ a proportion?

<u>Solution</u>: A proportion is an equation which states that two ratios are equal.

Since $\frac{12}{20} = \frac{3}{5}$ and $\frac{36}{60} = \frac{3}{5}$, the ratios $\frac{12}{20}$ and $\frac{36}{60}$ are equal, and, therefore, $\frac{12}{20} = \frac{36}{60}$ is a proportion.

● **PROBLEM** 240

Solve for the unknown, c, in the proportion 18 : 6 = c : 9.

<u>Solution</u>: A theorem tells us that, in a proportion, the product of the means is equal to the product of the extremes.

In this problem, 6 and c are the means and 18 and 9 are the extremes. Therefore,

$$6c = 18 \times 9$$

$$6c = 162$$

$$c = \frac{162}{6} = 27.$$

The answer c = 27 can be checked by substituting

back into the original proportion.

$$18 : 6 = 27 : 9$$

$$3 : 1 = 3 : 1$$

Since the two ratios are equal, c = 27.

● **PROBLEM** 241

Find the mean proportional between 4 and 16.

Solution: Let x = the mean proportional between 4 and 16.

By definition of the mean proportional, the proportion 4 : x = x : 16, must be satisfied.

The product of the means equals the product of the extremes in a proportion. Hence,

$$x^2 = 4 \times 16$$

$$x^2 = 64$$

$$x = \pm 8.$$

In geometry, we restrict our discussion to positive values. Therefore, x = 8.

Check: 4 : x = x : 16 (Substitute x = 8)

$$4 : 8 = 8 : 16$$

$$1 : 2 = 1 : 2$$

Therefore, the mean proportional between 4 and 16 is 8.

● **PROBLEM** 242

Find the fourth proportional to 3, 4, and 9.

Solution: Let x = the fourth proportional to 3, 4 and 9.

If x is the fourth proportional, then the equation 3 : 4 = 9 : x is a proportion. Since the product of the means equals the product of the extremes, we have

$$3x = 9 \times 4$$

$$3x = 36$$

Thus,   x = 12.

To check this, we substitute x = 12 into
3 : 4 = 9 : x and get 3 : 4 = 9 : 12, which reduces to
3 : 4 = 3 : 4, a proportion.

Therefore, the fourth proportional to 3, 4, and 9,
is 12.

PROVING TRIANGLES SIMILAR

● **PROBLEM** 243

Given the A.A.A. (Angle, Angle, Angle) Similarity Theorem,
prove the A.A. (Angle, Angle) Similarity Theorem.

Solution: The A.A.A. Similarity Theorem states: If there
exists a correspondence between $\triangle ABC$ and $\triangle DEF$ such that
corresponding angles are congruent, then $\triangle ABC \sim \triangle DEF$.

Suppose we are given 2 triangles and that the corresponding angles of 2 pairs of angles are congruent. Noting
that the sum of the angles of any triangle is 180°, it
follows that the corresponding angles of the third pair of
angles are also congruent and that the 2 triangles are
similar by the A.A.A. theorem. Hence, we obtain a generalization of the A.A.A. Similarity Theorem. The A.A.
Similarity Theorem: If there exists a correspondence be-
tween $\triangle ABC$ and $\triangle DEF$ such that 2 angles of $\triangle ABC$ are congruent to the corresponding angles of $\triangle DEF$, then $\triangle ABC \sim$
$\triangle DEF$.

● **PROBLEM** 244

(a) If 2 triangles are congruent, does it follow that they
are similar? Why? (b) If 2 triangles are similar, does it
follow that they are congruent? Why?

Solution: (a) Let us call the 2 triangles ABC and DEF,
and assume, according to the statement of the problem, that
$\triangle ABC \cong \triangle DEF$.

By definition, a one-to-one correspondence of the
vertices of triangles ABC and DEF is a congruence if and
only if the corresponding parts are congruent (i.e.,
$\angle A \cong \angle D$, $\angle B \cong \angle E$, $\angle C \cong \angle F$, $\overline{AB} \cong \overline{DE}$, $\overline{AC} \cong \overline{DF}$, $\overline{BC} \cong \overline{EF}$.)

$$\overline{AB} \cong \overline{DE} \quad \rightarrow \quad AB = DE$$

$$\overline{AC} \cong \overline{DF} \quad \rightarrow \quad AC = DF$$

$$\overline{BC} \cong \overline{EF} \quad \rightarrow \quad BC = EF$$

Hence,

$$\frac{AB}{DE} = \frac{AC}{DF} = \frac{BC}{EF} = 1$$

This means that corresponding sides of the 2 triangles are proportional. Furthermore, as was shown above,

$$\angle A \cong \angle D$$

$$\angle B \cong \angle E$$

$$\angle C \cong \angle F.$$

Hence, a correspondence between the vertices of triangles ABC and DEF exists in which corresponding angles are congruent, and corresponding sides have proportional measures. By definition, this means that $\triangle ABC \sim \triangle DEF$ ($\sim$ is read "is similar to"). Therefore, congruent triangles are similar.

(b) To show that similar triangles are not congruent, we provide a counter-example. Let the 2 triangles be ABC and DEF, and suppose a correspondence between the vertices of $\triangle ABC$ and $\triangle DEF$ exists such that

$$\frac{AB}{DE} = \frac{BC}{EF} = \frac{AC}{DF} = K$$

$$\angle A \cong \angle D, \quad \angle B \cong \angle E, \quad \angle C \cong \angle F.$$

If $K \neq 1$, then

$$AB \neq DE \quad \rightarrow \quad \overline{AB} \not\cong \overline{DE}$$

$$AC \neq DF \quad \rightarrow \quad \overline{AC} \not\cong \overline{DF}$$

$$BC \neq EF \quad \rightarrow \quad \overline{BC} \not\cong \overline{EF}.$$

Hence, $\triangle ABC \cong \triangle DEF$ unless $K = 1$.

● **PROBLEM** 245

In the accompanying figure, triangle ABC is similar to triangle A'B'C', and AC corresponds to A'C'. (a) Find the ratio of similitude. (b) Find A'B' and B'C'.

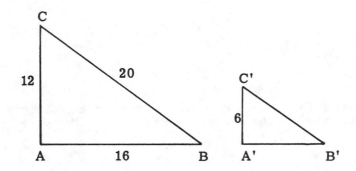

Solution: (a) The ratio of similitude of two similar polygons is defined as the ratio of the measures of any two corresponding sides. We are told AC corresponds to A'C'. Therefore,

Ratio of Similitude $= \frac{AC}{A'C'} = \frac{12}{6} = \frac{2}{1} = 2 : 1$.

(b) The ratio of similitude of $\triangle ABC$ to $\triangle A'B'C'$ being 2 : 1 tells us that each side of $\triangle ABC$ is twice as long as its corresponding part in $\triangle A'B'C'$. Conversely, each side of $\triangle A'B'C'$ is ½ as long as its corresponding side in $\triangle ABC$.

(i) $A'B' = ½ (AB) = ½(16) = 8$.

(ii) $B'C' = ½ (BC) = ½(20) = 10$.

● **PROBLEM** 246

The lengths of the sides of a triangle are 6, 8, and 12. The lengths of the sides of a second triangle are 1½, 2, and 3. Are the two triangles similar?

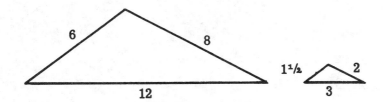

Solution: Since we are given three sides, it is best to use the SSS Similarity Theorem: two triangles are similar, if the corresponding sides are proportional.

The sides are proportional if,

(i) $\frac{6}{1\,½} = \frac{8}{2} = \frac{12}{3}$.

Dividing through, we have

(ii)   4 = 4 = 4.

    The proportion holds, therefore, the triangles are
similar.

Triangles ABC and A'B'C' have dimensions as indicated in the
figure. Angle A = angle A', angle B = angle B.'   Find the
measures of sides A'B' and C'B'.

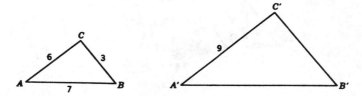

Solution: Two triangles are similar if, and only if, two
angles of one triangle are equal in measure to two angles
of the other triangle. Since m ∢A = m ∢A', and m ∢B =
m ∢B', the two triangles ΔABC and ΔA'B'C', are similar.
In similar triangles, corresponding sides are proportional.
Hence,

$$\frac{AC}{A'C'} = \frac{CB}{C'B'} = \frac{AB}{A'B'} \ .$$

Substituting the given,

$$\frac{6}{9} = \frac{3}{C'B'} = \frac{7}{A'B'}$$

    We can first solve for C'B', using the proportion:

$$\frac{6}{9} = \frac{3}{C'B'}$$

$$6 \cdot C'B' = 27$$

$$C'B' = \frac{9}{2} \ .$$

    We then solve for A'B', using the proportion:

$$\frac{6}{9} = \frac{7}{A'B'}$$

$$6 \cdot A'B' = 63 \qquad A'B' = \frac{21}{2}$$

Therefore, side A'B' = $\frac{21}{2}$, and side C'B' = $\frac{9}{2}$.

A boy knows that his height is 6 ft. and his shadow is 4 ft. long. At the same time of day, a tree's shadow is 24 ft. long. How high is the tree?

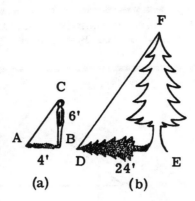

(a)   (b)

Solution: Show that △ABC ∿ △DEF, and then set up a proportion between the known sides AB and DE, and the sides BC and EF.

First, assume that both the boy and the tree are ⊥ to the earth. Then, $\overline{BC}$ ⊥ $\overline{BA}$ and $\overline{EF}$ ⊥ $\overline{ED}$. Hence,

$$∢ ABC \stackrel{\sim}{=} ∢ DEF.$$

Since it is the same time of day, the rays of light from the sun are incident on both the tree and the boy at the same angle, relative to the earth's surface. Therefore,

$$∢ BAC \stackrel{\sim}{=} ∢ EDF.$$

We have shown, so far, that 2 pairs of corresponding angles are congruent. Since the sum of the angles of any triangle is 180°, the third pair of corresponding angles is congruent (i.e. ∢ ACB $\stackrel{\sim}{=}$ ∢ DFE). By the Angle Angle Angle (A.A.A.) Theorem,

$$△ABC \sim △DEF.$$

By definition of similarity,

$$\frac{FE}{CB} = \frac{ED}{BA} .$$

CB = 6', ED = 24', and BA = 4'. Therefore,

$$FE = (6')(24'/4') = 36'.$$

Let ABC be a triangle where D is a point on $\overline{AB}$, and E is a point on $\overline{AC}$. Prove that if $\overline{DE} \parallel \overline{BC}$, then AB/AD = BC/DE. (See figure.)

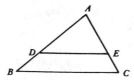

Solution: We will use the fact that $\overline{DE} \parallel \overline{BC}$ to prove that $\triangle ADE \sim \triangle ABC$. We can then set up a proportion and show that AB/AD = BC/DE.

Since the corresponding angles of 2 parallel lines cut by a transversal are congruent, $\angle ADE \cong \angle ABC$ and $\angle AED \cong \angle ACB$. Furthermore, $\angle DAE \cong \angle BAC$. Therefore, by the A.A.A. (angle angle angle) Similarity Theorem, $\triangle ADE \sim \triangle ABC$. Hence, corresponding sides of the 2 triangles must be proportional, and we may write

$$\frac{AD}{AB} = \frac{AE}{AC} = \frac{DE}{BC} \qquad \text{or} \qquad \frac{AD}{AB} = \frac{DE}{BC} \; .$$

Taking reciprocals of both sides:

$$\frac{AB}{AD} = \frac{BC}{DE} \; .$$

Given: $\triangle ABC$ with $\overline{ADB}$ and $\overline{AEC}$ such that $\angle EDB$ is supplementary to $\angle C$. Prove:

$$\frac{AE}{AB} = \frac{AD}{AC} \; .$$

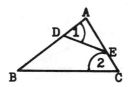

Solution: Whenever proportions are mentioned, it is wise to look for similar triangles. For AE to correspond to AB, and AD to correspond to AC, the similar triangles must be $\triangle ADE$ and $\triangle ACB$. Thus, it is sufficient to show $\triangle ADE \sim \triangle ACB$

to prove the proportion.

To prove two triangles similar, we can show that two pairs of corresponding angles of the two triangles are congruent (the A.A. Similarity Theorem).

STATEMENTS	REASONS
1. $\triangle ABC$; $\overline{ADB}$ and $\overline{AEC}$; $\sphericalangle EDB$ is supplementary to $\sphericalangle C$.	1. Given.
	(Now, we show $\sphericalangle 1 \overset{\sim}{=} \sphericalangle 2$.)
2. $\sphericalangle EDB$ and $\sphericalangle 1$ form a linear pair	2. Definition of a linear pair.
3. $\sphericalangle EDB$ is supplementary to $\sphericalangle 1$	3. Two angles that form a linear pair are supplementary.
4. $\sphericalangle 1 \overset{\sim}{=} \sphericalangle 2$	4. Supplements of the same angle are congruent.
5. $\sphericalangle A \overset{\sim}{=} \sphericalangle A$	5. Every angle is congruent to itself.
6. $\triangle ADE \sim \triangle ACB$	6. The A.A. Similarity Theorem.
7. $\dfrac{AE}{AB} = \dfrac{AD}{AC}$	7. The sides of similar triangles are proportional (from the definition of similar triangles).

● **PROBLEM** 251

(a) In $\triangle ABC$, if D is the midpoint of $\overline{AB}$, E is the midpoint of $\overline{AC}$, and F is the midpoint of $\overline{BC}$, then prove that $\overline{DE} \parallel \overline{BC}$, $\overline{EF} \parallel \overline{AB}$, and $\overline{DF} \parallel \overline{AC}$. (b) Prove that $\triangle DEF \sim \triangle CBA$.

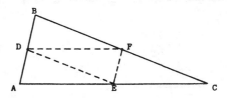

Solution: (a) The method is to show that $\triangle ADE \sim \triangle ABC$. We will then have $\sphericalangle ADE \overset{\sim}{=} \sphericalangle ABC$ and be able to conclude that $\overline{DE} \parallel \overline{BC}$. The same procedure is used for the pairs of triangles CFE and CBA, and BDF and BAC.

Focus attention on triangles ADE and ABC.

First, D and E are the midpoints of $\overline{BA}$ and $\overline{CA}$, respectively. Hence,

$$\frac{AD}{AB} = \frac{AE}{AC} = \frac{1}{2}$$

Also, $\angle DAE \stackrel{\sim}{=} \angle BAC$. Therefore, by the Side Angle Side (S.A.S.) Similarity Theorem, $\triangle ADE \sim \triangle ABC$. By definition then, $\angle ADE \stackrel{\sim}{=} \angle ABC$. Because these 2 angles are congruent corresponding angles of the segments $\overline{DE}$ and $\overline{BC}$, $\overline{DE} \parallel \overline{BC}$.

In an exactly analogous manner, we may prove that $\triangle CFE \sim \triangle CBA$, and $\triangle BDF \sim \triangle BAC$. Hence, $\angle CFE \stackrel{\sim}{=} \angle CBA$ and $\angle BDF \stackrel{\sim}{=} \angle BAC$. Therefore, $\overline{FE} \parallel \overline{BA}$ and $\overline{DF} \parallel \overline{AC}$.

(b) From part (a), we know that

$$\triangle ADE \sim \triangle ABC$$

$$\triangle CFE \sim \triangle CBA \qquad\qquad (1)$$

$$\triangle BDF \sim \triangle BAC.$$

Hence, from equations (1),

$$\frac{DE}{BC} = \frac{AD}{AB} = \frac{1}{2}$$

$$\frac{FE}{BA} = \frac{CF}{CB} = \frac{1}{2} \qquad\qquad (2)$$

$$\frac{DF}{AC} = \frac{BD}{BA} = \frac{1}{2}$$

Here, we have used the fact that D, E, and F are midpoints of $\overline{BA}$, $\overline{AC}$, and $\overline{CB}$, respectively. From (2), then,

$$\frac{DE}{BC} = \frac{FE}{BA} = \frac{DF}{AC} = \frac{1}{2}$$

By the S.S.S. (Side Side Side) Similarity Theorem,

$$\triangle DEF \sim \triangle CBA.$$

● **PROBLEM** 252

Given: $\overline{AD}$ is an angle bisector of $\triangle ABC$; point E is on $\overleftrightarrow{AD}$ such that AB · AC = AD · AE. Prove: $\angle B \stackrel{\sim}{=} \angle AEC$.

213

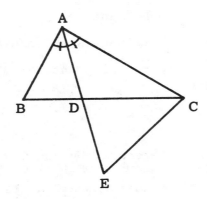

Figure 2

<u>Solution</u>: Whenever there are proportions, the best procedure is to find similar triangles. There are five possible triangles in the figure but only two, ΔABD and ΔAEC, allow us to use the given fact that ∢ BAD $\stackrel{\sim}{=}$ ∢ EAC. By using this and the information that AB · AC = AD · AE (which show that two pairs of sides are proportional) we can prove ΔABD ∿ ΔAEC by the A.S.A. Similarity Theorem. The congruence of ∢ B and ∢ AEC follows immediately.

STATEMENT	REASONS
1. $\overline{AD}$ bisects ∢ BAC	1. Given.
2. ∢ BAD $\stackrel{\sim}{=}$ ∢ DAC	2. Definition of an angle bisector.
3. AB · AC = AD · AE	3. Given.
4. $\frac{AB}{AE} = \frac{AD}{AC}$	4. $\frac{a}{b} = \frac{c}{d}$ if and only if ad = bc.
5. ΔABD ∿ ΔAEC	5. S.A.S. Similarity Theorem.
6. ∢ B $\stackrel{\sim}{=}$ ∢ AEC	6. Corresponding angles of similar triangles are congruent.

● **PROBLEM** 253

In the accompanying figure, ΔABC is isosceles with $\overline{AB} \stackrel{\sim}{=}$ $\overline{AC}$. Segment $\overline{AF}$ is the altitude on $\overline{BC}$. From a point on $\overline{AB}$, call it D, a perpendicular is drawn which is extended to meet $\overleftrightarrow{BC}$. It meets $\overline{BC}$ at point P. Prove that FC : DB = AC : PB.

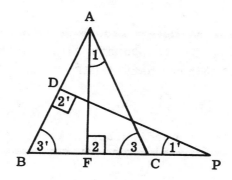

Solution: $\overline{FC}$ and $\overline{AC}$ are sides of $\triangle FCA$ and correspond to sides $\overline{DB}$ and $\overline{PB}$ of $\triangle DBP$. By proving $\triangle FCA \sim \triangle DBP$, we can conclude that FC : DB = AC : PB is a proportion, because there is a theorem which states that corresponding sides of similar triangles are proportional.

The corresponding angles of $\triangle FCA$ and $\triangle DBP$ have been numbered, in the diagram, by a "prime and no prime" system.

Similarity will be proved by showing two pairs of corresponding angles congruent.

STATEMENT	REASON
1. $\overline{AB} = \overline{AC}$	1. Given.
2. $\angle 3 \cong \angle 3'$	2. If two sides of a trignale are congruent, then the angles opposite these sides are congruent.
3. $\overline{AF} \perp \overline{BC}$	3. An altitude drawn to a side is perpendicular to that side.
4. $\overline{PD} \perp \overline{AB}$	4. Given.
5. $\angle 2$ and $\angle 2'$ are right angles	5. Perpendicular lines intersect forming right angles.
6. $\angle 2 \cong \angle 2'$	6. All right angles are congruent.
7. $\triangle FCA \sim \triangle DBP$	7. A.A. $\cong$ A.A.
8. $\dfrac{FC \ (opp. \angle 1)}{DB \ (opp. \angle 1')} = \dfrac{AC \ (opp. \ \angle 2)}{PB \ (opp. \ \angle 2')}$	8. Corresponding sides of similar triangles are proportional.

215

In the accompanying figure, points D, E, and F are such that $\overleftrightarrow{ABD}$, $\overleftrightarrow{BEC}$, $\overleftrightarrow{CFA}$. Show that if $\overleftrightarrow{DEF}$, then

$$\frac{AD}{BD} \cdot \frac{BE}{EC} \cdot \frac{CF}{FA} = 1.$$

(Hint: Consider the perpendiculars drawn to $\overline{FD}$ from A, B, C.)

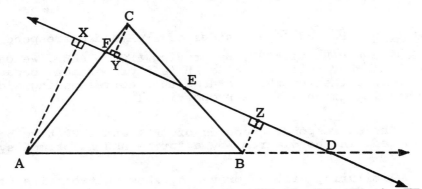

Solution: In the accompanying figure $\overline{AX} \perp \overline{FD}$, $\overline{CY} \perp \overline{FD}$, and $\overline{BZ} \perp \overline{FD}$. We wish to show

$$\frac{AD}{BD} \cdot \frac{BE}{EC} \cdot \frac{CF}{FA} = 1,$$

an equation involving three ratios. Whenever dealing with ratios, it is wise to look for similar triangles. We may then be able to substitute equivalent ratios for the given ones which may make the equality more apparent.

Consider the first ratio $\frac{AB}{BD}$. These are corresponding parts of $\triangle AXD$ and $\triangle BZD$. Note that, by reflexivity, $\angle BDZ \cong$

$\angle ADX$. Furthermore, right angles $\angle AXD$ and $\angle BZD$ are congruent. By the A-A Similarity Theorem, $\triangle AXD \sim \triangle BZD$. Therefore,

(i) $$\frac{AD}{BD} = \frac{AX}{BZ} = \frac{XD}{ZD}$$

Consider the ratio $\frac{BE}{EC}$. Because $\angle BZE \cong \angle CYE$ and $\angle CEY \cong \angle BEZ$, $\triangle CEY \sim \triangle BEZ$. Thus,

(ii) $$\frac{BE}{EC} = \frac{EZ}{EY} = \frac{BZ}{CY}$$

Finally, CF and FA are corresponding parts of △CFY and △AFX. Since ∡AFX ≅ ∡CFY, ∡AXF ≅ ∡CYF, then △CFY ∼ △AFX.

(iii) $$\frac{CF}{FA} = \frac{CY}{AX} = \frac{FY}{XF}$$

From these three sets of equations, we can find substitute ratios for each ratio in the to-be-proved equation that will make the result discernible. We find that by substituting $\frac{AD}{BD} = \frac{AX}{BZ}$, $\frac{BE}{EC} = \frac{BZ}{CY}$, and $\frac{CF}{FA} = \frac{CY}{AX}$, we obtain

(iv) $$\frac{AX}{BZ} \cdot \frac{BZ}{CY} \cdot \frac{CY}{AX} = 1$$

By cancelling out, we see that the desired result is true.

● **PROBLEM** 255

In the accompanying figure, points D, E, and F are points on the triangle △ABC such that $\overline{AD}$, $\overline{BE}$, and $\overline{CF}$ are concurrent at point P. Show that

$$\left(\frac{BD}{DC}\right)\left(\frac{CE}{EA}\right)\left(\frac{AF}{FB}\right) = 1.$$

(Hint: Consider $\overline{MAN} \parallel \overline{BDC}$ such that $\overline{MFPC}$ and $\overline{BPEN}$.)

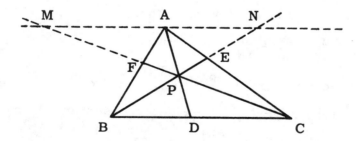

Solution: In the accompanying figure, M and N lie on the line passing through A parallel to the side $\overline{BDC}$. We wish to show

$$\left(\frac{BD}{DC}\right)\left(\frac{CE}{EA}\right)\left(\frac{AF}{FB}\right) = 1.$$

Whenever ratios are concerned, it is wise to look for similar triangles. From these triangles, equivalent ratios can be found which, when substituted in the equation to be proved, make the equality more readily apparent.

Note that the lengths of the third term, $\frac{AF}{FB}$, are corresponding sides of $\triangle FAM$ and $\triangle FBC$. Opposite angles $\sphericalangle\, MFA$ and $\sphericalangle\, CFB$ are congruent. Furthermore, because $\overline{MAN}\parallel\overline{BDC}$, alternate interior angles $\sphericalangle\, FMA$ and $\sphericalangle\, FCB$ are congruent. Thus, $\triangle FAM \sim \triangle FBC$ by the A.A. Similarity Theorem, This leads to the proportions,

(i) $\qquad\qquad \dfrac{AF}{FB} = \dfrac{MF}{FC} = \dfrac{MA}{BC}$.

For the second term, $\frac{CE}{EA}$, note that $\overline{CE}$ and $\overline{EA}$ are corresponding sides of $\triangle EAN$ and $\triangle ECB$. Note $\sphericalangle\, AEN \stackrel{\sim}{=} \sphericalangle\, CEB$ (opposite angles) and $\sphericalangle\, ANE \stackrel{\sim}{=} \sphericalangle\, CBP$ (alternate interior angles). By the A.A. Similarity Theorem, $\triangle EAN \sim \triangle ECB$. Thus,

(ii) $\qquad\qquad \dfrac{CE}{EA} = \dfrac{CB}{AN} = \dfrac{EB}{EN}$.

To find proportions related to the first term, $\frac{BD}{DC}$, we must use a more intricate method. Note that there are no similar triangles in which BD and DC are corresponding sides. We must derive a relation involving $\frac{BD}{DC}$ indirectly.

First we show that $\triangle PAM \sim \triangle PDC$ by a method similar to the above. Thus,

(iii) $\qquad\qquad \dfrac{PD}{PA} = \dfrac{PC}{PM} = \dfrac{DC}{AM}$.

Also we can show $\triangle PAN \sim \triangle PDB$. Thus,

(iv) $\qquad\qquad \dfrac{PD}{PA} = \dfrac{PB}{PN} = \dfrac{BD}{AN}$.

Combining from (iii) $\frac{PD}{PA} = \frac{DC}{AM}$ and from (iv) $\frac{PD}{PA} = \frac{BD}{AN}$, we have $\frac{PD}{PA} = \frac{DC}{AM} = \frac{BD}{AN}$; $\frac{DC}{AM} = \frac{BD}{AN}$ can be rearranged $\left(\text{by}\right.$ multiplying both sides by $\left.\frac{AN}{DC}\right)$ as $\frac{AN}{AM} = \frac{BD}{DC}$.

We gather our ratios and inspect them.

(i) $\qquad\qquad \dfrac{AF}{FB} = \dfrac{MF}{FC} = \dfrac{MA}{BC}$

(ii) $\qquad\qquad \dfrac{CE}{EA} = \dfrac{CB}{AN} = \dfrac{EB}{EN}$

(iii) $\qquad\qquad \dfrac{BD}{DC} = \dfrac{AN}{AM}$

Suppose we substitute $\frac{AF}{FB} = \frac{MA}{BC}$, $\frac{CE}{EA} = \frac{BC}{AN}$, and $\frac{BD}{DC} =$

$\frac{AN}{AM}$. Then, $\left(\frac{BD}{DC}\right)\left(\frac{CE}{EA}\right)\left(\frac{AF}{FB}\right) = 1$, becomes

(v) $\qquad \frac{AN}{AM} \cdot \frac{BC}{AN} \cdot \frac{MA}{BC} = 1$.

This equation is true. The similarities between the triangles are true; therefore, the substitutions are valid and the equation $\left(\frac{BD}{DC}\right)\left(\frac{CE}{EA}\right)\left(\frac{AF}{FB}\right) = 1$ is true.

● **PROBLEM** 256

The lengths of the radii of two circles are 15 in. and 5 in. Find the ratio of the circumferences of the two circles.

Solution: Let ⊙P be the circle of radius r = 15 in. and ⊙P' be the circle of r' = 5 in. Then the ratio of the circumferences is $\frac{C'}{C}$. Since the circumference, C, of a circle of radius r is $2\pi r$, for the circumference of ⊙P, we have C = $2\pi(15$ in.$) = 30\pi$ in. For the circumference of ⊙P' we have C' = $2\pi(5$ in.$) = 10\pi$ in. Then $\frac{C'}{C} = \frac{10\pi \text{ in.}}{30\pi \text{ in.}} = \frac{1}{3}$, or a ratio of 1:3.

A second method is to note that in the circumference expression $2\pi r$, 2π is a constant factor and therefore the ratio of circumferences $\frac{C'}{C} = \frac{2\pi r'}{2\pi r}$ equals $\frac{r'}{r}$, the ratio of the radii. Therefore, the ratio of the circumferences $\frac{C'}{C}$ equals $\frac{r'}{r} = \frac{5 \text{ in.}}{15 \text{ in.}} = \frac{1}{3}$, or 1:3.

● **PROBLEM** 257

Prove that any two regular polygons with the same number of sides are similar.

Solution: For any two polygons to be similar, their corresponding angles must be congruent and their corresponding sides proportional. It is necessary to show that these conditions always exist between regular polygons with the same number of sides.

Let us examine the corresponding angles first. For a regular polygon with n sides, the measure of each central angle is $\frac{360}{n}$ and each vertex angle is $\frac{(n-2)\ 180}{n}$. Therefore, two regular polygons with the same number of sides will have corresponding central angles and vertex angles that are all of the same measure and, hence, are all congruent. This fulfills our first condition for similarity.

We must now determine whether or not the corresponding sides are proportional. It will suffice to show that the ratios of the lengths of every pair of corresponding sides are the same.

Since the polygons are regular, the sides of each one will be equal. Call the length of the sides of one polygon ℓ_1 and the length of the sides of the other polygon ℓ_2. Hence, the ratio of the lengths of corresponding sides will be ℓ_1/ℓ_2. This will be a constant for any pair of corresponding sides and, hence, the corresponding sides are proportional.

Thus, any two regular polygons with the same number of sides are similar.

● **PROBLEM** 258

Show that a regular pentagon and the pentagon determined by joining all the vertices of the regular pentagon are similar.

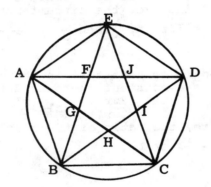

<u>Solution</u>: We must show that pentagon FGHIJ ∿ polygon ABCDE. Polygon ABCDE is a regular pentagon. Therefore, FGHIK must also be a regular pentagon. Since all regular pentagons are similar, it is sufficient to show that FGHIJ is a regular pentagon to complete the proof.

To show FGHIJ is a regular pentagon, we show that all angles are congruent and all sides are congruent.

The sides of pentagon FGHIJ are corresponding sides of ΔAGF, ΔEFJ, ΔDJI, ΔCIH, and ΔBGH. We show these triangles congruent by the SAS Postulate. The interior angles of the pentagon have opposite angles and, as such, angles congruent to them in ΔAFE, ΔEJD, ΔDIC, ΔCHB, and ΔBGA. We show these triangles congruent by ASA.

In our proof we (1) show triangles formed by three consecutive vertices are congruent - that is ΔAED, ΔEDC, ΔDCB, ΔCBA, and ΔBAE by SAS.

(2) Therefore, ∢ ABE ≅ ∢ EAD ≅ ∢ AEB ≅ ∢ DEC ≅ ...;and

(3) ΔBGA ≅ΔAFE ≅ ΔEJD ≅ΔDIC ≅ ΔCHB   by ASA.

By corresponding angles, the angles opposite the interior angles are congruent. Thus, all five interior angles are congruent.

To show the sides congruent, we (4) show $\overline{CI}$ ≅ $\overline{CH}$ ≅ $\overline{DJ}$ ≅ $\overline{DI}$ ≅ $\overline{BG}$ ≅ $\overline{BH}$ ≅ $\overline{AG}$ ≅ $\overline{AF}$ ≅ $\overline{EF}$ ≅ $\overline{EJ}$ by corresponding parts.

(5) Show ΔACD ≅ ΔEBC ≅ ΔDAB ≅ ΔCAE ≅ ΔBED by SSS;
(6) Therefore ∢ DAC ≅ ∢ BEC ≅ ∢ ADB ≅ ∢ ECA ≅ ∢ DBE; and
(7) ΔAGF ≅ ΔEFJ ≅ ΔDJI ≅ ...; and (8) by corresponding sides $\overline{GF}$ ≅ $\overline{FJ}$ ≅ $\overline{JI}$ ≅ $\overline{IH}$ ≅ $\overline{GH}$.

GIVEN: regular pentagon ABCDE.
　　　　$\overline{AD}$ ∩ $\overline{EB}$ = point F; $\overline{AD}$ ∩ $\overline{EJ}$ = point J.
　　　　$\overline{AC}$ ∩ $\overline{BE}$ = point G; $\overline{AC}$ ∩ $\overline{BD}$ = point H;
　　　　$\overline{BD}$ ∩ $\overline{CE}$ = point I.
PROVE: pentagon FGHIJ ∿ ABCDE.

STATEMENTS	REASONS
1. (see above)	1. Given.
2. $\overline{AE}$ ≅ $\overline{ED}$ ≅ $\overline{DC}$ ≅ $\overline{CB}$ ≅ $\overline{BA}$	2. The sides of a regular polygon are congruent.
3. ∢ EAB ≅ ∢ AED ≅ ∢ EDC ≅ ∢ DCB ≅ ∢ CBA	3. The angles of a regular polygon are congruent.
4. ΔBAE ≅ ΔAED ≅ ΔEDC ≅ ΔDCB ≅ ΔCBA	4. The SAS Postulate.
5. ∢ ABE ≅ ∢ EAD ≅ ∢ DEC ≅ ∢ CDB ≅ ∢ BCA Also, ∢ AEB ≅ ∢ EDA ≅ ∢ DCE ≅ ∢ CBD ≅∢ BAC	5. Corresponding parts of congruent triangles are congruent.
6. ΔBAE, ΔAED, ΔEDC, ΔDCB, ΔCBA are isosceles	6. If two sides of a triangle are congruent, then the triangle is isosceles.
7. ∢ ABE ≅ ∢ EAD ≅ ∢ DEC ≅ ∢ CDB ≅ ∢ BCA ≅ ∢ AEB ≅ ∢ EDA ≅ ∢ DCE≅∢ CBD≅∢ BAC	7. The base angles of an isosceles triangle are congruent. Also, step 5.
8. ΔAFE ≅ ΔEJD ≅ ΔDIC ≅	8. The ASA Postulate.

$\triangle CHB \cong \triangle BGA$

9. $\sphericalangle AFE \cong \sphericalangle EJD \cong \sphericalangle DIC \cong$ $\sphericalangle CHB \cong \sphericalangle BGA$

9. Corresponding angles of congruent triangles are congruent.

10. $\sphericalangle GFJ \cong \sphericalangle AFE$
$\sphericalangle FJI \cong \sphericalangle EJD$
$\sphericalangle JIH \cong \sphericalangle DIC$
$\sphericalangle GHI \cong \sphericalangle BHC$
$\sphericalangle FGH \cong \sphericalangle AGB$

10. Opposite angles are congruent.

11. $\sphericalangle GFJ \cong \sphericalangle FJI \cong \sphericalangle JIH \cong$ $\sphericalangle GHI \cong \sphericalangle FGH$

11. Transitivity Postulate
(Thus the interior angles are congruent.)

12. $\triangle AFE, \triangle EJD, \triangle DIC, \triangle CHB,$ and $\triangle BGH$ are isosceles triangles.

12. If two angles of a triangle are congruent, then the triangle is isosceles (see Step 7).

13. $\overline{AF} \cong \overline{EF} \cong \overline{EJ} \cong \overline{DJ} \cong \overline{DI} \cong$ $\overline{CI} \cong \overline{CH} \cong \overline{BH} \cong \overline{BG} \cong \overline{AG}$

13. In an isosceles triangle, the sides opposite the base angles are congruent.

14. $\overline{AD} \cong \overline{AC} \cong \overline{EB} \cong \overline{EC} \cong \overline{BD}$

14. Corresponding sides of congruent triangles are congruent (see Step 4).

15. $\triangle CAD \cong \triangle BEC \cong \triangle ADB \cong$ $\triangle ECA \cong \triangle EBD$

15. The SSS Postulate.

16. $\sphericalangle CAD \cong \sphericalangle BEC \cong \sphericalangle ADB \cong$ $\sphericalangle ECA \cong \sphericalangle EBD$

16. The corresponding angles of congruent triangles are congruent.

17. $\triangle GAF \cong \triangle FEJ \cong \triangle JDI \cong$ $\triangle ICH \cong \triangle HBG$

17. The SAS Postulate.

18. $\overline{GF} \cong \overline{FJ} \cong \overline{JI} \cong \overline{IH} \cong \overline{HG}$

18. Corresponding parts of congruent triangles are congruent.

19. Pentagon FGHIJ is a regular pentagon

19. If the sides of a polygon are all congruent and the angles are all congruent, then the polygon is regular.

20. Pentagon FGHIJ $\sim$ ABCDE

20. All regular polygons of the same number of sides are similar.

The lengths of two corresponding sides of two similar polygons are 4 and 7. If the perimeter of the smaller polygon is 20, find the perimeter of the larger polygon.

<u>Solution</u>: We know, by theorem, that the perimeter of two similar polygons have the same ratio as the measures of any pair of corresponding sides.

If we let s and p represent the side and perimeter of the smaller polygon and s' and p' the corresponding side and perimeter of the larger one, we can then write the proportion

$$p : p' = s : s'$$

By substituting the given, we can solve for p'.

$$20 : p' = 4 : 7$$

$$4p' = 140$$

$$p' = 35.$$

Therefore, the perimeter of the larger polygon is 35.

CHAPTER 14

COMPUTATIONS INVOLVING SIMILAR TRIANGLES

Basic Attacks and Strategies for Solving Problems in this Chapter. See pages 224 to 245 for step-by-step solutions to problems.

In the figure below $\triangle ABC \sim \triangle DEF$. This means that $\angle A \cong \angle D$, $\angle B \cong \angle E$, $\angle C \cong \angle F$, and $\dfrac{AB}{DE} = \dfrac{BC}{EF} = \dfrac{AC}{DF}$.

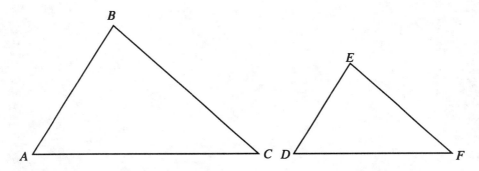

In many of the problems in this chapter, it must first be established that two given triangles are similar and then lengths of sides of triangles can be calculated using ratios of lengths of sides of the similar triangles. Problem 243 of the previous chapter establishes the fact that when two angles of one triangle are congruent to two angles of another triangle, the triangles are similar. This famous AA similarity theorem is applied to several problems in this chapter.

Here are some other important theorems which are used in various problems in this chapter.

(1) In a proportion, the product of the means equals the product of the

extremes: $\left(\dfrac{a}{b} = \dfrac{c}{d} \leftrightarrow ad = bc\right)$

(2) A line segment which joins the midpoints of two sides of a triangle is parallel to the third side of the triangle, and its length is one-half the length of the third side.

(3) If a line is parallel to one side of a triangle and intersects the other two sides of the triangle, then it divides these other two sides proportionally.

(4) If a line divides two sides of a triangle proportionally, then it is parallel to the third side.

(5) The bisector of an angle of a triangle divides the side opposite the angle into segments proportional to the sides adjacent to the angle.

(6) The altitude to a hypotenuse of a right triangle is the mean proportional between the two segments of the hypotenuse.

(7) The medians of a triangle meet in a point which is two-thirds of the way from any vertex to the midpoint of the opposite side. (This theorem is proved in Problem 277.)

LENGTHS

● **PROBLEM** 260

In an isosceles trapezoid, the length of the lower base is 15, the length of the upper base is 5, and each congruent side is of length 6 (see figure). By how many units must each nonparallel side be extended to form a triangle?

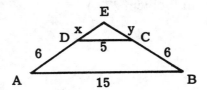

Solution: First, as is done in the accompanying figure, extend $\overline{AD}$ and $\overline{BC}$ to form a triangle. We are given that ABCD is a trapezoid. Therefore, we know that $\overline{DC} \parallel \overline{AB}$. There is a corollary which states that a line parallel to one side of a triangle, and intersecting the other two sides in different points, cuts off a triangle similar to the given triangle. We can therefore conclude that $\triangle DEC \sim \triangle AEB$ and that their corresponding sides are in proportion.

It seems likely that the large, as well as the small, triangle will be isosceles. However, we will not assume this, and later on check to see if this is true.

By similarity, the proportions $\dfrac{ED}{EA} = \dfrac{DC}{AB}$ and $\dfrac{EC}{EB} = \dfrac{DC}{AB}$ exist. In both expressions, everything is known except for ED and EC, the two extensions.

Let ED = x. Hence, from the figure, EA = x + 6. By substitution the first proportion becomes $\dfrac{x}{x+6} = \dfrac{5}{15}$. Solving, we get 15x = 5x + 30, or 10x = 30 and, finally, x = 3.

224

The second proportion can be solved in the same manner.

Let EC = y. Hence, EB = y + 6. By substitution, the second proportion becomes $\frac{y}{y + 6} = \frac{5}{15}$. This is the same calculation encountered above, and we can conclude y = 3.

Therefore, the extension of each nonparallel side will be 3 units and since ABCD is an isosceles trapezoid, the triangles will, indeed, be isosceles.

● **PROBLEM** 261

In the figure below, $\overline{AB}$ intersects $\overline{CD}$ at E. AE = 3, EB = 4 ½, CE = 4, and ED = 6. Find BD if AC = 3 ½.

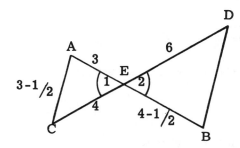

Figure 1

Solution: To find DB, we first prove that ΔAEC ∿ ΔBED. This enables us to write the proportion

$$\frac{AC}{BD} = \frac{CE}{DE} \cdot$$

Since we know AC, CE, and DE, the proportion can be solved for BD.

First, though, we must show that ΔAEC ∿ ΔBED. To show two triangles similar it is only necessary to prove that two sides of one triangle are proportional to two sides of the second triangle and that the included angle of the first triangle is congruent to the included angle of the second triangle (the S.A.S. Similarity Theorem).

The two sides, AE and CE, of triangle AEC are proportional to the two sides, EB and ED, of triangle BED, if and only if the proportion

(i) $\frac{AE}{EB} = \frac{CE}{ED}$

is true. Substituting in values, we have

(ii) $\frac{3}{4 \frac{1}{2}} = \frac{4}{6} \cdot$

Equation (ii) is true if and only if the product of the means equals the product of the extremes, i.e. cross-multiplying

(iii) $3(6) = (4\text{ }\frac{1}{2})\text{ }4$

(iv) $18 = 18.$

Therefore, the proportion of equation (i) is true and two pairs of sides of the triangles are proportional.

To complete the proof that $\triangle AEC \sim \triangle BED$, we need only show that the included angle of $\triangle AEC$, $\sphericalangle 1$, is congruent to the included angle of $\triangle BED$, $\sphericalangle 2$. $\sphericalangle 1$ and $\sphericalangle 2$ are vertical angles. Therefore, they are congruent and $\triangle AEC \sim \triangle BED$ by the S.A.S. Similarity Theorem.

Having proved the triangles similar, we can now use the proportion

$$\frac{AC}{BD} = \frac{CE}{DE}$$

to solve for BD. Substituting in values, we have

(v) $\frac{3\text{ }\frac{1}{2}}{BD} = \frac{4}{6}$ or $\frac{3\text{ }\frac{1}{2}}{BD} = \frac{2}{3}$.

Setting the product of the means equal to the product of the extremes, we obtain

(vi) $2 \cdot BD = 3 \cdot 3\text{ }\frac{1}{2}$

(vii) $2 \cdot BD = 10\text{ }\frac{1}{2}$

(viii) $BD = 5\text{ }\frac{1}{4}.$

● **PROBLEM** 262

Line segments $\overline{AC}$ and $\overline{BD}$ intersect at E, as shown in the accompanying diagram. $\overline{AB}$ is parallel to $\overline{CD}$. (a) Prove that $\triangle ABE \sim \triangle CDE$ and (b) if DE = 10, BE = 15 and CE = 20, find AE.

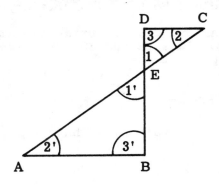

<u>Solution:</u> (a) For this proof, we will employ the theorem which tells us that two triangles are similar if two angles of one triangle are congruent to two corresponding angles of the other.

STATEMENT	REASON
1. $\overline{AC}$ and $\overline{BD}$ are straight line segments	1. Given.
2. ∢ 1 and ∢ 1' are vertical angles	2. Definition of vertical angles
3. ∢ 1 ≅ ∢ 1'	3. If two angles are vertical angles, then they are congruent.
4. $\overline{AB}$ || $\overline{CD}$	4. Given.
5. ∢ 2 ≅ ∢ 2'	5. If parallel lines are cut by a transversal, then the alternate interior angles are congruent.
6. ΔABE ∾ ΔCDE	6. A.A. Similarity Theorem.

(b) Corresponding sides of similar triangles must be proportional. Therefore, since DE corresponds to BE and CE corresponds to AE, the proportion DE : BE = CE : AE must hold.The lengths of all sides other than $\overline{AE}$ are known and can be substituted into the proportion, enabling us to calculate AE.

$$10 : 15 = 20 : AE$$

$$10 \ (AE) = 15 \cdot 20$$

$$AE = 30.$$

● **PROBLEM** 263

In the accompanying figure, the line segment,$\overline{KL}$, is drawn parallel to $\overline{ST}$, intersecting $\overline{RS}$ at K and $\overline{RT}$ at L in ΔRST. If RK = 5, KS = 10, and RT = 18, then find RL.

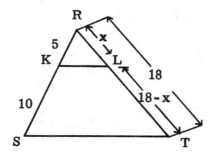

<u>Solution</u>: If a line is parallel to one side of a triangle, then it divides the other two sides proportionally. Since $\overline{KL} \parallel \overline{ST}$, $\overline{RT}$ and $\overline{RS}$ are divided proportionally.

Let x = RL. Then 18 - x = LT. Set up the proportion

$\frac{RK}{KS} = \frac{RL}{LT}$ and substitute to obtain

$\frac{5}{10} = \frac{x}{18 - x}$

or   10x = 90 - 5x

     15x = 90

       x = 6

Therefore, x = RL = 6.

Alternatively, instead of forming a ratio of upper to lower segment, we can form a ratio of the upper segment to the whole side.

We are given RT = 18, RS = 15, and RK = 5. If we let RL = x, the proportion becomes

$\frac{RK}{RS} = \frac{RL}{RT}$

$\frac{5}{15} = \frac{x}{18}$

15x = 90

  x = 6.

It is seen, then, that the same solution can be arrived at in several ways.

● **PROBLEM** 264

In triangle ABC, CD = 6, DA = 5, CE = 12, and EB = 10, as shown in the figure. Is $\overline{DE}$ parallel to $\overline{AB}$?

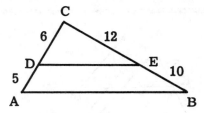

<u>Solution</u>: A postulate tells us that if a line divides two sides of a triangle proportionally, the line is parallel to the third side.

Set up the ratio of upper to lower segments for both $\overline{CA}$ and $\overline{CB}$. If these ratios are equal, the lines are proportional and we can then conclude that $\overline{DE} \parallel \overline{AB}$.

$$\frac{CD}{DA} = \frac{6}{5} \qquad \text{and} \qquad \frac{CE}{EB} = \frac{12}{10} = \frac{6}{5}$$

Therefore, $\overline{CA}$ and $\overline{CB}$ are divided proportionally.

We can thus conclude that $\overline{DE}$ is parallel to $\overline{AB}$.

PROPORTIONS INVOLVING ANGLE BISECTORS

● PROBLEM 265

The sides of a triangle have lengths 15, 20 and 28. Find the lengths of the segment into which the bisector of the angle with the greatest measure divides the opposite side.

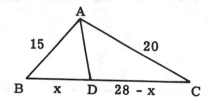

<u>Solution</u>: An angle bisector of any triangle divides the side of the triangle opposite the angle into segments proportional to the sides adjacent to the angle. Therefore,

(i) $$\frac{BD}{DC} = \frac{AB}{AC} \; .$$

If we let BD = x, then DC = 28 - x. Substituting in equation (i), we have

(ii) $$\frac{x}{28 - x} = \frac{15}{20} \; .$$

Reducing the fraction on the right to lowest terms,

(iii) $$\frac{x}{28 - x} = \frac{3}{4}$$

For real numbers a, b, c, d, it must be true that if $\frac{a}{b} = \frac{c}{d}$, then ad = cb. Therefore, from equation (iii), we have

(iv) $$4x = 3(28 - x).$$

Multiplying out, we have

(v)   4x = 84 - 3x

    Adding 3x to both sides,

(vi)   4x + 3x = 84 - 3x + 3x

(vii)   7x = 84

    Dividing by 7 leaves

(viii)   x = 12.

Then,

(ix)   BD = 12

(x)   DC = 28 - 12 = 16.

● **PROBLEM** 266

A right triangle has legs of length 6 and 8 inches. $\overline{CD}$ bisects the right angle. Find the lengths of $\overline{AD}$ and $\overline{DB}$.

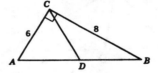

Solution: We must somehow relate the unknown segment lengths to the given data to derive the required results.

    Recall that the bisector of one angle of a triangle divides the opposite side so that the lengths of its segments are proportional to the lengths of the adjacent sides. Thus,

$$\frac{6}{8} = \frac{AD}{DB} .$$

We see that $\frac{8}{8} = \frac{DB}{DB}$ is always true. Hence, adding this to the above proportion, we obtain

$$\frac{6 + 8}{8} = \frac{AD + DB}{DB} .$$

But, AD + DB = AB, the hypotenuse of a right triangle. Applying the Pythagorean theorem, $a^2 + b^2 = c^2$, where a = 6, b = 8, we obtain

$$6^2 + 8^2 = c^2$$

$$36 + 64 = c^2$$

$$100 = c^2.$$

230

Thus, c = 10, or hypotenuse AB = 10. Substituting we obtain

$$\frac{6 + 8}{8} = \frac{AD + DB}{DB}$$

$$\frac{14}{8} = \frac{AB}{DB}$$

$$\frac{14}{8} = \frac{10}{DB} \; .$$

Since, in a proportion, the product of the means is equal to the product of the extremes, we obtain

$$14 \cdot DB = 8 \cdot 10,$$

$$DB = \frac{80}{14} = \frac{40}{7} \; .$$

To find AD, we notice that

$$AD = AB - DB$$

$$= 10 - \frac{40}{7} = \frac{30}{7} \; .$$

Therefore, $\qquad AD = \dfrac{30}{7} \;$ and $\; DB = \dfrac{40}{7} \; .$

● **PROBLEM** 267

In triangle ABC, in the accompanying diagram, D is the midpoint of $\overline{AB}$, and E is the midpoint of $\overline{AC}$. If BC = 7x + 1 and DE = 4x - 2, find x and calculate the lengths of $\overline{BC}$ and $\overline{DE}$.

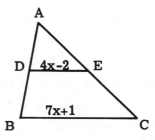

<u>Solution</u>: There is a theorem which states that if a line segment joins the midpoints of two sides of a triangle, the segment is parallel to the third side and its length is one-half the length of the third side. $\overline{DE}$ is just such a line segment and therefore its length equals ½ the length of $\overline{BC}$.

To solve for x, set DE = ½BC

$$4x - 2 = \tfrac{1}{2}(7x + 1)$$
$$2(4x - 2) = 7x + 1$$
$$8x - 4 = 7x + 1$$
$$8x - 7x = 1 + 4$$
$$x = 5$$

By substitution, the length of $\overline{DE}$ is $DE = 4x - 2 = 4(5) - 2 = 18$ and the length of $\overline{BC}$ is

$$BC = 7x + 1 = 7(5) + 1 = 36.$$

Therefore, the complete solution is

$$x = 5, DE = 18 \text{ and } BC = 36.$$

PROPORTIONS IN RIGHT TRIANGLES

● **PROBLEM** 268

Given ΔACB, as shown in the figure. Prove (a) (AD)(AB) = (AC)^2; (b) (DB)(AB) = (CB)^2; (c) (AD)(DB) = (DC)^2.

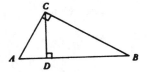

Solution: We will obtain the required results by proving, for each part, that the two triangles in the figure are similar. By setting up proportions between the lenghts of the sides of these triangles, the desired result is obtained.

(a) In order to show that (AD)(AB) = (AC)^2, examine triangles ACD and ABC. ∢CAD $\stackrel{\sim}{=}$ ∢BAC, by the reflexive property. ∢ADC $\stackrel{\sim}{=}$ ∢ACB, because they are both right angles. Hence, by the A.A. Similarity Theorem, ΔACD ∿ ΔABC. This implies that

$$\frac{AD}{AC} = \frac{AC}{AB}$$

or

$$(AD)(AB) = (AC)^2.$$

(b) To prove that (DB)(AB) = (CB)^2, focus attention on triangles ABC and CBD. ∢ABC $\stackrel{\sim}{=}$ ∢CBD by reflextivity. ∢ACB $\stackrel{\sim}{=}$ ∢CDB, because they are both right triangles. Hence, by the A.A. Similarity Theorem, ΔABC ∿ ΔCBD, which implies that

$$\frac{AB}{CB} = \frac{CB}{DB}$$

or

$$(AB)(DB) = (CB)^2.$$

(c) From (a) and (b), we recall that

$$\Delta ACD \sim \Delta ABC$$

$$\Delta CBD \sim \Delta ABC.$$

By the transitive property, then $\triangle ACD \sim \triangle CBD$ which implies

$$\frac{AD}{CD} = \frac{CD}{BD}$$

or

$$(AD)(BD) = (CD)^2.$$

The sides of triangle ABC measure 5, 7, and 9. The shortest side of a similar triangle, A'B'C', measures 10.

(a) Find the measure of the longest side of triangle A'B'C'. (b) Find the ratio of the measures of a pair of corresponding altitudes in triangles ABC and A'B'C'. (c) Find the perimeter of triangle A'B'C'.

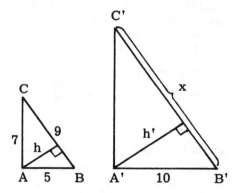

Solution: (a) As seen in the diagram, the longest side of $\triangle A'B'C'$ is $\overline{B'C'}$; it corresponds to $\overline{BC}$, the longest side of $\triangle ABC$. The shortest corresponding sides are $\overline{AB}$ and $\overline{A'B'}$.

Since $\triangle ABC \sim \triangle A'B'C'$, a proportion exists between corresponding sides. For these triangles, one proportion is

$$\frac{AB}{A'B'} = \frac{BC}{B'C'} .$$

The only length in this proportion which is unknown is B'C', and this can be determined by substitution and algebra.

Substituting:

$$\frac{5}{10} = \frac{9}{B'C'}$$

$$90 = 5 \ B'C'$$

$$B'C' = 18.$$

The longest side of $\triangle A'B'C'$, $\overline{B'C'}$ measures 18.

(b) The ratio of the measures of the pair of

233

altitudes,h and h' will be the same for these two similar triangles as will be the ratio of the measures of any other pair of corresponding linear parts.

Choose AB and A'B' as the determinants of the ratio. Therefore,

$$\frac{h}{h'} = \frac{AB}{A'B'} \ .$$

By substitution, we conclude that

$$\frac{h}{h'} = \frac{5}{10} \quad \text{or} \quad \frac{1}{2} \ .$$

(c) The perimeter of A'B'C' is a non-angular measure and, as with the altitudes, will have a ratio of similitude which is the same as that possessed by all other pairs of linear corresponding parts. If p is the perimeter of △ABC, and p' that of △A'B'C', we have

$$p : p' = 1 : 2.$$

By substitution, we obtain

$$21 : p' = 1 : 2, \text{ or } p' = 42.$$

● **PROBLEM** 270

In the figure below, △ABC is a right triangle. $\overline{AD}$ is an altitude on the hypotenuse $\overline{BC}$. $\overline{BD}$ = 2, DC = 8. Find AD, AB, and AC.

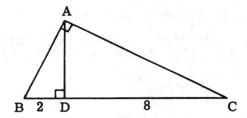

Solution: The geometric mean of two numbers x and z, is defined as that number, y, such that

$$\frac{x}{y} = \frac{y}{z} \ .$$

In the right triangle, the altitude to the hypotenuse is the geometric mean between the segments of the hypotenuse. Therefore,

(i) $$\frac{BD}{AD} = \frac{AD}{DC}$$

Substituting in the values, BD = 2 and DC = 8, the equation becomes

234

(ii) $$\frac{2}{AD} = \frac{AD}{8} \ .$$

For the proportionality to be true, the product of the means must equal the product of the extremes.

The equation then becomes

(iii) $$(AD)^2 = 2 \cdot 8 = 16$$

(iv) $$AD = \sqrt{16} = 4.$$

To find AB, we note that AB is the hypotenuse of right triangle $\triangle ABD$. Since we know the lengths of both legs, we can use the Pythagorean theorem,

(v) $$c^2 = a^2 + b^2.$$

Substituting, we obtain

(vi) $$(AB)^2 = (BD)^2 + (AD)^2$$

(vii) $$(AB)^2 = 2^2 + 4^2 = 4 + 16 = 20$$

(viii) $$AB = \sqrt{20} = \sqrt{4 \cdot 5} = \sqrt{4} \cdot \sqrt{5} = 2 \cdot \sqrt{5}.$$

Another method of solving for AB is to realize that there is another geometric mean implicit in the triangle. The altitude to the leg divides the hypotenuse so that either leg is the geometric mean between the hypotenuse and the segment of the hypotenuse adjacent to that leg. Therefore, it must be that

(ix) $$\frac{BD}{AB} = \frac{AB}{BC}$$

We substitute in the values BD = 2 and BC = 10,

(x) $$\frac{2}{AB} = \frac{AB}{10} \ ,$$

(xi) $$(AB)^2 = 2 \cdot 10 = 20,$$

(xii) $$AB = \sqrt{20} \quad \text{or} \quad 2 \sqrt{5}.$$

We find AC in a similar manner. AC is the geometric mean of DC and BC. Therefore,

(xiii) $$\frac{DC}{AC} = \frac{AC}{BC} \ ,$$

(xiv) $$\frac{8}{AC} = \frac{AC}{10} \ .$$

Remember: the product of the means must equal the product of the extremes,

(xv) $$(AC)^2 = 8 \cdot 10 = 80$$

(xvi) $$AC = \sqrt{80} = \sqrt{16 \cdot 5} = \sqrt{16} \cdot \sqrt{5} = 4 \sqrt{5}.$$

In right triangle ABC, altitude $\overline{CD}$ is drawn to hypotenuse $\overline{AB}$, as seen in the figure. If AD = 6 and DB = 24, find (a) CD and (b) AC.

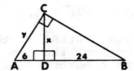

Solution: If the altitude is drawn to the hypotenuse of a right triangle, as in $\triangle ACB$, then the length of the altitude (CD) is the mean proportional between the lengths of the segments of the hypotenuse ($\overline{AD}$ and $\overline{DB}$). Also, the triangles thus formed ($\triangle ADC$ and $\triangle CDB$) are both similar to the given triangle ($\triangle ACB$).

(a) The first part of the above conclusion allows us to form the proportion

$$\frac{AD}{CD} = \frac{CD}{DB} .$$

Since we want to determine CD, let x = length of $\overline{CD}$ and substitute the given values of AD and DB into the proportion. This gives us $\frac{6}{x} = \frac{x}{24}$. The product of the means equals the product of the extremes, therefore, $x^2 = 144$ and $x = 12$; the length of altitude $\overline{CD}$ is 12.

(b) The second fact, that $\triangle ADC \sim \triangle ACB$, tells us that corresponding sides are in proportion. Therefore, to find AC, which is the hypotenuse of $\triangle ADC$ and the shorter leg of $\triangle ACB$, we set up the following proportion:

$$\frac{AD}{AC} = \frac{AC}{AB} .$$

If we let y = AC, and substitute in the given,we find that $\frac{6}{y} = \frac{y}{30}$. As stated in part (a), this expression is equal to $y^2 = 180$. Therefore,

$$y = \sqrt{180} = \sqrt{36} \cdot \sqrt{5} = 6\sqrt{5} \quad \text{which is the length of AC.}$$

In right triangle ABC, altitude $\overline{CD}$ is drawn to hypotenuse $\overline{AB}$, as shown in the diagram. If CD = 12 and AD exceeds DB by 7, find the lenghts of DB and AD.

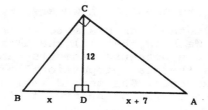

Solution: By theorem, we know that the length of the altitude to the hypotenuse will be the mean proportional between the lengths of the segments of the hypotenuse. In this case, AD : CD = CD : BD.

   If we let x = the length of $\overline{DB}$,

   then x + 7 = the length of $\overline{AD}$.

   By substituting this and CD = 12 we get (x + 7) : 12 = 12 : x. Since the product of the means equals the product of the extremes in a proportion, this equals $x^2 + 7x = 144$.

It follows then that $x^2 + 7x - 144 = 0$.

By factoring, this becomes (x - 9)(x + 16) = 0.

   Therefore, x = 9 or x = - 16. (The negative is rejected for lack of geometric significance.)

   If x = 9, then x + 7 = 16.

   Therefore, the length of $\overline{DB}$ = 9 and the length of $\overline{AD}$ = 16.

● **PROBLEM 273**

Given: $\overline{CD}$ is an altitude and $\overline{CE}$ is an angle bisector of △ABC; ∢ ACB is a right angle. Prove:

$$\frac{AD}{DB} = \frac{AE^2}{EB^2} .$$

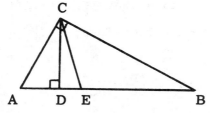

Solution: The best method to use in solving these problems is to express each side in the proportion in terms of lengths commonly related to all of them. In this case, both the altitude and the angle bisector can be related to the sides of the triangle.

   We have that $\overline{CD}$ is an altitude, and, thus, we can take advantage of the proportions that the altitude of a right triangle creates. The proportion we are trying to prove involves $\overline{AD}$ and $\overline{DB}$, the segments into which the altitude divides the base. We know that (1) the altitude is the mean proportional of the hypotenuse segments, and (2) each leg of the triangle is the mean proportional of the hypotenuse and the adjacent segment. The second fact will allow us to derive equations that will be helpful in this problem

(1) $\dfrac{AB}{AC} = \dfrac{AC}{AD}$, or $AC^2 = AB \cdot AD$ and

(2) $\dfrac{AB}{BC} = \dfrac{BC}{BD}$, or $BC^2 = AB \cdot BD$.

These are the equations we have involving AD and DB.

For AE and EB, we take advantage of the fact that $\overline{CE}$ is the angle bisector. We know that the angle bisector divides the side of the triangle opposite the angle into segments proportional to the adjacent sides. Thus,

(3) $\dfrac{AC}{BC} = \dfrac{AE}{EB}$.

We thus have equations involving all four lengths, in terms of AB, BC, and AC, the lengths of the sides of the triangle. We find $\dfrac{AD}{DB}$ in terms of AB, BC, and AC. Then we find $\dfrac{AE^2}{EB^2}$ in terms of AB, BC, and AC. If the two expressions are equal, then $\dfrac{AD}{DB} = \dfrac{AE^2}{EB^2}$.

To obtain $\dfrac{AD}{DB}$, first we solve (1) for AD. $AD = \dfrac{AC^2}{AB}$. From (2), we obtain $DB = \dfrac{BC^2}{AB}$. Therefore,

$$\dfrac{AD}{DB} = \dfrac{AC^2/AB}{BC^2/AB} = \dfrac{AC^2}{BC^2} \ .$$

To obtain $\dfrac{AE^2}{EB^2}$, note that $\dfrac{AE^2}{EB^2} = \left(\dfrac{AE}{EB}\right)^2$. From equation (3), $\dfrac{AE}{EB} = \dfrac{AC}{BC}$. Therefore, $\left(\dfrac{AE}{EB}\right)^2 = \left(\dfrac{AC}{BC}\right)^2$.

Thus, $\dfrac{AE^2}{EB^2} = \dfrac{AC^2}{BC^2}$. Also, $\dfrac{AD}{DB} = \dfrac{AC^2}{BC^2}$.

Therefore, $\dfrac{AE^2}{EB^2} = \dfrac{AD}{DB}$.

● **PROBLEM** 274

An altitude and an angle bisector are drawn to the hypotenuse of a right triangle whose sides are 8, 15, and

238

17 inches long. Find the length of the segment joining the points where the altitude and angle bisector intersect the hypotenuse (see figure). Express your answer to the nearest hundredth.

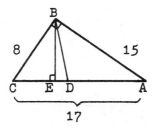

Solution: In the accompanying figure, $\triangle ABC$ is a right triangle with AB = 15, BC = 8, and AC = 17. $\overline{BD}$ is the bisector of the right angle and $\overline{BE}$ is the altitude drawn to the hypotenuse. We wish to find DE. With the given information, we can determine (1) DC and (2) EC. Since DE = DC - EC, we can then solve for DE.

(1) To solve for DC, note that the angle bisector divides the opposite side into segments proportional to the sides.

(i) $$\frac{AD}{DC} = \frac{AB}{BC}$$

Note AB = 15 and BC = 8. Since AD + DC = AC = 17, AD can be expressed as 17 - DC. Substituting these results, we have

(ii) $$\frac{17 - DC}{DC} = \frac{15}{8} \ .$$

We have one equation and one unknown. To solve for DC, multiply both sides by 8 · DC:

(iii) $$8\,(17 - DC) = 15 \cdot DC$$

(iv) $$136 - 8DC = 15\,DC$$

Adding 8DC to both sides and dividing by 23, we obtain

(v) $$DC = \frac{136}{23} \ .$$

(2) To solve for EC, note that $\overline{BE}$ is an altitude drawn to the hypotenuse of a right triangle and cuts segments $\overline{AE}$ and $\overline{EC}$ on the hypotenuse. Since the length of each leg of the given triangle is the mean proportional between the hypotenuse length and the adjacent segment, we have

(vi) $$BC^2 = EC \cdot AC.$$

239

We substitute BC = 8, AC = 17, and solve for EC:

(vii) $\qquad 8^2 = EC \cdot 17$

(viii) $\qquad EC = \dfrac{64}{17}$

Having found values for DC and EC, we can now solve for DE:

(ix) $\qquad DE = DC - EC$

(x) $\quad DE = \dfrac{136}{23} - \dfrac{64}{17} = \dfrac{17(136) - 64(23)}{23(17)}$

(xi) $\quad DE = \dfrac{840}{391} \overset{\sim}{=} 2.15.$

● **PROBLEM** 275

D and E are respective points of side $\overline{AB}$ and $\overline{BC}$ of $\triangle ABC$, so that $\dfrac{AD}{DB} = \dfrac{2}{3}$ and $\dfrac{BE}{EC} = \dfrac{1}{4}$. If $\overline{AE}$ and $\overline{DC}$ meet at P, find $\dfrac{PC}{DP}$.

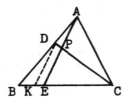

Solution: We are asked to find ratio $\dfrac{PC}{DP}$. To find the value of a ratio $\dfrac{a}{b}$, we can either (1) find exact values of a and b; (2) express the values of a and b in terms of a common length; or (3) form a proportion with $\dfrac{a}{b}$ as one side and a known ratio on the other side. Here, no values are given. Thus, PC and DP cannot be solved exactly. Neither is there a common length in which DP and PC can be expressed. The third method calls for a proportion. Proportions imply (1) similar triangles; (2) a triangle cut by a line parallel to one side; and (3) sets of parallel lines. To create proportional sides, draw segment $\overline{DK}$ parallel to $\overline{AE}$ such that K is between B and C. Note that in $\triangle CDK$, line $\overline{PE}$ is parallel to side DK. Therefore, $\dfrac{PC}{DP} = \dfrac{EC}{EK}$. Our proportional is formed. $\dfrac{PC}{DP}$ is on one side. The term on the other side, $\dfrac{EC}{EK}$, can be found since both EC and EK can be found in terms of BC.

To find EK in terms of BC, we show, (1) in $\triangle BAE$,

$\overleftrightarrow{DK} \parallel$ side $\overline{AE}$. Therefore, sides $\overline{BA}$ and $\overline{BE}$ are cut proportionally.

$$\frac{KE}{KB} = \frac{AD}{DB} \cdot \text{ Since } \frac{AD}{DB} = \frac{2}{3}, \quad \frac{KE}{KB} = \frac{2}{3} .$$

(2) From $\frac{KE}{KB}$, we can find KE in terms of BE. $\frac{KE}{KB} = \frac{2}{3}$. Then, $KE = \frac{2}{3} KB$, or $KB = \frac{3}{2} KE$. Since $BE = KE + BK$, $BE = KE + \frac{3}{2} KE = \frac{5}{2} KE$. Thus,

$$KE = \frac{2}{5} BE.$$

(3) We can find BE in terms of BC. From the given $\frac{BE}{EC} = \frac{1}{4}$ or $BE = \frac{1}{4} EC$ or $EC = 4BE$. Since $BC = BE + EC = BE + 4BE = 5 BE$, then

$$BE = \frac{1}{5} BC.$$

(4) We can combine the results of 2 and 3 to find KE in terms of BC. $KE = \frac{2}{5} BE$ and $BE = \frac{1}{5} BC$. Therefore, $KE = \frac{2}{5} \left[\frac{1}{5} BC \right] = \frac{2}{25} BC$.

To find EC in terms of BC, note that $\frac{BE}{EC} = \frac{1}{4}$ or $BE = \frac{1}{4} EC$. Since $BC = BE + EC = \frac{1}{4} EC + EC = \frac{5}{4} EC$, then $EC = \frac{4}{5} BC$.

Combining these two results, we have

$$\frac{EC}{KE} = \frac{\frac{4}{5} BC}{\frac{2}{25} BC} = 10.$$

Earlier, we found $\frac{PC}{DP} = \frac{EC}{KE}$. Therefore, $\frac{PC}{DP} = 10$.

PROPORTIONS INVOLVING MEDIANS

● PROBLEM 276

In triangle ABC, medians $\overline{AD}$, $\overline{BE}$, and $\overline{CF}$ intersect at P, as seen in the figure. If AD = 24 in., find the length of $\overline{AP}$.

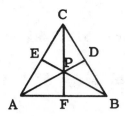

Solution: A theorem tells us that the medians of a triangle are con-current at a point whose distance from any vertex is two-thirds the distance from that vertex to the midpoint of the opposite side. The vertex in question is A, the point of concurrency is P, and the distance to the midpoint of the side opposite A is given by AD. Therefore, the length of $\overline{AP}$ equals $\frac{2}{3}$ the length of $\overline{AD}$. Algebrai-cally, this means

$$AP = \frac{2}{3} AD$$

$$AP = \frac{2}{3}(24") = 16" .$$

Therefore, the length of $\overline{AP}$ is 16".

● **PROBLEM** 277

Show that the medians of a triangle are concurrent at a point on each median located two-thirds of the way from each vertex to the op-posite side.

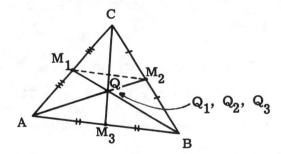

Solution: Referring to the figure above, we must show (1) Q is a point on $\overline{AM_2}$, $\overline{BM_1}$, and $\overline{CM_3}$; and (2) that $AQ = \frac{2}{3} AM_2$, $BQ = \frac{2}{3} BM_1$, and $CQ = \frac{2}{3} CM_3$. Since $AM_2 = AQ + QM_2$, $QM_2 = AM_2 - AQ = AM_2 - \frac{2}{3} AM_2 = \frac{1}{3} AM_2$, or $QM_2 = \frac{1}{3} AM_2$. Thus, to prove the second part, we need to show that the point of concurrency divides the median AM_2 into two segments such that the segment near the vertex is twice the length of the segment near the side. In algebraic notation, what we must show for the second part is that $\dfrac{AQ}{QM_2} = \dfrac{BQ}{QM_1} = \dfrac{CQ}{QM_3} = 2$.

Wherever there are proportions, it is wise to look for similar triangles. Suppose we first wish to show that $\dfrac{AQ}{QM_2} = \dfrac{2}{1}$. Then we find two triangles such that $\overline{AQ}$ and $\overline{QM_2}$ are corresponding sides. $\triangle AQB$ and $\triangle M_2QM_1$ are such triangles. We show similarity by the A-A Simi-larity Theorem. Then $\dfrac{AQ}{QM_2} = \dfrac{BQ}{QM_1} = \dfrac{AB}{M_1M_2}$. $\dfrac{AB}{M_1M_2}$ is known. $\overline{M_1M_2}$ is a midline of $\triangle ABC$. Therefore, $AB = 2 \cdot M_1M_2$, $\frac{AB}{M_1M_2} = 2$, and $\frac{AQ}{QM_2} = 2$, proving the second part. We can repeat the procedure for the two other sides.

NOTE: It may seem that we have proven the second part without actually showing that the lines are indeed concurrent, but we really don't need the concurrency to show the two-thirds division. In fact, the exact reverse is true. We show that the intersection of any two medians is a point two-thirds the length of each median from the respective vertex. Since there is only one point that is two-thirds the median length from the vertex, the points of intersection must be the same. Concurrency (part one) follows from the two-thirds division (part two).

Given: $\overline{AM_2}$, $\overline{BM_1}$, and $\overline{CM_3}$ are medians of $\triangle ABC$.

Prove: (1) Q is a point on $\overline{AM_2}$, $\overline{BM_1}$, and $\overline{CM_3}$.
      (2) $AQ = \frac{2}{3} AM_2$, $BQ = \frac{2}{3} BM_1$, $CQ = \frac{2}{3} CM_3$.

Statements	Reasons
1. $\overline{AM_2}$, $\overline{BM_1}$, and $\overline{CM_3}$ are medians of $\triangle ABC$	1. Given.
2. Q_1 is the intersection of $\overline{AM_2}$ and $\overline{BM_1}$.	2. Two non-parallel, non-co-incident, lines intersect in a point.
3. $\sphericalangle AQ_1B \cong \sphericalangle M_2Q_1M_1$	3. Opposite angles formed by two intersecting lines are congruent.
4. $\overline{M_1M_2}$ is a midline of $\triangle ABC$	4. The segment that connects the midpoints of two sides of a triangle is a midline of the triangle.
5. $\overline{M_1M_2} \parallel \overline{AB}$	5. The midline of a triangle is parallel to the third side.
6. $\sphericalangle M_1M_2Q \cong \sphericalangle QAB$	6. Alternate interior angles of a transversal are congruent.
7. $\triangle M_1QM_2 \sim \triangle AQB$	7. If two angles of one triangle are congruent with the corresponding two angles of a second triangle, then the two triangles are similar.
8. $\frac{M_2Q}{AQ} = \frac{M_1Q}{BQ} = \frac{M_1M_2}{AB}$	8. The corresponding sides of similar triangles are proportional.
9. $AB = 2 \cdot M_1M_2$ or $\frac{M_1M_2}{AB} = \frac{1}{2}$	9. The midline of a triangle is half the length of the third side.
10. $\frac{M_2Q}{AQ} = \frac{M_1Q}{BQ} = \frac{1}{2}$	10. Substitution Postulate.
11. $AM_2 = AQ + M_2Q$ $BM_1 = BQ + M_1Q$	11. Q_1 is the intersection of two segments $\overline{AM_2}$ and $\overline{BM_1}$. Therefore Q_1 must lie between the endpoints of the segment. The equations at left follow

	from the definition of between-ness.
12. $M_2Q = \frac{1}{2} AQ_1$ $M_1Q_1 = \frac{1}{2} BQ_1$	12. Multiplication Postulate and Step 10.
13. $AM_2 = 1\frac{1}{2} AQ_1$ $BM_1 = 1\frac{1}{2} BQ_1$	13. Substitution of Step 12 into 11.
14. $AQ_1 = \frac{2}{3} AM_2$ $BQ_1 = \frac{2}{3} BM_1$	14. Multiplication Postulate
15. Q_2 is the intersection of $\overline{AM_2}$ and $\overline{CM_3}$	15. Two non-parallel, non-coincident lines intersect in a unique point.
16. $AQ_2 = \frac{2}{3} AM_2$ $CQ_2 = \frac{2}{3} CM_3$	16. Obtained by repeating procedure used from Steps 3 through 13. The midline is now $\overline{M_1M_3}$ not $\overline{M_1M_2}$.
17. Q_3 is the intersection of medians $\overline{CM_3}$ and $\overline{BM_1}$	17. Two non-parallel non-coincident lines intersect in a unique point.
18. $CQ_3 = \frac{2}{3} CM_3$ $BQ_3 = \frac{2}{3} BM_1$	18. Obtained by repeating procedure used from Steps 3 through 13. The midline is now $\overline{M_1M_3}$.
19. Q_1 is a point on segment $\overline{AM_2}$ such that $AQ_1 = \frac{2}{3} AM_2$. Q_2 is a point on segment $\overline{AM_2}$ such that $AQ_2 = \frac{2}{3} AM_2$.	19. Repetition of results from Steps 14 and 16.
20. Point Q_1 is point Q_2	20. On a given segment, there is only one point on the segment that is a given distance from a given endpoint.
21. Q_2 is a point on segment $\overline{CM_3}$ such that $CQ_2 = \frac{2}{3} CM_3$. Q_3 is a point on segment $\overline{CM_3}$ such that $CQ_3 = \frac{2}{3} CM_3$.	21. Repetition of results from Steps 16 and 18.
22. Point Q_2 is point Q_3	22. Same reason as Step 20.
23. Let point $Q = Q_1 = Q_2 = Q_3$	23. From Steps 20 and 22, we have shown $Q_1 = Q_2 = Q_3$ and here we give this common point of intersection of the three medians a more general name.
24. Q is a point on $\overline{AM_2}$, $\overline{BM_1}$, and $\overline{CM_3}$	24. Follows from Step 23 and the definitions of C_1, C_2, and C_3.
25. $AQ = \frac{2}{3} AM_2$ $BQ = \frac{2}{3} BM_1$ $CQ = \frac{2}{3} CM_3$	25. Substitution Postulate (Substitute Q for Q_1, Q_2, and Q_3 in Steps 14, 16, and 18.)

CIRCLES

CHAPTER 15

CENTRAL ANGLES AND ARCS

Basic Attacks and Strategies for Solving Problems in this Chapter. See pages 246 to 257 for step-by-step solutions to problems.

A central angle of a circle is defined as an angle whose vertex is at the center of the circle. In the picture below, $\angle BOA$ is a central angle since O is the center of the circle. This angle can be measured in terms of the arc that is intercepted. The circle can be divided into 360 congruent parts of 1° each and, since the part of the circle from A to B makes up 30 of those parts, the measure of $\angle BOA$ is 30°. The symbol $\overset{\frown}{AB}$ is used to represent the portion of the circle from A to B and is called "arc AB." The degree measure of this arc is also said to be 30°:

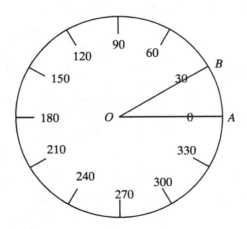

The arc measure is sometimes expressed in standard **linear units**, which is accomplished as follows. The formula for the circumference of the circle is $2\pi r$. In this case, let the radius, $\overline{OB}$, be 1 inch long, so $C = 2\pi$ inches.

246-A

Since $\dfrac{30}{360} = \dfrac{1}{12}$, the length of $\overset{\frown}{AB}$ (measured in inches) is $\dfrac{1}{12}$ the circumference of the circle.

$$\frac{1}{12} \cdot 2\pi \text{ inches} = \frac{1}{6}\pi \text{ inches}$$

Thus the linear measure of $\overset{\frown}{AB}$ is $\dfrac{1}{6}\pi$ inches. A more formal development of this process is included in Problem 278.

Consider the figure below. Suppose you know that $\angle AOB \cong \angle COD$ and O is the center of the circle. Then it is easy to establish that $\triangle AOB \cong \triangle COD$ and it follows that $\overline{AB} \cong \overline{CD}$. Also, the degree measures of $\overset{\frown}{AB}$ and $\overset{\frown}{CD}$ are equal since the degree measures of the corresponding central angles are the same. Now it is obvious that the linear measures of $\overset{\frown}{AB}$ and $\overset{\frown}{CD}$ are equal.

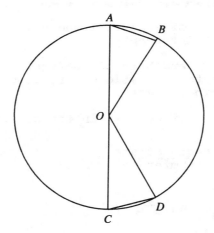

In this situation, $\overset{\frown}{AB}$ is said to be the intercepted arc for $\angle AOB$ and also the intercepted arc for the chord $\overline{AB}$. (A chord is a line segment whose endpoints lie on the circle.) For each central angle, there is a corresponding chord and an intercepted arc. In the paragraph above it was argued that when two central angles are congruent, the intercepted arcs are congruent and the corresponding chords are congruent. It is also the case that if the arcs are congruent, then the chords are congruent and the central angles are congruent; and if the chords are congruent, then the arcs are congruent and the central angles are congruent. This information will be used in several problems in this chapter.

Step-by-Step Solutions to Problems in this Chapter, "Central Angles and Arcs"

● **PROBLEM** 278

In a circle whose radius is 8 inches, find the number of degrees contained in the central angle whose arc length is 2π inches.

<u>Solution</u>: The measure of a central angle is equal to the measure of the arc it intercepts.

The ratio of arc length to circumference, in linear units, will be equal to the ratio of arc length to circumference as measured in degrees.

If n = the number of degrees in the arc 2π inches long,

Then
$$\frac{\text{length of arc}}{\text{circumference}} = \frac{n}{360°}$$

By substitution,
$$\frac{2\pi \text{in.}}{2\pi(8 \text{ in.})} = \frac{n}{360°}$$

$$\frac{1}{8} = \frac{n}{360°}$$

$$360° = 8n$$

$$n = 45°$$

Therefore, the central angle contains 45°.

● **PROBLEM** 279

Find the circumference of a circle whose radius is 21 in. $\left[\text{Use } \pi = \frac{22}{7}\right]$

Solution: By a theorem, we know that the circumference (C) of a circle is equal to twice the product of π and its radius (r), i.e. $C = 2\pi r$.

By substitution, $C = 2 \cdot \left[\frac{22}{7}\right] (21)$ in.

$$C = 2(22)(3) \text{ in.} = 132 \text{ in.}$$

Therefore, the circumference is 132 in.

● **PROBLEM** 280

Find the diameter of a circle whose circumference is 628 ft. [Use $\pi = 3.14$]

Solution: The circumference of a circle is given by the formula $C = \pi d$, where d is the diameter of the circle.

Using $\pi = 3.14$, and substituting the given information, we obtain

$$628 \text{ ft.} = (3.14)d$$

$$\frac{628}{3.14} \text{ft.} = d$$

$$200 \text{ ft.} = d$$

Therefore, the diameter is 200 ft.

● **PROBLEM** 281

The ratio of the circumference of two circles is 3:2. The smaller circle has a radius of 8. Find the length of a radius of the larger circle.

Solution: Since the circumference of a circle is a linear function of the radius, the ratio of circumferences will be equal to the ratio of the radii of the circles.

Let the ratio of circumferences $= \frac{C}{C'} = \frac{3}{2}$

  $r' = 8$, the radius of the smaller circle
  r = length of a radius of the larger circle.

Therefore, $\frac{C}{C'} = \frac{r}{r'}$

$$\frac{3}{2} = \frac{r}{8}$$

$$2r = 24$$

$$r = 12.$$

Therefore, the length of a radius of the larger circle is 12.

Find the number of degrees, to the nearest degree, in an angle subtended at the center of a circle by an arc 5 ft. 10 in. in length. The radius of the circle is 9 ft. 4 in.

<u>Solution</u>: Let us convert the mixed unit measures into single unit measure.

$$(12 \text{ in.} = 1 \text{ ft.})$$

$$5 \text{ ft. } 10 \text{ in.} = 5(12) + 10 = 70 \text{ in.}$$

$$9 \text{ ft. } 4 \text{ in. } = 9(12) + 4 = 112 \text{ in.}$$

We solve this problem by determining what fraction of the total circumference the arc occupies and then multiplying this by 360° to find the degree measure of the arc. The angle subtended at the center is a central angle and contains the same number of degrees as the arc.

Circumference $C = \pi 2 r$, where r = the radius length. By substitution, $C = 2\pi(112) = 224\pi$ in. The fraction of the total circumference is given by $\frac{70}{224\pi} = \frac{70}{(224)(3.14)} = \frac{70}{704}$.

Hence, there are $(360)\left(\frac{70}{704}\right)$ or 35° 48' contained in this arc and in its central angle. To the nearest degree, the answer is 36°.

Find, to the nearest tenth of an inch, the length of an arc of 60° in a circle whose radius is 12 in.

<u>Solution</u>: Since a circle contains 360°, an arc measuring 60° represents $\frac{60°}{360°}$, or $\frac{1}{6}$ of the total measure of the circle. As such, the arc must extend over $\frac{1}{6}$ of the circumference of the circle. Its length will equal, then, $\frac{1}{6}C$, where $C = 2\pi r$ is the circumference of the circle of radius r. By substitution, $C = 2\pi(12 \text{ in.}) = 24\pi$ in.

Using $\pi = 3.14$, $C = 24(3.14)$ in. = 75.36 in.

Therefore, the length of the arc in question equals $\frac{1}{6}$ (75.36) in., or 12.6 in.

Show that the diameter of a circle divides the circle into two congruent arcs.

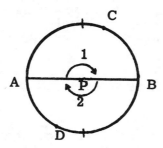

Solution: In the accompanying figure, $\overline{AB}$ is a diameter of circle P. We wish to show $\overarc{ACB} \cong \overarc{ADB}$. $\overarc{ACB}$ is intercepted by central angle ∢1; $\overarc{ADB}$ is intercepted by central angle ∢2. Since the measure of an arc is determined by the central angle, two arcs are congruent if and only if their central angles are congruent.

Therefore, to show $\overarc{ACB} \cong \overarc{ADB}$, we show ∢1 ≅ ∢2.

To show ∢1 ≅ ∢2, note that $\overline{AB}$ is a diameter, and therefore points A and B are collinear with center P. Thus, ∢1 and ∢2 are straight angles. All straight angles are congruent; therefore ∢1 ≅ ∢2, and the arcs intercepted by these angles are congruent. Therefore, $\overarc{ACB} \cong \overarc{ADB}$.

Hence, we have shown that the diameter of a circle divides the circle into two congruent arcs.

Let $\overarc{AB}$ be an arc of a circle whose center is 0. $\overarc{AB}$ is of length 11, and m ∢AOB = 10°. (The length of a circular arc is proportional to the central angle which cuts the arc.)

    a). Compute the circumference of the circle.
    b). Approximating π by 22/7, compute the diameter of the circle.

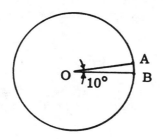

Solution: The circumference, C, of a circle of radius, r, is given by $C = 2\pi r$. If the diameter of the circle is d, we note that $2r = d$. Hence,

$$C = 2\pi r = \pi(2r) = \pi d.$$

a) Now, the circumference of a circle can be thought of as the length of an "arc" which wraps itself completely around the circle. Since arc-length is proportional to the angle which cuts the arc, we may set up the following proportions:

$$\frac{C}{360°} = \frac{\text{length of } \overset{\frown}{AB}}{10°} = \frac{11}{10°}.$$

The angle which "cuts" the "arc" represented by C is 360°. Solving for C by multiplying both sides of the last equation by 360°, we find

$$C = (36)(11) = 396.$$

b) Since $C = \pi d$ and $C = 396$ from part (a), we obtain

$$\pi d = 396.$$

Letting $\pi = \frac{22}{7}$, $\frac{22d}{7} = 396$ or $d = \frac{7}{22} \cdot 396 = 126$.

● **PROBLEM** 286

In the figure, show that isosceles triangles OAB and O'A'B' are similar.

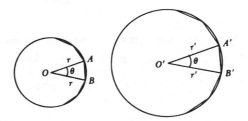

Solution: We will show that $\triangle OAB \sim \triangle O'A'B'$ by the S.A.S. (side angle side) Similarity Theorem.

Note that $\angle AOB \cong \angle A'O'B'$, from the figure. We also find that

$$\frac{AO}{A'O'} = \frac{r}{r'} \qquad \frac{BO}{B'O'} = \frac{r}{r'}$$

as shown in the figure. Hence, $\overline{AO} \sim \overline{A'O'}$ and $\overline{BO} \sim \overline{B'O'}$. Therefore, by the S.A.S. Similarity Theorem,

$$\triangle OAB \sim \triangle O'A'B'.$$

A and B are points on circle Q such that ΔAQB is equilateral. If AB = 12, find the length of AB͡.

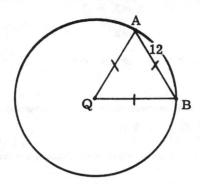

<u>Solution</u>: To find the arc length of AB͡, we must find measure of the central angle ∢AQB and the measure of the radius QA͞. ∢AQB is an interior angle of the equilateral triangle ΔAQB. Therefore, m∢AQB = 60°. Similarly, in the equilateral ΔAQB, AQ = AB = QB = 12. Given the radius, r, and the central angle, n, the arc length is given by $\frac{n}{360} \cdot 2\pi r$. Therefore, by substitution, ∢AQB = $\frac{60}{360} \cdot 2\pi \cdot 12 = \frac{1}{6} \cdot 2\pi \cdot 12 = 4\pi$. Therefore, length of arc AB͡ = 4π.

In circle O, the measure of AB͡ is 80° Find the measure of ∢A.

<u>Solution</u>: The accompanying figure shows that AB͡ is intercepted by central angle ∢AOB. By definition, we know that the measure of the central angle is the measure of its intercepted arc. In this case,

mAB͡ = m∢AOB or m∢AOB = 80°

Radius OA͞ and radius OB͞ are congruent and form two sides of ΔOAB. By a theorem, the angles opposite these two congruent sides must, themselves, be congruent. Therefore, m∢A = m∢B.

The sum of the measures of the angles of a triangle is 180°. Therefore,

$$m\sphericalangle A + m\sphericalangle B + m\sphericalangle AOB = 180°.$$

Since $m\sphericalangle A = m\sphericalangle B$, we can write

$$m\sphericalangle A + m\sphericalangle A + 80° = 180° \text{ or}$$

$$2m\sphericalangle A = 100° \text{ or}$$

$$m\sphericalangle A = 50°.$$

Therefore, the measure of $\sphericalangle A$ is 50°

● **PROBLEM** 289

Circle O is drawn in the figure, with point M as the midpoint of chord $\overline{AB}$. Prove that $\overline{OM}$ bisects $\sphericalangle AOB$.

Solution: To show $\overline{OM}$ bisects $\sphericalangle AOB$, we will prove $\triangle AOM \cong \triangle BOM$ by the S.S.S. $\cong$ S.S.S., method and then employ the corresponding parts rule to show the sufficient condition for bisection, namely that $\sphericalangle AOM \cong \sphericalangle BOM$.

The fact that all radii of a circle are congruent will be used in the proof.

Statement	Reason
1- $\overline{AM} \cong \overline{BM}$	1- Given.
2- $\overline{OM} \cong \overline{OM}$	2- Reflexive property of congruence.
3- $\overline{OA} \cong \overline{OB}$	3- Radii of a circle are congruent.
4- $\triangle AOM \cong \triangle BOM$	4- S.S.S. $\cong$ S.S.S.
5- $\sphericalangle AOM \cong \sphericalangle BOM$	5- Corresponding parts of congruent triangles are congruent.
6- $\overline{OM}$ is the bisector of $\sphericalangle AOB$.	6- If a segment divides an angle into two congruent parts, it is the angle bisector.

● **PROBLEM** 290

In circle O, in the figure shown, $\overline{OS} \perp \overline{RT}$. Prove $\overset{\frown}{RS} \cong \overset{\frown}{ST}$.

Solution: We can prove the two arcs congruent by proving that their central angles are congruent. Congruent central angles intercept congruent arcs. The central angles of $\overset{\frown}{RS}$ and $\overset{\frown}{ST}$ are ∢ROS and ∢SOT, respectively.

These two angles are contained, respectively, in ΔOER and ΔOET. By proving these two triangles to be congruent to each other, we can conclude that the required central angles are congruent to each other.

Statement	Reason
1. $\overline{OS} \perp \overline{RT}$	1. Given.
2. ∢OER and ∢OET are right angles.	2. Perpendicular lines intersect in right angles.
3. ∢OER ≅ ∢OET	3. All right angles are congruent.
4. $\overline{OR} \cong \overline{OT}$	4. Radii of the same circle are congruent.
5. $\overline{OE} \cong \overline{OE}$	5. Reflexive property of congruence.
6. ΔOER ≅ ΔOET	6. Hypotenuse-Leg Theorem.
7. ∢ROS ≅ ∢SOT	7. Corresponding angles of congruent triangles are congruent.
8. $\overset{\frown}{RS} \cong \overset{\frown}{ST}$	8. Congruent central angles of the same circle intercept congruent arcs.

● **PROBLEM** 291

In the accompanying figure, if ∢COD ≅ ∢FOE, prove that $\overline{CE} \cong \overline{DF}$.

Solution: In this problem, we will make use of the theorem that congruent chords always intercept congruent arcs in the same circle. The converse of this is also true: congruent arcs have congruent chords.

Statement	Reason
1- ∢COD ≅ ∢FOE	1- Given.
2- $\overgroup{CD}$ ≅ $\overgroup{FE}$	2- Congruent central angles intercept congruent arcs.
3- $\overgroup{DE}$ ≅ $\overgroup{DE}$	3- Reflexive property of congruence.
4- $\overgroup{CD}$ + $\overgroup{DE}$ ≅ $\overgroup{FE}$ + $\overgroup{DE}$ or $\overgroup{CE}$ ≅ $\overgroup{FD}$	4- Congruent quantities added to congruent quantities result in congruent quantities.
5- $\overline{CE}$ ≅ $\overline{FD}$	5- Congruent arcs have congruent chords.

● **PROBLEM** 292

If the grooves of a record were straightened out in a line, how many miles long would it be? Assume the grooves form concentric circles with outermost circle of radius 6 and innermost circle of radius 3, and that at 33 1/3 revolutions per minute, the playing time is 20 minutes.

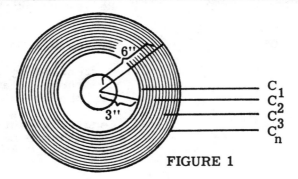

FIGURE 1

Solution: Let $C_1, C_2, C_3 \ldots C_n$ be the circumferences of the concentric circles that make up the grooves of the record. Then,

(i) Length of grooves = $C_1 + C_2 + C_3 + \ldots + C_n$.

Since the circumference of a circle = 2π × radius of the circle, we have

(ii) Length of groove = $2\pi r_1 + 2\pi r_2 \ldots + 2\pi r_n$ =
$2\pi(r_1 + r_2 + r_3 \ldots + r_n)$.

Thus, the length of the groove equals 2π times the sum of the radii, $r_1 + r_2 \ldots r_n$. To find the sum of the radii, we must first find (1) the number of radii, and, (2) the length of each.

(1.) To find the number of radii, find the number of circles. The number of circles equal the number of revolutions. We are given that the record makes 33 1/3 revolutions per minute and that there are 20 minutes of playing time. Thus, the number of radii = number of revolutions = (33 1/3 rev/min) × (20 min) = 666 2/3. Since we wish to deal with complete circles, we round this off to 667.

(2) Assuming that adjacent grooves are equally spaced apart, the difference between the radii of adjacent circles is constant. Let the difference between radii be d. Since there are 667 circles (or 666 spaces) between r = 6 and r = 3, we have that the space between r = 6 and r = 3 must be 666d. (Figure 2).

(iii) $666d = 6 - 3 = 3.$

(iv) $d = \dfrac{3}{666} = \dfrac{1}{222}$

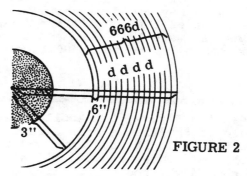

FIGURE 2

The radius of the inner circle is 3. The radius of the second innermost is $3 + d = 3 + \dfrac{1}{222}$. The radius of the mth innermost circle is $3 + \dfrac{m-1}{222}$. Thus,

(v) $r_1 + r_2 + r_3 + \ldots r_{667} = 3 + 3\dfrac{1}{222} + 3\dfrac{2}{222} + \ldots + 3\dfrac{666}{222}$

Rearranging, we obtain:

(vi) $r_1 + r_2 + \ldots r_{667} = 667(3) + \dfrac{1}{222}(1+2+3+\ldots+666).$

Recall, from algebra, that the sum of the first n numbers is $\dfrac{n(n+1)}{2}$. Thus, $1+2+3+\ldots 666$ equals $\dfrac{666(667)}{2}$. Substituting the result in (vi), we obtain:

(vii) $r_1 + r_2 + \ldots + r_{667} = 667(3) + \dfrac{1}{222}\dfrac{666(667)}{2}$

(viii) $r_1 + r_2 + \ldots + r_{667} = 667(3) + \dfrac{3}{2}(667) = 667(3 + \dfrac{3}{2})$

$$= 667 \left(\frac{9}{2}\right).$$

Substituting this in eq. (ii), we find the total length of the groove:

(ix)   Length of groove = $2\pi \cdot 667\left(\frac{9}{2}\right) = 9(667)\pi$.

The length is approximately 18,859 inches. To convert to feet, remember that 1 in = $\frac{1}{12}$ ft. Thus,

(x)   Length of groove = 18,859 in = $18,859\left(\frac{1}{12} \text{ ft}\right) =$ 1,571.6 ft.

To convert to miles, remember 1 ft = $\frac{1}{5280}$ miles.

(xi)   Length of groove = 1571.6 ft = $1571.6\left(\frac{1}{5280} \text{ miles}\right)$ = .3 miles.

If the grooves of one side of the long playing record are stretched out, they would reach .3 miles. (If every groove of every one of the 400 million long playing records sold by the Beatles were laid down end to end, it would reach from the earth to the moon 500 times.)

● **PROBLEM** 293

In the accompanying figure, the Apollo 11 rocket is launched making an angle of θ with the moon directly overhead. Given that the rocket travels at 25000 mph and the moon completes a circular orbit of radius 240,000 miles every 30 days, at what angle θ should the rocket be fired so that the rocket will intercept the moon?

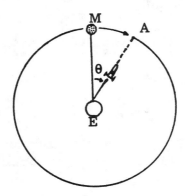

Solution: At the time of the launch, the moon is at position M, the rocket at point E. If they are to intersect at point A, the time it takes for the moon to travel to point A must equal the time it takes for the rocket to reach A.

To find the time required for the rocket to reach point A, remember that distance = rate × time and therefore time = $\frac{distance}{rate}$.

(i)   Time of rocket = $\frac{AE}{speed}$.

We are given that the moon's orbit is a circular orbit of radius 240,000 miles. Thus, radius AE = 240,000 miles. Also, we know that the speed of the rocket equals 25,000 mph.

(ii)   Time of rocket = $\frac{240,000 \text{ mi}}{25,000 \text{ mi/hr}}$ = 9.6 hr.

In 9.6 hr., the moon must travel an arc of θ degrees. We are given that the moon completes a revolution, or travels an arc of 360°, in 30 days. Given that the moon travels at constant speed, we can set up the following proportion:

(iii) $\frac{\theta}{360} = \frac{9.6 \text{ hr.}}{30 \text{ days}}$

To be consistent in our measurements, we convert the 30 days into hours. Note: 1 day = 24 hrs. Therefore, 30 days = 30(1 day) = 30(24 hrs.) = 720 hrs.

(iv) $\frac{\theta}{360} = \frac{9.6 \text{ hrs.}}{720 \text{ hrs.}}$.

Multiplying both sides by 360, we obtain:

(v) $\theta = 360(\frac{9.6 \text{ hrs.}}{720 \text{ hrs.}}) = 4.8°$.

Thus, the rocket must be launched at an angle of 4.8° with the moon in order to intercept it.

CHAPTER 16

INSCRIBED ANGLES AND ARCS

Basic Attacks and Strategies for Solving Problems in this Chapter. See pages 258 to 271 for step-by-step solutions to problems.

The previous chapter contained a discussion of central angles and related intercepted arcs. A brief review of the introduction to that chapter might be helpful in order to understand the problems in this chapter.

An arc of a circle can be measured two ways, in standard linear units (inches, feet, centimeters, meters, etc.) or in degrees. For example, in the picure below, A and B are on the circle and O is the center of the circle. Then $\overset{\frown}{AB}$ is used to denote the arc intercepted by $\angle AOB$ and $m\overset{\frown}{AB}$ is used to denote the degree measure of AB. Since $\angle AOB$ is a right angle, $m\overset{\frown}{AB} = 90°$.

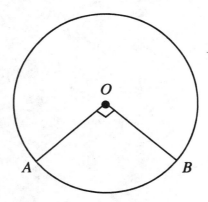

An inscribed angle is an angle whose vertex is on the circle and whose sides contain chords of the circle. In the figure on the next page, $\angle ACB$ is an inscribed angle. Also, $\overset{\frown}{AB}$ is the arc intercepted by that angle.

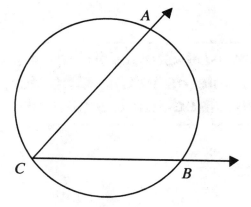

Most of the problems in this chapter involve finding and using the relationship between the measure of the inscribed angle and the measure of the intercepted arc.

The last problem in this chapter (Problem 308) is particularly important. In this case, a relationship is established between the measure of the angle formed by two chords and various intercepted arcs.

Step-by-Step Solutions to Problems in this Chapter, "Inscribed Angles and Arcs"

● **PROBLEM** 294

In the figure, a) is ∢B inscribed in circle C? Why? b) is ∢D? Why? and c) is ∢P? Why?

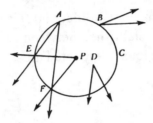

Solution: An angle is inscribed in a circle if its vertex lies on the circle, and if its sides contain chords of the circle.

a) Here, the vertex of B is a point of the circle, but its sides do not contain chords of the circle. ∢B is not inscribed in the circle.

b) ∢D is not inscribed in the circle because its vertex does not lie on the circle.

c) ∢P is not inscribed in the circle because its vertex does not lie on the circle nor do its sides contain chords of the circle.

● **PROBLEM** 295

In the circle shown, if m$\overarc{PR}$ = 70 and m$\overarc{QR}$ = 80, find m∢P, m∢Q, and m∢R.

Solution: By definition, the three angles whose measure we are asked to determine are inscribed angles. We will prove the theorem which states that the measure of an inscribed angle is equal to one-half the measure of the intercepted

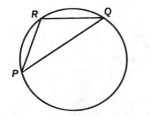

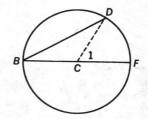

Figure 1   Figure 2

arc. This theorem can be proved in three cases. One where
the center of the circle lies on a side of the angle, one
where it lies in the interior of the angle, and one where it
is assumed to be on the exterior of the angle.

Case 1: We have a circle with C as the center and ∢DBF inscribed so that $\overline{BF}$ is a diameter (figure 2). We draw radius
$\overline{CD}$. Since they are radii, $\overline{BC} \cong \overline{CD}$. In △BCD, ∢B ≅ ∢D because when a triangle has two sides congruent, the angles op-
posite those sides are congruent. ∢1 and ∢C are, by definition, supplementary. Therefore, m∢1 + m∢C = 180°. Since they
are the three angles of △BCD,

$$m∢D + m∢B + m∢C = 180.$$

Therefore, by transitivity,

$$m∢D + m∢B = m∢1.$$

Since ∢1 is a central angle, $m∢1 = \overset{\frown}{mDF}$ and

$$2(m∢B) = \overset{\frown}{mDF} \text{ and}$$

$$m∢B = \frac{1}{2} \overset{\frown}{mDF}.$$

Case 1 is now established and will be instrumental in proving
Cases 2 and 3.

Case 2: Now C is in the interior of ∢DBF (figure 3). If we
draw diameter $\overline{AB}$, as shown, then m∢DBF = m∢1 + m∢2. Angles
1 and 2 are both inscribed angles with one side of each containing the center of the circle and, as such, their measures
can be found by referring to case 1.

$$m∢1 = \frac{1}{2}\overset{\frown}{mFA} \text{ and } m∢2 = \frac{1}{2} \overset{\frown}{mAD}$$

$$m∢1 + m∢2 = \frac{1}{2} \overset{\frown}{mFA} + \frac{1}{2} \overset{\frown}{mAD}$$

$$= \frac{1}{2} (\overset{\frown}{mFA} + \overset{\frown}{mAD}).$$

As the figure shows, $\overset{\frown}{FA}$ and $\overset{\frown}{AD}$ compose $\overset{\frown}{FD}$.

Therefore, $\overset{\frown}{mFA} + \overset{\frown}{mAD} = \overset{\frown}{mFD}$.

$$m∢1 + m∢2 = \frac{1}{2} \overset{\frown}{mFD}.$$

By substitution, then, m∢DBF = $\frac{1}{2}$ $\overset{\frown}{mFD}$.

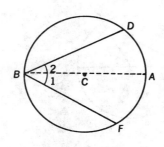

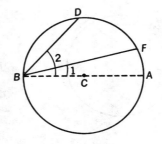

Figure 3   Figure 4

Case 3:   Here, C is in the exterior of the inscribed angle
(figure 4), but a technique similar to case 2 is used. Introduce the diameter to from an angle like that covered by
case 1.

$$m\angle DBF = m\angle 2 - m\angle 1 .$$

But, by the case 1 result, since $\angle 2$ and $\angle 1$ both have one side
that includes the circle's center,

$$m\angle 2 = \frac{1}{2}m\widehat{DA} \quad \text{and} \quad m\angle 1 = \frac{1}{2}m\widehat{FA}.$$

By substitution, $m\angle DBF = \frac{1}{2}m\widehat{DA} - \frac{1}{2}m\widehat{FA}$

$$= \frac{1}{2}(m\widehat{DA} - m\widehat{FA}).$$

Since $\widehat{DA}$ is composed of $\widehat{DF}$ and $\widehat{FA}$, the $m\widehat{DA} - m\widehat{FA} = m\widehat{DF}$.
By substitution, $m\angle DBF = \frac{1}{2}m\widehat{DF}$.
Now we can use the theorem that states if an angle is inscribed, its measure is equal to $\frac{1}{2}$ the measure of the intercepted
arc, to answer the original question.

We were told that $m\widehat{PR} = 70$ and $m\widehat{QR} = 80$. Therefore,
since a circle contains a measure of 360, $m\widehat{PQ} = 360 - (70+80)$
or $m\widehat{PQ} = 210$. $\angle P$, $\angle Q$, and $\angle R$ are all inscribed angles. This
allows us to find their measure by using the theorem we have
just proved. $\angle P$ intercepts $\widehat{QR}$, $\angle Q$ intercepts $\widehat{PR}$, and $\angle R$ intercepts $\widehat{PQ}$.

Therefore, $m\angle P = \frac{1}{2}m\widehat{QR} = \frac{1}{2}(80) = 40$

$$m\angle Q = \frac{1}{2}m\widehat{PR} = \frac{1}{2}(70) = 35$$

$$m\angle R = \frac{1}{2}m\widehat{PQ} = \frac{1}{2}(210) = 105.$$

● **PROBLEM** 296

Let $\angle A$ be inscribed in a circle, and let $m\angle A < 90°$. Let
$\angle P$ be the angle, with vertex at the center of the circle,
which intercepts the same arc as $\angle A$. (Then $\angle A$ and $\angle P$ are
said to be related). Prove: $m\angle A = \frac{1}{2}m\angle P$.

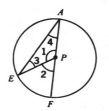 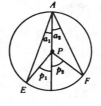 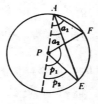

Figure (a)   Figure (b)   Figure (c)

<u>Solution:</u> For convenience, we consider the 3 possible sit-
uations corresponding to this theorem separately, as shown
in figures (a) through (c).

In figure (a), one side of the inscribed ⊰A contains the
center of the circle, P. Since ⊰2 is exterior to △AEP, we
can find a relationship between m⊰2 and m⊰3 + m⊰4. Then,
using the fact that $\overline{AP}$ and $\overline{EP}$ are radii of the same circle,
We will then know m⊰A in terms of m⊰P.

Now, ⊰2 is an exterior angle of △EPA. ⊰2 is supplementary to ⊰EPA as is ⊰3 + ⊰4. Therefore m⊰2 + m⊰EPA = 180 =
m⊰3 + m⊰4 + m⊰EPA,

Hence, m⊰2 = m⊰3 + m⊰4.   (1)

As was said above, EP = PA since both $\overline{EP}$ and $\overline{PA}$ are radii of
the same circle. This means that △EPA is an isosceles tri-
angle. Hence, by definition, m⊰3 = m⊰4. Using this in (1)
yields,

$$m⊰2 = 2m⊰4,$$

which yields,

$$m⊰4 = \frac{1}{2}m⊰2.$$

noting that m⊰A = m⊰4, and that m⊰P = m⊰2, hence,

$$m⊰A = \frac{1}{2}m⊰P$$

Figure (b) shows the case where the center of the circle,
P, is in the interior of ⊰A. Noting that the angles to the
left of $\overrightarrow{AP}$ satisfy the conditions for the case we just proved
(and similarly for the angle to the right of $\overrightarrow{AP}$), we may
write

$$m⊰a_1 = \frac{1}{2}m⊰p_1$$
$$m⊰a_2 = \frac{1}{2}m⊰p_2$$ (2)

or $$m⊰p_1 = 2m⊰a_1$$

$$m⊰p_2 = 2m⊰a_2$$ (3)

Figure (b) shows that

$$m⊰P = m⊰p_1 + m⊰p_2$$ (4)

Hence, using (3) in (4) yields

$$m\sphericalangle P = 2m\sphericalangle a_1 + 2m\sphericalangle a_2 = 2(m\sphericalangle a_1 + m\sphericalangle a_2).$$

But, the figure also shows that

$$m\sphericalangle A = m\sphericalangle a_1 + m\sphericalangle a_2.$$

Combining the last 2 equations, we obtain

$$m\sphericalangle P = 2m\sphericalangle A$$

or $$m\sphericalangle A = \frac{1}{2}m\sphericalangle P.$$

The last case is shown in figure (c). Here, the center of the circle is in the exterior of $\sphericalangle A$. We may use the same procedure as in parts (a) and (b), and note from figure (b) that

$$m\sphericalangle a_1 = \frac{1}{2}m\sphericalangle p_1$$

$$m\sphericalangle a_2 = \frac{1}{2}m\sphericalangle p_2.$$

(5)

But, from figure (c),

$$m\sphericalangle A = m\sphericalangle a_1 - m\sphericalangle a_2$$

$$m\sphericalangle P = m\sphericalangle p_1 - m\sphericalangle p_2.$$

(6)

Subtracting the equations listed in (5) and using (6), we obtain

$$m\sphericalangle a_1 - m\sphericalangle a_2 = \frac{1}{2}(m\sphericalangle p_1 - m\sphericalangle p_2)$$

or $$m\sphericalangle A = \frac{1}{2}m\sphericalangle P.$$

● PROBLEM 297

In the circle shown, if $m\sphericalangle P = 50$, and $m\sphericalangle R = 60$, find $m\overarc{PR}$, $m\overarc{QR}$, and $m\overarc{PQ}$.

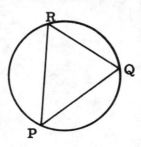

Solution: We will determine the arc measures by applying a

theorem already proved, which states that an inscribed angle has a measure equal to one-half the measure of the intercepted arc.

We are given the measures of two of the angles of $\triangle PRQ$. We can find the third by recalling that the measures of the angles of a triangle sum to 180.

$$m\sphericalangle Q + m\sphericalangle R = m\sphericalangle P + 180.$$

By sutstituting the given,

$$m\sphericalangle Q + 60 + 50 = 180$$

Accordingly,

$$m\sphericalangle Q = 70.$$

Each angle of $\triangle PRQ$ intercepts one of the arcs whose measure we wish to find. $\sphericalangle P$ intercepts $\overarc{QR}$, $\sphericalangle R$ intercepts $\overarc{PQ}$, and $\sphericalangle Q$ intercepts $\overarc{PR}$.

Therefore, by the theorem cited,

$$m\sphericalangle P = \tfrac{1}{2}m\overarc{QR}. \qquad\qquad m\sphericalangle P = 50.$$

By substitution, $50 = \tfrac{1}{2}m\overarc{QR}$. Therefore, $m\overarc{QR} = 100$.

$$m\sphericalangle R = \tfrac{1}{2}m\overarc{PQ} \qquad\qquad m\sphericalangle R = 60.$$

By substitution, $60 = \tfrac{1}{2}m\overarc{PQ}$. Therefore, $m\overarc{PQ} = 120$.

$$m\sphericalangle Q = \tfrac{1}{2}m\overarc{PR} \qquad\qquad m\sphericalangle Q = 70.$$

By substitution, $70 = \tfrac{1}{2}m\overarc{PR}$. Therefore, $m\overarc{PR} = 140$.

● **PROBLEM** 298

Prove that inscribed angles which intercept the same arc are congruent.

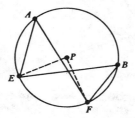

Solution: We must show that $\sphericalangle EAF \cong \sphericalangle EBF$. Construct the central angle, EPF, which intercepts the same arc ($\overarc{EF}$) as

do angles EAF and EBF. Then, by definition, ∢EAF and ∢EPF
are related and ∢EBF and∢EPF are related. Hence,

$$m\angle EAF = \frac{1}{2}m\angle EPF$$

(1)

$$m\angle EBF = \frac{1}{2}m\angle EPF$$

Using equations (1),

$$m\angle EAF = m\angle EBF$$

or   ∢EAF ≅ ∢EBF

● **PROBLEM** 299

In circle O, the measure of angle AOB is 80°. If point
C, as shown in the diagram, is on the minor arc ⌢AB, find the
measure of ∢ACB.

Solution: Since minor arc ⌢AB is intercepted by central angle
AOB, by definition, the measure of ⌢AB is also 80°. Knowing
that an entire circle measures 360°, and using the fact that
the whole must equal the sum of its parts, arc ⌢AMB must measure 360° - 80° or 280°. ∢ACB is inscribed in circle O and
intercepts ⌢AMB. By a theorem, the measure of ∢ACB must be
one-half the measure of its intercepted arc, ⌢AMB. Sine ⌢AMB
measures 280°, we conclude that ∢ACB measures 140°.

● **PROBLEM** 300

In the figure, let m∢C = 100°. What are the measures
of ∢a and ∢b? Given that ∢C is a central angle?

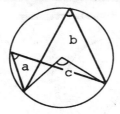

Solution: To solve this problem, we use the fact that ∢C
is related to both ∢a and ∢b, since the former is a central
angle of a circle, and subtends the same arc as the inscribed angles, a and b. Knowing this, we may write

$$m\angle b = \frac{1}{2}m\angle C = (\frac{1}{2})100° = 50°$$

$$m\angle a = \frac{1}{2}m\angle C = (\frac{1}{2})100° = 50°.$$

● **PROBLEM** 11

In a circle, as shown in the accompanying figure, congruent chords $\overline{AB}$ and $\overline{CD}$ are extended through B and D, respectively, until they intersect at P. Prove that triangle APC is an isosceles triangle.

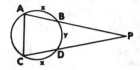

Solution: To show that $\triangle APC$ is isosceles, we must show $\overline{PA} \cong \overline{PC}$. This can be done by showing $\angle A \cong \angle C$.

The proof of this last point will involve using the theorem which states that congruent chords have arcs of equal measure. This, along with the reflexive property of arc measure, will show that $\angle A$ and $\angle C$ are inscribed in arcs of equal measure, which implies the necessary congruence condition.

Statement	Reason
1. Chord $\overline{AB} \cong$ Chord $\overline{CD}$.	1. Given.
2. $m\overparen{BD} = m\overparen{BD} = y$	2. Reflexive Property of Equality.
3. $m\overparen{AB} = m\overparen{CD} = x$	3. In a circle, congruent chords have equal arcs.
4. $m\overparen{ABD} = x + y$ and $m\overparen{BDC} = x + y$	4. The measure of an arc is equal to the sum of the measures of its parts.
5. $m\overparen{ABD} = m\overparen{BDC}$	5. Transitive Property of Equality
6. $\angle A \cong \angle C$	6. In a circle, if inscribed angles intercept equal arcs, the angles are congruent.
7. $\overline{PA} \cong \overline{PC}$	7. In a triangle, if two angles are congruent, the sides opposite these angles are congruent.
8. $\triangle APC$ is an isosceles triangle	8. An isosceles triangle is a triangle that has two congruent sides.

In circle O, $\overline{BD}$ is a diameter, $\overline{AB}$ and $\overline{BC}$ are chords, and AB > BC. Prove that m⊀ABD < m⊀CBD.

Solution: Since ⊀ABD and ⊀CBD are inscribed angles, their measures are equal to one-half the measure of the intercepted arcs. Therefore, the inequality can be proven by showing that intercepted arc $\overset{\frown}{AD}$ < intercepted $\overset{\frown}{CD}$.

We are given that $\overline{BD}$ is a diameter. As such, it divides the circle into two congruent arcs: m$\overset{\frown}{BAD}$ = m$\overset{\frown}{BCD}$.

In the same circle, unequal chords intercept unequal minor arcs, and the greater chord intercepts the greater minor arc. Since AB > BC, m$\overset{\frown}{AB}$ > m$\overset{\frown}{BC}$.

When unequal quantities are subtracted from equal quantities, the results are unequal in the opposite sense to the original inequality. Therefore, m$\overset{\frown}{BAD}$ − m$\overset{\frown}{AB}$ < m$\overset{\frown}{BCD}$ − m$\overset{\frown}{BC}$, or m$\overset{\frown}{AD}$ < m$\overset{\frown}{DC}$.

Halves of unequal quantities are unequal in the same sense. Therefore, since m⊀ABD = $\frac{1}{2}$m$\overset{\frown}{AD}$ and m⊀CBD = $\frac{1}{2}$m$\overset{\frown}{DC}$, m⊀ABD < m⊀CBD.

Let m⊀A = 90° in ΔABC. Let D, E, and F be the midpoints of $\overline{AB}$, $\overline{AC}$, and $\overline{BC}$, respectively. Prove that F is the center of a semicircle which contains B, A, and C.

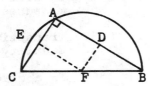

Solution: The figure shows ΔABC and part of the circle on which A, B, and C lie. To show that F is the center of the circle, we will prove that $\overline{FE}$ and $\overline{FD}$ are perpendicular bisectors of chords $\overline{CA}$ and $\overline{AB}$. Since the perpendicular bisectors of the chords of a circle meet at the center of the circle, we will be able to show that F is the center of the circle passing through A, B, and C. Since an angle inscribed in a semi-circle is a right angle, and ⊀A is a right angle inscribed in a circle, we will be able to conclude that $\overset{\frown}{CAB}$ is a semi-circle, and F is its center.

First, note that E and D are the midpoints of $\overline{CA}$ and $\overline{AB}$, as the problem states. Hence, $\overline{EF} \parallel \overline{AB}$ and $\overline{DF} \parallel \overline{AC}$. Since the corresponding angles of 2 parallel lines cut by a transversal are congruent, $\measuredangle CEF \cong \measuredangle CAD$ and $\measuredangle BDF \cong \measuredangle BAC$. Therefore, $\overline{FE} \perp \overline{CA}$ and $\overline{FD} \perp \overline{AB}$. Because E bisects $\overline{CA}$ and D bisects $\overline{AB}$, FE and FD are the perpendicular bisectors of chords $\overline{CA}$ and $\overline{AB}$. As explained above, the perpendicular bisectors of the chords of a circle intersect at the center of the circle. Hence, F is the center of the circle containing points A, B, and C. (See figure.) Furthermore, $\measuredangle A$ is inscribed in circle F and has measure 90°. Hence, $\measuredangle A$ must intercept a semi-circle i.e. CAB is a semi-circle, with center at F.

● **PROBLEM 304**

In the figure, let O be the center of the circle. What can be said about $\measuredangle 1$ and $\measuredangle 2$? About $\measuredangle 3$ and $\measuredangle 4$?

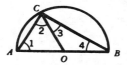

Solution: We will use the properties of circles and isosceles triangles to relate $\measuredangle 1$ and $\measuredangle 2$, and $\measuredangle 3$ and $\measuredangle 4$.

First, looking at the figure, note that $\overline{OA} = \overline{OC}$, and $\overline{OC} = \overline{OB}$, since all of these segments are radii of circle O. Hence, by definition, $\triangle OAC$ and $\triangle OBC$ are isosceles triangles. The base angles of $\triangle OAC$ are $\measuredangle 1$ and $\measuredangle 2$, and the base angles of $\triangle OBC$ are $\measuredangle 3$ and $\measuredangle 4$. Because the base angles of an isosceles triangle are congruent, $\measuredangle 1 \cong \measuredangle 2$ and $\measuredangle 3 \cong \measuredangle 4$.

● **PROBLEM 305**

Let $\measuredangle B$ be an angle inscribed in a circle and let it have measure greater than 90°. (See figure.) Prove that

$$m\measuredangle B = 180° - \frac{1}{2}m\measuredangle P.$$

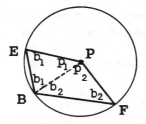

Solution: We use the fact that $\overline{PB} \cong \overline{PE} \cong \overline{PF}$, since all are radii of the same circle. This tells us that $\triangle PEB$ and $\triangle PBF$ are isosceles. Hence, by definition, $m\measuredangle BEP = m\measuredangle EBP = m\measuredangle b$, and $m\sphericalangle BFP = m\sphericalangle PBF = m\sphericalangle b_2$. Now, applying the fact that the sum of the measures of the angles of a triangle is 180° to triangles PEB and PBF, we may write

$$m\sphericalangle BEP + m\sphericalangle p_1 + m\sphericalangle b_1 = 180°$$

$$(1)$$

$$m\sphericalangle BFP + m\sphericalangle p_2 + m\sphericalangle b_2 = 180°.$$

Due to the fact that both triangles are isosceles,

$$m\sphericalangle BEP = m\sphericalangle b_1$$

$$(2)$$

$$m\sphericalangle BFP = m\sphericalangle b_2$$

Using (2) in (1) yields

$$2m\sphericalangle b_1 + m\sphericalangle p_1 = 180°$$

$$(3)$$

$$2m\sphericalangle b_2 + m\sphericalangle p_2 = 180°$$

Furthermore, from the figure,

$$m\sphericalangle P = m\sphericalangle p_1 + m\sphericalangle p_2$$

$$(4)$$

$$m\sphericalangle B = m\sphericalangle b_1 + m\sphericalangle b_2$$

Adding the 2 equations in (3), and using (4), we obtain

$$2(m\sphericalangle b_1 + m\sphericalangle b_2) + (m\sphericalangle p_1 + m\sphericalangle p_2) = 360°$$

or $2m\sphericalangle B + m\sphericalangle P = 360°$

or $2m\sphericalangle B = 360° - m\sphericalangle P$

$$m\sphericalangle B = 180° - \frac{1}{2}m\sphericalangle P.$$

● **PROBLEM** 306

Given: Points A, B, C, and D are in ⊙P; $\overset{\frown}{AB} \cong \overset{\frown}{AD}$; $\overset{\frown}{BC} \cong \overset{\frown}{DC}$.
Prove: $\sphericalangle B \cong \sphericalangle D$.

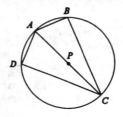

268

Solution: There are two ways of solving this problem. The first is to prove ΔADC ≅ ΔABC and then, ∢B ≅ ∢D follows by corresponding parts. The second method is to show ∢B and ∢D both intercept equal arcs.

Method 1: In the same circle, or in congruent circles, if two arcs are congruent, then their chords are congruent. Therefore, ÂD ≅ ÂB, which is given, implies $\overline{AD}$ ≅ $\overline{AB}$ and B̂C ≅ D̂C, which is given, implies $\overline{BC}$ ≅ $\overline{DC}$. ΔADC ≅ ΔABC follows by the SSS Postulate.

Statements	Reasons
1. Points A, B, C, and D are in ⊙P; ÂB ≅ ÂD; B̂C ≅ D̂C	1. Given.
2. $\overline{AB}$ ≅ $\overline{AD}$; $\overline{BC}$ ≅ $\overline{DC}$	2. In the same circle, congruent arcs intercept congruent chords.
3. $\overline{AC}$ ≅ $\overline{AC}$	3. Reflexivity Property.
4. ΔABC ≅ ΔADC	4. The SSS Postulate.
5. ∢B ≅ ∢D	5. Definition of congruenct triangles.

Method 2:

Statements	Reasons
1. Points A, B, C, and D in ⊙P; ÂB ≅ ÂD; B̂C ≅ D̂C	1. Given
2. mÂB + mB̂C = mÂD + mD̂C	2. Addition Postulate.
3. mÂB̂C = mÂD̂C	3. Substitution.
4. ÂB̂C ≅ ÂD̂C	4. Definition of Congruent Arcs.
5. ∢B ≅ ∢D	5. Inscribed angles that intercept congruent arcs are congruent.

● **PROBLEM** 307

Given, in the circle shown, that $\overline{BA}$ ⊥ $\overline{AD}$, $\overline{BC}$ ⊥ $\overline{CD}$ and $\overline{BD}$ bisects ∢ABC, prove ÂB ≅ B̂C.

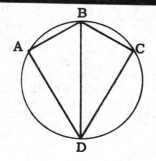

Figure 1

Solution: In this problem, we will employ the theorem that in the same circle congruent chords cut congruent arcs.

We can prove $\overset{\frown}{AB} \cong \overset{\frown}{BC}$ by proving $\overline{AB} \cong \overline{BC}$. We will do this by showing $\triangle ABD \cong \triangle CBD$ and applying the corresponding parts rule.

Statement	Reason
1. $\overline{BA} \perp \overline{AD}$ and $\overline{BC} \perp \overline{CD}$	1. Given.
2. ⦠BAD and ⦠BCD are right angles.	2. Perpendicular lines intersect to form right angles.
3. ⦠BAD ≅ ⦠BCD	3. All right angles are congruent.
4. $\overline{BD}$ bisects ⦠ABC.	4. Given.
5. ⦠ABD ≅ ⦠CBD	5. An angle bisector divides the angle into two congruent angles.
6. $\overline{BD} \cong \overline{BD}$	6. Reflexive property of congruence.
7. $\triangle ABD \cong \triangle CBD$.	7. A.A.S. = A.A.S.
8. $\overline{AB} \cong \overline{BC}$.	8. Corresponding sides of congruent triangles are congruent.
9. $\overset{\frown}{AB} \cong \overset{\frown}{BC}$.	9. Congruent chords intersect congruent arcs.

● **PROBLEM** 308

Given two intersecting chords of a circle, show that the measure of the angle formed by the intersection is one-half the sum of the measures of the arcs intercepted by the angle and its vertical angle.

Solution: We are required to relate ⦠AEM to $\overset{\frown}{AM}$ and $\overset{\frown}{NF}$. Here, by drawing $\overline{MF}$, we can express $\overset{\frown}{AM}$ and $\overset{\frown}{NF}$ in terms of inscribed angles ⦠NMF and ⦠AFM. ⦠NEF can also be expressed in terms of these two angles, since ⦠NEF is an exterior angle of $\triangle MEF$. Thus, we solve for ⦠NEF in terms of ⦠NMF and ⦠AFM. Then we solve for ⦠NMF in terms of $\overset{\frown}{NF}$ and for ⦠AFM in terms of $\overset{\frown}{AM}$. Substituting into the expression for ⦠NEF, we will then have an equation for ⦠NEF in terms of the intercepted arcs.

Given: $\overline{AF}$ and $\overline{MN}$ are chords of $\odot P$ that intersect at E.

Prove: $m\angle NEF = \frac{1}{2}(m\overset{\frown}{AM} + m\overset{\frown}{NF})$

Statements	Reasons
1. $\overline{AF}$ and $\overline{MN}$ are chords of $\odot P$ that intersect at E.	1. Given.
2. $m\angle AFM = \frac{1}{2}m\overset{\frown}{AM}$ $m\angle NMF = \frac{1}{2}m\overset{\frown}{NF}$	2. The measure of the inscribed angle is equal to one-half the measure of the intercepted arc.
3. $\angle NEF$ is an exterior angle of $\triangle MEF$	3. An exterior angle of a triangle is an angle that forms a linear pair with one of the interior angles of the angle.
4. $m\angle NEF = m\angle AFM + m\angle NMF$	4. The measure of the exterior angle of a triangle equals the sum of the measure of the two remote interior angles.
5. $m\angle NEF = \frac{1}{2}m\overset{\frown}{AM} + \frac{1}{2}m\overset{\frown}{NF}$ or $m\angle NEF = \frac{1}{2}(m\overset{\frown}{AM} + m\overset{\frown}{NF})$	5. Substitution Postulate (Steps 2 and 4)

CHAPTER 17

CHORDS

Basic Attacks and Strategies for Solving Problems in this Chapter. See pages 272 to 292 for step-by-step solutions to problems.

A **chord** is a segment with endpoints on the circle. For example, in the figure below, $\overline{AB}$ and $\overline{CD}$ are chords.

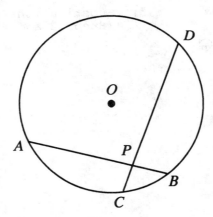

When two chords intersect as do $\overline{AB}$ and $\overline{CD}$, there is an interesting relationship among the lengths of the segments $\overline{AP}$, $\overline{PB}$, $\overline{CP}$, and $\overline{PD}$. This relationship, $\dfrac{AP}{DP} = \dfrac{CP}{BP}$, is established in Problem 312. Many of the other problems in this chapter involve application of this important theorem.

There are also several problems which involve the relationship between the lengths of two chords and their distance from the center of the circle. In general the distance from a line to a point is the length of a line segment which is perpendicular to the line, which has one endpoint on the line, and has the point itself as its other endpoint. In the following diagram, the distance from P to l is PQ:

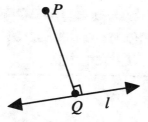

In the figure on the previous page, $\overline{AB}$ is farther from O (the center of the circle) than $\overline{CD}$. Problem 329 shows that this implies that $AB < CD$. On the other hand, if $\overline{AB}$ and $\overline{CD}$ were equally distant from O, then the chords would be congruent (Problem 327).

Below is an important theorem concerning the measure of an angle formed by two intersecting chords and various intercepted arcs.

For two intersecting chords, the measure of any angle formed is one-half the sum of the measures of the arcs formed by the angle and the corresponding vertical angle.

This important relationship is proved in Problem 308 of the previous chapter. Applying this theorem to the chords in the first figure, $m \sphericalangle APC = \frac{1}{2}(m\,\widehat{AC} + m\,\widehat{BD})$. Several of the problems in this chapter involve comparable application of this theorem.

● **PROBLEM** 309

If, from a point within a circle, more than two equal line segments can be drawn to the circumference, prove that such a point is the center of the circle.

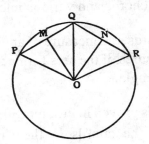

Solution: In the figure, O is given point. P,Q, and R are points on the circumference of the circle such that OP = OQ = OR. To show that point O is the center of the circle, we recall that the center is on the perpendicular bisectors of all chords of the circle. Thus, the perpendicular bisectors of any two chords must have the center of the circle as a common point. We will show that the perpendicular bisector of chords $\overline{PQ}$ and $\overline{QR}$ have only point O in common. Therefore, point O must be the center of the circle

We draw $\overline{PQ}$ and $\overline{QR}$. We are given OP = OQ = OR. Therefore, ΔOPQ and ΔQOR are isosceles.

Let M be the midpoint of $\overline{PQ}$ and N be the midpoint of $\overline{QR}$. From an earlier theorem, we know that the segment drawn from the vertex of an isosceles triangle to the midpoint of the base is perpendicular to the base. Therefore, $\overline{OM} \perp \overline{PQ}$ and $\overline{ON} \perp \overline{QR}$.

Since $\overline{OM}$ also intersects $\overline{PQ}$ at the midpoint, $\overline{OM}$ is the perpendicular bisector of chord $\overline{PQ}$. Similarily, $\overline{ON}$ is the perpendicular bisector of chord $\overline{QR}$.

Since point O is common to the perpendicular bisectors of two distinct chords of the circle, point O must be the center of the circle.

(Note: The following question may arise: how can we show O is the only common point? Remember that no three points on a circle can be collinear. Therefore, P, Q and R are not collinear, and $\overleftrightarrow{PQ}$ and $\overleftrightarrow{QR}$ are not coincident. Furthermore, $\overleftrightarrow{PQ}$ and $\overleftrightarrow{QR}$ share common point Q and thus cannot be parallel. Since $\overline{OM}$ and $\overline{ON}$ are perpendicular to lines that are neither coincident nor parallel, $\overleftrightarrow{OM}$ and $\overleftrightarrow{ON}$ cannot be parallel nor coincident. Thus, $\overleftrightarrow{OM}$ and $\overleftrightarrow{ON}$ can only intersect in one point.)

● **PROBLEM** 310

Show that if two chords of a circle bisect each other, the chords are diameters and the angles formed are central angles.

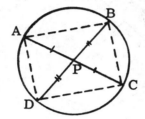

Solution: In the accompanying figure, chords $\overline{AC}$ and $\overline{DB}$ bisect each other and intersect at point P. We must show that $\overline{AC}$ and $\overline{DB}$ are diameters and that the angles are central angles.

Remember that a chord that divides the circle into two congruent arcs of 180° each is a diameter. To show that $\overline{AC}$ is a diameter, we show $\overset{\frown}{ABC} \cong \overset{\frown}{ADC}$.

To show $\overset{\frown}{ABC} \cong \overset{\frown}{ADC}$, we prove

(1) $\overline{AD} \cong \overline{BC}$. To see this, remember that $\overline{AC}$ and $\overline{BD}$ bisect each other. Thus $\overline{AP} \cong \overline{PC}$ and $\overline{DP} \cong \overline{PB}$. Furthermore, opposite angles ∢APD and ∢CPB are congruent. By the SAS Postulate, △APD ≅ △CPB. By corresponding parts, it follows that $\overline{AD} = \overline{BC}$.

(2) $\overline{AB} \cong \overline{DC}$. Using a procedure similar to (1), we prove △APB ≅ △DPC by the SAS Postulate. $\overline{AB} \cong \overline{DC}$ follows by corresponding parts.

(3) $\overset{\frown}{AD} \cong \overset{\frown}{BC}$ and $\overset{\frown}{AB} \cong \overset{\frown}{DC}$, since congruent chords intercept congruent arcs.

By the addition postulate, it follows that $m\overset{\frown}{AD} + m\overset{\frown}{CD} \cong m\overset{\frown}{BC} + m\overset{\frown}{AB}$. Since $m\overset{\frown}{ABC} \cong m\overset{\frown}{BC} + m\overset{\frown}{AB}$ and $m\overset{\frown}{ADC} \cong m\overset{\frown}{AD} + m\overset{\frown}{CD}$, we can conclude that $\overset{\frown}{ABC} \cong \overset{\frown}{ADC}$ and $\overline{AC}$ is a diameter.

From (3), we can also conclude that $m\overarc{AD} + m\overarc{AB} \cong m\overarc{DC} +$ $m\overarc{CB}$. Thus, $\overarc{DAB} \cong \overarc{DCB}$ and $\overline{DB}$ is also a diameter.

To show that the angles formed are central angles, remember that an angle is a central angle of a circle only if its vertex is the center of the circle. In this case, the vertex of each angle is point P, the intersection of diameters $\overline{AC}$ and $\overline{DB}$. Since diameters intersect only at the center of the circle, P is the center of the circle, and the angles formed are central angles.

● PROBLEM 311

> Prove that the line containing the midpoints of the major and minor arcs of a chord of a circle is the perpendicular bisector of the chord.

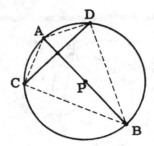

<u>Solution:</u> A perpendicular bisector is a line, all of whose points are equidistant from the endpoints of the given segment. We show that two points, the midpoints of the major and minor arcs, are equidistant from the endpoints of the chord and thus that the line determined by these points must be the perpendicular bisector. To show A and B, the midpoints of the arcs, are equidistant from chord endpoints C and D, we show $\overline{AC} \cong \overline{AD}$ and $\overline{CB} \cong \overline{DB}$. However, from the given $\overarc{AC} \cong \overarc{AD}$ and $\overarc{CB} \cong \overarc{BD}$, and the congruence of the segments follows.

Given: $\overline{CD}$ is a chord of ⊙P. A is the midpoint of the minor arc $\overarc{CD}$. B is the midpoint of major arc $\overarc{CD}$.

Prove: $\overleftrightarrow{AB}$ is the perpendicular bisector of $\overarc{CD}$.

Statements	Reasons
1. A is the midpoint of minor arc $\overarc{CD}$. B is the midpoint of major arc $\overarc{CD}$.	1. Given
2. $m\overarc{AC} = m\overarc{AD} = 1/2\ m\overarc{CD}$ $m\overarc{BC} = m\overarc{BD} = \frac{1}{2}m\overarc{CD}$	2. The midpoint of an arc divides the arc into two congruent arcs.

3. $\overline{AC} \cong \overline{AD}$; $\overline{CB} \cong \overline{BD}$	3. If two arcs are congruent, their chords are also congruent.
4. A is equidistant from C and D. B is equidistant from C and D.	4. Definition of equidistance.
5. $\overleftrightarrow{AB}$ is the perpendicular bisector of $\overline{BC}$.	5. If a line contains two points equidistant from the endpoints of the segment, then the line is the perpendicular bisector of the segment.

● **PROBLEM** 312

Two chords intersect in the interior of a circle, thus determining two segments in each chord. Show that the product of the length of the segments of one chord equals the product of the lengths of the segments of the other chord.

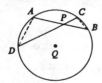

Solution: Restating the above problem, in terms of the figure, we obtain AP · PB = DP · PC. We can rewrite this (divide both sides by PB · DP) as $\frac{AP}{DP} = \frac{PC}{PB}$. To prove the proportionality, we show that two triangles are similar. In this case, ΔAPD ∿ ΔCPB by the A-A Similarity Theorem.

Given: Chords $\overline{AB}$ and $\overline{CD}$ of ⊙Q intersect at point P.
Prove: AP · PB = DP · PC.

Statements	Reasons
1. Chords $\overline{AB}$ and $\overline{CD}$ of ⊙Q intersect at point P.	1. Given.
2. ∡ADC ≅ ∡ABC.	2. Inscribed angles that cut the same arc are congruent.
3. ∡APD ≅ ∡CPB.	3. Vertical angles are congruent.
4. ΔAPD ∿ ΔCPB.	4. The A-A Similarity Theorem.

5. $\frac{AP}{PD} = \frac{PC}{PB}$

5. The sides of similar triangles are proportional.

6. $AP \cdot PB = PC \cdot PD$.

6. In a proportion, the product of the means must equal the product of the extremes.

● **PROBLEM** 313

In circle O, chord $\overline{CD}$ is bisected at S by chord $\overline{EF}$, as shown in the accompanying figure. If ES = 16 and SF = 4, find the length of chord $\overline{CD}$.

Solution: When two chords intersect, as in this problem, the product of the measures of the two segments of one chord equals the product of the measures of the segments of the other, i.e. CS × SD = ES × SF.

Since $\overline{EF}$ bisects $\overline{CD}$, $\overline{CS} \cong \overline{SD}$ and the measures of $\overline{CS}$ and $\overline{SD}$ are equal. Let x = the measure of $\overline{CS}$ and $\overline{SD}$. Now substitute the given. This gives us $x^2 = 16 \cdot 4 = 64$ or $x = 8$, which is $\frac{1}{2}$ the length of CD. Therefore, the length of chord $\overline{CD}$ is 2x = 16.

● **PROBLEM** 314

In circle O, chords $\overline{AB}$ and $\overline{CD}$ intersect at E, as shown in the diagram. If AE = 6, EB = 8, and CE = 12, find ED.

Solution: If two chords intersect inside a circle, the product of the measures of the segments of one chord is equal to the product of the measures of the segments of the other. We now see that AE × EB = CE × ED. Let ED = x. If we substitute the known data, we obtain

$$6 \times 8 = 12x$$
$$48 = 12x$$
$$4 = x$$

Therefore, the measure of $\overline{ED}$ is 4 units.

In circle O, the length of the chord of a minor arc is 8. The height of the arc, i.e. the length of the line segment joining the midpoint of the arc to the midpoint of its chord, is 2. Find the length of the radius of the circle. (See figure.

Solution: Since $\overline{CD}$ is the perpendicular bisector of $\overline{AB}$, $\overline{CD}$ will pass through the center of circle O, when it is extended, as shown in the accompanying diagram. $\overline{CE}$ will represent the diameter of circle O.

Chords AB and CE divide each other into segments AD, DB, CD, and DE. From an earlier theorem, we know that if two chords intersect inside a circle, the product of the measures of the segments of one chord is equal to the product of the measures of the segments of the other. Therefore, AD × DB = CD × DE. By substituting in the given information, we can determine the length of DE. AD = 4 and DB = 4 because of the given fact that AB, which measures 8, is bisected by chord $\overline{CE}$. Therefore, 4 × 4 = 2 × DE

$$16 = 2DE$$
$$8 = DE$$

The length of diameter CE = CD + DE = 2 + 8 = 10. Therefore, the length of radius $\overline{OE}$ (or any other radius) is $\frac{1}{2}$CE or 5 units.

In circle O, chords $\overline{AB}$ and $\overline{CD}$ intersect at E. If the measure of ∢CEB equals 50° and the measure of $\overset{\frown}{CB}$ equals 40°, find the measure of minor arc $\overset{\frown}{AD}$.

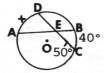

Solution: The measure of an angle formed by two chords intersecting within a circle is equal to one-half the sum of the measures of the intercepted arcs. From the figure, it can be seen that arcs $\overset{\frown}{AD}$ and $\overset{\frown}{BC}$ are intercepted by chords $\overline{AB}$ and $\overline{CD}$. Therefore, m∢CEB = $\frac{1}{2}$(m$\overset{\frown}{BC}$ + m$\overset{\frown}{AD}$). If we let x = the measure of minor arc $\overset{\frown}{AD}$, then, by substitution,

$$50° = \frac{1}{2}(40 + x)$$
$$100° = 40 + x$$
$$60° = x$$

Therefore, the measure of minor arc $\overset{\frown}{AD}$ is 60 .

● **PROBLEM** 317

In the figure below, if m$\overset{\frown}{BN}$ = 50 and m∢OFG = 35, find m$\overset{\frown}{OG}$.

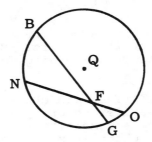

Solution: We know that given two intersecting chords in a circle, the measure of the angle formed by the intersection is equal to one-half the sum of the measures of the arcs intercepted by the angle and its vertical angle. Here, $\overset{\frown}{BN}$ and $\overset{\frown}{OG}$ are the intercepted arcs of ∢OFG and its vertical angle. Therefore,

(i)   m∢OFG = $\frac{1}{2}$(m$\overset{\frown}{BN}$ + m$\overset{\frown}{OG}$).

Substituting in the values m$\overset{\frown}{BN}$ = 50 and m∢OFG = 35, we obtain:

(ii)   35 = $\frac{1}{2}$(50 + m$\overset{\frown}{OG}$)

(iii)   (2)35 = 50 + m$\overset{\frown}{OG}$

(iv)   m$\overset{\frown}{OG}$ = (2)35 – 50 = 70 – 50 = 20

● **PROBLEM** 318

In circle 0, $\overline{BD} > \overline{AC}$. Prove that $\overline{CD} > \overline{AB}$.

Solution: In a circle, if two chords are unequal, then their minor arcs are unequal in measure and the greater chord has the greater minor arc.

Since $\overline{BD} > \overline{AC}$, m $\overset{\frown}{BCD}$ > m $\overset{\frown}{ABC}$. By reflexivity, $\overset{\frown}{BC} \cong \overset{\frown}{BC}$.

Hence, m $\overset{\frown}{BCD}$ - m $\overset{\frown}{BC}$ > m $\overset{\frown}{ABC}$ - m $\overset{\frown}{BC}$ or m $\overset{\frown}{CD}$ > $\overset{\frown}{AB}$. (This follows due to the fact when equal quantities are subtracted from unequal quantities, the results are unequal in the same order.)

If minor arcs of the same circle are unequal, then their chords are unequal and the greater minor arc has the greatest chord. Hence, since m $\overset{\frown}{CD}$ > m $\overset{\frown}{AB}$, $\overline{CD} > \overline{AB}$.

● **PROBLEM** 319

In the accompanying figure, diameter $\overline{BC}$ is drawn in circle O. Radius $\overline{OE}$ is drawn parallel to chord $\overline{CD}$. Prove that the measure of arc BE is equal to the measure of arc ED, i.e. $\overset{\frown}{BE} \cong \overset{\frown}{ED}$.

Solution: We know that in any given circle, arcs of equal measure have central angles of equal measure. Radius $\overline{OD}$, when drawn, becomes the transversal of parallel segments $\overline{OE}$ and $\overline{CD}$. Diameter $\overline{CB}$ is also a transversal of these lines. If we prove m⊀EOD $\cong$ m⊀BOE, then we have shown that the arcs which these two central angles intercept are also equal.

Statement	Reason
1. Draw radius $\overline{OD}$	1. One and only one straight line may be drawn through two points.
2. $\overline{OE} \parallel \overline{CD}$.	2. Given.
3. m⊀ EOD = m⊀CDO.	3. If two parallel lines are cut by a transversal, the alternate interior angles are congruent and of equal measure.
4. $\overline{OD} \cong \overline{OC}$	4. Radii of a circle are congruent.
5. m⊀BOE = m⊀OCD.	5. If two parallel lines are cut by a transversal, the corresponding angles are congruent and of equal measure.

279

6. m∢OCD = m∢CDO.

6. If two sides of a triangle are congruent, the angles opposite these sides are congruent and of equal measure.

7. m∢BOE = m∢CDO.

7. Transitive property of equality.

8. m∢BOE ≅ m∢EOD.

8. Same as reason 7.

9. $\overarc{BE}$ ≅ $\overarc{ED}$.

9. In a circle, central angles whose measures are equal intercept congruent arcs.

● **PROBLEM** 320

An arch is built in the form of an arc of a circle and is subtended by a chord 30 ft. long. If a chord 17 ft. long subtends half that arc, what is the radius of the circle?

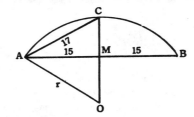

Solution: Since we do not know the radius, we cannot draw the arch. However, let us sketch an arch, as in the accompanying figure, to help us with our reasoning.

$\overline{AB}$ is 30 ft. long and $\overline{AC}$ is 17 ft. long. C is the midpoint of arc AB. Let O be the center of the circle. The line between the midpoint of an arc and the center of the circle will be the perpendicular bisector of the chord subtending that arc. Hence $\overline{OC} \perp \overline{AB}$ and AM = MB = 15. We wish to find the length of radius $\overline{OC}$ (or OA).

△ACM and △AMO are both right triangles and we can apply the Pythagorean Theorem to obtain the desired results. We will determine CM, represent OM in terms of this, and then find OA, the length of the radius.

In △ACM, $(AC)^2 = (CM)^2 + (AM)^2$

$$(17)^2 = (CM)^2 + (15)^2$$

$$289 = (CM)^2 + 225$$

$$64 = (CM)^2$$

$$8 = CM$$

Since OC = OA (they are both radii),

$$OM = OA - 8$$

In $\triangle AMO$, $(OA)^2 = (15)^2 + (OA-8)^2$

$$(OA)^2 = 225 + (OA)^2 - 16(OA) + 64$$

$$16(OA) = 289$$

$$OA = 18 \text{ ft. } \frac{3}{4} \text{ in.}$$

Hence, the radius of the circle is 18 ft. $\frac{3}{4}$ in.

● **PROBLEM** 321

Given: $\overline{AB}$ and $\overline{CD}$ are chords of ⊙P; $\overline{AB} \cong \overline{CD}$. Prove: $\overset{\frown}{AC} \cong \overset{\frown}{BD}$.

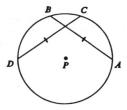

Solution: In the same circle or congruent circles, equal chords intercept equal arcs. Since $\overline{AB} \cong \overline{CD}$, $\overset{\frown}{AB} \cong \overset{\frown}{CD}$. If we now subtract $\overset{\frown}{BC}$ from both of them, the resulting arcs, $\overset{\frown}{AC}$ and $\overset{\frown}{BD}$, should be congruent.

Statements	Reasons
1. $\overline{AB}$ and $\overline{CD}$ are chords of ⊙P. $\overline{AB} \cong \overline{CD}$.	1. Given.
2. $\overset{\frown}{ACB} \cong \overset{\frown}{DBC}$.	2. Equal chords intercept equal arcs.
3. $m\overset{\frown}{ACB} = m\overset{\frown}{DBC}$.	3. Definition of congruent arcs.
4. $m\overset{\frown}{BC} = m\overset{\frown}{BC}$.	4. Reflexive Property.
5. $m\overset{\frown}{ACB} - m\overset{\frown}{BC} = m\overset{\frown}{DBC} - m\overset{\frown}{BC}$ or $m\overset{\frown}{AC} = m\overset{\frown}{BD}$.	5. If equals are subtracted from equals, the results are equal.
6. $\overset{\frown}{AC} \cong \overset{\frown}{BD}$.	6. Definition of congruent arcs.

In the circle shown, $\overline{BE} \perp \overline{AC}$ and B is the midpoint of arc $\overset{\frown}{AC}$. Prove $\overline{AE} \cong \overline{CE}$.

Solution: In this proof, we must use the theorem that congruent arcs have congruent chords. To show $\overline{AE} \cong \overline{CE}$, we first prove $\triangle ABE \cong \triangle CBE$. To show $\triangle ABE \cong \triangle CBE$, we use the Hypotenuse Leg Theorem. Right angles BEA and BEC are congruent. $\overline{BE} \cong \overline{BE}$. For the third congruence, we note that $\overline{AB}$ and $\overline{BC}$ are chords that intersect congruent arcs. From the above theorem, it follows that $\overline{AB} \cong \overline{BC}$.

Statement	Reason
1. $\overline{BE} \perp \overline{AC}$	1. Given.
2. ∢BEA and ∢BEC are right angles.	2. Perpendicular lines intersect to form right angles.
3. △ABE and △BCE are right triangles.	3. A triangle containing a right angle is a right triangle.
4. B is the midpoint of $\overset{\frown}{AC}$.	4. Given.
5. $\overset{\frown}{AB} \cong \overset{\frown}{BC}$.	5. A midpoint of an arc divides the arc into two congruent sections.
6. $\overline{AB} \cong \overline{BC}$	6. Congruent arcs have congruent chords.
7. $\overline{BE} \cong \overline{BE}$	7. Reflexive property of congruence.
8. $\triangle ABE \cong \triangle CBE$.	8. hy. leg $\cong$ hy. leg.
9. $\overline{AE} \cong \overline{CE}$.	9. Corresponding sides of congruent triangles are congruent.

Given: Two concentric circles with center O; chord $\overline{BC}$ is a subset of chord $\overline{AD}$. Prove: $\overline{AB} \cong \overline{CD}$. (Hint: Draw a perpendicular from O to $\overline{BC}$.)

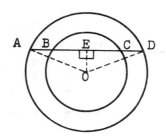

Solution: Employing the hint, we draw $\overline{OE} \perp \overline{BC}$. Since $\overline{OE}$ is perpendicular to a chord and passes through the center of the circle, $\overline{OE}$ bisects $\overline{BC}$, i.e. $\overline{BE} \cong \overline{EC}$.

$\overline{AB}$ and $\overline{CD}$ can be represented as $\overline{AE} - \overline{BE}$ and $\overline{DE} - \overline{CE}$, respectively. Therefore, if we prove $\overline{AE} \cong \overline{DE}$, then $\overline{AB} \cong \overline{CD}$, since the differences between two congruent pairs is a congruent pair.

We can prove $\overline{AE} \cong \overline{DE}$ by showing $\triangle AEO \cong \triangle DEO$ and applying corresponding parts.

$\overline{OE} \cong \overline{OE}$ by the reflexive property of congruence. $\overline{OA} \cong \overline{OD}$ because they are radii of the same circle. Since perpendicular lines intersect at right angles, ∢OEA and ∢OED are right angles and $\triangle AEO$ and $\triangle DEO$ are right triangles. Therefore, $\triangle AEO \cong \triangle DEO$ by hy leg ≅ hy leg.

Hence, $\overline{AE} \cong \overline{DE}$, because they are corresponding parts.

Therefore, $\overline{AE} - \overline{BE} \cong \overline{DE} - \overline{CE}$ or $\overline{AB} \cong \overline{CD}$ by the reasons mentioned previously.

● **PROBLEM 324**

In circle O, as shown in the accompanying figure, diameter $\overline{AB}$ has been constructed. Chords $\overline{AC}$ and $\overline{DB}$ have been drawn in such a way that $\overline{AC} \parallel \overline{DB}$. Prove chord $\overline{AC} \cong$ chord $\overline{DB}$.

Solution: A theorem states that if two chords are equidistant from the origin, then they are congruent. Therefore, if we construct the perpendiculars from the origin to the two chords (these now represent the distances from the origin to the chords), and prove them to be congruent (or of equal length), then we can conclude that the chords are congruent.

This will be done by proving $\triangle OEA \cong \triangle OFB$ by the S.A.A. ≅ S.A.A. method. Then we apply the corresponding parts rule to show that $\overline{OE} \cong \overline{OF}$.

Statement	Reason
1. Draw $\overline{OE} \perp \overline{AC}$, $\overline{OF} \perp \overline{DB}$	1. From a given point outside a line, one and only one perpendicular can be drawn to the line.
2. $\sphericalangle OEA \cong \sphericalangle OFB$	2. All right angles are congruent.
3. $\overline{AC} \parallel \overline{DB}$	3. Given.
4. $\sphericalangle EAO \cong \sphericalangle FBO$	4. If two parallel lines are cut by a transversal, the alternate interior angles are congruent.
5. $\overline{OA} \cong \overline{OB}$	5. Radii of a circle are congruent.
6. $\triangle OEA \cong \triangle OFB$	6. S.A.A. $\cong$ S.A.A.
7. $\overline{OE} \cong \overline{OF}$	7. Corresponding parts of congruent triangles are congruent.
8. Chord $\overline{AC} \cong$ Chord $\overline{DB}$	8. In a circle, if two chords are equidistant from the center, they are congruent.

● **PROBLEM** 325

Given that $\overline{POM}$ is a diameter and A is any point on it, prove that AM is shorter than any other line from A to the circumference and that AP is longer than any other line from A to the circumference.

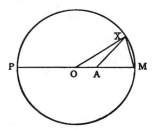

Solution: In the drawing, $\overline{AX}$ is any segment not on the diameter connecting A to the circumference. We must show $AX > AM$ and $AX < AP$. To show $AX > AM$, we find a triangle with sides $\overline{AX}$ and $\overline{AM}$, $\triangle$ AXM. From an earlier theorem, we know that if two angles of a triangle are unequal, the greater side is opposite the greater angle. We will show m $\sphericalangle$ AMX > m $\sphericalangle$ AXM. Thus, AX > AM.

To show m $\sphericalangle$ AMX > m $\sphericalangle$ AXM, consider $\triangle$ XOM. $\overline{OX}$ and $\overline{OM}$ are radii of $\odot O$. Therefore, $\overline{OX} \cong \overline{OM}$ and $\triangle$ XOM is isosceles. Thus, base angles $\sphericalangle$ OXM and $\sphericalangle$ AMX are congruent.

Since m ≯ AXM = m ≯ OXM - m ≯ OXA, m ≯ AXM < m ≯ OXM. By transitivity, m ≯ AXM < m ≯ AMX. Consequently, the side opposite ≯ AMX is longer than the side opposite ≯ AXM. Thus, any line segment connecting A to the circumference that is not on the diameter (such as $\overline{AX}$) is longer than $\overline{AM}$.

To show AP > AX, we divide $\overline{AP}$ into segments that can be compared with $\overline{AX}$. Consider △ OAX. Note that $\overline{OP}$ and $\overline{OX}$ are radii. Therefore, OP = OX. Since PA = OP + OA, $\overline{PA}$ is the sum of the lengths of two sides of the triangle. By the Triangle Inequality, we know the length of any one side of a triangle is less than the sum of the lengths of the other two. Thus, AX < OA + OX. Since OA + OX = OA + OP = PA, AX < PA.

● **PROBLEM** 326

Given: Point Q is in the interior of ⊙P; $\overrightarrow{QP}$ intersects ⊙P in R; S is any point on ⊙P.

Prove: RQ > SQ (Hint: Use $\overline{SP}$).

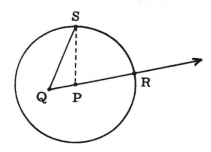

Solution: We must show RQ > SQ. Since RQ = RP + PQ, this is the same as proving RP + PQ > SQ.

To relate SQ and PQ, we use the Triangle Inequality Theorem - that is, in △ SPQ, SQ < QP + SP. Since $\overline{SP}$ and $\overline{PR}$ are radii, $\overline{SP} \cong \overline{PR}$ and QP + SP = QP + PR. The inequality then becomes SQ < QP + PR or SQ < RQ.

Statements	Reasons
1. (see problem statement)	1. Given
2. QR = QP + PR	2. Definition of betweenness.
3. SQ < QP + PS	3. Triangle Inequality Theorem.
4. SP = PR	4. All radii of a circle are congruent.
5. SQ < QP + PR	5. Substitution Postulate (Step 3,4)
6. SQ < QR	6. Substitution Postulate (Step 5,2)

● **PROBLEM** 327

Given: $\overline{AB}$ and $\overline{CD}$ are chords of circle P; $\overline{PM} \perp \overline{AB}$ at M; $\overline{PN} \perp \overline{CD}$ at N; $\overline{PM} \cong \overline{PN}$. Prove: $\overline{AB} \cong \overline{CD}$.

Solution: To prove congruence of lines, we must look for congruent triangles. Here, in drawing the radii $\overline{AP}$ and $\overline{DP}$,

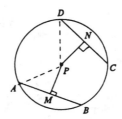

we make use of the fact that all radii of a given circle are congruent. After proving ΔAMP ≅ ΔDNP, we have $\overline{AM} \cong \overline{DN}$. We could then repeat the procedure, drawing radii $\overline{PC}$ and $\overline{PB}$, and prove $\overline{NC} \cong \overline{MB}$, but that is unnecessary. The line containing the center of the circle and perpendicular to any chord bisects that chord. Therefore, if $\overline{AM} \cong \overline{DN}$; $\overline{AB}$ (which equals 2·AM) must be congruent to $\overline{CD}$ (which equals 2·DN).

	Statements		Reason
1.	$\overline{AB}$ and $\overline{CD}$ are chords of circle P; $\overline{PM} \perp \overline{AB}$ at M; $\overline{PN} \perp \overline{CD}$ at N; $\overline{PM} \cong \overline{PN}$.	1.	Given
2.	AP = DP.	2.	All radii of a circle are equal in measure.
3.	ΔAMP ≅ ΔDNP.	3.	If the hypotenuse and a leg of a right triangle are congruent to the hypotenuse and leg of another right triangle, then the two triangles are congruent.
4.	AM = DN.	4.	Corresponding parts of congruent triangles are congruent.
5.	M is the midpoint of $\overline{AB}$; N is the midpoint of $\overline{CD}$.	5.	The line containing the center of the circle and perpendicular to the chord intersects the chord at its midpoint.
6.	AB = 2AM CD = 2DN	6.	Property of the midpoint.
7.	AB = CD.	7.	Multiplication Postulate and Substitution Postulate.
8.	$\overline{AB} \cong \overline{CD}$	8.	Two line segments are congruent if they are of the same length.

● **PROBLEM** 328

In the figure below, $\overline{QR} \perp \overline{EF}$ at R and $\overline{QS} \perp \overline{GH}$ at S. If RQ = 5 and QS = 4, is EF > GH?

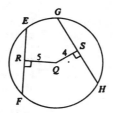

Solution: No. The closer a chord is to the center of the circle, the longer the chord becomes. In fact, the largest possible chord in the circle, the diameter, passes through the center. In general, in the same circle or in congruent circles, if two chords are not congruent, then the chord nearer the center of the circle is the longer of the two chords.

In this case, the distance RQ is greater than the distance QS, Therefore, EF < GH.

● **PROBLEM** 329

Show that in the same circle (or in congruent circles) if two chords are not congruent, then the longer chord is nearer the center of the circle than the shorter chord.

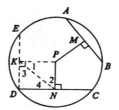

Solution: Consider chords $\overline{AB}$ and $\overline{CD}$ in the accompanying figure. AB < CD. We want to show PM > PN.

The inequality AB < CD could lead us to an inequality of angles, if $\overline{AB}$ and $\overline{CD}$ were sides of a triangle. Similarly, if $\overline{PM}$ and $\overline{PN}$ were sides of a triangle, then by showing the angle opposite $\overline{PM}$ is greater than the angle opposite $\overline{PN}$, we could reach the desired conclusion.

We can create the necessary triangles by constructing chord $\overline{ED} \cong \overline{AB}$. Consider point K, the midpoint of $\overline{ED}$. Then, KD = ½ED. Also, DN = ½DC.

Since DC > ED, then DN > KD. In triangle Δ DKN, m ∡ 3 > m ∡ 4. Because ∡ DKP and ∡ DNP are right angles, m ∡ 3 + m ∡ 1 = m ∡ 4 + m ∡ 2 = 90°. Since m ∡ 3 > m ∡ 4, m ∡ 1 < m ∡ 2. PK > PN and the distance of chord $\overline{AB}$ from the center, PM = PK, is greater than the distance of the longer chord $\overline{DC}$ from the center, $\overline{PN}$.

Given: $\overline{AB}$ and $\overline{CD}$ are chords of ⊙ P; $\overline{PM} \perp \overline{AB}$ at M;

   $\overline{PN} \perp \overline{CD}$ at N; CD > AB.
Prove: PN < PM.

Statements	Reasons
1. $\overline{AB}$ and $\overline{CD}$ are chords of ⊙ P; $\overline{PM} \perp \overline{AB}$ at M; $\overline{PN} \perp \overline{CD}$ at N; CD > AB	1. Given.

287

2. Construct chord $\overline{ED}$ such that ED = AB and E and P are on the same side of DC.	2. From any point on the circle, a chord can be constructed of a given length, provided the length is less than the diameter.
3. K is the midpoint of $\overline{ED}$	3. Every line segment $\overline{AB}$ has one and only one point C such that AC = CB.
4. KD = ½ED	4. Definition of the midpoint of a segment.
5. $\overline{PK} \perp \overline{ED}$	5. The line joining the center of the circle and the midpoint of a chord is the perpendicular bisector of the chord.
6. PK = PM	6. In the same circle (or congruent circles), chords are congruent if and only if they are equidistant from the center of the circle.
7. DN = ½DC	7. The perpendicular from the center of the circle to a chord bisects the chord.
8. DC > ED	8. Substitution Postulate
9. DN > KD	9. Halves of unequal quantities are unequal in the same sense.
10. m ⊰ 3 > m ⊰ 4	10. In a triangle, if the length of the side opposite the first angle is greater than the length of the side opposite the second angle, the first angle is greater in measure than the second angle.
11. m ⊰ 3 + m ⊰ 1 = $90°$ m ⊰ 4 + m ⊰ 2 = $90°$	11. If two angles form a right angle, their measures sum to $90°$.
12. m ⊰ 1 = $90°$ - m ⊰ 3 m ⊰ 2 = $90°$ - m ⊰ 4	12. Equals subtracted from equals leave equals.
13. $90°$ - m ⊰ 4 > $90°$ - m ⊰ 3	13. If a = b and c < d, then a - c > b - d. (The equality is $90° = 90°$; the inequality comes from Step 10.)
14. m ⊰ 2 > m ⊰ 3	14. Substitution Postulate
15. PK > PN	15. In a triangle, if the measure of the angle opposite the first side is greater than the measure of the angle opposite the second side, then the length of the first side is greater than the length of the second.
16. PM > PN	16. Substitution Postulate (Steps 15 and 6)

• **PROBLEM** 330

If two chords of a circle subtend different acute inscribed angles, the smaller angle belongs to the shorter chord. Prove the statement and its converse.

<u>Solution</u>: In the accompanying figure, $\overline{AB}$ and $\overline{CD}$ are chords of ⊙O. ⊰α and ⊰β are inscribed angles of ⊙O such that ⊰α intercepts chord $\overline{AB}$ and ⊰β intercepts chord $\overline{CD}$. To prove that the shorter chords subtend the smaller inscribed angles, we must show, given AB > CD, that m⊰α > m⊰β.   To prove

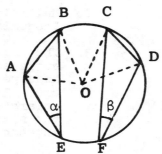

the converse, that smaller inscribed angles intercept shorter chords, we must show, given m∢α > m∢β, that AB > CD. We prove both by relating the measures of both chords and angles to the measure of the appropriate central angle.

Given: ⊙O with chords $\overline{AB}$ and $\overline{CD}$ such that AB > CD; inscribed ∢α intercepts $\overline{AB}$; inscribed ∢β intercepts $\overline{CD}$. Prove: m∢α > m∢β.

Statement	Reason
1. (See above)	1. Given.
2. $\overline{AO} \cong \overline{BO} \cong \overline{CO} \cong \overline{DO}$.	2. All radii of a circle are congruent.
3. AB > CD.	3. Given.
4. m∢AOB > m∢COD.	4. If two sides of one triangle (△AOB) are congruent to two sides of another triangle (△COD) and the third side of the first is greater than the third side of the second, then the included angle of the first is greater than the included angle of the second.
5. $\frac{1}{2}$m∢AOB > $\frac{1}{2}$m∢COD.	5. Division Property of Inequality.
6. m∢β = $\frac{1}{2}$m∢COD.	6. The measure of the inscribed angle equals one-half the measure of the intercepted arc. Also, the measure of the arc of a circle equals the measure of the subtending central angle.
7. m∢α > m∢β.	8. Transitive Property of Inequality.

Given: ⊙O with chords $\overline{AB}$ and $\overline{CD}$. Inscribed ∢α intercepts $\overline{AB}$; inscribed ∢β intercepts $\overline{CD}$. m∢α > m∢β. Prove: AB > CD.

Statement	Reason
1. (See above)	1. Given.

2. $2m\sphericalangle\alpha > 2m\sphericalangle\beta$.	2. Multiplication Property of Inequality.
3. $m\sphericalangle AOB = 2m\sphericalangle\alpha$ $m\sphericalangle COD = 2m\sphericalangle\beta$.	3. The measure of the inscribed angle equals one-half the measure of the intercepted arc. Also, the measure of an arc of a circle equals the measure of the subtending central angle.
4. $m\sphericalangle AOB > m\sphericalangle COD$.	4. Transitive Property of Inequality.
5. $\overline{AO} \cong \overline{BO} \cong \overline{CO} \cong \overline{DO}$.	5. All radii of a circle are congruent.
6. $AB > CD$.	6. If two sides of a triangle are congruent to two sides of another triangle and the included angle of the first is greater than the included angle of the second, then the third side of the first is longer than the third side of the second.

● **PROBLEM** 331

In circle O, $\overline{OE} \perp \overline{AB}$, $\overline{OF} \perp \overline{CD}$ and $m \measuredangle OEF > m \measuredangle OFE$. Prove that AB > CD.

Solution: Two chords are unequal if they are unequal distances from the center of the circle, and the longer chord is closer to the center of the circle. Therefore, if OE < OF, then AB > CD.

In any triangle, if two angles are unequal, the sides opposite them are unequal and the greater side is opposite the greater angle. In Δ OEF, we are given that $m \measuredangle OEF > m \measuredangle OFE$. Therefore, OF > OE.

Thus, CD is a greater perpendicular distance from the center of the circle than AB. As such, AB > CD.

● **PROBLEM** 332

If a,b,c, and d are four equidistant parallel chords in a circle, prove that $a^2 - d^2 = 3(b^2 - c^2)$.

290

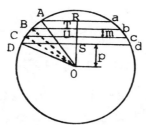

Solution: In the figure, a,b,c, and d are parallel equidistant chords of circle O. We are asked to relate the lengths of these chords; in particular, to show $a^2 - d^2 = 3(b^2-c^2)$. To do this, we express the lengths a,b,c, and d in terms of lengths common to all of them. Every chord can be expressed in terms of the radius and the distance of the chord from the center. Since the chords are equidistant, we can relate the distance from the center in terms of the distance between any two adjacent chords.

Draw $\overline{OR}$ perpendicular to a. Because a,b,c, and d are parallel, it must hold that $\overline{OR} \perp b$, $\overline{OR} \perp c$, and $\overline{OR} \perp d$. Furthermore, (since the perpendicular drawn from the center of a circle to a chord bisects the chord), $\overline{OR}$ bisects a,b, c, and d.

Let $\overline{OS}$, the distance between the center and chord d, be p, and let the distance between chords be m.

We now express each chord length in terms of r, the radius; and combinations of m and p, the distances from the center.

In right $\triangle OAR$, $AR^2 = OA^2 - OR^2$

$$(1/2a)^2 = r^2 - (3m + p)^2$$

$$a^2 = 4r^2 - 36m^2 - 24mp - 4p^2$$

In right $\triangle OBT$, $BT^2 = OB^2 - OT^2$

$$(1/2b)^2 = r^2 - (2m + p)^2$$

$$b^2 = 4r^2 - 16m^2 - 16mp - 4p^2$$

In right $\triangle OCU$, $CU^2 = OC^2 - OU^2$

$$(1/2c)^2 = r^2 - (m + p)^2$$

$$c^2 = 4r^2 - 4m^2 - 8mp - 4p^2$$

In right $\triangle ODS$, $DS^2 = OD^2 - OS^2$

$$(1/2d)^2 = r^2 - (p)^2$$

$$d^2 = 4r^2 - 4p^2$$

$(a^2 - d^2)$ equals $(4r^2 - 36m^2 - 24mp - 4p^2) -$

$(4r^2 - 4p^2)$ or $-36m^2 - 24mp$.

Also, $3(b^2-c^2) = 3\left[(4r^2-16m^2-16mp-4p^2) - (4r^2-4m^2-8mp-4p^2)\right]$

$$= 3\left[-12m^2-8mp\right] = -36m^2-24mp$$

Thus, $a^2-d^2 = -36m^2-24mp = 3(b^2-c^2)$

or $(a^2-d^2) = 3(b^2-c^2)$.

CHAPTER 18

TANGENTS AND INTERSECTING CIRCLES

Basic Attacks and Strategies for Solving Problems in this Chapter. See pages 293 to 312 for step-by-step solutions to problems.

Note that in the figure below, $\overleftrightarrow{AB}$ intersects the circle only at B. $\overleftrightarrow{AB}$ is said to be tangent to the circle, and $\overline{AB}$ is said to be a **tangent** to the circle.

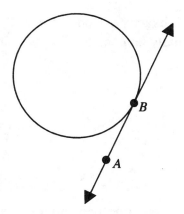

An important theorem which is applied repeatedly in this chapter follows:

The lengths of two tangents drawn to a circle from an exterior point are congruent.

This means that for the tangents $\overline{AB}$ and $\overline{AC}$ pictured below, $\overline{AB} \cong \overline{AC}$.

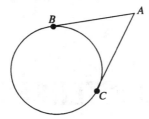

It is possible for two circles to have common tangents. For example, $\overline{AB}$ is a common tangent for the first two circles, and $\overline{CD}$ is a common tangent for the second pair of circles.

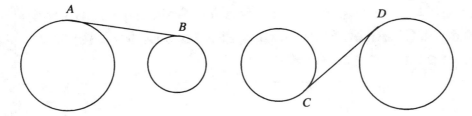

$\overline{AB}$ is called a **common external tangent** and $\overline{CD}$ is called a **common internal tangent**.

Circles which intersect in exactly one point are said to be tangent. The two circles on the left are internally tangent and the two circles on the right are externally tangent.

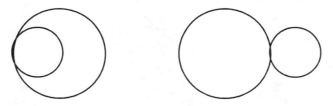

It is also possible for two non-congruent circles to intersect in two points, and in the figure below $\overline{AB}$ is said to be a common chord.

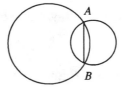

This chapter contains several problems which involve tangents, circles, and intersecting circles.

Step-by-Step Solutions to Problems in this Chapter, "Tangents and Intersecting Circles"

TANGENTS

Prove that if two circles are tangent externally, tangents drawn to the circles from any point on their common internal tangent are equal in length. (See figure)

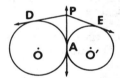

Solution: A common internal tangent is a line which is tangent to both circles and intersects their line of center. The line of center has endpoints O and O'. $\overleftrightarrow{PA}$ is the common internal tangent of circles O and O', and $\overrightarrow{PD}$ and $\overrightarrow{PE}$ are external tangents drawn from a point P on the common internal tangent. If both $\overrightarrow{PE}$ and $\overrightarrow{PD}$ can be shown to be equal in length to the same line segment, then we can conclude that they are equal to each other.

Statement	Reason
1. $\overleftrightarrow{PD}$ and $\overleftrightarrow{PA}$ are tangents to circle O.	1. Given
2. PD = PA.	2. The lengths of tangents drawn to a circle from an external point are equal.
3. $\overleftrightarrow{PE}$ and $\overleftrightarrow{PA}$ are tangents to circle O'.	3. Given.
4. PA = PE.	4. Same as Reason 2.
5. If PD = PA and PE = PA, then PD = PE.	5. Transitive Property of Equality.

Given: $\overline{AB}$ and $\overline{CD}$ are common tangents to circles O and O'.
Prove: AB = CD.

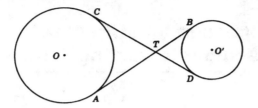

Solution: The Addition Property of Equality states that if equal quantities are added to equal quantities, then the results are equal. In this problem, we are proving that AT = CT and BT = DT. By the Addition Property of Equality, it then follows that AT + BT = CT + DT. Since AB = AT + BT and CD = CT + DT, we will have proved AB = CD.

Since $\overline{AB}$ and $\overline{CD}$ are common tangents of the circles, any subsegments of $\overline{AB}$ and $\overline{CD}$, including the points of tangency, will also be tangent to their respective circles. Therefore, $\overline{BT}$ and $\overline{DT}$ are tangent to circle O' and $\overline{AT}$ and $\overline{CT}$ are tangent to circle O.

BT = DT and AT = CT because tangents to a circle from the same external point are of equal length.

Since the length of the whole segment is equal to the sum of the lengths of all subsegments, AB = AT + BT and CD = CT + DT.

By the addition property of equality (as stated in the first paragraph), AB = CD.

Given that $\overline{AB}$ is a common tangent to circles O and O'. Prove that ∢AOC ≅ ∢BO'C. (See figure.)

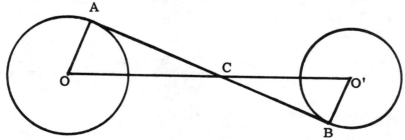

Solution: Since, if ΔAOC ∿ ΔBO'C, then ∢AOC ≅ ∢BO'C. Therefore we shall show that ΔAOC ∿ ΔBO'C by the Angle-Angle Similarity Theorem.

First, $\overline{AB}$ is tangent to circles O and O' at points A and B, respectively. Since the radius of a circle is perpendicular to the tangent line at the point of tangency, $\overline{OA} \perp \overline{AB}$ and $\overline{O'B} \perp \overline{AB}$. Hence, ∢ OAC and ∢ O'BC are right angles, i.e. ∢OAC ≅ ∢O'BC. Since opposite angles of two intersecting lines are equal, ∢ACO ≅ ∢BCO'. By the Angle-Angle Similarity Theorem, ΔAOC ∼ ΔBO'C, which implies that ∢AOC ≅ ∢BO'C.

● **PROBLEM** 336

A belt moves over two wheels which are the same size and crosses itself at right angles as shown in the figure. A,B,C, and D are points of tangency; E is the intersection of the tangents; and O is the center of one of the circles. The radius of each wheel is 7 inches.

(a)   Show that m∢AOE = 45°.

(b)   Find

(1)   the length of $\overline{AE}$

(2)   the length of major arc $\overarc{AC}$. $\left[\text{Use } \pi = \dfrac{22}{7} \right]$

(3)   the length of the entire belt.

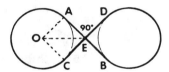

Solution: (a)   We will first show that AOCE is a square in order to prove that ∢AOC is a right angle. Then we will show that diagonal $\overline{OE}$ bisects ∢AOC. If m∢AOC = 90°, then ∢AOE must equal 1/2 of 90° or 45°.

Statement	Reason
1. $\overline{OA} \cong \overline{OC}$	1. Radii of a circle are congruent.
2. $\overline{AE}$ and $\overline{CE}$ are tangents at points A and C respectively.	2. Given.
3. $\overline{OA} \perp \overline{AE}$ and $\overline{OC} \perp \overline{CE}$	3. If a line is tangent to a circle, the line is perpendicular to the radius drawn to the point of contact.
4. ∢OAE and ∢OCE are right angles.	4. Perpendicular lines intersect to form right angles.
5. ∢AEC is a right angle.	5. Given.
6. ∢OAE is supplementary to ∢AEC.	6. Any two right angles are supplementary.
7. $\overline{OA} \mid\mid \overline{EC}$	7. If interior angles on the

same side of a transversal
are supplementary, then the
lines are parallel.

8. ∡OCE is supplementary
to ∡AEC.

8. Any two right angles are
supplementary.

9. $\overline{OC} \parallel \overline{AE}$

9. Same as reason 7.

10. AOCE is a square.

10. A quadrilateral in which
(1) all opposite sides are
parallel; (2) there exists
two adjacent sides which
are congruent; and (3) at
least one vertex angle is
a right angle is a square.

11. ∡AOC is a right angle.

11. All vertex angles of a
square are right angles.

12. m∡AOC = 90°

12. Right angles measure 90°.

13. $\overline{OE}$ bisects ∡AOC

13. The diagonal of a square
bisects the vertex angles
through which it is drawn.

14. m∡AOC = 2m∡AOE

14. An angle bisector divides
the angle into two angles
of equal measure.

| Statement | Reason |

15. 2(m∡AOE) = 90°.

15. Substitution Postulate.

16. m∡AOE = 45°.

16. Division Property of
Equality.

(b) 1. Since all four sides of a square are of equal measure, AE = OC. $\overline{OC}$ is a radius (r) given to be 7 in. in length.

Therefore, AE = 7 in.

2. Since central angle AOC measures 90°, minor arc $\overset{\frown}{AC}$ = 90°. Therefore, major arc $\overset{\frown}{AC}$ = 360 - 90 or 270.
arc length = $\dfrac{\text{degree measure of the arc}}{360}$ x Circumference.
Circumference = C = 2πr, where r is the radius. By substitution, C = 2π(7) = 14π. Using π = $\dfrac{22}{7}$, we obtain C = $\dfrac{14 \times 22}{7}$ = 44 in. Therefore, length of major arc $\overset{\frown}{AC}$ = $\dfrac{270}{360}$(44) = $\dfrac{3}{4}$(44) = 33. Therefore, length of major arc $\overset{\frown}{AC}$ = 33 in.

3. From the diagram, we find that length of entire belt = length of major arc $\overset{\frown}{AC}$ + $\overline{AE}$ + $\overline{EC}$ + $\overline{ED}$ + $\overline{EB}$ + length of major arc $\overset{\frown}{BD}$. We know that $\overline{AE}$ = $\overline{EC}$ = 7 inches , and that the length of major arc $\overset{\frown}{AC}$ = 33 inches.
Using the same manner we employed to find the above measurements, we can derive $\overline{ED}$, $\overline{EB}$ and length of major arc $\overset{\frown}{DB}$.

If we drew the radii from D and B to the center of the circle on the right, they would form a square symmetric to AOCE and, as such, $\overline{ED}$ = $\overline{EB}$ = radius of that circle. We are given the radius = 7 in. for both circles. Therefore, $\overline{ED} = \overline{EB} = 7$ in.

By a similar symmetry argument, we prove major arc $\overset{\frown}{BD}$ has the same length of major arc $\overset{\frown}{AC}$. Therefore, major arc $\overset{\frown}{BD} = 33$ in.

By substitution, length of entire belt = 33 + 7 + 7 + 7 + 7 + 33 = 94

Therefore, the length of the entire belt = 94 in.

● PROBLEM 337

Given: Circles E and F; $\overset{\leftrightarrow}{AB}$ and $\overset{\leftrightarrow}{CD}$ tangent to both circles at A, B, C, and D. Prove: $\overline{AB} \cong \overline{CD}$.

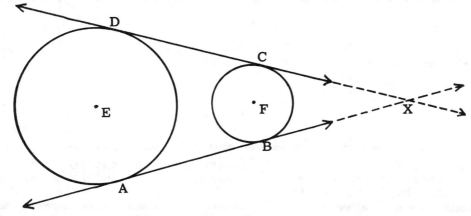

Solution: Extend $\overset{\leftrightarrow}{CD}$ and $\overset{\leftrightarrow}{AB}$ to the right until they intersect at a point X.

Since A, B, C, and D are points of tangency, $\overset{\leftrightarrow}{AX}$ and $\overset{\leftrightarrow}{DX}$ will be tangent to ⊙E and $\overset{\leftrightarrow}{BX}$ and $\overset{\leftrightarrow}{CX}$ will be tangents to ⊙F. Two tangents drawn to a circle from an external point are of equal length. Therefore, $\overline{AX} \cong \overline{DX}$ and $\overline{BX} \cong \overline{CX}$.

We know that the length of the whole segment is equal to the sum of the lengths of its parts. Hence, $\overline{AX} = \overline{AB} + \overline{BX}$ and $\overline{DX} = \overline{DC} + \overline{CX}$. Accordingly, $\overline{AB} = \overline{AX} - \overline{BX}$ and $\overline{DC} = \overline{DX} - \overline{CX}$. Since $\overline{AX} \cong \overline{DX}$ and $\overline{BX} \cong \overline{CX}$, we can conclude that $\overline{AB} \cong \overline{DC}$.

● PROBLEM 338

A triangle, ABC, is formed by the intersection of three tangents to a circle. Two of the tangents, AM and AN, are fixed in position, while BC touches the circle at a variable point, P. Show that the perimeter of triangle ABC is a constant no matter where P is located and is equal to 2 · AN.

Solution: Remember that from an external point, the tangents drawn to the circle are equal in length. Therefore, AN = AM; CP = CN; and BP = BM. We use these equalities to relate the sides of ΔABC to the tangent lengths.

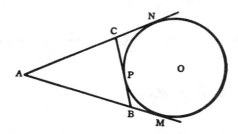

(i) Perimeter of ΔABC = AC + AB + CB. Since the whole is equal to the sum of its parts, this reduces to AC + AB + CP + BP. From above, we have CP = CN and BP = BM. Then,

(ii) Perimeter of ΔABC = AC + AB + CN + BM = (AC + CN) + (AB + BM). Points C and B are points on $\overline{AN}$ and $\overline{AM}$. Therefore, AN = AC + CN and AM = AB + BM. Substituting, we have

(iii) Perimeter of ΔABC = AN + AM.

Finally, $\overline{AN}$ and $\overline{AM}$ are tangents drawn to a circle from an external point. Therefore, AM = AN.

(iv) Perimeter of ΔABC = AN + AN = 2 · AN.

● **PROBLEM** 339

Line segments $\overline{AB}$, $\overline{BC}$, and $\overline{CA}$ are drawn tangent to circle O, as shown in the figure. $\overline{AD}$ measures 5 in. and $\overline{BE}$ measures 4 in. Find the length of $\overline{AB}$.

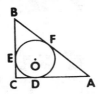

Solution: A theorem states that if two tangents are drawn to a circle from an external point, these tangents are of equal length. Therefore, AF = AD = 5 in. and BF = BE = 4 in. Since AB = AF + BF, by substitution, we know that the length of AB = (5 + 4)in. = 9 in.

● **PROBLEM** 340

Two tangents to a circle are at right angles with each other. A point on the smaller arc between the two points of tangency is 4½ in. from one tangent line and 4 in. from the other. Find the radius of the circle.

Solution: We will solve this problem by applying several properties of tangent lines, parallel lines and parallelograms.

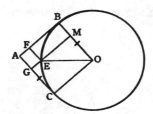

The accompanying figure represents the situation described in the problem. First we prove that quadrilateral
ACOB is a square.

We are given that $\overline{CA}\perp\overline{AB}$. If a radius intersects a tangent line at the point of tangency, then it is perpendicular
to the tangent. Hence, $\overline{OB}\perp\overline{AB}$. Lines perpendicular to the
same line are parallel and therefore, $\overline{OB}\parallel\overline{AC}$. Again, since
a radius and a tangent meet at the point of tangency, $\overline{OC}\perp\overline{AC}$.
It follows then that $\overline{OC}\parallel\overline{AB}$.

Since both pairs of opposite sides of ACOB are parallel
lines, ACOB is a parallelogram.

Tangents to a circle drawn from the same external point
are congruent. Hence, $\overline{AB} = \overline{AC}$. Also $\overline{OE} = \overline{OC}$, since radii
of the same circle are congruent.

Since ∡BAC is a right angle, ACOB is a square.

Draw $\overline{EG} \perp \overline{AC}$ and extend it to point M on line $\overline{BO}$. Also
draw $\overline{EF} \perp \overline{AB}$. We are given $\overline{EG}$ = 4 in. and $\overline{EF}$ = 4½ in. Since
$\overline{OB} \parallel \overline{AC}$ and $\overline{MG} \perp \overline{AC}$, then $\overline{MG} \perp \overline{OB}$. ∡EMO will be a right angle
making ΔOME a right triangle.

The perpendicular distance between parallel lines is
constant at all places between the lines. Hence, GM = AB =
OC = radius of ⊙O = r.

Since the lengths of all sides of ΔOME can be represented in terms of r, we can use the Pythagorean theorem to solve
for r.

EM = MG - EG = r - 4.

MO = OB - BM. OB is a radius, and equals r. Since
$\overline{AB} \perp \overline{AC}$ and $\overline{MG} \perp \overline{AC}$, $\overline{AB} \parallel \overline{MG}$. As such, EF = BM = 4½ in.
Hence, MO = r - 4½. In right ΔOME,

$$(EM)^2 + (MO)^2 = r^2$$
$$(r-4)^2 + (r-4½)^2 = r^2$$
$$(r^2-8r+16) + (r^2-9r+20.25) - r^2 = 0$$
$$r^2 - 17r + 36.25 = 0.$$

We multiply by 4 to remove decimals: $4r^2 - 68r + 145 = 0$.
Using the factoring method to solve for r, we obtain
(2r-5)(2r-29) = 0.

$$2r - 5 = 0 \quad \text{or} \quad 2r - 29 = 0$$
$$r = \frac{5}{2} \quad \text{or} \quad r = \frac{29}{2}.$$

Therefore, the radius of the circle described can be of length $\frac{5}{2}$ in. or $\frac{29}{2}$ in.

Since $\overline{FE}$ is the same length as $\overline{BM}$ and $\overline{FE}$ is $4\frac{1}{2}$ in., $\overline{BM}$ must also equal $4\frac{1}{2}$ in.. The length of the radius of the circle is equal to the length of $\overline{BO}$, which is equal to the sum of the lengths of $\overline{BM}$ and $\overline{MO}$. Therefore, the length of the radius must be greater than the length of $\overline{BM}$, $4\frac{1}{2}$ in.. This rules out 5/2 inch as a solution and the length of r is 29/2 inch.

● PROBLEM 341

Given two concentric circles with centers at O. Let $\overline{AB}$ and $\overline{CD}$ be chords of the larger circle which are both tangent to the smaller circle. Prove that AB = CD. (See figure.)

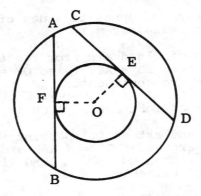

Solution: We will use the fact that if the perpendicular distances between the center of a circle and two chords of the circle are equal, the chords have equal length.

In the figure, both $\overline{CD}$ and $\overline{AB}$ are tangent to the smaller circle. Hence, $\overline{OE} \perp \overline{CD}$ and $\overline{OF} \perp \overline{AB}$, where $\overline{OE}$ is the distance from O to $\overline{CD}$, and $\overline{OF}$ is the distance from O to $\overline{AB}$. Since $\overline{OF}$ and $\overline{OE}$ are the radii of the smaller circle, $\overline{OF} \cong \overline{OE}$. By the theorem stated in the first paragraph, $\overline{CD} \cong \overline{AB}$.

CIRCLES THAT ARE TANGENT TO OTHER CIRCLES

● PROBLEM 342

Circles A and B are externally tangent at C. Prove that A, C, and B are collinear.

Solution: If two adjacent angles are right angles, then their non-common sides form a straight angle. Since the rays of a

straight angle form a straight line, any points on these rays
are collinear.

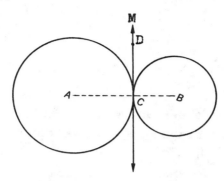

      C is the point of tangency of line m and circle B. Since
radii of a circle are perpendicular to tangent lines at the
points of tangency, therefore, radius $\overline{CB}$ is perpendicular to
line m. Accordingly, radius $\overline{AC}$ is also perpendicular to line
m.

      Since perpendicular lines intersect to form right angles,
adjacent angles DCB and DCA are right angles.

      Hence, the non-common sides of ∢DCB and ∢DCA form a
straight angle, and the points A, C, and B are collinear.

● PROBLEM 343

Circles P and Q are internally tangent at point T, chord $\overline{TA}$
of ⊙Q meets ⊙P at B, and chord $\overline{TC}$ of ⊙Q meets ⊙P at D. If
$m\overset{\frown}{AC} = 106$, find $m\overset{\frown}{BD}$.

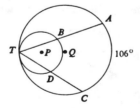

Solution: $\overset{\frown}{BD}$ is an arc intersected by the inscribed angle
∢ATC. Therefore, $m\overset{\frown}{BD} = 2 \cdot m\angle ATC$. To find m∢ATC, note that
∢ATC is also an inscribed angle of circle ⊙Q. Thus, m∢ATC =
$\frac{1}{2}m\overset{\frown}{AC}$. $m\overset{\frown}{AC} = 106$. Therefore, m∢ATC = $\frac{1}{2}(106) = 53$, and $m\overset{\frown}{BD}$ =
2 × m∢ATC = 2(53) = 106°.

● PROBLEM 344

Prove that $\overline{B_1C_1} \parallel \overline{B_2C_2}$, given that C_1 and C_2 are the centers
of the circles shown in the figure.

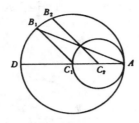

Solution: Let DA be the transversal cutting the 2 lines $\overleftrightarrow{B_1C_1}$ and $\overleftrightarrow{B_2C_2}$. If the corresponding angles $\angle DC_1B_1$ and $\angle DC_2B_2$ are congruent, then $\overline{B_1C_1} \parallel \overline{B_2C_2}$. In order to show the necessary congruence; of the two angles we use some facts about related angles.

$\angle DAB_1$ and $\angle DC_1B_1$ are related, relative to the large circle. Hence, $m\angle DAB_1 = \frac{1}{2}m\angle DC_1B_1$. Furthermore, $\angle DC_2B_2$ and $\angle DAB_1$ are related, relative to the small circle. Hence, $m\angle DAB_1 = \frac{1}{2}m\angle DC_2B_2$. Combining the last 2 facts,

$$\frac{1}{2}m\angle DC_1B_1 = \frac{1}{2}m\angle DC_2B_2 \text{ or } \angle DC_1B_1 \cong \angle DC_2B_2.$$

Hence, $\overline{B_1C_1} \parallel \overline{B_2C_2}$.

● **PROBLEM** 345

Given two circles with a common tangent at a point A such that the second circle passes through the center of the first. Show that every chord of the first circle that passes through A is bisected by the second circle. (See figure.)

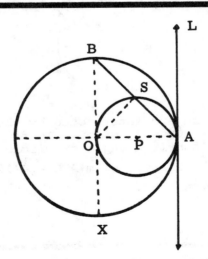

Solution: We are asked to show that if circles P and O are tangent at A, and O lies on circle P, then $\overline{AB}$ is bisected by circle P at point S. We will do this by showing that $\overline{OS}$ is the perpendicular bisector of $\overline{BA}$.

First, note that $\overline{OPA}$ connects the centers of 2 circles tangent at the same point (See figure.) Hence, $\overline{OPA}$ is a straight line passing through the center of circle P. $\overline{OPA}$ is, therefore, a diameter of P, which implies that ⊀OSA is inscribed in a semicircle ($\overline{OSA}$). This means that ⊀OSA is a right angle, and that $\overline{OS} \perp \overline{BA}$. However, any line segment drawn from the center of a circle perpendicular to a chord of the same circle bisects the chord. Since $\overline{OS} \perp \overline{BA}$ at S, $\overline{BS} \cong \overline{SA}$.

● **PROBLEM** 346

Circles A, B, and C are tangent to one another. Find the radii of the three circles if AB = 7, AC = 5, and BC = 9.

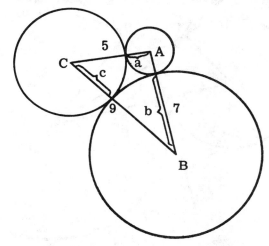

Solution: The segment connecting the centers of tangent circles passes through the point of tangency on the circumference of the circles. Thus, the distance to the point of tangency from the center of ⊙C is the radius, c, from the center of ⊙A is the radius, a, and from the center of ⊙B is the radius, b. Therefore, the length of the segments between the centers is the sum of the radii. Hence,

$$AC = c + a$$
$$AB = a + b$$
$$BC = b + c$$

Since AC = 5, AB = 7, and BC = 9, we have

(i)   5 = c + a
(ii) 7 = a + b
(iii) 9 = b + c

Solving for a in terms of c 5 = c + a, a = 5 - c. Substituting this result in the other equations, we obtain:

(ii) 7 = (5-c) + b = 5 + b - c or 2 = b - c
(iii) 9 = b + c

Adding both equations, we eliminate c.

$$11 = 2b \quad \text{or} \quad b = 5.5$$

Since $9 = b + c$, $c = 9 - 5.5 = 3.5$ Hence $a = 5 - c = 5 - 3.5 = 1.5$.   Thus, $a = 1.5$, $b = 5.5$, and $c = 3.5$.

Prove that, if three circles are tangent to one another, the tangents at the points of contact are concurrent.

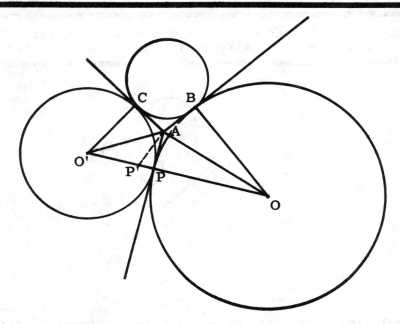

Solution: In the accompanying figure, the points of tangency are P, B, and C. The tangents drawn at C and B meet at A. We must prove that A is also a point on the tangent at P; that is, that $\overleftrightarrow{AP}$ is the tangent to circles O and O'. To show this, we prove $\overline{AP} \perp \overline{O'P}$ and $\overline{AP} \perp \overline{OP}$, or $\overline{AP} \perp \overline{OO'}$. (Remember that the line passing through the center of two tangent circles also passes through the point of tangency.) We will prove this result indirectly.

First, we will find a relationship between $\overline{OA}$, $\overline{O'A}$, $\overline{OP}$, and $\overline{O'P}$. Second, we will assume that $\overline{AP}$ is not tangent to the circles; that, instead, there is some other point on $\overline{O'O}$, P', such that $\overline{AP'} \perp \overline{OO'}$. We then find a relationship between $\overline{OA}$, $\overline{O'A}$, $\overline{OP'}$, and $\overline{O'P'}$. We use our two relationships to find a relation involving $\overline{OP}$, $\overline{O'P}$, $\overline{O'P'}$, and $\overline{OP'}$. We will show this relation to be impossible. Thus, our assumption that "$\overline{AP}$ not $\perp \overline{OO'}$" will have to be false; $\overline{AP} \perp \overline{OO'}$ and, consequently, the three tangents drawn at the points of intersection, are concurrent at point A.

Part (1). First, we find an expression involving OA, O'A, OP, and O'P. In right $\triangle OBA$, $OA^2 = OB^2 + BA^2$ or $BA^2 = OA^2 - OB^2$. In right $\triangle O'CA$, $O'A^2 = O'C^2 + AC^2$ or $AC^2 = O'A^2 - O'C^2$.

$\overline{AB}$ and $\overline{AC}$ are tangents to a circle drawn from external point A. Therefore, they are congruent, and AB = AC. Hence, we can equate the above expressions to obtain

$$OA^2 - OB^2 = O'A^2 - O'C^2 \text{ or } OA^2 - O'A^2 = OB^2 - O'C^2.$$

Note that $\overline{OB}$ and $\overline{OP}$ are radii of $\odot O$. Thus, OB = OP. Furthermore, $\overline{O'C}$ and $\overline{O'P}$ are radii of $\odot O'$. Thus, O'C = O'P. Substituting, we obtain an expression involving $\overline{OA}$, $\overline{O'A}$, $\overline{OP}$, and $\overline{O'P}$.

$$OA^2 - O'A^2 = OP^2 - O'P^2$$

Part (2). Now we assume $\overline{AP'} \perp \overline{OO'}$ and find an expression involving $\overline{OA}$, $\overline{O'A}$, $\overline{O'P'}$, and $\overline{OP'}$. In right $\triangle AO'P'$, $P'A^2 = O'A^2 - O'P'^2$. In right $\triangle AOP'$, $P'A^2 = OA^2 - OP'^2$.

Since $P'A^2 = P'A^2$, we have

$$O'A^2 - O'P'^2 = OA^2 - OP'^2 \text{ or } OA^2 - O'A^2 = OP'^2 - O'P'^2.$$

This is our equation involving $\overline{OA}$, $\overline{O'A}$, $\overline{O'P'}$, and $\overline{OP'}$. Note that the left side of the equation is the same as the left side of the equation in Part I. We equate the right sides of the equations to obtain an expression in $\overline{OP}$, $\overline{O'P}$ $\overline{OP'}$, and $\overline{O'P'}$:

$$OP^2 - O'P^2 = OP'^2 - O'P'^2$$

(a) $(OP+O'P)(OP-O'P) = (OP'+O'P')(OP'-O'P').$

Note that:

$$CO'=OP+O'P=OP'+O'P'$$

Substituting OP+O'P for OP'+O'P' into equation (a) one obtains:

$$(OP+O'P)(OP-O'P)=(OP+O'P)(OP'-O'P')$$

Then cancelling the equal quantities, we obtain:

$$OP-O'P=OP'-O'P' \text{ or } OP+O'P'=O'P+OP'.$$

Examining the diagram, we see that this equation cannot be true if $P \neq P'$. Thus, our assumption, $\overline{AP'} \perp \overline{OO'}$, is wrong. Thus, $\overline{AP} \perp \overline{OO'}$ and the tangents from the three points of intersection are concurrent.

INTERSECTING CIRCLES

● **PROBLEM** 348

Circles F and D intersect at points B and E. $\overline{AE}$ and $\overline{CE}$ are the respective diameters. Prove A, B, and C are collinear.

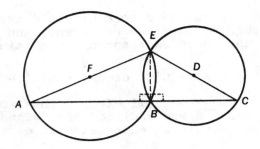

Solution: Begin this problem by drawing segment $\overline{BE}$. If we prove adjacent angles ∢EBC and ∢EBA are right angles, then their non-common sides form a straight line and A, B, and C will be collinear.

Since $\overline{EC}$ is a diameter of circle D, arc $\overparen{EBC}$ is a semi-circle. Any angle inscribed in a semi-circle is a right angle. Therefore, ∢EBC is a right angle. $\overline{AE}$ is a diameter of circle F and, by the same logic, ∢EBA is a right angle.

Since both adjacent angles have been shown to be right angles, we can conclude that their non-adjacent sides form a straight line and, as such, that A, B, and C are collinear.

● **PROBLEM** 349

Given: Circles P and Q intersect at points A and C; chord $\overline{AB}$ of ⊙P is tangent to ⊙Q; chord $\overline{AD}$ of ⊙Q is tangent to ⊙P; E is a point on ⊙Q; F is a point on ⊙P. **Prove:** ΔABC ∿ ΔDAC and AC is the mean proportional between BC and DC.

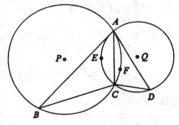

Solution: To show similar triangles, we use either (1) A-A Similarity Theorem, (2) SAS Similarity Theorem, or (3) SSS Similarity Theorem. There are no obvious proportions between the sides. Therefore, we try to find an A-A Similarity. We are to prove ΔABC ∿ ΔDAC, therefore, ∢ABC corresponds to ∢CAD. ∢B is an inscribed angle that intercepts arc AFC and, thus, m∢B = $\frac{1}{2}$m$\overparen{AFC}$. ∢CAD is an angle formed by a chord and a tangent of ⊙P and has a measure equal to $\frac{1}{2}$ the intercepted arc or $\frac{1}{2}$m$\overparen{AFC}$. Therefore, both m∢B and m∢CAD equal $\frac{1}{2}$m$\overparen{AFC}$ and ∢B ≅ ∢CAD.

In a similar manner, we show ∢BAC ∿ ∢CDA. Thus, ΔABC ∿ ΔDAC. By definition of similar triangles, we show AC is the mean proportional between BC and DC.

Statements	Reasons
1. (See problem statement)	1. Given.
2. $m \angle B = \frac{1}{2} m\overarc{AFC}$ in $\odot P$. $m \angle D = \frac{1}{2} m\overarc{AEC}$ in $\odot Q$.	2. The measure of the inscribed angle equals one-half the measure of the intercepted arc.
3. $m \angle CAD = \frac{1}{2} m\overarc{AFC}$ in $\odot P$. $m \angle BAC = \frac{1}{2} m\overarc{AEC}$ in $\odot Q$.	3. The measure of an angle formed by a chord and a tangent of a circle equals one-half the intercepted arc.
4. $m \angle B = m \angle CAD$, or, $\angle B \cong \angle CAD$ $m \angle D = m \angle BAC$, or, $\angle D \cong \angle BAC$.	4. Transitivity Postulate and Definition of Congruence of Angles.
5. $\triangle ABC \sim \triangle DAC$.	5. The A-A Similarity Theorem.
6. $\dfrac{BC}{AC} = \dfrac{AC}{DC}$	6. The sides of similar triangles are proportional.
7. $AC^2 = BC \cdot DC$.	7. For a proportion to hold, the product of the means equal the product of the extremes.
8. AC is the mean proportional between BC and DC.	8. Defintion of the mean proportional.

● **PROBLEM** 350

Given: $\odot R$ intersects $\odot S$ at points A and B; E, F, D, and C are points of $\odot R$; P and Q are points of $\odot S$; $\overline{EAP}$, $\overline{FAQ}$, $\overline{DBQ}$, and $\overline{CBP}$ are straight lines. Prove: $\overarc{EC} \cong \overarc{FD}$.

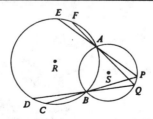

Solution: We must show $\overarc{EC} \cong \overarc{FD}$. $\overarc{EC} = \overarc{EF} + \overarc{FC}$ and $\overarc{FD} = \overarc{FC} + \overarc{CD}$. Since $\overarc{FC} \cong \overarc{FC}$, we need only show $\overarc{EF} \cong \overarc{CD}$. $\overarc{EC} \cong \overarc{FD}$ follows from this. $\overarc{EF}$ is intercepted by inscribed angle $\angle EAF$. $\overarc{CD}$ is intercepted by inscribed angle $\angle DBC$. Therefore, $\overarc{EF} = \overarc{CD}$ only if $\angle EAF \cong \angle DBC$. To relate these two angles, note that the angles opposite them, $\angle PAQ$ and $\angle PBQ$, are inscribed angles of $\odot S$. Since $\angle PAQ$ and $\angle PBQ$ intercept the same arc, $\angle PAQ \cong \angle PBQ$. Thus, $\angle EAF \cong \angle DBC$, and the arcs are congruent.

Statements	Reasons
1. (see problem statement)	1. Given
2. ∢PAQ ≅ ∢PBQ.	2. Inscribed angles that inter-cept congruent arcs are con-gruent.
3. ∢EAF ≅ ∢PAQ ∢DBC ≅ ∢PBQ.	3. Opposite angles formed by intersecting lines are con-gruent.
4. ∢EAF ≅ ∢DBC.	4. Transitivity Postulate.
5. m$\widehat{DC}$ = 2 · m∢DBC m$\widehat{EF}$ = 2 · m∢EAF.	5. The measure of an inscribed angle equals one-half the intercepted arc.
6. m$\widehat{DC}$ = m$\widehat{EF}$.	6. Transitivity Postulate.
7. m$\widehat{FC}$ = m$\widehat{FC}$.	7. An arc is congruent to it-self.
8. m$\widehat{FC}$ + m$\widehat{DC}$ = m$\widehat{FC}$ + m$\widehat{EF}$.	8. Addition Postulate.
9. m$\widehat{EC}$ = m$\widehat{FD}$.	9. An arc is specified by its endpoints.
10. $\widehat{EC}$ ≅ $\widehat{FD}$.	10. Definition of congruent arcs.

● **PROBLEM** 351

Two parallel tangents are drawn to a circle whose center is
O. A third tangent is drawn intersecting the first two at
A and B. Prove that the circle constructed with diameter $\overline{AB}$
passes through O.

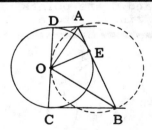

Solution: In the figure, E is the point of tangency of $\overline{AB}$
and ⊙O. $\overline{AB}$ intersects parallel tangents $\overleftrightarrow{AD}$ and $\overleftrightarrow{BC}$ at A and
B, respectively. We must show that the circle of diameter
$\overline{AB}$ passes through point O.

If $\overline{AB}$ is a diameter, then $\widehat{AB}$ is a semicircle. By an
earlier theorem, we know that for point O to be on circle of
diameter $\overline{AB}$, then ∢AOB must be a right angle.

Therefore, if we show m∢AOB = 90°, then point O is on
the circle. To show m∢AOB = 90°, we will have to show

308

(1)   m∡EAO = $\frac{1}{2}$m∡DAE and m∡OBE = $\frac{1}{2}$m∡CBE,

(2)   m∡DAE + m∡CBE = 180°, (3) m∡EAO + m∡OBE = $\frac{1}{2}$·180° = 90°,
and (4) m∡OAE + m∡OBE + m∡AOB = 180° or
m∡AOB = 180 - m∡OAE - m∡OBE = 180° - 90° = 90°.

To show $\overline{AO}$ and $\overline{BO}$ are angle bisectors, we show ΔAEO ≅
ΔADO. Since tangent segments to a given circle from an ex-
terior point are congruent, $\overline{AE}$ ≅ $\overline{AD}$. Because radii are congruent, $\overline{EO}$ ≅ $\overline{OD}$. Finally, $\overline{AO}$ ≅ $\overline{AO}$. By the SSS Theorem,
ΔAEO ≅ ΔADO. By corresponding parts, ∡EAO ≅ ∡DAO or
m∡EAO = m∡DAO = $\frac{1}{2}$m∡DAE.

By showing ΔBEO ≅ ΔBCO, we can analogously show
m∡OBE ≅ m∡OBC = $\frac{1}{2}$m∡CBE.

To show (2), note that $\overleftrightarrow{AD}$ and $\overleftrightarrow{BC}$ are given to be parallel
tangents. Therefore, the interior angles on the same side of
the transversal, $\overleftrightarrow{AB}$, are supplementary. Thus,
m∡DAE + m∡CBE = 180°.

Note that m∡OAE + m∡OBE = $\frac{1}{2}$m∡DAE + $\frac{1}{2}$m∡CBE =

$\frac{1}{2}$(m∡DAE + m∡CBE) = $\frac{1}{2}$(180) = 90°.

Finally, ∡OAE, ∡OBE, and ∡AOB are the angles of ΔAOB
and sum to 180°. Thus, m∡AOB = 180 - (m∡OAE + m∡OBE) =
180 - 90 = 90. ΔAOB is a right triangle with hypotenuse $\overline{AB}$.
Therefore, O must be a point on the semicircle whose diameter
is $\overline{AB}$.

● **PROBLEM** 352

Given: Circles P and Q intersect at points A and C; $\overline{PQ}$
meets $\overrightarrow{BA}$ and $\overrightarrow{DC}$ at T, a point of circle P. Prove that
$\overline{AB}$ ≅ $\overline{CD}$.

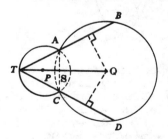

Solution: Construct segment $\overline{AC}$ so that it intersects $\overline{PQ}$
at point S. Draw perpendicular line segments from Q to $\overline{AB}$
and $\overline{CD}$. The perpendicular to $\overline{AB}$ intersects $\overline{AB}$ at E, and
the perependicular to $\overline{CD}$ intersects $\overline{CD}$ at F. (See figure.)

We first prove that $\overline{TPSQ}$ is the perpendicular bisector
of $\overline{AC}$. This will enable us to show that ΔATS ≅ ΔCTS, which

309

implies that ∢ATS ≅ ∢CTS. We will then have enough data to prove that ΔQET ≅ ΔQFT or that $\overline{QE}$ ≅ $\overline{QF}$. Since chords which are equidistant from the center of a circle are congruent, we will have shown that $\overline{AB}$ ≅ $\overline{CD}$.

Circles P and Q intersect at points A and C. Hence, $\overline{QA}$ and $\overline{QC}$ are radii of circle Q, and points A and C are equidistant from point Q. Similarly, A and C are equidistant from point P. Since the locus of points equidistant from two points is the perpendicular bisector of the line segment joining the two points, P and Q must lie on the perpendicular bisector of $\overline{AC}$. Furthermore, since two points determine a line, $\overleftrightarrow{PQ}$ is the perpendicular bisector of $\overline{AC}$. By construction, S and T lie on $\overleftrightarrow{PQ}$. $\overline{TPSQ}$ is the perpendicular bisector of $\overline{AC}$. We use this fact to show that ΔATS ≅ ΔCTS.

$\overline{AS}$ ≅ $\overline{CS}$ and ∢AST ≅ ∢CST because $\overline{TPSQ}$ is the perpendicular bisector of $\overline{AC}$. $\overline{TS}$ ≅ $\overline{TS}$. By the S.A.S. Postulate, ΔATS ≅ ΔCTS, and ∢ATS ≅ ∢CTS. A lies on $\overline{TE}$, C lies on TF, and $\overline{TPSQ}$ is a straight line. Hence, the last statement can be rewritten as ∢ETQ ≅ ∢FTQ. By construction, ∢TEQ ≅ ∢TFQ. $\overline{TQ}$ ≅ $\overline{TQ}$ and, by the A.A.S. Postulate, ΔQET ≅ ΔQFT. Hence, $\overline{QE}$ ≅ $\overline{QF}$ or QE = QF. This means that $\overline{AB}$ and $\overline{CD}$ are equally distant from Q (because $\overline{QE}$ ⊥ $\overline{AB}$ and $\overline{QF}$ ⊥ $\overline{CD}$). Therefore, since chords drawn equidistantly from the center of a circle are congruent, $\overline{AB}$ ≅ $\overline{CD}$.

● **PROBLEM** 353

Two equal circles are drawn so that the center of each is on the circumference of the other. Their intersection points are A and B. Prove that if, from A, any line is drawn cutting the circles at D and C, then ΔBCD is equilateral.

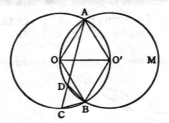

Solution: In the accompanying figure, ⊙O and ⊙O' intersect at A and B. $\overleftrightarrow{AD}$ is an arbitrary line that intersects ⊙O' at D and ⊙O at C. To show ΔBCD is an equilateral triangle, we must either show (1) sides DC = CB = BD or (2) ∢DCB ≅ ∢CBD ≅ ∢BDC. We will show that m∢DCB = 60° and m∢ BDC = 60°. Therefore, m∢CBD = 180 - m∢DCB - m∢BDC = 60. Thus, we will be able to conclude that ΔBCD is an equilateral triangle.

To show m∢DCB = 60°, note that ∢DCB is the same angle as ∢ACB which is an inscribed angle of ⊙O. Thus, m∢ACB = m∢DCB = $\frac{1}{2}$m $\overarc{AO'B}$.

To find m $\overarc{AO'B}$, we consider ΔOAO' and ΔOBO'. Because

310

equal circles have equal radii, OA = OO' = O'A and OO' =
OB = BO'. Thus △OAO' and △OBO' are equilateral triangles
and m∢AOO' = m∢BOO' = 60°. m∢AOB = m∢AOO' + m∢BOO' =
60 + 60 = 120°. Thus, m $\widehat{AO'B}$ = m∢AOB = 120°. Since
m∢DCB = $\frac{1}{2}$m $\widehat{AO'B}$, m∢DCB = $\frac{1}{2}$(120°) = 60°.

To show m∢CDB = 60°, we show that its supplement, exterior angle ∢ADB, has a measure of 120°. Note that ∢ADB
is an inscribed angle of ⊙O'. Thus, m∢ADB = $\frac{1}{2}$ $\widehat{AMB}$. Since

m$\widehat{AMB}$ + m$\widehat{AOB}$ = 360, then m$\widehat{AMB}$ = 360 - m$\widehat{AOB}$ = 360 - 120 = 240.
Substituting, we have m∢ADB = $\frac{1}{2}$(240) = 120.

Because exterior angle ∢ADB forms a linear pair with
∢CDB, m∢ADB + m∢CDB = 180. Thus, m∢CDB = 180 - m∢ADB = 180
- 120 = 60.

Finally, in △BCD, if m∢DCB = 60° and m∢CDB = 60°, then
m∢DBC = 180 - m∢DCB - m∢CDB = 180 - 60 - 60 = 60°.

Thus, all angles of △BCD are equal and △BCD is an equilateral triangle.

● **PROBLEM** 354

Three unequal circles ⊙A, ⊙B, and ⊙C, intersect at point O.
Lines $\overleftrightarrow{OA}$, $\overleftrightarrow{OB}$, and $\overleftrightarrow{OC}$ intersect the circles at points A', B',
and C'. Show that the sides of △A'B'C' pass through the
other points of intersection of the circles and are, respectively, parallel to the sides of △ABC.

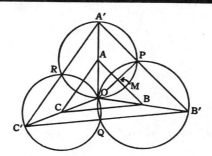

Solution: In the figure, point P is the intersection of ⊙A
and ⊙B. We must show $\overleftrightarrow{A'B'}$ ‖ $\overline{AB}$ and $\overleftrightarrow{A'P B'}$.

To show that point P is between A' and B', we show
$\overleftrightarrow{A'P}$ ‖ $\overline{AB}$ and $\overleftrightarrow{PB'}$ ‖ $\overline{AB}$. Both $\overleftrightarrow{A'P}$ and $\overleftrightarrow{PB'}$ will be lines parallel to $\overline{AB}$ passing through the given point P. Since only
one line can be drawn through a given point parallel to a
given line, $\overleftrightarrow{A'P}$ and $\overleftrightarrow{PB'}$ are on the same line, $\overleftrightarrow{A'PB'}$.

To show $\overleftrightarrow{A'P}$ ‖ $\overline{AB}$, we show corresponding angles ∢OA'P
and ∢OAM are congruent. Consider the common chord, $\overline{OP}$.
The line passing through the centers of the intersecting
circles is the perpendicular bisector of the common chord.
Thus, $\overleftrightarrow{AM}$ is the perpendicular bisector of $\overline{OP}$ and

$OM = MP = \frac{1}{2}OP$ (or $\frac{OM}{OP} = \frac{1}{2}$). Furthermore, OA' is a diameter and OA a radius of ⊙A. Thus, $OA' = 2 \cdot OA$ (or $\frac{OA}{OA'} = \frac{1}{2}$).

Hence, $\frac{OM}{OP} = \frac{OA}{OA'}$. Since two sides of ΔOAM are proportional to two sides of ΔOA'P, by reflexivity, ∢AOM ≅ ∢A'OP. The SAS Similarity Theorem allows us to conclude ΔOAM ∿ ΔOA'P. Consequently, by corresponding parts, ∢OAM ≅ ∢OA'P. Since ∢OAM and ∢OA'P are corresponding angles of lines $\overleftrightarrow{AB}$ and $\overrightarrow{A'P}$, $\overleftrightarrow{AB} \parallel \overrightarrow{A'P}$.

By proving ΔOBM ∿ ΔOB'P, we can show ∢OBM ≅ ∢OB'P and $\overleftrightarrow{AB} \parallel \overrightarrow{PB'}$. Since $\overrightarrow{A'P}$ and $\overrightarrow{B'P}$ are both parallel to $\overleftrightarrow{AB}$ and both pass through P, $\overrightarrow{A'P} = \overrightarrow{B'P}$ or $\overleftrightarrow{A'PB'}$. The second part of the proof, $\overleftrightarrow{A'B'} \parallel \overleftrightarrow{AB}$, follows. Since $\overleftrightarrow{A'PB'}$ and $\overrightarrow{A'P} \parallel \overleftrightarrow{AB}$, then $\overleftrightarrow{A'B'} \parallel \overleftrightarrow{AB}$.

The same procedure can be used to show $\overleftrightarrow{B'QC'} \parallel \overleftrightarrow{BC}$ and $\overleftrightarrow{C'RA'} \parallel \overleftrightarrow{AC}$.

CHAPTER 19

ANGLES FORMED BY TANGENTS, SECANTS, AND CHORDS

Basic Attacks and Strategies for Solving Problems in this Chapter. See pages 313 to 336 for step-by-step solutions to problems.

In previous chapters, arcs and the measures of arcs were discussed. This was done somewhat ambiguously. For example, in the figure below, it is not clear whether $\overset{\frown}{AB}$ refers to the part of the circle which contains C or the part of the circle which contains D.

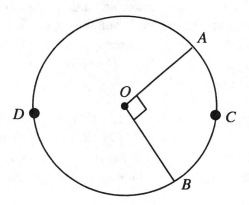

In previous situations, when an arc was mentioned, the reference was to the short arc. In this chapter, three letters will sometimes be used to avoid ambiguity. In the figure above, O is the center of the circle and $\angle AOB$ is a right angle. Then,

$m \overset{\frown}{ACB} = 90°$ and $m \overset{\frown}{ADB} = 270°$.

A line which intersects a circle in two points is a secant. In the figure below, $\overleftrightarrow{BC}$ and $\overleftrightarrow{DE}$ are secants with a common point A, and this point is exterior to the circle. The measure of the angle formed at A is equal to the difference between the measures of the intercepted arcs, $\overarc{BFD}$ and $\overarc{CGE}$.

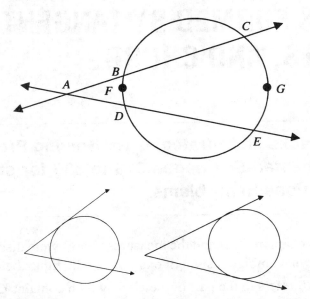

Similarly, both in the case of an angle formed by the intersection of a secant and a tangent, as well as the case of an angle formed by the intersection of two tangents, the measure of the angle outside the circle is equal to half the difference of the measures of the intercepted arcs. These three valuable relationships are established in Problem 371 and are then used to solve several other problems in this chapter.

In the figures below, intersecting chords, secants, and tangents are shown in various ways. In each case the lengths of various segments denoted by x, y, z, and w are related and the relationships are established in problems in this chapter. There are several problems in which application of these relationships is made.

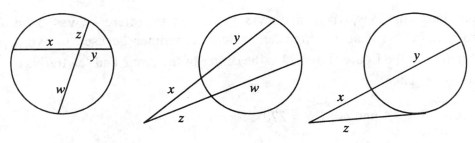

Step-by-Step Solutions to Problems in this Chapter, "Angles Formed by Tangents, Secants, and Chords"

TANGENTS

● **PROBLEM** 355

Prove that the measure of the angle formed by a tangent and a chord of a circle is one-half the measure of its intercepted arc.

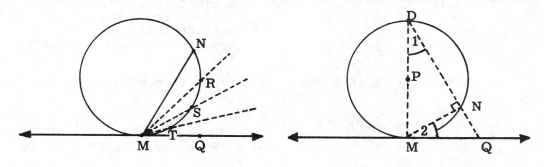

Figure 1   Figure 2

<u>Solution:</u> To see why the statement must be true, consider the inscribed angle ⊀NMR. We know that m⊀NMR = $\frac{1}{2}$N͡R. Similarly, for the larger angle ⊀NMS, it is true that m⊀NMS = $\frac{1}{2}$mN͡RS. Even as we pick points on the circle closer to M, the statement continues to be true. At the limit, the series of lines M⃡R, M⃡S, M⃡T ... approach the tangent M͞Q, and what we shall prove here is that the relationship still holds: m⊀NMQ = $\frac{1}{2}$mN͡SM.

There are two methods for proving the relation and both involve relating the angle ⊀NMQ with some inscribed angle of the circle that intercepts the same arc.

Method 1: Let D be the point on the circle such that D͞M is a diameter. Define Q such that Q is the intersection of D⃡N and the tangent. Inscribed angle ⊀1 = $\frac{1}{2}$mM͡N. If we prove ⊀2 ≅ ⊀1, then ⊀2 = $\frac{1}{2}$mM͡N and the proof is complete. We prove

313

∡2 ≅ ∡1 by showing ΔNMQ ∿ ΔMDQ by the A-A Similarity Theorem.

Given: $\overleftrightarrow{MQ}$ is tangent to ⊙P;
$\overline{DM}$ is a diameter of ⊙P;
$\overline{MN}$ is a chord of ⊙P.

Prove: m∡NMQ = $\frac{1}{2}$m$\overarc{MN}$

Statement	Reason
1. $\overleftrightarrow{MQ}$ is tangent to ⊙P; $\overline{DM}$ is a diameter of ⊙P; $\overline{MN}$ is a chord of ⊙P.	1. Given.
2. $\overline{PM} \perp \overline{MQ}$.	2. The tangent to a circle is perpendicular to the radius of the circle drawn to the point of intersection.
3. ∡DMQ is a right angle.	3. Definition of perpendicularity.
4. ΔDMN is a right triangle.	4. If one side of a triangle inscribed in circle is a diameter, then the triangle is a right triangle with the hypotenuse being the diameter.
5. ∡DNM is a right angle.	5. Definition of a right triangle.
6. ∡MNQ is a right angle	6. If one angle of a linear pair is a right angle, then the other angle is also a right angle.
7. ∡MNQ ≅ ∡DMQ	7. All right angles are congruent.
8. ∡NQM ≅ ∡NQM	8. A-A Similarity Theorem.
9. ∡1 ≅ ∡2	9. Corresponding angles of similar triangles are congruent.
10. m∡1 = $\frac{1}{2}$m$\overarc{MN}$	10. The measure of an inscribed angle equals one-half the measure of the intercepted arc.
11. m∡2 = $\frac{1}{2}$m$\overarc{MN}$	11. Substitution Postulate

Method 2: Here, again, we show ∡1 ≅ ∡2. By showing ∡1 and ∡DMN are complementary, and ∡2 and ∡DMN are complementary, we can equate ∡1 and ∡2 and, therefore, m∡1 = m∡2 = $\frac{1}{2}$m$\overarc{MN}$.

Statement	Reason

The first 5 steps are the same as the last proof.

Statement	Reason
6. ∡1 and ∡DMN are complementary.	6. The acute angles of a right triangle are complementary.
7. ∡2 and ∡DMN are complementary	7. If the sum of two angles is 90°, then they are complementary.
8. ∡1 ≅ ∡2	8. Two angles that are complementary to the same angle are congruent.
9. m∡1 ≅ $\frac{1}{2}$m⌢MN	9. The measure of an inscribed angle equals one-half the measure of the intercepted arc.
10. m∡2 = $\frac{1}{2}$m⌢MN	10. Substitution Postulate.

● **PROBLEM** 356

$\overline{NM}$ is a chord of ⊙P. $\overleftrightarrow{LMQ}$ is tangent to ⊙P at M. m$\overset{\frown}{MDN}$ = 310. Find m∡NMQ.

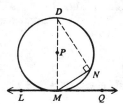

Solution: The measure of the angle formed by a tangent and a chord of a circle is one-half the measure of its intercepted arc. Therefore m∡NMQ = $\frac{1}{2}$m$\overset{\frown}{MN}$. To find m$\overset{\frown}{MN}$, note that m$\overset{\frown}{MDN}$ + m$\overset{\frown}{MN}$ = 360. Since m$\overset{\frown}{MDN}$ = 310, m$\overset{\frown}{MN}$ = 360 - 310 or m$\overset{\frown}{MN}$ = 50. By substitution, m∡NMQ = $\frac{1}{2}$(50). Therefore, m∡NMQ = 25°.

● **PROBLEM** 357

The angle formed by two tangents drawn to a circle from the same external point measures 80°. Find the measure of the minor intercepted arc.

Solution: The measure of an angle formed by two tangents drawn to a circle from an outside point is equal to one-half the difference of the measures of the intercepted arcs.

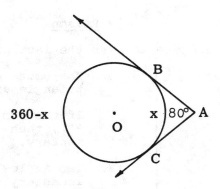

If we let x = the measure of minor arc $\overset{\frown}{BC}$, then 360°-x = the measure of the major arc $\overset{\frown}{BC}$. Applying the above rule,

$$m\measuredangle BAC = \frac{1}{2}(m(\text{major } \overset{\frown}{BC}) - m(\text{minor } \overset{\frown}{BC})).$$

By substitution,

$$80° = \frac{1}{2}(360° - 2x).$$

80° = 180° - x, which implies x = 100°. Therefore, the measure of the minor intercepted arc is 100°.

An alternative solution for this problem is as follows. Draw $\overline{BC}$. Since tangents to a circle from the same external point are congruent, triangle ABC is an isosceles triangle. Hence, m∡ABC = m∡ACB. But the sum of the measures of the angles of any triangle is 180°. Therefore,

m∡BAC + m∡ABC + m∡ACB = 180°.

Since m∡ABC = m∡ACB and m∡BAC = 80°, we obtain

m∡ABC = m∡ACB = 50°.

Angle ABC is formed by a tangent and a chord. Therefore, its measure equals $\frac{1}{2}$ the measure of the intercepted arc; i.e., m∡ABC = $\frac{1}{2}$m$\overset{\frown}{BC}$. By substituting 50° for the m∡ABC, we can calculate the m$\overset{\frown}{BC}$ to be 100°. This method is less direct than the first, but is a valid option.

● **PROBLEM** 358

Circles O and O' are tangent internally at B, with $\overleftrightarrow{BF}$ as the common tangent line. $\overline{BD}$ is a chord of circle O. C is a point on $\overline{BD}$. Prove: m$\overset{\frown}{BC}$ = m$\overset{\frown}{BD}$.

Solution: We will show that both arcs must have the same measure by relating both to the m∡DBF.

∡DBF intercepts arc $\overset{\frown}{BD}$ in circle O and arc $\overset{\frown}{BC}$ in circle O'. In circle O, ∡DBF is formed by a tangent line and a

316

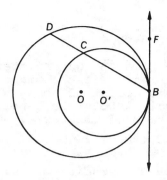

chord drawn from the point of contact. In circle O', the angle is formed by a secant ray and a tangent ray with its vertex on the circle.

In either case, the angle measure is equal to one-half the measure of the intercepted arc. Hence,

$$m\angle DBF = \frac{1}{2}m\overset{\frown}{BD} \quad \text{and} \quad m\angle DBF = \frac{1}{2}m\overset{\frown}{BC}.$$

It follows that $m\overset{\frown}{BD} = 2m\angle DBF$ and $m\overset{\frown}{BC} = 2m\angle DBF$. Therefore, by transitivity, $m\overset{\frown}{BD} = m\overset{\frown}{BC}$.

● **PROBLEM** 359

If perpendiculars from the ends of a diameter of any circle are drawn to a tangent of that circle, prove that the sum of the lengths of the perpendiculars is equal to the length of the diameter.

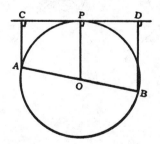

<u>Solution:</u> In the accompanying figure, tangent $\overline{CD}$ intersects ⊙O at P. From the endpoints of diameter $\overline{AB}$, $\overline{AC}$ and $\overline{BD}$ are drawn to meet $\overline{CD}$ perpendicularly. We must show that AC + BD = AB. We do this by showing (1) $\overline{OP}$ is the median of trapezoid ABDC and, as such, AC + DB = 2 · OP and (2) diameter AB = 2 · radius OP. Thus, AC + BD = AB.

From the given, we have $\overline{AC} \perp \overline{CD}$ and $\overline{DB} \perp \overline{CD}$. Since $\overline{OP}$ is the radius to the point of tangency, we have $\overline{OP} \perp \overline{CD}$. Since $\overline{AC}$, $\overline{DB}$, and $\overline{OP}$ are perpendicular to the same line, $\overline{AC} \parallel \overline{DB} \parallel \overline{OP}$. Parallel lines $\overline{AC}$, $\overline{DB}$, and $\overline{OP}$ cut $\overline{AB}$ into equal segments $\overline{AO}$ and $\overline{OB}$ (because all radii of ⊙O are congruent). Therefore, $\overline{AC}$, $\overline{DB}$, and $\overline{OP}$ cut equal segments in

every other transversal. $\overleftrightarrow{CD}$ is a transversal of the parallel
lines. Therefore, CP = PD.

Note that since $\overline{AC} \parallel \overline{DB}$, quadrilateral ABDC is a trapezoid. Furthermore, since P is the midpoint of nonparallel
side $\overline{CD}$ and O is the midpoint of nonparallel side $\overline{AB}$, $\overline{OP}$ is
the median of trapezoid ABDC. The length of the median of a
trapezoid is the average of the bases. That is,

$$OP = \frac{1}{2}(AC + BD) \text{ or } AC + BD = 2 \cdot OP.$$

Remember that OP is a radius of ⊙O. Therefore, diameter AB = 2 · OP. Since 2 · OP = AC + BD, AB = AC + BD. Thus,
the lengths of the perpendiculars, $\overline{AC}$ and $\overline{BD}$, sum to the diameter.

● **PROBLEM** 360

Given: ⊙P is externally tangent to ⊙Q at T; $\overline{DTC}$ (that is,
D, T, and C are collinear); D is a point of ⊙P; C is a point
of ⊙Q; T is the point of intersection of ⊙Q and ⊙P. $\overleftrightarrow{ACE}$ is
tangent to ⊙Q at C; $\overleftrightarrow{BDF}$ is tangent ⊙P at D. Let points A
and B be the intersection of the tangents with the common
internal tangent of ⊙P and ⊙Q.
Prove: $\overleftrightarrow{ACE} \parallel \overleftrightarrow{BDF}$
(Hint: Consider the common internal tangent.)

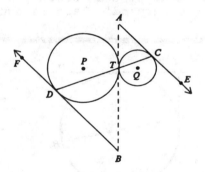

Solution: By our definitions, $\overline{DB}$ and $\overline{BT}$ are tangents drawn
to ⊙P from external point B. Therefore, DB = BT. Similar-
ly, the tangents, $\overline{AC}$ and $\overline{AT}$, drawn from external point A to
⊙Q are congruent. Thus, ΔBDT and ΔATC are isosceles tri-
angles. Since base angles of isosceles triangles are congruent, ∢BDT ≅ ∢BTD and ∢ATC ≅ ∢ACT.

To show $\overleftrightarrow{ACE} \parallel \overleftrightarrow{BDF}$, we show that a pair of alternate
interior angles, ∢BDT and ∢ACT, are congruent. Note that
∢BTD and ∢ATC are vertical angles. Thus, ∢BTD ≅ ∢ATC. From
the above, we know that ∢BDT ≅ ∢BTD and ∢ATC ≅ ∢ACT. By the
Transitive Property, we obtain ∢BDT ≅ ∢ACT. Since alternate
interior angles are congruent, $\overleftrightarrow{ACE} \parallel \overleftrightarrow{BDF}$.

If P and Q are points on the circumference of two concentric circles, prove that the angle included between the tangents at P and Q is congruent to that subtended at the center by $\overline{PQ}$.

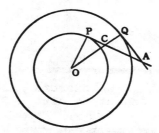

Solution: In the figure, the tangents drawn from P and Q intersect at A. C is the intersection of $\overline{OQ}$ and $\overline{AP}$. We are asked to prove ∡PAQ ≅ ∡POQ.

We will prove the required congruency by showing that both angles are complements of the same angle.

Statement	Reason
1. $\overline{QA}$ tangent to external circle O.	1. Given.
2. $\overline{OQ} \perp \overline{QA}$	2. The radius of a circle drawn to a point of tangency is perpendicular to the tangent line.
3. ∡CQA is a right angle.	3. Perpendiculars intersect to form a right angle.
4. ΔCQA is a right triangle.	4. A triangle containing a right angle is a right triangle.
5. ∡PAQ and ∡ACQ are complements.	5. The acute angles of a right triangle are complements.
6. ∡ACQ ≅ ∡OCP	6. Vertical angles are congruent.
7. ∡PAQ and ∡OCP are complements.	7. If one angle is complementary to a given angle, then it is complementary to any angle congruent to the given angle.
8. $\overline{OP} \perp \overline{PA}$	8. Reason 2.
9. ∡OPC is a right angle.	9. Reason 3.

10. ΔCOP is a right tri-angle.	10. Reason 4.
11. ∢POQ and ∢OCP are complements.	11. Reason 5.
12. ∢POQ ≅ ∢PAQ.	12. Complements of the same angle are congruent to each other.

● **PROBLEM** 362

As shown in the diagram, $\overline{BD}$ bisects ∢ABC, which is inscribed in circle O. $\overleftrightarrow{EC}$ is tangent to circle O at point C. Prove that ∢ABD ≅ ∢DCE.

Solution: This proof will involve recalling that both the measure of an inscribed angle in a circle and the measure of an angle formed by a tangent and a chord to that tangency point are equal to one-half the measure of the intercepted arc. With this in mind, we can prove ∢DCE ≅ ∢DBC. By applying the definition of an angle bisector and the transitive property of congruence, we can conclude that ∢ABD = ∢DCE.

Statement	Reason
1. $\overline{BD}$ bisects ∢ABC	1. Given.
2. ∢ABD ≅ ∢DBC	2. A bisector divides an angle into two congruent angles.
3. $\overleftrightarrow{EC}$ is tangent to the circle.	3. Given.
4. m∢DCE = $\frac{1}{2}$m$\overparen{DC}$	4. The measure of an angle formed by a tangent and a chord at the point of tangency is equal to one-half the measure of the intercepted arc.
5. m∢DBC = $\frac{1}{2}$m$\overparen{DC}$	5. The measure of an inscribed angle is equal to one-half the measure of the intercepted arc.
6. m∢DBC = m∢DCE	6. Transitive property of equality.

7. ∢DBC ≅ ∢DCE	7. Two angles are congruent if their measures are equal.
8. ∢ABD ≅ ∢DCE	8. Transitive property of congruence.

● **PROBLEM** 363

In the accompanying diagram, △ABC is inscribed in the circle. The ratio of the mBĈ: mCÂ: mÂB is 2:3:5. Find the measure of the acute angle formed by side BC̄ and the tangent to the circle at B(BD̂).

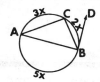

Solution: To solve this problem, determine the measure of each arc drawn, and then apply the theorem stating that the measure of an angle formed by a tangent to a circle and a chord drawn from the point of contact is equal to one-half the measure of its intercepted arc. ∢DBC is formed in just such a way, and its measure is to be determined.

Represent the measures of arcs BĈ, CÂ, and ÂB by 2x, 3x, and 5x, respectively. To solve for x, sum the measures of the arcs and set the sum equal to 360°, the measure of the whole circle. We have

$$2x + 3x + 5x = 360°$$
$$10x = 360°$$
$$x = 36°$$
$$mB\widehat{C} = = 2x° = 72°$$

Therefore, by the theorem previously mentioned, m∢DBC = $\frac{1}{2}$mBĈ = $\frac{1}{2}$(72°) = 36°. Therefore, acute angle DBC measures 36°.

SECANTS

● **PROBLEM** 364

Given: N is an interior point of ⊙P; M is an exterior point of ⊙P.
Prove: m ∢ N > m ∢ M.

Solution: When two angles of a triangle are not equal in measure, the angle of greater measure lies opposite the longer side. ∢ N and ∢ M are both angles of △ PMN. By showing PM > PN, it then follows that m ∢ N > m ∢ M.

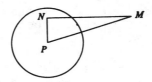

To show PM > PN, we relate both to the radius of ⊙P. Note that
$\overline{PM}$ is a segment drawn from the center of the circle to a point outside
the circle. The exterior of the circle contains all points at a distance
greater than the radius. Thus, PM > radius of ⊙P. Also, note that
$\overline{PN}$ is a segment drawn from the center of the circle to a point inside
the circle. The interior of the circle contains all points at a distance
less than the radius. Thus, PN < radius of ⊙P.

By the Transitive Property of Inequality, it then follows that PM >
radius of ⊙P > PN or PM > PN. Since ∡ N lies opposite the longer
side, m ∡ N > m ∡ M.

● **PROBLEM** 365

Given: A circle with tangent $\overline{TP}$ and secant SP intersecting
at point P. Prove:

$$\frac{PS}{PT} = \frac{PT}{PR} \qquad \text{or} \quad (PT)^2 = PS \times PR.$$

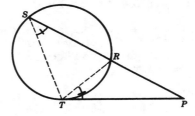

Solution: When solving a segment proportion problem, it
is necessary to locate similar triangles which contain
the segments as corresponding parts.

We will use the A.A. similarity theorem.

∡ TSR is an inscribed angle of the given circle and,
as such, its measure is equal to one-half the measure of
the intercepted arc. Consequently,

m ∡ TSR = ½ m $\overarc{TR}$.

∡ RTP is an angle formed by a tangent and a chord
drawn from the point of tangency. Its measure is one-half
the measure of its intercepted arc. Accordingly,

m ∡ RTP = ½ m $\overarc{TR}$.

Hence, by transitivity, m ∡ TSR = m ∡ RTP, or ∡ TSR ≅ ∡ RTP.

By the reflexive property of congruence, ∡ SPT ≅ ∡ RPT.
By the A.A. similarity theorem, ΔPST ∼ ΔPTR.

322

Since corresponding sides of similar triangles have the same ratio of similitude,

$$\frac{PS}{PT} = \frac{PT}{PR}.$$

In a proportion, the product of the means equals the product of the extremes and, as such,

$$(PT)^2 = PS \times PR.$$

● **PROBLEM** 366

Two secant lines of the same circle share an endpoint in the exterior of the circle. Show that the product of the lengths of one secant segment and its external segment equal the product of the lengths of the other secant segment and its external segment.

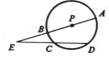 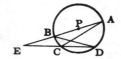

Solution: The secant segment is the portion of the secant line bounded by the common endpoint and the point of intersection with the circle farthest from this point.

Restating the problem in terms of the above figure, we must show that EB · EA = EC · ED. We can rewrite this as the proportion $\frac{EB}{ED} = \frac{EC}{EA}$. To prove the proportion, we show ΔEBD ∿ ΔECA by the A–A Similarity Theorem.

Given: $\overline{ABE}$ and $\overline{DCE}$ are secant segments of ⊙P. Point E is exterior to OP.
Prove: EB · EA = EC · ED.

Statements	Reasons
1. $\overline{ABE}$ and $\overline{DCE}$ are secant segments of ⊙P. Point E is exterior to ⊙P.	1. Given.
2. m∢BAC = $\frac{1}{2}$m$\overset{\frown}{BC}$ m∢BDC = $\frac{1}{2}$m$\overset{\frown}{BC}$	2. The measure of an inscribed angle equals one-half the measure of the subtended arc.
3. m∢BAC = m∢BDC	3. Transitivity Postulate.
4. ∢BAC ≅ ∢BDC	4. Definition of congruence of angles.
5. ∢E ≅ ∢E	5. An angle is congruent to itself.

323

6. △EBD ∼ △ECA

6. The A-A Similarity Theorem.

7. $\dfrac{EB}{ED} = \dfrac{EC}{EA}$

7. The sides of similar triangles are proportional.

8. EB · EA = EC · ED.

8. In a proportion, product of the means must equal the product of the extremes.

● **PROBLEM** 367

$\overline{PB}$ and $\overline{PD}$, which are secants drawn to circle O, intersect the circle in points A and C, respectively. In the figure shown, if PA = 4, AB = 5, and PD = 12, find PC.

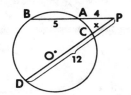

Solution: When two secants from a point outside the circle are drawn to a circle, as in this example, the product of the length of one secant and the length of its external segment is equal to the product of the length of the other secant and the length of its external segment. (In the figure, $\overline{PA}$ and $\overline{PC}$ are external segments.)

The length of $\overline{PB}$ equals the length of $\overline{PA}$ plus the length of $\overline{AB}$; i.e. PB = 4 + 5 = 9.

The above theorem allows us to state that PB × PA = PD × PC If we let PC = x, then, by substitution of the given, we obtain 9 × 4 = 12x. Solving for x, x = 3.

Therefore, the length of PC is 3.

● **PROBLEM** 368

From a point outside a circle, a tangent and a secant are drawn to the circle. The point, C, at which the secant intersects the circle divides the secant into an external segment of length 4 ($\overline{PC}$) and an internal segment of length 12 ($\overline{AC}$). Find the length of the tangent ($\overline{PD}$). (See figure.)

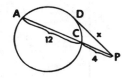

<u>Solution:</u> In a situation of this type, a theorem tells us
that the length of the tangent is the mean proportional between
the length of the secant and the length of its external segment. By consulting the accompanying figure, we see that
this statement can be presented as PA : PD = PD : PC. If we
let x = the length of tangent $\overline{PD}$, and then substitute the
given information in the proportion, we obtain: 16:x = x:4.
Since the product of the means equals the product of the extremes in a proportion, we can calculate x, the tangent
length, in the following way:

$$x^2 = 16 \times 4 = 64$$
$$x = 8.$$

Therefore, the length of tangent $\overline{PD}$ is 8.

● **PROBLEM** 369

In the accompanying figure, the length of a radius of circle
O is 4. From a point P outside circle O, tangent $\overline{PA}$ is
drawn. If the length of $\overline{PA}$ is 3, find the distance from P
to the circle.

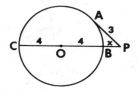

<u>Solution:</u> Let B be the point of intersection of the circle
with the segment connecting point P to the center of ⊙O. $\overline{PB}$
is the shortest segment that can be drawn from P to any point
on the circle. Therefore, PB is the distance from P to the
circle.

Now locate point C such that $\overline{PC}$ intersects the circle
at B and C and passes through both O and P.

From this, we can determine that $\overline{PC}$ is a secant to
circle O and, as such, the tangent $\overline{PA}$ will be, by theorem,
the mean proportional between the length of the secant, $\overline{PC}$, and
the length of its external segment, $\overline{PB}$. Thus, PC:PA = PA:PB
is the proportion.

Diameter CB = 8, since we are given the radius of the
circle as 4.

If we let x = the length of $\overline{PB}$, then x + 8 = the length
of $\overline{PC}$. By substitution, (x+8):3 = 3:x. As stated in prior
examples, the product of the means equals the product of the
extremes. Therefore, x(x+8) = 9. By the distributive property, this becomes $x^2 + 8x = 9$ or $x^2 + 8x - 9 = 0$. To solve
for x, factor the left side of the equation and set each
factor equal to zero. (x-1) and (x+9) are the factors.
x-1 = 0 implies x = 1 and x+9 = 0 implies x = -9. The nega-

tive value has no geometric significance and is rejected. Therefore, the distance from P to the circle, PB, equals 1.

$\overline{EBA}$ and $\overline{EGD}$ are secant segments of ⊙P. Chord $\overline{AF}$ intersects $\overline{EGD}$ at point G. If EB = 5, BA = 7, EC = 4, GD = 3, and AG = 6, find GF.

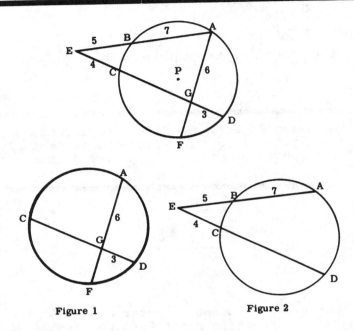

Figure 1   Figure 2

<u>Solution:</u> We can look at this problem as two separate problems.

From Fig. 2, we can find the length of CD. From the lengths of CD and GD in Fig. 1, we obtain CG. Given CG, GD, and GA, GF can be obtained.

If two secant segments of the same circle share an endpoint in the exterior of the circle, then the product of the lengths of one secant segment and its external segment must equal the product of the lengths of the other secant segment and its external segment. From Fig. 2, therefore,

(i)   EB · EA = EC · ED.

Substituting in values EB = 5, EA = EB + BA = 5 + 7 = 12, and EC = 4, we obtain

(ii)   5 · 12 = 4 · ED

(iii)   ED = $\frac{1}{4}$ · 60 = 15.

Since the whole equals the sum of its parts, ED = EC + CD, hence CD = ED - EC.

326

Therefore, CD = 15 - 4 = 11.

If two chords intersect in the interior of a circle, thus determining two segments in each chord, then the product of the lengths of the segment of one chord equals the product of the lengths of the segments of the other cord. In Fig. 1, therefore,

    (iv) AG · GF = CG · GD.

AG = 6, and GD = 3. CG = CD - GD = 11 - 3 = 8. Substituting these values, we obtain

    (v) 6 · GF = 8 · 3 = 24.
    (vi) GF = $\frac{1}{6}$ · 24 = 4.

● **PROBLEM** 371

Prove: Whether an angle that is exterior to the circle is formed by two secants to the circle, a secant and a tangent to the circle, or two tangents to the circle, the measure of the angle equals the difference in the measures of the intercepted arcs.

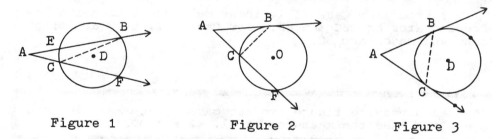

Figure 1   Figure 2   Figure 3

<u>Solution:</u> Though three separate proofs are needed, the reasoning for all three is the same. The cases of Fig. 2 and Fig. 3 are actually special cases of Fig. 1. To see this, imagine that ray AB is slowly rotating upward with A fixed. As AB rotates, the points where it intersects the circle, B' and E', get closer and closer. At the point where they meet, secant AB becomes tangent AB, or Fig. 2. If we now rotate AF downward, we obtain Fig. 3.

Proof: To prove that m∢A equals the difference of the intercepted arcs, we draw line segment $\overline{BC}$. ∢A and ∢ABC are thus interior angles of ∆ABC, and ∢BCF is the remote exterior angle. We then proceed to (1) find ∢A in terms of ∢ABC and ∢BCF, (2) use the fact that ∢ABC and ∢BCF are inscribed angles to express them in terms of the intercepted arcs, and (3) substitute the expressions derived in (2) into (1) to arrive at an expression for m∢A in terms of the intercepted arcs.

Case I. Given $\overleftrightarrow{AEB}$ and $\overleftrightarrow{ACF}$ are secants of ⊙D. A is a point exterior to ⊙D.

327

Prove: $m\angle A = \frac{1}{2}(m\overparen{BF} - m\overparen{EC})$.

Statements	Reasons
1. $\overrightarrow{AEB}$ and $\overrightarrow{ACF}$ are secants of $\odot D$. A is a point exterior to $\odot D$.	1. Given
2. $m\angle BCF = \frac{1}{2}m\overparen{BF}$ $m\angle EBC = \frac{1}{2}m\overparen{EC}$	2. The measure of an inscribed angle equals one-half the measure of the intercepted arc.
3. $m\angle BCF = m\angle A + m\angle EBC$	3. The measure of an exterior angle of a triangle equals the sum of the measures of the remote interior angles.
4. $\frac{1}{2}m\overparen{BF} = m\angle A + \frac{1}{2}m\overparen{EC}$	4. Substitution Postulate.
5. $m\angle A = \frac{1}{2}m\overparen{BF} - \frac{1}{2}m\overparen{EC}$	5. Subtraction Property of Equality.
6. $m\angle A = \frac{1}{2}(m\overparen{BF} - m\overparen{EC})$	6. Distributive Property.

The proof for Case II is identical except that the intercepted arcs are $\overparen{BC}$ and $\overparen{BF}$. For Case III, they are minor arc $\overparen{BC}$ and major arc $\overparen{BC}$.

● **PROBLEM** 372

In the accompanying figure, we are given that $m\overparen{AE} = 80°$ and $m\angle C = 20°$. Find the measures of $\angle 1$, $\angle 2$ and $\overparen{BD}$.

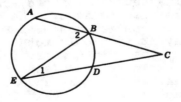

Solution: $\angle 2$ is an inscribed angle which intercepts an arc, $\overparen{AE}$, of $80°$. Recall that the measure of an inscribed angle is equal to one-half the measure of its intercepted arc. Hence, $\angle 2 = 40$.

We can derive $m\overparen{BD}$ by noticing that it is intercepted by $\angle C$. $\angle C$ is formed by the intersection of two secants outside the circle. Recall that an angle formed in this way is equal to one-half the difference of its intercepted arcs. The intercepted arcs of $\angle C$ are $\overparen{AE}$ and $\overparen{BD}$. Therefore, $\angle C = \frac{1}{2}(\overparen{AE} - \overparen{BD})$. By substituting the given values for $\angle C$ and $\overparen{AE}$, we can solve for $m\overparen{BD}$.

$$20° = \frac{1}{2}(80° - \overset{\frown}{BD})$$

$$40° = 80° - \overset{\frown}{BD}$$

$$\overset{\frown}{BD} = 40°.$$

Since ∢1 is an inscribed angle which intercepts this 40° arc, ∢1 must measure one-half of 40°, or 20°.

Therefore, m∢1 = 20°, m∢2 = 40°, and m$\overset{\frown}{BD}$ = 40°.

● **PROBLEM** 373

In circle O, diameter $\overline{AB}$ is extended to point C. Line $\overleftrightarrow{CF}$ intersects the circle at points D and E. If $\overline{DC} \overset{\sim}{=} \overline{OE}$ show that m∢EOA is three times as large as m∢ACE. [Hint: Draw radius $\overline{OD}$].

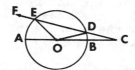

Solution: The procedure followed in this problem is to let x = m∢ACE. Then, by a series of applications of rules for isosceles triangles, the measures of supplements and the sum of measures of angles in a triangle, we will deduce that m∢EOA = 3x.

Since they are radii of the same circle, $\overline{OD} \overset{\sim}{=} \overline{OE}$. We are given $\overline{OE} \overset{\sim}{=} \overline{DC}$. Therefore, by transitivity, $\overline{OD} \overset{\sim}{=} \overline{DC}$. As such, ΔODC is an isosceles triangle and has base angles, m∢DOB = m∢DCO. We let x equal m∢DCO.

As the third angle in ΔODC,

m∢ODC = 180 - (x + x) = 180 - 2x.

Since $\overleftrightarrow{FC}$ is a straight line, ∢ODE and ∢ODC are supplements. Therefore,

m∢ODE + m∢ODC = 180.

By substitution, we obtained

m∢ODE + (180 - 2x) = 180

m∢ODE = 180 - (180 - 2x) = 2x.

Triangle ODE is isosceles, since two sides are congruent. Its base angles, ∢OED and ∢ODE, are congruent. m∢OED + m∢ODE = 2(2x) = 4x. The third angle of ΔODE, ∢EOD, has a measure equal to 180 - (m∢OED + m∢ODE) = 180 - 4x.

329

∢EOC is the supplement of the angle whose measure we wish to determine, ∢EOA.

m∢EOC = m∢EOD + m∢DOB.

By substitution m∢EOC = (180 − 4x) + (x)

m∢EOC = 180 − 3x.

Since they are supplements,

m∢EOC + m∢EOA = 180.

Again, substituting, 180 − 3x + m∢EOA = 180.

Therefore, m∢EOA = 3x. Hence, we have shown m∢EOA = 3(m∢ACE).

● **PROBLEM** 374

In the accompanying figure, $\overleftrightarrow{AB}$ is tangent to circle O at point B, $\overline{AEC}$ is a secant and $\overline{DE}$, $\overline{FC}$, and $\overline{BC}$ are chords. The measure of $\overparen{EB}$ is 50°, the measure of $\overparen{BC}$ is (4x − 50°), the measure of $\overparen{CD}$ is x, the measure of $\overparen{DF}$ is (x + 25°), and the measure of $\overparen{FE}$ is (x − 15°).

a) Find the measures of $\overparen{BC}$, $\overparen{CD}$, $\overparen{DF}$, and $\overparen{FE}$

b) Find the measures of ∢a, ∢b, ∢c, and ∢d.

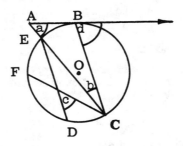

Solution: (a) To determine the unknown arc measures, find the sum of the given representations of the measures, set this equal to 360°, and then solve for x.

$$m\overparen{EB} + m\overparen{BC} + m\overparen{CD} + m\overparen{DF} + m\overparen{FE} = 360°$$

Substituting the given representations for these arcs,

50° + 4x − 50° + x + x + 25° + x − 15° = 360°

Combining and solving,   7x + 10° = 360°

7x = 350°

x = 50°

Therefore,

330

$$m\overset{\frown}{BC} = 4(50°) - 50° = 150°$$

$$m\overset{\frown}{CD} = x = 50°$$

$$m\overset{\frown}{DF} = x + 25°=50° + 25° = 75°$$

$$m\overset{\frown}{FE} = x - 15° = 50° - 15° = 35°$$

(b) ∢a is formed by a tangent and a secant, and the measure of this angle is one-half the difference of the measures of the arcs it intercepts, i.e. 1/2 (major arc-minor arc). m∢a = 1/2 (m$\overset{\frown}{BC}$ - m$\overset{\frown}{EB}$). By substitution,

$$m∢a = 1/2 \ (150° - 50°) = 1/2 \ (100°) = 50°$$

Therefore, the measure of ∢a is 50°.

∢b is inscribed, and its measure is equal to 1/2 the measure of the intercepted arc.

$$m∢b = 1/2 \ (m\overset{\frown}{EB}) = 1/2 \ (50°) = 25°$$

Therefore, the measure of ∢b is 25°.

∢c is formed by the intersection of two chords, and its measure is equal to one-half the sum of the measures of the intercepted arcs.

$$m∢c = 1/2 \ (m\overset{\frown}{CD} + m\overset{\frown}{FE})$$

$$= 1/2 \ (50° + 35°)$$

$$= 1/2 \ (85°) = 42.5°$$

Therefore, the measure of ∢c is 42.5°.

∢d is formed by a tangent and a chord to the point of tangency. Its measure is equal to one-half the measure of the intercepted arc.

$$m∢d = 1/2 \ (m\overset{\frown}{BC}) = 1/2 \ (150°) = 75°$$

Therefore, the measure of ∢d is 75°.

● **PROBLEM** 375

As seen in the figure, $\overleftrightarrow{PA}$ is a tangent and $\overleftrightarrow{PCB}$ is a secant drawn to the circle. The measure of angle P is represented by x and the measures of $\overset{\frown}{AB}$, major arc $\overset{\frown}{BC}$ and $\overset{\frown}{CA}$ are (3x - 20°), 6x, and y, respectively.

    a)   In terms of x and y, write a set of equations that can be used to solve for x and y.
    b)   Solve the equations written in the answer to part (a).
    c)   Find the number of degrees contained in $\overset{\frown}{BAC}$.

Solution: a) ∢P is formed by a tangent and a secant and, as such, its measure is equal to one-half the difference be-

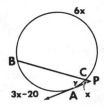

tween the major and minor arcs they intercept. This fact is
used to derive the first equation.

$$m \angle P = \frac{1}{2}(m\overset{\frown}{BA} - m\overset{\frown}{AC}).$$

Substituting, $x = \frac{1}{2}(3x - 20° - y)$.

$$2x = 3x - 20° - y$$
$$20° = x - y.$$

This is the first equation of the set. The second equation
is derived by setting the sum of the measures of each arc
equal to the measure of the whole circle, 360°.

$$m\overset{\frown}{AC} + m\overset{\frown}{CB} + m\overset{\frown}{BA} \qquad = 360°$$
$$y + 6x \ + 3x - 20° = 360°$$
$$9x + y - 20° = 360°$$
$$9x + y \ = 380°.$$

This is the second equation of the set. Therefore, the set
of equations to solve for x and y is

$$x - y = 20°$$
$$9x + y = 380°.$$

b) To determine a solution for the equations formed in part
(a), add them, solve for x and substitute back again to solve
for y.

$\left. \begin{array}{l} x - y = 20° \\ 9x + y = 380° \end{array} \right\}$ add these to obtain $10x = 400°$ or $x = 40°$

Substitute $x = 40°$ into $x - y = 20°$, and obtain $40° - y = 20°$
or $y = 20°$.

Therefore, the solution of the set of equations found
in part (a) is $x = 40°$, $y = 20°$.

c) The measure of $\overset{\frown}{BAC}$ can be found by adding the expressions
for the measure of $\overset{\frown}{BA}$ and $\overset{\frown}{AC}$ and then substituting the values
of x and y found in part (b).

$$m\overset{\frown}{BAC} = m\overset{\frown}{BA} + m\overset{\frown}{AC}$$
$$m\overset{\frown}{BAC} = 3x - 20° + y$$
$$m\overset{\frown}{BAC} = 3(40°) - 20° + 20° = 120° - 20° + 20° = 120°.$$

Therefore, the measure of arc $\overset{\frown}{BAC}$ is 120°.

332

Line segment $\overline{CD}$ is the diameter of circle O, as seen in the diagram, and $\overline{AD}$ is tangent to the same circle. Given that $\overline{ABC}$ is a straight line, prove that ΔABD and ΔDBC are mutually equiangular.

Solution: To show that two triangles are mutually equiangular, we must prove that all three angles in one triangle are congruent to the corresponding angles of the other triangle. In familiar notation, this states that A.A.A. ≅ A.A.A. for mutually equiangular triangles to exist.

The proof will proceed in three stages, each one dealing with a different angle.

Statement	Reason
1. $\overline{CD}$ is a diameter in circle O.	1. Given.
2. ∢CBD is a right angle	2. An angle inscribed in a semi-circle is a right angle.
3. $\overleftrightarrow{ABC}$ is a straight line.	3. Given.
4. ∢ABD is supplementary to ∢CBD.	4. If two adjacent angles have their non-common sides lying on a straight line, the angles are supplementary.
5. ∢ABD is a right angle.	5. The supplement of a right angle is a right angle.
6. ∢CBD ≅ ∢ABD	6. All right angles are congruent.
7. $\overleftrightarrow{AD}$ is tangent to circle O.	7. Given.
8. ∢CDA is a right angle.	8. A tangent to a circle is perpendicular to a radius at the point of contact.
9. ∢ADB is complementary to ∢BDC.	9. Two angles are complementary if the sum of their measures is 90°.
10. ΔBCD is a right triangle.	10. Any triangle that contains a right angle is a right triangle.

11. ∢BCD is complementary to ∢BDC.	11. The acute angles of a right triangle are complementary.
12. ∢BCD ≅ ∢ADB	12. If two angles are complements of the same angle, they are congruent.
13. ∢BDC ≅ ∢BAD	13. If two angles in one triangle are congruent to two angles in another triangle, the third angle in these triangles are congruent.
14. △ABD and △DBC are mutually equiangular.	14. If three angles in one triangle are congruent to three angles in another triangle, the triangles are mutually equiangular.

● **PROBLEM** 377

In the accompanying figure, △MBC is formed by the intersection of three tangents, two of which, $\overline{MN}$ and $\overline{MP}$, are fixed lines. The third tangent, $\overline{BC}$, is a variable tangent. Prove that no matter where point A lies, the angle, BOC, is constant in measure.

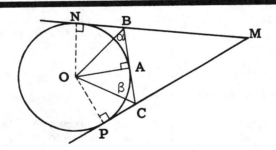

Solution: We wish to show m∢BOC does not change as the tangency point, A, changes. We do this by expressing m∢BOC in terms of m∢M. Since m∢M does not change as point A changes, m<BOC is constant.

Given: ⊙O with fixed tangents $\overline{MN}$ and $\overline{MP}$. The tangent at point A intersects $\overline{MN}$ at B and $\overline{MP}$ at C. Prove: m∢BOC is constant regardless of the position of A.

Statements	Reasons
1. (See given)	1. Given.
2. m∢M is constant.	2. If the sides of an angle are fixed, the angle does not change.
3. m∢M + m∢MBC + m∢MCB = 180°.	3. The angle sum of a triangle equals 180°.

4. $m\angle MBC + m\angle MCB = 180 - m\angle M$.	4. Subtraction Property of Equality.
5. $m\angle NBC + m\angle MBC = 180$ $m\angle BCP + m\angle MCB = 180$	5. Pairs of angles that form a linear pair sum to 180°.
6. $m\angle NBC + m\angle BCP + m\angle MBC + m\angle MCB = 360$	6. Addition Property of Equality.
7. $m\angle NBC + m\angle BCP = 360 - (m\angle MBC + m\angle MCB) = 360 - (180 - m\angle M) = 180 + m\angle M$.	7. The Subtraction Property of Equality. The property that the sum of the interior angles of a triangle equals 180, together with the substitution postulate.
8. $\angle ONB$ is a right angle $\angle OPC$ is a right angle	8. The intersection of a tangent with a radius at point of tangency forms a right angle.
9. $\overline{OB} \cong \overline{OB}$	9. Reflexive property.
10. $\overline{OB}$ is a hypotenuse	10. A sides of a triangle opposite a right angle is the hypotenuse.
11. $\overline{ON} \cong \overline{OP} \cong \overline{OA}$	11. All radii of a circle are equal.
12. $\triangle ONB \cong \triangle OAB$	12. The Hypotenuse-Leg Theorem.
13. $\triangle OAC \cong \triangle OPC$	13. The Hypotenuse-Leg Theorem.
14. $\angle NBO \cong \angle ABO$ ($\overline{OB}$ bisects $\angle AON$ $\angle NBA$) $\angle OCA \cong \angle OCP$ ($\overline{OC}$ bisects $\angle AOP$ $\angle ACP$)	14. Corresponding parts of congruent triangles are congruent, and definition of angle bisector.
15. $m\angle NBO = m\angle ABO = \frac{1}{2}m\angle NBC$. $m\angle OCA = m\angle OCP = \frac{1}{2}m\angle BCP$.	15. The Angle Addition Postulate,
16. $m\angle ABO + m\angle OCA = \frac{1}{2}(m\angle NBC + m\angle BCP) = \frac{1}{2}(180 + m\angle M)$.	16. Angle Addition Postulate, and Substitution Postulate from step 7.
17. $m\angle BOC + m\angle ABO + m\angle OCA = 180$.	17. The angle sum of a triangle is 180°.
18. $m\angle BOC + \frac{1}{2}(180 + m\angle M) = 180°$.	18. Substitution Postulate.

19. m∢BOC = 180 −
$\frac{1}{2}$(180 + m∢M) =
90 − $\frac{1}{2}$m∢M.

19. Subtraction Property of Equality.

20. 90 − $\frac{1}{2}$m∢M is a constant.

20. The subtraction of a constant from a constant is a constant.

21. m∢BOC is a constant.

21. If a quantity equals a constant, then the quantity is a constant.

CHAPTER 20

AREAS

Basic Attacks and Strategies for Solving Problems in this Chapter. See pages 337 to 349 for step-by-step solutions to problems.

All of the problems in this chapter concern circles. Two equivalent circumference formulas follow:

$$C = \pi d$$
$$C = 2\pi r$$

The formula for the area of a circle is

$$A = \pi r^2.$$

As mentioned above, the problems in this chapter involve circles, but often it is desired to find the area of a certain portion of a circle. The shaded region in the figure on the next page is called a **sector** of a circle. Since O is the center of the circle, it is not very difficult to find the area of the sector if the area of the circle is known and the measure of the central angle is known. For the example below, let $m \angle COB = 60°$. Then, since the circle "has" 360°, the area of the sector is 60/360 times the area of the circle. In general, if the measure of the central angle is $x°$, and if A is the area of the circle, then the area

of the sector is $\dfrac{x}{360} A$.

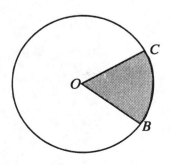

The area of the shaded region below could be found by subtracting the area of ∡ *AOB* from the area of the sector.

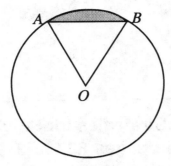

This technique of adding or subtracting areas of familiar regions to find areas of irregular regions is used in several problems in this chapter.

Step-by-Step Solutions to Problems in this Chapter, "Areas"

● **PROBLEM** 378

The circumference of a tree trunk is 6.6 ft.

   a) What is the diameter of the trunk?

   b) What is the area of a cross-section of the
      tree trunk?

Solution: To a first approximation, we may assume that the tree trunk is cylindrical, and that its cross-section is circular.

   a) The circumference, C, of a circle of diameter, d, is

$$C = \pi d.$$

In our case, we want d and we know C. Hence,

$$d = \frac{C}{\pi} = \frac{6.6}{3.14} \text{ ft.} = 2.10 \text{ ft.}$$

   b) The area, A, of a circle of radius, r, is

$$A = \pi r^2 \tag{1}$$

Since 2r = d, where d is the circle's diameter, we may rewrite (1) as

$$A = \pi (d/2)^2 = \frac{\pi d^2}{4}$$

Using the value of d found in part (a), we have

$$A = \frac{(3.14)}{(4)} (2.10 \text{ ft})^2$$

$$A = 3.46 \text{ ft.}^2$$

Find the area of a circle whose radius is 7 in. [Use $\pi = \frac{22}{7}$].

Solution: The area (A) of a circle is equal to π times the square of the length of the radius (r), i.e., $A = \pi r^2$.

By substitution, $A = \pi(7 \text{ in.})^2 = 49 \pi \text{ in.}^2$.

Use $\pi = \frac{22}{7}$. $A = 49(\frac{22}{7}) \text{ in.}^2 = 154 \text{ in.}^2$.

Therefore, the area = 154 square inches.

Find the area of a circle of radius 4. What is the area of the sector subtended by a central angle of $45°$.

Solution: The area of the circle is πr^2 . Therefore, for a circle of radius 4, the area is $\pi(4)^2 = 16\pi$.

The area of a sector with radius r and a central angle of measure n equals

$$\frac{n}{360} \cdot \pi r^2 .$$

In this case, $n = 45°$ and area equals $\frac{45}{360} \cdot \pi(4)^2 = \frac{1}{8} \cdot \pi 16 = 2\pi$.

The area of a circular region is known to be 154 sq. in.
a) What is the diameter of the circle?
b) What is its circumference?
(Assume $\pi = 22/7$.)

Solution: a) The area, A, of a circle of radius r is

$$A = \pi r^2 . \tag{1}$$

If the circle has diameter $d = 2r$, then we can rewrite (1) as

$$A = \pi(\frac{d}{2})^2 = \frac{\pi d^2}{4} \tag{2}$$

$$d = \sqrt{\frac{4A}{\pi}}$$

Using the given information, and $\pi = 22/7$

$$d = \sqrt{\frac{(4)(154 \text{ sq. in.})}{22/7}}$$

$$d = \sqrt{196 \text{ sq. in.}} = 14 \text{ in.}$$

b) The circumference, c, of a circle of diameter d is

$$c = \pi d .$$

Using $\pi = 22/7$ and the result of part (a),

$$c = (\frac{22}{7})(14 \text{ in.}) = 44 \text{ in.}$$

If the circumference of a circle is 88 ft., find the area of the circle. [Use π = 22/7 .]

Solution: Since the area, A, of a circle of radius r is $A = \pi r^2$, we must determine the length of the radius before we can calculate the area of the circle.

We are given the circumference, C, and knowing $C = 2\pi r$, we substitute C = 88 ft. and solve to find the value of r. Therefore,

$$88 \text{ ft.} = 2\pi r$$

and

$$r = \frac{88 \text{ ft.}}{2\pi} = \frac{44 \text{ ft.}}{\pi} .$$

Using π = 22/7, we find

$$r = 44\left(\frac{7}{22}\right) \text{ft.} = 2 \times 7 \text{ ft.} = 14 \text{ ft.}$$

Knowing r = 14 ft. and using π = 22/7 again, by substitution we can determine the area,

$$A = \left(\frac{22}{7}\right)(14)^2 \text{ft.}^2 = \left(\frac{22}{7}\right)(196)\text{ft.}^2 = 616 \text{ ft.}^2 .$$

Therefore, the area is 616 sq. ft.

The area of circle O is 36 sq. in. If there are 40° contained in central angle COD, find the number of square inches in the area of sector COD.

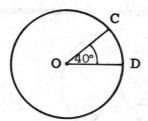

Solution: As seen in the accompanying figure, the rays of central angle COD, along with minor arc $\overset{\frown}{CD}$, form the boundaries of sector COD.

Since there are 360° in an entire circle, the area of a given sector equals

$$\frac{\text{the measure of the central angle of the sector}}{360°} \times A ,$$

where A = the area of the whole circle. In this example,

$$\text{Area of sector} = \frac{40°}{360°} (36 \text{ in.}^2) = \frac{1}{9}(36)\text{in.}^2 = 4 \text{ in.}^2 .$$

Therefore, the area of sector is 4 sq. in.

In circle 0, shown in the figure, a sector whose angle contains
120° has an area of 3π sq. in. Find the radius of the circle.

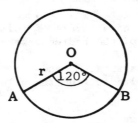

Solution: Since the central angle of the sector, ∡ AOB, contains 120°,
or $\frac{1}{3}$ the total measure of the circle, the area of the sector is $\frac{1}{3}$ the
area of the circle. We are given the area of the sector, 3π sq. in.
Therefore, the area of the entire circle is 3(3π) sq. in. = 9π sq. in.

By the area formula, $A = \pi r^2$, where A is the area of the circle
and r is its radius, we can calculate the radius by setting
$\pi r^2 = 9\pi$ sq. in. and solve for r.

$$\pi r^2 = 9\pi \text{ sq. in.}$$
$$r^2 = 9 \text{ sq. in.}$$
$$r = 3 \text{ in.}$$

Therefore, the radius of the circle is 3 in.

In a circle whose radius is 12, find the area of a minor segment
whose arc has a central angle of 60°. [Leave the answer in terms of
π, and in radical form.]

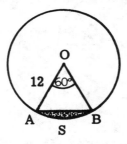

Solution: A segment of a circle is the union of an arc of the circle,
its chord and the region bounded by the arc and the chord. While there
is no formula for the area of a segment, a general rule that can be
applied is to subtract the area of the triangle formed by the rays of
the central angle and chord of the arc from the area of the sector
bounded by the rays of the central angle and the arc. In the accompanying figure, Area of segment ASB = Area of sector OASB
                                   - area of triangle AOB.

Area of sector OASB = $\dfrac{\text{measure of central angle}}{360°}$ x πr^2 ,

where r is the radius of the circle O. By substitution,

Area of sector $OASB = \frac{60°}{360°} \times \pi(12)^2 = \frac{1}{6}(144)\pi = 24\pi$.

Since $\overline{OA}$ and $\overline{OB}$ are radii of a circle, $\overline{OA} \cong \overline{OB}$, and, since the angle included between them is given to be 60°, $\triangle AOB$ is equilateral. As such, its area equals $\frac{s^2}{4}\sqrt{3}$, where s is the length of 1 side of the $\triangle$. We are told s = 12 and, by substitution,

$$\text{Area of } \triangle AOB = \frac{(12)^2}{4}\sqrt{3} = \frac{144}{4}\sqrt{3} = 36\sqrt{3} .$$

Therefore,

Area of segment ASB = Area of sector OASB - Area of triangle AOB

$$= 24\pi - 36\sqrt{3} .$$

Therefore, Area of segment ASB = $24\pi - 36\sqrt{3}$.

● **PROBLEM** 386

In a circle whose radius is 6 in., find, to the nearest square inch, the area of a segment whose chord has an arc which contains 80°.

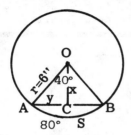

Solution: A segment of a circle is the region bounded by an arc and its chord . The area of such a region can only be found indirectly. The approach applicable to this problem is to find the area of sector OASB, as shown in the figure, and subtracting from it the area of triangle OAB, to find the area of segment ASB.

Since sector OASB has an arc that measures 80°, we can conclude that it occupies 80°/360°, or 2/9 , of the area of the whole circle.

Area of the whole circle = $\pi r^2 = \pi(6 \text{ in.})^2 = 36\pi \text{ in.}^2$

Area of sector OASB = $\frac{2}{9}(36\pi)$ in.^{2} = 8π in.^{2}

Use π = 3.14. Then Area of sector OASB = 8(3.14) in.^{2} = 25.1 in.^{2} .

To find the area of $\triangle OAB$ we will employ the formula A = ½bh, where b and h are the lengths of the base and altitude of the triangle. We draw the altitude, $\overline{OC}$, to base $\overline{AB}$. $\overline{OC} \perp \overline{AB}$. Since $\triangle AOB$ is isosceles, $\overline{OC}$ bisects $\angle$ AOB and, as such, m $\angle$ AOC = 40° . Additionally, $\overline{OC}$ bisects $\overline{AB}$, such that $\overline{AC} \cong \overline{BC}$.

Triangle AOC is a right triangle, and the sine and cosine ratios can be used to determine the legs of the triangle, which are necessary to find the original base and height.

$$\text{Cos } 40° = \frac{\text{length of adjacent leg}}{\text{length of hypotenuse}} = \frac{x}{6 \text{ in.}} .$$

But cos 40° = 0.7660, according to a standard cosine table. Hence,

$$0.7660 = \frac{x}{6 \text{ in.}}$$

$$x = 6(.7660) \text{ in.} = 4.5960 \text{ in.}$$

$$\text{Sin } 40^\circ = \frac{\text{length of opposite leg}}{\text{length of hypotenuse}} = \frac{y}{6 \text{ in.}}$$

Sin 40° = 0.6428, according to a standard sine table. Hence,

$$0.6428 = \frac{y}{6 \text{ in.}}$$

$$y = 6(0.6428)\text{in.} = 3.8568 \text{ in.}$$

Since we are looking for an answer correct to the nearest whole number, we can round off to the nearest tenth in intermediate results, i.e., x = 4.6 in., y = 3.9 in.

Returning to the isosceles triangle, we now know its height, x, is 4.6 in., and its base, $\overline{AB}$, equals 2AC = 2y = 2(3.9)in. = 7.8 in.

Area of $\triangle AOB = \frac{1}{2}bh = \frac{1}{2}(7.8)(4.6)\text{in.}^2 = 17.94 \text{ in.}^2$.

Therefore,

Area of segment ASB = area of sector OASB - area of triangle AOB.

By substitution,   = (25.1 - 17.9)in.^{2} = 7.2 in.^{2}

Area of segment ASB = 7 sq. in., to the nearest square inch.

● **PROBLEM** 387

In the figure shown, $\triangle PQR$ is equilateral and the radius of circle Q is 6. Find the area of the shaded region.

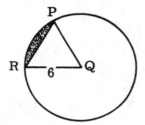

<u>Solution</u>:   The area of the shaded region equals the area of the sector PRQ minus the area of triangle $\triangle PQR$. The area of sector PQR is given by

$$\frac{n}{360} \cdot \pi r^2$$

where n is the measure of the central angle, $\sphericalangle$ PQR. Since $\sphericalangle$ PQR is also an angle of the equilateral triangle $\triangle PRQ$, $\sphericalangle$ PQR = 60° .

Therefore, the area of the sector equals $\frac{60}{360} \cdot \pi \cdot (6)^2 = 6\pi$. The area of the equilateral triangle with side of measure 6 equals

$$\frac{s^2 \sqrt{3}}{4} = \frac{6^2 \cdot \sqrt{3}}{4} = 9\sqrt{3} \quad .$$

The area of the shaded region, therefore, equals (area of sector PRQ) - (area of $\triangle PRQ$) or $6\pi - 9\sqrt{3}$.

then if we let $\pi = 3.14159$ and $\sqrt{3} = 1.73205$,

$$6\pi - 9\sqrt{3} \cong 3.26108$$

● **PROBLEM** 388

The ratio of the area of two circles is 16:1. If the diameter of the smaller circle is 3, find the diameter of the larger circle.

<u>Solution</u>:   The area of a circle equals πr^2 and, from this, we can conclude the ratio of areas is equal to the square of the ratio of radii length. However, we know that $r = \frac{1}{2}d$, where d = diameter, therefore, by substitution,

$$A = \pi(\tfrac{1}{2}d)^2 = \tfrac{1}{4}\pi d^2 .$$

This shows the ratio of any two areas will be equal to the square of the ratio of the diameters. Let A/A' = the ratio of areas. Also, d' = the smaller diameter, and d = the larger diameter. Then,

$$\frac{A}{A'} = \left(\frac{d}{d'}\right)^2 .$$

By substitution,

$$\frac{16}{1} = \left(\frac{d}{3}\right)^2$$

$$\frac{16}{1} = \frac{d^2}{9}$$

$$144 = d^2 \implies d = 12 .$$

Therefore, the diameter of the larger circle is 12.

● **PROBLEM** 389

If the areas of the two circles are 48 and 75, find the ratio of their circumferences.

<u>Solution</u>:   This problem could be solved by using the formula for a circle's area, πr^2, and solving for the two radii. Once we have the two radii, the circumferences can be calculated and the ratio of the circumferences obtained. However, a more direct method is to recognize that all circles are similar. The ratio of their circumferences is the ratio of similitude and can be related directly to the ratio of the areas.

From an earlier theorem, we know that the ratio of the areas of two similar triangles equals the square of their ratio of similitude. Conversely, the ratio of similitude of two triangles equals the square root of the ratios of their areas. The square relationship of ratios of similitude and the ratio of areas is a general result. It holds not only for triangles, but for all polygons and circles. Thus, for circles C_1 and C_2 with radii r_1 and r_2, it must be that

$$\frac{\text{Area of } C_1}{\text{Area of } C_2} = \frac{r_1^2}{r_2^2}$$

or

$$\frac{r_1}{r_2} = \sqrt{\frac{\text{Area } C_1}{\text{Area } C_2}} .$$

The ratio of the circumferences,

$$\frac{2\pi r_1}{2\pi r_2} = \frac{r_1}{r_2} ,$$

therefore, equals

$$\sqrt{\frac{\text{Area } C_1}{\text{Area } C_2}} = \sqrt{\frac{48}{75}} = \sqrt{\frac{16 \cdot 3}{25 \cdot 3}} = \sqrt{\frac{16}{25}} = \frac{4}{5}$$

The ratio of the areas of two circles is 9:4. The length of the radius of the larger circle is how many times greater than the length of the radius of the smaller circle?

<u>Solution</u>:   The ratio of the areas of any two circles, say A and A_1, which have radii of lengths represented by r and r_1 will equal

$$\frac{A}{A_1} = \frac{\pi r^2}{\pi r_1^2} = \frac{r^2}{r_1^2} .$$

We are given that $\frac{A}{A_1} = \frac{9}{4}$. Therefore, if we let $\frac{r}{r_1}$ = the ratio of the length of the radii, then,

$$\frac{9}{4} = \left(\frac{r}{r_1}\right)^2$$

$$\frac{r}{r_1} = \frac{3}{2}$$

Since the ratio of the length of the radius of the larger circle to the length of the smaller circle is $3/2$, the radius of the larger circle is $1\frac{1}{2}$ times as long as the radius of the smaller one.

The ancient Greeks worked in vain to "square the circle", to find the square that has the same area as the circle. Modern mathematicians have shown that the problem is impossible. Does this seem obvious? In the accompanying figure, find the area of △ABC, in terms of $\overline{AC}$ and $\overline{BC}$. Find the area of R_1 plus the area of R_2 . Compare.

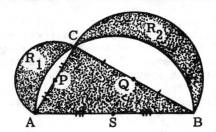

<u>Solution</u>: Since △ABC is a right triangle, the area is one-half the product of the legs.

$$\text{Area of } \triangle ABC = \tfrac{1}{2} AC \cdot BC .$$

The area of $R_1 + R_2$ is the sum of two semicircles of diameters $\overline{AC}$ and $\overline{BC}$ minus the white areas APC and CQB. (1) The sum of two semicircles of diameters $\overline{AC}$ and $\overline{BC}$ is

$$\tfrac{1}{2}\pi r_1^2 + \tfrac{1}{2}\pi r_2^2 = \tfrac{1}{2}\pi(AP)^2 + \tfrac{1}{2}\pi(CQ)^2 , \text{ or since}$$

$AP = \tfrac{1}{2} AC$ and $CQ = \tfrac{1}{2} CB$; $\tfrac{1}{2}\pi(\tfrac{1}{2} AC)^2 + \tfrac{1}{2}\pi(\tfrac{1}{2} CB)^2 = \frac{1}{8}\pi \, AC^2 +$

$$+ \frac{1}{8}\pi \, CB^2 .$$

(2) The area of the white regions equals the area of the semicircle of

diameter $\overline{AB}$ minus the area of the triangle $\triangle ABC$. (i) The area of the semicircle of diameter AB or radius $\frac{1}{2}$ AB equals $\frac{1}{2}\pi(\frac{1}{2}AB)^2 = \frac{1}{8}\pi AB^2$; (ii) the area of triangle $\triangle ABC$ was found above to be $\frac{1}{2}$ AC·BC.

Then, the area of the white regions APC and CQB equals $\frac{1}{8}\pi AB^2 - \frac{1}{2}$ AC·BC. The area of $R_1 + R_2$ equals (sum of the areas of semicircles of diameters $\overline{AC}$ and $\overline{BC}$) - (area of the white regions) =
$(\frac{1}{8}\pi AC^2 + \frac{1}{8}\pi CB^2) - (\frac{1}{8}\pi AB^2 - \frac{1}{2}$ AC·BC$) = \frac{1}{8}\pi(AC^2 + CB^2 - AB^2) + \frac{1}{2}$ AC·BC.

To reduce the expression in the parenthesis, note that $\overline{AC}$ and $\overline{CB}$ are the legs of right triangle $\triangle ABC$ with hypotenuse $\overline{AB}$. By the Pythagorean Theorem, $AC^2 + CB^2 = AB^2$ and the expression in parenthesis equals 0.

Therefore, the area of $R_1 + R_2$ reduces to $\frac{1}{2}$ AC·BC, the same area as $\triangle ABC$.

The area of two semicircular regions has been shown to equal the area of a triangle. The ancient Greeks knew of this example. Is it any wonder that they thought a circle could be squared?

● **PROBLEM** 392

Given: P is a point on $\overset{\frown}{AB}$ of circle Q, so that $\overset{\frown}{AP} \cong \overset{\frown}{BP}$; $\overset{\frown}{ANB}$ is an arc of circle P; $\overline{AQB}$.

Prove: The area of the shaded region = $(PQ)^2$.

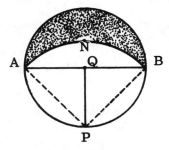

Solution: To find the area of any irregular region, look for areas that can be calculated and that either sum or form, by their difference, the desired region. Here, the shaded region is the difference of semicircle Q minus the region ANBQ. The area of semicircle Q can be found with the formula $\frac{1}{2}\pi r^2$. The region ANBQ is the difference of sector PANB and $\triangle APB$. Therefore, the area of region ANBQ equals the area of the sector PANB minus the area of $\triangle APB$.

Statements	Reasons
1. P is a point on $\overset{\frown}{AB}$ of circle Q, so that $\overset{\frown}{AP} \cong \overset{\frown}{BP}$; $\overset{\frown}{ANB}$ is an arc of circle P; $\overline{AQB}$.	1. Given.
2. AB is a diameter.	2. Any chord that passes through the center of the circle is a diameter.
3. m ∡ AQB = 180°.	3. The measure of a straight angle is 180°.

345

4. $\angle$ AQP $\cong$ $\angle$ PQB .

5. m $\angle$ AQP + m $\angle$ PQB = m $\angle$ AQB.
6. 2m $\angle$ AQP = 180 or
m $\angle$ AQP = m $\angle$ PQB = $90°$.
7. AQ = QB = PQ .
8. $PB^2 = BQ^2 + PQ^2 = 2 \cdot PQ^2$.

9. $\triangle$ APB is a right triangle.

10. $\angle$APB is a right angle.
11. Area of sector PANB =
$\dfrac{90}{360} \cdot \pi \cdot PB^2$.

12. Area of sector PANB =
$\frac{1}{4} \pi(2PQ^2) = \frac{1}{2} \pi PQ^2$.
13. Area of $\triangle$APB = $\frac{1}{2}AB \cdot QP$.

14. Area of $\triangle$APB = $\frac{1}{2}PQ \cdot (2PQ)$
= $(PQ)^2$.
15. Area of region ANBQ = (Area
of sector PANB) - (Area of
$\triangle$APB) = $\frac{1}{2}\pi PQ^2 - PQ^2$.

16. Area of semicircle with dia-
meter $\overline{AQB}$ = $\frac{1}{2}\pi(AQ)^2$.
17. Area of shaded area = (Area
of semicircle $\overline{AQB}$) - (Area of
region ANBQ).

18. Area of shaded region =
$\frac{1}{2}\pi(AQ)^2 - (\frac{1}{2}\pi PQ^2 - PQ^2)$.
19. $(\frac{1}{2}\pi - \frac{1}{2}\pi + 1)PQ^2 = PQ^2$.

4. Central angles that intercept
congruent arcs are congruent.
5. Angle Addition Postulate.
6. Substitution Postulate.

7. All radii are congruent.
8. Pythagorean Theorem and Substitution Postulate.
9. Any triangle inscribed in a semicircle with the diameter as a side
is a right triangle, with the
hypotenuse being the diameter.
10. Definition of right triangle.
11. The area of a sector with radius
r and central angle n is
$\dfrac{n}{360} \cdot \pi r^2$.

12. Substitution Postulate and Step 8.

13. The area of a triangle with base
b and altitude h is $\frac{1}{2}$bh.
14. Substitution Postulate.

15. The area of non-overlapping regions is the sum of the areas of
each region. Also, Substitution
Postulate.
16. Area of a semicircle of radius r
equals $\frac{1}{2}\pi r^2$.
17. The area of non-overlapping regions
is the sum of the areas of each
region.
18. Substitution Postulate.

19. Substitution Postulate and Factoring.

● **PROBLEM** 393

Using a compass, draw a circle of radius 1". Superimpose a
1/8" grid, and estimate the area of the circle.

Solution: The figure shows the circle drawn on a 1/8" grid. If we
know the area of a 1/8" on a side square, we can, by counting the
number of squares enclosed by the circle, estimate the area of the
circle.

As is shown in the figure, the area occupied by the circle lies
between the areas of the shaded and unshaded regions. That is

Area (shaded) < Area (circle) < Area (unshaded)   (1)

The shaded region contains 180 boxes, 1/8" on a side. The area of
a 1/8" square is 1/64 sq. in. Hence,

Area (shaded) = (1/64 $\underline{\text{sq. in.}}$)(180 boxes)
             box

Area (shaded) = 2.8125 sq. in.   (2)

Similarly, the unshaded region contains 216 boxes, 1/8" on a side.

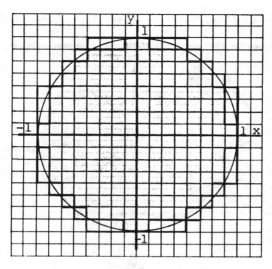

Hence,

  Area (unshaded) = (1/64 <u>sq. in.</u>)(216 boxes)
         box

  Area (unshaded) = 3.375 sq. in.   (3)

Using (3) and (2) in (1) yields

  2.8125 sq. in. < Area (circle) < 3.375 sq. in. .

If we required a more accurate estimate, we could repeat the procedure using a smaller grid. In fact, as the size of the squares become infinitesimally small, the estimate we make comes closer and closer to the exact value of the area of the circle.

               ● **PROBLEM** 394

   The lengths of the radii of two circles are in the ratio of 1:4. Find the ratio of the areas of the circles.

<u>Solution</u>: Since the area of a circle is a function of the square of the length of the radius, the ratio between any two areas will be equal to the ratio of the squares of the lengths of their radii.

  Let A/A' = ratio of areas. r/r' = ratio of radii.
then,

$$\frac{A}{A'} = \left(\frac{r}{r'}\right)^2$$

$$\frac{A}{A'} = \left(\frac{1}{4}\right)^2$$

By substitution,

$$\frac{A}{A'} = \frac{1}{16} .$$

Therefore, the ratio of the areas is 1:16.

               ● **PROBLEM** 395

   A 1-acre field in the shape of a right triangle has a post at the midpoint of each side. A sheep is tethered to each of the side posts and a goat to the post on the hypotenuse. The ropes are just long enough to let each animal reach the two adjacent vertices. What is the total area the two sheep have to themselves, i.e., the area the goat cannot reach?

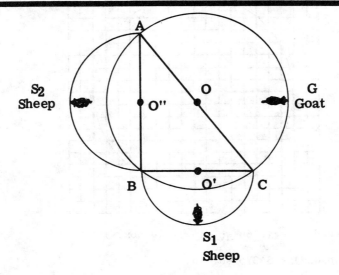

Solution: The goat is tethered at point O and, therefore, has free access to any of the land within the circle of radius OA. Each animal can move in a circle around its post. However, the only area that the sheep can move in apart from the goat is crescents BS_1C and AS_2B.

The area of these crescents is the area we wish to find. Notice that the area of the crescents BS_1C and AS_2B = area of right $\triangle ABC$ + area of semicircle O' + area of semicircle O'' - area of semicircle O. We are given that the area of right $\triangle ABC$ = 1 acre. The area of a semicircle of radius r is $\frac{1}{2}\pi r^2$. The radius of semicircle O' = BO' = $\frac{1}{2}$BC; radius of semicircle O'' = AO'' = $\frac{1}{2}$AB, and radius of semicircle O = CO = $\frac{1}{2}$AC. By substitution,

$$\text{Area of crescents } (BS_1C + AS_2B) = 1 \text{ acre} + \tfrac{1}{2}\pi(\tfrac{1}{2}BC)^2 + \tfrac{1}{2}\pi(\tfrac{1}{2}AB)^2$$
$$- \tfrac{1}{2}\pi(\tfrac{1}{2}AC)^2$$
$$= 1 \text{ acre} + \frac{1}{8}(BC^2 + AB^2 - AC^2)\pi.$$

But, by the Pythagorean Theorem $AC^2 = BC^2 + AB^2$. Therefore,

$$BC^2 + AB^2 - AC^2 = 0 .$$

Hence, the area of the two crescents that the sheep have completely to themselves is 1 acre.

● **PROBLEM** 396

A farmer is cutting a field of oats with a machine which takes a 5 ft. cut. The field he is cutting is circular and when he has been round it 11½ times (starting from the perimeter) he calculates that he has cut half the area of the field. How large is the field? (Answer to the nearest 100 sq. yds.)

Solution: In the accompanying figure, the shaded area is the area already cut; r is the radius of the field. Originally the entire field was uncut and the uncut area was a circle of radius r. The

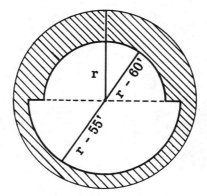

first trip around, he reduces the uncut area from a circle of radius r to a circle of radius r - 5. On the nth trip, he reduces the uncut area from a circle of radius r - 5(n-1) to a circle of radius r - 5n. Thus, on the 11th trip, the uncut area is a circle of radius r - 55. Because the farmer made 11 and ½ trips, the uncut area consists of a semicircle of radius r - 55 and a semicircle of radius r - 60.

We are given that the area of the uncut portion equals the area of the cut portion. Thus, the area of the unplowed area is one-half the area of the field. The area of the uncut portion is the sum of the areas of the semicircle of radius r - 55 and the area of semicircle of radius r - 60. Given that the formula of the semicircle of radius x is $\frac{1}{2}\pi x^2$, we have

(i)   area of semicircle of r - 55 = $\frac{1}{2}\pi(r-55)^2$
(ii) area of semicircle of r - 60 = $\frac{1}{2}\pi(r-60)^2$
(iii) area of unplowed area = $\frac{1}{2}\pi(r-55)^2 + \frac{1}{2}\pi(r-60)^2$

$$= \frac{1}{2}\pi((r-55)^2 + (r-60)^2)$$

(vi) area of field = 2 · area of unplowed area.

(v) $\pi r^2 = 2 \cdot \frac{1}{2}\pi((r-55)^2 + (r-60)^2)$
(vi) $r^2 = r^2 - 110r + 3025 + r^2 - 120r + 3600$
(vii) $r^2 - 230r + 6625 = 0$

Using the quadratic formula,

$$ax^2 + bx + c = 0, \quad x = \frac{-b \pm \sqrt{b^2 - 4ac}}{2a} \quad ,$$

we obtain for the radius of the field,

(viii) $r = \dfrac{230 \pm \sqrt{230^2 - 4(6625)}}{2} = \dfrac{230 \pm \sqrt{26400}}{2}$

(ix) $r = \dfrac{230 \pm 162.48}{2}$ = 196.24 or 33.76.

Note that the answer r = 33.76 makes no sense since this would imply that the radius of the unplowed portion would be r = 33.76 - 55 = -21.24. Thus, r = 196.24 ≃ 196 ft. is the only solution.

To find the area of the circular field, remember that the area of the circle is πr^2 .

(x) Area of field = $\pi(196)^2 = 38416\pi$

$$38416 \, \pi \approx 120687 \text{ sq. ft.}$$

Since 1 sq. ft. = $\frac{1}{9}$ sq. yd.,

(xi) Area of field = $120687(\frac{1}{9}$ sq. yd.) = 13,400 sq. yd.

CHAPTER 21

PYTHAGOREAN THEOREM AND APPLICATIONS

Basic Attacks and Strategies for Solving Problems in this Chapter. See pages 350 to 370 for step-by-step solutions to problems.

Perhaps the most famous of all mathematical theorems is the **Pythagorean Theorem** which states:

If, in a right triangle, a and b are the lengths of the legs, and c is the length of the hypotenuse, then $a^2 + b^2 = c^2$.

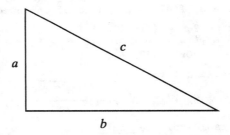

The converse of the Pythagorean Theorem is also true.

If a, b, and c are the lengths of the sides of a triangle, and if $a^2 + b^2 = c^2$, then the triangle is a right triangle.

The problems in this chapter are of two types:

(1) Find the lengths of the side(s) of a triangle by showing it is a right triangle and applying the Pythagorean Theorem.

(2) Determine whether a triangle is a right triangle or not, based on information about the lengths of its sides, and application of the converse of the Pythagorean Theorem.

Several problems involve algebra skills, including applying the square root generalization $\sqrt{ab} = \sqrt{a}\sqrt{b}$, and equation-solving procedures utilizing the quadratic formula.

Step-by-Step Solutions to Problems in this Chapter, "Pythagorean Theorem and Applications"

● **PROBLEM** 397

The legs of a right triangle are 3 feet and 4 feet in length. What is the length of the hypotenuse of the triangle?

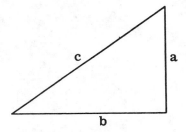

<u>Solution</u>: Apply the Pythagorean Theorem, which states that the sum of the squares of the lengths of two legs, a and b, of a right triangle is equal to the square of the length of the hypotenuse, c: $a^2 + b^2 = c^2$.

We are given that a = 3, b = 4, and substituting in the above formula,

$$3^2 + 4^2 = c^2$$

$$9 + 16 = c^2$$

$$25 = c^2$$

$$5 = c$$

Therefore, the hypotenuse of the triangle is 5 feet.

● **PROBLEM** 398

The legs of a certain right triangle are equal and the hypotenuse is $\sqrt{8}$. What is the length of either leg of the triangle?

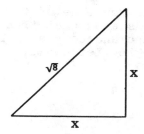

<u>Solution</u>: Recall the Pythagorean Theorem, $a^2 + b^2 = c^2$, where a and b are the lengths of the two legs of a right triangle, and c is the length of the hypotenuse. We can use this to solve for the length of the legs. Let x = the length of each of the equal legs. Substituting in the above formula, x = a = b and $\sqrt{8}$ for c,

$$x^2 + x^2 = (\sqrt{8})^2$$

$$2x^2 = 8$$

$$x^2 = 4$$

$$x = 2$$

Therefore, each leg is 2 units in lengths.

● **PROBLEM** 399

The lengths of the sides of a triangle are 8, 15, and 17. Show that the triangle is a right triangle.

<u>Solution</u>: This problem requires the use of the converse of the Pythagorean Theorem, which states that if the square of the length of one side of a triangle is equal to the sum of the squares of the lengths of the other two sides, then the triangle is a right triangle.

Let x = 17, the longest side of the triangle and let y = 8 and z = 15.

Then, $x^2 = (17)^2 = 289$

and $y^2 + z^2 = (8)^2 + (15)^2 = 64 + 225 = 289.$

Since $17^2 = 8^2 + 15^2$, the triangle is a right triangle.

● **PROBLEM** 400

Is a triangle with sides 3, 7, and 11 inches a right triangle?

Solution: Recall that the converse of the Pythagorean Theorem is also true. It states that if a triangle has sides of length a, b, and c and $c^2 = a^2 + b^2$, then the triangle is a right triangle. Let a = 3, b = 7, and c = 11. We compute the squares:

$$a^2 = 3^2 = 9$$

$$b^2 = 7^2 = 49$$

$$c^2 = 11^2 = 121$$

$$121 \neq 49 + 9.$$

Since the sum of any two of these squares does not equal the third square, the triangle is not a right triangle.

● **PROBLEM** 401

A triangle has sides that measure 20, 21, and 29 inches. Determine whether the triangle is a right triangle.

Solution: In this problem the converse of the Pythagorean Theorem will be applied. It states that if a triangle has sides of length a, b, and c, and $c^2 = a^2 + b^2$, then the triangle is a right triangle. Let a = 20, b = 21, and c = 29 and compute their squares,

$$a^2 = 20^2 = 400$$

$$b^2 = 21^2 = 441$$

$$c^2 = 29^2 = 841.$$

$$841 = 441 + 400.$$

Therefore, the triangle is a right triangle.

● **PROBLEM** 402

In an isosceles triangle, the length of each of the congruent sides is 10 and the length of the base is 12. Find the length of the altitude drawn to the base.

Solution: In the figure shown, altitude $\overline{BD}$ has been drawn. By a theorem, we know that $\overline{BD} \perp \overline{AC}$ and that $\overline{BD}$ bisects $\overline{AC}$. Therefore, $\overline{AD} \cong \overline{DC}$ and they both measure 6.

Since $\overline{BD} \perp \overline{AC}$, we can conclude that $\triangle BDC$ is a right triangle and then apply the Pythagorean Theorem. As applied here, it states that

$$(DC)^2 + (BD)^2 = (BC)^2.$$

We were given that BC = 10, and we have found that DC = 6. If we let x = BD and substitute into the equation above, we arrive at

$$6^2 + x^2 = 10^2$$

$$36 + x^2 = 100$$

$$x^2 = 64$$

$$x = 8.$$

Therefore, the length of the altitude drawn to the base is 8.

● **PROBLEM** 403

A rectangle has dimensions 5 × 12 ft. What is the length of one of its diagonals?

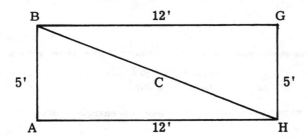

Solution: A rectangle of the given dimensions with diagonal of length c, is shown. Since the figure is a rectangle, m ∢ BGH = 90°, and $\triangle BGH$ is a right triangle. According to the Pythagorean Theorem, if a triangle has a right angle, then the square of the measure of the hypotenuse equals the sum of the squares of the measures of the legs. In right $\triangle BGH$, $\overline{BH}$ is the hypotenuse and $\overline{BG}$ and $\overline{GH}$ are the legs. Hence,

$$(BH)^2 = (BG)^2 + (GH)^2$$

$$(BH)^2 = (12 \text{ ft})^2 + (5 \text{ ft})^2 = 169 \text{ ft}^2$$

$$BH = 13 \text{ ft}.$$

● **PROBLEM** 404

Let $\triangle ABC$ be an equilateral triangle of side length s.
Let D be the midpoint of $\overline{BC}$. Compute AD. (See figure.)

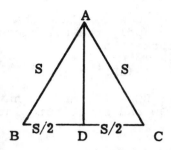

Solution: We will apply the Pythagorean Theorem to ∆ABD to obtain AD.

$$(AB)^2 = (AD)^2 + (BD)^2$$

Hence,

$$(AD)^2 = (AB)^2 - (BD)^2$$

$$AD = \sqrt{(AB)^2 - (BD)^2}$$

Since BD = DC = s/2, and AB = s, therefore,

$$AD = \sqrt{s^2 - s^2/4}$$

$$AD = \sqrt{3s^2/4} = s\frac{\sqrt{3}}{2}$$

● **PROBLEM** 405

Find, in radical form, the length of the hypotenuse of an isosceles right triangle, each of whose legs is 4 units long.

Solution: We let h = the length of the hypotenuse. Then, we can apply the Pythagorean Theorem to obtain

$$h^2 = 4^2 + 4^2 = 16 + 16 = 32.$$

Therefore, $h = \sqrt{32} = \sqrt{16}\sqrt{2} = 4\sqrt{2}$

Therefore, the length of the hypotenuse is $4\sqrt{2}$.

Note that this problem verifies a general rule stating that in an isosceles right triangle, the length of the hypotenuse equals the length of a leg times $\sqrt{2}$.

Find, in radical form, the length of the altitude of an equilateral triangle whose sides measure 12 inches (see figure).

Solution: When the altitude of an equilateral triangle is drawn, two 30-60-90 triangles are formed. The altitude is opposite the 60° base angle, and its length (h) is equal to one-half the length of the side $\overline{BC}$ times $\sqrt{3}$. Therefore,

$$h = \tfrac{1}{2}(\overline{BC})\ \sqrt{3} \quad \text{and by substitution,}$$

$$h = \tfrac{1}{2}\ (12 \text{ in.})\ \sqrt{3} = 6\ \sqrt{3} \text{ in.}$$

Alternatively, the Pythagorean Theorem could have been employed directly for right triangle CDB. We know that $\overline{AB}$ is bisected by the altitude and, accordingly, DB = 6 in. Therefore,

$$h^2 + 6^2 \text{ in.}^2 = 12^2 \text{ in.}^2$$

$$h^2 + 36 \text{ in.}^2 = 144 \text{ in.}^2,$$

$$h^2 = 108 \text{ in.}^2 \text{ and}$$

$$h = \sqrt{108 \text{ in.}} = \sqrt{36}\ \sqrt{3} \text{ in.} = 6\ \sqrt{3} \text{ in.}$$

Therefore, the length of the altitude of △ABC is 6 $\sqrt{3}$ inches.

The lengths of the diagonals of the rhombus shown in the figure are 30 and 40. Find the perimeter of the rhombus.

Solution: Since the diagonals of a rhombus are perpendicular bisectors of each other, △AEB is a right triangle in which EB = $\tfrac{1}{2}$(30),or 15, and AE = $\tfrac{1}{2}$ (40),or 20. With this in mind, we can calculate the length of the hypotenuse $\overline{AB}$, of △AEB by using the Pythagorean Theorem:

$$(AB)^2 = (EB)^2 + (AE)^2.$$

$\overline{AB}$ is one side of a rhombus. All sides of a rhombus are congruent, thus, the perimeter can be calculated by multiplying AB by 4. If we let AB = x, by substitution,

$$x^2 = (15)^2 + (20)^2 = 225 + 400$$

or $$x^2 = 625; \quad \text{solving } x = 25.$$

Then, the perimeter = 4x or 4 (25) = 100.

● **PROBLEM** 408

Find, to the nearest tenth of an inch, the length of a side of a square whose diagonal measures 8 inches.

Solution: When a diagonal is drawn in a square, an isosceles right triangle is formed. (This can be seen in the accompanying figure.)

We let s = the length of the side of a square, then, applying Pythagoras' Theorem to ΔABC, we obtain,

$$s^2 + s^2 = 8^2$$

$$2s^2 = 64$$

$$s^2 = 32$$

$$s = \sqrt{32} = \sqrt{16} \ \sqrt{2} = 4 \ \sqrt{2}.$$

Approximating, $\sqrt{2} = 1.41$, we get s = 4 (1.41) = 5.64, and, to the nearest tenth, s = 5.6.

Therefore, the length of the side of the square is 5.6 inches, to the nearest tenth.

● **PROBLEM** 409

Suppose that the area of a rectangle is 300 sq. in., and that the length of its diagonal is 25 in. Find the lengths of the sides of the rectangle.

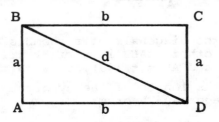

356

<u>Solution</u>: Let the sides of the rectangle be labeled as in the figure. Then the area of the rectangle A(R) is

$$A(R) = ba = 300 \text{ sq. in.} \tag{1}$$

Since $\triangle ABD$ is a right triangle,

$$d^2 = a^2 + b^2 \tag{2}$$

by Pythagoras' Theorem. Noting that d = 25 in., we can write:

$$625 \text{ sq. in.} = a^2 + b^2 \tag{3}$$

We now have 2 equations ((1) and (2)) and 2 unknowns (a and b). We may, therefore, solve explicitly for a and b. First solve (1) for a:

$$a = \frac{300 \text{ sq. in.}}{b} \tag{4}$$

Inserting (4) into (3),

$$625 \text{ sq. in.} = \left(\frac{300 \text{ sq. in.}}{b}\right)^2 + b^2$$

$$625 \text{ sq. in.} = \frac{90000 \text{ in}^4}{b^2} + \frac{b^4}{b^2}$$

$$625 \text{ sq. in.} = \frac{b^4 + 90000 \text{ in}^4}{b^2}$$

$$b^4 - (625 \text{ sq. in.}) b^2 + 90000 \text{ in}^4 = 0$$

Let
$$x = b^2 \tag{5}$$

then,
$$x^2 - (625 \text{ sq. in.}) x + 90000 \text{ in}^4 = 0 \tag{6}$$

We solve this by means of the quadratic formula. If an equation is of the form

$$ax^2 + bx + c = 0,$$

its roots are

$$x = \frac{-b \pm \sqrt{b^2 - 4ac}}{2a} \tag{7}$$

In equation (6), a = 1, b = - 625 sq. in., and c = 90,000 in^4. Using these in (7) yields

$$x = \frac{625 \text{ sq. in.} \pm \sqrt{(-625 \text{ sq. in.})^2 - (4)(1)(90,000 \text{ in}^4)}}{(2)(1)}$$

$$x = \frac{625 \text{ sq. in.} \pm \sqrt{30625 \text{ in}^4}}{2}$$

$$x = \frac{625 \text{ sq. in.} \pm 175 \text{ sq. in.}}{2}$$

x = 400 sq. in.   or 225 sq. in.

But, from (5), $x = b^2$; hence,

b^2 = 400 sq. in. or 225 sq. in.

Therefore,   b = 20 in. or 15 in.   (8)

(Negative roots are eliminated because b is a distance and must be positive.)

Using (8) in (4), we find

$a = \frac{300}{20}$ in.,   or $\frac{300}{15}$ in.

a = 15 in., or 20 in.

Hence, the lengths are

a = 15 in., b = 20 in.

or,   a = 20 in., b = 15 in.

● **PROBLEM** 410

A chord, 16 inches long, is 6 inches from the center of a circle. Find the length of the radius of the circle.

Solution: The distance of a chord from the center of a circle is measured on a line perpendicular to the chord from the center of the circle. Therefore, as labeled in the accompanying figure, $\overline{OC} \perp \overline{AB}$ and $\triangle OCB$ is a right triangle.

In addition to being perpendicular to $\overline{AB}$, the segment $\overline{OC}$ also intersects chord $\overline{AB}$ at its midpoint. Hence, $\overline{OC}$ bisects $\overline{AB}$. Therefore, $\overline{AC} \cong \overline{BC}$, and they both measure 8 in. We are given that OC = 6 in.

Since $\triangle OCB$ is a right triangle, we can apply the Pythagorean Theorem to calculate the length of the hypotenuse of the triangle, which is also the radius of the circle.

In this problem,

$$(OB)^2 = (OC)^2 + (CB)^2.$$

If we let r = the length of radius $\overline{OB}$, by substitution

$$r^2 = 6^2 + 8^2$$

$$r^2 = 36 + 64$$

$$r^2 = 100$$

$$r = 10.$$

Therefore, the length of the radius of the circle is 10 inches.

● **PROBLEM** 411

In the figure, EC = s is called the sagitta. $\overline{AC}$ is the radius (of length r) of arc $\overset{\frown}{BCD}$. Derive the sagitta relationship, $s = \dfrac{x^2}{2r}$, assuming that s is much smaller than all other relevant lengths.

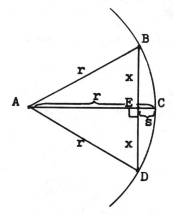

Solution: The fact that a radius perpendicular to a chord also bisects the chord is illustrated in the figure. We shall apply the Pythagorean Theorem to right triangle AEB, and thereby derive the sagitta relationship (after taking necessary approximations into account).

By the Pythagorean Theorem,

$$(AE)^2 + (EB)^2 = (AB)^2.$$

Since AB = r, AE = r - s, and EB = x, this becomes

$$(r - s)^2 + x^2 = r^2$$

or $\qquad r^2 - 2sr + s^2 + x^2 = r^2$

or $\qquad s^2 - 2sr + x^2 = 0.$ $\qquad\qquad$ (1)

But, because s is much smaller than all other relevant lengths (i.e., r and x), we may neglect s^2 in comparison with 2sr and x^2. Hence, from (1)

$$x^2 - 2sr = 0$$

or,
$$s = \frac{x^2}{2r} .$$

● **PROBLEM 412**

In a right triangle, the length of the hypotenuse is 20 and the length of one leg is 16. Find the length of the other leg.

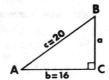

Solution: The Pythagorean Theorem tells us that in a right triangle, the square of the length of the hypotenuse is equal to the sum of the squares of the lengths of the legs.

Referring to the figure, the above statement can be written as

$$c^2 = a^2 + b^2 .$$

By substitution, we obtain

$$(20)^2 = a^2 + (16)^2$$

$$400 = a^2 + 256$$

$$144 = a^2 \quad \text{and} \quad a = 12.$$

Therefore, the length of the other leg is 12.

● **PROBLEM 413**

Find the length of the diagonal of a rectangle whose sides are $\sqrt{5}$ inches and 2 inches, respectively.

Solution: The diagonal of a rectangle acts as the hypotenuse forming, along with two sides of the rectangle, a right triangle, such as ΔABC in the given figure. The Pythagorean Theorem states that the length of the hypotenuse squared equals the sum of the squares of the lengths of the legs.

Let x = the length of the diagonal.

Therefore, by substitution,

$$x^2 = (2 \text{ in.})^2 + (\sqrt{5} \text{ in.})^2 = 4 \text{ in.}^2 + 5 \text{ in.}^2 = 9 \text{ in.}^2$$

which implies x = 3 in.

Therefore, the length of the diagonal of the rectangle is 3 inches.

● **PROBLEM** 414

A square has sides of $\sqrt{8}$ feet. How long is its diagonal?

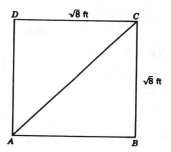

Solution: We will see that when the diagonal $\overline{AC}$ is drawn, the given square is divided into two right triangles.

Since all sides of a square are equal, both AB and BC have a length of $\sqrt{8}$ ft.; hence, triangle ABC is an isosceles right triangle.

Recall that in this special type of triangle, the hypotenuse is equal to the length of one of its arms multiplied by $\sqrt{2}$. Thus, the hypotenuse, or diagonal $\overline{AC}$, equals

$$\sqrt{8} \cdot \sqrt{2} = \sqrt{16} = 4 \text{ feet.}$$

This problem could also be solved by using the Pythagorean theorem from which the above results were derived,

$$(\sqrt{8})^2 + (\sqrt{8})^2 = c^2. \qquad \text{Again AC} = 4.$$

● **PROBLEM** 415

In a circle, angle ABC, formed by diameter $\overline{AB}$ end chord $\overline{BC}$, is 30°. If the length of the diameter of the circle is 20, find the length of chord $\overline{AC}$ and, in radical form, the length of chord $\overline{BC}$.

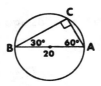

Solution: From the figure shown, we see that ∢ BCA is inscribed in a semi-circle, therefore is 90°. Angle CAB is the third angle of ΔABC, and it measures 180° - (m ∢ ABC + m ∢ BCA), or 180° - (30° + 90°) = 60°. There-fore, ΔABC is a 30-60-90 triangle.

Chord $\overline{AC}$ is opposite the 30° angle, and, accordingly, measures one-half of the length of hypotenuse $\overline{AB}$. Therefore,

$$\overline{AC} = \tfrac{1}{2}\ \overline{AB} = \tfrac{1}{2}(20) = 10.$$

Chord $\overline{BC}$ is opposite the 60° angle, and its length is given by $\tfrac{1}{2}$ the length of the hypotenuse times $\sqrt{3}$. Thus,

$$\overline{BC} = \tfrac{1}{2}\ \overline{AB}\ \sqrt{3} = \tfrac{1}{2}\ (20)\ \sqrt{3} = 10\ \sqrt{3}.$$

Therefore, the length of $\overline{AC}$ is 10 and the length of $\overline{BC}$ is 10 $\sqrt{3}$.

● **PROBLEM** 416

In triangle ABC, m ∢ A = 30°, m ∢ B = 60°, m ∢ C = 90° and $\overline{AC}$ measures 6 inches. (Refer to the accompanying figure as a visual aid.) Find BC to the nearest tenth of an inch.

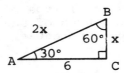

Solution: $\overline{BC}$ is opposite the 30° angle in a 30° - 60° right triangle and measures $\dfrac{1}{\sqrt{3}}$ times the length of the side opposite the 60° angle, $\overline{AC}$ = 6 inches. Therefore, $\overline{BC}$ measures $\dfrac{6}{\sqrt{3}}$. Approximately, $\sqrt{3} \approx 1.73$ and, to the nearest tenth, $\dfrac{6}{\sqrt{3}}$ = 3.5.

However, because the hypotenuse is twice the length of leg $\overline{BC}$ for a 30 - 60 - 90 triangle, the Pythagorean Theorem can be utilized in an alternative method.

If we let x = the length of $\overline{BC}$,
then 2x = the length of hypotenuse $\overline{AB}$.

Therefore,

$$(2x)^2 = x^2 + 6^2$$

$$4x^2 = x^2 + 36$$

$$3x^2 = 36$$

$$x^2 = 12$$

$$x = \sqrt{12} = \sqrt{4} \cdot \sqrt{3} = 2\sqrt{3}.$$

Again, approximating $\sqrt{3} \approx 1.73$, x = 3.5.

Thefore, chord $\overline{BC}$ measures 3.5 inches to the nearest tenth of an inch.

● **PROBLEM** 417

An equilateral triangle has sides of 8 inches. What is its height?

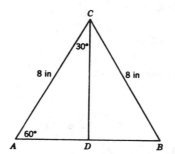

Solution: Since the triangle is equilateral, its three angles are 60°. Draw the bisector of angle C. (In an equilateral triangle this is also the altitude.) It divides angle C into two equal angles of 30°. Hence, triangle ADC is a 30°-60°-90° triangle. Recall the relations in this special right triangle: the side opposite the 30° angle is half the length of the hypotenuse, and the side opposite the 60° angle is equal to the length of the side opposite the 30° angle multiplied by $\sqrt{3}$. Since AC = 8 = the hypotenuse length, and m ∢ ACD = 30°, AD = 4.

Therefore, CD = the height = $4\sqrt{3}$ inches.

● **PROBLEM** 418

Which of the following are Pythagorean Triples?
(a) 1, 2, 3; (b) 3, 4, 5 ; (c) 5, 6, 7; (d) 5, 12, 13;

(e) 11, 60, 61; (f) 84, 187, 205.

Solution: A set of 3 positive integers, a, b, and c, (where c is the largest of the set) is a Pythagorean Triple if and only if

$$c^2 = a^2 + b^2$$

(i.e. they satisfy Pythagorean Theorem).

(a) For (1, 2, 3) to be a Pythagorean triple, it must be true that

$$3^2 = 2^2 + 1^2 \qquad \text{or} \qquad 9 = 4 + 1 = 5.$$

Since $9 \neq 5$, therefore, (1, 2, 3) is not a Pythagorean Triple.

(b) c = 5, a = 3, b = 4.

$$c^2 \overset{?}{=} a^2 + b^2$$

$$25 \overset{?}{=} 9 + 16$$

Since $\qquad$ 25 = 25,

$\qquad$ (3, 4, 5) is a Pythagorean Triple.

(c) c = 7, a = 5, b = 6

$$c^2 \overset{?}{=} a^2 + b^2$$

$$49 \overset{?}{=} 25 + 36$$

$$49 \neq 61$$

Therefore, (5, 6, 7) is not a Pythagorean Triple.

(d) c = 13, a = 5, b = 12.

$$c^2 \overset{?}{=} a^2 + b^2$$

$$169 \overset{?}{=} 25 + 144$$

$$169 = 169$$

Hence, (5, 12, 13) is a Pythagorean Triple.

(e) c = 61, a = 11, b = 60

$$c^2 \overset{?}{=} a^2 + b^2$$

$$3721 \overset{?}{=} 121 + 3600$$

$$3721 = 3721$$

Therefore, (11, 60, 61) is a Pythagorean Triple.

(f) c = 205, a = 84, b = 187

$$c^2 \overset{?}{=} a^2 + b^2$$

$$42025 \overset{?}{=} 7056 + 34969$$

$$42025 = 42025$$

Therefore, (84, 187, 205) is a Pythagorean Triple.

● **PROBLEM** 419

A 25-foot ladder leans against the wall such that the base of the ladder is 20 feet from the wall. The ladder slides until the base is 24 feet from the wall. Locate the point X that is common to both ladder positions (see diagram).

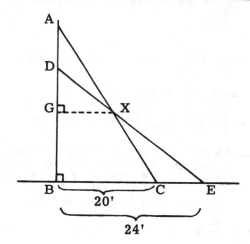

Solution: $\overline{AC}$ is the original position of the ladder; $\overline{DE}$, the final position. The length of the ladder is constant. Hence, AC = DE = 25 ft. The original distance from the base of the ladder to the wall, BC = 20; the final distance BE = 24. To find the height of the ladder top in both cases, by using the Pythagorean Theorem:

The original height

$$AB = \sqrt{(AC)^2 - (BC)^2} = \sqrt{25^2 - 20^2} = \sqrt{225} = 15 \text{ ft.}$$

The final height

$$DB = \sqrt{(DE)^2 - (BE)^2} = \sqrt{25^2 - 24^2} = \sqrt{49} = 7 \text{ ft.}$$

We wish to find the point X common to $\overline{DE}$ and $\overline{AC}$.

We draw the line $\overline{GX} \perp \overline{AB}$. Remember that a line drawn through two sides of a triangle and parallel to the third side forms 2 similar triangles. Thus, $\triangle AGX \sim \triangle ABC$ and also $\triangle DGX \sim \triangle DBE$. We can relate these two similarities to find GX and AG. By the Pythagorean Theorem, we can find AX. Thus, we will have the location of X: a distance AX from the top of the ladder in the original position.

$\triangle AGX \sim \triangle ABC$ by the A.A. Similarity Theorem ($\angle A \stackrel{\sim}{=} \angle A$, and $\angle AGX \stackrel{\sim}{=} \angle ABC$ = right angles). Similarly, $\triangle DGX \sim \triangle DBE$, since $\angle D \stackrel{\sim}{=} \angle D$ and $\angle AGX \stackrel{\sim}{=} \angle ABC$. Since corresponding sides of similar triangles are proportional, we have

(i) $\qquad \dfrac{AG}{GX} = \dfrac{AB}{BC} = \dfrac{15}{20} = \dfrac{3}{4}$

(ii) $\qquad \dfrac{DG}{GX} = \dfrac{DB}{BE} = \dfrac{7}{24}$.

We can relate AG and DG by expressing AG in terms of GX and GX in terms of DG. From (i) and (ii), we obtain

(iii) $\qquad AG = \dfrac{3}{4} GX$

(iv) $\qquad DG = \dfrac{7}{24} GX \qquad$ or $\qquad GX = \dfrac{24}{7} DG.$

By substitution, we have,

(v) $\qquad AG = \dfrac{3}{4} GX = \dfrac{3}{4} \left(\dfrac{24}{7} DG \right) = \dfrac{18}{7} DG.$

To solve for AG and DG, we need only one more relation. Note that AG = AD + DG. AD is the difference in ladder heights, AB - DB = 15 - 7 = 8. Then, by substitution,

(vi) $\qquad AG = AD + DG = 8 + DG, \qquad$ or $\qquad DG = AG - 8.$

Combining (v) and (vi), we obtain:

(vii) $\qquad AG = \dfrac{18}{7} (AG - 8) = \dfrac{18}{7} AG - \dfrac{144}{7}$

(viii) $\qquad \dfrac{18}{7} AG - AG = \dfrac{144}{7}$

(ix) $\qquad \dfrac{11}{7} AG = \dfrac{144}{7} \qquad$ or $\qquad AG = \left(\dfrac{7}{11} \right) \left(\dfrac{144}{7} \right) = \dfrac{144}{11}$.

Thus, $AG = \dfrac{144}{11}$. To find GX, remember that by (iii), $AG = \dfrac{3}{4} GX$. Then, $GX = \dfrac{4}{3} \left(\dfrac{144}{11} \right) = \dfrac{192}{11}$.

We could stop here and say that point X, in terms of the original ladder position, is $\dfrac{144}{11}$ ft. down from the top of the wall and $\dfrac{192}{11}$ ft. from the wall; but it may be easier to measure distances along the ladder. Therefore, we can find AX. Note that AX is the hypotenuse of right $\triangle AGX$. By the Pythagorean Theorem,

$$(AX)^2 = (AG)^2 + (GX)^2 = \left(\frac{144}{11}\right)^2 + \left(\frac{192}{11}\right)^2.$$

DIGRESSION: Whenever cumbersome calculations present themselves, it is best to look for easier methods. We are asked to square two larger numbers and then to find a square root. We wonder, therefore, whether or not we are squaring unnecessary factors. Note that

$$\frac{144}{11} = \frac{48}{11} \, (3) \quad \text{and} \quad \frac{192}{11} = \frac{48}{11} \, (4).$$

Then, $(AX)^2 = \left(\frac{48}{11} \times 3\right)^2 + \left(\frac{48}{11} \times 4\right)^2$

$$= \left(\frac{48}{11}\right)^2 \times 3^2 + \left(\frac{48}{11}\right)^2 \times 4^2 = \left(\frac{48}{11}\right)^2 (3^2 + 4^2)$$

$$= \left(\frac{48}{11}\right)^2 (9 + 16) = \left(\frac{48}{11}\right)^2 (25).$$

Then, $AX = \sqrt{\left(\frac{48}{11}\right)^2 (25)} = \sqrt{\left(\frac{48}{11}\right)^2} \sqrt{25} = \frac{48}{11} \, (5) = \frac{240}{11}$

Therefore, the point X is $\frac{240}{11}$ feet down from the top end of the ladder.

Thus, the point in space that is occupied by the ladder in the original position <u>and</u> the final position is $\frac{240}{11}$ or 21.8 ft. from the top of the ladder along the original line of the ladder.

● **PROBLEM** 420

A 25-foot ladder leans against a wall such that the top of the ladder is y feet from the ground, and the base of the ladder is x feet from the wall. The base of the ladder slides another 17 feet from the wall so that the base is now y feet from the wall and the top is x feet from the ground. Find y, the original height that the ladder reached.

Solution: In the accompanying figure, $\overline{AC}$ is the original position of the ladder; $\overline{DE}$ is the final position. Since the length of the ladder does not change, AC = DE = 25 ft.

Let AB, the original height = y and DB, the final

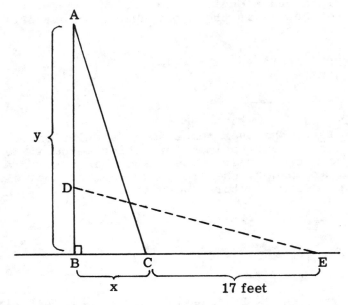

height = x. Also, the original distance from the base
BC = x; and the final distance BE = y. From the problem
statement, we know that

$$BC + 17 = BE, \quad \text{or} \quad x + 17 = y.$$

To relate the various distances, we assume that the
wall is perpendicular to the ground, and we use the
Pythagorean Theorem:

(i) $(AB)^2 + (BC)^2 = (AC)^2$

(ii) $y^2 + x^2 = 25^2$

Since x + 17 = y or x = y - 17, we can substitute
this for x. We then obtain an equation in y and thus can
solve for y.

(iii) $y^2 + (y - 17)^2 = 25^2$

(iv) $y^2 + (y^2 - 34y + 289) = 625$

(v) $2y^2 - 34y + 289 - 625 = 0$

(vi) $2y^2 - 34y - 336 = 0$

(v) $y^2 - 17y - 168 = 0.$

To solve, we can either factor, or use the quadratic
equation. In this case, note that 7(- 24) = - 168 and
7 + (- 24) = - 17. Therefore,

(vi) $(y + 7)(y - 24) = 0.$

For this equation to be true, one factor must be 0.
Either y + 7 = 0 or y - 24 = 0. Thus, either y = - 7 or
y = 24. Remember, though, that y is the height of the ladder.

368

We discard y = - 7 as being physically meaningless. Thus, the original height of the ladder is y = 24 feet.

We can determine x, by substituting y = 24 into equation (ii)

$$x^2 + y^2 = 25^2.$$

Hence, $x^2 = 25^2 - 24^2 = 49$ and x = 7,

the original distance from the foot of the ladder to the base of the wall.

After the 17 ft. slip, the base of the ladder was (7 + 17) ft. or 24 ft. from the wall.

● **PROBLEM** 421

Show that if d is the length of a diagonal of a rectangular solid in which every pair of intersecting edges is perpendicular, then $d^2 = a^2 + b^2 + c^2$, where a, b, and c are the lengths of the edges.

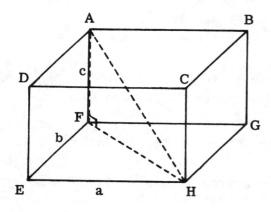

Figure 1

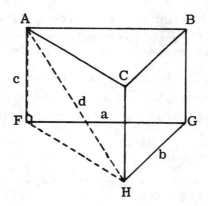

Figure 2a

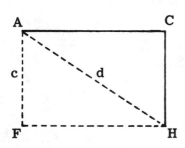

Figure 2b

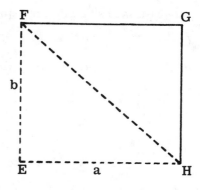

Figure 3

369

<u>Solution</u>: Let the width EH = a; the length EF = b; and the height AF = c. If we cut along the diagonal AC straight down, we obtain two triangular solids, one of which is shown below in Fig. 2a.

In Fig. 2b, quadrilateral ACFH is redrawn. If ΔAFH were a right triangle, we could state, using the Pythagorean Theorem,

(i) $\qquad d^2 = c^2 + (FH)^2.$

To prove that ∢ AFH is a right angle, we show that $\overline{AF}$ is perpendicular to the plane determined by quadrilateral EFGH, and therefore it is perpendicular to every line in the plane passing through the foot of $\overline{AF}$. Since $\overline{FH}$ is such a line, $\overline{AF}$ must be perpendicular to it.

PROOF: We are given that the figure in Fig. 1 is a rectangular solid. Therefore, the face EFGH must be a rectangle. A rectangle is a plane figure. Therefore, all sides of rectangle EFGH ($\overline{EF}$, $\overline{GH}$, $\overline{EH}$ and $\overline{FG}$) must lie on the same plane - let's call it P.

We are given that all intersecting edges must be perpendicular. Therefore, $\overline{AF} \perp \overline{FE}$ and $\overline{AF} \perp \overline{FG}$. Since $\overline{AF}$ is perpendicular to two distinct lines in plane P, $\overline{AF}$ is perpendicular to plane P.

Points F and H lie in plane P; therefore, line segment $\overline{FH}$ lies in plane P. $\overline{AF}$ is perpendicular to all lines in plane P; therefore, $\overline{AF} \perp \overline{FH}$.

Thus equation (i) is valid. To solve for d, though, we must still find FH.

Redrawing the base rectangle EFGH from Fig. 1, we note that FH can be found, using the Pythagorean theorem.

(ii) $\qquad FH^2 = a^2 + b^2$

Substituting in equation (i), we obtain:

(iii) $\qquad d^2 = c^2 + (a^2 + b^2)$

(iv) $\qquad d^2 = a^2 + b^2 + c^2.$

370

CHAPTER 22

TRIGONOMETRIC RATIOS

Basic Attacks and Strategies for Solving Problems in this Chapter. See pages 371 to 385 for step-by-step solutions to problems.

Although trigonometry has many varied applications, the problems in this chapter concern the trigonometry of right triangles. In that case, the trigonometric functions are applied to acute angles in right triangles. The three most commonly used trigonometric functions are the **sine**, **cosine**, and **tangent**. As applied to angle A in the right triangle pictured below, the definitions follow:

$$\text{sine (sin)} \, A \; = \; \frac{\text{length of side opposite} \; \angle A}{\text{length of hypotenuse}}$$

$$\text{cosine (cos)} \, A \; = \; \frac{\text{length of side adjacent to} \; \angle A}{\text{length of hypotenuse}}$$

$$\text{tangent (tan)} \, A \; = \; \frac{\text{length of side opposite} \; \angle A}{\text{length of side adjacent to} \; \angle A}$$

These formulas are commonly remembered with the aid of the following acronym: SOHCAHTOA: Sine = Opposite/Hypotenuse; Cosine = Adjacent/Hypotenuse; Tangent = Opposite/Adjacent.

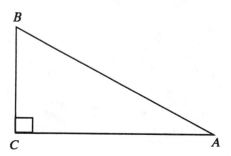

In the figure on the previous page, $\overline{AB}$ is the hypotenuse, and $\overline{BC}$ and $\overline{AC}$ are opposite and adjacent to $\angle A$, respectively. Thus the trigonometric ratios can be expressed in this way:

$$\sin A = \frac{BC}{AB}$$

$$\cos A = \frac{AC}{AB}$$

$$\tan A = \frac{BC}{AC}$$

Note that with respect to angle B, $\overline{AC}$ is the opposite side, $\overline{BC}$ is the adjacent side, and $\overline{AB}$ is the hypotenuse. Also, note that these formulas only make sense when applied to the acute angles of a right triangle.

Trigonometric functions are routinely listed in tables, and several problems in this chapter suggest referring to the table to find appropriate ratios. Most scientific calculators also have the capability of giving trigonometric ratios, and in most instances a calculator can be used instead of a table.

In several problems, questions are asked about triangles which are not right triangles. In those cases it is possible to make some constructions to get a right triangle "related" to the original triangle, and then to apply the trigonometry to the right triangle.

Step-by-Step Solutions to Problems in this Chapter, "Trigonometric Ratios"

A road is inclined 8° to the horizontal. Find, to the nearest hundred feet, the distance one must drive to increase one's altitude 1,000 ft.

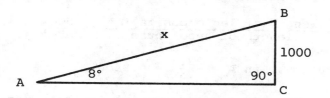

Solution: If we begin driving at point A in the accompanying diagram, then the upward sloping road, $\overline{AB}$, the imaginary altitude line, $\overline{BC}$, and the "flat" ground upon which the road is built form a right triangle. We know one acute angle, the length of the side opposite it, and we wish to find the length of the hypotenuse (the road length to be traveled). The sine ratio can be employed here.

$$\text{Sin A} = \frac{\text{length of leg opposite } \angle A}{\text{length of hypotenuse}} = \frac{BC}{AB}$$

Let x = AB, and substitute the given facts.

$$\text{Sin } 8° = \frac{1000}{x} \quad \text{and} \quad x = \frac{1000}{\sin 8°}.$$

According to a standard sine table, Sin 8° = 0.1392. Therefore,

$$x = \frac{1000}{.1392} = 7184$$

One must drive 7200 ft. (correct to the nearest hundred feet).

A ladder 25 feet long leans against a building and reaches a point 23.5 feet above the ground. Find, to the nearest degree, the angle which the ladder makes with the ground.

Solution: As seen in the accompanying drawing, the building forms a right triangle with the ground. Angle A is the angle the question asks us to determine. We are given the length of the side opposite ∢ A, 23.5, and the length of the hypotenuse (the ladder) 25. With this information, we can calculate the measure of the angle by using the sine ratio.

$$\text{Sin A} = \frac{\text{length of leg opposite} \angle A}{\text{length of hypotenuse}} = \frac{BC}{AB} .$$

By substitution, $\text{Sin A} = \frac{23.5}{25} = 0.9400$.

Now, we consult a standard sine table to determine which angles have sines near 0.9400.

We find that Sin 70° = 0.9397 and Sin 71° = 0.9455. Since 0.9400 is closer to Sin 70°, we can conclude that m ∢ A is closer to 70°.

Therefore, the ladder makes a 70° angle with the ground.

A boy who is flying a kite lets out 300 feet of string which makes an angle of 38° with the ground. Assuming that the string is perfectly taut and forms a straight line, how high above the ground is the kite? (Give the answer to the nearest foot.)

<u>Solution</u>: The boy, his kite, and the point on the ground immediately below the kite form a right triangle. (See figure.) The string represents the hypotenuse, and its angle with the ground is an acute angle of the right triangle. The unknown height of the kite is the length of the side opposite ⊿ B. Given an angle and the hypotenuse of a right triangle, we can find the length of the side opposite the angle by an application of the sine ratio. In our case,

$$\text{Sin B} = \frac{\text{length of leg opposite} \angle \text{B}}{\text{length of hypotenuse}} = \frac{\text{KG}}{\text{BK}}$$

If we let x = KG, then, by substitution,

$$\text{Sin } 38° = \frac{x}{300} \cdot$$

From a standard sine table we find that Sin 38° = 0.6157. Therefore,

$$0.6157 = \frac{x}{300}$$

$$x = 0.6157 \ (300) = 184.71.$$

Therefore, the kite is, to the nearest foot, 185 ft. above the ground.

● **PROBLEM** 425

A guy wire reaches from the top of a pole to a stake in the ground. The stake is 10 feet from the foot of the pole. The wire makes an angle of 65° with the ground. Find, to the nearest foot, the length of the wire.

<u>Solution</u>: A right triangle is formed by the pole, the wire and the ground. The wire acts as the hypotenuse of this triangle, and its length is the unknown to be determined. An acute angle and the length of the side adjacent to it are given and, to find the length of the hypotenuse, the cosine ratio can be applied.

$$\text{Cos S} = \frac{\text{length of leg adjacent to} \angle \text{S}}{\text{length of hypotenuse}} = \frac{\text{SB}}{\text{ST}} \cdot$$

Let x = ST and, by substitution, Cos 65° = $\frac{10}{x}$.
In a standard cosine table, we find Cos 65° = 0.4226.

Substitute this value into the equation

$$0.4226 = \frac{10}{x}$$

$$x = \frac{10}{0.4226} = 23.6.$$

Therefore, the guy wire is 24 feet long, correct to the nearest foot.

● **PROBLEM** 426

A plane took off from a field and rose at an angle of 8° with the horizontal ground. Find, to the nearest ten feet, the horizontal distance the plane had covered when it had flown 2,000 feet in air.

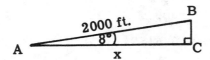

Solution: The line of flight, $\overline{AB}$, the altitude, $\overline{BC}$, and the horizontal distance, $\overline{AC}$, form a right triangle, $\triangle ABC$ (as shown in the figure). We are given the angle of take-off, i.e. ∢ A and the length of the hypotenuse of the right triangle. We wish to find the horizontal distance, which is the length of the side adjacent to ∢ A. This implies an application of the cosine ratio.

$$\text{Cos A} = \frac{\text{length of leg adjacent } \sphericalangle A}{\text{length of hypotenuse}} = \frac{AC}{AB}.$$

By substitution, $\cos 8° = \frac{x}{2000}$. From a standard cosine table, we find $\cos 8° = 0.9903$. Substituting this number into the equation, we obtain

$$0.9903 = \frac{x}{2000},$$

$$x = (2000)\ 0.9903 = 1980.6.$$

Therefore, the horizontal distance, to the nearest ten feet is 1980.

● **PROBLEM** 427

At a point on the ground 40 feet from the foot of a tree, the angle of elevation to the top of the tree is 42°. Find the height of the tree to the nearest foot.

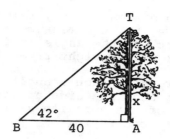

Solution: The geometric figure formed is a right triangle (see figure). Since the unknown height of the tree is opposite the given angle of elevation, and we are given the side adjacent to this angle, we can solve the problem using the tangent ratio. The tangent is the ratio of the length of the leg opposite the acute angle to the length of the leg adjacent to the acute angle in any right triangle. In this example,

$$\tan B = \frac{\text{length of leg opposite} \sphericalangle B}{\text{length of leg adjacent} \sphericalangle B},$$

$$\tan B = \frac{AT}{BA}.$$

Let $x = AT$, and consult a standard table of tangents to find that $\tan 42° = 0.9004$. Since $BA = 40$, we obtain

$$0.9004 = \frac{x}{40}.$$

Therefore, $x = 40 (0.9004) = 36.016.$

Therefore, the height of the tree, to the nearest foot, is 36 feet.

● **PROBLEM** 428

From the top of a lighthouse 160 ft. above sea level, the angle of depression of a boat at sea is 35°. Find, to the nearest foot, the distance from the boat to the foot of the lighthouse.

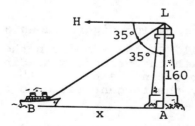

Solution: The distance from the foot of the lighthouse to the boat is represented by BA, (see figure), which is the unknown we wish to find. We will determine BA by an application of the tangent ratio.

The angle of depression, ∢ BLH, is complementary to
∢ BLA. Since m ∢ BLH = 35°, the measure of ∢ BLA is 90° -
35°, or 55°. We cannot use the angle of depression because
it is on the exterior of the right triangle.

Given the measure of an acute angle of a right
triangle and the length of the side adjacent to it, we can
determine the length of the opposite side by a direct
application of the tangent ratio.

$$\tan ∢ BLA = \frac{\text{length of leg opposite } ∢ \text{ BLA}}{\text{length of leg adjacent } ∢ \text{ BLA}} = \frac{BA}{LA}$$

Let x = BA, thus $\tan 55° = \frac{x}{160}$. Consult a standard
tangent table to find tan 55° = 1.4281, and substitute
this to obtain

$$1.4281 = \frac{x}{160}$$

$$x = (1.4281)\, 160$$

$$x = 228.496$$

Therefore, the boat is 228 ft. away from the foot of the
lighthouse (to the nearest foot).

● **PROBLEM** 429

A ladder, which is leaning against a building, makes an
angle of 75° with the ground. If the top of the ladder
reaches a point which is 20 feet above the ground, find,
to the nearest foot, the distance from the foot of the
ladder to the base of the building.

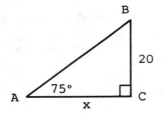

Solution: The accompanying figure shows a ladder, $\overline{AB}$,
leaning against wall, $\overline{BC}$. We are given the angle at which
the ladder meets with the ground, ∢ A. Since we want to
find the distance between the base of the ladder and the
wall, we can employ the tangent ratio to determine the
unknown.

$$\tan ∢ A = \frac{\text{length of leg opposite } ∢ \text{ A}}{\text{length of leg adjacent } ∢ \text{ A}} = \frac{BC}{AC} \; .$$

If we let x = AC, then $\tan 75° = \frac{20}{x}$; $x = \frac{20}{\tan 75°}$.

Since tan 75° = 3.7321, the final result

$$x = \frac{20}{3.7321} = 5.3580.$$

● **PROBLEM** 430

Find to the nearest degree, the measure of the angle of elevation of the sun when a vertical pole 6 feet high casts a shadow 8 feet long.

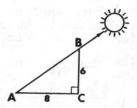

Solution: When a line is extended from the farthest end of the shadow, point A in the figure shown, toward the sun to form the angle of elevation, the line will intersect the pole at its uppermost tip (point B). This will form a right triangle in which the length of the sides both opposite and adjacent to angle A are known. In order to calculate m∢A, the tangent ratio can be applied.

$$\text{Tan A} = \frac{\text{length of leg opposite } ∢ A}{\text{length of leg adjacent } ∢ A} = \frac{BC}{AC}.$$

We are given BC = 6, and AC = 8, therefore, by substitution,

$$\tan A = \frac{6}{8} = .7500.$$

By inspection of a standard tangent table, it is seen that tan 36° = .7265 and tan 37° = .7536. Since .7536 is closed to .7500 than it is to .7265, we conclude that m ∢ A is closer to 37°. (When we calculate the tangent of an unknown angle, we always compare the calculated value of the tangent to the value immediately less than it and the value immediately greater than it on the tangent table. The table value which is closer to the calculated value is accepted as the unknown angle's measure.)

● **PROBLEM** 431

The diagonals of a rhombus are 10 and 24. Find, to the nearest degree, the sum of the angles of the rhombus.

Solution: As drawn in the accompanying diagram, BD = 10 and AC = 24.

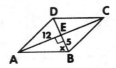

Since the diagonals of a rhombus are perpendicular bisectors of each other, ∢ AEB is a right angle, EB = ½ (DB) = ½(10) = 5 and AE = ½(AC) = ½(24) = 12.

Since ABCD is a rhombus, diagonal $\overline{DB}$ bisects ∢ ABC and, as such m ∢ ABC = 2(m ∢ ABE). Therefore, we will find the measure of ∢ ABE in right triangle AEB and multiply the result by 2 to find m ∢ ABC.

∢ ABE is one of the acute angles of right △AEB. Since we know the lengths of both the opposite and adjacent sides, we can use the tangent ratio to determine its measure.

Let x = the measure of ∢ ABE, thus,

tan x = $\dfrac{\text{length of leg opposite ∢ ABE}}{\text{length of leg adjacent ∢ ABE}}$.

By substitution, tan x = $\dfrac{12}{5}$ = 2.400.

From a standard tangent table, we see that the tan 67° = 2.3559 and the tan 68° = 2.4751. Since 2.4000 is closer to 2.3559 than it is to 2.4751, 67° is the best degree measure of x.

Therefore, m ∢ ABC = 2(m ∢ ABE) = 2 (67°) = 134°.

Since the consecutive angles of a rhombus are supplementary, ∢ BCD is supplementary to ∢ ABC.

As such, m ∢ BCD = 180° - m ∢ ABC or

m ∢ BCD = 180° - 134° = 46°.

Since opposite angles of a rhombus are congruent, m ∢ CDA = m ∢ ABC = 134°, and m ∢ DAB = m ∢ BCD = 46°.

Therefore, the angles of the rhombus contain 46°, 134°, 46°, and 134°, respectively. The sum of these angles is

46 + 134 + 46 + 134 = 360°.

● **PROBLEM** 432

In isosceles triangle ABC, AC = CB = 20 and m ∢ A = m ∢ B = 68°. $\overline{CD}$ is an altitude, thus $\overline{CD} \perp \overline{AB}$. (a) Find the length of altitude $\overline{CD}$ to the nearest tenth; (b) Find the length of $\overline{AB}$ to the nearest tenth.

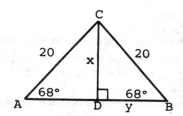

Solution: (a) Altitude $\overline{CD}$ divides $\triangle ABC$ into two right triangles. In right $\triangle BDO$, an acute angle and the length of the hypotenuse are known and, therefore, the length of the altitude, which is opposite ∢ B, can be found by applying the sine ratio:

$$\sin B = \frac{\text{length of leg opposite } ∢ B}{\text{length of hypotenuse}} = \frac{CD}{DB}$$

Let x = CD and substitute the given facts:

$$\sin 68° = \frac{x}{20} \ .$$

According to standard sine tables, sin 68° = .9272. Substitute this number into the equation above to calculate x.

$$.9272 = \frac{x}{20}$$

$$x = (.9272)(20) = 18.5440.$$

To the nearest tenth, the length of the altitude is 18.5.

  (b) Since the altitude to the base of an isosceles triangle bisects that base, to find $\overline{AB}$ we need only find $\overline{DB}$ and multiply it by two. $\overline{DB}$ is adjacent to angle B and can be found through use of the cosine ratio:

$$\cos B = \frac{\text{length of leg adjacent } ∢ B}{\text{length of hypotenuse}} = \frac{DB}{BC}$$

Let y = DB. Then, by substitution,

$$\cos 68° = \frac{y}{20} \ .$$

A standard cosine table shows that cos 68° = .3746. Substituting this into the previous equation yields

$$.3746 = \frac{y}{20}$$

$$y = (20)(.3746) = 7.4920.$$

$$AB = 2DB. \text{ Therefore, } AB ≒ 2y = 2(7.4920)$$

$$AB = 14.9840$$

The length of base $\overline{AB}$ is 15.0, to the nearest tenth.

In a circle whose radius is 4 in., find, to the nearest inch, the length of the minor arc intercepted by a chord 6 inches in length.

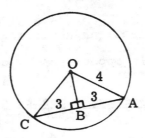

Solution: To find the length of minor arc $\overset{\frown}{AC}$, we must first find the measure of the arc in degrees. Then we will apply the arc length formula which states that

$$\text{arc length} = \frac{\text{degree measure of arc}}{360} \times \text{Circumference}.$$

Circumference = C = $2\pi r$, where r = radius of circle. By substitution, C = $2\pi(4)$ = 8π. $\angle AOC$ is the central angle of the arc in question, $\overset{\frown}{AC}$. m$\angle AOC$ is what we wish to determine. $\triangle AOC$ is isosceles because, as radii of circle O, $\overline{OC} \cong \overline{OA}$. Draw the altitude to $\overline{AC}$. By a theorem, we know that altitude $\overline{OB}$ bisects vertex angle $\angle AOC$. Therefore, m$\angle AOC$ = m$\angle AOB$ + m$\angle COB$ and, since $\overline{OB}$ is the bisector of $\angle AOC$, m$\angle AOB$ = m$\angle COB$ = 1/2 m$\angle AOC$.

Hence, m$\angle AOC$ = 2(m$\angle AOB$). $\triangle AOB$ is a right triangle, since $OB \perp AC$.

Since we know the hypotenuse and the side opposite $\angle AOB$, we can use the sine ratio to determine the measure of $\angle AOB$.

$$\text{Sin }(\angle AOB) = \frac{\text{leg opposite }\angle AOB}{\text{hypotenuse}} = \frac{AB}{OA}$$

$\overline{OA}$ is a radius given to be 4 in. Since a line drawn from the center of a circle perpendicular to a chord bisects that chord, AB = 1/2(AC) = 1/2(6) = 3

Therefore, sin ($\angle AOC$) = $\frac{3}{4}$ = 0.75

The measure of the angle whose sine is closest to 0.75 is 49.

Hence, by substitution

$$\text{m}\angle AOC = 2(\text{m}\angle AOB) = 2(49) = 98°$$

Then, length of minor arc $\overset{\frown}{AC}$ = $\frac{98}{360} \times 8\pi$ = 2.2π

Using π = 3.14, 2.2π = 6.9

Therefore, to the nearest inch, length of minor arc $\overset{\frown}{AC}$ = 7in.

Show for △ABC, with radius of the circumscribed circle = R, that:

$$\frac{a}{\sin A} = \frac{b}{\sin B} = \frac{c}{\sin C} = 2R.$$

(Hint: Draw inscribed right △BCJ.)

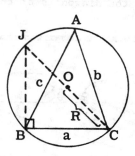

Solution: In the accompanying figure, ⊙ O circumscribes △ABC. $\overline{JC}$ is a diameter of ⊙ C. The lengths of the sides of △ABC are a, b, and c. To show $\frac{a}{\sin A}$ = 2R, we will show (a) sin∢BJC = $\frac{a}{2R}$; (b) ∢ BJC $\stackrel{\sim}{=}$ ∢ A.

(a) Since $\overline{JC}$ is a diameter, △JBC is a triangle inscribed in a semicircle. Thus, △JBC is a right triangle and ∢ BJC is an acute angle of a right triangle.

The sine of an angle of a right triangle

$$= \frac{\text{length of the opposite side}}{\text{length of the hypotenuse}}$$

Thus,   sin∢BJC = $\frac{BC}{JC}$ = $\frac{a}{2R}$.

(b) Inscribed ∢ BJC and inscribed ∢ A both intercept the same arc $\overparen{BC}$. Thus, ∢ BJC $\stackrel{\sim}{=}$ ∢ A; and

(i)   sin A = $\frac{a}{2R}$.

Multiplying both sides by $\frac{2R}{\sin A}$, we obtain

(ii)   2R = $\frac{a}{\sin A}$ or $\frac{a}{\sin A}$ = 2R.

By drawing other right triangles, we can also show that $\frac{b}{\sin B}$ = 2R,   and $\frac{c}{\sin C}$ = 2R. Thus,

(iii) $\quad \dfrac{a}{\sin A} = \dfrac{b}{\sin B} = \dfrac{c}{\sin C} = 2R.$

● **PROBLEM** 435

A clock hangs on the wall of the Pitman Security Agency. The wall is 71 ft. 9 in. long and 10 ft. 4 in. high. The night watchman notices that the hands of the clock are pointing in opposite directions and are parallel to one of the diagonals of the wall. What was the exact time?

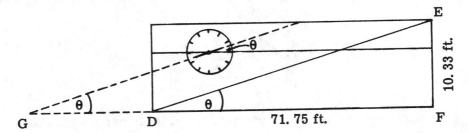

Solution: Assuming that the clock is circular and hanging perfectly straight on the wall, the line segment formed by the hands the watchman is reading intersects the line of the floor $(\overline{GDF}$, when extended) in an angle congruent to that of the diagonal. This is the case because the hands are parallel to the diagonal and corresponding angles formed by a transversal $\overline{GDF}$ of parallel lines are congruent.

The segment joining the 9 and the 3 on the clock will be parallel to the floor, according to how the clock is hanging, hence, the acute angle between the segment and the clock hands will be equal in measure to the acute angle between the clock hands and the floor. The clock hands (extended to G) now act as a transversal for the floor line and the segment joining 9 and 3.

We can find the measure of this angle, θ, using the tangent ratio since we know the length of the sides opposite and adjacent to θ in right $\triangle$ DEF.

$$\tan \theta = \frac{\text{length of opposite side}}{\text{length of adjacent side}} = \frac{10.33}{71.75} = .14 .$$

Hence, according to a standard tangent table, $\theta = 8.17°$ for $\tan \theta = .14$.

We will now relate 8.17 to 360, the total number of degrees in a circle, to find the percentage of the circle the angle occupies. Multiplying 60 minutes by this percentage will give you the angle in (time) minutes and allow you to determine the time.

$$\frac{8.17}{360} = 2.3\% \text{ of the circular clock}$$

$$(2.3\%)(60 \text{ min.}) = 1 \text{ min. } 22 \text{ sec. of time.}$$

During the shift of a night watchman, the hands will form the segment joining 9 to 3 at 2:45 am. (as a good approximation).

The angle θ = 1 min. 22 sec., as drawn, tells us that the time the watchman saw was 1 min. 22 sec. before 2:45 or 2:43 and 38 seconds, exactly.

● **PROBLEM** 436

In circle O, $\overline{AB}$ is a diameter. Radius $\overline{OC}$ is extended to meet tangent $\overline{BD}$ at D. The measures of arc $\overset{\frown}{AC}$ and arc $\overset{\frown}{CB}$ are in the ratio 2:1 and $\overline{AB}$ = 32.

(a) Find the number of degrees contained in arc $\overset{\frown}{CB}$.
(b) Find, to the nearest integer, the perimeter of the figure bounded by $\overline{BD}$, $\overline{DC}$ and arc $\overset{\frown}{CB}$. [π = 3.14]
(See figure)

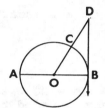

Solution: (a) To find the measure of arc $\overset{\frown}{CB}$, we must make use of two known facts. First, arc $\overset{\frown}{ACB}$ is a semi-circle, and, as such, contains 180°. Second, the composite arcs of $\overset{\frown}{ACB}$; $\overset{\frown}{AC}$ and $\overset{\frown}{CB}$, have measures in a ratio of 2:1. If we let x = the measure of arc $\overset{\frown}{CB}$, then 2x = the measure of arc $\overset{\frown}{AC}$. Since the measure of $\overset{\frown}{ACB}$ = measure of $\overset{\frown}{AC}$ + measure of $\overset{\frown}{CB}$, and $\overset{\frown}{ACB}$ is a semicircle, we may write 180° = 2x + x = 3x or x = 60°. Therefore, the number of degrees contained in arc $\overset{\frown}{CB}$ is 60°.

(b) To find the perimeter of the figure in question, we must determine the lengths of arc $\overset{\frown}{CB}$ and segments $\overline{DC}$ and $\overline{DB}$, and then sum these quantities. We will first calculate the length of arc $\overset{\frown}{CB}$.

$$\text{length of arc } \overset{\frown}{CB} = \frac{\text{measure of central angle COB}}{360°} \times \text{Circumference}$$

of circle O. Arc $\overset{\frown}{CB}$ has a measure of 60°, and its central angle, ∢COB, by theorem, has the same measure. Circumference of circle O = $2\pi r$, where r is the radius of circle O. We are given diameter AB = 32. Therefore, any radius = $\frac{1}{2}$(32) or 16.

Circumference of circle O = $2\pi(16)$ = 32π = 32(3.14) = 100.5.

Therefore, length of arc $\overset{\frown}{CB}$ = $\frac{60°}{360°} \times 100.5$ = 16.75.

Next, we need to find BD and DC. ΔODB is a right tri-
angle, since tangent $\overline{BD}$ is perpendicular to radius $\overline{OB}$. (A
tangent to a circle is always perpendicular to a radius
drawn to the point of tangency.) Hence, the various trigonometric ratios can be employed to find the lengths of the
unknown segments. We know m≯DOB = 60°, and the leg adjacent
to it, $\overline{OB}$, has a length of 16. We can find the length of
the opposite leg, $\overline{DB}$, by using the tangent ratio, and the
length of the hypotenuse, $\overline{OD}$, by using the cosine ratio.

$$\tan(\text{≯DOB}) = \frac{\text{length of leg opposite ≯DOB}}{\text{length of leg adjacent ≯DOB}} = \frac{DB}{OB} .$$

By substitution, $\tan 60° = \frac{DB}{16}$. According to a standard tangent table, $\tan 60° = 1.7321$, hence $1.7321 = \frac{DB}{16}$, DB =
(16)(1.7321). Therefore, BD = 27.7. The final segment,
whose length we need, is DC. DC = OD - OC. We know OC = 16,
because it is a radius of O. $\overline{OD}$ is the hypotenuse of rt.
ΔODB. We will use the cosine ratio to find its length.

$$\text{cosine}(\text{≯DOB}) = \frac{\text{length of leg adjacent ≯COB}}{\text{length of hypotenuse}} = \frac{OB}{OD}$$

OB = 16 and m≯DOB = 60°. By substitution, $\cos 60° = \frac{16}{OD}$

OD = $\frac{16}{\cos 60°}$. According to a standard cosine table,
$\cos 60° = .5$.

$$OD = \frac{16}{5} = 32$$

Therefore, length of $\overline{DC}$ = OD - OC = 32 - 16 = 16. Hence,
the perimeter of the figure bounded by $\overline{BD}$, $\overline{DC}$, and arc $\overline{CB}$
is the sum of the measures of these individual parts, or
27.7 + 16 + 16.75 = 60.45. The perimeter of the required
figure is 60, to the nearest integer.

● **PROBLEM** 437

Given a triangle with sides of length a, b, and c, show
that $\cos C = \dfrac{a^2 + b^2 - c^2}{2ab}$.

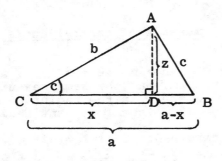

<u>Solution</u>: In the accompanying figure $\triangle ABC$ has sides of length a, b, and c. The altitude $\overline{AD}$ has length z and intersects side $\overline{BC}$ at point D. The length of $\overline{CD}$ = x. Because DB = CB - CD, then **DB** = a - x.

We are asked to find the cosine of $\angle$ C. Note that $\angle$ C is also an angle of right triangle ACD. By definition, the cosine of $\angle$ C in right $\triangle ACD$ equals

(length of adjacent side)/(length of the hypotenuse)

$$= \frac{CD}{CA} = \frac{x}{b} \; .$$

Since the three sides of $\triangle ABC$ are given, b is known. To find x, we use the fact that AD = z is a leg of two right triangles ACD and ABD. By the Pythagorean Theorem, we have

(i) $AD^2 = AC^2 - CD^2$ or $z^2 = b^2 - x^2$

(ii) $AD^2 = AB^2 - BD^2$ or $z^2 = c^2 - (a - x)^2$.

Combining these two expressions, we have a single equation with only one unknown - x. Combining (i) and (ii) by the Transitive Property of Equality, we obtain:

(iii) $b^2 - x^2 = c^2 - (a - x)^2$

Multiplying out and simplifying, we obtain

(iv) $b^2 - x^2 = c^2 - a^2 + 2ax - x^2$

(v) $b^2 = c^2 - a^2 + 2ax$

(vi) $x = \dfrac{a^2 + b^2 - c^2}{2a}$

With the expression for x, we can now solve for cos C.

(vii) $\cos C = \dfrac{x}{b} = \dfrac{a^2 + b^2 - c^2}{2ab}$.

385

CHAPTER 23

AREAS OF QUADRILATERALS AND TRIANGLES

> **Basic Attacks and Strategies for Solving Problems in this Chapter. See pages 386 to 419 for step-by-step solutions to problems.**

Most of the problems in this chapter involve the development or application of standard area formulas for squares, rectangles, parallelograms, rhombuses, trapezoids, and triangles. The formulas for a rectangle and square are listed below.

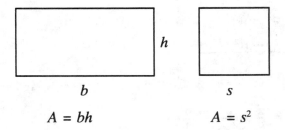

$$A = bh \qquad\qquad A = s^2$$

In the case of a rectangle formula, b is the length of one side and h is the length of an adjacent side. Often the side whose length is b is called the **base**, and the side whose length is h is called the **altitude**. Note that the formula for the area of a square is really the formula for a rectangle with both its height and base equal to s.

For a triangle, it is common to classify one side as the base. For the triangle pictured on the next page, AB is the base. The area of a triangle is one-half the product of the length of the base and the length of the altitude to that base.

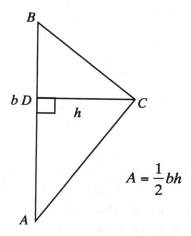

$$A = \frac{1}{2}bh$$

Note that $b = AB$ and $h = DC$. It should also be noted that selection of the base side is arbitrary. It is somewhat common to call the horizontal side the base, but the orientation of the triangle in the illustration is also arbitrary, so "horizontal" has no real geometric meaning.

It is worth noting that although $A = \frac{1}{2}bh$ is the most common way of representing the area of a triangle, there are other ways of expressing that same value. Depending on what information is given, it may be more convenient to calculate the area of the triangle using a different form of the formula, as in Problems 458-460.

The area formula for a parallelogram is given below. Again, it is common to classify one side of the parallelogram as the base and to call its length b, and to find an altitude to that base and call its length h.

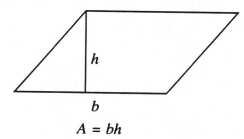

$$A = bh$$

A **trapezoid** is defined as a quadrilateral with exactly two parallel sides. These parallel sides are called bases. In the trapezoid on the next page, the lengths of these bases are b_1 and b_2, and the length of an altitude is h. The area formula is given below.

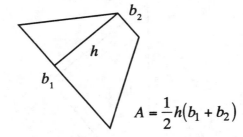

$$A = \frac{1}{2}h(b_1 + b_2)$$

There is an interesting relationship between the areas of similar figures. If the ratio of lengths of "corresponding" parts of any pair of similar figures is a/b, then the ratio of the areas of those figures is $(a/b)^2$, regardless of the type of figure involved. That generalization is illustrated in several problems in this chapter.

SQUARES & RECTANGLES

● **PROBLEM** 438

Find the area of a rectangle whose base is 1½ feet and whose altitude is 6 inches.

Solution: The area of a rectangle is equal to the product of the length of its base (b) and the length of its altitude, (h), (i.e., A = bh).

However, before we can use this formula, the base and altitude must be converted to the same units. Let us convert the base, given in feet, into inches. Since 1 ft. = 12 in., 1½ ft. = 18 in. Hence,

$$A = bh \; ; \; b = 18 \text{ in.}; \; h = 6 \text{ in.}$$

By substitution, $A = (18 \text{ in.} \times 6 \text{ in.}) = 108 \text{ in.}^2$

Therefore, the area is 108 sq. in.

● **PROBLEM** 439

A rectangle, the lengths of whose base and altitude are in the ratio 4:1, is equivalent to a square, the length of whose side is 6 inches. Find the dimensions of the rectangle.

Solution: The fact that the square and the rectangle are equivalent implies that they have the same area.

Let x = the altitude of the rectangle in inches.

Then 4x = the base of the rectangle in inches. Therefore, the area of the rectangle = bh, where b is the base length, and h is the height of the altitude. By substitution, $A_r = (4x)x = 4x^2$.

If we let s = the side of a square in inches, then, the area of the square, $A_s = s^2$ and, by substitution, $A_s = (6 \text{ in.})^2 = 36 \text{ in.}^2$.

Since we are told that the square and rectangle are equivalent, we can set $A_r = A_s$ to solve for x.

$$4x^2 = 36 \text{ in.}^2$$
$$x^2 = 9 \text{ in.}^2$$
$$x = 3 \text{ in.}$$
Thus, $\qquad 4x = 12 \text{ in.}$

Therefore, the base of the rectangle = 12 in. and the altitude of the rectangle = 3 in.

Using the fact that the area of a square of side a is a^2, prove that the area of a rectangle, of side lengths h and w, is hw .

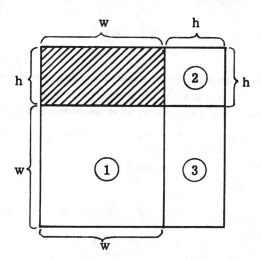

Solution: In the figure, we must find the area of the shaded region. We will exploit our knowledge of the expression for the area of a square in order to find the area of a rectangle. First, construct squares of edge length w and h adjacent to the shaded rectangle (regions (1) and (2).). Region (3) is a rectangle of side lengths h and w. Hence, the shaded region and region (3) are congruent. Finally, the entire figure is a square of edge length w + h .

We may now write

Total area = area of (1) + area of (2)
+ area of (3) + shaded area .

Denoting the area of (3) and the area of the shaded region as area of rectangle, we obtain

Total area = area of (1) + area of (2)
+ 2 area of rectangle   (1)

Since the area of a square of edge length a is a^2 , we know that

area of (1) = w^2

area of (2) = h^2 .   (2)

Total area = $(h+w)^2$ = h^2 + 2hw + w^2 .

Hence, using (2) in (1),

$h^2 + 2hw + w^2$ = h^2 + w^2 + 2 (area of rectangle).

Cancelling like terms,

2hw = 2$($area of rectangle$)$

or

hw = area of rectangle.

What is the effect on the area of a square if we change its side lengths by a factor of t.

Solution: By definition, the original square having edge length a, has an area a^2. The new square has edge length ta, and area $(ta)^2 = t^2a^2$. Hence, if we change the edge length by a factor of t, we change the area of the square by a factor of t^2

● **PROBLEM** 442

Find the area of a square whose perimeter is 20 ft.

Solution: The perimeter, p, of a square is given by 4 times the length of one side, s; that is, p = 4s.
But, by substitution,

$$p = 20 \text{ ft.}$$
$$20 \text{ ft.} = 4s$$
$$5 \text{ ft.} = s \quad .$$

The area of a square is given by s^2 . Hence, A = s^2 and, by substitution,

$$A = (5 \text{ ft.})^2 = 25 \text{ ft.}^2$$

Therefore, the area of the square is 25 sq. ft.

● **PROBLEM** 443

Find the area of square ABCD where AC = d .

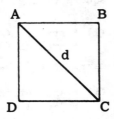

Solution: The area of a square equals s^2 where s is the length of a side. We are given only the length of the diagonal d and must relate this to the length of the side. By the Pythagorean Theorem, for right triangle $\triangle ACD$, $\overline{AC}^2 = \overline{CD}^2 + \overline{AD}^2$. The sides of a square are equal. Therefore, $\overline{CD} = \overline{AD} = s$, and the formula becomes

$$d^2 = s^2 + s^2 \quad \text{or} \quad s^2 = \frac{d^2}{2} \quad .$$

Taking square roots of both sides, we obtain $s = \sqrt{\frac{d^2}{2}} = \frac{d}{\sqrt{2}} = \frac{d}{\sqrt{2}} \cdot \frac{\sqrt{2}}{\sqrt{2}} = \frac{d\sqrt{2}}{2}$. Since we now have s, we can use the area formula.

$$\text{area} = s^2 = \left(\frac{d\sqrt{2}}{2}\right)^2 = \frac{d^2}{2} .$$

● **PROBLEM** 444

A man has a rectangular piece of property measuring 200 x 300 ft.

388

He plans to put a concrete sidewalk 3 ft. wide around the edge. What will the area of the sidewalk be?

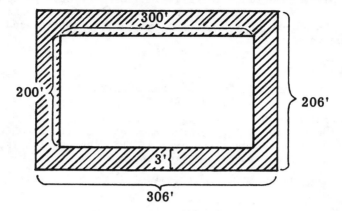

<u>Solution</u>: We assume that the man puts the sidewalk down as shown in the figure. We must find the area of the shaded region. First, find the area of the figure bounded by the outer border (the large rectangle). Next, find the area of the small rectangle and subtract this from the area of the large rectangle. We will then have the area of the side-walk.

Area of large rectangle = (206 ft.)(306 ft.)
                        = 63036 sq. ft.

Area of small rectangle = (300 ft.)(200 ft.)
                        = 60000 sq. ft.

Area of sidewalk   = (63036 - 60000) sq. ft.
                        = 3036 sq. ft.

This is much more efficient than breaking up the sidewalk region into rectangles, calculating their areas, and summing them up.

● **PROBLEM** 445

A man wishes to put linoleum tile on a rectangular floor which measures 5 x 7 yd. If each piece of linoleum tile is a 9 in. square, how many pieces of tile will be needed? If a quart of tile adhesive covers 10 sq. ft., how many quarts will he need?

<u>Solution</u>:   The actual problem is to find out how many tiles (9" per side) are needed to cover the surface of a rectangle 5 x 7 yd.
    The area of the rectangle (the floor) is made up of the sum of the areas of the square tiles. Let us represent the number of tiles needed by x. Then the previous statement may be written, mathematically, as

$$x(\text{Area of 1 tile}) = \text{Area of floor}. \tag{1}$$

The area of 1 tile is (9 in.)^2 = 81 sq. in., since each tile is a square. The area of the floor is (5 yd.)(7 yd.) = 35 sq. yd., since the floor is rectangular. Using these results in (1), we obtain

$$x(81 \text{ sq. in.}) = 35 \text{ sq. yd.}$$

Dividing both sides by 81 sq. in.,

$$x = \frac{35 \text{ sq. yd.}}{81 \text{ sq. in.}} \qquad (2)$$

Now, we have mixed units in (2), In order to get a pure number in (2) (which we must, since the number of tiles (x) is dimensionless) we must change either sq. yds. to sq. ins., or vice versa. We take the former course.

$$1 \text{ sq. yd. } (1 \text{ yd.})^2 = (36 \text{ in.})^2 = 1296 \text{ sq. in.}$$

Using this in (2),

$$x = \frac{(35 \text{ sq. yd.})(1296 \text{ sq. in./sq. yd.})}{(81 \text{ sq. in.})} = 560 \quad.$$

The man will need 560 tiles to cover his floor.

As for the adhesive, we know that each quart of adhesive covers 10 sq. ft. We must cover the area of the floor, which is, from the above calculations, 35 sq. yd. Since

$$1 \text{ sq. yd. } = (1 \text{ yd.})^2 = (3 \text{ ft.})^2 = 9 \text{ sq. ft.}$$

the floor area is $(35)(9)$ sq. ft. = 315 sq. ft.

The man will therefore need

$$\frac{315}{10} = 31.5 \quad \text{quarts of adhesive.}$$

TRIANGLES

● **PROBLEM** 446

Calculate the area of a right triangle by using the fact that the area of a rectangle, of sides h and w, is hw.

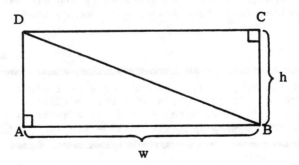

Solution: We use our knowledge of the area of a rectangle to find the area of a right triangle. Looking at the figure, the area of the rectangle is the sum of the areas of $\triangle ABD$ and $\triangle CDB$. Hence,

$$\text{Area } \triangle CDB + \text{Area } \triangle ABD = hw \qquad (1)$$

where we have used the fact that the area of the rectangle is hw.

Now, by definition of a rectangle, $\overline{CD} \cong \overline{AB}$, $\overline{CB} \cong \overline{AD}$. Furthermore, $\overline{DB} \cong \overline{DB}$. Hence, by the S.S.S. (side-side-side) Postulate, $\triangle CDB \cong \triangle ABD$. Hence, Area $\triangle ABD$ = Area $\triangle CDB$. Therefore, (1) becomes

$$2 \text{ Area } \triangle CDB = 2 \text{ Area } \triangle ABD = hw$$

or

$$\text{Area } \triangle CDB = \text{Area } \triangle ABD = \frac{hw}{2} \quad.$$

In right triangle ABC, ∡ C measures 90°, $\overline{AB}$ is of length 20 in., and the length of $\overline{AC}$ is 16 in. Find the area of triangle ABC.

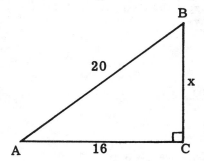

Solution: Assume, as in the accompanying figure, that $\overline{AC}$ is the base of triangle ABC. Since ∡ C is a right angle, $\overline{BC}$ is the altitude to the base of the triangle.

The area (A) of a right triangle is given by one-half the product of the length of the base (b) times the length of the corresponding altitude (h). $A = \frac{1}{2}bh$.

We are not given the length of the altitude, but can calculate it by applying the Pythagorean Theorem to △ABC. If we let x = the length of the altitude, then,

$$x^2 + (16 \text{ in.})^2 = (20 \text{ in.})^2$$
$$x^2 + 256 \text{ in.}^2 = 400 \text{ in.}^2$$
$$x^2 = 144 \text{ in.}^2$$

which implies $\quad\quad\quad\quad\quad$ x = 12 in.

The area of △ABC = $\frac{1}{2}bh$. By substitution,

$$\text{area of } \triangle ABC = \frac{1}{2}(16 \text{ in.})(12 \text{ in.}) = 96 \text{ in.}^2$$

Therefore, the area of △ABC is 96 sq. in.

Given that the area of a right triangle of base length b and perpendicular height h is bh/2, find an expression for the area of an arbitrary triangle.

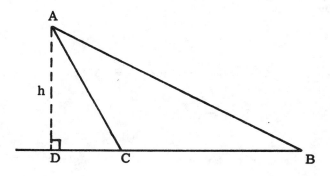

Solution: The idea here is to use our knowledge of a right triangle and extrapolate to any arbitrary triangle.

We must calculate the area of triangle ABC. Since the union of triangles ACB and ADC is triangle ADB, we may write

$$\text{Area } \triangle \text{ ADB} = \text{Area } \triangle \text{ACB} + \text{Area } \triangle \text{ADC} . \tag{1}$$

Now, both triangles ADC and ADB are right triangles. Therefore,

$$\text{Area } \triangle \text{ADB} = (\text{ DB })h/2 \tag{2}$$

$$\text{Area } \triangle \text{ADC} = (\text{ DC })h/2 . \tag{3}$$

Hence, using (2) and (3) in (1),

$$\frac{h \text{ DB}}{2} = \frac{h \text{ DC}}{2} + \text{Area } \triangle \text{ACB} .$$

Subtracting $\dfrac{h \text{ DC}}{2}$ from both sides,

$$\text{Area } \triangle \text{ACB} = (h \text{ DB } - h \text{ DC })/2$$

$$= h(\text{ DB } - \text{ DC })/2 .$$

But   DB - DC = CB , as the figure shows,   Hence,

$$\text{Area } \triangle \text{ACB} = \frac{h \text{ CB}}{2} .$$

The area of an arbitrary triangle is ½ the product of its base $\left(\text{ CB } \right)$ and its altitude (h).

● **PROBLEM** 449

The area of a certain triangle is 52 square feet and the height is 13 feet. What is the measure of the base of the triangle?

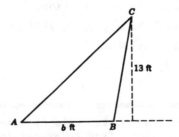

Solution: Recall the formula for the area of a triangle, A = ½bh, where b = length of a base and h = corresponding height of the triangle. Since we wish to find the length of the base, b, let us solve the equation for b:

$$A = \tfrac{1}{2}bh$$

$$\frac{2A}{h} = b .$$

We are given that A = 52 sq. ft. and h = 13 ft. Substituting into the above equation we have:

$$\frac{2(52 \text{ ft.}^2)}{13 \text{ ft.}} = b$$

$$8 \text{ ft.} = b .$$

Therefore, the measure of the base of the triangle is 8 feet.

Find the area of a triangle with a base of 5 inches and a height of 8 inches.

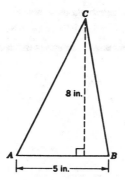

<u>Solution</u>: Recall that the area of a triangle is given by the formula,

Area = ½bh,

where b = length of a base and h = the corresponding height of the triangle.

We are given b = 5 inches and h = 8 inches; therefore,

Area = (½)(5 in.)(8 in.) = 20 square inches.

Find the area of the equilateral triangle in terms of the length of its side s.

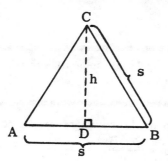

<u>Solution</u>: The formula of the area of any triangle is ½bh where
b = base length and h = corresponding height. We are given the length
of base $\overline{AB}$ = s . To find the height or altitude $\overline{CD}$, we use the
symmetry of the equilateral triangle. Because △ABC is equilateral,
it is also isosceles. Therefore, the altitude to the base, $\overline{CD}$, is the
perpendicular bisector of the base $\overline{AB}$, and, thus (i) AD = DB = ½AB =
½s; and (ii) ∢ CDB is a right angle.

In right triangle △CDB, we have a hypotenuse of length s, and
a leg of length ½s. We wish to find the length of the other leg $\overline{CD}$.
By the Pythagorean Theorem, $CB^2 = CD^2 + DB^2$ or

$$CD = \sqrt{CB^2 - DB^2} = \sqrt{s^2 - (\tfrac{s}{2})^2} = \sqrt{s^2 - \frac{s^2}{4}} = \sqrt{\frac{3s^2}{4}} = \frac{\sqrt{3}}{2}\,s \ .$$

Substituting in the area formula for a triangle, we obtain,

Area of an equilateral triangle = $\frac{1}{2}bh = \frac{1}{2}(s)(\frac{\sqrt{3}}{2}s) = \frac{\sqrt{3}}{4}s^2$.

● **PROBLEM 452**

Find the area of an equilateral triangle whose perimeter is 24.

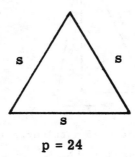

p = 24

<u>Solution</u>: The perimeter of an equilateral triangle, whose sides are represented by s, is given by 3s. In this case, given the perimeter equal to 24, we have 3s = 24, or s = 8.

The area formula for an equilateral triangle is given by the expression

$$A = \frac{s^2\sqrt{3}}{4} .$$

By substitution,

$$A = \frac{(8)^2\sqrt{3}}{4} = \frac{64\sqrt{3}}{4} = 16\sqrt{3} .$$

Therefore, the area of the triangle is $16\sqrt{3}$ sq. units.

● **PROBLEM 453**

Find the length of a side of an equilateral triangle whose area is $4\sqrt{3}$.

<u>Solution</u>: The area of an equilateral triangle, of side length s, is given by the formula

$$Area = \frac{s^2\sqrt{3}}{4} .$$

Substituting the given area into this formula, we obtain,

$$4\sqrt{3} = \frac{s^2\sqrt{3}}{4}$$

$$4 = \frac{s^2}{4}$$

$$16 = s^2$$

$$4 = s .$$

Therefore, the length of the side is 4.

● **PROBLEM 454**

Find the area of triangle ABC, to the nearest square inch, if

AB = 8 in., AC = 4 in., and the measure of ∡ A is 135°. (See figure.)

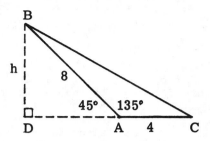

<u>Solution</u>: To find the area of △ABC, we need to know the length of its base (b) and altitude (h), because A = ½bh. The base length, $\overline{AC}$, is given as 4 in. The altitude, however, is not given and must be calculated.

Since $\overline{AC}$ is being used as the base, it must be extended outward so that the altitude to it can be drawn. $\overline{BD}$ is this altitude, and its length can be found in the following manner.

Noticing that ∡ BAC and ∡ BAD are supplementary and, because we are told m ∡ BAC = 135°, we can conclude that ∡ BAD measures 45°. This implies △BDA is an isosceles right triangle. Its hypotenuse, $\overline{BA}$, measures 8 in.

Let h = the length of the congruent sides, i.e., $\overline{BD}$ and $\overline{DA}$. The length of $\overline{BD}$ is the altitude we need for the area calculation. By the Pythagorean Theorem, as applied to △BDA,

$$h^2 + h^2 = 8^2 \text{ in.}^2$$
$$2h^2 = 64 \text{ in.}^2$$
$$h^2 = 32 \text{ in.}^2$$
$$h = \sqrt{32} \text{ in.} = \sqrt{16}\sqrt{2} \text{ in.} = 4\sqrt{2} \text{ in.}$$

The area of △ABC = ½bh. By substitution,

$$A = \tfrac{1}{2}(4 \text{ in.})(4\sqrt{2} \text{ in.})$$
$$= 8\sqrt{2} \text{ in.}^2$$

Letting $\sqrt{2}$ = 1.4, A = 8(1.4) in.² = 11.2 in.² . Therefore, the area of △ABC is 11 sq. in., to the nearest square inch.

● **PROBLEM** 455

Show that the median of any triangle separates the triangle into two regions of equal area.

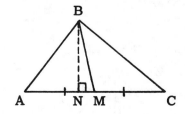

Solution: The area of a triangle with base length b_1 and corresponding height h_1 is $\frac{1}{2}b_1h_1$. For any triangles with the same base length, b_1, and measured height, h_1, the expressions for their areas are the same, hence, their areas must be equal. Therefore, to show that two triangles have the same area, it is sufficient to show that their base lengths and their heights are equal.

Consider as the base the side of the triangle to which the median is drawn. Then, the base of the triangle is divided into two equal segments by the median. The bases of the two triangular regions created by the median are equal; i.e., $\overline{AM} = \overline{MC}$. The median does not affect the height $\overline{BN}$; both regions have the same height to the bases $\overline{AM}$ and $\overline{MC}$, respectively. Since the two triangular regions have the same height and base length, they have the same area.

Given: $\overline{BM}$ is a median of $\triangle ABC$.

Prove: Area of $\triangle ABM$ = Area of $\triangle CBM$.

Statements	Reasons
1. $\overline{BM}$ is the median of $\triangle ABC$.	1. Given.
2. M is the midpoint of $\overline{AC}$.	2. Definition of the median of a triangle.
3. AM = MC.	3. Definition of the midpoint of a line segment.
4. Locate N in $\overleftrightarrow{AC}$ so that $\overline{BN} \perp \overleftrightarrow{AC}$.	4. Through a point external to a line, there is one and only one line perpendicular to the given line.
5. $\overline{BN}$ is an altitude of both $\triangle ABM$ and $\triangle CBM$.	5. Definition of an altitude of a triangle.
6. Area of $\triangle ABM$ = Area of $\triangle CBM$.	6. Two triangles have equal areas if their bases have the same length and the altitudes to their bases have the same length.

● **PROBLEM** 456

In the figure, $\overline{BC} \parallel \overleftrightarrow{AA'}$. Compute the area of the region bounded by triangle (a) ABC; (b) A'BC.

Solution: (a) + (b) We observe that both triangles have the same base, $\overline{BC}$ (BC = 6). Furthermore, since $\overleftrightarrow{AA'} \parallel \overline{BC}$, all $\perp$ distances from $\overleftrightarrow{AA'}$ to $\overline{BC}$ are equal in length. Hence, both triangles have the same height (h = 5). Using the formula for the area of a triangle,

$$\text{Area } \triangle ABC = \frac{(\text{base})(\text{height})}{2} = \frac{(6)(5)}{2} = 15 .$$

From the discussion above,

$$\text{Area } \triangle A'BC = 15.$$

For the figure below, show that $h_1 b_1 = h_2 b_2 = h_3 b_3$.

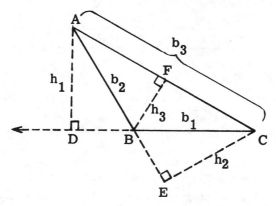

Solution: The area of any triangle equals one-half the product of the lengths of its base and the altitude to that base.
b_1, b_2, and b_3 designate sides $\overline{BC}$, $\overline{AB}$, and $\overline{AC}$, respectively
in $\triangle ABC$. h_1, h_2, and h_3 are altitudes of $\triangle ABC$ drawn from
vertices A, C, and B, respectively.

If we consider each side as a base, we obtain:

$$\text{Area of } \triangle ABC = \tfrac{1}{2} h_1 b_1 ,$$

$$\text{Area of } \triangle ABC = \tfrac{1}{2} h_2 , b_2 ,$$

$$\text{Area of } \triangle ABC = \tfrac{1}{2} h_3 , b_3 ,$$

or

$$\tfrac{1}{2} h_1 b_1 = \tfrac{1}{2} h_2 b_2 = \tfrac{1}{2} h_3 b_3$$

Multiplying by two, we obtain the desired result:

$$h_1 b_1 = h_2 b_2 = h_3 b_3 .$$

● **PROBLEM** 458

Given the lengths of three sides of a triangle, the most efficient
method for finding the area is to use Heron's Formula: given sides of
length a,b, and c; let the semiperimeter,
$$s = \frac{a+b+c}{2} .$$
Then, the area is given by
$$\sqrt{s(s-a)(s-b)(s-c)} .$$

Consider the accompanying figure. $\triangle ABC$ has sides of length
a,b, and c. $\overline{CD}$ is the altitude to side $\overline{AB}$, dividing $\overline{AB}$ into
segments of length x and c-x . Find x .

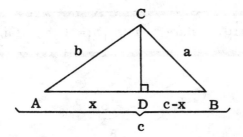

Solution: The length x is determined by the point D. The restriction on D is that $\overline{CD} \perp \overline{AB}$, and thus $\triangle ACD$ and $\triangle CDB$ must be right triangles. Since all right triangles must obey the Pythagorean Theorem, $b^2 = x^2 + CD^2$ and $a^2 = CD^2 + (c-x)^2$. We thus have two simultaneous equations with two unknowns, CD and x. Since we are solving for x, we eliminate $\overline{CD}$. From the first equation $b^2 = x^2 + CD^2$, we obtain $CD^2 = b^2 - x^2$. Substituting this result in the second equation, we obtain:

$$a^2 = (b^2 - x^2) + (c - x)^2 = b^2 - x^2 + c^2 - 2cx + x^2$$
$$a^2 = b^2 + c^2 - 2cx .$$

We isolate x by adding $2cx$ to both sides, subtracting a^2 from both sides, and dividing by $2c$.

$$x = \frac{b^2 + c^2 - a^2}{2c} .$$

● **PROBLEM** 459

Given the value of x in the previous problem, find the altitude CD. Do not perform any unnecessary multiplications. Factor whenever possible. The final answer should contain four terms. Note that

(1) $x^2 - y^2 = (x+y)(x-y)$; and (2) $(x+y)^2 = x^2 + 2xy + y^2$.

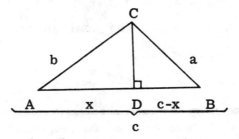

Solution: $\overline{CD}$ is a leg of right triangle $\triangle ACD$, Therefore,

$$AC^2 = AD^2 + CD^2 \quad \text{or} \quad CD^2 = AC^2 - AD^2 .$$

$AC = b$ and $AD = x$. Therefore, $CD^2 = b^2 - x^2$. This is of the form $x^2 - y^2$. Therefore, $CD^2 = (b+x)(b-x)$. Since

$$x = \frac{b^2 + c^2 - a^2}{2c}, \quad CD^2 = \left(b + \frac{b^2 + c^2 - a^2}{2c} \right)\left(b - \frac{b^2 + c^2 - a^2}{2c} \right) .$$

To proceed with the simplification, we must bring the b's into the denominators of the fractions. Note that

398

$$b = b \cdot 1 = b \cdot \frac{2c}{2c} = \frac{2bc}{2c} \quad .$$

Substituting in, we obtain

$$\left(\frac{2bc + b^2 + c^2 - a^2}{2c}\right)\left(\frac{2bc - b^2 - c^2 + a^2}{2c}\right) \quad .$$

Note that $2bc + b^2 + c^2 = b^2 + 2bc + c^2 = (b+c)^2$. Similarly $+2bc - b^2 - c^2 = -(b^2 - 2bc + c^2) = -(b-c)^2$. Subsequently,

$$CD^2 = \left(\frac{(b+c)^2 - a^2}{2c}\right)\left(\frac{a^2 - (b-c)^2}{2c}\right) \quad .$$

Once again, the numerators are of the form $x^2 - y^2$. Thus, $(b+c)^2 - a^2 = ((b+c)+a)((b+c)-a)$, and

$$CD^2 = \left[\frac{(b+c+a)(b+c-a)}{2c}\right]\left[\frac{(a-b+c)(a+b-c)}{2c}\right]$$

$$CD = \sqrt{\frac{(b+c+a)(b+c-a)(a+c-b)(a+b-c)}{(2c)(2c)}}$$

$$= \frac{\sqrt{(b+c+a)(b+c-a)(a+c-b)(b+a-c)}}{\sqrt{(2c)(2c)}}$$

$$= \frac{1}{2c}\sqrt{(b+c+a)(b+c-a)(a+c-b)(b+a-c)} \quad .$$

● **PROBLEM** 460

The semiperimeter s of a triangle of sides $a, b,$ and c is defined as half the perimeter, or,

$$\frac{a+b+c}{2} \quad .$$

Find an expression for the area of $\triangle ABC$ and show that it is equivalent to $\sqrt{s(s-a)(s-b)(s-c)}$.

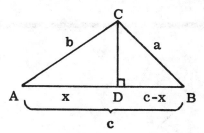

Solution: The area of a triangle equals $\frac{1}{2}Q \cdot H$ where Q is the length of the base and H is the altitude. If, in $\triangle ABC$, we choose $\overline{AB}$ as the base, then $Q = AB = c$. Then the altitude H must be $\overline{CD}$, or

$$H = \overline{CD} = \frac{1}{2c}\sqrt{(b+a+c)(b+c-a)(a+c-b)(a+b-c)} \quad .$$

The area, therefore, equals $\frac{1}{2}Q \cdot H$ or

$$\frac{1}{2}(c)\left[\frac{1}{2c}\sqrt{(a+b+c)(b+c-a)(a+c-b)(a+b-c)}\right]$$

or

$$\frac{1}{4} \sqrt{(a+b+c)(b+c-a)(a+c-b)(a+b-c)} \quad .$$

Consider the expression $\sqrt{s(s-a)(s-b)(s-c)}$:

if $s = \dfrac{a+b+c}{2}$; $s-a = \dfrac{a+b+c}{2} - a = \dfrac{b+c-a}{2}$;

$s-b = \dfrac{a+b+c}{2} - b = \dfrac{a+c-b}{2}$; and $s-c = \dfrac{a+b+c}{2} - c = \dfrac{a+b-c}{2}$.

Then, $\sqrt{s(s-a)(s-b)(s-c)} = \sqrt{\left(\dfrac{a+b+c}{2}\right)\left(\dfrac{b+c-a}{2}\right)\left(\dfrac{a+c-b}{2}\right)\left(\dfrac{a+b-c}{2}\right)}$

Factoring out the $\frac{1}{2}$'s, we have

$$\sqrt{\frac{1}{16}(a+b+c)(a+b-c)(a+c-b)(b+c-a)}$$

or

$$\frac{1}{4}\sqrt{(a+b+c)(b+c-a)(a+c-b)(a+b-c)} \quad .$$

Thus, the two <u>expressions we h</u>ave found are equivalent. Therefore, Heron's Formula, $\sqrt{s(s-a)(s-b)(s-c)}$ is a valid formula for the area of a triangle, if

$$s = \frac{a+b+c}{2}$$

where a, b, and c are the lengths of the sides of the triangle.

SIMILAR TRIANGLES

● **PROBLEM** 461

Show that the ratio of the areas of two similar triangles equals the square of their ratio of similitude.

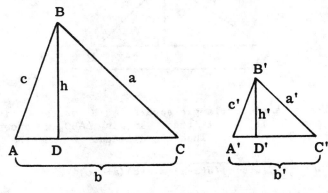

<u>Solution</u>: Suppose each side of Δ1 is a times larger than the corresponding sides of Δ2. Then, Area of Δ2 = $\frac{1}{2}b_2h_2$ and the area of Δ1 = $\frac{1}{2}b_1h_1$. Since $b_1 = ab_2$ and $h_1 = ah_2$, area of Δ1 = $\frac{1}{2}(ab_2)(ah_2) = a^2(\frac{1}{2}b_2h_2)$ or a^2 times larger

400

than the area of $\triangle 2$.

Given: $\triangle ABC \sim \triangle A'B'C'$; $\overline{BD}$ and $\overline{B'D'}$ are the altitudes of their respective triangles. The lengths are indicated in the figure.

Prove: $\dfrac{\text{Area of } \triangle ABC}{\text{Area of } \triangle A'B'C'} = \left(\dfrac{b}{b'}\right)^2$ where $\dfrac{b}{b'}$ is the ratio of similitude.

Statements	Reasons
1. $\triangle ABC \sim \triangle A'B'C'$; $\overline{BD}$ and $\overline{B'D'}$ are the altitudes of their respective triangles. $\angle BDC$ and $\angle B'D'C'$ are right angles.	1. Given.
2. $\dfrac{a}{a'} = \dfrac{b}{b'}$	2. The sides of similar triangles are proportional.
3. Area of $\triangle ABC = \frac{1}{2}bh$ Area of $\triangle A'B'C' = \frac{1}{2}b'h'$	3. The area of a triangle of base length b and altitude h equals $\frac{1}{2}bh$.
4. $\dfrac{\text{Area of } \triangle ABC}{\text{Area of } \triangle A'B'C'} = \dfrac{\frac{1}{2}bh}{\frac{1}{2}b'h'} = \dfrac{bh}{b'h'}$	4. Multiplicative Property.
5. $\angle BDC \cong \angle B'D'C'$.	5. All right angles are congruent.
6. $\angle C \cong \angle C'$.	6. Corresponding angles of similar triangles are congruent.
7. $\triangle BDC \sim \triangle B'D'C'$.	7. If two pairs of corresponding angles of two triangles are congruent, then the triangles are similar.
8. $\dfrac{a}{a'} = \dfrac{h}{h'}$.	8. The sides of similar triangles are proportional.
9. $\dfrac{h}{h'} = \dfrac{b}{b'}$.	9. Transitive Property (Step 2).
10. $\dfrac{\text{Area of } \triangle ABC}{\text{Area of } \triangle A'B'C'} = \dfrac{b^2}{(b')^2} = \left(\dfrac{b}{b'}\right)^2$	10. Substitution Postulate.

● **PROBLEM** 462

If two triangles have congruent bases, then the ratio of their areas equals the ratio of the lengths of their altitudes.

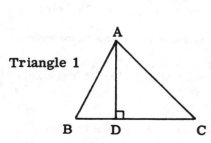

Triangle 1

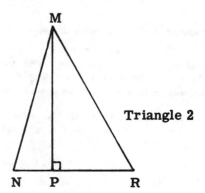

Triangle 2

<u>Solution</u>: The area of triangle 1 is $\frac{1}{2}b_1 h_1$. The area of triangle 2, a triangle whose base is congruent to the base of triangle 1, is

401

$\frac{1}{2}b_1h_2$. The ratio is therefore,

$$\frac{\text{area of } \Delta 1}{\text{area of } \Delta 2} = \frac{\frac{1}{2}b_1h_1}{\frac{1}{2}b_1h_2} = \frac{h_1}{h_2} .$$

Given: $\overline{AD}$ and $\overline{MP}$ are altitudes of $\triangle ABC$ and $\triangle MNR$ respectively;

$\overline{BC} \cong \overline{NR}$.

Prove: $\dfrac{\text{Area of } \triangle ABC}{\text{Area of } \triangle NMR} = \dfrac{AD}{MP}$.

Statements	Reasons
1. $\overline{AD}$ and $\overline{MP}$ are altitudes of $\triangle ABC$ and $\triangle MNR$ respectively; base $\overline{BC} \cong$ base $\overline{NR}$.	1. Given
2. Area of $\triangle ABC = \frac{1}{2}BC \cdot AD$ Area of $\triangle NMR = \frac{1}{2}NR \cdot MP$	2. Area of triangle = $\frac{1}{2}$base·altitude.
3. $\dfrac{\text{Area of } \triangle ABC}{\text{Area of } \triangle NMR} = \dfrac{AD}{MP}$	3. Division of both sides of an equation by equals does not change the equality.

● **PROBLEM** 463

Triangle ABC is similar to triangle A'B'C'. If BC = 4 and B'C' = 12, find the ratio of the areas of the triangles.

Solution: A theorem tells us that the ratio of the areas of any two similar triangles is equal to the ratio of the square of the lengths of any two corresponding parts.

$\overline{BC}$ and $\overline{B'C'}$ are corresponding parts of $\triangle ABC$ and $\triangle A'B'C'$. Therefore, the ratio of the areas is

$$\frac{\text{Area of } \triangle ABC}{\text{Area of } \triangle A'B'C'} = \frac{(BC)^2}{(B'C')^2} .$$

By substitution, $\left(\dfrac{BC}{B'C'}\right)^2 = \left(\dfrac{4}{12}\right)^2 = \left(\dfrac{1}{3}\right)^2 = \dfrac{1}{9}$

Therefore, the ratio of the areas of the triangles if 1:9 .

● **PROBLEM** 464

If the ratio of a pair of corresponding altitudes of two similar triangles is 6/7, and the area of the larger triangle is 98, find the area of the smaller triangle.

Solution: The ratio of the similitude (the measure of a side of a triangle divided by the measure of the corresponding side of the similar triangle) is 6/7. Therefore, the ratio of the areas must be

$$\left(\frac{6}{7}\right)^2 \text{ of } \frac{36}{49} .$$

If K is the area of the smaller triangle, then the proportion can be written as

$$\frac{K}{98} = \frac{36}{49} .$$

Since the product of the means must equal the product of the extremes for the proportion to hold, 49K = 36(98), and

$$K = \frac{36(98)}{49} = 72 .$$

The area of the smaller triangle is 72.

● **PROBLEM** 465

The areas of two similar triangles are in the ratio of 4:1. The length of a side of the smaller triangle is 5. Find the length of the corresponding side in the larger triangle.

<u>Solution</u>: We know, by theorem, that the ratio of the areas of two similar triangles is the square of the ratio of the lengths of any corresponding sides. It follows then, that if 4:1 is the ratio of areas, 2:1 must be the ratio of corresponding sides.

Therefore, if the length of a side in the smaller triangle is 5, its corresponding side in the larger triangle must be twice as long, i.e., its length must be 10.

● **PROBLEM** 466

In trapezoid ABCD, as in the accompanying diagram, the larger base, $\overline{AB}$, measures 24 in., the smaller base, $\overline{DC}$, measures 8 in., and the altitude, $\overline{FG}$, measures 6 in.. The nonparallel sides $\overline{AD}$ and $\overline{BC}$ are extended to meet at E.
a) In triangle DEC: (a) Find EF, the measure of the altitude from E to $\overline{DC}$;
b) Find the area of △DEC.

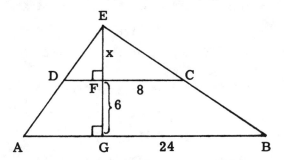

<u>Solution</u>: In the trapezoid ABCD, $\overline{DC} \parallel \overline{AB}$. Then, △DEC ~ △AEB, because a line parallel to one side of a triangle, which intersects the other two sides at different points, creates a triangle similar to the given triangle. Therefore, by corresponding parts, the proportion

$$\frac{\text{length of altitude } \overline{EF}}{\text{length of altitude } \overline{EG}} = \frac{\text{length of base } \overline{DC}}{\text{length of base } \overline{AB}}$$

is valid. Let x = EF, then x + 6" = EG. By substitution,

$$\frac{x}{x+6"} = \frac{8"}{24"} = \frac{8}{24} .$$

Since, in a proportion, the product of the means equals the product of the extremes,

$$24x = 8x + 48''$$
$$16x = 48''$$
$$x = 3'' \ .$$

Therefore, the altitude of $\triangle DEC$ to $\overline{DC}$, $\overline{EF}$, measures 3 in.

b) The area, A, of $\triangle DEC = \frac{1}{2}bh = \frac{1}{2}DC \times EF$. By substitution,

$$A = \frac{1}{2}(8 \text{ in.} \times 3 \text{ in.})^2 = 12 \text{ in.}^2 \ .$$

Therefore, the area of $\triangle DEC = 12$ sq. in.

● **PROBLEM** 467

In the figure, $\overline{DE} \parallel \overline{BC}$.
a) Write the formula for the area of trapezoidal region DECB.
b) Let S be the triangular region bounded by $\triangle ABC$ and let T be the triangular region bounded by $\triangle ADE$. Write the formulas for the area of S, A(S), and the area of T, A(T). What can be said about A(S) - A(T)?
c) Use (a) and (b) to prove that $h_2/h_1 = b_2/b_1$.

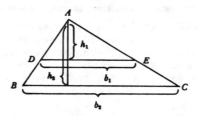

Solution: a) The area of a trapezoid is $\frac{1}{2}$ times the product of its perpendicular height and the sum of the lengths of its bases. Using the figure,

$$\text{Area DECB} = \frac{1}{2}(h_2 - h_1)(b_1 + b_2)$$

Note that we have used the fact that the perpendicular height of the trapezoid is $h_2 - h_1$.

b) By the definition of the area of a triangle, and using the notation of the figure,

$$A(S) = \frac{1}{2}h_2 b_2$$
$$A(T) = \frac{1}{2}h_1 b_1$$

Since $A(S) = A(T) + \text{Area DECB}$.

$$A(S) - A(T) = \text{Area DECB} \ .$$

c) Using the results of (a) and (b),

$$A(S) - A(T) = \text{Area DECB}$$
$$\frac{1}{2}h_2 b_2 - \frac{1}{2}h_1 b_1 = \frac{1}{2}(h_2 - h_1)(b_1 + b_2)$$

Expanding the right side of the preceding equation,

$$\frac{1}{2}h_2 b_2 - \frac{1}{2}h_1 b_1 = \frac{1}{2}h_2 b_1 + \frac{1}{2}h_2 b_2 - \frac{1}{2}h_1 b_1 - \frac{1}{2}h_1 b_2$$

Cancelling like terms,

$$\tfrac{1}{2}h_2b_1 - \tfrac{1}{2}h_1b_2 = 0 \ .$$

Adding $\tfrac{1}{2}h_1b_2$ to both sides, and multiplying by 2,

$$h_2b_1 \ = \ h_1b_2 \ .$$

Dividing both sides by b_1h_1 ,

$$\frac{h_2}{h_1} = \frac{b_2}{b_1} \ .$$

● **PROBLEM** 468

Let the two congruent sides of an isosceles triangle have lengths a, and let the included angle have measure θ. Choosing the third side as a base, let the corresponding altitude be of length r. Prove that the area enclosed by the triangle is given by the formula

$$A = r^2 \tan \theta/2 \ .$$

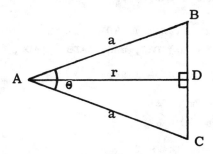

<u>Solution</u>:   The figure shows isosceles ΔABC, with altitude $\overline{AD}$ of length r. By definition of an altitude, $\overline{AD} \perp \overline{BC}$, as indicated. We will use the general formula for the area of a triangle, and "tailor" it to the special properties of an isosceles triangle to prove that $A = r^2 \tan \theta/2$.

The area, A, of any triangle with base length b and corresponding altitude length h is

$$A = \tfrac{1}{2}bh.$$

In the ΔABC, the length of the base is BC, and the length of the altitude $\overline{AD}$ is r. Hence,

$$A = \tfrac{1}{2}(BC)r \ .$$

It is now necessary to find BC in terms of the known quantities r and θ. In order to do this, we must digress for a moment.

We note that

$$BC = BD + DC \tag{1}$$

Therefore, if we can find BD and DC, in terms of r and θ, our problem will be solved. We will do this by proving ΔADB ≅ ΔADC, and concluding that BD = DC. Finding either BD or DC in terms of r and θ, and using (1), we will have finished the problem.

First, note that ΔABC is an isosceles triangle. By definition, this means that AB = AC . Furthermore, AD = AD. Now, the sum of the measures of the angles of any triangle is $180°$. Applying this fact to ΔABD and ΔACD,

$$m \measuredangle BAD + m \measuredangle ADB + m \measuredangle DBA = 180°$$
$$m \measuredangle CAD + m \measuredangle ADC + m \measuredangle DCA = 180° \ . \tag{2}$$

But m ∡ ADB = m ∡ ADC = 90°, and m ∡ DBA = m ∡ DCA. (The last équality
follows from the fact that ΔABC is an isosceles triangle.) Using
these facts, (2) may be written

$$m ∡ BAD + 90° + m ∡ DBA = 180°$$ (3)

$$m ∡ CAD + 90° + m ∡ DBA = 180° .$$

Comparing the 2 equations in (3), we conclude that m ∡ BAD = m ∡ CAD,
or ∡ BAD ≅ ∡ CAD. By the S.A.S. (side-angle-side) Postulate,
ΔADB ≅ ΔADC. Hence, $\overline{BD} ≅ \overline{DC}$ and BD = DC, upon which (1) becomes

$$BC = 2BD .$$ (4)

Using (4) in the formula $A = \frac{1}{2}(BC)r$ yields

$$A = (BD)r.$$ (5)

We may write BD in terms of θ and r by introducing the concept
of the tangent of an angle (abbreviated tan). The tangent of an
angle of a right triangle is defined as the ratio of the length of
the side of the triangle opposite the angle and the length of the
side of the triangle adjacent to the angle, so long as neither of the
sides are the hypotenuse. For instance, in the figure,

$$\tan(m ∡ BAD) = \frac{BD}{DA} = \frac{BD}{r}$$

or

$$BD = r \tan(m ∡ BAD) .$$ (6)

Since m ∡ BAC = m ∡ BAD + m ∡ DAC, and ΔADB ≅ ΔADC, (which means that
m ∡ BAD = m ∡ DAC), we have

$$m ∡ BAC = 2m ∡ BAD.$$

But, m ∡ BAC = θ,

$$θ = 2m ∡ BAD$$

or

$$m ∡ BAD = θ/2 .$$ (7)

Using (7) in (6),

$$BD = r \tan θ/2 .$$

Inserting this in (5) yields

$$A = r^2 \tan θ/2 .$$

This is a formula for the area of ΔABC in terms of r and θ, as
defined by the figure.

● PROBLEM 469

Let the congruent sides of an isosceles triangle have lengths r,
and let the included angle have measure θ. Prove that the area enclosed by the triangle is given by $A = \frac{1}{2}r^2 \sin θ$.

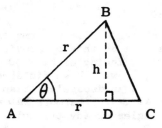

Solution: An isosceles triangle is a triangle which has 2 sides of
equal length. The figure shows an isosceles triangle ABC, with
$\overline{BD} ⊥ \overline{AC}$. Hence, $\overline{BD}$ defines an altitude (of length h) of ΔABC.

In order to find the area of triangle ABC, we will use the general formula for the area of a triangle, and "tailor" it to the specific properties of an isosceles triangle.

In general, the area A of a triangle of base length b, and its corresponding altitude length h is

$$A = \tfrac{1}{2}bh.$$

For △ABC,

$$A = \tfrac{1}{2}(AC)(BD) .$$

But AC = r, and BD = h ; hence,

$$A = \tfrac{1}{2}rh .$$ (1)

However, we do not know the value of h. We are only given θ and r. Our task, then, is to find an expression relating h to r and θ. To do this, we must introduce the concept of the sine of an angle (abbreviated sin). By definition, the sine of an angle of a right triangle is the ratio of the length of the side opposite the angle and the length of the hypotenuse.

For instance, (see figure)

$$\sin \theta = \frac{BD}{AB} = \frac{h}{r} .$$

Multiplying both sides of this equation by r, we obtain

$$h = r \sin \theta ,$$

which relates h to r and θ, as required. Using the last equation in (1) yields

$$A = \tfrac{1}{2}(r)(r \sin \theta) = \tfrac{1}{2}(r^2 \sin \theta),$$

which gives the area of an isosceles triangle in terms of the length of its congruent sides and the angle enclosed by these sides.

PARALLELOGRAMS

● **PROBLEM** 470

Find the area of a parallelogram whose base has a length of 8 ft. and whose altitude has a length of 18 in.

Solution: The area of a parallelogram is equal to the product of the length of any base and the length of its corresponding altitude.

To apply the above, both the base length and the altitude length must be in the same units. Therefore, convert 18 in. to $1\tfrac{1}{2}$ ft.

Now, A = bh. Since b = 8 ft., h = $1\tfrac{1}{2}$ ft., by substitution, A = (8)($1\tfrac{1}{2}$) ft.^{2} = 12 ft.^{2} .

Therefore, the area of the parallelogram is 12 sq. ft.

● **PROBLEM** 471

A parallelogram whose base is represented by x + 4 and whose altitude is represented by x - 1 is equivalent to a square whose side is 6. Find the base and altitude of the parallelogram.

Solution: Since they are equivalent, the area of the square is equal to the area of the parallelogram. This holds for all polygons.

The area of the square equals the square of the length of any side s. Therefore, $A_s = s^2$. By substitution, s = 6, and $A_s = 6^2 = 36$.

The area of a parallelogram, A_p, is equal to the product of lengths of any base and its corresponding altitude. In this example, $A_p = (x+4)(x-1) = x^2 + 3x - 4$.

The above rule for equivalent polygons allows us to set $A_p = A_s$ and solve for x. Hence,

$$x^2 + 3x - 4 = 36$$
$$x^2 + 3x - 40 = 0$$

By factoring,
$$(x+8)(x-5) = 0 .$$

Then, either x + 8 = 0 or x - 5 = 0, which implies x = -8 or x = 5. The negative answer is rejected because it has no geometric significance. Therefore, Base = x + 4 = 5 + 4 = 9 .

$$\text{Altitude} = x-1 = 5-1 = 4.$$

Therefore, base of parallelogram = 9, altitude = 4.

● **PROBLEM** 472

The lengths of two consecutive sides of a parallelogram are 10 inches and 15 inches, and these sides include an angle of 63°. (As shown in the figure). a) Find, to the nearest tenth of an inch, the length of the altitude drawn to the longer side of the parallelogram. b) Find, to the nearest square inch, the area of the parallelogram.

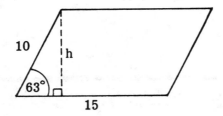

<u>Solution</u>: Let h = the length of the altitude.
a) When the altitude is drawn, a right triangle is formed (see figure). We know an acute angle, the length of the hypotenuse, and we wish to determine the length of the leg opposite the given angle. Therefore, we can apply the sine ratio:
$$\sin 63° = \frac{h}{10}$$
and, according to a standard sine table, sin 63° = 0.8910 . By substitution, $0.8910 = \frac{h}{10}$ in.

$$h = 8.910 \text{ in.}$$

Therefore, the altitude is 8.9 in., to the nearest tenth of an inch.

b) The area of a parallelogram, A, equals the product of the lengths of the base (b) and altitude (h). A = bh. By substituting,

$$b = 15 \text{ in. and } h = 8.9 \text{ in.}$$

$$A = (15)(8.9) \text{ in.}^2 = 133.5 \text{ in.}^2$$

Therefore, the area equals 134 sq. in., to the nearest square inch.

Find the area of a rhombus, one of whose angles measures 60°, and whose shorter diagonal is 6.

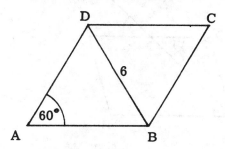

Solution: ABCD, in the accompanying figure, is a rhombus and, as such, $\overline{BA} \cong \overline{AD}$. Since the angles opposite the two congruent sides of a triangle are themselves congruent, $\angle ADB \cong \angle ABD$. Therefore, m $\angle$ ADB = m $\angle$ ABD.

We are told m $\angle$ DAB = 60°. Therefore, since the angle-sum of a triangle is 180°, m $\angle$ ADB + m $\angle$ ABD = 180° - 60° = 120°. Since $\angle$ ABD and $\angle$ ADB are of equal measure, we conclude that m $\angle$ ABD = m $\angle$ ADB = 60° and, thus, that $\triangle$DAB is equilateral.

Area, A, of $\triangle$DAB = $\dfrac{s^2\sqrt{3}}{4}$, where s is the length of one side of the equilateral triangle. We are given s = 6. By substitution,

$$A = \frac{6^2\sqrt{3}}{4} = \frac{36\sqrt{3}}{4} = 9\sqrt{3} \ .$$

Knowing that opposite angles of a rhombus are congruent, and by using reasoning similar to that employed before, it can be shown that $\triangle$DCB is also equilateral with sides of the same length as the sides of $\triangle$DAB. Therefore we conclude that $\overline{BD}$ bisects rhombus ABCD into 2 equivalent equilateral triangles.

Therefore, the Area of ABCD = 2 area of $\triangle$DAB. By substitution, Areas of ABCD = 2(9$\sqrt{3}$) = 18$\sqrt{3}$. The area of rhombus ABCD is 18$\sqrt{3}$.

In parallelogram ABCD, M is the midpoint of side $\overline{DC}$, as shown in the figure. Line segment $\overline{AM}$, when extended, intersects $\overleftrightarrow{BC}$ at K.

a) Prove that triangle ADM is congruent to triangle KCM.
b) Prove that triangle AKB is equal in area to parallelogram ABCD.

Solution: a) In this proof, we will prove congruence of the two triangles by the A.S.A. $\cong$ A.S.A. method.

Statements	Reasons
1. ABCD is a $\square$	1. Given.
2. $\overleftrightarrow{KA}$ and $\overleftrightarrow{KB}$ are straight lines	2. Given.
3. $\angle 1 \cong \angle 2$	3. If two angles are vertical angles

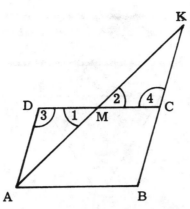

4. $\overline{AD} \parallel \overline{BC}$.	then they are congruent.
	4. The opposite sides of a parallel-ogram are parallel.
5. $\angle 3 \cong \angle 4$	5. If two parallel lines are cut by a transversal, then the alternate in-terior angles are congruent.
6. $\overline{DM} \cong \overline{CM}$	6. A midpoint divides a line into two congruent segments.
7. $\triangle ADM \cong \triangle KCM$.	7. A.S.A. $\cong$ A.S.A.

b) To prove that the area of $\triangle AKB$ is equal to the area of $\square$ ABCD, we must show that the sum of the areas of quadrilateral ABCM and $\triangle KCM$ is equal to the sum of the areas of quadrilateral ABCM and $\triangle ADM$. The first sum is precisely the area of $\triangle AKB$; the second sum is the area of $\square$ ABCD.

Statements	Reasons
1. $\triangle KCM \cong \triangle ADM$.	1. Proved in part (a).
2. Area of $\triangle KCM$ = Area of $\triangle ADM$.	2. If two triangles are congruent, then they are equal in area.
3. Area of ABCM = Area of ABCM.	3. Reflexive Property of Equality.
4. Area of $\triangle KCM$ + Area of ABCM = Area of $\triangle ADM$ + Area of ABCM.	4. Equal quantities added to equal quantities sum to equal quanti-ties (see step (2)).
5. Area of $\triangle KCM$ + Area of ABCM = Area of $\triangle AKB$ and Area of $\triangle ADM$ + Area of ABCM = Area of $\square$ ABCD.	5. If a polygon which encloses a region is separated into several polygons which do not overlap, then its area is the sum of the areas of these several polygons. (Area-Addition Postulate).
6. Area of $\triangle AKB$ = Area of $\square$ ABCD.	6. Substitution Postulate.

● **PROBLEM** 475

Find the altitude AF of a rhombus if the lengths of the diagonals are 10 and 24.

Solution: We are given the diagonals of the rhombus and must relate them to the altitude. Few equations involve the altitude of a rhombus. One of those is that the area of the rhombus equals the base times the altitude. If we can find the area of the rhombus and the length of the base (the length of a side), then we can solve for the altitude.

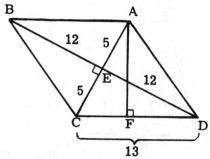

Given the lengths of the diagonals, d_1, d_2, the area of the rhombus can be found by the formula:

$$\text{Area} = \tfrac{1}{2}(d_1 \cdot d_2) .$$

This can be derived by considering the figures that comprise a rhombus. Assume the rhombus shown has diagonals of length d_1 and d_2, where $AC = d_1$ and $BD = d_2$.

Area of rhombus ABCD = area of $\triangle ABC$ + area of $\triangle ADC$. The two triangles share a common base of length d_1. Since the diagonals of a rhombus bisect each other, they are perpendicular to each other. Segments $\overline{BE}$ and $\overline{DE}$ are the respective altitudes of $\triangle ABC$ and $\triangle ADC$ and have lengths equal to $\tfrac{1}{2}$ the length of $\overline{BD}$, or $\tfrac{1}{2}d_2$.

Area of a triangle = $\tfrac{1}{2}bh$. By substitution,

$$\text{Area of rhombus ABCD} = \tfrac{1}{2}d_1(\tfrac{1}{2}d_2) + \tfrac{1}{2}d_1(\tfrac{1}{2}d_2)$$

$$= \tfrac{1}{2}d_1[\tfrac{1}{2}d_2 + \tfrac{1}{2}d_2] = \tfrac{1}{2}d_1 d_2 .$$

In this case, $d_1 = 10$ and $d_2 = 24$. Therefore,

area of rhombus ABCD = $\tfrac{1}{2}(10)(24) = 120$.

To find the length of a side (the base), we note:
(1) the diagonals of a rhombus bisect each other. Therefore, $AE = EC = \tfrac{1}{2}AC = 5$ and $BE = ED = \tfrac{1}{2}BD = 12$;

(2) the diagonals of a rhombus are perpendicular to each other. Thus, $\angle CED$ is a right angle.

In the right triangle $\triangle CED$, the lengths of the two legs are known. We wish to find the length of side $\overline{CD}$, the hypotenuse. By the Pythagorean Theorem

$$CD^2 = CE^2 + ED^2 = 5^2 + 12^2 = 25 + 144 = 169.$$

Therefore, $CD = \sqrt{169} = 13$. Substituting this result into an alternative area equation, we obtain:

$$\text{Area of rhombus} = b \cdot h$$

$$120 = 13 \cdot AF$$

$$\text{`}AF = \frac{120}{13} .$$

Therefore, length of altitude $\overline{AF} = 9.2$.

● PROBLEM 476

Find the area of a rhombus, each of whose sides is 10 in., and one of whose diagonals is 16 in.

411

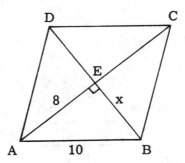

Solution: Since the area of a rhombus is equal to one-half the product of the lengths of the diagonals, we must determine the length of the second diagonal before the area can be calculated. (It has been represented in the accompanying figure).

Since the diagonals of a rhombus bisect each other at right angles, we can conclude that $\overline{AE} \perp \overline{BD}$, and that $AE = EC = \frac{1}{2}AC$. $\overline{AC}$ is assumed to be the given 16 in. diagonal. Therefore,

$$AE = \frac{1}{2}(16 \text{ in.}) = 8 \text{ in.}$$

Because $\triangle AEB$ is a right triangle we can apply the Pythagorean Theorem to determine the length of half-diagonal $\overline{EB}$. Let $x = EB$. Then,

$$x^2 + (8 \text{ in.})^2 = (10 \text{ in.})^2$$
$$x^2 + 64 \text{ in.}^2 = 100 \text{ in.}^2$$
$$x^2 = 36 \text{ in.}^2$$
$$x = 6 \text{ in.}$$

Therefore, the whole diagonal $\overline{BD}$ measures twice EB or $BD = 2EB = 2(6 \text{ in.}) = 12 \text{ in.}$

Area of a rhombus $= \frac{1}{2}d_1 d_2$, where d_1 and d_2 are the lengths of the rhombus diagonals. By substitution,

$$\text{Area} = \frac{1}{2}(12)(16)\text{in.}^2$$
$$= 6(16)\text{in.}^2 = 96 \text{ in.}^2$$

Therefore, the area of rhombus $ABCD = 96$ sq. in.

● **PROBLEM** 477

The area of a rhombus is 90, and one diagonal is 10. Find the length of the other diagonal.

Solution: We can solve the problem by direct substitution into the area formula for a rhombus: $A = \frac{1}{2} d_1 d_2$. Here, A is the area of the rhombus, and d_1 and d_2 are the lengths of the diagonals.

Let d_2 = the unknown diagonal. Then, by substitution,

$$90 = \frac{1}{2}(10) d_2$$
$$90 = 5 d_2$$
$$18 = d_2 .$$

Therefore, the length of the second diagonal is 18.

Derive a formula for the area of the trapezoidal region shown in the figure.

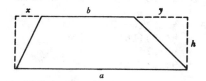

Solution: The procedure to be used is one in which we employ the formulas for the area of a triangle and of a rectangle, and to extrapolate from these a formula for the area of a trapezoid.

The rectangle shown in the figure is composed of 3 regions: the trapezoid, and 2 triangles. The area of the rectangle is equal to the sum of the areas of each of these sub-regions. That is

$$\text{Area of rectangle} = \text{Area of } \triangle I + \text{Area of } \triangle II \tag{1}$$
$$+ \text{Area of trapezoid.}$$

Using the formulas for the area of a triangle and of a rectangle, we obtain

$$\text{Area of rectangle} = (x+y+b)(h)$$
$$\text{Area of } \triangle I = (x)(h)/2 \tag{2}$$
$$\text{Area of } \triangle II = (y)(h)/2$$

Here, we have used the fact that the long side of the rectangle has length x+y+b (see figure). We have also called the triangle on the left of the figure $\triangle I$, and the one on the right of the figure $\triangle II$. Using (2) in (1),

$$(x+y+b)h = \frac{xh}{2} + \frac{yh}{2} + \text{Area of trapezoid.}$$

Subtracting $\frac{xh}{2} + \frac{yh}{2}$ from both sides of the previous equation,

$$\text{Area of trapezoid} = bh + \frac{xh}{2} + \frac{yh}{2}$$

$$\text{Area of trapezoid} = \frac{(x+y+2b)h}{2}$$

$$\text{Area of trapezoid} = \frac{[(x+y+b) + b]h}{2} = \frac{(a+b)h}{2} .$$

The area of a trapezoid equals ½ the product of its height and the sum of the lengths of its bases.

A trapezoid has two bases, of 6 and 10 inches, and a height of 1 foot. What is its area?

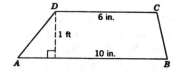

<u>Solution</u>: Recall that the area of a trapezoid is given by the formula,

$$\text{Area} = \tfrac{1}{2}h(b+b'),$$

where h is the height and b and b' are the lengths of the bases of the trapezoid. We are given that b = 6 in., b' = 10 in., and h = 1 ft. Converting 1 ft. to 12 in., and substituting in the equation we obtain:

$$\begin{aligned}
A &= \tfrac{1}{2}(12)(6+10) \\
&= \tfrac{1}{2}(12)(16) \\
&= 6 \cdot 16 \\
&= 96 \text{ sq. in.}
\end{aligned}$$

Therefore, the area of the trapezoid is 96 square inches.

● **PROBLEM** 480

The bases of a trapezoid are 12 in. and 20 in. If the area of the trapezoid is 128 sq. in., find the length of its altitude.

<u>Solution</u>: The area of a trapezoid is given by the formula $A = \tfrac{1}{2}h(b+b')$, where h is the height of the trapezoid, and b' and b are the lengths of its bases. We are given $A = 128$ in.^{2}, b = 20 in., and b' = 12 in. By substitution, we can solve for the length of the unknown altitude.

$$\begin{aligned}
A &= \tfrac{1}{2}h(b+b') \\
128 \text{ in.}^2 &= \tfrac{1}{2}h(20+12)\text{in.} \\
128 \text{ in.}^2 &= \tfrac{1}{2}h(32)\text{in.} \\
128 \text{ in.}^2 &= h(16 \text{ in.}) \\
8 \text{ in.} &= h
\end{aligned}$$

Therefore, the altitude is 8 inches.

● **PROBLEM** 481

Find the area of trapezoid ABCD , as shown in the diagram, if the length of base $\overline{AB}$ = 9 in., the length of base $\overline{DC}$ = 5 in., and the length of altitude $\overline{DE}$ = 6 in.

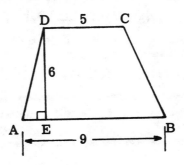

<u>Solution</u>: A theorem tells us that the area of a trapezoid is equal to one-half the product of the length of its altitude, h, and the sum of the lengths of its bases, b_1 and b_2 . $A = \tfrac{1}{2}h(b+b'))$.

We are given b = 9 in., b' = 5 in., and h = 6 in.

414

Therefore, by substitution, A = ½(6)(9+5) in.² = (3)(14) in.² = 42 in.²
Therefore, the area of trapezoid ABCD = 42 sq. in.

The wall at one end of an attic takes on the shape of a trapezoid
because of a slanted ceiling. The wall is 8 feet high at one end, 10
feet wide, and only 3 feet high at the other end. What is the wall's
area?

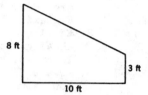

8 ft

3 ft

10 ft

Solution: The wall forms the shape of a trapezoid with bases 8 and
3, and a height of 10. Recall the formula for the area of a trapezoid, A = ½h(b+b'), where h is the height and b and b' are the
lengths of the bases. This formula can be applied to find the required area. We are given that b = 8, b' = 3, and h = 10. Substituting
in the equation we obtain:

$$A = \tfrac{1}{2}(10)(8+3)$$
$$= 55$$

Therefore, the area of the wall is 55 square feet.

In the accompanying figure, $\overline{AB}$ and $\overline{DC}$ are the bases of trapezoid
ABCD. Diagonals $\overline{AC}$ and $\overline{BD}$ intersect in E. Prove that triangle
ADE is equal in area to triangle BCE.

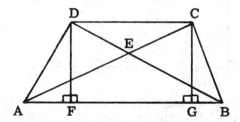

Solution: The purpose of this proof is to show that the area of △ADB
is equal to the area of △ACB. We will then prove that the difference
between the areas of △ADB and △AEB (which is the area of △ADE),
is equal to the difference between the areas of △BCA and △AEB
(which is the area of △BCE.)
        We will construct the altitudes of △ADB and △BCA and prove
equality of the two areas by showing that two triangles having congruent bases and altitudes are equivalent.
This is derived from the formula for the area of a triangle, A = ½bh,
where b and h are the base length and height, respectively.

Statements	Reasons
1. Draw $\overline{DF} \perp \overline{AB}$ and $\overline{CG} \perp \overline{AB}$.	1. An altitude may be drawn to a side of a triangle.
2. $\overline{AB}$ and $\overline{DC}$ are bases of trapezoid ABCD.	2. Given.
3. $\overline{DC} \parallel \overline{AB}$.	3. The bases of a trapezoid are parallel.
4. DF = CG or $\overline{DF} \cong \overline{CG}$.	4. Parallel lines are everywhere equidistant.
5. $\overline{AB} \cong \overline{AB}$.	5. Reflexive Property of Congruence.
6. Area of $\triangle ADB$ = area of $\triangle BCA$.	6. Two triangles which have congruent bases and congruent altitudes are equal in area.
7. Area of $\triangle AEB$ = area of $\triangle AEB$.	7. Reflexive Property of Equality.
8. Area of $\triangle ADB$ - area of $\triangle AEB$ = area of $\triangle BCA$ - area of $\triangle AEB$	8. Equal quantities subtracted from equal quantities yield equal differences.
9. Area of $\triangle ADB$ - area of $\triangle AEB$ = area of $\triangle ADE$ and, Area of $\triangle BCA$ - area of $\triangle AEB$ = area of $\triangle BCE$.	9. Area-addition postulate: If a polygon is made up of several non-overlapping polygons, then the area of any one of the composite polygons is equal to the area of the larger polygon minus the sum of the areas of the remaining composite polygons. (See figure.)
10. Area of $\triangle ADE$ = area of $\triangle CEB$.	10. Substitution Postulate.

● **PROBLEM** 484

In isosceles trapezoid ABCD , $\overline{CF}$ and $\overline{DE}$ are altitudes drawn to base $\overline{AB}$, and the measure of ∡ B is 60° (as shown in the figure). If CD exceeds AD by 5 and the perimeter of ABCD is 110, find, in radical form, the area of the trapezoid.

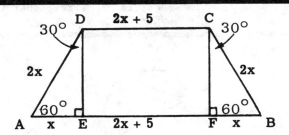

Solution: To find the area, we must first determine the length of both bases and the length of the altitude of trapezoid ABCD.

Triangle AED is a right triangle; and since m ∡ DAE = 60°, then the m ∡ ADE = 30°.

In a 30-60-90 triangle, the length of the side opposite the 30° angle equals ½ the length of the hypotenuse.

Therefore, AD = 2AE . Let AE = x. Then, AD = 2x. We are told that $\overline{CD}$ exceeds $\overline{AD}$ by 5. Therefore, CD = 2x + 5.

Since DCFE is a rectangle, EF = CD = 2x + 5.

Because ABCD is isosceles, $\triangle DAE \cong \triangle CBF$, by A.S.A. $\cong$ A.S.A.

(AD = BC = 2x; m $\angle$ ADE = m $\angle$ BCF = 30°; and m $\angle$ DAE = m $\angle$ CBF = 60°.)

By corresponding parts, AE = FB = x.

Referring to the diagram, we see that $\overline{AE} + \overline{EF} + \overline{FB} + \overline{BC} + \overline{CD} + \overline{DA}$ = perimeter of ABCD. By substitution,

$$x + (2x+5) + x + 2x + (2x+5) + 2x = 110$$
$$10x + 10 = 110$$
$$10x = 100$$
$$x = 10 .$$

Therefore, the length of base $\overline{CD}$ = 2(10) + 5, or 25. The length of base $\overline{AB}$ = 10 + (2(10) + 5) + 10 = 45.

To calculate the altitude DE of trapezoid ABCD we apply the Pythagorean Theorem to right triangle AED. Let h = length of altitude $\overline{DE}$.

We have found that AD = 20 and AE = 10. By substitution,

$$(20)^2 = h^2 + (10)^2$$
$$400 = h^2 + 100$$
$$h^2 = 300$$
$$h = \sqrt{300} = \sqrt{100}\sqrt{3} = 10\sqrt{3} .$$

Therefore, since the area of a trapezoid = $\frac{1}{2}h(b+b')$, by plugging in the values we have determined, we obtain

$$A = \frac{1}{2}(10\sqrt{3})(25+45)$$
$$= \frac{1}{2}(10\sqrt{3})(70)$$
$$= 35(10\sqrt{3})$$
$$= 350\sqrt{3} .$$

Therefore, the area of the trapezoid is $350\sqrt{3}$.

● **PROBLEM** 485

Find the area of an isosceles trapezoid if the measure of one angle is 135 and the lengths of the bases are 10 and 18.

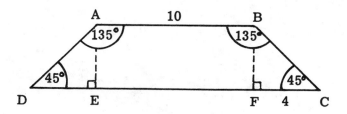

Figure 1

Solution: The area of a trapezoid equals one-half the product of the height and the sum of the bases.

We have b_1 = 10 and b_2 = 18. To find the altitude AE or BF, we show:
(1)   quadrilateral ABEF is a rectangle, therefore, m $\angle$ EAB = m $\angle$ FBA = 90°.
(2)   m $\angle$ DAE = m $\angle$ DAB - m $\angle$ EAB = 135° - 90° = 45°.
(3)   m $\angle$ ADE = 90° - m $\angle$ DAE = 90° - 45° = 45°.

417

(4)   △DAE is isosceles; therefore, $\overline{AE} = DE$.
(5)   Similarly, m ∢ FCB = $45°$, and BF = FC .
(6)   △DAE $\cong$ △CBF . Therefore, DE = FC .
(7)   By properties of the rectangle, EF = AB = 10 .
(8)   DC = DE + EF + FC .
(9)   18 = 2·DE + 10 or DE = 4 .
(10)   Since AE = DE, then AE = 4, or h = 4.

Then, Area = $\frac{1}{2}h(b_1+b_2)$ = $\frac{1}{2}(4)(10+18)$ = 56.

To show (1) quadrilateral ABFE is a rectangle: $\overline{AE}$ and $\overline{BF}$ are altitudes. Therefore, $\overline{AE} \perp \overline{EF}$ and $\overline{BF} \perp \overline{EF}$. Since $\overline{AE}$ and $\overline{BF}$ are perpendicular to the same line, $\overline{AE} \parallel \overline{BF}$. Because ABFE is a quadrilateral, AB $\parallel$ EF . Since both pairs of sides are parallel, ABFE is a parallelogram. Since one of the angles is a right angle, ABFE is a rectangle.

To show (6) △DAE $\cong$ △CBF .

Since $\overline{AE}$ and $\overline{BF}$ are altitudes, ∢ DEA and ∢ CFB are right angles. $\overline{AE}$ and $\overline{BF}$ are opposite sides of a parallelogram; therefore, $\overline{AE} \cong \overline{BF}$; finally, $\overline{DA} \cong \overline{BC}$, because the nonparallel sides of an isosceles trapezoid are congruent. Since the hypotenuse and one leg of right triangle △DAE are congruent to the hypotenuse and one leg of right triangle △CBF, △DAE $\cong$ △CBF .

POLYGONS

CHAPTER 24

INTERIOR AND EXTERIOR ANGLES

Basic Attacks and Strategies for Solving Problems in this Chapter. See pages 420 to 431 for step-by-step solutions to problems.

For $\triangle ABC$ pictured below, $\angle ABC$, $\angle ACB$, and $\angle CAB$ are all interior angles, while $\angle CAD$ is an exterior angle. For quadrilateral $EFGH$, $\angle EFG$, $\angle FGH$, $\angle GHE$, and $\angle HEF$ are all interior angles, while $\angle HGI$ and $\angle JEF$ are both exterior angles.

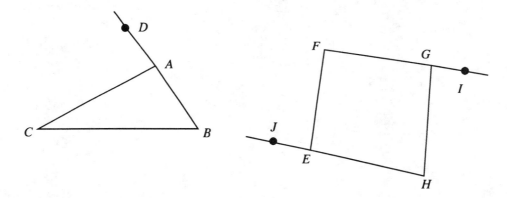

The problems in this chapter involve finding and applying relationships between the measures of the interior angles and the number of sides of a polygon, the measures of the exterior angles and the number of sides of a polygon, and the measures of the interior angles and the exterior angles of a polygon.

One important theorem is used repeatedly in this chapter

The sum of the measures of the angles of a triangle is 180°.

Convex and non-convex polygons are mentioned in Problem 492. As pictured below, quadrilateral *IJKL* is said to be convex because, for any segment drawn between any two points in the polygon, all the points of the segment are within the polygon; quadrilateral *MNOP* is non-convex because $\overline{NP}$ goes outside of the polygon.

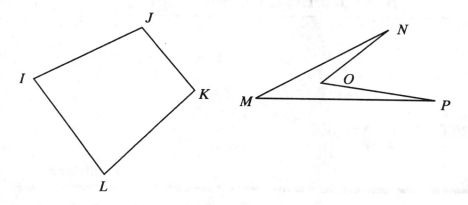

The angle measure agreements in non-convex polygons are a little tricky. For example, under normal circumstances, the measure of ∡ *NOP* is said to be about 60°, but when considered an angle in the quadrilateral, it has a measure of about 300°.

Regular polygons are mentioned in several problems. In a regular polygon, all the sides are congruent and all the angles are congruent. Several regular polygons are pictured below:

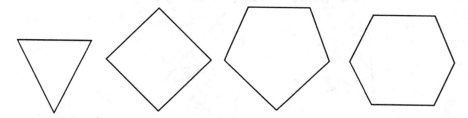

A regular triangle is an equilateral triangle while a regular quadrilateral is a square.

Step-by-Step Solutions to
Problems in this Chapter,
"Interior and Exterior Angles"

EXTERIOR ANGLES

● **PROBLEM** 486

What relation holds for an angle of a polygon and its exterior angle?

Solution: In the figure, ABCDE is an arbitrary polygon. ∢BCD is an (interior) angle of ABCDE, and ∢LCD is its exterior angle. Note that both of these angles share a common ray ($\overrightarrow{CD}$), and also, $\overrightarrow{CB}$ is directly opposite $\overrightarrow{CL}$. Hence, ∢BCD and ∢LCD form a linear pair of angles, and they are therefore supplementary angles (i.e. m∢BCD + m∢LCD = 180°).

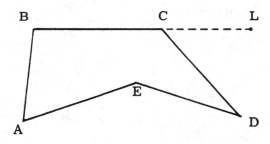

● **PROBLEM** 487

If an exterior angle of a polygon is obtuse, what can one say about the corresponding interior angle?

Solution: The exterior angle of a polygon and the corresponding interior angle are supplementary. This means that the sum of the measures of these angles is 180°. If one of the 2 angles is obtuse (i.e. has measure greater than 90°), then the other angle must have a measure less than 90°(i.e., it is acute).

Find the number of degrees in the measure of each exterior angle of
a regular polygon which has 12 sides.

<u>Solution</u>: By a theorem, we know that each exterior angle of a regular
polygon of n sides contains $\frac{360}{n}$ degrees. With this in mind, the
problem is one of substitution.

Since $n = 12$, $\frac{360°}{n} = \frac{360°}{12} = 30°$.

Therefore, the measure of each exterior angle of a regular 12-sided
polygon is $30°$.

Show that the measure of the exterior angle of a regular n-sided
polygon is given by the formula $\frac{360}{n}$.

<u>Solution</u>: An exterior angle of a polygon is defined to be an angle
that forms a linear pair (i.e., supplementary) with an interior angle
of a polygon. Thus,

   (i)   m ∡ exterior + m ∡ interior = $180°$.

   To find m ∡ interior, recall that the sum of the interior angles
of an n-sided polygon is $(n-2) \cdot 180°$. Since all interior angles of a
regular polygon are congruent, each of the n interior angles has
measure $\frac{(n-2) \cdot 180°}{n}$. Substituting m ∡ interior = $\frac{(n-2)}{n} \cdot 180°$ in
equation (i), we obtain

   (ii)   m ∡ exterior + $\frac{n-2}{n} \cdot 180°$ = $180°$.

   Solving for m ∡ exterior, we have:

   (iii)   m ∡ exterior = $180° - \frac{n-2}{n} \cdot 180°$

$$= 180°(1 - \frac{n-2}{n}) = 180°(\frac{2}{n})$$
$$= \frac{360°}{n} .$$

INTERIOR ANGLES

In the accompanying figure, ABCDE is a regular pentagon -
that is, all angles are congruent and all sides congruent
-M is the midpoint of side $\overline{CD}$. Show that AM is the perpendicular bisector of $\overline{CD}$.

<u>Solution</u>: To show that a line is the perpendicular bisector
of a segment, we prove that two distinct points on the line

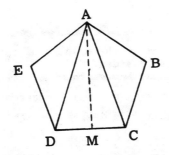

are equidistant from the endpoints of the segment. Here, we show that points A and M are each equidistant from the endpoints of $\overline{DC}$.

Note, M is the midpoint of $\overline{DC}$. Thus, MD = MC and M is equidistant from endpoints D and C.

To show A is equidistant from the endpoints - that is, AD = AC, we first prove $\triangle DEA \cong \triangle CBA$. Because all the sides of the regular pentagon are congruent, $\overline{EA} \cong \overline{AB}$ and $\overline{ED} \cong \overline{BC}$. Furthermore, because all angles of the regular pentagon are congruent, $\angle E \cong \angle B$. By the SAS Postulate, $\triangle DEA \cong \triangle CBA$. Because corresponding parts of congruent triangles are congruent, then $\overline{AD} \cong \overline{AC}$. Thus, AD = AC, and A is equidistant from the endpoints.

Because A and M are equidistant from the endpoints of $\overline{CD}$, the line determined by them, $\overleftrightarrow{AM}$, must be the perpendicular bisector of $\overline{CD}$.

● **PROBLEM** 491

Show that the sum of the measures of the interior angles of a convex polygon of n sides equals (n-2)·180. (It is given that the sum of the measures of the angles of a triangle is 180˚.)

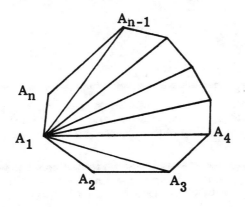

<u>Solution</u>: We look ahead at the result "$(n-2) \cdot 180$" and the given "sum of the angles of a triangle equals $180°$." In other words, the sum of the interior angles of an n-gon is the sum of the angle measures of n-2 triangles. Therefore, we can prove the result by dividing the n-gon into n-2 triangles such that every interior angle of the polygon is included once and only once in the angles of the n-2 triangles.

Polygon $A_1 A_2 \ldots A_n$ is an arbitrary polygon. Choose an arbitrary vertex, A_1, for example. Draw line segments from A_1 to the other n-1 vertices. Thus, n-2 triangles are formed. We express each interior angle of the polygon in terms of the angles of the triangles. $\measuredangle A_1$, the angle at the arbitrary vertex, is divided into n-2 angles, one in each triangle.

$$m \measuredangle A_1 = m \measuredangle A_2 A_1 A_3 + m \measuredangle A_3 A_1 A_4 + m \measuredangle A_4 A_1 A_5 + \ldots m \measuredangle A_{n-1} A_1 A_n$$

$\measuredangle A_2$ and $\measuredangle A_n$, the angles adjacent to A_1, are not split by the triangle.

$$m \measuredangle A_2 = m \measuredangle A_1 A_2 A_3$$
$$m \measuredangle A_n = m \measuredangle A_1 A_n A_{n-1}$$

Every other angle $A_3, A_4, \ldots, A_{n-1}$ has a segment drawn to it from A_1 and is, thus, split into two angles.

$$m \measuredangle A_3 = m \measuredangle A_1 A_3 A_2 + m \measuredangle A_1 A_3 A_4$$

$$\cdot$$
$$\cdot$$

$$m \measuredangle A_{n-1} = m \measuredangle A_1 A_{n-1} A_{n-2} + m \measuredangle A_1 A_{n-1} A_n \; .$$

Then for the sum of the interior angles of the polygon, we obtain

$$\begin{aligned}
\text{sum} &= m \measuredangle A_1 + m \measuredangle A_2 + \ldots m \measuredangle A_n \\
&= (m \measuredangle A_2 A_1 A_3 + m \measuredangle A_3 A_1 A_4 + \ldots m \measuredangle A_{n-1} A_1 A_n) + m \measuredangle A_1 A_2 A \\
&\quad + (m \measuredangle A_1 A_3 A_2 + m \measuredangle A_1 A_3 A_4) \ldots + \\
&\quad + m \measuredangle A_1 A_n A_{n-1} \; .
\end{aligned}$$

Rearranging and substituting, we obtain,

$$\begin{aligned}
\text{Sum} &= (m \measuredangle A_2 A_1 A_3 + m \measuredangle A_1 A_2 A_3 + m \measuredangle A_2 A_3 A_1) \\
&\quad + (m \measuredangle A_3 A_1 A_4 + m \measuredangle A_1 A_3 A_4 + m \measuredangle A_1 A_4 A_3) + \ldots \\
&\quad (m \measuredangle A_n A_1 A_{n-1} + m \measuredangle A_1 A_n A_{n-1} + m \measuredangle A_1 A_{n-1} A_n) \\
&= (\text{sum of the angles of } \triangle A_1 A_2 A_3) + \\
&\quad (\text{sum of the angles of } \triangle A_1 A_3 A_4) + \ldots \\
&\quad (\text{sum of the angles of } \triangle A_1 A_{n-1} A_n) \\
&= 180 + 180 + 180 \ldots \\
&= (n-2) \cdot 180 .
\end{aligned}$$

For convex polygons, the sum of the interior angles equals (n-2)·180, where n is the number of sides of the polygon. Show that the formula is also valid for the nonconvex polygon in Figure 1.

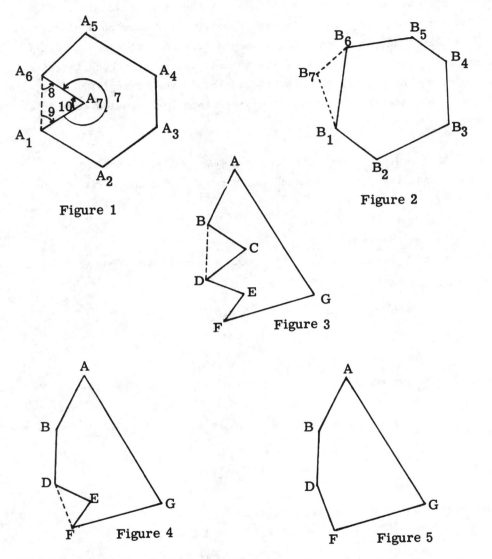

Figure 1

Figure 2

Figure 3

Figure 4

Figure 5

Solution: One way of thinking of the formula is the following: For each additional side, the sum of the interior angles increases by 180°. For a convex polygon, this can be seen more easily. Consider Figure 2. The difference between the angle sum of $B_1 B_2 \ldots B_7$ and the angle sum of $B_1 B_2 \ldots B_6$ is the angle sum of $\triangle B_1 B_6 B_7$. Since the angle sum of a triangle equals 180°, the sum of the angles of a seven sided polygon exceeds the sum of the angles of a six sided polygon by 180°.

To solve the problem, we show that the same behavior (i.e., each additional side adds $180°$ to the angle sum) also applies to nonconvex polygons. In other words, we prove:

(i) Angle sum of nonconvex $A_1A_2 \ldots A_7$ = Angle sum of convex
$$A_1A_2 \ldots A_6 + 180°$$

or, by subtracting,

(ii) Angle sum of $A_1A_2 \ldots A_7$ - Angle sum of $A_1A_2 \ldots A_6$ = 180.

Note that

(iii) Angle sum of $A_1A_2 \ldots A_7$ = $m \angle A_7A_1A_2 + m \angle A_2 + m \angle A_3$
$$+ m \angle A_4 + m \angle A_5 + m \angle A_5A_6A_7$$
$$+ m \angle 7 \ .$$

(iv) Angle sum of $A_1A_2 \ldots A_6$ = $m \angle A_6A_1A_2 + m \angle A_2 + m \angle A_3$
$$+ m \angle A_4 + m \angle A_5 + m \angle A_5A_6A_1 \ .$$

From (iii) and (iv), it follows that

(v) Angle sum of $A_1A_2 \ldots A_7$ - Angle sum of $A_1A_2 \ldots A_6$
$$= m \angle 7 + m \angle A_5A_6A_7 + m \angle A_7A_1A_2 - m \angle A_5A_6A_1 - m \angle A_6A_1A_2 .$$

To simplify this expression, note that $m \angle 8 =$
$m \angle A_5A_6A_1 - m \angle A_5A_6A_7$ and $m \angle 9 = m \angle A_6A_1A_2 - m \angle A_7A_1A_2$.

(vi) Angle sum $A_1A_2 \ldots A_7$ - Angle sum $A_1A_2 \ldots A_6$ = $m \angle 7 - m \angle 8$
$- m \angle 9$.

To further simplify, we relate these three angles to $\angle 10$. Since the angle sum of a triangle is 180, it follows that

(vii) $m \angle 8 + m \angle 9 + m \angle 10 = 180°$.

Also, because $\angle 10$ and $\angle 7$ form a circle, we have

(viii) $m \angle 10 + m \angle 7 = 360°$.

From (vii), we obtain $m \angle 10 = 180° - m \angle 8 - m \angle 9$. Substituting this result in (viii) and simplifying, we obtain

(ix) $m \angle 7 = m \angle 8 + m \angle 9 + 180°$.

Thus, equation (vi) becomes

(x) Angle sum $A_1A_2 \ldots A_7$ - Angle sum $A_1 \ldots A_6$ = $180°$.

or

(xi) Angle sum $A_1A_2 \ldots A_7$ = Angle sum $A_1 \ldots A_6 + 180°$.

Since $A_1A_2 \ldots A_6$ is convex, its angle sum is $(6-2) \cdot 180$.

(xii) Angle sum $A_1A_2 \ldots A_7$ = $(6-2) \cdot 180 + 180 = (7-2) \cdot 180$.

Thus, for the nonconvex figure $A_1A_2 \ldots A_7$, the formula $(n-2) \cdot 180$ still applies.

Note: One way of showing a polygon of n sides is convex is to consider the n possible $(n-1)$-gons, formed by omitting one of the vertices. If in every case the omitted vertex is outside the $(n-1)$-

gon, then the original n-gon is convex.

We can extend the method of proof in this problem to show that the formula (n-2)·180 holds for all n-gons, convex and nonconvex.

Consider the nonconvex polygon in Figure 3. ABCDEFG is not convex because point C is in the interior of ABDEFG. Using the reasoning of the proof, though, we know that angle sum of ABCDEFG = angle sum ABDEFG + 180. Similarly ABDEFG is not convex because E is in the interior of ABDFG, and angle sum of ABDEFG = Angle sum of ABDFG + 180. ABDFG is convex, so angle sum of ABDFG = (5-2)·180.

Thus, angle sum of ABDEFG = (5-2)·180 = (6-2)·180, and angle sum of ABCDEFG = (6-2)·180 + 180 = (7-2)·180 = (number of sides -2)·180.

This reasoning can be extended to any number of sides, and thus, in general, Sum of the Interior Angles of a Polygon equals (number of sides -2)·180.

● **PROBLEM** 493

What is the measure of the interior angle of a regular triangle? A regular quadrilateral? A regular 10-gon? A regular 2000-gon?

Solution: The sum of the measures of the interior angles of an n-gon is (n-2)·180. Since the n-angles of a regular n-gon are all equal in measure, each angle must have a measure of $[(n-2)\cdot 180]/n$ or $\frac{n-2}{n}\cdot 180$. For a regular triangle, the formula yields $\frac{3-2}{3}\cdot 180$ or $60°$, an equilateral triangle. For a quadrilateral, the formula yields $\frac{4-2}{4}\cdot 180 = 90°$. For 10 sides, the angle measure is $\frac{10-2}{10}\cdot 180 = \frac{8}{10}\cdot 180 = 144°$. For a 2000-gon, $\frac{2000-2}{2000}\cdot 180 = 179.82$. Note that as the number of sides increase, $\frac{n-2}{n}$ approaches 1 and the interior angles approach $180°$.

● **PROBLEM** 494

Find the number of degrees contained in each interior angle of a regular hexagon.

Solution: A regular hexagon is a 6 sided polygon in which all sides and all interior angles are of equal measure. A theorem tells us that the measure of each interior angle of a regular polygon of n sides is $\frac{180°(n-2)}{n}$. In this case, n equals 6.

Therefore, each angle's measure is calculated as follows:

$$\frac{180°(n-2)}{n} = \frac{180°(6-2)}{6} = \frac{180°(4)}{6} = 120°.$$

Therefore, each angle of a regular hexagon measures $120°$.

If each interior angle of a regular polygon contains 135°, find the number of sides that the polygon has.

Solution: We know that each interior angle of a regular polygon of n sides contains $\frac{180(n-2)}{n}$ degrees. Since we are given that each interior angle measures 135°, we can determine the number of sides of the polygon by setting the above expression equal to 135°, and then solving for the unknown, n. This is done as follows:

$$135° = \frac{180°(n-2)}{n}$$
$$135°n = 180°n - 360°$$
$$45°n = 360°$$
$$n = 8 .$$

Therefore, the polygon with interior angles measuring 135° has 8 sides, and is called an octagon.

Each interior angle of a regular polygon contains 120°. How many sides does the polygon have?

120°   60°

Solution: At each vertex of a polygon, the exterior angle is supplementary to the interior angle, as shown in the diagram.

Since we are told that the interior angles measure 120 degrees, we can deduce that the exterior angle measures 60°.

Each exterior angle of a regular polygon of n sides measures $\frac{360}{n}$ degrees. We know that each exterior angle measures 60°, and, therefore, by setting $\frac{360°}{n}$ equal to 60°, we can determine the number of sides in the polygon. The calculation is as follows:

$$\frac{360°}{n} = 60°$$
$$60°n = 360°$$
$$n = 6 .$$

Therefore, the regular polygon, with interior angles of 120°, has 6 sides and is called a hexagon.

Find the greatest number of sides that a regular polygon can have and yet still have an integral number of degrees in each interior angle.

__Solution__: For a regular polygon of n sides, the number of degrees in each interior angle is given by the formula,

(i)   m ∢ interior = $\frac{n-2}{n} \cdot 180°$.

Note that as the number of sides increase, $\frac{n-2}{n}$ increases and, thus, the interior angle measure increases. Therefore, the polygon of integral angle measure that has the greatest number of sides must also be the polygon with the greatest integral angle measure.

To find the polygon of greatest integral angle measure, note that n-2 is always less than n. Therefore, $\frac{n-2}{n}$ is always less than 1 and $\frac{n-2}{n} \cdot 180$ is always less than 180°. Therefore, the greatest integral number of degrees is 179°.

To solve for the number of sides in an 179° polygon, we substitute into equation (i):

(ii) $179 = \frac{n-2}{n} \cdot 180$

(iii) $\frac{179}{180} n = n-2$

(iv) $n - \frac{179}{180} n = 2$

(v) $n(\frac{1}{180}) = 2$

(vi) n = 360 sides.

$\overline{AX}$ and $\overline{BX}$ are two adjacent sides of a regular polygon. If the measure of angle ABX equals $\frac{1}{3}$ the measure of angle AXB, how many sides has the regular polygon?

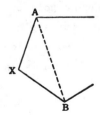

__Solution__: We first determine the measure of vertex ∢ AXB. Since we are told the polygon is regular, the number of sides will be uniquely determined. The measure of each vertex angle of a regular polygon is given by the expression $\frac{180(n-2)}{n}$, where n is the number of sides.

We will set m $\not\subset$ AXB = $\frac{180(n-2)}{n}$ and solve for n.

Draw $\overline{AB}$ to complete $\triangle$ AXB. Since $\overline{AX}$ and $\overline{BX}$ are sides of a
regular polygon, $\overline{AX} \cong \overline{BX}$. Therefore, $\triangle$ AXB is isosceles.

The measures of the base angles of an isosceles triangle are equal.
As such, m $\not\subset$ ABX = m $\not\subset$ XAB. We are given that m $\not\subset$ AXB = 3 m $\not\subset$ ABX.
The measure of the angles of a triangle sum to 180. Therefore,
m $\not\subset$ AXB + m $\not\subset$ ABX + m $\not\subset$ XAB = 180. By substitution,

$$3m \not\subset ABX + m \not\subset ABX + m \not\subset ABX = 180$$

$$5m \not\subset ABX = 180$$

$$m \not\subset ABX = \frac{180}{5} = 36°.$$

Therefore, m $\not\subset$ AXB = 3(36) = 108. Set $108 = \frac{180(n-2)}{n}$ and solve for
n.

$$108n = 180n - 360$$

$$-72n = -360$$

$$n = \frac{360}{72} = 5.$$

Therefore, the regular polygon has 5 sides, i.e. it is a pentagon.

● **PROBLEM** 499

The interior angles of a polygon are in arithmetic progression; the
least angle is 120° and the common difference is 5°. Find the number
of sides of the polygon.

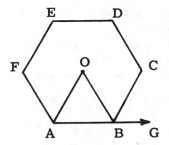

Solution: We know that the sum of the measures of the angles of a polygon is 180(n-2), where n is the number of sides.

We could take the sum of the measures of the angles described in the
problem until a number divisible by 180 without a remainder is reached.
The number of sides will be two more than $\frac{sum}{180}$.

However, a more elegant approach is available. We can express the
sum as a function of n by applying the formula

$$sum = \frac{n}{2}(2a + (n-1)d).$$

This formula is for the general sum of an arithmetic progression beginning at the number a and with the common difference between terms being given by d.

In this problem, a = 120 and d = 5. Hence, by substitution,

$$\text{Sum of the angles} = \frac{n}{2}(2(120) + (n-1)5)$$
$$= \frac{n}{2}(240 + 5n - 5)$$
$$= \frac{n}{2}(235 + 5n).$$

Therefore, to solve for n, set

$$\frac{n}{2}(235 + 5n) = 180(n-2).$$

Then,
$$2[\frac{235n}{2} + \frac{5n^2}{2} = 180n - 360]$$

or
$$\frac{1}{5}[235n + 5n^2 = 360n - 720]$$

or
$$47n + n^2 = 72n - 144$$

or
$$n^2 - 25n + 144 = (n-9)(n-16) = 0.$$

From this equation, we see n = 9 or 16. However, n = 16 has no meaning. In that case, and under the assumptions made about the angle measures, one angle of the polygon would measure 180, which could never be an angle of a polygon. The two adjacent sides would be indistinguishable.

Therefore, the polygon described in this problem has 9 sides.

● **PROBLEM** 500

For a regular polygon of six sides, find the number of degrees contained in (a) each central angle, (b) each interior angle (c) each exterior angle. (See figure.)

Solution: (a) The central angle of a regular polygon is an angle formed by two radii of the polygon drawn to consecutive vertices of the polygon. Its measure is equal to $\frac{360°}{n}$, where n = the number of sides in the polygon. In the figure shown, ∢AOB is a central angle. Hence, m∢AOB $= \frac{360°}{6} = 60°$.

Therefore, each central angle contains 60°.

(b) An interior angle of a regular polygon is formed by two consecutive sides of the polygon meeting at a vertex. ∢ABC is such an angle. Its measure is given by $\frac{(n - 2) \, 180°}{n}$, where n = the number of sides. In this example, n = 6 and, therefore, by substitution, each angle measures $\dfrac{(6 - 2) \ 180°}{6}$, or 120°.

Therefore, each interior angle contains 120°.

(c) An exterior angle of a regular polygon is formed outside the regular polygon by one side and the outward extension of an adjacent side. ∡CBG is an exterior angle. (See figure.) Its measure equals $\dfrac{360}{n}$ degrees. In this problem $\dfrac{360°}{n} = \dfrac{360°}{6} = 60°$.

Therefore, each exterior angle contains 60°.

CHAPTER 25

CYCLIC QUADRILATERALS

Basic Attacks and Strategies for Solving Problems in this Chapter. See pages 432 to 441 for step-by-step solutions to problems.

A quadrilateral which can be inscribed in a circle is called a **cyclic quadrilateral**. For example, quadrilateral *ABCD* is cyclic. Any set of points, all of which are on a circle, is said to be **concyclic**.

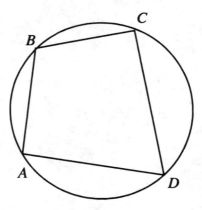

The prevailing question in several problems in this chapter is, "Is this set of points concyclic?" In the case of three noncollinear points, the answer is always "Yes." That fact is established in Problem 501. In the case of four points, the situation is more complex. The usual way to show that four points are concyclic is to show that the related quadrilateral is cyclic. From this point of view, there are two well-known ways to establish that four points are concyclic:

(1) Consider ∡B and ∡D in the quadrilateral pictured on the previous page. The intercepted arc for ∡B is $\overarc{CDA}$ and the intercepted arc for ∡D is $\overarc{ABC}$. Since the measure of an inscribed angle is half the measure of the intercepted arc, and since $\overarc{ABC}$ and $\overarc{CDA}$ make up the entire circle, ∡B and ∡D are supplementary. Similarly, ∡A and ∡C are supplementary. Conversely, if the opposite angles of a quadrilateral are supplementary, then the quadrilateral can be inscribed in a circle, and the points which are the vertices of the quadrilateral are concyclic.

(2) Suppose diagonals $\overline{BD}$ and $\overline{AC}$ are drawn in. Note that ∡ACD and ∡ABD both intercept the same arc and are congruent.

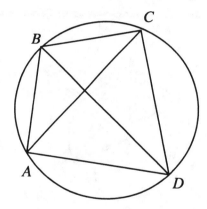

The converse also holds true, and this means that when a side of a quadrilateral, $\overline{AD}$, subtends angles at B and C which are congruent, the quadrilateral is cyclic and the four points are concyclic. This fact is established in Problem 506.

Also, Problem 505 establishes another way to prove that four points are concyclic. This proof is related to the theorem concerning intersecting chords.

● **PROBLEM** 501

Show that any three noncollinear points are concyclic.

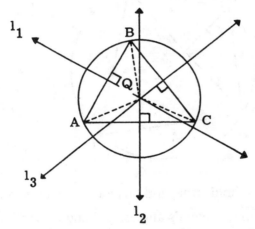

Solution: All points on a circle must be equidistant from the center. Therefore, for A, B, C, D, etc. to be concyclic, in the same plane as A, B, C, D, etc, there must exist a point P equidistant from A, B, C, D, etc. If such a point exists, then P is the center of the circle and PA = PB = PC = ... = radius of the circle.

For the particular case of three noncollinear points, we show that such a point P always exists and any set of three noncollinear points are concyclic.

Consider the three noncollinear points, A, B, and C. Draw △ABC. Let ℓ_1 be the perpendicular bisector of $\overline{AB}$ and ℓ_2 be the perpendicular bisector of $\overline{BC}$.

We now show that $\ell_1 \cap \ell_2$ is a point. Either ℓ_1 and ℓ_2 are coincident, parallel, or intersect in a point. We show that ℓ_1 and ℓ_2 are not coincident by indirect proof. Suppose ℓ_1 and ℓ_2 were coincident in the same line ℓ. It would then follow that $\ell \perp \overline{AB}$ and $\ell \perp \overline{AC}$. But this would imply that $\overline{AB}$ and $\overline{AC}$ are perpendicular to the same line and, as such $\overleftrightarrow{AB} \parallel \overleftrightarrow{AC}$ or $\overleftrightarrow{AB}$ and $\overleftrightarrow{AC}$ are coincident. Since the $\overleftrightarrow{AB}$ and $\overleftrightarrow{AC}$ share common point A, they cannot be parallel. Thus, $\overleftrightarrow{AB}$ and $\overleftrightarrow{AC}$ are coincident or $\overleftrightarrow{BAC}$. This contradicts our assumption that A, B, and C are noncollinear. Therefore, it is impossible that ℓ_1 and ℓ_2 are coincident. A similar argument shows that $\ell_1 \parallel \ell_2$ is impossible. Thus ℓ_1 and ℓ_2 intersect at a point Q.

Remember that the perpendicular bisector is a segment consisting of all points equidistant from the endpoints of the segment. Since Q is a point on ℓ_1, Q is equidistant from A and B. Since Q is a point on ℓ_2, QB = QC.

Combining these two facts, we obtain QA = QB = QC or Q is equidistant from points A, B, and C. Thus Q is the center of the circle that contains points A, B, and C with radius QA.

Hence, any three noncollinear points are concyclic.

● **PROBLEM** 502

In the figure below, the measures of angles are as indicated. Are the points B, C, D, and E concyclic? What is m∢DEB?

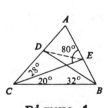

Figure 1

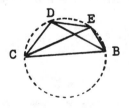

Figure 2

Solution: Inscribed angles of the same circle must be congruent if they intercept the same arc. Therefore, if points B, C, D, and E lie on the same circle, the inscribed angles ∢CDB and ∢CEB must be congruent. In fact, the reverse is true. If we show that inscribed angles ∢CDB and ∢CEB are congruent, B, C, D, and E must be concyclic. (The theorem we use here is usually stated in terms of the quadrilateral determined by the four points. If one side of a quadrilateral subtends congruent angles at the two nonadjacent vertices, then the quadrilateral is cyclic.)

Proof: ∢CEB is supplementary to ∢CEA. Therefore

m∢CEB = 180 − m∢CEA = 180 − 80 = 100.

In triangle $\triangle$CDB, m∢CDB + m∢DCB = m∢DBC = 180. Since the measure of the whole is equal to the sum of its parts, m∢DCB = m∢DCE + m∢ECB. Therefore by substitution, m∢CDB =

433

180 - 28 - 20 - 32 = 100.

m∢CEB = m∢CDB. Therefore, inscribed angles ∢CEB and ∢CDB are congruent, and points B, C, D, and E are concyclic.

To find m∢DEB, we use two facts: (1) B, C, D, and E are concyclic, which means ∢DEB is an inscribed angle. (2) D͡CB, D͡E, and E͡B comprise the entire circle and therefore

(i)   mD͡CB + mD͡E + mE͡B = 360.

Because D͡E is the intercepted arc of inscribed ∢DBE and ∢DCE, it must be true that mD͡E = 2 · m∢DCE. Since m∢DCE = 28, mD͡E = 2 · 28 = 56.

Similarly, E͡B is the intercepted arc of inscribed angle ∢ECB and thus mE͡B = 2 · m∢ECB = 2·20 = 40.

Substituting these results in equation (i), we obtain:

(ii)   mD͡CB + 56 + 40 = 360

(iii)   mD͡CB + 360 - 56 - 40 = 264.

Since the unknown, ∢DEB, is the inscribed angle that intercepts D͡CB, then

(iv)   m∢DEB = $\frac{1}{2}$mD͡CB = $\frac{1}{2}$ · 264 = 132.

A shorter method would have been to note that the opposite angles in an inscribed quadrilateral are supplementary. Therefore,

m∢DEB = 180 - m∢DCB = 180 - (m∢DCE + m∢ECB) = 180 - (20 + 28) = 180 - 48 = 132.

● **PROBLEM** 503

Given: Isosceles △DEC with $\overline{DE} \cong \overline{DC}$; $\overline{ADC}$; $\overline{AB}$ intersects $\overline{DE}$ in F; m∢A = 1/2 m∢ABE. Prove: Points A, B, D, and E are concyclic.

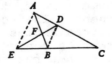

Solution: We show points A, B, D, and E are concyclic by showing quadrilateral ABDE cyclic. Since the vertices of a cyclic quadrilateral lie on a common circle, we will have shown that A, B, D, and E are concyclic.

To show a quadrilateral cyclic, we show that one side of a quadrilateral subtends congruent angles at the two non-adjacent vertices. The side we choose is $\overline{BD}$. The angles we must show congruent are ∢A (∢BAD) and ∢E (∢BED).

To show $\sphericalangle A \cong \sphericalangle E$, we relate the triangles $\triangle AFD$ and $\triangle EFB$. Note: Since the angle sum of a triangle is 180°, $m\sphericalangle A + m\sphericalangle AFD + m\sphericalangle FDA = 180°$ and $m\sphericalangle E + m\sphericalangle EBF + m\sphericalangle EFB = 180°$. Thus,

(1) $m\sphericalangle A + m\sphericalangle AFD + m\sphericalangle FDA = m\sphericalangle E + m\sphericalangle EBF + m\sphericalangle EFB$. But $\sphericalangle AFD$ and $\sphericalangle EFB$ are opposite angles. Thus, $m\sphericalangle AFD = m\sphericalangle EFB$, and we delete them from the equation.

(2) $m\sphericalangle A + m\sphericalangle FDA = m\sphericalangle E + m\sphericalangle EBF$. We are given that $m\sphericalangle A = \frac{1}{2}m\sphericalangle ABE = \frac{1}{2}m\sphericalangle FBE$ or $m\sphericalangle FBE = 2 \cdot m\sphericalangle A$

(3) $m\sphericalangle A + m\sphericalangle FDA = m\sphericalangle E + 2m\sphericalangle A$.

We have an equation relating $\sphericalangle A$, $\sphericalangle E$, and the unwanted $\sphericalangle FDA$. We rid ourselves of $\sphericalangle FDA$ by showing $m\sphericalangle FDA = 2 \cdot m\sphericalangle E$. Note that $\sphericalangle FDA$ is an exterior angle of $\triangle EDC$. Its measure equals the sum of remote interior angles $\sphericalangle E$ and $\sphericalangle C$. But $\triangle EDC$ is isosceles. Thus, $\sphericalangle E \cong \sphericalangle C$. Since $m\sphericalangle FDA = m\sphericalangle E + m\sphericalangle C$, $m\sphericalangle FDA = 2 \cdot m\sphericalangle E$. Using this result in (3), we have:

(4) $m\sphericalangle A + 2 \cdot m\sphericalangle E = m\sphericalangle E + 2 \cdot m\sphericalangle A$

(5) $m\sphericalangle A = m\sphericalangle E$ or $\sphericalangle A = \sphericalangle E$.

Since side $\overline{DB}$ intercepts congruent angles from the non-adjacent vertices, quadrilateral ABDE is cyclic.

Statements	Reasons
1. (See the problem statement.)	1. Given.
2. $\sphericalangle E \cong \sphericalangle BCD$	2. The base angles of an isoceles triangle are congruent.
3. $m\sphericalangle ADE = m\sphericalangle E + m\sphericalangle DCE$ $= 2 \cdot m\sphericalangle E$.	3. The measure of an exterior angle of the triangle equals the sum of the measures of the remote interior angles.
4. $m\sphericalangle AFD \cong m\sphericalangle EFB$.	4. Opposite angles formed by intersecting lines are congruent.
5. $m\sphericalangle AFD + m\sphericalangle A + m\sphericalangle ADF = 180$ $m\sphericalangle E + m\sphericalangle EFB + m\sphericalangle FBE = 180$	5. The sum of the angles of a triangle equals 180°.
6. $m\sphericalangle AFD + m\sphericalangle A + 2 \cdot m\sphericalangle E = 180$ $m\sphericalangle AFD + m\sphericalangle E + 2 \cdot m\sphericalangle A = 180$.	6. Substitution Postulate. (Steps 5, 4, 3, and 1).
7. $m\sphericalangle E - m\sphericalangle A = 0$.	7. Subtraction Postulate.
8. $m\sphericalangle E = m\sphericalangle A$.	8. Addition Postulate.
9. Quadrilateral ABDE is concyclic.	9. If a side of a quadrilateral subtends congruent angles

with non-adjacent vertices, then the quadrilateral is cyclic.

10. A, B, D, and E are concyclic.

10. The vertices of cyclic polygons are cyclic (Definition of concyclic polygon.)

Given: ΔABC is isosceles with $\overline{AB} \cong \overline{AC}$; $\overline{ADB}$ and $\overline{AEC}$; $\overline{DE} \parallel \overline{BC}$.
Prove: Quadrilateral EDBC is cyclic.

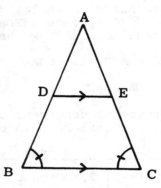

Solution: To show that a quadrilateral is cyclic, we can prove (1) a pair of opposite angles of the quadrilateral are supplementary or (2) one side of the quadrilateral subtends congruent angles at the two non-adjacent vertices.

Here, we show that the opposite angles, ∢DEC and ∢DBC, are supplementary by showing ∢DEC is supplementary to ∢ECB and ∢ECB ≅ ∢ABC.

Statements	Reasons
1. ΔABC is isosceles with $\overline{AB} \cong \overline{AC}$; $\overline{ADB}$ and $\overline{AEC}$; $\overline{DE} \parallel \overline{BC}$.	1. Given
2. ∢C is supplementary to ∢DEC.	2. The interior angles on the same side of the transversal are supplementary.
3. ∢C = ∢B.	3. The base angles of an isoceles triangle are congruent.
4. ∢B is supplementary to ∢DEC.	4. Substitution Postulate.
5. Quadrilateral EDBC is cyclic.	5. If a pair of opposite angles of a quadrilateral are supplementary, the quadrilateral is cyclic.

Given: $\overline{AB}$ and $\overline{CD}$ intersect at P; AP · BP = CP · DP. Prove: Points A, B, C, and D are concyclic.

Solution: From the accompanying figure, we will show points A, B, C, and D are concyclic by showing that they are the vertices of cyclic quadrilateral ABCD. To prove quadrilateral ABCD is cyclic, we use the theorem that states "a quadrilateral (ABCD) is cyclic if one side of the quadrilateral ($\overline{BC}$) subtends congruent angles (∢CAB and ∢CDB) with the nonadjacent vertices (points A and D)."

Thus, to show quadrilateral ABCD is cyclic, we show ∢CAB ≅ ∢CDB. To show ∢CAB ≅ ∢CDB, which are corresponding angles of similar triangles △CAP and △BDP, we prove △CAP ∿ △BDP.

We are given AP · BP = CP · DP. Dividing through by PC · PB, we obtain $\frac{AP}{PC} = \frac{DP}{BP}$. Thus, sides $\overline{AP}$ and $\overline{PC}$ are proportional to sides $\overline{DP}$ and $\overline{PB}$. Also, ∢APC ≅ ∢DPB because they are opposite angles formed by intersecting lines. Hence, △CAP ∿ △BDP by the SAS Similarity Theorem. Since corresponding angles of similar triangles are congruent, ∢CAP ≅ ∢BDP. Side $\overline{BC}$ intercepts congruent angles with the nonadjacent vertices A and D, and quadrilateral ADBC is cyclic. The vertices of a cyclic figure are concyclic. Thus, points A, B, C, and D are concyclic.

In a circle, all inscribed angles that intercept the same arc are congruent. The vertices of the inscribed angles and the endpoints are all on the same circle, i.e. concyclic. Show that the reverse is true. In the figure below, segment $\overline{BA}$ subtends equal angles with points C and D. Show that A, B, C, and D are concyclic.

Solution: A circle can always be found to contain three noncollinear points. Four points are concyclic only in special cases. Here, we prove one of the special cases.

Consider the circle determined by the three points, A, B, and C. What we wish to show is that point D must be on the circle. We shall use an indirect proof. Point D is either (1) in the interior of the circle, (2) outside the circle, or (3) on the circle. We shall show that the first two cases lead to a contradiction.

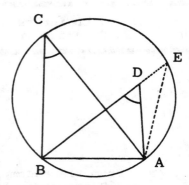

Figure 1   Figure 2

Only the third possibility, i.e. when D is concyclic
with A, B, and C, can be true. We shall prove this by finding an inscribed angle that is congruent to ⊰BCA, and then
showing that subtended angle ⊰BDA cannot be congruent to it.

Given: $\overline{AB}$. Points C and D are coplanar. Points C
and D are on the same halfplane created by $\overleftrightarrow{AB}$.
⊰ACB ≅ ⊰ADB. Points A, B, and C are points on
circle Q.

Prove: D is on circle Q.

Proof: Point D can either be (1) in the interior of
⊙Q, (2) outside ⊙Q, or (3) on circle ⊙Q.

Case (I).   D is in the interior of ⊙Q. Extend $\overleftrightarrow{BD}$ so that
it intersects ⊙Q at E. This leads to two lines
of reasonings: (a) ⊰BEA is an inscribed angle
that intercepts arc $\overarc{BA}$. Since inscribed ⊰BCA
also intercepts $\overarc{BA}$, ⊰BCA ≅ ⊰BEA. From the
given, we know that ⊰BCA ≅ ⊰BDA. Therefore,
⊰BDA ≅ ⊰BEA. (b) ⊰BDA is an exterior angle of
△DEA.   DEA is a remote interior angle
of △DEA. Since the exterior angle is greater
than either remote interior angles, m⊰BDA >
⊰BEA. The two lines of reasoning lead to a
contradiction. Therefore, D cannot be in the
interior of ⊙Q.

Case (II).   D is outside ⊙Q. Let point E be the intersection of $\overline{BD}$ with ⊙Q. This leads to two lines of
reasoning: (a) ⊰BEA is an inscribed angle that
intercepts arc $\overarc{BA}$. Since inscribed ⊰BCA also
intercepts $\overarc{BA}$, ⊰BCA ≅ ⊰BDA. From the given, we
know ⊰BCA ≅ ⊰BDA. Therefore, ⊰BDA ≅ ⊰BEA. (b)
⊰BEA is an exterior angle of △DEA. ⊰BDA is a
remote interior angle of △DEA. Therefore,
m⊰BEA > m⊰BDA.

These lines of reasoning also lead to a
contradiction. Therefore, point D cannot be
outside the circle.

Since D is neither outside nor inside the circle, it lies on the circle and A, B, C, and D are concyclic points.

Note: Given three noncollinear points, a circle can always be found to circumscribe it. Thus, a triangle can always be circumscribed since this requires that only three points, the vertices, be concyclic. What we have shown for the four points, A, B, C, and D, can be used to prove certain quadrilaterals inscribable. This theorem, although rather poorly worded, is used frequently in the proof of inscribable or cyclic quadrilaterals. It states the following: If one side of a quadrilateral (in our proof the side was AB, the quadrilateral was ABCD) subtends congruent angles (∢ACB and ∢ADB) at the two nonadjacent vertices (the vertices nonadjacent to the side AB), then the quadrilateral is concyclic.

● **PROBLEM** 507

Given three noncollinear points, a circle can always be found that passes through each of them. That is to say, three non-collinear points are concyclic. Given four points, it is only in special cases that all four are concyclic. In the accompanying figures, find four sets of four concyclic points. (All quadrilaterals are rectangles, all closed curves are circles, and the points chosen must be labelled points.)

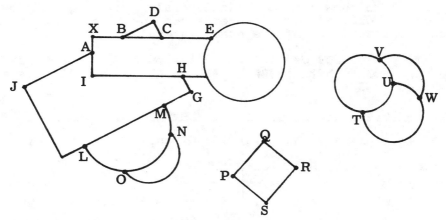

Solution: The four sets of points are as follows:

(1) P, Q, R, and S. These are the vertices of a rectangle, and a rectangle can always be inscribed in a circle. The diagonals of the parallelogram bisect each other and are congruent. Therefore, the center of the circumscribed circle is the intersection of the diagonals and the diameter is a diagonal.

(2) L, M, N, and O. These are points on a circle.

(3) A, X, D, and C. Let $\overline{AC}$ be the diameter of a circle. Remember that any angle inscribed in a semicircle is a right angle. Similarly, any point, say point Z, that determines a

right angle, say ∢AZC, must be a point on the semicircle.
Since m∢AXC = 90°, X is a point on the semicircle ÂC. Also,
m∢ADC = 90°, thus D is a point on the semicircle ÂC. Therefore, X and D are points on semicircle ÂC, and all four points
are concyclic.

(4) A, H, I, D: Similarly, consider $\overline{AH}$ the diameter.
Since ∢ADH and ∢AIH are right angles, points I and D are
each on one of the semicircles with endpoints A and H,
and the four points are concyclic.

● **PROBLEM** 508

Given: $\overline{AB}$ is a diameter of ⊙Q; $\overline{CQ} \perp \overline{AB}$ at Q; $\overline{AC}$ intersects
⊙Q at D; $\overline{BC}$ intersects ⊙Q at F. Prove: Quadrilateral CDQB
is cyclic; ∢QDB ≅ ∢QCA.

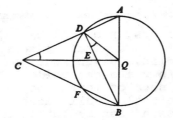

Solution: To show that a quadrilateral is cyclic, we show a
side of the quadrilateral subtends congruent angles with the
nonadjacent vertices of the quadrilateral.

To prove CDQB cyclic, we must show side QB subtends
∢BDQ and ∢BCQ such that ∢BDQ ≅ ∢BCQ. Also, we will have to
show ∢QDB ≅ ∢QCA.

	Statements		Reasons
1.	AB is a diameter of ⊙Q; $\overline{CQ} \perp \overline{AB}$ at Q; $\overline{AC}$ intersects ⊙Q at D; $\overline{BC}$ intersects ⊙Q at F.	1.	Given.
2.	$\overline{DQ} = \overline{QB}$.	2.	All radii of a circle are congruent.
3.	∢QDB ≅ ∢QBD.	3.	If two sides of a triangle are congruent, the angles opposite them are congruent.
4.	∢ADB is a right angle.	4.	Any angle inscribed in a semicircle is a right angle.
5.	∢DAB and ∢QBD are complementary.	5.	The acute angles of right triangle ADB are complementary.
6.	∢AQC and ∢BQC are right angles.	6.	Perpendicular lines intersect at right angles.

440

7. ∢ACQ and ∢DAB are complementary.	7. The acute angles of right triangles △CAQ are complementary.
8. ∢ACQ ≅ ∢QBD.	8. Angles complementary to congruent angles are congruent.
9. $\overline{CQ} \cong \overline{CQ}$.	9. A segment is congruent to itself.
10. $\overline{AQ} \cong \overline{QB}$.	10. The radii of a circle are congruent.
11. ∢AQC ≅ ∢BQC.	11. All right angles are congruent.
12. △AQC ≅ △BQC.	12. The SAS Postulate.
13. ∢ACQ ≅ ∢BCQ.	13. Corresponding parts of congruent triangles are congruent.
14. ∢BCQ ≅ ∢QBD.	14. Substitution Postulate. (Steps 13, 8).
15. ∢BCQ ≅ ∢QDB.	15. Substitution Postulate. (Steps 14, 3)
16. Quadrilateral CDQB is cyclic.	16. If a side of a quadrilateral subtends equal angles with the two nonadjacent vertices, then the quadrilateral is cyclic.

CHAPTER 26

CIRCLES INSCRIBED IN POLYGONS

Basic Attacks and Strategies for Solving Problems in this Chapter. See pages 442 to 450 for step-by-step solutions to problems.

In the figures below, a circle is inscribed in a triangle and another circle is inscribed in a quadrilateral. Another way of saying the same thing is to say that a triangle is **circumscribed** about a circle and a quadrilateral is circumscribed about another circle.

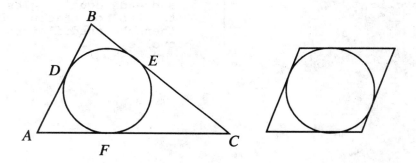

In the case of $\triangle ABC$, $\overleftrightarrow{AB}$ is tangent to the circle at D, $\overleftrightarrow{BC}$ is tangent to the circle at E, and $\overleftrightarrow{AC}$ is tangent to the circle at F. Since $\overline{AD}$ and $\overline{AF}$ are tangents to the circle from the exterior point A, a theorem established in Chapter 18 confirms that $\overline{AF} \cong \overline{AD}$. In the same way it follows that $\overline{BD} \cong \overline{BE}$ and $\overline{CE} \cong \overline{CF}$. This generalization is used in several problems in this chapter. Specifically, this generalization can be used to find an expression for the length of the radius in terms of the lengths of the sides of the triangle.

● **PROBLEM** 509

Show that in a quadrilateral circumscribed about a circle, the sum of the lengths of a pair of opposite sides equals the sum of the lengths of the remaining pair of opposite sides.

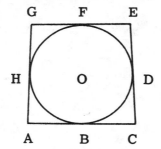

<u>Solution</u>: In the accompanying figure, quadrilateral AGEC is circumscribed about ⊙O and is tangent to ⊙O at points H, F, D, and B. We are asked to show

(i)   GE + AC = AG + CE.

We can accomplish this by rewriting each side of the equation in terms of segments such that each segment length on the left side of the equation has a segment equal in length to it on the opposite side.

To find these congruent segments, we use the result that the two tangent segments to a circle from an external point are congruent. Consider a vertex G and the two segments drawn to the circle, $\overline{GF}$ and $\overline{GH}$. Since $\overline{GH}$ is tangent to ⊙O at H and $\overline{GF}$ is tangent to ⊙O at F, $\overline{GF}$ and $\overline{GH}$ are tangent segments drawn from external point G and, thus, $\overline{GF} = \overline{GH}$. If we consider E, C, and A as external points, then we relate the tangent segments by the congruencies $\overline{EF} \cong \overline{ED}$, $\overline{CD} \cong \overline{CB}$, and $\overline{AB} \cong \overline{AH}$.

We can rewrite the equation to be proved in terms of the tangent segments. Note that GE = GF + FE, AC = AB + BC, GA =

GH + HA, and EC = ED + DC. Substituting these results in (i), we obtain:

(ii) GF + FE + AB + BC = GH + HA + ED + DC.

Since GF = GH, FE = ED, AH = AB, and CB = CD, the equation holds and, thus, GE + AC = AG + CE.

● **PROBLEM** 510

Find the radius of a circle inscribed in a triangle whose sides have lengths 3, 4 and 5.

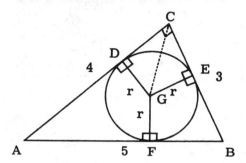

Solution: In the accompanying figure, △ABC circumscribes circle G. AC = 4, BC = 3, and AB = 5. Points D, E, and F are the points of tangency of ⊙G with the sides of △ABC. We are asked to find GE - the length of the radius of the inscribed circle. We will do this by (1) finding the length of tangent segment, CE, and (2) showing △CEG is isosceles with $\overline{GE} \cong$ tangent segment $\overline{CE}$. Thus, the radius, r, will equal CE.

(1) To find the lengths of the tangent segments, we use the theorem which states that, from an external point (such as C), the tangent segments (DC and CE) drawn to the circle (G) are congruent ($\overline{DC} \cong \overline{CE}$). Thus, if we consider C, A, and B respectively as the external point, we obtain:

(i)   DC = CE   (C is the external point)
(ii) AD = AF   (A is the external point)
(iii) BF = BE   (B is the external point)

But from the figure, we know that

(iv) AD + DC = AC = 4
(v)   CE + EB = CB = 3
(iv) AF + FB = AB = 5.

We have 6 equations with 6 variables. To solve for tangent segment CE, we first eliminate AD, BE, and DC by substituting in equations (iv), (v), and (vi) the values AD = AF, BE = BF, and DC = CE.

(vii) AF + CE = 4
(viii) CE + BF = 3
(ix)   AF + FB = 5.

We now have 3 equations in 3 unknowns. Eliminate AF by

substituting (ix) in (vii).

    (x)   CE - FB = 4 - 5 = -1
  (xi)   CE + FB = 3

We eliminate FB by adding (x) and (xi).

  (xii) 2 · CE = 2 or CE = 1.

  (2)   Thus, we have a value for CE.

To show ΔCEG is isosceles, we will show m∡ECG = m∡EGC = 45°. In order to show m∡ECG = 45°, we prove ∡DCG ≅ ∡ECG. By the Hypotenuse Leg Theorem, ΔDCG ≅ ΔECG, since points D and E are points of tangency, radii $\overline{DG} \perp \overline{DC}$ and $\overline{GE} \perp \overline{CE}$ and, thus, m∡CDG = m∡CEG = 90°. Since $\overline{DG}$ and $\overline{GE}$ are radii, $\overline{DG} \cong \overline{GE}$ and $\overline{CG} \cong \overline{CG}$. Thus, ∡DCG ≅ ∡ECG because they are corresponding angles of congruent triangles. Because a 3-4-5 triangle is a right triangle, ∡ACB is a right angle. Thus, we have

    m∡DCG + m∡ECG = m∡ACB.

By substitution, 2·m∡ECG = 90°
               m∡ECG = 45°.

We will now show m∡CGE also equals 45°. Since m∡ECG = 45° and m∡GEC = 90°, then to find m∡CGE, we have the following:

    m∡GCE + m∡GEC + m∡CGE = 180°
       45° + 90° + m∡CGE = 180°
                 m∡CGE = 45°.

Since m∡CGE ≅ m∡GCE, ΔCGE is isosceles. Therefore, GE ≅ CE, and r = GE = CE = 1. The radius of the inscribed circle equals 1.

● **PROBLEM** 511

Given: ΔABC with sides of length a, b, and c. The inscribed circle, Q, intersects $\overline{AB}$ at D, $\overline{BC}$ at E, and $\overline{CA}$ at F. Find the lengths AD and DB in terms of a, b, and c.

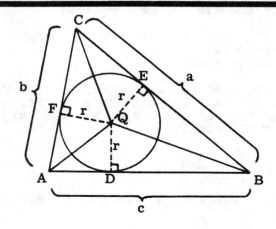

444

<u>Solution</u>: We are asked to find AD and DB, the lengths of the segments into which the tangency point, D, divides side AB of ΔABC, in terms of the sides of the triangle. We know therefore, that we must (1) use the properties of the inscribed circle Q; and (2) relate segments $\overline{AD}$ and $\overline{DB}$ to the other sides of ΔABC.

Because D, E, and F are the points of tangency, the radii $\overline{QD}$, $\overline{QE}$, and $\overline{QF}$ are all radii to the points of tangency and, therefore, perpendicular to the sides.

Thus, from the fact that ⊙Q is the inscribed circle, we know that (1) QD = QE = QF (because all radii are congruent); and (2) $\overline{QD} \perp \overline{AB}$, $\overline{QE} \perp \overline{CB}$, and $\overline{QF} \perp \overline{AC}$.

We now relate segments $\overline{AD}$ and $\overline{DB}$ to the corresponding sides. Note, from the problem statement, that $\overline{AD}$ and $\overline{DB}$ are segments divided by the point of tangency. We can use the same procedure as we are using now to find AF and FC, or CE and EB. Therefore, it should not be surprising that the expressions for these segments be, in some way, similar to the expressions for AD and DB. Nor should it be unexpected that, when we relate segments AD and DB to side AC, those segments correspond to segments AF and FC. In fact, we can relate segment $\overline{AD}$ to $\overline{AF}$, they are congruent. Consider $\overline{\Delta AQD}$ and ΔAQF. QF = QD because they are radii of ⊙Q. ∡QFA and ∡QDA are both right angles and therefore congruent. $\overline{AQ} \cong \overline{AQ}$ is true by the reflexive property. Since the hypotenuse and the leg of right triangle ΔAQD are congruent to the hypotenuse and the leg of right triangle ΔAQF, ΔAQD ≅ ΔAQF. By corresponding parts, $\overline{AD} \cong \overline{AF}$.

In relating segments $\overline{AD}$ and $\overline{DB}$ to side $\overline{CB}$, we note that ΔQDB ≅ ΔQEB and, therefore, $\overline{DB} \cong \overline{EB}$. Segments CF and CE cannot be related to $\overline{AD}$ and $\overline{DB}$, but they can be related to each other. By showing ΔQFC ≅ ΔQEC, we know by corresponding parts that $\overline{CF} \cong \overline{CE}$.

We have (1) used the properties of inscribed circle Q and (2) related $\overline{AD}$ and $\overline{DB}$ to the sides of ΔABC. We collect our data, and solve for AD and DB.

Because F, E, and D are points on $\overline{AC}$, $\overline{BC}$, and $\overline{AB}$, respectively, we have from the definition of "betweenness" that:

$$AB = AD + DB$$
$$AC = AF + FC$$
$$BC = BE + EC.$$

From the properties of the inscribed circles we have shown that

$$\overline{AD} \cong \overline{AF}$$
$$\overline{DB} \cong \overline{EB}$$
$$\overline{FC} \cong \overline{EC}$$

From the given, we know that BC = a
AC = b
AB = c.

We substitute a, b, c for BC, AC, and AB and substitute AD, DB, and FC for AF, EB, and EC in the above "betweenness" equations to obtain the following:

$$c = AD + DB$$
$$b = AD + FC$$
$$a = DB + FC$$

Here, we have three equations in three unknowns. FC is a variable for which we do not wish to solve. Therefore, we eliminate that variable first. Since b = AD + FC, FC = b - AD. Substituting this result in the equation, we obtain the following:

$$c = AD + DB$$
$$a = DB + (b-AD) = -AD + DB + b$$

Adding the equations, we eliminate AD.

$$c + a = (AD-AD) + DB + DB + b = 2DB + b$$

$$DB = \frac{c + a - b}{2}$$

To solve for AD, remember that c = AD + DB and, therefore, AD = c - DB.

$$AD = c - \frac{c + a - b}{2} = \frac{2c - (c+a-b)}{2} = \frac{c + b - a}{2}$$

Remember the earlier supposition that the expressions would probably be similar? Segment $\overline{DB}$ touches vertex B, and its formula can be written as

$$\frac{\left(\begin{array}{l}\text{sum of the length of the}\\\text{sides of }\triangle ABC\text{ that touch B}\end{array}\right) - \left(\begin{array}{l}\text{length of the}\\\text{third side}\end{array}\right)}{2}.$$

The same formula can be written for AD. For any segment that touches an arbitrary vertex X, the formula for the length is

$$\frac{\left(\begin{array}{l}\text{sum of the lengths of}\\\text{sides that touch x}\end{array}\right) - \left(\begin{array}{l}\text{length of the}\\\text{third side}\end{array}\right)}{2}$$

● **PROBLEM** 512

Find the radius, r, of the inscribed circle of right triangle ABC in terms of leg lengths a and b and hypotenuse length c.

Solution: In the accompanying figure, circle O of radius r is inscribed in right $\triangle ABC$ with AB = c, BC = a, and AC = b. Points P, Q, and S are points of tangency of the circle with the triangle. Point R is the intersection of $\overline{AB}$ and the extension of $\overline{OP}$. g is the length $\overline{OR}$. We must find r in terms of a, b, and c.

First, we will show $\triangle ARP \sim \triangle ABC$, which will give a relationship involving (known) a, b, and c and (unknown) g and r. Second, we will show $\triangle ORS \sim \triangle ABC$, which will give us a

446

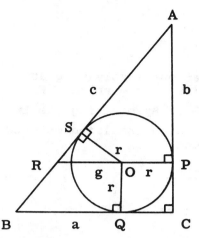

second relationship involving unknown g and r. With two
equations in two unknowns, we can then solve for r in terms
of a, b, and c.

To show ΔARP ∿ ΔABC, we use the A-A Similarity Theorem.
Because P is a point of tangency, $\overline{OP} \perp \overline{AC}$ and ∢OPA is a
right angle. ∢BCA is given to be a right angle. Thus ∢OPA
= ∢BCA. ∢RAP ≅ ∢BAC by the reflexive property. Thus,
ΔARP ∿ ΔABC. Because corresponding sides of similar tri-
angles are proportional, $\frac{RP}{BC} = \frac{AP}{AC}$. Note: RP = RO + OP =
g + r, BC = a and AC = b. Furthermore, quadrilateral OPCQ
is a parallelogram. (Since corresponding angles ∢BQO and
∢BCP are congruent, $\overline{QO} \parallel \overline{PC}$. Similarly, $\overline{OP} \parallel \overline{QC}$. Since both
pairs of opposite sides are parallel, OPCQ is a parallel-
ogram.) Opposite sides of a parallelogram are congruent.
Thus, QC = OP = r, and PC = OQ = r. Thus, AP = AC - PC =
b - r. Substituting these values, we obtain the following:

(i) $\frac{g+r}{a} = \frac{b-r}{b}$

Crossmultiplying, we obtain

(ii) gb + rb = ab - ar.

Solving for g, we obtain

(iii) g = (ab - ar - rb)/b.

Now we find a second relationship by showing ΔORS ∿
ΔABC. Note: S is the point of tangency. Therefore, the
radius $\overline{OS} \perp \overline{AB}$ and ∢OSR is a right angle. Since ∢ACB is a
right angle, ∢OSR ≅ ∢ACB. Furthermore, note that OPCQ is a
parallelogram and therefore $\overline{OP} \parallel \overline{QC}$. Since $\overline{RO}$ is an extension of $\overline{OP}$ and $\overline{BC}$ an extension of $\overline{QC}$, $\overline{RO} \parallel \overline{BC}$. Since corresponding angles of parallel lines are congruent, ∢SRO ≅
∢CBA. By the A-A Similarity Theorem, we have ΔORS ∿ ΔABC.
Because corresponding sides of similar triangles are proportional, $\frac{OR}{SO} = \frac{AB}{AC}$. Note: OR = g, OS = r, AB = c, and AC = b.
Substituting these values in the proportion, we obtain
    (iv) $\frac{g}{r} = \frac{c}{b}$.

447

Solving for g, we obtain

(v) $g = \dfrac{rc}{b}$.

Now we have two relations involving g and r. From (iii), we know that $g = (ab - ar - rb)/b$. From (v), we know that g also equals $\dfrac{rc}{b}$. By equating the two expressions for g, we obtain an expression involving only the unknown, r, and the known, a, b, and c.

(vi) $\dfrac{rc}{b} = \dfrac{(ab-ar-rb)}{b}$

Solving for r, we obtain

(vii) $rc = ab-ar-rb$
(viii) $ra+rb+rc = ab$
(ix) $r(a+b+c) = ab$
(x) $r = \dfrac{ab}{a+b+c}$

● **PROBLEM** 513

In the accompanying figure, $\odot P_1$ is inscribed in equilateral $\triangle ABC$. $\odot P_2$ is inscribed such that it is tangent to $\overline{AB}$, $\overline{AC}$, and $\odot P_1$. $\odot P_3$ is tangent to the same two sides ($\overline{AB}$, $\overline{AC}$) and $\odot P_2$. This process continues indefinitely. Given that the radius of $\odot P_1$ equals 1, find

(a) the radius of P_2 and
(b) the sum of the circumferences of all the circles in the figure.

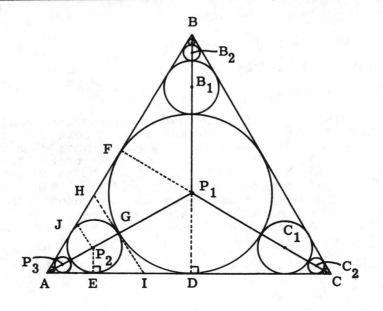

448

<u>Solution</u>: In the figure, let points E and D be the points of tangency of $\odot P_2$ and $\odot P_1$ with side $\overline{AC}$.

(a) We relate the radius of $\odot P_2$, $\overline{P_2E}$, and the radius of $\odot P_1$, $\overline{P_1D}$, by noting that $\overline{P_2E}$ and $\overline{P_1D}$ are corresponding parts of similar triangles $\triangle AP_2E$ and $\triangle AP_1D$. We will prove that $\triangle AP_2E \sim \triangle AP_1D$ by the A-A Similarity Theorem. From this, we will find a proportion that involves P_2E and P_1D. Since P_1D is known, we can solve for P_2E.

To show $\triangle AP_2E \sim \triangle AP_1D$, note that $\overline{P_2E}$ and $\overline{P_1D}$ are radii drawn to the points of tangency. Thus, $\angle P_2EA$ and $\angle P_1DA$ are right angles. Since all right angles are congruent, $\angle P_2EA \cong \angle P_1DA$. Note also that points P_1 and P_2 are on the angle bisector of $\angle FAD$ and thus $\overleftrightarrow{AP_1P_2}$. (To see this, note that circle P_1 is tangent to $\overline{AF}$ and $\overline{AD}$. Thus, the distances $\overline{P_1F}$ and $\overline{P_1D}$, from the center point, P_1, to the sides $\overline{AB}$ and $\overline{AC}$, are radii. Point P_1 is therefore equidistant from the two sides of $\angle FAD$ and must lie on the angle bisector of $\angle FAD$. Similar reasoning shows P_2 is on the bisector.) $\angle AP_2 = \angle AP_1$ and $\angle P_1AD \cong \angle P_2AE$.

By the A-A Similarity Theorem, $\triangle AP_2E \sim \triangle AP_1D$ and

(i) $\dfrac{AP_2}{P_2E} = \dfrac{AP_1}{P_1D}$

We wish to find a relation involving P_2E and P_1D. Therefore, we solve for AP_1 and AP_2 in terms of P_2E and P_1D.

To find AP_1, note that $\triangle ABC$ is an equilateral triangle and $\overline{AP_1}$ is an angle bisector. Then, $m\angle P_1AD = \frac{1}{2}m\angle FAD = \frac{1}{2}(60°) = 30°$. In right $\triangle AP_1D$, $\sin P_1AD = \dfrac{P_1D}{AP_1}$ or $AP_1 = \dfrac{P_1D}{\sin P_1AD}$ $= \dfrac{P_1D}{1/2} = 2P_1D$. Finally, $AP_2 = AP_1 - GP_1 - GP_2$. Because P_2G and P_2E and radii of $\odot P_2$, $P_2G = P_2E$. Substituting, we obtain $AP_2 = 2P_2D - P_1D - P_2E = P_1D - P_2E$.

We substitute our results in (i)

(ii) $\dfrac{P_1D - P_2E}{P_2E} = \dfrac{2 \cdot P_1D}{1 \cdot P_1D}$

(iii) $P_1D - P_2E = 2P_2E$.

(iv) $P_2E = \frac{1}{3}P_1D$.

We now have a general relationship between the radii of $\odot P_1$ and $\odot P_2$. By similar reasoning, we could show that any circle has radius one-third the length of the radius of the larger adjacent circle.

To find the radius in this particular case, we substitute $P_1D = 1$ in equation (iv).

(v) $P_2E = \frac{1}{3}(1) = \frac{1}{3}$.

(b). (vi) Sum of circumferences in the figure = (circum. of P_1) + (circum. of P_2 + circum. of P_3 + ...) + (circum. of B_1 + circum. of B_2...) + (circum of C_1 + circum. of C_2...). Note that since $\triangle ABC$ is equilateral, the clusters of circles, P_2, P_3, P_4, P_5..., B_1, B_2, B_3..., and C_1, C_2, C_3..., are all congruent. Thus, (circum. of P_2 + circum. of P_3 +...) = (circum. of B_1 + ...) = (circum. of C_1 + ...).

Substituting in our result, we have:

(vii) Sum of circum. = (circum. of P_1) + 3 · (circum. of P_2 + circum. of P_3 + ...).

Since the circumference of a circle of radius r equals $2\pi r$, we have:

(viii) Sum of circum. = (2π) + 3 · $(2\pi r_2 + 2\pi r_3 + 2\pi r_4...)$

Factoring, we obtain:

(ix) Sum of circum. = $2\pi + 3 \cdot 2\pi \cdot (r_2 + r_3 + r_4...)$.

From part (a), we found out that each succeeding circle has radius one-third the length of the previous circle. Thus, $r_2 = \frac{1}{3}r_1$, $r_3 = (\frac{1}{3})r_2 = (\frac{1}{3})(\frac{1}{3}r) = (\frac{1}{3})^2 r$, $r_n = (\frac{1}{3})^{n-1}r_1$. Hence,

(x) $r_2 + r_3 + r_4... = \frac{1}{3} + \frac{1}{9} + \frac{1}{27} + ...$

The right side of the equation is the sum of a geometric series with first term $\frac{1}{3}$ and ratio $\frac{1}{3}$. From algebra, we know that for a geometric series of first term a and ratio-between-terms of r, the sum of the terms equals $\frac{a}{1-r}$. Thus,

(xi) $r_2 + r_3 + r_4... = \frac{1/3}{1 - 1/3} = \frac{1}{2}$.

Substituting this result in (viii), we have:

(xii) Sum of circum. = $2\pi + 3 \cdot 2\pi(\frac{1}{2}) = 5\pi$.

450

CHAPTER 27

CIRCLES CIRCUMSCRIBING POLYGONS

> **Basic Attacks and Strategies for Solving Problems in this Chapter. See pages 451 to 471 for step-by-step solutions to problems.**

In the figure below the circle circumscribes pentagon *ABCDE*. This means that pentagon *ABCDE* is inscribed in the circle.

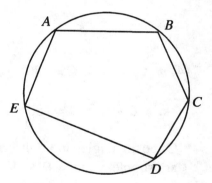

The problems in this chapter concern various relationships between measures of angles and lengths of sides of polygons which have circles circumscribing them.

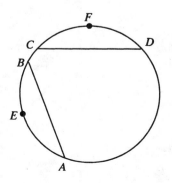

There are several theorems which are used in this chapter and which are illustrated below.

(1) $\overline{AB} \cong \overline{CD}$ if and only if $\overset{\frown}{AEB} \cong \overset{\frown}{CFD}$. In a circle, congruent chords subtend congruent arcs and the chords corresponding to congruent arcs are congruent. (Problem 516)

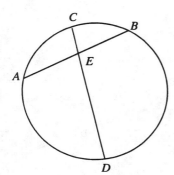

(2) $AE \cdot EB = CE \cdot ED$ (Problem 526)

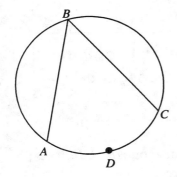

(3) $m \not\angle B = 1/2\ m\ \overset{\frown}{ADC}$ The measure of an inscribed angle is one-half the measure of the intercepted arc. (Problem 527)

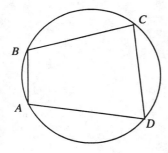

(4) $m \not\angle A + m \not\angle C = 180°$

$m \not\angle B + m \not\angle D = 180°$

Opposite angles in an inscribed quadrilateral are supplementary. (Problem 515)

Careful consideration of these theorems prior to the investigation of the problems in this chapter is recommended.

● **PROBLEM** 514

If the length of the apothem of a regular polygon is 5 inches, find the length of the diameter of the circle that is inscribed in the regular polygon.

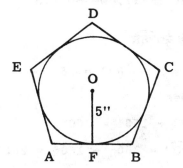

Solution: The apothem of a regular polygon is defined to be the radius of the inscribed circle. In polygon ABCDE, as shown in the figure, $\overline{OF}$ is the apothem and, as such, the radius of inscribed circle O.

The length of the diameter of a circle is twice the length of the radius. In this case, the diameter is 2 × 5 in., or 10 in.

Therefore, the length of the diameter of the inscribed circle is 10 inches.

● **PROBLEM** 515

If quadrilateral PQRS is inscribed in ⊙N and m∢Q = 80°, find m∢S.

Solution: Inscribed angle, ∢S, is the unknown. Inscribed angle ∢Q is the known (m∢Q = 80°) ∢S and ∢Q are related by the fact that the intercepted arcs of ∢S and ∢Q comprise the entire circle.

To solve for m⊀S in terms of m⊀Q, remember that (1) the measure of an inscribed angle equals one half the measure of the intercepted arc, and (2) the arc measure of a complete circle is 360°.

Thus,

(i) $m⊀S = \frac{1}{2}m\overset{\frown}{PQR}$

(ii) $m⊀Q = \frac{1}{2}m\overset{\frown}{PSR}$.

(iii) $m\overset{\frown}{PQR} + m\overset{\frown}{PSR} = 360°$

Furthermore, we are given that

(iv) $m⊀Q = 80°$

Substituting (iv) into (ii), we obtain

(v) $80° = \frac{1}{2}m\overset{\frown}{PSR}$.

(vi) $m\overset{\frown}{PSR} = 2 \times 80° = 160°$

Equation (iii) becomes

(vii) $m\overset{\frown}{PQR} + 160° = 360°$

(viii) $m\overset{\frown}{PQR} = 360° - 160° = 200°$

Finally, we substitute (viii) into equation (i) to obtain

(ix) $m⊀S = \frac{1}{2}(200°) = 100°$

Note that ⊀S and ⊀Q are supplementary. In general, the opposite angles of a cyclic quadrilateral are supplementary.

● **PROBLEM** 516

In the figure shown, $\overset{\frown}{AB} \cong \overset{\frown}{DC}$ and $\overset{\frown}{AD} \cong \overset{\frown}{BC}$. Prove ABCD is a parallelogram.

<u>Solution</u>: We will prove ABCD is a parallelogram by proving that each set of opposite sides consists of a congruent pair of segments.

The theorem stating that congruent arcs have congruent chords will help us in this problem.

Statement	Reason
1. $\overarc{AB} \cong \overarc{DC}$	1. Given.
2. $\overline{AB} \cong \overline{DC}$	2. Congruent arcs have congruent chords.
3. $\overarc{AD} \cong \overarc{BC}$	3. Given.
4. $\overline{AD} \cong \overline{BC}$	4. Same as reason 2.
5. ABCD is a parallelogram	5. A quadrilateral in which both pairs of opposite sides are congruent is a parrallelogram.

● **PROBLEM** 517

Prove that the sum of the alternate angles of any hexagon inscribed in a circle is equal to four right angles.

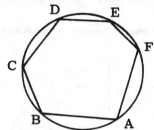

<u>Solution</u>: We see from the figure that each angle is an inscribed angle of the circle. As such, each angle has a measure equal to $\frac{1}{2}$ the measure of its intercepted arc. If the alternate angles of the hexagon sum to 360°, the measure of four right angles, then we have proved the desired results.

We choose ∢A, ∢E, and ∢C as the alternate angles.

$$m\angle A = \frac{1}{2}m\overarc{BCDEF}$$

$$m\angle E = \frac{1}{2}m\overarc{FABCD}$$

$$m\angle C = \frac{1}{2}m\overarc{DEFAB}$$

Hence, $m\angle A + m\angle E + m\angle C = \frac{1}{2}(m\overarc{BCDEF} + m\overarc{FABCD} + m\overarc{DEFAB})$

We now show that these three arcs form two entire circular angles. Note:

$m\overarc{BCDEF} = m\overarc{BC} + m\overarc{CD} + m\overarc{DE} + m\overarc{EF}$
$m\overarc{FABCD} = m\overarc{FA} + m\overarc{AB} + m\overarc{BC} + m\overarc{CD}$
$m\overarc{DEFAB} = m\overarc{DE} + m\overarc{EF} + m\overarc{FA} + m\overarc{AB}$

Adding these three equations, we obtain

$$\overset{\frown}{mBCDEF} + \overset{\frown}{mFABCD} + \overset{\frown}{mDEFAB} = (\overset{\frown}{mBC} + \overset{\frown}{mCD} + \overset{\frown}{mDE} + \overset{\frown}{mEF} + \overset{\frown}{mFA} + \overset{\frown}{mAB}) + (\overset{\frown}{mBC} + \overset{\frown}{mCD} + \overset{\frown}{mDE} + \overset{\frown}{mEF} + \overset{\frown}{mFA} + \overset{\frown}{mAB})$$

Note, though, that $\overset{\frown}{mBC} + \overset{\frown}{mCD} + \overset{\frown}{mDE} + \overset{\frown}{mEF} + \overset{\frown}{mFA} + \overset{\frown}{mAB}$ contains the whole circle or 360°. Thus, $\overset{\frown}{mBCDEF} + \overset{\frown}{mFABCD} + \overset{\frown}{mDEFAB} = 2 \cdot 360 = 720$.

In other words, if we start at point B and trace the three arcs in the sum, then we will find that we have made two entire trips around the circle. Therefore, the sum of the arcs is 2(360°) or 720°. By substitution

$$m\angle A + m\angle E + m\angle C = \frac{1}{2}(720) = 360 \ .$$

With similar reasoning, it can be shown $m\angle B + m\angle F + m\angle D = 360°$. Hence, we conclude that the sum of the alternate angles of a hexagon will always sum to the measure of four right angles.

● **PROBLEM** 518

Given: Regular hexagon ABCDEF. Prove: Quadrilateral ABDE is a rectangle.

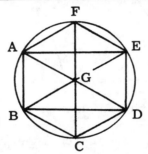

Solution: To show quadrilateral ABDE is a rectangle, we can either show (1) it has four right angles, (2) it is a parallelogram with one right angle, or (3) it is a parallelogram with congruent diagonals. Since there is not any given data concerning right angles, the third method is suggested.

We show ABDE is a parallelogram by showing both pairs of opposite sides are congruent. Then, we show that, since the diagonals of the parallelogram pass through the center of the hexagon, their lengths are both equal to twice the length of the radii of the hexagon and, as such, the diagonals are congruent.

STATEMENTS	REASONS
1. Regular hexagon ABCDEF	1. Given.
2. $\overline{AB} \cong \overline{DE} \cong \overline{AF} \cong \overline{FE} \cong \overline{BC} \cong \overline{CD}$	2. The sides of a regular polygon are congruent.
3. $\angle F \cong \angle C$	3. The interior angles of a regular polygon are congruent.

4. $\triangle AFE \cong \triangle BCD$	4. The SAS Postulate.
5. $\overline{AE} \cong \overline{BD}$	5. Corresponding parts of congruent triangles are congruent.
6. Quadrilateral ABDE is a parallelogram	6. If both pairs of opposite sides of a quadrilateral are congruent, then the quadrilateral is a parallelogram.
7. G is the center of ABCDEF	7. Every regular polygon has a center that is concurrent with the centers of the inscribed and circumscribed circles.
8. $\overline{AG} \cong \overline{BG} \cong \overline{CG} \cong \overline{DG} \cong \overline{EG} \cong \overline{FG}$	8. The radii of a regular polygon are congruent.
9. $\triangle AGB \cong \triangle AGF \cong \triangle FGE \cong \triangle EGD \cong \triangle DGC \cong \triangle CGB$	9. The SSS Postulate.
10. $\angle AGF \cong \angle FGE \cong \angle EGD \cong \angle DGC \cong \angle CGB \cong \angle BGA$	10. Corresponding parts of congruent triangles are congruent.
11. $m\angle AGF + m\angle FGE + m\angle EGD + m\angle DGC + m\angle CGB + m\angle BGA = 360°$	11. Angles that form a complete circle sum to 360°.
12. $6 \cdot m\angle AGF = 360°$	12. Substitution Postulate.
13. $m\angle AGF = 60°$	13. Division Property of Equality.
14. $m\angle AGD = m\angle AGF + m\angle FGE + m\angle EGD$ $m\angle BGE = m\angle BGA + m\angle AGF + m\angle FGE$	14. Angle Addition Postulate.
15. $m\angle AGD = 60° + 60° + 60° = 180°$ $m\angle BGE = 60° + 60° + 60° = 180°$	15. Substitution Postulate.
16. $\angle AGD$ and $\angle BGE$ are straight angles and $\overline{AGD}$ and $\overline{BGE}$	16. A straight angle has measure 180°.
17. $AD = AG + DG = 2 \cdot AG$ $BE = BG + EG = 2 \cdot AG$	17. Definition of "Betweenness" and Substitution
18. $AD = BE$	18. Transitivity Postulate.
19. $\overline{AD} \cong \overline{BE}$	19. Definition of congruence of segments.

20. Quadrilateral ABDE is
    a rectangle

20. A parallelogram whose
    diagonals are congruent
    is a rectangle.

● **PROBLEM** 519

Find the length of an arc intercepted by a side of a regular
hexagon inscribed in a circle whose radius is 18 in.

6 sides

Solution: Since a regular hexagon consists of 6 sides of
equal measure, the degree measure of each arc intercepted by
a side must equal $\frac{1}{6}$ the total degree measure of the circle,
or $\frac{1}{6}(360°) = 60°$.

$$\text{arc length} = \frac{\text{degree measure of arc}}{360°} \times \text{Circumference}$$

The circumference (C) of the circle of radius r is $C = 2\pi r$.
By substitution, $C = 2\pi(18 \text{ in.}) = 36\pi$ in. Therefore,

$$\text{arc length} = \frac{60°}{360°}(36\pi)\text{in.} = \frac{60}{10}\pi \text{ in} = 6\pi \text{ in.}$$

Using $\pi = 3.14$, arc length = 6(3.14 in.) = 18.84 in. Therefore, the length of the arc intercepted by a side of a regular hexagon in the circle of radius 18 in. is 18.84 in.

● **PROBLEM** 520

In the figure, Δ ABC is inscribed in a circle. If m ✗ A = 70°,
m ✗ B = 60°, and m ✗ C = 50°, which side of Δ ABC is nearest the
center of the circle.

Solution: Each side of Δ ABC is also a chord of the circle. By a
theorem, we know that the longer chord is closer to the center of the
circle. Therefore, the longest side of Δ ABC is nearest to the circle
center.
    To find the longest side, we recall that, given two unequal angles
of a triangle, the longer side is opposite the angle of greatest measure.
In Δ ABC, ✗ A has the greatest measure. Therefore, side $\overline{BC}$ is the
longest side, and thus, the chord nearest the center of the circle.

If an equilateral triangle is inscribed in a circle, it divides the circle into three equal arcs. Prove this statement.

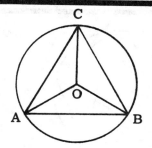

Solution: Each side of the equilateral triangle will be a chord of the circle in which it is inscribed. If we draw the radii of circle O to the vertices of equilateral triangle ABC, then we can show triangles AOB, AOC, and BOC are congruent. This will allow us to conclude ∢AOB ≅ ∢AOC ≅ ∢BOC. Since congruent central angles intercept arcs of equal measure, we will be able to draw the desired conclusion.

Statement	Reason
1. △ABC is equilateral and intercepts circle O at A, B. and C.	1. Given
2. Draw radii $\overline{OC}$, $\overline{OA}$ and $\overline{OB}$.	2. Radii can be drawn from the center to any point on a circle.
3. $\overline{AC} \cong \overline{CB} \cong \overline{AB}$	3. All three sides of an equilateral triangle are congruent.
4. $\overline{OA} \cong \overline{OC} \cong \overline{OB}$	4. Radii of a given circle are congruent.
5. △AOB ≅ △AOC ≅ △BOC	5. S.S.S. ≅ S.S.S.
6. ∢AOB ≅ ∢AOC ≅ ∢BOC	6. Corresponding angles of congruent triangles are congruent.
7. ∢AOB intercepts arc $\overset{\frown}{AB}$ ∢AOC intercepts arc $\overset{\frown}{AC}$ ∢BOC intercepts arc $\overset{\frown}{BC}$	7. An angle intercepts an arc if the endpoints of the arc are the intersection of the sides of the angle and the circle.
8. $\overset{\frown}{AB} \cong \overset{\frown}{AC} \cong \overset{\frown}{BC}$	8. Arcs which have congruent central angles are congruent to each other.

Note that in step 4, since two sides of each triangle are radii, the three congruent radii provide for two corresponding congruent sides in each triangle.

Prove that if two diagonals of a regular pentagon intersect each other, the longer segments will be equal in
length to the sides of the pentagon.

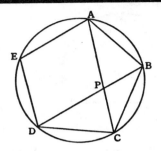

Solution: In regular pentagon ABCDE, as the figure shows,
diagonals $\overline{AC}$ and $\overline{BD}$ intersect at P. Since ABCDE is a
regular polygon, a circle can be circumscribed about it.

This will allow us to prove ⦨ABP $\overset{\sim}{=}$ ⦨APB which then gives
us justification to conclude AB = AP, or that the longer
segment of one diagonal is equal to a side of the
pentagon.

⦨ABP is an inscribed angle. The measure of an inscribed angle is one-half the measure of its intercepted
arc. Therefore, m⦨ABP = ½ m$\overset{\frown}{AED}$ = ½(m$\overset{\frown}{ED}$ + m$\overset{\frown}{EA}$).

⦨APB is an angle formed by two chords intersecting
inside a circle. Its measure is equal to one-half the sum
of the measures of the arcs intercepted by the chords.
Hence, m⦨APB = ½(m$\overset{\frown}{AB}$ + m$\overset{\frown}{DC}$).

Since the sides of the pentagon are equal, the
vertices of ABCDE divide the circle into 5 congruent arcs.
As such, m$\overset{\frown}{AB}$ = m$\overset{\frown}{BC}$ = m$\overset{\frown}{DC}$ = m$\overset{\frown}{ED}$ = m$\overset{\frown}{EA}$.

Since they are both equal to equal quantities,
m⦨ABP = m⦨APB, or ⦨ABP $\overset{\sim}{=}$ ⦨APB.

We then conclude that $\overline{AP}$ $\overset{\sim}{=}$ $\overline{PA}$, or AP = AB, because
if two angles of a triangle (△APB) are congruent, the
sides opposite them are also congruent.

By similar logic, we can prove PD = DC.

An equilateral triangle and a regular hexagon are inscribed in the
same circle. Prove that the length of an apothem of the hexagon is
greater than the length of an apothem of the equilateral triangle.

Solution: All sides of the triangle and hexagon are chords of the
circle in which they are inscribed. The apothem is the perpendicular

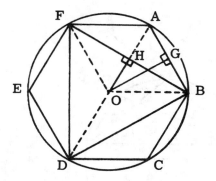

distance from the center of the circumscribing circle to the chords that form the sides of the inscribed regular polygon. Since, in the same circle, longer chords lie at a smaller distance from the center of the circle, the figure with the longer apothem is the figure with the shorter sides. We will show that any side, say $\overline{BF}$, of equilateral triangle BFD is greater than any side, say $\overline{AB}$, of regular hexagon ABCDEF.

We know that when two triangles have two pairs of congruent corresponding sides, the triangle with the greater included angle will have the greater third side. Therefore, we can derive our results from analyzing △'s OBF and OAB and applying the above theorem.

$\overline{OF} \cong \overline{OB} \cong \overline{OA}$ because all radii of a circle are congruent.

The measure of the central angles of a regular polygon is given by $\frac{360^\circ}{n}$, where n is the number of sides. Therefore, the central angles △ BFD measure $\frac{360^\circ}{3}$ or 120° and those of regular hexagon ABCDEF measure $\frac{360}{6}$ or 60°. Hence, m ⦨ FOB = 120° and m ⦨ AOB = 60°. Since two sides of △ OBF are congruent to two corresponding sides of △ OAB and the included angle in △ OBF is greater than that in △ OAB, by the theorem stated above, BF > AB.

Since BF > AB, the perpendicular distances OH and OG, from the center to each chord (the apothem) will be related by the inequality OG OH.

Therefore, we have shown the apothem of the regular hexagon will be greater than the apothem of the equilateral triangle.

● **PROBLEM** 524

Prove: (a) If a triangle has two unequal angles, the smaller angle has the longer angle bisector (measured from the vertex to the opposite side). (b) Any triangle that has two equal angle bisectors is isosceles. (The Steiner-Lehmus Theorem).

Solution: (a) Let ABC be the triangle with m⦨NBC = B and m⦨MCB = C. Let BM and CN bisect the angles B and C. Let M' be the point on BM such that m⦨M'CN = 1/2 B. We wish to show BM > CN, given B > C. We will do this by showing BM > BM' > CN in steps:

(1) Show points B, N, M', and C are concyclic and therefore are chords of a circle.

(2) Because m⦨M'CB > m⦨NCB, and the greater inscribed angle intersects the greater chord, BM' > CN.

(3) Because BM > BM', BM > CN.

459

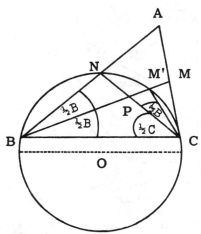

(1) To show that points B, N, M', and C are concyclic, we show that the quadrilateral, BNM'C,
is cyclic. To show a quadrilateral is cyclic,
we show that side $\overline{BC}$ subtends congruent angles
∢BNC and ∢BM'C with the nonadjacent vertices
N and M'. Because the angle sum of a triangle
equals 180°,

    (i)   m∢NBM' + m∢BNC + m∢NPB = 180°

    (ii)   m∢M'CP + m∢CM'P + m∢M'PC = 180°, using
        the Transitive Property of Equality,
        we have

    (iii)   m∢NBM' + m∢BNC + m∢NPB = m∢M'CP +
         m∢CM'P + m∢M'PC.

Note, though, that m∢NBM' = m∢M'CP = $\frac{1}{2}$B. Also, opposite
angles ∢NPB and ∢M'PC must also be congruent. Subtracting
these equal measures from both sides of equation (iii), we obtain:

    (iv)   m∢BNC = m∢CM'P or ∢BNC ≅ ∢CM'P.

Since side BC of quadrilateral BNM'C subtends congruent
angles with nonadjacent vertices N and M', quadrilateral BNM'C
is cyclic, and there exists a circle O such that B, N, M', and
C are points on ⊙O. Thus, BN and BM' are chords of ⊙O.

(2) We will now show that m∢M'CB > m∢NBC.

Since m∢ACB > m∢NBC, $\frac{1}{2}$m∢ACB > $\frac{1}{2}$m∢NBC or m∢NCB > m∢M'BC.

Hence, m∢NCB + $\frac{1}{2}$B > m∢M'BC + $\frac{1}{2}$B or m∢M'CB > m∢NBC. Remember
from an earlier proof, that if two inscribed angles intercept
unequal chords, the greater angle intercepts the greater chord.
∢M'CB intercepts chord $\overline{M'B}$ and ∢NBC intercepts chord $\overline{CN}$. Since
m∢M'CB > m∢NBC, M'B > CN.

    (3) By the transitive property, since BM > BM' and
        BM' > CN, BM > BN. Thus, the bisector of
        smaller ∢B is shorter than the bisector of ∢C.

(b)   For this part, we prove the contrapositive of the theorem. Since a statement is true if and only if its contrapositive is true, proving the contrapositive is sufficient to prove the statement.

Here, we are asked to show any triangle that has two equal angle bisectors is isosceles. The contrapositive of this statement is as follows: if a triangle is not isosceles, the bisectors are not equal.

From Part (a), we know that if two angles of a triangle are not congruent, the smaller angle has the longer angle bisector. That is, if two angles of a triangle are not congruent, then the angle bisectors are not congruent. Since no two angles of a non-isosceles △ are congruent, no two angle bisectors can be congruent. Thus, the contrapositive has been proved and the proof is complete: Any triangle with two equal angle bisectors is isosceles.

● **PROBLEM** 525

Prove that if a quadrilateral is inscribed in a circle, the product of the diagonals is equal to the sum of the products of both pairs of opposite sides. (Hint: Consider $\overline{AE}$ such that ∢DAE ≅ ∢CAB.)

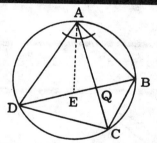

Solution: In the accompanying figure, cyclic quadrilateral ABCD has diagonals $\overline{AC}$ and $\overline{BD}$. We must prove

$$AC \cdot BD = AD \cdot BC + AB \cdot CD.$$

The hint is to draw $\overline{AE}$ such that ∢DAE ≅ ∢CAB. By doing this, two pairs of similar triangles involving the two diagonals and the four sides are created.

Note, one of terms above involves AD and BC. The pair of triangles, whose similarity will relate these sides, is ∢ABC and ∢ADE. According to the definition of $\overline{AE}$, ∢DAE ≅ ∢CAB. Furthermore, inscribed angles ∢ADE and ∢ACB intercept the same arc, $\overset{\frown}{AB}$. Therefore, ∢ADE ≅ ∢ACB, and by the A-A Similarity Theorem, △ABC ∿ △ADE. The proportions that follow from this can be written as

(i) $\dfrac{AD}{AC} = \dfrac{AE}{AB} = \dfrac{DE}{BC}$

By performing the cross multiplication, we obtain

(ii) $AD \cdot BC = AC \cdot DE.$

461

The other term on the right side of the equation to be proved is AB × CD. This term can be derived by showing ΔABE ∿ ΔACD. Inscribed angles ∢DCA and ∢DBA intercept the same arc, AD. Then, ∢DCA ≅ ∢DBA. By construction, ∢DAE ≅ ∢QAB. Therefore,

m∢DAE + m∢EAQ = m∢QAB + m∢EAQ, or ∢DAQ ≅ ∢BAE.

By the A-A Similarity Theorem, ΔABE ∿ ΔACD. The proportions that follow from this are:

(iii) $\dfrac{AB}{AC} = \dfrac{AE}{AD} = \dfrac{BE}{CD}$

By performing the crossmultiplication to derive a AB× CD term, we obtain

(iv)   AB · CD = AC · BE.

The entire right side AD × BC + AB × CD therefore equals AC × DE + AC × BE. Factoring out AC, we have AC · (DE + BE). Since DE + BE = DB, we have AC · DB, the product of two diagonals.

Thus the product of the diagonals equals the sum of the product of the opposite sides. This important result is better known as Ptolemy's Theorem.

● **PROBLEM** 526

In regular polygon ABCDE, diagonals $\overline{AC}$ and $\overline{BD}$ intersect at F. Prove AF × FC = BF × FD.

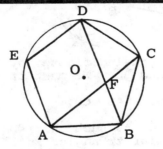

Solution: In this problem we employ the fact that a circle may be circumscribed about any regular polygon. This will change the problem from one involving a polygon to one involving a circle, about which more theorems exist. (The circle has been circumscribed in the diagram accompanying the problem).

STATEMENT

REASON

1. ABCDE is a regular polygon with diagonals $\overline{AC}$ and $\overline{BD}$ intersecting at F.

1. Given.

2. Circumscribe circle O about regular polygon ABCDE

2. A circle may be circumscribed about any regular polygon.

3. In circle O,
   AF × FC = BF × FD

3. If two chords intersect inside a circle, then the product of the lengths of the segments of one chord equals the product of the lengths of the segments of the other chord.

● **PROBLEM** 527

Prove that the two diagonals of a regular pentagon, not drawn from a common vertex, divide each other in extreme and mean ratio.

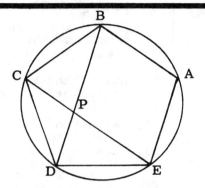

Solution: In the accompanying figure, diagonals $\overline{EC}$ and $\overline{BD}$ of regular pentagon ABCDE intersect at point P.

If P divides $\overline{EC}$ and $\overline{BD}$ into mean and extreme ratio, then $\dfrac{EC}{EP} = \dfrac{EP}{PC}$ and $\dfrac{BD}{BP} = \dfrac{BP}{PD}$. It will be sufficient to prove either of these proportions. Hence, we will show that $\dfrac{BD}{BP} = \dfrac{BP}{PD}$.

To prove this proportion, we show $\triangle BCD \sim \triangle CPD$ and, thus, $\dfrac{BD}{CD} = \dfrac{CD}{PD}$. We will then show $CD = BP$, to verify that $\dfrac{BD}{BP} = \dfrac{BP}{PD}$.

STATEMENT	REASON
1. ABCDE is a regular pentagon	1. Given.
2. $\overline{BC} \cong \overline{CD} \cong \overline{DE}$	2. The sides of a regular polygon are congruent.
3. ⊀BCD $\cong$ ⊀CDE	3. The vertex angles of a regular polygon are congruent.
4. $\triangle BCD \cong \triangle CDE$	4. SAS Postulate.
5. ⊀CDB $\cong$ ⊀DCE	5. Corresponding parts of congru-

ent triangles are congruent.

6. $\angle CDB \cong \angle CBD$

6. Reason 10.

7. $\triangle BCD \sim \triangle CPD$

7. A.A. Similarity Theorem.

8. $\dfrac{BD}{CD} = \dfrac{CD}{PD}$

8. Corresponding parts of similar triangles are proportional.

To show CD = BP we will prove $\triangle BCD$ and $\triangle BCP$ are isosceles.

9. $m\angle BCD = 108°$

9. The vertex angle of a regular pentagon has a measure equal to $\dfrac{5-2}{5}\,180°$.

10. $\triangle BCD$ is isosceles

10. If two sides of a triangle are congruent, then the angles oppisite them are also congruent.

11. $m\angle BCD + m\angle CDB + m\angle CBD = 180°$

11. Sum of the measures of the angles of a triangle is 180°.

12. $m\angle CDB = m\angle CBD$

12. Definition of isosceles triangles.

13. $180° - m\angle BCD = 2m\angle CDB$

13. Substitition.

14. $m\angle CDB = m\angle CBD = \frac{1}{2}(180° - 108°) = 36°$.

14. Substitution.

15. $m\angle BCP = m\angle BCD - m\angle PCD$

15. Angle addition postulate.

16. $\angle PCD$ is an inscribed angle

16. Definition of inscribed angle.

17. $m\angle PCD = \frac{1}{2}m\overset{\frown}{DE}$

17. The measure of an inscribed angle is equal to one-half the measure of the arc it intercepts.

18. $m\overset{\frown}{DE} = 1/5(360°) = 72°$

18. The vertices of a regular pentagon divide the circumscribed circle into 5 congruent arcs.

19. $m\angle PCD = \frac{1}{2}(72°) = 36°$

19. Substitution postulate.

20. $m\angle BCP = 108° - 36° = 72°$

20. Same as 19.

21. $\angle BPC$ is an exterior angle of $\triangle PCD$

21. Definition of exterior angle.

22. $m\angle BPC = m\angle PCD + m\angle PDC = 36° + 36° = 72°$

22. The measure of an exterior angle is equal to the sum of the measures of the remote interior angles. Substitution.

23. △BCP is isosceles	23. A triangle whose base angles have equal measure is isosceles.
24. BP = BC	24. The sides opposite the equal angles of an isosceles triangle are equal in measure.
25. BC = CD	25. All sides of a regular polygon are congruent.
26. BP = CD	26. Transitive property.
27. Therefore, $\frac{BD}{BP} = \frac{BP}{PD}$	27. Substitution of 20 in 8.

Thus, P divides diagonal $\overline{BD}$ into mean and extreme ratio. The same result can be derived for $\overline{EC}$ in an analogous manner.

● **PROBLEM** 528

The product of two sides of a triangle is equal to the product of the altitude to the third side and the diameter of the circumscribed circle. Prove this. (Hint: consider the diameter that passes through the included vertex of the first two sides.)

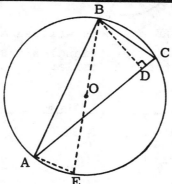

Solution: In the figure above, ⊙ O is circumscribed about △ABC. $\overline{BD}$ is an altitude. We are asked to show that AB · BC = BE · BD.

Whenever there are products, proportions are also present; and whenever proportions are present, we are well advised to search for similar triangles. If AB · BC = BE · BD, then, dividing both sides by BE · BC, we obtain, $\frac{AB}{BE} = \frac{BD}{BC}$. It is sufficient for this proof to prove this proportion.

These are proportions that correspond to △ABE and △DBC. Thus, if we prove that △ABE ∿ △DBC, then the proof is almost complete.
Note that △ABE is a triangle inscribed in a semicircle (since $\overline{BE}$ is a diameter). Therefore, △ABE is a right triangle

465

and ∢BAE is a right angle. Since $\overline{BD}$ is an altitude, ∢ BDC is also a right angle and ∢ BAE $\overset{\sim}{=}$ ∢ BDC. Thus, one pair of corresponding angles are congruent. Note also that ∢ AEB and ∢ DCB are inscribed angles that intercept the same arc, $\overset{\frown}{AB}$. Therefore, ∢ AEB $\overset{\sim}{=}$ ∢ DCB. By the A.A. Similarity Theorem, △ABE ∿ △DBC. Since corresponding sides of similar triangles are proportional,

$$\frac{AB}{BE} = \frac{BD}{BC} \quad \text{or } AB \cdot BC = BE \cdot BD.$$

Thus, the product of two sides of a triangle is equal to the product of the altitude on the third side and the diameter of the circumscribed circle.

● **PROBLEM** 529

Given: △ABC is inscribed in ⊙P; $\overline{AD}$ is an altitude of △ABC; $\overline{APE}$ is a diameter of ⊙P. Prove: AB · AC = AD · AE.

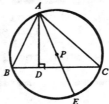

Solution: AB · AC = AD · AE can be rewritten as $\frac{AB}{AD} = \frac{AE}{AC}$. This is a proportion between similar triangles △ABD and △AEC. We can show △ABD ∿ △AEC by the A-A Similarity Theorem, and thus the proportion holds.

	Statements		Reasons
1.	△ABC is inscribed in ⊙P; $\overline{AD}$ is an altitude of △ABC; $\overline{APE}$ is a diameter of ⊙P.	1.	Given
2.	m∢ABC $\cong \frac{1}{2}m\overset{\frown}{AC}$ m∢AEC $\cong \frac{1}{2}m\overset{\frown}{AC}$	2.	The measure of an inscribed angle equals one-half the intercepted arc.
3.	m∢ABC $\cong$ m∢AEC or ∢ABC $\cong$ ∢AEC	3.	Transitivity and Definition of Congruent Angles.
4.	∢ADB is a right angle.	4.	The altitude to a given side is perpendicular to the side
5.	∢ACE is a right angle.	5.	An angle inscribed in a semicircle is a right angle.
6.	∢ADB $\cong$ ∢ACE	6.	All right angles are congruent.
7.	△ABD ∿ △AEC	7.	The A.A. Similarity Theorem

8. $\dfrac{AB}{AD} = \dfrac{AE}{AC}$

8. The sides of similar triangles are proportional.

9. $AB \cdot AC = AE \cdot AD$

9. The product of the means equals the product of the extremes.

● **PROBLEM** 530

Triangle ABC is inscribed in a circle, as seen in the figure. $\overline{BD}$ is the altitude to $\overline{AC}$, and $\overline{BR}$ is a diameter of circle O. Prove that $BA \times BC = BR \times BD$.

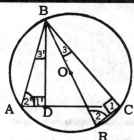

Solution: In order to prove that $BA \times BC = BR \times BD$, it is necessary to show that $BR : BA = BC : BD$ is a proportion. (The last statement is equivalent to the first statement because the product of the means = the product of the extremes.) This can be accomplished by showing $\triangle BRC \sim \triangle BAD$ by A.A. $\overset{\sim}{=}$ A.A. and employing the corresponding parts rule for similar triangles to obtain the proportion. The corresponding angles in the two triangles have been numbered with a "prime-no prime" notation. (Consult the diagram.)

To form $\triangle BRC$ (this must be done before the proof proceeds), chord $\overline{RC}$ had to be added to the figure.

STATEMENT	REASON
1. Draw $\overline{RC}$ to form $\triangle BRC$	1. One and only one straight line may be drawn between two points.
2. $\overline{BD} \perp \overline{AC}$	2. Definition of altitude.
3. $\overline{BR}$ is a diameter	3. Given.
4. $\angle 1$ is a right angle	4. An angle inscribed in a semi-circle is a right angle.
5. $\angle 1'$ is a right angle	5. Perpendicular lines intersect forming right angles.
6. $\angle 1 \overset{\sim}{=} \angle 1'$ (A. $\overset{\sim}{=}$ A.)	6. All right angles are congruent.
7. $m \angle 2 = \frac{1}{2} m \overset{\frown}{BC}$ and $m \angle 2' = \frac{1}{2} m \overset{\frown}{BC}$	7. The measure of an inscribed angle is equal to

	½ the measure of the intercepted arc.
8. m ∢ 2 = m ∢ 2'	8. Transitive Property of Equality.
9. ∢ 2 ≅ ∢ 2'	9. Definition of congruent angles.
10. ΔBRC ∿ ΔDAB	10. A.A. ≅ A.A.
11. $\dfrac{BR \ (opp. \ ∢ 1)}{BA \ (opp. \ ∢ 1')} = \dfrac{BC \ (opp. \ ∢ 2)}{BD \ (opp. \ ∢ 2')}$	11. Corresponding sides of similar triangles are in proportion.
12. BA × BC = BR × BD	12. In a proportion, the product of the means is equal to the product of the extremes.

● **PROBLEM** 531

Using Ptolemy's Theorem show that if a and b, with a ≥ b, are chords of two arcs of a circle of unit radius, then
$d = (a/2)(4-b^2)^{½} - (b/2)(4-a^2)^{½}$ is the chord of the difference of the two arcs.

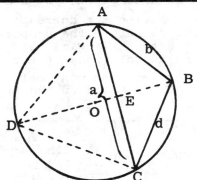

Solution: In circle O, chord AC = a, AB = b. DC is a diameter. The difference AC – AB = BC. We must find the length of chord BC. The hint in this problem is to use Ptolemy's Theorem which reads: The product of the diagonals of a cyclic quadrilateral equals the sum of the products of the opposite sides. Thus, we search for some convenient cyclic quadrilateral, each of whose diagonals and sides are known except for BC. Using Ptolemy's Theorem, BC can then be solved.

Consider the quadrilateral determined by diameter DC, cyclic quadrilateral ABCD. Side AB has known length b. Diagonal AC has known length a. Diagonal DB is a diameter and has known length 2 · radius or 2. Furthermore, because DB is a diameter, ΔDAB and ΔDCB are triangles inscribed in a semicircle and therefore are right triangles. Side DA is thus a leg of right ΔDAB and can be expressed in terms of known quantities using the Pythagorean Theorem. $DA^2 = DB^2 - AB^2$ or, substitu-

468

ting the given, $DA^2 = 2^2 - b^2$ or $DA = \sqrt{4-b^2}$.

Of the six elements (4 sides, 2 diagonals) used in Ptolemy's Theorem, we have solved for 4, one element is the unknown, and DC is an element that is also unknown but for which we are not solving. We recap our information up to this point:

$$AB = b, \quad AC = a, \quad DB = 2, \quad AD = \sqrt{4-b^2}.$$

(i) By Ptolemy's Theorem, $DB \times AC = AD \times BC + AB \times DC$.

(ii) Because of rt. $\triangle DCB$, $DC^2 + BC^2 = DB^2$.

Substituting our known values in (i) and (ii), we obtain:

(iii) $2 \cdot a = \sqrt{4-b^2} \cdot BC + b \cdot DC$

(iv) $DC^2 + BC^2 = 4$.

We then solve (iv) for DC in terms of BC. Then we substitute for DC in (iii) to obtain an expression totally in BC: From (iv), we have $DC = \sqrt{4-BC^2}$.

(v) $2 \cdot a = \sqrt{4-b^2} \cdot BC + b\sqrt{4-BC^2}$.

We now have an expression for BC, but we have a long algebraic road between (v) and a solution for BC. First, we eliminate all BC's in radicals by subtracting $\sqrt{4-b^2} \cdot BC$ from both sides and squaring.

(va) $(2a - \sqrt{4-b^2}BC)^2 = (\sqrt{4-b^2}BC + b\sqrt{4-BC^2} - \sqrt{4-b^2}BC)^2$

(vi) $(2a - \sqrt{4-b^2} \cdot BC)^2 = (b\sqrt{4-BC^2})^2$

(vii) $4a^2 - 4a \cdot BC \cdot \sqrt{4-b^2} + 4BC^2 - BC^2 \cdot b^2 = 4b^2 - b^2BC^2$

Gathering terms on the left side, we have:

(viii) $4BC^2 - (4a\sqrt{4-b^2})BC + (4a^2 - 4b^2) = 0$

We now divide each side by 4 and use the quadratic formula.

(ix) $BC^2 + (-a\sqrt{4-b^2})BC + (a^2-b^2) = 0$.

From the quadratic equation, we know that in the equation $qx^2 + px + r = 0$, $x = \dfrac{-p \pm \sqrt{p^2-4qr}}{2q}$. In this equation (ix), $q = 1$; $p = -a\sqrt{4-b^2}$; $r = (a^2-b^2)$.

(x) $BC = \dfrac{+a\sqrt{4-b^2} \pm \sqrt{(4a^2-a^2b^2) - 4(a^2-b^2)}}{2}$

(xi) $BC = \dfrac{a\sqrt{4-b^2} \pm \sqrt{4b^2 - a^2b^2}}{2}$

(xii) $BC = \frac{a}{2}\sqrt{4-b^2} \pm \frac{b}{2}\sqrt{4-a^2}$.

There would seem to be two answers, $BC = \frac{a}{2}\sqrt{4-b^2} + \frac{b}{2}\sqrt{4-a^2}$ and $BC = \frac{a}{2}\sqrt{4-b^2} - \frac{b}{2}\sqrt{4-a^2}$. But $\frac{a}{2}\sqrt{4-b^2} + \frac{b}{2}\sqrt{4-a^2}$ equals the sum of arcs a and b and therefore cannot be their difference. (Consult the previous problem.) Therefore the difference

$$d = \frac{a}{2}\sqrt{4-b^2} - \frac{b}{2}\sqrt{4-a^2}.$$

Note: To see where the extra answer crept in, remember that in step (vi), we squared each side, a procedure that sometimes leads to an extra root in the equation.

● **PROBLEM** 532

If a and b are the chords of two arcs of a circle of unit radius, show by Ptolemy's Theorem that the chord of the sum of the two arcs, S, equals in length $(a/2)(4-b^2)^{\frac{1}{2}} +$ $(b/2)(4-a^2)^{\frac{1}{2}}$.

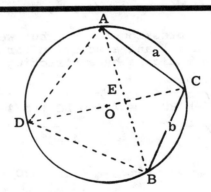

Solution: In circle O, chord AC = a, BC = b. $\overline{DC}$ is a diameter. We wish to find AB.

The hint of this problem is to use Ptolemy's Theorem which reads: The product of the diagonals of a cyclic quadrilateral equals the sum of the products of the opposite sides. Thus, we search for some convenient cyclic quadrilateral, whose diagonals or sides are known. The best quadrilateral is formed by the diameter $\overline{DC}$. Quadrilateral ACBD is thus formed.

Note sides $\overline{AC}$ and $\overline{BC}$ have known lengths a and b. DC, the diagonal, has the same length as the diameter, 2. Furthermore consider the triangles $\triangle ADC$, and $\triangle BCD$. They are triangles inscribed in semicircles. Therefore, they are right triangles and sides $\overline{DA}$ and $\overline{DB}$ can be found using the Pythagorean Theorem. Thus, we have 4 known sides and a known diagonal.

Using Ptolemy's Theorem, we can find the other diagonal, $\overline{AB}$, which is the desired length.
Let AC = a; BC = b, DC = diameter = 2 · radius = 2·1 = 2.

470

$AD^2 = DC^2 - AC^2 = 4 - a^2$; then $AD = \sqrt{4 - a^2}$

$DB^2 = DC^2 - BC^2 = 4 - b^2$; then $DB = \sqrt{4 - b^2}$.

According to Ptolemy's Theorem

$DC \cdot AB = DA \cdot BC + DB \cdot AC$. Then $AB = \dfrac{DA \cdot BC + DB \cdot AC}{DC}$

Substituting values, we obtain:

$$AB = \frac{(4-a^2)^{\frac{1}{2}} \cdot b + (4-b^2)^{\frac{1}{2}} \cdot a}{2}$$

$$= \frac{a}{2} \cdot (4-b^2)^{\frac{1}{2}} + \frac{b}{2} \cdot (4-a^2)^{\frac{1}{2}}$$

CHAPTER 28

PERIMETER

Basic Attacks and Strategies for Solving Problems in this Chapter. See pages 472 to 482 for step-by-step solutions to problems.

Most of the problems in this chapter are concerned with the perimeter of a polygon which is either inscribed in, or circumscribed about, a circle. This perimeter is often expressed as a function of the radius of the corresponding circle. Several of the polygons mentioned are regular polygons. The angles of a regular polygon are all congruent and the sides are also all congruent. This means that the perimeter of an n-sided regular polygon is na, where a is the length of one side.

For the regular hexagon $ABCDEF$ pictured below, O is the center of the circumscribed circle, and segment $\overline{OG}$ is called the apothem related to $\triangle ABO$:

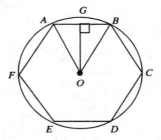

Several problems involve the interrelationships between the area and perimeter of a regular polygon.

Problem 540 illustrates the relationship between the perimeter of inscribed regular polygons and the circumference of the circle. The generalization illustrated in that problem is that as the number of sides is increased, the perimeter becomes closer to the circumference.

● **PROBLEM** 533

Show that the perimeter of a square which circumscribes a circle of radius of length r is 8r.

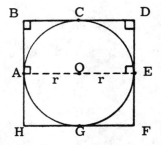

Solution: As indicated in the figure, points A, C, E, and G are the points at which segments HB, BD, DF, and HF are tangent to circle O. The perimeter of square BDFH is P = BD + DF + FH + HB. Since the figure is a square, BD = DF = FH = HB and, by substitution, P = 4BD. Hence, if we find BD, we can find P.

Since BH is tangent to O at point A, and DF is tangent to O at point E, we know that BH ⊥ OA and DF ⊥ OE.

Since OA and OE are perpendicular to parallel lines, OA || OE. Furthermore, since OA and OE share a common point O, OA and OE are coincident and AOE, or AE. It follows that AE ⊥ DF. Being that BDFH is a square, BD ⊥ DF. Hence,

BD || AE. Furthermore, as sides of a square BA || DE. Therefore, BAED is a rectangle, and BD = AE = 2r (since OA and OE are radii of circle O). As a result, P = 4BD = 8r.

● **PROBLEM** 534

Show that the perimeter of a square that is inscribed in a circle of radius length r is $4\sqrt{2}r$.

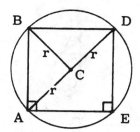

Solution: The figure shows the given situation. The perimeter of the square is AE + ED + BD + BA. Since AEDB is
a square, AE = ED = BD = BA and we may write

perimeter = 4BD.   (1)

If we can show that ⦨BCD is a right angle, then we can
apply Pythagorean Theorem to triangle BDC, and find BD
in terms of r, and, therefore, the perimeter of BDEA in
terms of r.

Since the square is inscribed in a circle, AC and
DC are radii of the circle. Hence, AC = DC. Furthermore,
BC = BC. Lastly, since BDEA is a square, BA = BD. By the
SSS Postulate, △BCD ≅ △BCA. This means that m⦨BCA =
m⦨BCD. Since m⦨BCA + m⦨BCD = 180°, m⦨BCA = 90° and
m⦨BCD = 90°. Hence △BCD is a right triangle.

Now, we may apply Pythagorean Theorem to △BCD.
The theorem states that

$BD^2 = BC^2 + CD^2$.

But, BC = r, CD = r. Hence,

$BD^2 = 2r^2$

$BD = r\sqrt{2}$.

Using this in expression (1)

perimeter = $(4\sqrt{2})r$.

● **PROBLEM** 535

Show that the perimeter of an equilateral triangle which
is inscribed in a circle with radius of length r is
$3\sqrt{3}r$.

Solution: The situation described is shown in the figure.
We will use several of the theorems of plane geometry to
show that △OBE is a 30 - 60 right triangle (that is, its
angles measure 30°, 60°, and 90°). Once that is done, we
will be able to find the length of $\overline{BC}$.

Since the triangle ABC is equilateral, $\overline{BC} \cong \overline{CA} \cong \overline{AB}$,
and the perimeter, P, of △ABC is P = AB + BC + CA = 3BC.

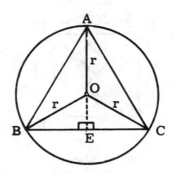

Focus attention on ΔABO and ΔCBO. $\overline{AB} \cong \overline{BC}$ since
ΔABC is equilateral. $\overline{AO} \cong \overline{CO}$, since $\overline{AO}$ and $\overline{CO}$ are radii of
the same circle. Lastly, $\overline{BO} \cong \overline{BO}$. Hence, by the SSS Postu-
late, ΔABO $\cong$ ΔCBO. It follows, therefore, that ∢ABO $\cong$ ∢OBC.
Since ΔABC is equilateral, m∢ABO + m∢OBC = 60°. Hence,
m∢OBC = 30°. Constructing the perpendicular $\overline{OE}$ to $\overline{BC}$ at
point E, we realize that m∢OEB = 90° and m∢OEC = 90°.
Applying the fact that the sum of the angles of a triangle
is 180° to triangles OBE and OCE, we conclude that m∢BOE
= 60° and m∢COE = 60°. We have, therefore, shown that
ΔOBE and ΔOCE are 30° - 60° - 90° triangles.

We next show that ΔOCE $\cong$ ΔOBE. First, OB = OC, since
they are both radii of the same circle. Also, ∢EBO $\cong$ ∢ECO,
and ∢BOE $\cong$ ∢COE, from what was said above. By the ASA
Postulate, ΔOCE $\cong$ ΔOBE. Hence, BE = CE. Since ΔBOE is a
30° - 60° - 90° triangle, BE (the long side opposite
hypotenuse $\overline{OB}$) has length $\sqrt{3}/2$ times the length of the
hypotenuse, r. Therefore, BE = $(\sqrt{3}/2)$r From the discussion
above, CE = BE = $\sqrt{3}/2$ r. Furthermore, BC = BE + EC (see
figure); hence, BC = $\sqrt{3}$r. The perimeter of ΔABC is then
P = 3BC = 3$\sqrt{3}$r.

● **PROBLEM** 536

Find, in radical form, the length of the radius of a circle circumscribed about an equilateral triangle, the length of whose side is
24.

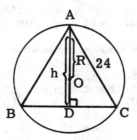

Solution: By a theorem, we know that the length of the radius of a
circle circumscribed about an equilateral triangle is equal to two-

thirds of the length of the altitude of the triangle. Therefore, we must calculate the length of altitude $\overline{AD}$, as shown in the figure.

$\overline{AD} \perp \overline{BC}$, and bisects $\overline{BC}$. Therefore, $\triangle ADB$ is a right triangle, and the Pythagorean Theorem can be invoked to determine the unknown altitude, h,

$$(AB)^2 = (BD)^2 + h^2 .$$

But $AB = 24$, and $BD = \frac{1}{2}BC = \frac{1}{2}(24) = 12$. By substitution,

$$(24)^2 = (12)^2 + h^2$$
$$576 = 144 + h^2$$
$$432 = h^2$$

$$h = \sqrt{432} = \sqrt{144}\sqrt{3} = 12\sqrt{3} .$$

Therefore, since the radius length, $R = \frac{2}{3}h$, then, by substitution,

$$R = \frac{2}{3}(12\sqrt{3}) = 8\sqrt{3} .$$

Therefore, length of the radius is $8\sqrt{3}$.

● **PROBLEM** 537

Show that for a triangle of area A, and perimeter P, the radius of the inscribed circle, r, equals 2A/P.

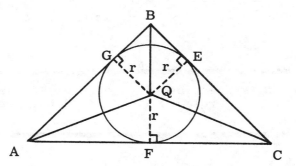

Solution: Let Q be the center of the inscribed circle. Triangle ABC can be divided into three triangles AQB, AQC, and BQC. Therefore, the area of $\triangle ABC$ equals the sum of the areas of the three triangles. We show that the radii of the inscribed circle are altitudes of these triangles, and, as such, the radius can be related to the area of the large triangle $\triangle ABC$.

Given: $\triangle ABC$ with perimeter P and area A; and r, radius of inscribed circle ⊙Q.

Prove: r = 2A/P .

Statements	Reasons
1. $\triangle ABC$ with perimeter P and area A; r, radius of the inscribed circle ⊙Q.	1. Given.
2. Let G,E, and F be the points of tangency with sides $\overline{AB}$, $\overline{BC}$, and $\overline{CA}$ respectively.	2. A line that is tangent to a circle intersects the circle in exactly one point.
3. $\overline{GQ}$, $\overline{QE}$, and $\overline{QF}$ are radii of the inscribed circle.	3. A line segment whose endpoints are the center of the circle and a point on the circle is a radius of the circle.

4. $\overline{GQ} \perp \overline{AB}$, $\overline{QE} \perp \overline{BC}$, $\overline{QF} \perp \overline{AC}$.

5. $\overline{GQ}$ is an altitude of $\triangle ABQ$.
   $\overline{QE}$ is an altitude of $\triangle CQB$
   $\overline{QF}$ is an altitude of $\triangle AQC$.

6. Area of $\triangle ABQ = \frac{1}{2}$ AB·GQ
   Area of $\triangle QBC = \frac{1}{2}$ QE·BC
   Area of $\triangle AQC = \frac{1}{2}$ QF·AC

7. Area of $\triangle ABC$ = Area of $\triangle ABQ$ +
   Area of $\triangle QBC$ + Area of $\triangle AQC$.

8. Area of $\triangle ABC = \frac{1}{2}$ AB·GQ + $\frac{1}{2}$QE·BC
   +$\frac{1}{2}$ QF·AC .

9. Area of $\triangle ABC = \frac{1}{2}$r·AB+$\frac{1}{2}$r·BC +
   $\frac{1}{2}$r·AC .

10. Area of $\triangle ABC = \frac{1}{2}$r·(AB+BC+CA).

11. P = AB+BC+CA.

12. A = $\frac{1}{2}$r·P .

13. r = 2A/P .

4. The radii drawn to the point of tangency are perpendicular to the tangent.

5. Definition of an altitude.

6. The area of a triangle equals one-half the base times the altitude to the base.

7. The area of a triangle divided into several non-overlapping parts equals the sum of the areas of the parts.

8. Substitution Postulate.

9. Substitution Postulate.

10. Distributive Property.

11. The perimeter of a polygon equals the sum of the lengths of the side.

12. Substitution Postulate.

13. Multiplication and Division Postulate.

● **PROBLEM** 538

A regular hexagon is inscribed in a circle. If the length of the radius of the circle is 7 in., find the perimeter of the hexagon.

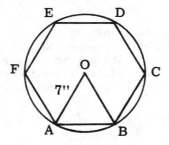

Solution: The sides of a central angle of the inscribed polygon form a triangle with the side whose endpoints they intercept. Since all 6 sides of a regular hexagon are equal, we can determine the perimeter by finding the length of one side of the polygon and multiplying by 6.

The measure of a central angle of a regular hexagon measures 360/6, or 60 degrees. In $\triangle OAB$, ∡AOB measures 60°.

$\overline{OA} \cong \overline{OB}$ because they are radii of a circle and, therefore, the angles opposite these sides in $\triangle OAB$ must also be congruent. Therefore, m∡OAB = m∡OBA. These measures must sum to 180° - m∡AOB, or 120°. We conclude that since the two measures are equal, m∡OAB = m∡OBA = 60°. Therefore, ΔOAB is equiangular and equilateral.

Since the length of $\overline{OA}$ is 7 in., the length of side $\overline{AB}$ of the triangle and the regular hexagon is also 7 in.

Therefore, the perimeter of regular hexagon ABCDEF is 6 × AB = 6 × 7 in. = 42 inches.

● **PROBLEM** 539

Prove that the perimeter of a regular polygon of n sides which is inscribed in a circle of radius r is given by P = 2nr sin(π/n).

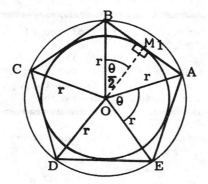

Solution: The figure shows the situation for n = 5. In general, the n-sided polygon can be considered to be composed of n triangles (such as ΔAOB, ΔBOC, etc.) Note that each of these n triangles is isosceles.

The perimeter of the polygon is

$$P = AB + BC + \ldots \ .$$

But, since the polygon is regular, each side is equal, and, since there are n sides, we may write

$$P = n(AB) \ . \tag{1}$$

To find AB, we inscribe a circle in the polygon. Then, by definition of inscription, $\overline{AB} \perp \overline{OM}$, at M_1 . We now prove that ΔOM_1B ≅ ΔOM_1A . $\overline{OB} \cong \overline{OA}$, because they are both radii of the large circle, $\overline{OM_1} \cong \overline{OM_1}$. Both ΔOM_1B and ΔOM_1A are, therefore, right triangles with 2 pairs of congruent sides. By Pythagorean Theorem, the third pair of sides are congruent and $\Delta OM_1B \cong \Delta OM_1A$ by the S.S.S. (side-side-side) Postulate. It follows that

$$\overline{M_1B} \cong \overline{M_1A}$$

and

$$M_1A = BM_1 \ .$$

But $AB = BM_1 + M_1A$, as the figure shows. Hence,

$$AB = 2M_1B \ . \tag{2}$$

But, by definition of the sine of an angle,

$$\sin \frac{\theta}{2} = \frac{M_1B}{r}$$

477

or
$$M_1B = r \sin \frac{\theta}{2} \qquad (3)$$

Also,
$$n\theta = 2\pi$$

or
$$\frac{\theta}{2} = \frac{\pi}{n} \qquad (4)$$

Using (4) in (3),
$$M_1B = r \sin \frac{\pi}{n} \qquad (5)$$

Inserting (5) in (2),
$$AB = 2r \sin \frac{\pi}{n} \qquad (6)$$

Substituting (6) in (1),
$$P = 2nr \sin \frac{\pi}{n} .$$

● **PROBLEM** 540

Find the perimeter of a regular polygon of n sides inscribed in a circle of radius $r = 10$ when n equals:
   a) 3; b) 4; c) 6; d) 9; e) 12; f) 18; g) 36;
   h) Find the circumference of the circle using $\pi = 3.14$;
   i) Comparing the answers to parts (a) through (g) with the answer to (h), what conclusion do you reach?

Solution: The perimeter, P, of an n-sided polygon inscribed in a circle of radius r is given by

$$P = 2nr \sin \frac{\pi}{n}$$

a) $n = 3,$ $P = (2)(3)(10) \sin \frac{\pi}{3} = 51.96 .$

b) $n = 4,$ $P = (2)(4)(10) \sin \frac{\pi}{4} = 56.57 .$

c) $n = 6,$ $P = (2)(6)(10) \sin \frac{\pi}{6} = 60.00 .$

d) $n = 9,$ $P = (2)(9)(10) \sin \frac{\pi}{9} = 61.56 .$

e) $n = 12,$ $P = (2)(12)(10) \sin \frac{\pi}{12} = 62.12 .$

f) $n = 18,$ $P = (2)(18)(10) \sin \frac{\pi}{18} = 62.51 .$

g) $n = 36,$ $P = (2)(36)(10) \sin \frac{\pi}{36} = 62.78 .$

h) The circumference, C, of a circle of radius r is given by

$$C = 2\pi r .$$
Hence, for our circle, $C = (2)(\pi)(10) = 62.80 .$

i) We conclude that as n gets larger, the circumference of the polygon inscribed in the circle approaches the actual circumference of the circle. (Compare parts (g) and (h). Note that we have rounded off to 2 decimal places, so that the perimeter of a circle of radius 10 is not exactly the same as the perimeter of a 36-sided polygon inscribed in the circle.)

● **PROBLEM** 541

In the figure, let ABCDE be a regular pentagon touching the circle at points M_1, M_2, M_3, M_4, and M_5. Assume that $OA = OB = OC = OD = OE$. (a) Prove that $\triangle AM_1O \cong \triangle BM_1O$.

(b) Prove that $\triangle AOB \cong \triangle BOC \cong \ldots \triangle EOA$. (c) Show that $\theta/2 = 36°$. (d) Prove that $AM_1 = r \tan 36°$. (e) Prove that the perimeter equals $10\, r \tan 36°$.

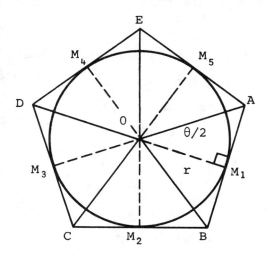

Solution: (a) We will prove that $\triangle AM_1O \cong \triangle BM_1O$ by using the SSS Postulate.

First, note that $\overline{OA} \cong \overline{OB}$, as we are told in the statement of the problem. Also, $\overline{OM_1} \cong \overline{OM_1}$. Since radii drawn to the point of tangency are perpendicular to the tangent segments, $OM_1 \perp AB$. Therefore, both $\triangle AM_1O$ and $\triangle BM_1O$ are right triangles, with 2 sides congruent. Hence, their third sides must also be congruent, by Pythagorean Theorem. Therefore, $\triangle AM_1O \cong \triangle BM_1O$, by the SSS Postulate.

(b) We will prove $\triangle AOB \cong \triangle BOC \cong \ldots \triangle EOA$ by using the properties of a regular polygon and the SSS Postulate.

Focus attention on $\triangle AOB$ and $\triangle BOC$. Since ABCDE is a regular polygon, $\overline{BC} \cong \overline{AB}$. By the statement of the problem, $\overline{OA} \cong \overline{OC}$. Lastly, $\overline{OB} \cong \overline{OB}$. Hence, $\triangle AOB \cong \triangle COB$, by the SSS Postulate. We may proceed in a similar matter, proving every adjacent set of triangles congruent. As a result, $\triangle AOB \cong \triangle BOC \cong \ldots \triangle EOA$.

(c) First, recall from part (a) that $\triangle AM_1O \cong \triangle BM_1O$. Hence,

$\qquad m\sphericalangle AOM_1 = m\sphericalangle BOM_1.$ \hfill (1)

But, $m\sphericalangle AOM_1 = \theta/2$ \hfill (2)

$\qquad$ Using (2) in (1),

$\qquad m\sphericalangle BOM_1 = \theta/2.$ \hfill (3)

Looking at the figure,

$$m \angle BOA = m \angle BOM_1 + m \angle AOM_1. \qquad (4)$$

Using (2) and (3) in (4),

$$m \angle BOA = \theta. \qquad (5)$$

It is also observed, from the figure, that

$$m \angle BOA + m \angle AOE + \ldots m \angle COB = 360°. \qquad (6)$$

Since ABCDE is a regular polygon,

$$m \angle BOA = m \angle AOE = \ldots m \angle COB. \qquad (7)$$

Using (7) in (6),

$$5m \angle BOA = 360°$$

or $m \angle BOA = 72°. \qquad (8)$

Using (5) in (8),

$$\theta = 72° \qquad \text{or} \qquad \frac{\theta}{2} = 36°.$$

(d) The tangent of an angle, α, is defined in terms of the right triangle of which α is one acute angle. It is the ratio of the length of the side of the right triangle opposite α, to the length of the side of the right triangle adjacent to α. For instance, in the figure,

$$\tan \frac{\theta}{2} = \frac{AM_1}{OM_1} = \frac{AM_1}{r}$$

Multiplying both sides by r, and recalling that $\theta/2 = 36°$, we obtain

$$AM_1 = r \tan 36°.$$

(e) The perimeter of ABCDE is

$$P = AB + BC + CD + DE + EA. \qquad (9)$$

But ABCDE is a regular polygon. Hence, AB = BC = ... EA, and (9) becomes

$$P = 5AB. \qquad (10)$$

From part (a), $\triangle AOM_1 \cong \triangle BOM_1$, which means that $AM_1 = M_1B$. Since $AB = AM_1 + M_1B$, (10) becomes

$$P = 5(AM_1 + M_1B) = 5(2AM_1) = 10 \ AM_1. \qquad (11)$$

From part (d), $AM_1 = r \tan 36°$, and (11) becomes

$$P = 10 \ r \tan 36°.$$

Show that the perimeter of an n-sided polygon circumscribed about a circle whose radius has length r is 2nr tan π/n .

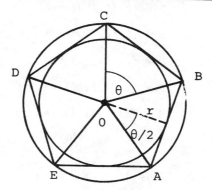

Solution: The case for n = 5 is illustrated in the figure. As shown, we may consider our 5-sided polygon to be composed of 5 triangles. Since the pentagon is regular, all its sides are equal. That is,

$$AB = BC = CD = DE = EA . \qquad (1)$$

The perimeter of the pentagon is

$$P = AB + BC + CD + DE + EA . \qquad (2)$$

Substituting (1) in (2), we obtain

$$P = 5(AB) . \qquad (3)$$

In the n-sided case, (3) is altered to

$$P = n(AB) \qquad (4)$$

where AB is the length of one side of the regular, circumscribed polygon. Our problem, therefore, is to calculate AB .

To do this, we circumscribe a circle about the polygon. By definition, the center of a polygon is the center of the circles inscribed in and circumscribed about the polygon. Since $\overline{OB}$ and $\overline{OA}$ are radii of the larger circle, $\overline{OA} \cong \overline{OB}$. Hence, $\triangle AOB$ is isosceles, and, for this type of triangle, we may write

$$AB = 2r \tan \frac{\theta}{2} .$$

Since nθ = 2π , θ/2 = π/n , and

$$AB = 2r \tan \frac{\pi}{n} .$$

Using this equation in (4) yields

$$P = 2nr \tan \frac{\pi}{n} .$$

Find the perimeter of a regular polygon of n sides which is circumscribed about a circle of radius length 10 when n equals a) 3; b) 4; c) 6; d) 9; e) 12; f) 18; g) 36.

Solution: The perimeter, P, of an n-sided polygon circumscribed about a circle of radius r is given by

$$P = 2nr \tan \frac{\pi}{n} \ .$$

a) If n = 3, $\qquad P = (2)(3)(10) \tan \frac{\pi}{3} = 103.92$

b) If n = 4, $\qquad P = (2)(4)(10) \tan \frac{\pi}{4} = 80.00$

c) If n = 6, $\qquad P = (2)(6)(10) \tan \frac{\pi}{6} = 69.28$

d) If n = 9, $\qquad P = (2)(9)(10) \tan \frac{\pi}{9} = 65.51$

e) If n = 12, $\qquad P = (2)(12)(10) \tan \frac{\pi}{12} = 64.31$

f) If n = 18, $\qquad P = (2)(18)(10) \tan \frac{\pi}{18} = 63.48$

g) If n = 36, $\qquad P = (2)(36)(10) \tan \frac{\pi}{36} = 62.99$

● **PROBLEM** 544

> Show that the perimeter of an equilateral triangle which circumscribes a circle of radius length r is $6\sqrt{3}\ r$.

Solution: The perimeter, P, of a polygon of n-sides inscribed in a circle of radius length r is

$$P = 2nr \tan \frac{\pi}{n} \ .$$

If n = 3, as in our problem,

$$P = 6r \tan \frac{\pi}{3}$$
$$P = 6r \tan 60$$
$$P = 6r \ \frac{\sqrt{3}/2}{\frac{1}{2}} = 6\sqrt{3}\ r \ .$$

CHAPTER 29

AREAS

Basic Attacks and Strategies for Solving Problems in this Chapter. See pages 483 to 505 for step-by-step solutions to problems.

Many of the problems in this chapter involve finding the areas of regular polygons. Four regular polygons are pictured below. Notice that in each case, all the angles are congruent and all the sides are congruent:

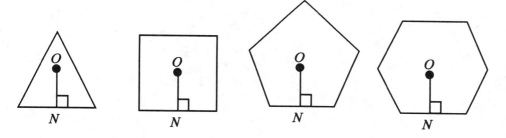

In each case, O is the center of the polygon and $\overline{ON}$ is perpendicular to the bottom side of the polygon. $\overline{ON}$ is called an **apothem**. Several problems in this chapter give a relationship between the length of this apothem and the area of the polygon.

The last six problems in the chapter require the use of trigonometric functions; a short review of the introduction to Chapter 22 might be desirable before those problems are addressed.

Step-by-Step Solutions to
Problems in this Chapter,
"Areas"

SOLVING FOR THE AREA USING THE PYTHAGOREAN THEOREM

● **PROBLEM** 545

Draw a circle with a circumscribed square. If the radius
length of the circle is r, prove that the area of the
square region is $4r^2$.

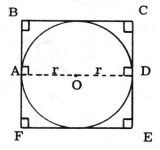

Solution: The formula for the area of a square is $A = s^2$
where s is the length of the side. Since all sides of a
square are congruent, let s = BC. We first show that the
length of side $\overline{BC}$ equals the diameter, d, of the inscribed circle. Since d = 2r, $A = d^2 = (2r)^2 = 4r^2$.

To show that s = d, note that, because the square
circumscribes the circle, the sides $\overline{BF}$ and $\overline{CE}$ are tangent
to the circle at points A and D. Therefore, the radii
drawn to the points of tangency are perpendicular to the
sides of the square. Thus, $\overline{AO} \perp \overline{BF}$ and $\overline{DO} \perp \overline{CE}$. We now
show (1) $\overline{AO}$ and $\overline{OD}$ are collinear; and (2) side $\overline{BC} \cong \overline{AD}$.
Note that $\overline{BF}$ and $\overline{CE}$ are parallel (opposite sides of
a parallelogram are parallel).

Because $\overline{AO}$ and $\overline{OD}$ are perpendicular to parallel lines,
$\overline{AO} \parallel \overline{OD}$. Because $\overline{AO}$ and $\overline{OD}$ share a common point O, $\overleftrightarrow{AO}$ and
$\overleftrightarrow{OD}$ must be coincident. Thus, $\overline{AOD}$.

Since adjacent sides of a square are perpendicular to each other, $\overline{BC} \perp \overline{BF}$. Due to the fact that $\overline{BC} \perp \overline{BF}$ and and $\overline{AOD} \perp BF$, $\overline{BC} \parallel \overline{AOD}$. Because $\overline{AB} \parallel \overline{CD}$ (opposite sides of a parallelogram are parallel), quadrilateral ABCD has two pairs of opposite sides parallel. Thus, ABCD is a parallelogram and $\overline{BC} \stackrel{\sim}{=} \overline{AD}$ because opposite sides of a parallelogram are congruent. Since AD = AO + OD = r + r = 2r, BC = 2r.

The area of square BFEC = $(BC)^2 = (2r)^2 = 4r^2$.

● **PROBLEM** 546

A rectangle is inscribed in a circle whose radius is 5 inches. The base of the rectangle is 8 inches. Find the area of the rectangle. (See figure.)

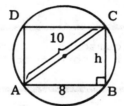

Solution: The diagonal of the rectangle is a diameter of the circle and as such, each of the two triangles shown in the figure is inscribed in a semi-circle. Therefore, they are right triangles.

Since △BAC is a right triangle, we can use the Pythagorean Theorem to determine the altitude, h, of ABCD, as indicated in the figure. We will need this to calculate the area of the rectangle. Therefore,

$$h^2 + (8 \text{ in.})^2 = (10 \text{ in.})^2$$
$$h^2 + 64 \text{ in.}^2 = 100 \text{ in.}^2$$
$$h^2 = 36 \text{ in.}^2 \quad \text{and}$$
$$h = 6 \text{ in.}$$

Now, Area = bh. Let b = 8 in., h = 6 in. By substitution,

$$A = (8 \times 6) \text{in.}^2 = 48 \text{ in.}^2$$

Therefore, the area of the inscribed rectangle is 48 sq. in.

● **PROBLEM** 547

Draw a circle with an inscribed square. If the radius length of the circle is r, prove that the area of the square region is $2r^2$.

Solution: We will first show that ∡AOB is a right angle (see figure), and then use Pythagorean Theorem to find AB.

$\overline{OB}$, $\overline{OD}$ and $\overline{OA}$ are all radii of circle O, and, hence, all have the same length, r. Therefore, OB = OD, and OA = OA. In addition, AD = AB since ABCD is a square. By the

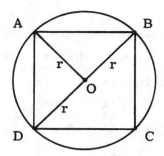

SSS Postulate, △AOD $\cong$ △AOB. As a result, ⦨DAO $\cong$ ⦨BAO.
But m⦨DAO + m⦨BAO = 90°, since ABCD is a square. This
means that m⦨DAO = m⦨BAO = 45°.

Now, focus attention on △AOB. Because AO = BO,
△AOB is isosceles. Therefore, its base angles are congruent and m⦨OAB = m⦨DBA. Since m⦨OAB = 45°, m⦨DBA = 45°.
Using the fact that the sum of the angles of a triangle
is 180°, m⦨AOB = 90°.

Using the Pythagorean Theorem on △AOB,

$(AO)^2 + (OB)^2 = (AB)^2$.

Because AO = OB = r, $(AB)^2 = 2r^2$ or $AB = r\sqrt{2}$.

The area of a square is equal to the square of the
length of one of its sides. For our square, the area is
$(AB)^2 = 2r^2$.

● **PROBLEM** 548

Find, in terms of π, the area of the shaded region in the diagram.

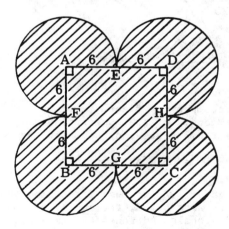

<u>Solution</u>: The shaded region consists of square ABCD, whose side is 12,
and four major sectors whose areas are equal because their radii and
central angles are equal. We will calculate the area of the square and
the major sectors, and then sum them, to determine the area of the

shaded region.

Area of the square = (AB)² = (12)² = 144.

Each sector spans 270°, since the corners of the square cut off arc lengths that measure 90° from each circle. Since the area of a sector is proportional to the central angle, the area of each sector is

$$\frac{270°}{360°},$$

or 3/4 of the area of each entire circle.

$$\text{Area of each sector} = \frac{3}{4}\pi r^2$$

$$= \frac{3}{4}\pi(6)^2$$

$$= \frac{3}{4}(36\pi)$$

$$= 27\pi.$$

Area of shaded region = area of square + area of 4 sectors

$$= 144 + 4(27\pi)$$

$$= 144 + 108\pi$$

Therefore, area of shaded region = 144 + 108π.

● **PROBLEM** 549

Find the area of a circle inscribed in a rhombus whose perimeter is 100 in. and whose longer diagonal is 40 in.

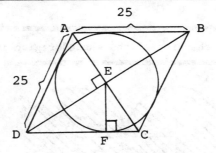

Solution: In the accompanying figure, rhombus ABCD circumscribes circle E. DB, the length of the longer diagonal, equals 40. Since all four sides of a rhombus are congruent and the perimeter equals 100, then

$$AD = DC = CB = AB = \frac{100}{4} = 25.$$

To find the area of ⊙E, we must first find the radius. To find the radius, we (1) show point E, the center of the circle is the intersection of the diagonals and thus $\overline{EF}$ is a radius of ⊙E;

(2) Show ΔDEC is a right triangle;
(3) We take advantage of the proportions involving the altitude of a right triangle to find EF, since $\overline{EF}$ is the altitude drawn to

the hypotenuse.

To show point E is the intersection of the diagonals, we recall that E, the center of the circle, is equidistant from all four sides. Consider the intersection of the diagonals E . In a rhombus, the diagonals bisect the angles of the rhombus. Therefore, E is on the angle bisector of each vertex angle. Because E is on the bisector of ∡ DAB, E is equidistant from sides $\overline{AD}$ and $\overline{AB}$. Because E is on the bisector of ∡ ABC, E is equidistant from sides $\overline{AB}$ and $\overline{BC}$. Because E is on the bisector of ∡ BCD, E is equidistant from sides $\overline{BC}$ and $\overline{DC}$. Combining these three facts and using the transitive property, we have that E is equidistant from all four sides. Therefore, E, the center of the inscribed circle, is the same point as E , the intersection of the diagonals.

To show the second step, that ΔDEC is a right triangle, we remember that the diagonals of a rhombus are perpendicular to each other. Therefore, ∡ DEC is a right angle and ΔDEC is a right triangle.

In right ΔDEC, we have the length of the hypotenuse, DC = 25. Because the diagonals of a parallelogram bisect each other, we have that the measure of the leg DE = ½DB = ½(40) = 20. By the Pythagorean Theorem,
$$EC^2 = DC^2 - DE^2 = 25^2 - 20^2 = 225;$$
or
$$EC = 15.$$

We wish to find the radius EF. EF, as the altitude to the hypotenuse, is the mean proportional of the hypotenuse segments, DF and FC. To find DF and FC, remember that the adjacent leg is the mean proportional of the segment and the hypotenuse. Thus,
$$\frac{DF}{DE} = \frac{DE}{DC} \quad \text{or} \quad DF = \frac{DE^2}{DC} = \frac{20^2}{25} = 16.$$

Then, DF = 16 and FC = DC - DF = 25 - 16 = 9.

Since altitude $\overline{EF}$ is the mean proportional of the base segments, we obtain
$$\frac{DF}{EF} = \frac{EF}{FC} \quad \text{or} \quad EF^2 = DF \cdot FC = 16 \cdot 9 = 144.$$

Then EF = $\sqrt{144}$ = 12.

Since the radius of the inscribed circle, EF, equals 12, the area of the circle = $\pi r^2 = \pi(12)^2 = 144\pi$.

● **PROBLEM** 550

A circle is inscribed in an equilateral triangle, whose side is 12. Find, to the nearest integer, the difference between the area of the triangle and the area of the circle. (Use π = 3.14 and $\sqrt{3}$ = 1.73.)

<u>Solution</u>: By determining the area of the triangle, and then subtracting the area of the inner circle, the required area can be found.

ΔABC is an equilateral triangle and its area is equal to
$$\frac{s^2}{4} \sqrt{3} \, ,$$
where s is the length of a side of the triangle. Therefore, the area of
$$\Delta ABC = \frac{s^2}{4} \sqrt{3} = \frac{(12)^2}{4} \sqrt{3} = 36\sqrt{3} \, .$$

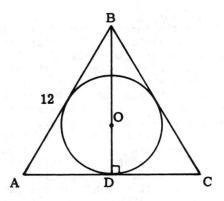

Using $\sqrt{3} = 1.73$, area of

$$\triangle ABC = 36(1.73) = 62.28$$

In order to calculate the area of the circle, remember that its radius equals $\frac{1}{3}$ the altitude of the circumscribing equilateral triangle. We must first determine the length of the altitude $\overline{BD}$. Let the length of $\overline{BD}$ be h. Since altitude $\overline{BD} \perp \overline{AC}$ and bisects it, $\triangle ABD$ is a right triangle and h can be found by applying the Pythagorean Theorem.

$$(AB)^2 = (AD)^2 + h^2 .$$

But AB = 12, and AD = $\frac{1}{2}$AC = $\frac{1}{2}$(12) = 6. By substitution,

$$(12)^2 = (6)^2 + h^2$$
$$144 = 36 + h^2$$
$$108 = h^2$$
$$h = \sqrt{108} = \sqrt{36}\sqrt{3} = 6\sqrt{3} .$$

In an equilateral triangle, the radius of the inscribed circle equals one third the altitude. Therefore, the length of radius $\overline{OD} = r = \frac{1}{3}(h)$. By substitution, $r = \frac{1}{3}(6\sqrt{3}) = 2\sqrt{3}$.

$$\text{Area of circle } O = \pi r^2 = \pi(2\sqrt{3})^2 = 12\pi .$$

Using $\pi = 3.14$,
$$\text{Area of circle } O = 12(3.14) = 37.68.$$
Difference in area = area of triangle ABC - area of circle O

$$= 62.28 - 37.68 = 24.6.$$

Therefore, the difference in area, to the nearest integer, is 25.

● **PROBLEM** 551

A side of a regular hexagon is 8 inches in length.
(a) Find the length of the apothem of the hexagon.
(b) Find the area of the hexagon. [Answers may be left in radical form.]

Solution: (a) In the accompanying figure, a representative regular hexagon has been drawn. The triangle formed by the sides of central angle AOB and $\overline{AB}$ is equilateral. This is the case since $\overline{OA} \cong \overline{OB}$ (as radii of the polygon) and the included angle measures 60° (as the central angle of a hexagon). We now know OA = AB = OB = 8 in.

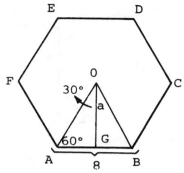

The apothem, a, is perpendicular to $\overline{AB}$ and forms a right triangle, $\triangle OAG$. Pythagoras' Theorem tells us that $(OA)^2 = (OG)^2 + (AG)^2$.

We found that OA = 8 inches, and, since the apothem bisects any side of a regular hexagon, AG = 4 in. Then, by substitution,

$(8 \text{ in})^2 = a^2 + (4 \text{ in.})^2$

$64 \text{ in.}^2 = a^2 + 16 \text{ in.}^2$

$a^2 = 48 \text{ in.}^2$

$a = \sqrt{48} \text{ in} = \sqrt{16}\sqrt{3} \text{ in.} = 4\sqrt{3} \text{ in.}$

Therefore, the length of the apothem of a regular hexagon is $4\sqrt{3}$ in.

(b) The area of a regular polygon is given by the equation $A = \frac{1}{2} ap$, where a = the length of the apothem and p = the perimeter.

From part (a) we know, $a = 4\sqrt{3}$ in.

The perimeter of a regular hexagon = 6 × length of any side. By substitution, $p = 6 \times 8 \text{ in.} = 48 \text{ in.}$

Therefore, the area $A = \frac{1}{2} (4\sqrt{3} \text{ in.}) (48 \text{ in.})$
$= 24 (4\sqrt{3}) \text{ in.}^2 = 96\sqrt{3} \text{ in.}^2$.

● **PROBLEM** 552

Consider a circle inscribed in a square. Let the circle have radius r. What is the area of the region outside the circle, but inside the square.

Solution: Let the area of the region be A(R), the area of the square A(S) and the area of the circle A(C). Then, by the Area-addition postulate, we may write

$A(R) = A(S) - A(C)$. (1)

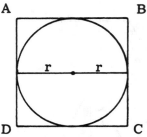

One side of the square has length 2r. Hence,

$$A(S) = (2r)^2 = 4r^2. \qquad (2)$$

The area of a circle of radius r is πr^2. Hence, $A(C) = \pi r^2$. Therefore, using (1), (2) and the last statement

$$A(R) = 4r^2 - \pi r^2 = (4 - \pi)r^2.$$

● **PROBLEM** 553

By how much does the area of the circumscribed circle exceed the area of the inscribed circle of a square of side 8.

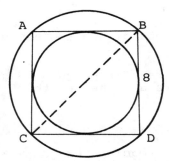

<u>Solution</u>:   The diameter of the inscribed circle equals the side of the square.   Therefore, the area equals $\pi r^2 = \pi(4)^2 = 16\pi$.
The diameter of the circumscribed circle is the diagonal of the square.   The diagonal of a square is the hypotenuse of the right triangle formed by any two sides of the square adjacent at a vertex not intersected by the diagonal.   Its length can be found by applying the Pythagorean Theorem, since we know the length of two sides of right $\triangle ABC$.

$$(AC)^2 + (AB)^2 = (BC)^2$$

AC = AB = 8 and accordingly, by substitution,

$$(8)^2 + (8)^2 = (BC)^2$$
$$2(64) = (BC)^2$$
$$BC = \sqrt{128} = \sqrt{64}\,\sqrt{2} = 8\sqrt{2}$$

The diagonal equals $8\sqrt{2}$, therefore, $r = 4\sqrt{2}$.   The area of the circle equals $\pi(4\sqrt{2})^2 = 32\pi$.   The area of the circumscribed circle exceeds the area of the inscribed circle by $32\pi - 16\pi = 16\pi$.   In general, for a given square, the area of the circumscribing circle is twice the area of the inscribed circle.

Show that the area of a regular polygon equals one-half the product of the lengths of the apothem and the perimeter.

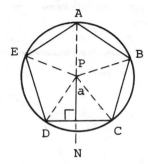

Solution: Every regular polygon can be circumscribed in some circle, and the line segment connecting the center of the circumscribing circle P to the vertices are radii and, therefore, all congruent. In the accompanying figure, for example PE = PA = PB = PC = PD. Since the polygon is regular, the sides are congruent or ED = DC = CB = BA = EA. By the SSS Postulate, all the triangles created by the radii and the sides of the polygon are congruent.

Therefore, the area of the n-sided polygon is the sum of n triangles, or n times the area of one triangle.

The area of a triangle = ½bh.

b = length of the side of the n-gon
h = length of the altitude to the side. By definition, this is the apothem, a.

Area of n-gon = n·½bh = ½(nb)a .

But nb is the perimeter of the n-gon. (Given a figure with n sides of length b, the perimeter equals b+b+b... = n·b). Therefore, by substitution,

Area of n-gon = ½ p·a

where p is the perimeter, and a is the length of the apothem.

SIDENOTE: This holds true even for the circle. Consider a circle of radius r as a regular polygon of infinitely many sides. Then the apothem is the radius r; the perimeter is the circumference $2\pi r$. Thus, the area of the circle, according to the formula is

$$\tfrac{1}{2}(r)(2\pi r) = \pi r^2 \ .$$

● **PROBLEM** 555

If the length of a side of a regular polygon of 12 sides is represented by s, and the length of its apothem is represented by a , find the area of the polygon, A, in terms of a and s.

Solution: In order to solve this problem, we must first determine the perimeter of the polygon, since the area is given by one-half the product of the perimeter and the length of the apothem. Recall that the apothem is the perpendicular distance from the center to any side.

Since, the regular polygon has 12 sides, the perimeter equals 12s.

Area = ½ap = ½a(12s) = 6as .

Therefore, the area of the 12 sided regular polygon is 6as .

● **PROBLEM** 556

Find the area of the regular polygon whose perimeter is 40 and whose apothem is 6.

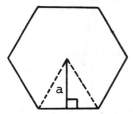

Solution: From the previous problem, we know that the area of the regular polygon equals one-half the product of the apothem and the perimeter.

Area = ½ a·p = ½(6)(40) = 120.

● **PROBLEM** 557

Find the area of a regular hexagon if one side has the length of 6.

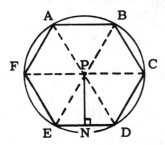

Solution: The area of a regular polygon equals one-half the product of the perimeter and the apothem. The perimeter of an n-gon equals nb where b is the length of the side. Therefore, perimeter of the hexagon equals 6(6) = 36. We find the apothem by (see the accompanying figure):

(1) Let P be the center of the hexagon; $\overline{ED}$ is the side of the triangle; then $\overline{PN}$, the perpendicular to $\overline{ED}$, is an apothem.
(2) $\overline{PN}$ is the altitude of triangle △PED .
(3) △PED is an equilateral triangle. (Sides $\overline{PE}$ and $\overline{PD}$ are radii of a regular polygon, and, thus, are congruent. The vertex angle ∡ EPD equals the central angle of the hexagon, 360/6 or 60° . △PED is thus an isosceles triangle with vertex angle of 60° . Therefore, △PED is equilateral.)
(4) Then, PE = PD = ED = 6.
(5) EN = ND = ½ED = 3 . (The altitude of an isosceles triangle bisects the base.)

(6) We thus have right triangle $\triangle PEN$ with hypotenuse = 6, one leg = 3, and the other leg unknown. Using Pythagorean Theorem, $PN^2 = PE^2 - EN^2 = 6^2 - 3^2 = 36 - 9 = 27$. Thus, $PN = \sqrt{27} = \sqrt{9} \cdot \sqrt{3} = 3\sqrt{3}$.

(7) $\overline{PN}$ is an apothem. Therefore, the length of the apothem of the hexagon must equal $3\sqrt{3}$.

With the apothem and the perimeter known, the area can be obtained.

$$\text{Area} = \tfrac{1}{2} \, a \cdot p = \tfrac{1}{2}(3\sqrt{3})(36) = 54\sqrt{3} \ .$$

● **PROBLEM** 558

The side of a regular pentagon is 20 inches in length. (a) Find, to the nearest tenth of an inch, the length of the apothem of the pentagon. (b) Using the result obtained in part (a), find, to the nearest ten square inches, the area of the pentagon.

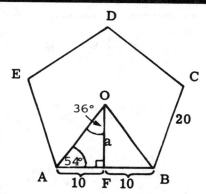

Solution: (a) When the central angle AOB is drawn, as shown in the figure, its rays, along with side $\overline{AB}$ of pentagon ABCDE form a triangle. The central-angle theorem states that the measure of a central angle of a regular polygon equals $360°$ divided by the number of sides of the polygon. Therefore, the measure of $\angle$ AOB is $360°/5$, or $72°$, where 5 is the number of sides in the pentagon.

By theorem, apothem $\overline{OF}$ bisects central angle AOF and pentagon side $\overline{AB}$. Since $\overline{OF} \perp \overline{AB}$, therefore, $\triangle OFA$ is a right triangle.

$$m \angle \text{AOF} = \tfrac{1}{2}(m \angle \text{AOB}) = \tfrac{1}{2}(72°) = 36° \ .$$

Therefore, $m \angle \text{FAO} = 180° - (m \angle \text{AOF} + m \angle \text{OFA})$.

By substitution, $m \angle \text{FAO} = 180° - (36° + 90°) = 54°$. We also know, $\overline{AF} = \tfrac{1}{2}\overline{AB} = \tfrac{1}{2}(20 \text{ in.}) = 10 \text{ in.}$

Since $\triangle OAF$ is a right triangle, knowing an angle measure and the length of its adjacent side, we can calculate the length of $\overline{OF}$ by using the tangent ratio.

$$\tan \angle \text{OAF} = \frac{\text{length of leg opposite } \angle \text{ OAF}}{\text{length of leg adjacent } \angle \text{ OAF}} \ .$$

Let a = length of leg opposite $\angle$ OAF (see figure.) Then, by substitution,

$$\tan 54° = \frac{a}{10 \text{ in.}} \ .$$

From a standard tangent table, we find $\tan 54° = 1.3764$. Therefore,

$$1.3764 = \frac{a}{10 \text{ in.}}$$

493

$$a = (10 \text{ in.})(1.3764) = 13.764 \text{ in.}$$

The length of apothem $\overline{OF}$ is 13.8 in.

(b) The area of a regular polygon is given by $A = \frac{1}{2}$(apothem length)(perimeter) $= \frac{1}{2}ap$. The perimeter, P, of pentagon ABCDE $= 5(20)$ in. $= 100$ in. Therefore, by substitution,

$$A = \frac{1}{2}(13.8)(100) \text{ in.}^2 = (50)(13.8) \text{ in.}^2 = 690 \text{ in.}^2 .$$

Therefore, the area of pentagon ABCDE is 690 sq. in.

● **PROBLEM** 559

Prove that the area bounded by a regular polygon of n sides which is inscribed in a circle with radius of length r is given by the formula

$$A = \frac{nr^2}{2} \sin \frac{2\pi}{n} .$$

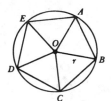

Solution: The figure shows the situation for the case where n = 5. Note that the polygon is composed of n triangles. Each of these triangles is isosceles (e.g. $\triangle AOB$, AO = BO). But the area, A, of an isosceles triangle, whose congruent sides have length r and include an angle θ , is

$$A = \frac{1}{2}r^2 \sin \theta ,$$

Each of the n triangles composing the polygon are congruent. Look at triangles AOB and BOC. $\overline{BO} \cong \overline{AO} \cong \overline{CO}$ because all are radii of the same circle. $\measuredangle AOB \cong \measuredangle BOC$ because all central angles of a regular polygon are congruent. By the S.A.S. (side-angle-side) Postulate, $\triangle AOB \cong \triangle BOC$. This procedure may be used repeatedly to prove that each of the n triangles are congruent to each other. Since they are all congruent, they all have the same area.

Since $n\theta = 2\pi$, then $\theta = \frac{2\pi}{n}$. Thus the total area of the polygon is $A = \frac{nr^2}{2} \sin \frac{2\pi}{n}$.

● **PROBLEM** 560

Find the area of a regular hexagon inscribed in a circle of radius r. Calculate the area explicitly when

a) r = 4, b) r = 9, c) r = 16, d) r = 25.

Solution: The area of a regular n-sided polygon inscribed in a circle of radius r is

$$A = \frac{nr^2}{2} \sin \frac{2\pi}{n} .$$

For a hexagon, n = 6 and

$$A = 3r^2 \sin \frac{\pi}{3} = 3r^2 \frac{\sqrt{3}}{2} \ .$$

a) If r = 4, $\qquad A = (3)(16)\left(\frac{\sqrt{3}}{2}\right) = 24\sqrt{3} \ .$

b) If r = 9, $\qquad A = (3)(81)\left(\frac{\sqrt{3}}{2}\right) = \frac{243}{2}\sqrt{3}$

c) If r = 16, $\qquad A = (3)(256)\left(\frac{\sqrt{3}}{2}\right) = \frac{768}{2}\sqrt{3}$

d) If r = 25, $\qquad A = (3)(625)\left(\frac{\sqrt{3}}{2}\right) = \frac{1875}{2}\sqrt{3} \ .$

● **PROBLEM** 561

Draw a circle circumscribed about a square of edge length
s. What is the area of the region outside the square but
inside the circle?

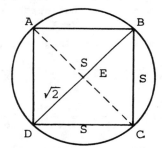

Solution: Let the area of the region inside the circle
but outside the square be A(R). Furthermore, let the area
of the circle be A(C) and the area of the square A(s).
From the figure, we have

(1) $\qquad$ A(R) = A(C) - A(s).

Since the square is of edge length s, A(s) = s^2.
To find A(C) = πr^2, we must first find the radius r.

We first show that the diameter of the circle equals
the diagonal of the square. Let E be the center of the
circumscribing circle. To show $\overline{DB}$ is a diameter, we must
show $\overline{DEB}$ is a straight line. We will show this by showing
that ∢DEA and ∢BEA form a straight angle. From an earlier
theorem, we know that the central angles of a regular
polygon are congruent. Therefore, m∢DEA = m∢AEB = m∢BEC
= m∢CED = $\frac{360°}{4}$ = 90°. m∢DEA + m∢BEA = 90° + 90° = 180°.
Since m∢DEA + m∢BEA = m∢DEB, m∢DEB = 180°, then ∢DEB is
a straight angle and D, E, and B are collinear. Thus
$\overline{DB}$, the diagonal, is also a diameter.

To find DB, note that ΔDBC is a right triangle.

Using the Pythagorean Theorem, we have $DB^2 = DC^2 + CB^2 = s^2 + s^2 = 2s^2$. Thus, $DB = \sqrt{2s^2} = \sqrt{2}s$.

The radius of the circle equals $\frac{1}{2}$ the diameter DB or $\frac{1}{2}(\sqrt{2}s)$. The area of the circle equals

$$\pi r^2 = \pi \left[\frac{\sqrt{2}}{2}\, s\right]^2 = \frac{\pi s^2}{2}.$$

Returning to equation (1), we obtain

$$A(R) = \frac{\pi s^2}{2} - s^2 = s^2 \left(\frac{\pi}{2} - 1\right)$$

● **PROBLEM** 562

Polygon WXYZ is similar to polygon W'X'Y'Z'. If WX = 3 and W'X' = 12, find the ratio of the areas of the two polygons.

<u>Solution</u>: The ratio of the areas of any two similar polygons is equal to the ratio of the squares of the lengths of any two corresponding sides.

$\overline{WX}$ and $\overline{W'X'}$ are corresponding sides and, therefore,

$$\frac{\text{Area of polygon WXYZ}}{\text{Area of polygon W'X'Y'Z'}} = \left(\frac{WX}{W'X'}\right)^2.$$

By substitution,
$$\left(\frac{WX}{W'X'}\right)^2 = \left(\frac{3}{12}\right)^2 = \left(\frac{1}{4}\right)^2 = \frac{1}{16}.$$

Therefore, the ratio of the areas of the two polygons is 1:16.

● **PROBLEM** 563

The lengths of two corresponding sides of two similar polygons are 4 and 6. If the area of the smaller polygon is 20, find the area of the larger polygon.

<u>Solution</u>: The ratio of the area of two similar polygons is equal to the square of the ratio of the lengths of two correspondings sides. Let A' = the larger polygon's area, then

$$\frac{20}{A'} = \frac{(4)^2}{(6)^2}$$

$$\frac{20}{A'} = \frac{16}{36}$$

$$16A' = 720$$

$$A' = \frac{720}{16}$$

$$= 45.$$

Therefore, the area of the larger polygon is 45.

The area of pentagon ABCDE is 18 sq. in.; the area of a similar pentagon A'B'C'D'E' is 32 sq. in. The diagonal AC is 6 in.; find the length of A'C' .

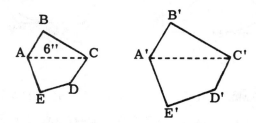

Solution: For two polygons to be similar, all corresponding angles are congruent, and all corresponding sides must be proportional. We will first show that the ratio of similitude of the diagonals, A'C'/AC, is the same as the ratio of similitude of the sides, A'B'/AB. Then, we find this ratio of similitude of the two polygons using the fact that the ratio of the area of similar polygons equals the square of the ratio of similitude.

To show that A'C'/AC = ratio of similitude, we show

$$\triangle ABC \sim \triangle A'B'C' .$$

Note $\angle B \cong \angle B'$ and $\dfrac{A'B'}{AB} = \dfrac{B'C'}{BC}$ because ABCDE and A'B'C'D'E' are similar polygons, By the S.A.S. Similarity Theorem, $\triangle ABC \sim \triangle A'B'C'$, and thus $\dfrac{A'C'}{AC} = \dfrac{A'B'}{AB}$.

Since $\dfrac{A'B'}{AB}$ is the ratio of similitude, $\dfrac{A'C'}{AC}$ = ratio of similitude.

To find the ratio of similitude, remember that

$$(\text{ratio of similitude})^2 = \frac{\text{area of A'B'C'D'E'}}{\text{area of ABCDE}} = \frac{32}{18} = \frac{16}{9}$$

$$\text{ratio of similitude} = \sqrt{\frac{16}{9}} = \frac{4}{3} .$$

Thus, $\dfrac{A'C'}{AC} = \dfrac{4}{3}$. Since AC = 6, then $\dfrac{A'C'}{6} = \dfrac{4}{3}$ or A'C' = $\dfrac{4}{3}$(6) = 8 in.

Prove that the ratio of the areas of any polygon circumscribed about a given circle is equal to the ratio of their perimeters.

Solution: Let ABCD and A'B'C'D'E'F' be any two polygons circumscribed about circle R. We wish to show

$$\frac{\text{Area ABCD}}{\text{Area A'B'C'D'E'F'}} = \frac{\text{Perimeter ABCD}}{\text{Perimeter A'B'C'D'E'F'}} .$$

First, we draw the segments connecting each vertex to the center of the inscribed circle, thus dividing the polygons into triangles. This will help us to determine an expression for area.

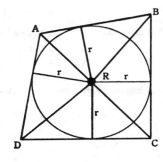

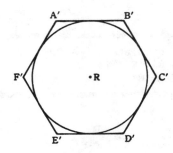

Area of ABCD =

      area of $\triangle$ARB + area of $\triangle$BRC + area of $\triangle$DRC + area of $\triangle$ARD.

The area of a triangle is $\frac{1}{2}$(base)(altitude). To find the altitude, note that each side is tangent to $\odot$R. Therefore, the altitude to any side will be represented by a radius of $\odot$R since a radius intersecting the tangent line at the point of tangency is perpendicular to the tangent.

    Let r = the length of the radius of $\odot$R. Then,

    Area of ABCD = $\frac{1}{2}$r(AB) + $\frac{1}{2}$r(BC) + $\frac{1}{2}$r(CD) + $\frac{1}{2}$r(AD)

                = $\frac{1}{2}$r(AB + BC + CD + AD)

                = $\frac{1}{2}$r(perimeter ABCD) = $\frac{1}{2}$rp.

By a similar process we can determine that

    Area of A'B'C'D'E'F' = $\frac{1}{2}$r(perimeter A'B'C'D'E'F') = $\frac{1}{2}$rp' .

Hence,

$$\frac{\text{Area of ABCD}}{\text{Area of A'B'C'D'E'F'}} \;=\; \frac{\frac{1}{2}rp}{\frac{1}{2}rp'} \;=\; \frac{p}{p'}$$

The ratio of the areas = the ratio of the perimeters.

● **PROBLEM** 566

    Upon the sides of a right triangle, three corresponding similar polygons are constructed. Show that the polygon upon the hypotenuse has an area equal to the sum of the areas of the polygons upon the legs.

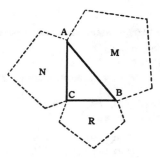

<u>Solution</u>: Let the areas of the polygons on $\overline{AB}$, $\overline{AC}$, and $\overline{BC}$ be represented by M, N, and R, respectively.

    We will derive the desired results by summing the ratios of the areas of the smaller polygons to the large one and proving that this sum is equal to one. In other words, we want to prove

$$\frac{N}{M} + \frac{R}{M} = \frac{N+R}{M} = 1 \; ,$$

which implies $N + R = M.$

The ratio of the areas of any two similar polygons is equal to the ratio of the squares of the lengths of corresponding sides. Therefore,

$$\frac{N}{M} = \frac{AC^2}{AB^2}$$

and

$$\frac{R}{M} = \frac{BC^2}{AB^2} \; .$$

Hence,

$$\frac{N}{M} + \frac{R}{M} = \frac{N+R}{M} = \frac{AC^2 + BC^2}{AB^2} \; .$$

Since $\triangle ABC$ is a right triangle, the Pythagorean Theorem tells us that $AB^2 = AC^2 + BC^2$. By substitution,

$$\frac{N+R}{M} = \frac{AB^2}{AB^2} = 1.$$

It therefore follows that $N + R = M$.

Prove that the area of an inscribed regular octagon is equivalent to that of a rectangle whose dimensions are the sides of the inscribed and circumscribed squares.

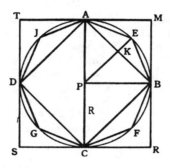

<u>Solution</u>: In the accompanying figure, AEBFCGDJ is the inscribed octagon; ABCD, the inscribed square; MRST, the circumscribed square. We are given circle P with radius of length r.

We will proceed by first determining the area of the octagon. Second, we determine the sides of the inscribed and circumscribed squares, which are the width and length of the rectangle. From this, we calculate the area of the rectangle. If the two areas are equal our work is done.

The area of octagon AEBFCGDJ = area of ABCD + area of $\triangle AEB$ + area of $\triangle BFC$ + area of $\triangle DGC$ + area of $\triangle AJD$.

The area of ABCD = AB^2. To find AB, note that it is the hypotenuse of isosceles right triangle APB. The legs are formed by radii of $\odot P$. Hence, by the Pythagorean Theorem,

$$AB^2 = r^2 + r^2 = 2r^2 \quad \text{and} \quad AB = r\sqrt{2} \; .$$

The area of ABCD = $(r\sqrt{2})^2 = 2r^2$. By the S.S.S. Postulate, all the triangles are congruent and the expression reduces to area of ABCD + 4(area of $\triangle$AEB), area of $\triangle$AEB = $\frac{1}{2}$(base)(altitude) = $\frac{1}{2}$(AB)(EK). We know AB = $r\sqrt{2}$.

EK = r - KP . KP can be found and, hence, EK will be known (in terms of r).

Since the octagon is regular, AE = EB. In the same circle congruent chords intercept congruent arcs which have congruent central angles. Hence, $\measuredangle$ APK $\cong$ $\measuredangle$ BPK or $\overline{PK}$ is the angle bisector of the vertex angle of isosceles triangle APB. As such, $\overline{PK}$ is the perpendicular bisector of $\overline{AB}$. Therefore, AK = KB = $\frac{1}{2}$(AB) = $\frac{1}{2}r\sqrt{2}$. $\triangle$AKP is a right triangle. Hence, we can apply the Pythagorean Theorem to find KP, given that we know AP = r and AK = $\frac{1}{2}r\sqrt{2}$,

$$(AP)^2 = (AK)^2 + (KP)^2$$
$$r^2 = (\tfrac{1}{2}r\sqrt{2})^2 + (KP)^2$$
$$r^2 - \tfrac{1}{4}2r^2 = (KP)^2$$
$$\tfrac{1}{2}r^2 = (KP)^2$$
$$\frac{\sqrt{2}}{\sqrt{2}} \cdot \frac{1}{\sqrt{2}}r = KP$$
$$\tfrac{1}{2}r\sqrt{2} = KP$$

As such, EK = r - $\frac{1}{2}r\sqrt{2}$.

$$\text{Area of } \triangle AEB = \tfrac{1}{2}(r\sqrt{2})(r - \tfrac{1}{2}r\sqrt{2})$$
$$= \tfrac{1}{2}(r^2\sqrt{2} - \tfrac{1}{2}2r^2) = \tfrac{1}{2}r^2(\sqrt{2} - 1) .$$

Recall, area of AEBFCGDJ = area of ABCD + 4(area of $\triangle$AEB). By substitution,
$$= 2r^2 + 4(\tfrac{1}{2}r^2(\sqrt{2} - 1))$$
$$= 2r^2 + 2r^2(\sqrt{2} - 1)$$

Therefore, Area of octagon = $2r^2\sqrt{2}$. We have one side of the required rectangle, AB = $r\sqrt{2}$. We will now find MR, the other side.

Since MRST is a square, $\overline{MT} \parallel \overline{SR}$ and MR is the perpendicular distance between $\overline{MT}$ and $\overline{SR}$. A and C are points of tangency. As such, $\overline{CA} \perp \overline{MT}$ and $\overline{CA} \perp \overline{SR}$ and CA is also a $\perp$ distance between $\overline{MT}$ and $\overline{SR}$. Since parallel lines are everywhere equidistant, MR = CA. CA is a diameter of OP. Hence, CA = 2r and MR = 2r.

The proposed rectangle has sides 2r and $r\sqrt{2}$. Its area then equals $2r^2\sqrt{2}$. Since this is the area of the octagon, we have shown the desired results.

SOLVING FOR THE AREA USING TRIGONOMETRIC FUNCTIONS

● **PROBLEM** 568

Derive a formula for the area of the region bounded by a regular n-sided polygon, where the length of the altitude of each isosceles triangle is r, in terms of r and n.

<u>Solution</u>: We will use a "building-up" procedure to solve this problem. First, we find the area of one triangle of which the polygon is composed (see figure (a)). Then, we prove that all the tri-

Case where
n = 12

Figure (a)

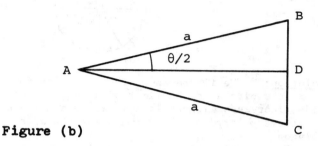

Figure (b)

angles are congruent, and, therefore, have the same area. Multiply-ing the area of one triangle by n (the total number of triangles composing the polygon), we obtain the area of the polygon.

Now, we are told that each triangle of which the polygon is com-posed (see figure (b)) is an isosceles triangle. Furthermore, the length of the altitude from the vertex formed by the intersection of the 2 congruent sides of the triangle, to the opposite side, is r. We recognize that the area of this specific type of triangle is

$$A = r^2 \tan \theta/2 \ .$$

We now prove that all the triangles of the polygon are congruent. Since we are told that all the triangles are isosceles, $OA = OB$, $OB = OC$, $OC = OD$, etc., and all the radial lines from the center of the polygon to its n vertices are congruent. Look at $\triangle AOB$ and $\triangle COB$. $\overline{AO} \cong \overline{CO}$. $\angle AOB \cong \angle COB$, because all the central angles of a regular polygon are congruent. Lastly, $\overline{BO} \cong \overline{BO}$. Hence, $\triangle AOB \cong \triangle COB$. The same thing can be proven for all the n triangles comprising the polygon. Hence, each triangle has the same area. By this dis-cussion,

$$\text{Area (polygon, n sides)} = n(r^2 \tan \theta/2) \qquad (1)$$

This is still not in terms of r and n. We must eliminate θ. First, note that each central angle (e.g., $\angle BAC$ in figure (b)) is twice the angle between one of the congruent sides and the altitude. Hence, each central angle has measure $(2)(\theta/2) = \theta$. Furthermore, there are n central angles, and their sum must be 2π . (This is true because in traversing all n central angles we would have gone around a circle one time.) Therefore,

$$n\theta = 2\pi$$

or

$$\frac{\theta}{2} = \frac{\pi}{n} \qquad (2)$$

Using (2) in (1),

$$\text{Area (polygon; n sides)} = nr^2 \tan \frac{\pi}{n}$$

Using the fact that the area of an n sided regular polygon is $nr^2 \tan \pi/n$, find the area of a circle. (Here, the polygon is composed of n isosceles triangles, with height r.   See figure.)

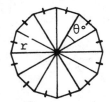

Solution: The figure shows the situation for n = 12. The idea here is that a circle is the limit of a regular polygon as the number of its sides increases without limit. We obtain the desired result by letting n get very large in the formula for the area of an n-sided regular polygon.

First, note that as n gets larger, tangent of π/n gets very small. Now, the tangent of any small angle is approximately equal to the angle itself.   Hence,

as n → ∞, tan π/n → π/n.

Then, using this in the formula given in the problem

as n → ∞ , $nr^2 \tan \pi/n → nr^2 \pi/n = \pi r^2$.

Hence, the area of a circle of radius r is πr^2 .

Prove that the area bounded by a regular polygon of n sides circumscribed about a circle with a radius of length r is given by the formula $A = nr^2 \tan \pi/n$.

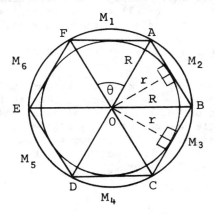

Solution: The figure shows the situation for the case where n = 6. In general, we see that we may consider the polygon of n sides to be composed of n triangles (for example, ΔAOB, ΔBOC, etc.). Since the polygon is circumscribed about circle O, $\overline{OM_2} \perp \overline{AB}$. This makes

$\overline{OM}_2$, of length r, an altitude of $\triangle AOB$. If $\triangle AOB$ were isosceles, we could find its area by using the formula $A = r^2 \tan \theta/2$. We now prove that $\triangle AOB$ is, indeed, isosceles.

In the figure, we have circumscribed a circle about the polygon, By definition, the center of a regular polygon is also the center of the inscribed and circumscribed circles. Now, look at $\triangle AOB$. $\overline{AO} \cong \overline{BO}$, since both are radii of the larger circle. Hence, $\triangle AOB$ is isosceles. Its area is, then,

$$A = r^2 \tan \theta/2$$

where r and θ are as shown.

Now we prove that each triangle (of which there are n) is congruent to every other triangle of the polygon. Consider $\triangle AOB$ and $\triangle BOC$, $\overline{AO} \cong \overline{CO}$, since both are radii of the large circle. $\overline{OB} \cong \overline{OB}$. $\overline{AB} \cong \overline{CB}$, because the sides of a regular polygon are all congruent. Therefore, by the S.S.S. (side-side-side) Postulate, $\triangle AOB \cong \triangle BOC$. This may be done n times to prove that all the triangles comprising the polygon are congruent. Hence, they all have the same area. Since the n-sided polygon is composed of the n adjacent triangles (see figure), the area of the polygon is the sum of the areas of each of the n triangles. Hence,

$$A_p = nr^2 \tan \theta/2 \ .$$

But

$$n\theta = 2\pi$$

or

$$\theta/2 = \pi/n \ .$$

Hence,

$$A_p = nr^2 \tan \pi/n \ .$$

● **PROBLEM** 571

Find the area of a regular hexagon circumscribed about a circle of radius r. Calculate the area explicitly if
a) r = 4; b) r = 9; c) r = 16; d) r = 25.

Solution: The area of a regular n-sided polygon circumscribed about a circle of radius r is given by

$$A = nr^2 \tan \pi/n \ .$$

In a hexagon, n = 6. $\tan \pi/6 = \sqrt{3}/3$
Thus

$$A = 6r^2 \tan \pi/6 = 2r^2 \sqrt{3}$$

a) For r = 4, $A = 2(16)\sqrt{3} = 32\sqrt{3}$

b) For r = 9, $A = 2(81)\sqrt{3} = 162\sqrt{3}$

c) For r = 16, $A = 2(256)\sqrt{3} = 512\sqrt{3}$

d) For r = 25, $A = 2(625)\sqrt{3} = 1250\sqrt{3}$.

Compute $\tan \pi/n$ and compare with π/n for

a) n = 9; b) n = 12; c) n = 18; d) n = 36; e) n = 180.

<u>Solution:</u> In order to compare $\tan \pi/n$ with π/n , we calculate

$$\frac{1}{\pi/n} \tan \frac{\pi}{n} = \frac{n}{\pi} \tan \frac{\pi}{n} ,$$

which tells us how close $\tan \pi/n$ is to π/n .

a) n = 9, $\qquad \frac{n}{\pi} \tan \frac{\pi}{n} = \frac{9}{\pi} \tan \frac{\pi}{9} = 1.04260$

b) n = 12, $\qquad \frac{n}{\pi} \tan \frac{\pi}{n} = \frac{12}{\pi} \tan \frac{\pi}{12} = 1.02349$

c) n = 18, $\qquad \frac{n}{\pi} \tan \frac{\pi}{n} = \frac{18}{\pi} \tan \frac{\pi}{18} = 1.01028$

d) n = 36, $\qquad \frac{n}{\pi} \tan \frac{\pi}{n} = \frac{36}{\pi} \tan \frac{\pi}{36} = 1.00255$

e) n = 180, $\qquad \frac{n}{\pi} \tan \frac{\pi}{n} = \frac{180}{\pi} \tan \frac{\pi}{180} = 1.00010$

As can be seen from the above calculations, $\frac{n}{\pi} \tan \frac{\pi}{n}$ gets very close
to 1 as n gets larger. This means that $\tan \frac{\pi}{n}$ gets closer to $\frac{\pi}{n}$
as n gets larger. Put in another way, the tangent of an angle gets
arbitrarily closer to the value of the angle in radians as the angle
gets smaller.

CONSTRUCTIONS

CHAPTER 30

LINES AND ANGLES

> **Basic Attacks and Strategies for Solving Problems in this Chapter. See pages 506 to 522 for step-by-step solutions to problems.**

The ultimate goal of all construction problems for a given geometric figure is to formulate a set of instructions such that any person in the world can execute them and construct exactly the same figure. Because the drawing ability of people varies widely, we cannot make our instructions too general and expect everyone to produce exactly the same figure. We cannot, for example, ask in one of our instructions, "Draw an angle so that it is smaller than a right angle, but only a little smaller." Similarly, we cannot state, "Draw a curved line." If we were to do that we would obtain very different results, indeed. What we must do is limit our instructions so that very few skills are required of the people following our instructions.

Euclid, the Father of Geometry, required the readers of his instructions to be able to do only three basic constructions. These were:

(1) given two points, draw the line segment between them;

(2) given a segment, extend it to form a straight line; and

(3) given a center point and a radius, construct the circle.

To this day, sets of instructions, called **constructions**, must be written so nothing more than these three skills is required.

In executing these three skills, we are allowed two instruments: the **straightedge** and the **compass**. The straightedge, as its name implies, is any object with a straight edge. The straightedge is used for drawing line segments and lines. By placing the straightedge on the given points A and B, line segment $\overline{AB}$ can be drawn. Similarly, by placing the straightedge so that its edge coincides with $\overline{AB}$, the line segment can be extended.

The second instrument is the compass. It has two legs, one of which has a pencil point at the tip. The compass is used for drawing circles. The leg with the sharp metal tip is placed at the center of the circle. The other leg, containing the pencil, is rotated until the pencil has drawn a complete circle. The compass is adjustable so that circles of different radii can be drawn.

In **solving** construction problems, the best procedure is to recognize the special properties of the figure and to remember that every complicated construction problem can be broken down into several simpler locus problems, whose loci can be drawn with just a straightedge and a compass. (For a discussion of locus problems see the introduction to Chapter 39.) It should be noted that in reality, many of the problems in this chapter require the construction of congruent triangles or similar triangles.

● **PROBLEM** 573

Draw a segment on $\overleftrightarrow{AB}$ congruent to $\overline{CD}$.

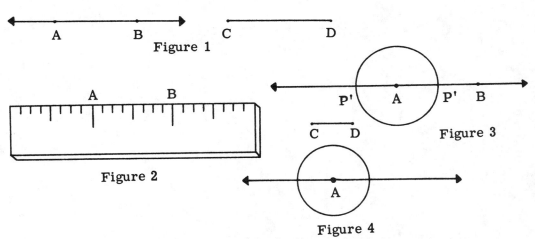

Figure 1

Figure 2

Figure 3

Figure 4

Solution: The ultimate goal of all construction problems for a given geometric figure is to formulate a set of instructions such that any person in the world can execute them and construct exactly the same figure. Because the drawing ability of people vary widely, we cannot make our instructions too general, and expect everyone to produce exactly the same figure. We cannot, for example, ask in one of our instructions, "draw an angle so that it is smaller than a right∢ but only a little smaller." Similarly, we cannot state, "draw a curved line". If we were to do that, we would obtain very different results, indeed. What we must do is limit our instructions so that very few skills are required of the people following our instructions.

Euclid, the Father of Geometry, required the readers of his instructions to be able to do only three basic constructions.
The three basic constructions were (1) given two points, draw the line segment between them; (2) given a segment, extend it to form a straight line; and (3) given a center point and a radius, construct the

circle. To this day, sets of instructions, called constructions, must be written so nothing more than these three skills is required.

In executing these three skills, we are allowed two instruments: the straightedge and the compass. The straightedge, as its name implies, is any object with a straight edge. The straightedge is used for drawing line segments and lines. By placing the straight edge on the given points A and B, line segment $\overline{AB}$ can be drawn. (Figure 2). Similarly, by placing the straightedge so that the straight edge coincides with $\overline{AB}$, the line segment can be extended.

The second instrument is the compass. It has two metal legs, one of which has a pencil point at the tip. The compass is used for drawing circles. The leg with the sharp metal tip is placed at the center of the circle. The other leg, containing the pencil, is rotated until the pencil has drawn a complete circle. The compass is adjustable so that circles of different radii can be drawn.

So much for the basic overview of constructions. In _solving_ construction problems, the best procedure is to recognize the special properties of the figure and to remember that every complicated construction problem can be broken down into several simpler locus problems, whose loci can be drawn with just a straightedge and a compass.

In _this_ example, we are required to draw a segment on $\overleftrightarrow{AB}$ congruent to $\overline{CD}$. Let A be one endpoint of the segment and P be the other endpoint. Since a segment is determined by its endpoints, we need only find point P.

This is our locus problem: find all points P such that (i) AP = CD and (ii) P is on $\overleftrightarrow{AB}$.

In any locus problem with several conditions, we draw the locus of each condition, and find their intersection. Condition (ii) is $\overleftrightarrow{AB}$ itself. The locus of condition (i) is a circle with center A and radius CD. To draw the locus, adjust the legs of the compass so that the metal tip rests on C and the pencil point on D. We are now assured that the radius of all circles drawn by the compass equals CD. Resting the metal tip on A, draw the circle of radius CD. (Figure 3).

The intersection of the two loci give the possible locations of P. Either $\overline{AP'}$ or $\overline{AP}$ is the solution.

NOTE: Often the intersection of the loci is evident. Therefore, it is not necessary to draw the whole circle, but only the arcs of the circle near the intersection, as shown below. In more complicated constructions where several equidistant points must be found, the drawing of arcs, as opposed to circles, saves not only pencils but also prevents a good deal of confusion. (Figure 4).

● **PROBLEM** 574

Construct a line segment whose length equals WX + YZ .

Solution: We can construct a segment $\overline{AB}$ such that AB = WX . Similarly, we can construct $\overline{BC}$ such that BC = YZ . Then if B is between A and C, we are assured by the definition of betweenness that AC = AB + BC = WX + YZ . To insure that B is between A and C, we draw with our straightedge line $\overleftrightarrow{AQ}$. Using the compass, we draw the circle centered about point A with radius WX . If we label the intersection of circle A and $\overleftrightarrow{AQ}$ as point B, then AB = WX . Similarly, if C is the intersection of $\overleftrightarrow{AQ}$ and the circle centered at

B with radius YZ, then we know BC = YZ . Since A,B, and C are
collinear AC = WX + YZ . (Figure 2).

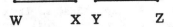

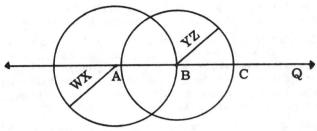

● **PROBLEM** 575

Construct a line perpendicular to a given line through a given
point on the given line.

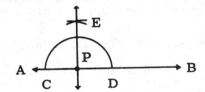

<u>Solution</u>: Let line $\overleftrightarrow{AB}$ and point P be the given line and the given
point, respectively.

We notice that ∡ APB is a straight angle. A line perpendicular
to $\overleftrightarrow{AB}$ from point P will form adjacent congruent angles with $\overleftrightarrow{AB}$,
by the definition of a perpendicular. Since ∡ APB is a straight
angle, the adjacent angles will be right angles. As such, the required perpendicular is the angle bisector of ∡ APB.

We can complete our construction by bisecting ∡ APB.

1. Using P as the center and any convenient radius, construct an
arc which intersects $\overleftrightarrow{AB}$ at points C and D.

2. With C and D as centers and with a radius greater in length
than the one used in Step 1, construct arcs that intersect. The
intersection point of these two arcs is point E.

3. Draw $\overleftrightarrow{EP}$.
$\overleftrightarrow{EP}$ is the required angle bisector and, as such, $\overleftrightarrow{EP} \perp \overleftrightarrow{AB}$.

● **PROBLEM** 576

Construct a line perpendicular to a given line through a given

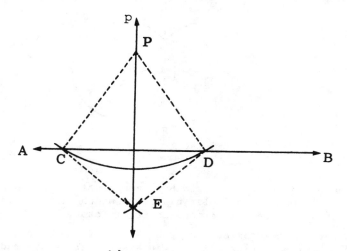

Solution: We are given line $\overleftrightarrow{AB}$ and point P, not on $\overleftrightarrow{AB}$, as in the figure shown. We wish to construct a line through P, perpendicular to $\overleftrightarrow{AB}$. Call this line p.

There is a segment $\overline{CD}$ on $\overleftrightarrow{AB}$ such that line p is the perpendicular bisector of $\overline{CD}$. To perform the construction we will (1) locate $\overline{CD}$; (2) construct the perpendicular bisector of $\overline{CD}$.

(1) Using P as a center, and any convenient radius, construct an arc which intersects $\overleftrightarrow{AB}$ at C and D. Since $\overline{PC}$ and $\overline{PD}$ are radii of the same arc, we have PC = PD. Since point P is equidistant from points C and D, point P must be on the perpendicular bisector of $\overline{CD}$.

(2) Construct the perpendicular bisector of $\overline{CD}$.

The perpendicular bisector contains point P and is perpendicular to given line $\overleftrightarrow{AB}$. Thus, the perpendicular bisector is the required line p.

● **PROBLEM** 577

Construct the perpendicular bisector of a given segment.

Solution: Given: $\overline{XY}$.

Wanted: Perpendicular bisector of $\overline{XY}$.

Construction: The perpendicular bisector is the locus of all points equidistant from the endpoints of the segment. Therefore, we construct the circle, which is by definition all points equidistant from a given point, centered at X with radius r, and the circle centered at Y with radius r. The intersection consists of two

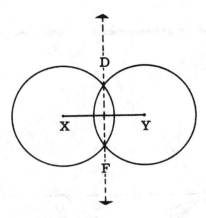

points each a distance r from both endpoints. The two points are therefore on the perpendicular bisector and the line drawn through these points must be the perpendicular bisector.

1. Adjust the compass so that the radius of the compass is more than half $\overline{XY}$. (The circles will not intersect unless the radii are greater than $\frac{1}{2}XY$.)
2. With X as center construct an arc of the circle extending to each side of $\overline{XY}$ (It is not necessary to draw the full circle.)
3. With Y as center construct an arc of the circle extending to each side of $\overline{XY}$. Label the points of intersection of the arcs, D and F.
4. Draw line $\overleftrightarrow{DF}$, the perpendicular bisector of $\overline{XY}$.

● **PROBLEM** 578

Bisect a given angle.

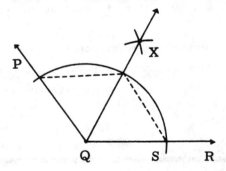

Solution: $\overrightarrow{QX}$ bisects ∢PQR if and only if ∢PQX ≅ ∢XQR . Let T be a point on $\overrightarrow{QP}$ and S be a point on $\overrightarrow{QR}$ such that △QTX ≅ △QSX . Then ∢PQX and ∢XQR are corresponding parts of congruent triangles and ∢PQX ≅ ∢XQR . Therefore, if we construct △QTX and △QSX such that △QTX ≅ △QSX , then $\overrightarrow{QX}$ bisects ∢PQR. We construct the triangles by the SSS Postulate.

Construction:
    1. Using Q as the center and any radius, construct an arc of a circle that intersects $\overrightarrow{QP}$ at T and $\overrightarrow{QR}$ at S. Since radii of a circle are equal, QS = QT . By reflexivity, QX = QX .

2. Since SX must equal TX for ΔQTX ≅ ΔQSX , as required by
the earlier reasoning, we make certain of this by construction.
Using S as a center and a radius of more than half ST, make an arc
in the interior of ∢PQR.

Using T as a center and the same radius, make an arc in the
interior of ∢PQR.

Since the radii are the same, SX = TX.

3. Draw Q⃗X . Since $\overline{QX} \cong \overline{QX}, \overline{TX} \cong \overline{SX}, \overline{QS} \cong \overline{QT}$, then ΔQTX ≅
ΔQSX by the SSS Postulate. Therefore, ∢TQX ≅ ∢XQS, and Q⃗X bisects
∢PQR .

● **PROBLEM** 579

Construct the bisector of an angle of a given triangle. (See
figure).

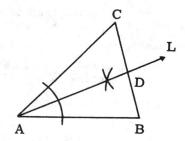

<u>Solution</u>: By referring to the bisection of an angle construction,
done previously, we can readily do the construction required in this
problem. If we terminate the line bisecting the angle at the side of
the triangle opposite the angle, then the line segment inside the triangle will be the bisector of an angle of a given triangle.

Follow the steps in the accompanying sketch.

1. Construct A⃗L, the bisector of ∢ A. (Again, we note that
this construction has been done previously.)

2. A⃗L will intersect side $\overline{BC}$, the side opposite ∢ A. Call
the point of intersection D.

3. Line segment $\overline{AD}$, as drawn, is the bisector of ∢ A in
ΔABC. $\overline{AD}$ is the bisector of ∢ A in ΔABC because, by definition,
the bisector of an angle of a triangle is the line segment, drawn from
a vertex, which bisects the angle at that vertex and which terminates
in the opposite side.

● **PROBLEM** 580

Construct an angle congruent to a given angle.

<u>Solution</u>: The principle behind this construction is that congruent
central angles, with arcs of equal radii, intercept congruent arcs which
have congruent chords. Therefore, by drawing two arcs with equal
radii, one centered on the vertex of the given angle and one centered
at an arbitrary point on a line, we can determine the chord length
intercepted by the given angle and then duplicate this chord length

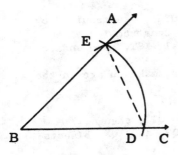
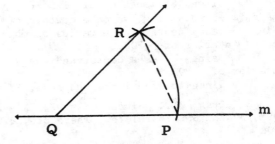

on the second arc. This chord uniquely determines an arc length and, as such, a central angle congruent to the given angle, since the arcs have the same radii.

Given: ∡ ABC
Wanted;   An angle congruent to ∡ ABC .

Construction:
    1. Using B as center and any radius that is convenient, con-struct an arc intersecting $\overrightarrow{BC}$ at D and $\overrightarrow{BA}$ at E.

    2. Construct line m. Label an arbitrary point Q.

    3. Using Q as center and same radius as in Step 1, construct an arc above and through $\overleftrightarrow{m}$, intersecting $\overleftrightarrow{m}$ at P. ($\overline{BD} \cong \overline{QP}$, $\overline{BE} \cong \overline{QR}$) .

    4. Adjust the radius the compass to equal ED.

    5. Using this radius and P as the center, construct an arc above m, intersecting the first arc. Label the point of intersection R. ($\overline{RP} \cong \overline{DE}$).

    6. Draw $\overrightarrow{QR}$. Since they are both central angles that intercept arcs of equal circles that have congruent chords,

$$\angle \text{ ABC} \cong \angle \text{ RQP} .$$

● **PROBLEM** 581

> To construct an angle whose measure is equal to the sum of the measures of two given angles.

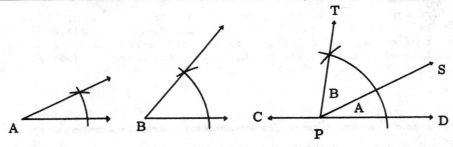

Solution: To construct an angle equal to the sum of the measures of two given angles, we must invoke the theorem which states that the whole is equal to the sum of the parts. The construction, then, will duplicate the given angles in such a way as to form one larger angle equal in measure to the sum of the measures of the two given angles.

    The two given angles, ∡ A and ∡ B, are shown in the figure.

    1. Construct any line $\overleftrightarrow{CD}$, and mark a point P on it.

2. At P, using $\overrightarrow{PD}$ as the base, construct ∡ DPS ≅ ∡ A .

3. Now, using $\overrightarrow{PS}$ as the base, construct ∡ SPT ≅ ∡ B at point P.

4. ∡ DPT is the desired angle, equal in measure to m ∡ A + m ∡ B. This follows because the measure of the whole, ∡ DPT, is equal to the sum of the measure of the parts, ∡ A and ∡ B .

● **PROBLEM** 582

A point is given within a given angle. Construct a line segment which shall be bisected by this point and shall terminate at the sides of the angle.

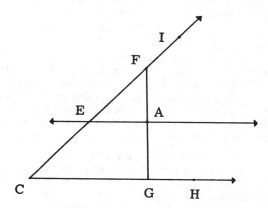

<u>Solution</u>: We are given an ∡ FCH and a point A in its interior. We are asked to find the segment $\overline{FAG}$ such that it is bisected by A and endpoints F and G lie on the sides of the angle.

Note that points C,F, and G determine a triangle, ΔCFG, with point A, a midpoint of unknown side $\overline{FG}$. Remember that a line drawn through two sides of a triangle parallel to the third divides the sides proportionally. Therefore, we draw $\overline{EA} \parallel \overline{CH}$ such that points C,E, and F are collinear. Since A is to be the midpoint of $\overline{FG}$,then E must be the midpoint of $\overline{CF}$. Knowing that E is the midpoint of $\overline{CF}$, we can locate point F.

The segment drawn through points A and F is the required segment for this construction.

Given: ∡ FCH and interior point A.

Construct: $\overline{FG}$ to yield $\overline{CEF}$, $\overline{CGH}$, $\overline{FAG}$ and FA = AG .

Construction:

1. Construct $\overline{AE} \parallel \overline{CH}$ such that $\overline{CEI}$.

2. Locate F such that $\overline{CEF}$ and CE = EF .

3. Draw $\overleftrightarrow{FA}$ such that it intersects $\overleftrightarrow{CH}$ at G. Then we have $\overleftrightarrow{EA}$, a line that passes through two sides of ΔFCG such that $\overline{EA} \parallel \overline{CG}$. Since $\overline{EA}$ bisects $\overline{FC}$, $\overline{EA}$ also bisects $\overline{FG}$. Thus, A is the midpoint of $\overline{FG}$.

513

Construct an angle containing $60°$, whose vertex is a given point. (See figure).

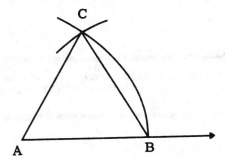

Solution: Since all angles in an equilateral triangle measure $60°$ (or $\frac{1}{3}(180)°$), by constructing such a triangle a $60°$ angle can be readily determined. This may seem like a "back door" method; however, there exists a fairly straightforward way to construct an equilateral triangle. This can be accomplished by drawing two radii of a circle whose points of intersection with the circle are a linear distance apart equal to the length of the radius.

1. Given point A, construct any line segment $\overline{AB}$.
2. Using A as the center, and a radius whose length is equal to AB, construct an arc of a circle.
3. Using B as the center, and the same radius as before, mark off another arc which intersects the first arc at C.
4. Construct $\overline{CA}$ and $\overline{CB}$, forming equilateral triangle ABC. The triangle will be equilateral because all three sides are radii of arcs which have been constructed with radii of the same length.

Since $\triangle ABC$ is equilateral, and, as such, equiangular, m $\angle$ CAB = $\frac{1}{3}$ of $180°$, the total of the measures of the angles of a triangle. Hence, m $\angle$ CAB = $60°$, the required angle to be constructed.

Divide a given line segment into parts proportional to given line segments.

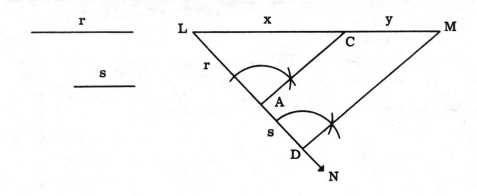

Solution: We know that a line intersecting two sides of a triangle, and parallel to the third side, will divide the two sides proportionally. First, we construct a triangle with (1) the first side equal in length to the given segment; and (2) a second side that can be easily divided into segments proportional to the given segments. Then we construct a line parallel to the third side that divides the second side into parts proportional to the given segment. By the above, we know that this line must also cut the first side, the given length, proportionally.

To find the second side that can be easily divided, note that there is no restriction on the length on the second side. Therefore, we construct the second side to be the length of the given segments combined.

1. Given segment $\overline{LM}$ and line segments of lengths r and s, draw $\overleftrightarrow{LN}$ making any convenient angle MLN.

2. On $\overleftrightarrow{LN}$, construct $\overline{LA}$ so that $LA = r$ and construct $\overline{AD}$ so that $AD = s$.

3. Construct $\overleftrightarrow{DM}$ to complete the triangle, $\triangle DML$.

4. Through A, construct $\overleftrightarrow{AC}$ parallel to $\overleftrightarrow{DM}$. C is the intersection of $\overleftrightarrow{AC}$ and $\overleftrightarrow{LM}$.

5. $\overline{LC}$ and $\overline{CM}$ are the required segments of length x and y.

The construction makes $\overleftrightarrow{AC} \parallel \overleftrightarrow{DM}$. Therefore, $x:y = r:s$ because a line parallel to one side of a triangle and intersecting the other two sides divides the other two sides proportionally.

● **PROBLEM** 585

Construct the mean proportional between two given segments.

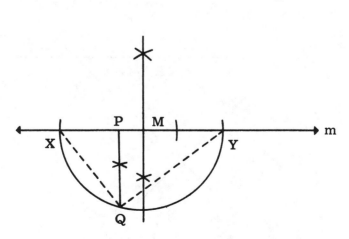

Solution: The mean proportional of two numbers a and b is defined as that number c such that $a/c = c/b$. The key here is "mean proportional". What geometric figures have properties involving the mean proportional?

In the right triangle, the altitude to the hypotenuse is the mean proportional of the segments of the hypotenuse.

Therefore, we construct a right triangle whose altitude divides the hypotenuse into segments of length AB and CD. The altitude will then be the mean proportional.

Given: $\overline{AB}$ and $\overline{CD}$.

Wanted: $\overline{PQ}$ such that $\dfrac{AB}{PQ} = \dfrac{PQ}{CD}$.

Construction:

(1) Construction of the hypotenuse: Choose a point P on any line. (P will be the intersection of the altitude and hypotenuse).

(2) On each side of P, construct segments $\overline{PX}$ and $\overline{PY}$ such that PX = AB and PY = CD. ($\overline{XY}$ is the hypotenuse.)

(3) Construction of the right triangle: (Any triangle inscribed in a semicircle is a right triangle and its hypotenuse is the diameter. We now construct the semicircle.) Bisect $\overline{XY}$ in order to find the midpoint M.

(4) Construct a semicircle above m using M as center and MX as radius.

(5) At P, construct a line perpendicular to m, intersecting the semicircle at Q.

PQ is the mean proportional between AB and CD since $\overline{PQ}$ is the altitude to hypotenuse $\overline{XY}$ of right $\triangle XQY$.

● **PROBLEM** 586

Construct the fourth proportional of three given segments. (In the proportion a/b = c/d, d is considered the fourth proportional, where a,b,c, and d are the lengths of the segments.)

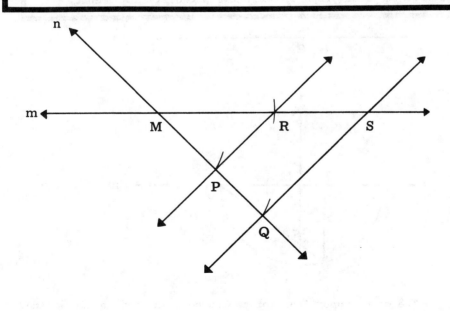

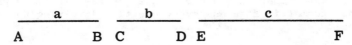

<u>Solution</u>: We are given segments of length AB, CD, and EF, and we are asked to find a segment $\overline{RS}$, such that AB/CD = EF/RS. By an earlier theorem, we have that, in a triangle, a line through two sides and parallel to the third side divides the two sides proportionally. Suppose in $\triangle MSQ$ (see figure above)

$$\text{line } \overleftrightarrow{PR} \parallel \text{side } \overleftrightarrow{QS} \text{ .}$$

Then
$$\frac{MP}{PQ} = \frac{MR}{RS} \text{ .}$$

This is a one-to-one correspondence with the proportion stated earlier in the solution. Thus, if we construct $\triangle MSQ$, with $\overleftrightarrow{PR} \parallel \overleftrightarrow{QS}$ such that MP = AB, PQ = CD, and MR = EF, then RS = RS.

Given: $\overline{AB}$, $\overline{CD}$, and $\overline{EF}$.

Wanted: $\overline{RS}$ such that $\dfrac{AB}{CD} = \dfrac{EF}{RS}$.

Construction:

(1) Construct any two intersecting lines m and n intersecting at M.

(2) On n, construct $\overline{MP}$ and $\overline{PQ}$ congruent to $\overline{AB}$ and $\overline{CD}$, respectively.

(3) On m, construct $\overline{MR}$ congruent to $\overline{EF}$.

(4) Draw $\overline{RP}$, the line through two sides and parallel to the third side of $\triangle MSQ$.

(5) Through S, now construct the third side $\overline{SQ} \parallel \overline{RP}$. By the theorem above, $\dfrac{MP}{PQ} = \dfrac{MR}{RS}$ or $\dfrac{AB}{CD} = \dfrac{EF}{RS}$. $\overline{RS}$ is the fourth proportional.

● **PROBLEM** 587

Construct the segment of length $\dfrac{a}{b}$ where a and b are real numbers. (Unit length shown below.)

<u>Solution</u>: Wherever proportions are involved, the theorems from similarity are usually helpful. From the theorem of the above problem, we know that, in $\triangle OUQ$, if $\overline{XP} \parallel \overline{UQ}$, then

$$\frac{PQ}{OP} = \frac{XU}{OX} \text{ .}$$

We are searching for c , such that $c = \dfrac{a}{b}$ or $\dfrac{c}{1} = \dfrac{a}{b}$. Therefore, if XU = a, OX = b, and OP = 1, then PQ will be the desired length c.

Given: AB = a, CD = b, EF = 1.

Wanted: segment of length $\dfrac{a}{b}$.

Construction:

(1) Construct lines m and n intersecting at point P.

(2) Construct $\overline{OP} \cong \overline{EF}$ on line m.

(3) Construct $\overline{OX} \cong \overline{CD}$ and $\overline{XU} \cong \overline{AB}$ on line n.

(4) Draw $\overleftrightarrow{XP}$, the line parallel to third side $\overleftrightarrow{UQ}$.

(5) Construct $\overleftrightarrow{UQ} \parallel \overleftrightarrow{XP}$. By the above theorem, $\dfrac{PQ}{OP} = \dfrac{XU}{OX}$ or $\dfrac{PQ}{1} = \dfrac{a}{b}$ or $PQ = \dfrac{a}{b}$.

Therefore PQ is the desired segment.

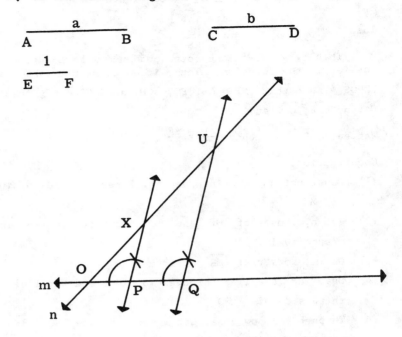

● **PROBLEM** 588

Divide any segment into three congruent segments.

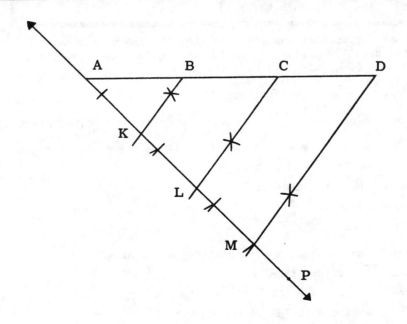

Solution: Given a line $\overleftrightarrow{AP}$ and a segment $\overline{AK}$ on the line, we can construct a segment $\overline{KL}$ on the line such that KL = AK . Similarly, we could construct another segment $\overline{LM}$ such that LM = AK. Thus, AK = $\frac{1}{3}$ AM . Hence, we can construct a segment $\overline{AM}$ that is cut into thirds by points K and L. Unfortunately, we cannot decide ahead of time how long the final segment $\overline{AM}$ will be. In this problem, we wish to divide the given segment $\overline{AD}$ into three congruent segments. We know $\overline{AM}$ is divided into congruent segments. If three or more parallel lines intercept congruent segments on one transversal, then they intercept congruent segments on any other transversal. Therefore, the set of parallel lines that intercept segment $\overline{AM}$ and $\overline{AD}$ must also cut congruent segments on $\overline{AD}$. The construction consists of: (1) Constructing a segment $\overline{AM}$ such that it is divided into three congruent segments; (2) drawing $\overleftrightarrow{MD}$ (since the parallel lines must intercept the non-mutual endpoints of $\overline{AM}$ and $\overline{AD}$, the parallel line that passes through M must pass through D); and (3) constructing lines through K and L parallel to $\overleftrightarrow{MD}$.

Given: $\overline{AD}$.

Wanted: The partition of $\overline{AD}$ into three congruent segments.

Construction:

(1) Draw $\overleftrightarrow{AP}$ through A at any angle to $\overline{AD}$.

(2) Using any appropriate radius of your compass, construct three consecutive congruent segments on $\overleftrightarrow{AP}$ starting at A, such that AK = KL = LM.

(3) Draw $\overleftrightarrow{MD}$.

(4) Construct $\overleftrightarrow{KB} \parallel \overleftrightarrow{MD}$ and $\overleftrightarrow{LC} \parallel \overleftrightarrow{MD}$.

(5) Since parallel lines $\overleftrightarrow{KB}$, $\overleftrightarrow{MD}$, and $\overleftrightarrow{LC}$ intercept congruent segments on $\overline{AM}$, they must intercept congruent segments on $\overline{AD}$. AB = BC = CD and $\overline{AD}$ has been divided into three congruent segments.

● **PROBLEM** 589

Partition a given segment into parts proportional to given segments.

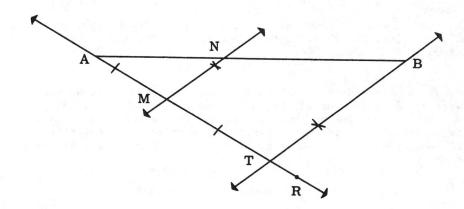

<u>Solution</u>: This problem is very similar to the division of a segment into congruent segments. Here, though, the divisions are not congruent but proportional. The major problem remains the same: the length of the segment to be divided is fixed. On arbitrary line $\overleftrightarrow{AR}$, we can construct segment $\overline{AM}$ such that AM = XY, one of the given segments, and segment $\overline{MT}$ such that MT = PQ. Then, segment $\overline{AT}$ is divided into the desired proportionals though it is not of the length of given segment AB. However, by an earlier theorem, a set of parallel lines cuts proportional segments on all transversals. Therefore, the set of parallel lines that intercept $\overline{AT}$ and $\overline{AB}$ must divide $\overline{AB}$ into parts proportional to the given segments. The construction consists of

(1) Constructing $\overline{AMT}$ such that AM = XY and MT = PQ;

(2) Drawing $\overleftrightarrow{TB}$ (if the parallel lines intercept $\overline{AT}$ and $\overline{AB}$, then the parallel line that passes through T must intersect $\overline{AB}$ at B);

(3) Constructing $\overleftrightarrow{MN} \parallel \overleftrightarrow{BT}$.

Given: $\overline{XY}$, $\overline{PQ}$, $\overline{AB}$.

Wanted: Segments proportional to $\overline{XY}$ and $\overline{PQ}$ on $\overline{AB}$.

Construction:

(1) Draw $\overleftrightarrow{AR}$ through A at any convenient angle to $\overline{AB}$.

(2) On $\overleftrightarrow{AR}$, construct $\overline{AM} \cong \overline{XY}$ and $\overline{MT} \cong \overline{PQ}$ so that we have $\overline{AMT}$.

(3) Draw $\overline{TB}$

(4) Construct $\overline{MN} \parallel \overline{TB}$ through M.

(5) A set of parallel lines cuts proportional segments on all transversals. Therefore, $\dfrac{AN}{NB} = \dfrac{AM}{MT}$ or $\dfrac{AN}{NB} = \dfrac{XY}{PQ}$.

● **PROBLEM** 590

Construct the segment of length equal to a·b where a and b are the lengths of the segments below. (Unit length is shown below.)

<u>Solution</u>: Let AB = a, CD = b. In the above triangle $\angle ORX$, $\overline{UP} \parallel \overline{RX}$.

From an earlier theorem (in a triangle, a line through two sides and parallel to the third side divides the two sides proportionally), we know that
$$\frac{PX}{UR} = \frac{OP}{OU} .$$

We can take advantage of this proportion. Note that we are searching for a segment of length c where c = a·b . Dividing both sides of the equation by b , we obtain c/b = a or c/b = a/1 . This is the exact form of the earlier proportion. If we construct $\triangle ORX$ such that OU = 1, OP = a, UR = b, then it must be that PX = c = a·b .

Given: AB = a , CD = b , EF = 1 .

Wanted: a segment of length a·b .

UNIT LENGTH

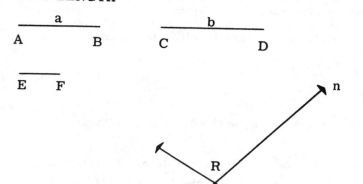

Construction:
(1) Construct two intersecting lines m and n. Label the intersection point O.
(2) Construct $\overline{OU}$ on n such that $\overline{OU} \cong \overline{EF}$.
(3) Construct $\overline{OP}$ on m such that $\overline{OP} \cong \overline{AB}$.
(4) Construct $\overline{UR}$ on n such that $\overline{UR} \cong \overline{CD}$.
(5) Draw $\overleftrightarrow{UP}$, the line to be parallel to the third side.
(6) Draw $\overleftrightarrow{RX} \parallel \overleftrightarrow{UP}$. By the above theorem, PX = OP·UR = AB·CD = a·b.

● **PROBLEM** 591

Given a line and a point not on the line, construct the line through the point that is parallel to the given line.

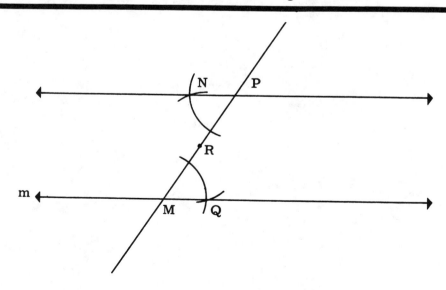

521

<u>Solution</u>: Two lines are parallel if their alternate interior angles are equal. Therefore, by constructing a transversal to m through point P, and constructing an interior angle on the "left" side congruent to the interior angle on the "right" side, the lines must be parallel.

Given: line m and point P .

Wanted: line $\overleftrightarrow{NP} \parallel$ m .

Construction:

(1) Let M and Q be points on m, with Q on the right of M. Draw $\overleftrightarrow{MP}$.

(2) Let R be any point of $\overleftrightarrow{MRP}$ on the same side of P from M. On $\overleftrightarrow{PR}$ construct $\sphericalangle$ RPN with N to the left of $\overleftrightarrow{MPR}$ such that $\sphericalangle$ RPN $\cong$ PMQ .

(3) Draw $\overleftrightarrow{NP}$. $\sphericalangle$ NPR $\cong$ $\sphericalangle$ RMQ by construction. By congruence of alternate interior angles, $\overleftrightarrow{NP} \parallel$ m .

CHAPTER 31

TRIANGLES

Basic Attacks and Strategies for Solving Problems in this Chapter. See pages 523 to 529 for step-by-step solutions to problems.

It will be useful in geometry to be able to construct triangles reliably with the various congruence postulates in Chapters 6, 7, and 8, it is sufficient to specify only a few of the desired triangle's measurements, and still guarantee a unique construction. For example, the solution of Problem 592 illustrates how to construct two triangles when two sides and the included angle of the triangle are given. In reality, this means that a triangle must be constructed in such a way that two sides and the included angle of the constructed triangle are congruent to the two angles and one segment pictured. Any two triangles correctly constructed to these specifications must be congruent by the SAS postulate. Similarly, if two people make the construction required in Problem 593, their triangles will be congruent by the ASA postulate. It should be noted that if two people are required to construct a triangle with three angles congruent to three given angles, the constructed triangles would not necessarily be congruent. For example, one person might construct triangle △ *ABC,* and the other might construct triangle △ *DEF:*

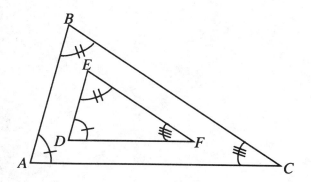

Many problems of this chapter require the following basic constructions, which are illustrated in Chapter 30:

(1) construct a segment congruent to a given segment (Problem 573);

(2) construct an angle congruent to a given angle (Problem 580); and

(3) construct the perpendicular bisector of a given segment (Problem 577).

Note that Problem 598 involves the construction of a triangle similar to a given triangle. Knowledge of the AA triangle similarity postulate (Problem 243) is a prerequisite for this construction.

Construct a triangle when two sides and the included angle are given. (See figure).

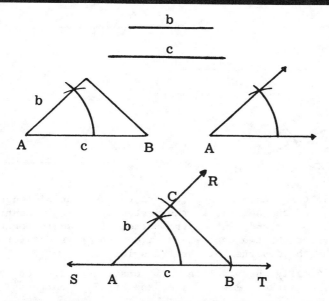

Solution: We are given two sides of lengths b and c, as shown in the figure, and one angle, ∢ A. We will duplicate ∢ A, and then construct segments congruent to b and c along the sides of ∢ A. Then, by connecting the endpoints of the segments farthest away from the vertex of ∢ A, a unique triangle can be drawn.

Follow the construction shown in the sketch.

1. On any line ST̷, use point A as the vertex to construct ∢ RAT ≅ ∢ A, the given angle. (The construction of an angle congruent to another angle was done previously.)

2. Using point A as a center, and c as the radius length construct AB̅, along side AT̷ of ∢ A, congruent to segment c.

3. Using A as the center again, and b as the radius length, construct AC̅, along side AR̷ of ∢ A, congruent to segment b.

4. Construct $\overline{BC}$. The required triangle is $\triangle ABC$.

$\triangle ABC$ is unique because, given any other triangle with side 1 ≅ b, side 2 ≅ c, and the included angle ≅ ∡ A , the two triangles would be congruent by S.A.S. ≅ S.A.S. By corresponding parts, side 3 = $\overline{BC}$. Therefore, all parts of the alternative triangle would be congruent to the corresponding parts of $\triangle ABC$ and the alternative triangle would be indistinguishable from $\triangle ABC$. This shows that, given two sides and the included angle, a unique triangle can be constructed.

● **PROBLEM**

Construct the triangle with ∡ A and ∡ B and the included side of length c.

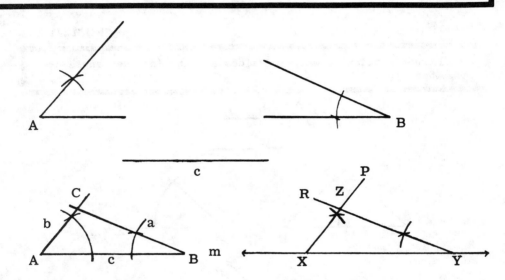

Solution: To do the construction, construct a segment of length c. Use one endpoint of the segment as the vertex of an angle congruent to ∡ A and the other endpoint as the vertex of an angle congruent to ∡ B. Construct the two angles and extend the rays until they join to form the third vertex of the triangle.

More specifically, draw a line m. Construct $\overline{XY}$ such that XY = c. At X, construct ∡ YXP ≅ ∡ A, and at Y construct ∡ XYR ≅ ∡ B.

Extend $\overrightarrow{XP}$ and $\overrightarrow{YR}$ until they intersect, and call the intersection Z. $\triangle XYZ$ is the required triangle.

● **PROBLEM** 594

Construct a triangle, given two of its sides, p and q, and the median, m, to the third side.

Solution: Let $\triangle COB$ be the triangle to be constructed, and point D, the midpoint of side $\overline{CB}$. We are given CO = q, OB = p and median OD = m.

The construction will have to be done in an indirect manner. We

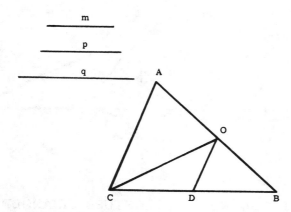

will have to construct ΔABC with one side = 2p, a median to that side equal to q and a midline equal in length to m. Then we will have the angle between the two given sides and can readily complete the required triangle, ΔCOB.

Given: three segments of length p,q, and m.

Construct: a triangle with sides of length p and q and a median to the third side of length m.

Construction:
1. Construct $\overline{AOB}$ with AOB = 2·p and point O the midpoint. Then, OB = p.
2. With O as the center, draw an arc with radius q.
3. With A as the center, draw an arc of radius 2m that intersects the first arc at point C. Connect O and A to C. (By construction, OC = q.)
4. Draw the line $\overline{DO}$ parallel to $\overline{AC}$.

Any line passing through the midpoint of a side of a triangle and parallel to another side is a midline of the triangle and intersects the third side at its midpoint. Thus, D is the midpoint of $\overline{CB}$. Furthermore, the midline length is one-half the length of the parallel side. Thus, OD = ½AC = ½(2m) = m. Hence, $\overline{OD}$ is the given median of length m.

ΔCOB has side CO = q, side OB = p, and median length OD = m. Thus, ΔCOB is the required triangle.

● **PROBLEM 595**

Construct an altitude of a given triangle.

Solution: This construction will vary slightly, depending upon whether the triangle is obtuse, acute, or right.

The basic idea when constructing an altitude of any triangle is to select a vertex, and draw a line through that vertex perpendicular to the opposite side. In the case of an obtuse triangle, the opposite side will have to be extended, and the point of intersection between the side and the perpendicular will be outside the triangle.

The construction can be followed in the accompanying sketch:
1. Through point A, construct $\overleftrightarrow{AT}$, a line perpendicular to $\overleftrightarrow{CB}$. Extend $\overline{CB}$ if necessary. (The construction of a perpendicular

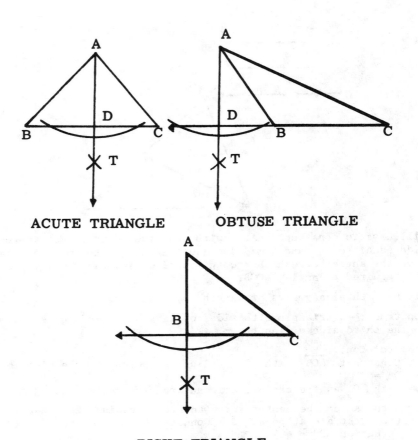

ACUTE TRIANGLE   OBTUSE TRIANGLE

RIGHT TRIANGLE

to a line through a given point outside the line has been described previously.)

2. In acute triangle ABC , $\overleftrightarrow{AT}$ intersects $\overline{CB}$ at point D. $\overline{AD}$ is the altitude from vertex A to side $\overline{CB}$.

3. In obtuse triangle ABC, $\overleftrightarrow{AT}$ intersects $\overline{CB}$ extended at point D. Line segment $\overline{AD}$ is the altitude from vertex A to side $\overline{CB}$.

4. In right triangle ABC, the point of intersection of $\overleftrightarrow{AT}$ and side $\overline{BC}$ is a vertex of the triangle, namely vertex B. Therefore, in addition to being one side of the triangle, $\overline{BC}$ is the altitude of right triangle ABC. This is so because ∢ B is a right angle.

● **PROBLEM** 596

Construct an isosceles triangle which has the same base as a given scalene triangle, and which is equal in area to it. (See figure).

Solution: The key to this construction lies in the fact that the area of a triangle is equal to ½ × base × height. The isosceles triangle to be constructed is to have the same base as the given scalene triangle. To insure that the isosceles triangle has the same area as the scalene triangle, we must construct it in such a way that both triangles have the same altitude.

If we draw line $\overleftrightarrow{RS}$ parallel to base $\overline{AB}$, the distance between the two lines will be the altitude of the scalene triangle (△ABC).

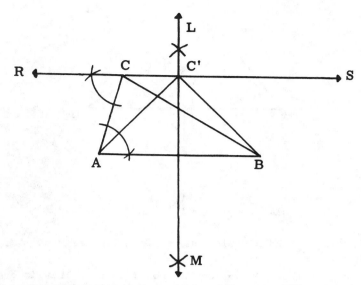

Any other perpendicular distance between R̂Ŝ and $\overline{AB}$ will be equal to the altitude of the scalene triangle. The altitude of the isosceles triangle will be one such perpendicular distance.

The construction follows:

1. Through vertex C, construct R̂Ŝ parallel to $\overline{AB}$. (This parallel line construction was done previously.)

2. Construct L̂M̂, the perpendicular bisector of base $\overline{AB}$. (This was also shown earlier).

3. R̂Ŝ and L̂M̂ intersect at the point C'.

4. Construct $\overline{AC'}$ and $\overline{BC'}$ to form isosceles triangle ABC'.

Because $\overline{AC'}$ and $\overline{BC'}$ are drawn to the same point on perpendicular bisector L̂M̂, $\overline{AC'} \cong \overline{BC'}$. It therefore follows that $\triangle ABC'$ must be isosceles.

We have constructed R̂Ŝ $\parallel$ ÂB. Since parallel lines are everywhere equidistant, the heights of the triangles, which equal the distance between the parallel lines, must be equal. With common bases and equal heights, the area of the isosceles triangle must be the same as the area of the given scalene triangle.

● **PROBLEM** 597

Construct a median of a given triangle.

Solution: The median of a triangle is that line segment which joins a vertex and the midpoint of the opposite side. We need two points to determine this line segment. One is the given vertex. The other point, the midpoint, will be on the side opposite the vertex intersected by its perpendicular bisector.

Follow the construction in the sketch shown.

1. Construct R̂Ŝ, the perpendicular bisector of B̂Ĉ. (Refer to earlier problems in this section for this construction).

2. R̂Ŝ intersects, and bisects, side $\overline{BC}$ at point M. As such, point M is the midpoint of $\overline{BC}$.

527

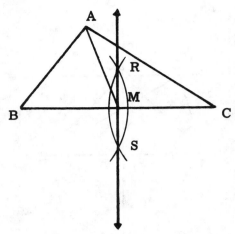

3. Join vertex A and midpoint M to form line segment $\overline{AM}$, the median desired.

$\overline{AM}$ is the median from vertex A to side $\overline{BC}$. This is because a median of a triangle is the line segment which joins a vertex and the midpoint of the opposite side.

● **PROBLEM** 598

Construct a triangle similar to a given triangle, using a given line segment as a base. (See figure.)

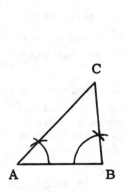
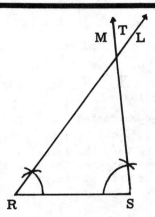

<u>Solution</u>: When given △ABC, we know that any triangle similar to △ABC must have two angles congruent to two corresponding angles of △ABC. Given line segment, RS we will construct at its endpoints two angles congruent to two angles of the given triangle.

By the A.A Similarity Theorem, the segment and the rays of the 2 angles will enclose a triangle similar to the given one.

1. Given $\overline{RS}$, construct ⦨ SRL congruent to ⦨ BAC at R.

2. At S, construct ⦨ RSM congruent to ⦨ ABC.

3. Let point T be the point of intersection of rays $\overrightarrow{RL}$ and $\overrightarrow{SM}$.

4. △RST is similar to △ABC.

The construction has made ⦨ R ≅ ⦨ A and ⦨ S ≅ ⦨ B. Therefore, △ABC ∼ △RST, because two triangles are similar if two angles of one triangle are congruent, respectively, to two angles of the other triangle.

Inscribe an equilateral triangle in a given circle.

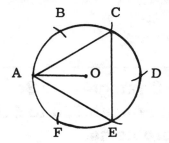

<u>Solution</u>: In the previous problem of inscribing a hexagon in a circle, we showed that an arc whose chord is equal in length to the radius of a circle, has measure $60°$. Two such arcs contain $120°$, and an angle inscribed in this compound arc will measure $60°$. Each angle of an equilateral triangle will be inscribed in this manner.

Therefore, we will mark off six arcs of the type described above, and join every other point to form a figure with 3 equal sides that is inscribed in the circle.

1.   Given circle O , draw any radius $\overline{OA}$.

2.   Using A as a center, and a radius of length OA , construct an arc which intersects the circle at B .

3.   Similarly, using B as a center and a radius of length OA, construct an arc which intersects the circle at C . In like manner, obtain points D, E and F on the circle.

4.   Construct chords $\overline{AC}$, $\overline{CE}$, and $\overline{EA}$.

5.   △ACE is an equilateral triangle. By the construction,

$\overset{\frown}{AC} \cong \overset{\frown}{CE} \cong \overset{\frown}{EA}$. Since the circle measures $360°$, each arc must measure $120°$ (the whole equals the sum of its parts). Therefore, ⊀ E , ⊀ A, and ⊀ C , which are inscribed in these three arcs, and are the angles of triangle △ACE must by a theorem, measure ½ the measure of their intercepted arcs. Therefore, m ⊀ E = m ⊀ A = m ⊀ C = $60°$.

△ACE is equiangular, and consequently is the desired inscribed equilateral triangle.

CHAPTER 32

CIRCLES

Basic Attacks and Strategies for Solving Problems in this Chapter. See pages 530 to 535 for step-by-step solutions to problems.

Understanding the constructions given in this chapter is dependent upon knowledge of several theorems about circles. Here is a list of those theorems, and problems which rely on them.

1. Any point on the perpendicular bisector of a segment is equidistant from the endpoints of that segment. (Problems 601, 602, and 604)

2. In a circle, congruent chords subtend congruent arcs. (Problem 602)

3. If a line is tangent to a circle at point A, then this line is perpendicular to the radius $\overline{OA}$, where O is at the center of the circle. (Problems 603 and 604)

4. Any point on the bisector of an angle is equidistant from the sides of the angle. (Problem 606)

5. The perpendicular bisectors of the sides of a triangle intersect in a point. (Problem 605)

6. The bisectors of the angles of a triangle intersect in a point. (Problem 606)

7. An angle inscribed in a semicircle is a right angle. (Problem 604)

Step-by-Step Solutions to Problems in this Chapter, "Circles"

● **PROBLEM** 600

Draw a circle using the length RS as a radius and R as a center.

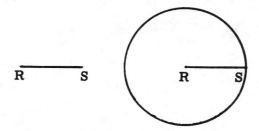

Solution: Adjust the compass so that the metal tip rests on point R and the pencil tip rests on S. Now, all circles drawn with the compass will have radius RS. Since R is to be the center of the circle, it is not necessary to move the metal tip. Keeping the metal tip stationary, rotate the pencil leg around until a full circle is drawn.

● **PROBLEM** 601

Locate the center of a given circle.

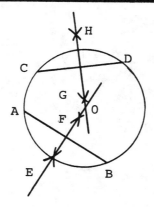

Solution: In any given construction, the special properties of the figure must be exploited. The center of the circle has many special properties that belong to no other points in the circle. Several of the properties, such as, "the center is equidistant from every point on the circle," are not useful to our problem because we do not know the radius, nor can we measure distances except in the crudest manner. We can, however, construct perpendicular bisectors of various chords. By earlier theorems, we know that the line connecting the center of the circle and the midpoint of a chord is the perpendicular bisector of the chord. This holds for every chord in the circle. Therefore, the center of the circle is on the perpendicular bisector of every chord in the circle. By choosing two non-parallel chords (so that their perpendicular bisectors do not coincide) and constructing their perpendicular bisectors, we know that only the point that lies on both bisectors can be the center.

Given: Any circle.

Wanted: The center, O, of the circle.

Construction:

(1) Draw any two nonparallel chords and label them $\overline{AB}$ and $\overline{CD}$.

(2) Construct the perpendicular bisector of each chord. Name them $\overleftrightarrow{EF}$ and $\overleftrightarrow{GH}$, respectively.

(3) The intersection of $\overleftrightarrow{EF}$ and $\overleftrightarrow{GH}$ is O, the center of the circle.

● **PROBLEM** 602

Bisect arc AB .

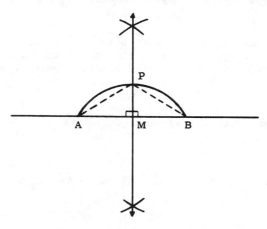

Solution: To bisect $\overset{\frown}{AB}$, we find the point P on $\overset{\frown}{AB}$ such that m $\overset{\frown}{AP}$ = m $\overset{\frown}{BP}$. If arcs of the same circle are congruent, their chords are congruent. As such, $\overline{AP} \cong \overline{PB}$, and ΔAPB would have to be isosceles. Therefore, if we construct the only isosceles triangle with base $\overline{AB}$ and vertex on the arc $\overset{\frown}{AB}$, then that vertex will be the point that bisects the arc.

    It is more straightforward than that though. We know that the median to the base in an isosceles triangle is perpendicular to the base, and thus the vertex P is on the perpendicular bisector of base $\overline{AB}$. Thus, P is the intersection of $\overset{\frown}{AB}$ and the perpendicular bisector of $\overline{AB}$.

Given: $\overset{\frown}{AB}$
Wanted: Bisect $\overset{\frown}{AB}$
Construction:
  (1)   Draw $\overline{AB}$.
  (2)   Construct the perpendicular bisector of $\overline{AB}$. The perpendicular bisector and $\overset{\frown}{AB}$ intersect at P, the bisector of $\overset{\frown}{AB}$.

<u>Proof</u>: Note that, because $\overleftrightarrow{PM}$ is the perpendicular bisector, $\overline{AM} \cong \overline{MB}$, right angles $\sphericalangle$ AMP $\cong \sphericalangle$ BMP, and $\overline{PM} \cong \overline{PM}$. By the SAS Postulate, $\triangle$APM $\cong \triangle$BPM and corresponding sides $\overline{AP} \cong \overline{PB}$. Since their chords are congruent $\overset{\frown}{AP} \cong \overset{\frown}{PB}$.

● **PROBLEM 603**

Construct a line tangent to a circle at any point of the circle.

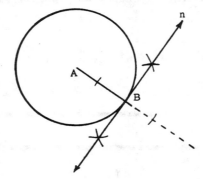

<u>Solution</u>: The tangent to a circle at point X is defined as the line passing through X perpendicular to the radius to point X. Thus, the problem of drawing the tangent reduces to drawing the perpendicular to a given line.

Given: Point B of $\odot$A.

Wanted: A tangent to $\odot$A at B.

Construction:

(1) Draw radius $\overline{AB}$.

(2) Construct line n, the perpendicular to $\overleftrightarrow{AB}$ at point B.

● **PROBLEM 604**

Construct the lines tangent to a circle through a point external to the circle.

<u>Solution</u>: Let P be the point of tangency. A line is tangent to a circle at point X, if the line is perpendicular to the radius drawn to point X. Therefore, in the figure shown, $\overline{BP} \perp \overline{PA}$ or $\sphericalangle$ BPA is a right angle and $\triangle$BPA is a right triangle. Thus, there are two requirements for point P: (i) P, the point of tangency, must be on circle A; and (ii) $\triangle$BPA must be a right triangle. The locus of

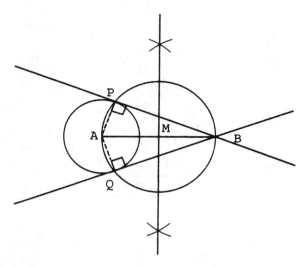

points satisfying condition (i) is the circle A. The locus of points satisfying condition (ii) is the circle (not the semicircle) with $\overline{BA}$ as the diameter. (To see this, remember that any triangle inscribed in a semicircle is a right triangle with the hypotenuse as the diameter. The converse is also true. All right triangles with given hypotenuse length can be inscribed in a semicircle with diameter equal to the hypotenuse.)

The intersection of the loci are thus the possible points of tangency, point P and Q .

Given: Point B in the exterior of ⊙A .

Wanted: The lines through B tangent to ⊙A .

Construction:

(1)   Draw and bisect $\overline{AB}$. Label the midpoint M .

(2)   With M as center and MB as the radius construct the circle through A and B. Circle M, with diameter $\overline{AB}$, intersects ⊙A at P and Q .

(3)   Draw $\overrightarrow{BP}$ and $\overrightarrow{BQ}$, the required tangents.

● **PROBLEM** 605

Construct a circle circumscribed around a triangle.

<u>Solution</u>: Given the center of the circle and a point on the circle, we can construct the circle. Because the circle circumscribes the triangle, the vertices of the triangle are points on the circle. Similarly, the sides of the triangle are chords of the circle. The center of the circle must, therefore, be on the perpendicular bisectors of any two sides.

Another way of viewing the perpendicular bisectors: the center of the circle must be equidistant from the vertices of the triangle. The perpendicular bisectors of the sides of the triangles intersect at a point characterized by this.

Given: ΔABC .

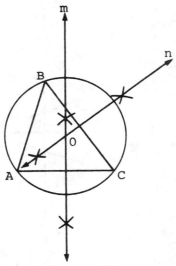

Wanted: A circle circumscribed about △ABC.

Construction:

(1) Construct n, the perpendicular bisector of $\overline{CB}$.

(2) Construct m, the perpendicular bisector of $\overline{AC}$.

(3) Lines n and m intersect at O .

(4) Construct the circle, using O as the center and either OC , OA or OB as the radius.

(5) Circle O is the desired circle.

● **PROBLEM** 606

Construct a circle inscribed in a triangle.

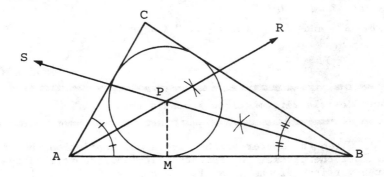

Solution: The circle inscribed in the triangle must be tangent to all three sides. The radius drawn to the point of tangency is, therefore, perpendicular to that side of the triangle. But the length of a segment drawn perpendicular to another segment from a point is the distance from the point to the segment. Consequently, r, the radius of the inscribed circle, is the distance from the center of the inscribed circle to each side. The center of the inscribed circle is therefore, equidistant from the sides of the triangle. By the Angle Bisector Theorem for concurrence, we know that the angle bisectors of the triangle are concurrent at the point equidistant from the three sides. Therefore, the point at which the angle bisectors of the triangle intersect is the center of the inscribed circle. By constructing a perpendicular from that point to a side, we construct a radius. Given the radius and the center of the circle, the circle can be constructed.

Given: $\triangle ABC$.

Wanted: A circle inscribed in $\triangle ABC$.

Construction:

(1) Construct $\overrightarrow{AR}$, the bisector of $\angle A$.

(2) Construct $\overrightarrow{BS}$, the bisector of $\angle B$.

(3) Label the intersection of $\overrightarrow{AR}$ and $\overrightarrow{BS}$, point P .

(4) Construct the perpendicular $\overline{PM}$ from P to $\overline{AB}$.

(5) Construct a circle of center P and radius PM .

(6) Circle P is the desired circle.

CHAPTER 33

POLYGONS

Basic Attacks and Strategies for Solving Problems in this Chapter. See pages 536 to 541 for step-by-step solutions to problems.

This short chapter involves the construction of polygons with certain properties. Most of these problems require the ability to construct a line perpendicular to a given line through either a point on the line or a point not on the line. These procedures are illustrated in Problems 575 and 576 of Chapter 30. In the solution to Problem 608, there is a construction of the fourth proportional to three given segments. That procedure is illustrated in Problem 586 of Chapter 30. A review of these three constructions is recommended.

Step-by-Step Solutions to Problems in this Chapter, "Polygons"

● **PROBLEM** 607

Given two squares, construct a square equivalent to their sum.

<u>Solution</u>: Two polygons are equivalent if their areas are equal. If the squares have sides of length a and b, then their areas are a^2 and b^2 . We must construct a square of side c such that its area, c^2, is the sum of these two areas, or $c^2 = a^2 + b^2$. We recognize this relationship as the Pythagorean Formula. Thus the sides of the two squares and the side of the equivalent square form a Pythagorean triplet. To construct the equivalent square, we find c, the length of the side. To find c, we construct a right triangle with legs a and b . The hypotenuse is then c.

Construction:
(1)   Construct segment $\overline{MN}$ equal in length to a.

(2)   Construct segment $\overline{PN} \perp \overline{MN}$ and PN = b.

(3)   Construct a square with a side of length PM.

● **PROBLEM** 608

Construct a rectangle that has a given base length and is equal in area to the area of a given parallelogram. (See figure).

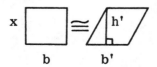

Area of Rectangle = Area of Parallogram

$$b\,x = b'h'$$

$$b : b' = h' : x$$

536

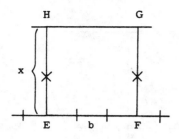
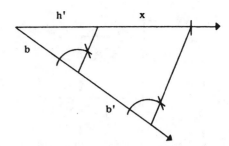

Solution: We are given parallelogram ABCD, with base of length b' and altitude of length h'. Accordingly, area of ⧜ ABCD = b'h'.

The rectangle we want to construct has a known base length of b, and an unknown altitude, say x. Then, area of rectangle to be constructed = bx.

Since we want the area of the rectangle to equal the area of the parallelogram, it must be that bx = b'h'. Therefore,

$$\frac{b}{b'} = \frac{h'}{x}$$

or b: b' = h':x is a proportion. We must find the fourth proportional of b,b' and h, (i.e., x), and then construct the rectangle with base b and altitude x.

Follow the construction in the accompanying sketch:

(1) Construct $\overline{DR}$, the altitude of ⧜ ABCD, whose length has been represented by h'.

(2) Construct the fourth proportional to b,b', and h', (i.e., x). (This construction is vitally important to the problem. It was done previously.)

(3) The segment we have just determined, whose length is x, is congruent to the altitude of the required rectangle.

(4) Let $\overline{EF}$ represent the base of the rectangle of length b. At points E and F, construct perpendiculars to $\overline{EF}$ of length x; these are the altitudes of the rectangle. (This construction has also appeared previously.)

(5) Through points H and G, construct $\overline{HG}$. Since $\overline{HE} \cong \overline{FG}$, $\overline{HE} \parallel \overline{FG}$ and ⦣ HEF is a right angle. EFGH is the required rectangle, whose base is a segment of length b, and whose area is equal to the area of ⧜ ABCD.

Since the construction makes x the fourth proportional of b,b' and h', then b: b' = h': x or bx = b'h'. Since the area of ⧜ABCD = b'h', and the area of rectangle EFHG = bx, the area of rectangle EFGH = area of ⧜ABCD. Therefore, EFGH is the rectangle we set out to construct.

● **PROBLEM** 609

Circumscribe a square about a given circle.

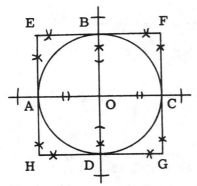

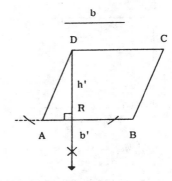

This construction is rather involved. It can best be understood when it is first done, and then justified. The steps of the construction will be presented, and then the logic will be reviewed in the proof.

1. Given circle O, draw diameter $\overline{AC}$ in circle O.

2. At O, construct diameter $\overline{BD}$ perpendicular to diameter $\overline{AC}$. (Review the earlier construction of a line perpendicular to a given line through a point on the line.)

3. At points A,B,C and D, construct tangents to circle O which intersect at points E, F, G and H. This is done by first extending diameters $\overline{AC}$ and $\overline{BD}$. (Recall that the tangent to a circle, at the point of intersection of the diameter, will be perpendicular to the diameter.) Draw the perpendicular to $\overline{AC}$ and $\overline{BD}$ at the points A,C,B,D, respectively.

4. EFGH is a square.

In the proof, we will show EBOA, FBOC, HDOA, and GDOC are all squares. This will allow us to conclude that the vertex angles of EFGH are right angles and that EFGH is a rectangle. By also showing a pair of adjacent sides of EFGH are congruent, we will prove EFGH is a square.

Statement	Reason
1. $\overline{EB} \cong \overline{EA}$	1. If two tangents are drawn to the same circle from a given point outside the circle, the two tangents are congruent.
2. $\angle$ EBO and $\angle$ EAO are right angles.	2. Tangents to a circle that intersect a diameter of that circle at the point of tangency, are perpendicular to the diameters and, as such, form right angles.
3. $\angle$ AOB is a right angle.	3. By the construction, $\overline{BD} \perp \overline{AC}$. Hence, $\overline{BD}$ and $\overline{AC}$ intersect to form right angles.
4. $\angle$ AOB is supplementary to $\angle$ EBO, and $\angle$ AOB is supplementary to $\angle$ EAO.	4. Any pair of right angles are supplementary to each other.
5. $\overline{AE} \parallel \overline{OB}$ and $\overline{AO} \parallel \overline{EB}$.	5. When two lines are cut by a transversal forming interior angles on the same side of the transversal that are supplementary, then the lines are parallel.
6. EBOA is a square.	6. Because opposite sides of EBOA are parallel, EBOA is a

538

parallelogram. Because ∡ AOB of parallelogram EBOA is a right angle, EBOA is a rectangle. Because adjacent sides are congruent, EBOA is a square.

7. FBOC, HDOA and GDOC are squares.

7. Apply the logic of steps 1 to 6 to each quadrilateral individually.

8. $\overline{EB} \cong \overline{AO}$

$\overline{BF} \cong \overline{OC}$

$\overline{AO} \cong \overline{HD}$

$\overline{OC} \cong \overline{DG}$

8. Opposites sides of a square are congruent.

9. $\overline{EB} + \overline{BF} \cong \overline{AO} + \overline{OC}$

$\overline{AO} + \overline{OC} \cong \overline{HD} + \overline{DG}$ or

$\overline{EF} \cong \overline{AC}$ and

$\overline{AC} \cong \overline{HG}$

9. Addition Postulate for congruence

10. $\overline{EF} \cong \overline{HG}$

10. Transitivity Postulate for congruence.

11. $\overline{EA} \cong \overline{OB}$

$\overline{AH} \cong \overline{OD}$

$\overline{OB} \cong \overline{FC}$

$\overline{OD} \cong \overline{CG}$

11. Same as reason 8.

12. $\overline{EA} + \overline{AH} \cong \overline{OB} + \overline{OD}$

$\overline{OB} + \overline{OD} \cong \overline{FC} + \overline{CG}$ or

$\overline{EH} \cong \overline{BD}$

$\overline{BD} \cong \overline{FG}$

12. Same as reason 9.

13. $\overline{EH} \cong \overline{FG}$

13. Same as reason 10.

14. $\overline{AC} \cong \overline{BD}$

14. All diameters of a given circle are congruent.

15. $\overline{EF} \cong \overline{AC}$ and

$\overline{EH} \cong \overline{BD}$

15. See steps 9 and 12.

16. $\overline{EF} \cong \overline{EH}$

16. Segments congruent to congruent segments must be congruent to each other.

17. In square HDOA, ∡ AHD is a right angle.

17. All angles formed by the intersection of two sides of a square are right angles.

18. EFGH is a square.

18. Because opposite sides of EFGH are congruent, EFGH is a parallelogram. Because ∡ AHD is a right angle, EFGH is a rectangle. Because adjacent sides $\overline{EF}$ and $\overline{EH}$ are congruent, EFGH is a square.

Therefore, the construction carried out results in a square. Since $\overline{EF}$, $\overline{EH}$, $\overline{HG}$, and $\overline{GF}$ are tangent to O at B,A,D, and C, respectively, EFGH is circumscribed about O, by definition.

Inscribe a square in a circle.

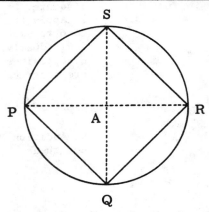

Solution: Given a side or a diagonal, we can construct a square. In the case of a circle, it is simpler to find the diagonal. To construct the diagonal, we use two facts: (1) Because the square is to be inscribed in the circle, all vertices of the square lie on the circle. Therefore, the diagonals of the square must be chords of the circle. (2) The center of a regular polygon is the center of the circumscribed circle. Since the diagonal of a square must pass through the center of the square, the diagonal must pass through the center of the circle. Any chord that passes through the center of the circle is a diameter. Therefore, the diagonals of the inscribed square are diameters of the circle, and any diameter can be constructed as the _first_ diagonal. The diagonals of any rhombus, of which the square is a special case, are perpendicular, therefore, the second diagonal must be the diameter perpendicular to the first diagonal. The points where the diagonals intercept the circle are the vertices of the inscribed square.

Given: Circle A.

Wanted: A square inscribed in ⊙A .

Construction:

(1) Locate the center of the circle. (The center of the circle is a point on the perpendicular bisector of every chord. Draw any two nonparallel chords. Construct their perpendicular bisectors. The point of intersection must be the center.)

(2) Draw diameter $\overline{PR}$.

(3) Construct the diameter $\overline{QS}$ perpendicular to $\overline{PR}$. Because they are diameters, $\overline{QS} \cong \overline{PR}$.

(4) Draw $\overline{PQ}$, $\overline{QR}$, $\overline{RS}$, and $\overline{SP}$. Because the diagonals of the square are congruent and perpendicular to each other, quadrilateral PQRS is a square.

Inscribe a regular hexagon in a circle.

Solution: Consider the n-sided regular polygon in the figure above (n, in this case, equals 6). From the center of the circumscribed circle, draw the radii to the vertices. Because all radii are congruent, two sides of every triangle must be congruent to two sides of every other triangle. Because this is a regular polygon, all the sides, $\overline{VT}$, $\overline{VP}$, $\overline{PQ}$, etc., must be congruent. By the SSS Postulate, all the triangles must be congruent, and, by corresponding parts, all the central angles must be congruent. Since the central angles must sum to 360°, the measure of the central angle of each side of an n-sided regular polygon must be 360/n .

This is the crux of every problem that involves inscribing a regular polygon in a circle. If we are able to construct an angle of measure 360/n , then the polygon can be constructed by constructing n isosceles triangles (each with two sides of length r and vertex angle of measure 360/n) side by side. In the case of the square, this critical angle is 360/4 or 90°— a right angle. Therefore, the inscription of a square in a circle is possible. (Note that the inscribed square is usually constructed by constructing two perpendicular diameters - or four right angles.) In the case of a heptagon, the critical angle is 360/7 or $51\frac{3}{7}°$. We cannot construct this

angle. Therefore this construction is impossible by our method.

In this problem, we are asked to inscribe a hexagon. The central angle is 360/6 = 60. We can construct an angle of 60° , and a hexagon can be inscribed in a circle by constructing 6 isosceles triangles with vertex angle = 60° and the vertex at the center of the circle.

However, much of the work can be omitted. In this case, the triangle to be constructed is an isosceles triangle with vertex angle = 60° . This is an equilateral triangle. Therefore, the side of the triangle that also forms a chord with the circle, must be equal to the radius. Consequently, for a hexagon inscribed in a circle, the measure of a side of the hexagon equals the measure of the radius.

This is extremely useful. We know the vertices are points on the circle and that they are spaced a distance r apart. There is no need for central triangles at all.

Given: Circle A.

Wanted: A regular hexagon inscribed in ⊙A .

Construction:
(1)   Find the center of ⊙A .

(2)   Choose any point P on the circle to be the first vertex. Using P as the center, and the radius of ⊙A as the radius, construct an arc intersecting the circle at Q. Since Q is on the circle and also a distance r from vertex P, Q is also a vertex.

(3)   Repeat this process to locate points R, S, T, and V .

(4)   Join these points to form $\overline{PQ}$, $\overline{QR}$, $\overline{RS}$, $\overline{ST}$, $\overline{TV}$, and $\overline{VP}$. PQRSTV is the inscribed hexagon.

CHAPTER 34

COMPLEX CONSTRUCTIONS

Basic Attacks and Strategies for Solving Problems in this Chapter. See pages 542 to 557 for step-by-step solutions to problems.

The problems in this chapter are quite difficult. In most situations, a complex construction is accomplished by a number of simple constructions. These simple constructions are illustrated in Chapter 30. Here is a list of some of them, together with the corresponding Chapter 30 problem number.

1. Construct the perpendicular bisector of a segment. (Problem 577)
2. Construct a line perpendicular to a given line through a point not on the given line. (Problem 576)
3. Construct a line perpendicular to a given line through a point on the given line. (Problem 575)
4. Construct a line parallel to a given line through a given point. (Problem 591)
5. Divide a given line segment into parts proportional to given segments. (Problem 584)
6. Find the mean proportional between two given segments. (Problem 585)
7. Construct a segment of length ab where a and b are lengths of given segments. (Problem 590)

There is one general technique illustrated in this chapter and also in the four previous chapters. When you don't immediately know how to make the construction, draw a rough sketch of the "completed figure" and assume that the construction has been completed. Look at the sketch for geometric relationships which can be more simply constructed. Then, try to make the appropriate constructions. This procedure of working backwards is a good problem-solving technique.

Step-by-Step Solutions to Problems in this Chapter, "Complex Constructions"

● **PROBLEM** 612

Construct △ABC given sides with the lengths of the following three segments (shown in Figure 1).

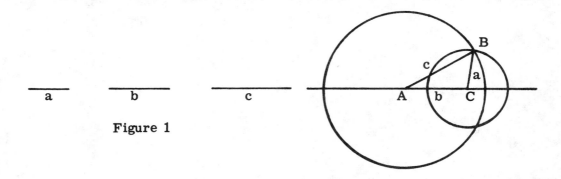

Figure 1

Figure 2

Solution: To find two vertices of the triangle, draw a line. Construct on it a segment of length b. Then the endpoints must be A and C.

To find B, you must solve a locus problem. The locus of points that vertex B can be chosen from, must satisfy two conditions: (1) AB = c and (2) BC = a. The locus of condition (i) is a circle. Construct a circle with center A and radius c. The locus of condition (ii) is also a circle. Construct a circle with center C and radius a. The intersection of the two loci yields two points. Either one satisfies all the requirements for B. Label either one of them B and ignore the other. To complete the construction, draw AB and BC.

● **PROBLEM** 613

Given the unit length below, construct a segment of length $\sqrt{3}$.

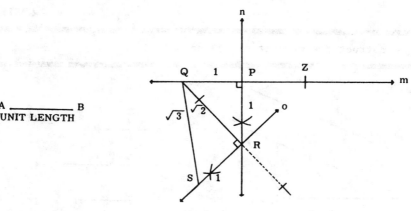

A ——————— B
UNIT LENGTH

Solution: By the Pythagorean Theorem, we know that $c^2 = a^2 + b^2$ or $c = \sqrt{a^2 + b^2}$. By constructing a right triangle with legs of length 1, we know that the length of the hypotenuse must be $\sqrt{1^2 + 1^2}$ or $\sqrt{2}$. Constructing a right triangle with sides $\sqrt{2}$ and 1, we obtain a hypotenuse of $\sqrt{\sqrt{2}^2 + 1^2} = \sqrt{2+1} = \sqrt{3}$. By continually taking the last hypotenuse of length $\sqrt{R}$, we can construct a right triangle with legs $\sqrt{n}$ and 1, which yields a hypotenuse of length $\sqrt{\sqrt{n}^2 + 1}$ or $\sqrt{n + 1}$. Thus, by proceeding in this manner, the square root of any integer can be constructed as the length of a segment.

Given: AB = 1

Wanted: a segment of length $\sqrt{3}$.

Construction:

(1) Draw line m.

(2) Construct segment $\overline{QP} \cong \overline{AB}$, the unit length.

(3) Construct line n perpendicular to $\overleftrightarrow{m}$ through point P. (To construct n, construct segment $\overline{PZ} \cong \overline{AB}$. Construct the perpendicular bisector of $\overline{QZ}$. Since PZ = QP = AB, P is the midpoint of $\overline{QZ}$ and the perpendicular bisector of $\overline{QZ}$ is the line perpendicular to $\overleftrightarrow{m}$ at point P.)

(4) Construct segment $\overline{PR} \cong \overline{AB}$ on n.

(5) Draw $\overleftrightarrow{QR}$. By the Pythagorean Theorem, QR = $\sqrt{2}$.

(6) Construct line O, the perpendicular to $\overleftrightarrow{QR}$ through point R.

(7) Construct segment $\overline{SR} \cong \overline{AB}$ on line O.

(8) Draw $\overline{QS}$. $QS^2 = QR^2 + SR^2 = (\sqrt{2})^2 + 1^2 = 3$ and, therefore, QS = $\sqrt{3}$.

Construct the product of 2 and 4½ .

Figure (a)

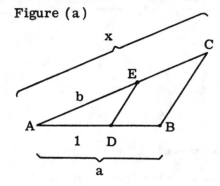

Figure (b)

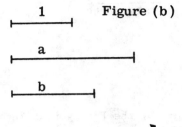

Figure (c)

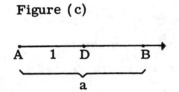

Figure (d)

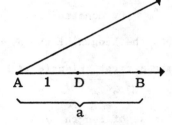

Figure (f)

Figure (e)

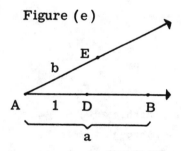

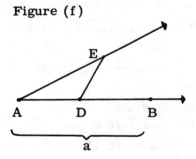

Figure (h)

Figure (g)

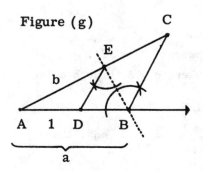

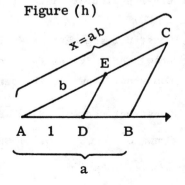

Figure (i)

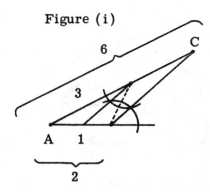

Solution: If we wish to construct the product of 2 numbers, a and b , then we really want to construct a segment whose measure is the product of the measures of two given segments (a and b). A method for doing this, which uses the properties of similar triangles, is indicated below.

In figure (a), $\triangle AED \sim \triangle ACB$ by the A.A.A. Similarity Theorem, since $\overline{DE} \parallel \overline{BC}$. Hence, $\frac{AE}{AC} = \frac{AD}{AB}$ or, with the notation of the figure

$$\frac{b}{x} = \frac{1}{a}$$

or
$$x = ab .$$

Hence, if we are to construct the product (x) of 2 given numbers (a and b), we can follow the program detailed below:

(1) Draw segments of length 1,a, and b, in arbitrary units. (Fig.(b)).

(2) Draw a ray, $\overrightarrow{AB}$, and lay off lengths of 1 and a along the ray, as measured from its endpoint, A. (Fig. (c)).

(3) Draw a second ray which passes through the endpoint, A, of the first ray, and makes an arbitrary angle with the latter. (Fig.(d)).

(4) Lay off a segment of length b along the second ray, $\overrightarrow{AE}$, as measured from the point A. (Fig. (e)).

(5) Draw the segment, $\overline{ED}$, connecting the ends of the segments of length 1 and b. (Fig. (f)).

(6) Construct a segment, $\overline{BC}$, such that it passes through the end, B, of segment $\overline{AB}$, and such that $\overline{BC} \parallel \overline{ED}$. (Fig. (g)).

(7) The distance, measured from point A, along ray $\overrightarrow{AE}$, to the point of intersection of $\overline{BC}$ and $\overleftrightarrow{AE}$ is x = ab . (Fig. (h) , since figure (h) is, exactly, figure (a).)

Following the procedure outlined above for the case when a = 2, b = 3 we obtain figure (i). (Here, we let our arbitrary unit be ½" .) Measurement of $\overline{AC}$ shows that it is, indeed, equal to 6 units of length.

● **PROBLEM** 615

Construct a triangle equal in area to a given pentagon. (See figure).

Solution: We have already shown that two triangles with a common base which both have a vertex on a line parallel to that base will be of equal area. This fact will guide our construction.

545

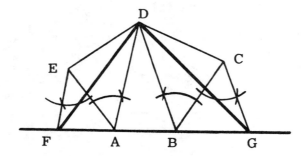

We can divide the pentagon ABCDE into three triangles by drawing diagonals $\overline{DB}$ and $\overline{DA}$. If we can find 2 triangles equal in area to △AED and △BCD which, when placed adjacent to △ABD form a triangle, our task if complete.

1. Given pentagon ABCDE, extend $\overline{AB}$ both to the left and to the right.

2. Construct diagonal $\overline{DB}$.

3. Through vertex C, construct $\overleftrightarrow{CG}$ parallel to $\overline{DB}$. [The construction of a line parallel to a fixed line through a given point was done previously.]
   $\overleftrightarrow{CG}$ intersects $\overleftrightarrow{AB}$ at G.

4. Construct $\overline{DG}$. At this point, △BCD and △BGD have been constructed.

5. Construct diagonal $\overline{DA}$.

6. Through vertex E, construct $\overleftrightarrow{EF}$ parallel to $\overline{DA}$. $\overleftrightarrow{EF}$ intersects $\overleftrightarrow{BA}$ at F.

7. Construct $\overline{DF}$. Now, △'s AED and AFD have been completed.

8. △FDG is the required triangle, which is equal in area to pentagon ABCDE.

   The construction makes $\overline{EF} \parallel \overline{DA}$. Therefore, the area of △AFD = the area of △AED because the two triangles have a common base, $\overline{DA}$, and vertices F and E lie on a line parallel to the base. A previously solved problem gives proof of this.

   Similarly, since $\overline{CG} \parallel \overline{DB}$, the area of △BGD = the area of △BCD.

   The area of △ABD = the area of △ABD, by reflexivity.

   Pentagon ABCDE is composed of △'s AED, BCD and ABD. Triangle FDG is composed of △'s AFD, BGD and ABD.

   Therefore, area of triangle DFG = area of pentagon ABCDE since the sums of equal quantities will be equal quantities. The area of △AFD + the area of △ABD + the area of △BGD = the area of △AED + the area of △ABD + the area of △BCD, and the sums of the areas of the respective components of triangle DFG and pentagon ABCDE are equal.

● **PROBLEM** 616

To outwit the land assessor, Farmer Cornfield bought an irregularly shaped piece of land. (Map shown below). Land assessor Coxeter, though, did geometric constructions in his spare time and thus was able to make an excellent approximation by reducing the irregular polygon to an equivalent triangle. Using a straight edge, compass, and ruler find the area of Cornfield's wheatfield. (You are allowed only two measurements with the ruler).

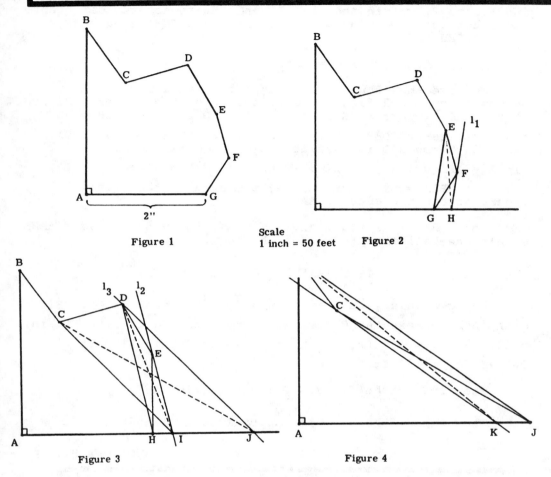

Figure 1

Scale
1 inch = 50 feet

Figure 2

Figure 3

Figure 4

Solution: The key to this problem is to reduce the seven sided figure into a six sided figure with the same area. From this, we reduce the figure to a five sided figure with the same area. We continue in this manner until we have a three sided figure. We then estimate the altitude and base and from this find the area of the wheatfield.

To reduce the seven sided figure to a six sided figure of the same area, we note that (see Figure 2)

    (i)   Area ABCDEFG = Area ABCDEG + Area ΔEFG.

Remember that triangles with the same altitude and the same base have equal areas. Draw $\ell_1 \parallel \overline{EG}$, and let point H be the intersection of $\overleftrightarrow{AG}$ and ℓ_1. Note ΔFEG and ΔHEG have the same base $\overline{EG}$. Furthermore, since parallel lines are everywhere equidistant, the altitude from F to $\overline{EG}$ and from H to $\overline{EG}$ must be equal. Since the triangles have equal base and altitude measures, Area ΔFEG = Area ΔHEG. Thus, substituting in equation (i), we have

(ii)   Area ABCDEFG = Area ABCDEG + Area ΔHEG.

This may not seem to be a great improvement, but consulting Figure 2, we can see that $\overline{AG}$ and $\overline{GH}$ are not two different sides but the same side $\overline{AH}$. Thus Area ABCDEG + Area ΔHEG = Area ABCDEH and

(iii)   Area ABCDEFG = Area ABCDEH.

We have thus found a six sided figure with the same area as the seven sided figure. Continuing in this manner, (Figure 3) we find ℓ_2 through E such that $\ell_2 \parallel \overline{DH}$. Area ΔDEH = ΔDIH and thus Area ABCDEH = Area ABCDH + Area ΔDEH = Area ABCDI. Next, ℓ_3 through D is parallel to $\overline{CI}$ and Area ABCDI = ABCJ. Finally (Figure 4), we find ℓ_4 through C such that $\ell_4 \parallel \overline{BJ}$, and Area ABCJ = Area ABK. Thus, the area of the original figure equals the area of ΔABK.

Note ⊄BAK is a right angle. Thus, the area of right ΔABK equals one-half the product of the legs.

(iv)   Area = $\frac{1}{2}$(AB)(AK).

Using a ruler, we note that AB = $2\frac{7}{8}$" and AK = $3\frac{3}{8}$". Since each inch represents 50 ft., $\overline{AB}$ represents ($2\frac{7}{8}$ in.)(50 ft./in.) = 143.75 ft. $\sim$ 140 ft. and $\overline{AK}$ represents ($3\frac{3}{8}$ in.)(50 ft./in.) = 168.75 ft. $\sim$ 170 ft.

The area of the field = $\frac{1}{2}$(140 ft.)(170 ft.).

The area of the field is approximately 12,000 sq. ft.

● **PROBLEM** 617

Construct a square equal in area to a given triangle (see figure).

Solution: We are given ΔABC with base length b and altitude length h . Therefore, the area of ΔABC = ½bh.

The construction asks us to form a square, the length of whose side is unknown, equal in area to ΔABC.
Let x = the length of the side of the square, therefore, the area of the square = x^2. The problem requires that the two areas be equal. Therefore, ½bh = x^2 = x·x . It follows that the proportion

$$\frac{b}{x} = \frac{x}{\frac{1}{2}h}$$

or b:x = x:½h must be true.
The construction will involve bisecting h, finding the mean proportional between b and ½h, and then using that length to construct the square.

1. In ΔABC, construct altitude $\overline{CD}$. This amounts to drawing a line $\overleftrightarrow{CD}$ perpendicular to $\overline{AB}$ from a point outside the latter line.

2. Bisect altitude $\overline{CD}$, whose length is h, so that CE = ½h .

3. On line $\overleftrightarrow{RT}$, construct x, the mean proportional between b

548

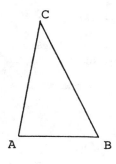

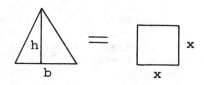

Triangle = Square

$$\tfrac{1}{2}bh = x^2$$

$$b : x = x : \tfrac{1}{2}h$$

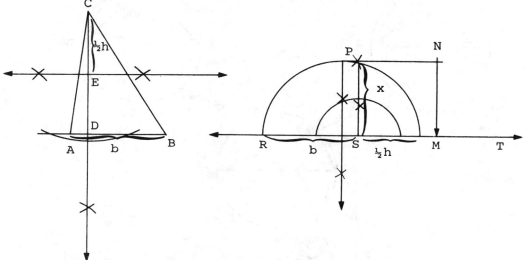

and $\tfrac{1}{2}h$. (For this type of construction, refer to previous problems in this section.)

4. $\overline{SP}$, whose length we call x, is the side of the required square.

5. Since $\overline{SP}$ has been drawn perpendicular to $\overleftrightarrow{ST}$, there is a quick way to finish constructing the square. On $\overleftrightarrow{ST}$, construct $\overline{SM}$ whose length is x. Using M and P as centers, and a radius of length x, construct two arcs which intersect at N. Draw $\overline{PN}$ and $\overline{NM}$.

6. SMNP is now the required square, which is equal in area to $\triangle ABC$.

By construction, we have made x the mean proportional between b and $\tfrac{1}{2}h$. Therefore, b: x = x:$\tfrac{1}{2}h$ or $x^2 = \tfrac{1}{2}bh$.
Therefore, square SMNP is the required square we were to construct.

● PROBLEM 618

Using a straight edge only, construct the perpendicular to a given diameter of a circle from any point off the diameter.

<u>Solution</u>: This construction is quite different from most others in this section in that it restricts you to only using a straight edge without a compass. Hence, the usual ways of dropping a perpendicular from a point to a line cannot be used here. We are given point P and circle O with diameter AB and asked to drop the perpendicular

from P to $\overline{AB}$.

Complete triangle PAB and note the points of intersection of PA and PB with the circle, points C and D, respectively.

Join point C to vertex B and point D to vertex B. Angles ACB and BDA are angles inscribed in a semi-circle and, as such, are right angles. Therefore, $\overline{BC} \perp \overline{AP}$ and $\overline{AD} \perp \overline{PB}$.

By the definition of an altitude of a triangle, $\overline{BC}$ is the altitude to $\overline{AP}$ and $\overline{AD}$ is altitude $\overline{PB}$.

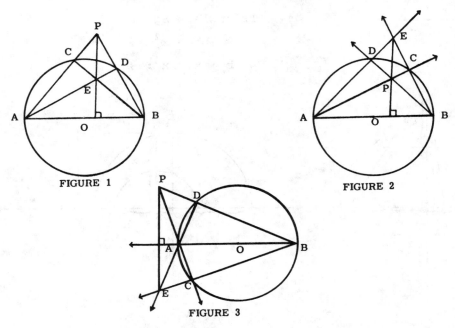

FIGURE 1

FIGURE 2

FIGURE 3

We have a theorem that tells us the three altitudes of a triangle are always concurrent at a point. Since $\overline{AD}$ and $\overline{BC}$ intersect at E, the altitude from P to $\overline{AB}$ must also pass through E. Two points determine a line. Therefore, by extending $\overline{PE}$ to intersect $\overline{AB}$ we will have drawn the third altitude and, hence, the perpendicular to $\overline{AB}$ from P.

The drawing in figure 1 has two restrictions. The first is that P is outside the circle and the second is that $\triangle PAE$ is acute. Either of these restrictions can be lifted and the construction will still work.

In figure 2, P is located in the interior of the circle. Draw $\overline{AP}$ and $\overline{BP}$ extended to C and D, respectively. By extending $\overline{AD}$ and $\overline{BC}$ complete triangle EAB. $\overline{AC}$ and $\overline{BD}$ are the respective altitudes to $\overline{BE}$ and $\overline{EA}$. Again $\overline{PE}$ extended to $\overline{AB}$ is the required perpendicular.

The altitude in figure 3 from P to $\overline{AB}$ will lie outside $\triangle PAB$ and this will show that our construction works when $\triangle PAB$ is obtuse.

Draw $\overline{PA}$ and $\overline{PB}$ and extend $\overline{PA}$ so that it intersects the circle at C. Draw $\overline{BC}$ and $\overline{AD}$ and continue them until they intersect at point E. Once again the line from E to P will be perpendicular to diameter $\overline{AB}$.

● **PROBLEM** 619

Given two circles at fixed locations, find the line that cuts equal chords in both of them.

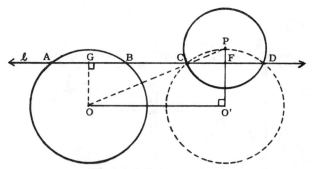

Solution: In the figure, we wish to find the line ℓ that cuts congruent chords $\overline{AB}$ and $\overline{CD}$ in circles O and P. Consider the point O' such that $\overline{OO'} \perp \overline{PO'}$. Now construct the circle at O' congruent to ⊙O . The line containing the common chord of ⊙P and ⊙O' will be the desired line.

To see this, remember that the line connecting the centers of intersecting circles is perpendicular to the common chord. Thus, $\overline{PO'} \perp \ell$. Furthermore, by construction, $\overline{PO'} \perp \overline{OO'}$. Thus, ℓ and $\overleftrightarrow{OO'}$ are parallel lines and everywhere equidistant. The distances FO' and GO must be equal. Recall that chords in the same circle or congruent circles that are the same distance from the center are congruent. Hence, the chord cut by line ℓ in ⊙O is, thus, congruent to the common chord of ⊙P and ⊙O' .

Given: Two circles O and P at fixed locations.

Construct: line ℓ cutting congruent chords in ⊙O and ⊙P .

Construction:

1. Construct a semicircle with diameter OP . (Every point on the semicircle forms a right triangle with $\overline{OP}$ as the hypotenuse.)

2. Select a point O' on the semicircle close enough to ⊙P so that a circle with center O' congruent to ⊙O will intersect ⊙P in two points, C and D. Construct ⊙O' congruent to ⊙O .

3. Construct line $\overleftrightarrow{CD}$. $\overleftrightarrow{CD}$ is the desired line ℓ .

● **PROBLEM** 620

Inscribe a square in a given semicircle.

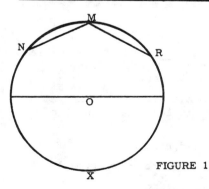

FIGURE 1

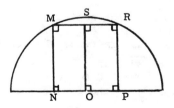

FIGURE 2

Solution: We are given semicircle O and asked to inscribe a square in it. Intuitively, one would envision one side of the square to be on the diameter, with point O being the midpoint of that side. In fact, this is the case and it can be shown by indirect proof.

Suppose less than two vertices lay on the diameter. (See Fig. 1).

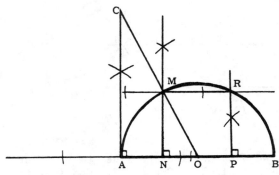

FIGURE 3

Since the square is inscribed, that would imply that at least three vertices N, M, and R lie on the semicircle. Consider angle $\angle$ NMR inscribed in $\odot$ O . Since points N and R are on the same side of the semicircle, major arc $\widehat{NXR}$ is of measure $> 180°$. Then, m $\angle$ NMR = $\frac{1}{2}$ $\widehat{NXR}$ or m $\angle$ NMR > $\frac{1}{2}(180)$ or m $\angle$ NMR > 90. But this contradicts the fact that $\angle$ NMR is a vertex angle of square MNPR and, therefore, is a right angle. Thus, if MNPR is to be a square, at least two vertices must lie on the diameter. Furthermore, since no three vertices of a polygon can be collinear, there are only two vertices on the diameter.

Suppose quadrilateral MNPR is the inscribed square with side $\overline{NP}$ on the diameter. (See Fig. 2). Draw the line $\overline{OS}$ through the center of the semicircle O parallel to sides $\overline{MN}$ and $\overline{PR}$, intersecting side $\overline{MR}$ at S. Since $\overline{MN} \perp \overline{MR}$, $\overline{OS} \perp MR$. Since the perpendicular drawn from the center of the circle to the chord bisects the chord, $\overline{OS}$ bisects $\overline{MSR}$. Since parallel lines $\overline{NM}$, $\overline{SO}$, and $\overline{RP}$ intercept congruent segments on $\overline{MSR}$, they also intercept congruent segments on $\overline{NP}$. Thus, NO = OP, and O is the midpoint of side $\overline{NP}$.

Using these two facts, we can formulate a method for constructing the square. Since the sides of a square are congruent, NP = NM. Since NO = $\frac{1}{2}$NP, NO = $\frac{1}{2}$MN. Thus, $\triangle$MNO is a right triangle ($\angle$MNO is the right angle) with one leg twice the length of the other. We can construct this right triangle by first constructing one similar to it and then drawing a line $\overline{MN} \perp \overline{AB}$ cutting off $\triangle$MNO with required side $\overline{MN}$.

Given: Semicircle O with $\overline{AB}$ as diameter.

Construct: Square MNPR inscribed in the semicircle.

Construction:

1.   Construct $\overline{AC} \perp \overline{AB}$ and AC = AB .

2.   Draw $\overline{CO}$ cutting semicircle O at M . $\triangle$CAO has right angle $\angle$ CAO. Side $\overline{CA}$ is the length of the diameter and $\overline{AO}$ the radius. Thus, leg $\overline{CA}$ is twice the length of the leg $\overline{AO}$.

3.   Construct $\overline{MN} \perp \overline{AB}$. $\triangle$MNO $\sim$ $\triangle$CAO by the A.A Similarity Theorem ($\angle$COA $\cong$ $\angle$ MON; $\angle$ CAO $\cong$ $\angle$ MNO). Thus, $\overline{MN}$ is twice the length of $\overline{NO}$ and $\overline{MN}$ is the desired side.

4.   Mark off OP = ON . Remember, O is the midpoint of side $\overline{NP}$.

5.   Construct $\overline{PR} \perp \overline{AB}$. MNPR is the desired square. To formally prove MNPR is a square, note that
     (1)   MN $\parallel$ RP, since both are perpendicular to $\overline{NP}$.

     (2) $\overline{MN} \cong \overline{RP}$, since they are corresponding sides of congruent triangles $\triangle$MNO and $\triangle$RPO . Thus, MNPR is a parallelogram.

     (3) $\angle$ MNO is a right angle since $\overline{MN} \perp \overline{NP}$. Thus, MNPR is a

552

rectangle.

(4) MN = NP, since NP = 2·NO = 2·(½ MN) = MN. Thus, MNPR is
a square.

Given two parallel lines and a transversal $\overleftrightarrow{FG}$, construct the
circle tangent to both the parallel lines and the transversal.

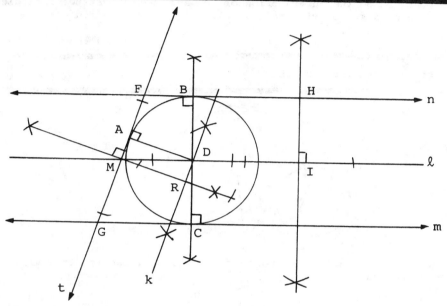

Solution: To construct a circle, we must find (a) the center; and
(b) the radius.

(a) The center: Consider the radii drawn to the tangent points,
$\overline{AD}$, $\overline{BD}$, and $\overline{CD}$. Because these are radii drawn to tangent points,
$\overline{AD}$ ⊥ line t, $\overline{BD}$ ⊥ line n, and $\overline{CD}$ ⊥ line m. Then, by definition,
AD, BD, and CD are the distances from lines t,n, and m to point
D. Therefore, D, the center of the circle, which must be equidistant
from t,m, and n.

The locus of points that are equidistant from n and m is the
line parallel to n and m and equidistant from both. Find the midpoint of the transversal segment cut by n and m and construct
line ℓ parallel to n. The distance BD between n and ℓ must
be the radius of the circle. The center must also be a distance HI
from t. The locus of points a distance HI from t is two parallel
lines a distance HI from t . The center of the circle is the intersection of ℓ and the lines parallel to t.

(b) The radius is the length of $\overline{HI}$ where $\overleftrightarrow{BD}$ is a line perpendicular to n,m, and ℓ .

Given: Two parallel lines n and m, and a transversal t. Let F
and G be the points of intersection of the transversal
with n and m.

Wanted: Construct the circle tangent to both the parallel lines and
the transversal.

Construction:

(1) Locate the midpoint of $\overline{FG}$, label it M.

553

(2) Construct line ℓ through M parallel to n.

(3) Construct line $\overset{\leftrightarrow}{HI}$ perpendicular to n such that point H is on line n and point I is on line ℓ .

(4) Construct $\overset{\leftrightarrow}{MR}$ perpendicular to t, such that $\overline{MR} \cong \overline{HI}$.

(5) Construct line k through point R such that k $\parallel$ t. (By constructing another parallel line on the other side of t, another centerpoint may be found. Since you are only required to find one circle, location of the other point is unnecessary.)

(6) Let D be the intersection of k and ℓ . Construct a circle centered at D with radius $\overline{HI}$. ⊙D is the desired circle.

● PROBLEM 622

Construct a regular decagon.

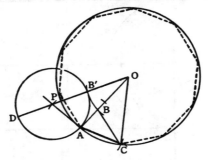

<u>Solution</u>: The radii of a regular decagon form isosceles triangles with vertex angles of

$$\frac{360^{\circ}}{10} = 36^{\circ} .$$

Thus, if we can construct central angles of 36° , then we can form 10 congruent arcs. The 10 congruent chords thus determined are the 10 congruent sides of the regular decagon. What the problem amounts to is constructing an isosceles triangle whose base angles are twice the measure of the vertex angle. If we let the measure of the vertex angle be x, then the base angle measure is 2x. Since the angle sum of a triangle is 180° , we have

$$x + (2x) + (2x) = 180$$
$$5x = 180 \quad \text{or}$$

$$x = 36^{\circ} .$$

Thus we can construct a 36° angle by constructing an isosceles triangle whose base angle measure is twice the vertex angle measure.

We first outline the steps and then justify the construction.

1. In ⊙O, draw any radius $\overline{OA}$.

2. Locate point B on $\overline{OA}$ such that OB is the mean proportional of OA and BA, by:
   (a) At A, construct $PA = \frac{1}{2}OA$ and $\overline{PA} \perp \overline{OA}$.

   (b) Draw circle with P as center and radius = PA.

   (c) Draw secant $\overline{OB'PD}$ and mark off $OB = OB'$ on $\overline{OA}$.

3. Draw chord AC = OB. $\overline{AC}$ is the side of the required decagon.

   To justify the construction, we show that $\triangle AOC$ is an isosceles triangle with vertex angle measure = 36°.

First, we justify step (2)—show that OB is actually the mean proportional of AO and AB, or

$$\frac{AO}{OB} = \frac{OB}{BA} .$$

By construction, $\overline{PA} \perp \overline{OA}$, and thus $\overline{OA}$ is a tangent segment from point O to $\odot P$. From an earlier theorem, we know that the tangent segment $(\overline{OA})$ drawn from an exterior point (O) is the mean proportional of any secant $(\overline{OD})$ and its external segment $(\overline{OB'})$. Thus,

$$\frac{OD}{OA} = \frac{OA}{OB'}, \text{ or}$$

$OD \cdot OB' = OA^2$. Since $OB = OB'$, $OD \cdot OB = OA^2$.

We now have two ways of proceeding; we can either prove $\frac{AD}{OB} = \frac{OB}{BA}$ by showing it is equivalent to the true statement $OD \cdot OB = OA^2$ or by showing the true statement $OD \cdot OB = OA^2$ equivalent to

$$\frac{AO}{OB} = \frac{OB}{BA} .$$

Since the to-be-proved

$$\frac{AO}{OB} = \frac{OB}{BA}$$

involves shorter segments than the true statement and it is usually easier to break down larger segments into smaller ones rather than the reverse, we choose the first method: converting $OD \cdot OB = OA^2$ in a series of algebraic steps to

$$\frac{AO}{OB} = \frac{OB}{AB} .$$

Note $OD = DB' + B'O$; $DB' = AO$; $B'O = BO$; then $OD = AO + BO$, and $OD \cdot OB = OA^2$ is equivalent to $(AO + OB) \cdot OB = OA^2$ or $AO \cdot OB + OB^2 = OA^2$ or $OB^2 = OA^2 - OA \cdot OB = OA \cdot (OA - OB)$. But $OA - OB = AB$. Therefore, $OA \cdot AB = OB^2$. Dividing by $OB \cdot AB$, we obtain the desired result

$$\frac{OA}{OB} = \frac{OB}{AB} .$$

Having justified step (2), we can now find the angle measure of $\angle AOC$ (or $\angle BOC$).

We know that $AC = OB$ and $\frac{OA}{OB} = \frac{OB}{AB}$. Hence, by substitution, we obtain $\frac{OA}{AC} = \frac{AC}{AB}$. With this in mind, $\triangle CBA \sim \triangle OAC$ by the SAS similarity theorem. (By reflexivity, $\angle OAC \cong \angle BAC$.)

Since $\triangle OAC$ is isosceles, $\triangle CBA$ must also be isosceles and $AC = OB = BC$ must be true. As such, $\triangle CBO$ is isosceles and $\angle BCO \cong \angle COB$.

By the rules for measures of exterior angles, $m \angle ABC = m \angle BCO + m \angle COB$. This reduces to $m \angle ABC = 2m \angle BCO$.

As base angles of $\triangle ABC$ and $\triangle AOC$, respectively, $m \angle ABC = m \angle BAC$ and $m \angle BAC = m \angle ACO$. By transitivity and substitution this becomes $m \angle ACO = m \angle BAC = m \angle ABC = 2 \angle BCO$.

The angles of a triangle sum to $180°$. Therefore,

$$m \angle BOC + m \angle BAC + m \angle ACO = 180.$$

By substitution, $m \angle BOC + 2m \angle BOC + 2m \angle BOC = 180$

$$5m \angle BOC = 180$$

$$m \angle BOC = 36° .$$

Thus, $\angle BOC$ intercepts an arc, $\overset{\frown}{AC}$, of $36°$. Ten such arcs can be formed from the circle with the chord subtending, each forming one side of the regular decagon.

COORDINATE/ANALYTIC GEOMETRY

CHAPTER 35

DISTANCE

Basic Attacks and Strategies for Solving Problems in this Chapter. See pages 558 to 570 for step-by-step solutions to problems.

In mathematics it is common to describe the positions of points in terms of numbers. This can be done in terms of positions of points on a line. In the picture below, A is the point which is labelled -2, and this indicates that A is two units to the left of the point labelled 0. Note also that the distance between A and B is 5 units.

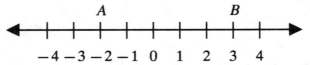

A more common way of describing the position of points is to make the description in terms of an entire plane, rather than just in terms of a line. C is associated with the ordered pair $(3, 2)$, while D is associated with $(-3, -2)$. These correspondences are often described just by writing "$C(3, 2)$" and "$D(-3, -2)$."

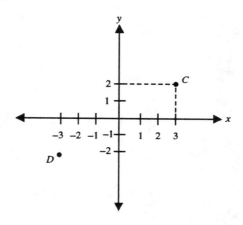

In many problems it will be necessary to find the distance between two points in the plane. The distance between $P_1(x_1, y_1)$ and $P_2(x_2, y_2)$ is given by $P_1P_2 = \sqrt{(x_1 - x_2)^2 + (y_1 - y_2)^2}$ and this is called the distance formula. Also, the midpoint of P_1P_2 is $P_0\left(\dfrac{x_1 + x_2}{2}, \dfrac{y_1 + y_2}{2}\right)$, and this is called the **midpoint formula**. Note that the x-coordinate of the midpoint is the average of the x coordinates of P_1 and P_2, and the y coordinate of the midpoint is the average of the y coordinates of P_1 and P_2. These formulas can both be established using the Pythagorean Theorem.

It is possible to use coordinate geometry methods to prove geometry theorems which are usually proven using more conventional methods. This approach is illustrated in Problems 630 and 635.

Step-by-Step Solutions to Problems in this Chapter, "Distance"

PLOTTING POINTS

● **PROBLEM** 623

Describe the location of the point (-1,-2) and then plot it on coordinate graph paper.

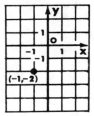

<u>Solution</u>: The first number in the point, the abscissa, is -1 and represents the number of horizontal units along the x-axis that will be moved left of the origin of the Cartesian plane when plotting this point. The second number, the ordinate, is -2 which represents the vertical distance to be moved downward from the origin.

With this in mind, we can plot the required point on the graph. We find that (-1,-2) will lie in the lower left region of the graph, Quadrant III, 2 units below the x-axis and 1 unit to the left of the y-axis.

● **PROBLEM** 624

Give the coordinates of points A,B,C,D,E,and F in the figure below and determine the quadrants in which they lie.

<u>Solution</u>: To find the coordinates of a point, z for example, imagine a line parallel to the y axis passing through z. Where that line intersects the x-axis, gives the x-coordinate. Similarly, the imaginary line parallel to the x-axis passing through z, gives the y coordinate. For example, the y-co-ordinate of A is 2. Similarly, the x-coordinate of A is 2.

558

Point	Coordinate	Quadrant
A	(2, 2)	I
B	(-3, -3)	III
C	(-2, 0)	on the x-axis
D	(-4, 3)	II
E	(0, -2)	on the y-axis
F	(4, -3)	IV

Repeating the procedure for points B,C,D,E, and F, we obtain the table below.

In determining the quadrant of a point, note that any point above the x-axis and to the right of the y-axis is in the first quadrant. A more common way of expressing first quadrant is to use the Roman numeral notation. Thus the first quadrant becomes Quadrant I. Similarly, if the point is above the x-axis but to the left of the y-axis, it is said to be in the second quadrant, or Quadrant II. Points below the x-axis and to the left of the y-axis are in Quadrant III; and points below the x-axis but to the right of the y-axis are in Quadrant IV. Points that lie on the x or y axis are said to lie on the x or y axis and are not in any quadrant.

There is a second method of determining the quadrant of the point. Note that all points above the x-axis have positive y-coordinates. Similarly, all points below the x-axis have negative y-coordinates. Points to the right of the y-axis have positive x-coordinates. Points to the left of the y-axis have negative x-coordinates. Since all points in Quadrant I are above the x-axis and to the right of the y-axis, all points in Quadrant I have the form (positive, positive). Repeating the procedure for the other five cases, we obtain:

>Quadrant II: (Negative, positive)
>Quadrant III: (negative, negative)
>Quadrant IV: (positive, negative)
>On the x-axis: (positive or negative, 0)
>On the y-axis: (0, positive or negative).

Given the graph and asked to find the quadrant of the point, method 1 is easier. If we are given the coordinates, method 2 is better.

DETERMINING DISTANCE

● PROBLEM 625

Find the distance between the points on the real number line whose coordinates are -4 and 3.

Solution: We know that the distance between any two points, call them A and B, on the real number line is equal to the absolute value of the difference of their respective coordinates, a and b.

Therefore:

(1) Let A be the point on the number line whose coordinate is -4 (i.e., a = -4);

(2) Call B that point on the number line whose coordinate is 3 (i.e., b = 3), and

(3) Let AB be the distance between points A and B.

Then, $AB = |a - b|$

$AB = |(-4) - 3|$

$AB = |-7| = 7$

Therefore, the distance between the two points given in this problem is 7.

Find the distance between the point A(1,3) and B(5,3).

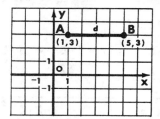

Solution: In this case, where the ordinate of both points is the same, the distance between the two points is given by the absolute value of the difference between the two abscissas. In fact, this case reduces to merely counting boxes as the figure shows.

Let, x_1 = abscissa of A y_1 = ordinate of A
x_2 = abscissa of B y_1 = ordinate of B
d = the distance.

Therefore, $d = |x_1 - x_2|$. By substitution, $d = |1 - 5| = |-4| = 4$. This answer can also be obtained by applying the general formula for distance between any two points

$$d = \sqrt{(x_1-x_2)^2 + (y_1-y_2)^2}$$

By substitution, $d = \sqrt{(1-5)^2 + (3-3)^2} = \sqrt{(-4)^2 + (0)^2} = 16 = 4$. The distance is 4.

Find the distance between the point C(-3,-2) and the point D(-3,4).

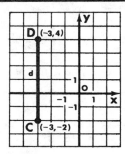

Solution: Since the abscissas of points C and D are the same, we have a choice of two techniques. In this special case, the distance is equal to the absolute value of the difference between the ordinates and can be found simply, without calculation, by counting the boxes traversed by $\overline{DC}$. The general formula for distance between two points could also be used. Both methods will be shown here. Let

x_1 = abscissa of C, y_1 = ordinate of C

x_2 = abscissa of D, y_2 = ordinate of D

d = distance.

<u>Method 1</u>: $d = |y_1 - y_2|$

By substitution, $d = |-2 - 4| = |-6| = 6$.

<u>Method 2</u>: $d = \sqrt{(x_1-x_2)^2 + (y_1-y_2)^2}$.

By substitution, $d = \sqrt{(-3-(-3))^2 + (-2-4)^2} = \sqrt{0^2+(-6)^2} = \sqrt{36}$ = 6.

Therefore, the distance is 6.

● **PROBLEM** 628

Problem: Show that the triangle with vertices A(6,7), B(-11,0), and C(1,-5) is isosceles.

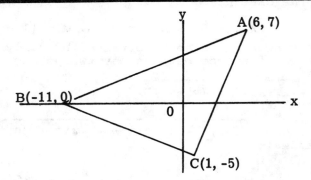

A(6, 7)

B(-11, 0)

0

x

y

C(1, -5)

<u>Solution</u>: To prove that triangle ABC is isosceles, we must find two equal sides. Calculate the length of the sides using the formula for distance in the plane: The length of the line segment determined by the two points $P_1(x_1,y_1)$, $P_2(x_2,y_2)$ is given by

$$P_1P_2 = \sqrt{(x_1-x_2)^2 + (y_1-y_2)^2}$$

Plugging in the values for A, B, C, we have

$$AB = \sqrt{(6-(-11))^2 + (7-0)^2}$$

$$= \sqrt{17^2 + 7^2} = \sqrt{338}$$

$$AC = \sqrt{(6-1)^2 + (7-(-5))^2} = \sqrt{5^2 + 12^2}$$

$$= \sqrt{25 + 144} = \sqrt{169} = 13.$$

$$BC = \sqrt{((-11)-1)^2 + (0-(-5))^2} = \sqrt{(-12)^2 + (+5)^2}$$

= 13.

Since AC = BC , triangle ABC is isosceles.

● **PROBLEM** 629

A circle whose center is at C(-4,2) passes through the point D(-3,5). Find R, the length of the radius, in radical form.

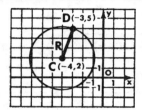

Solution: Since the length of the radius is, by definition, the length of a line segment drawn from the center of the circle to any point on the circle, R can be found by determining the length of $\overline{CD}$. On the Cartesian plane the length of a line segment is governed by the distance between its endpoints. The general formula for the distance between two points is d = $\sqrt{(x_1-x_2)^2 + (y_1-y_2)^2}$. Let

$$C = (x_1,y_1), \quad \text{then} \quad x_1 = -4 \quad y_1 = 2$$
$$D = (x_2,y_2), \quad \text{then} \quad x_2 = -3 \quad y_2 = 5$$

and   R = distance (length of the radius).

By substitution, R = $\sqrt{(-4-(-3))^2 + (2-5)^2}$ = $\sqrt{(-1)^2 + (-3)^2}$
$$= \sqrt{1 + 9} = \sqrt{10}$$

The length of the radius is $\sqrt{10}$.

● **PROBLEM** 630

Prove, using coordinate geometry, that the diagonals of a rectangle are congruent.

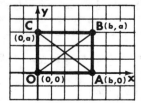

Solution: Prior to getting into the proof, let us draw a rectangle on the Cartesian plane. (See figure). Place one vertex at the origin and side $\overline{OA}$ along the x-axis, as shown. Extend side $\overline{AB}$ into quadrant I, so that vertex B is the point (b,a). The coordinates of the vertices of the rectangle are

O(0,0), A(b,0), B(b,a), and C(0,a).

Since ABCO is given to be a rectangle, with $\overline{OB}$ and $\overline{CA}$ as the diagonals, proving OB = CA will be sufficient to prove that the diagonals of a rectangle are congruent.

We can use the general formula for the distance between two points, $d = \sqrt{(x_1-x_2)^2 + (y_1-y_2)^2}$, to obtain the desired results. Here, (x_1,y_1) and (x_2,y_2) are the coordinates of the 2 points. Distance OB is given by

$$OB = \sqrt{(0-b)^2 + (0-a)^2} = \sqrt{b^2 + a^2}.$$

Distance AC is given by

$$AC = \sqrt{(b-0)^2 + (0-a)^2} = \sqrt{b^2 + a^2}.$$

Therefore, since OB = AC we can conclude that $\overline{OB} \cong \overline{AC}$.

MIDPOINTS

● **PROBLEM** 631

Find the coordinates of the midpoint of the line segment which joins the point R(4,6) and the point S(8,-2). (See figure)

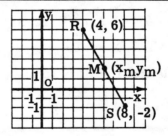

Solution: By a theorem, we know that the coordinates of the midpoint of a line segment are equal to one-half the sum of the corresponding x and y coordinates of the endpoints of the line segment. (These results can be checked by applying the general formula for the distance between two points). Let $M = (x_m, y_m)$, the midpoint of segment $\overline{RS}$.

Also let $R = (x_1, y_1)$ where $x_1 = 4$, $y_1 = 6$

and $S = (x_2, y_2)$ where $x_2 = 8$, $y_2 = -2$.

Then, $x_m = \frac{1}{2}(x_1+x_2) = \frac{1}{2}(4+8) = \frac{1}{2}(12) = 6$

and $y_m = \frac{1}{2}(y_1+y_2) = \frac{1}{2}(6-2) = \frac{1}{2}(4) = 2$.

564

To check that $M = (6,2)$ is the midpoint of $\overline{RS}$, calculate RM and MS by using the general distance formula to verify that they are equal.

$$RM = \sqrt{(x_1 - x_m)^2 + (y_1 - y_m)^2}$$

By substitution, $RM = \sqrt{(4-6)^2(6-2)^2} = \sqrt{(-2)^2 + (4)^2} = \sqrt{4+16}$

$$RM = \sqrt{20}$$

$$MS = \sqrt{(x_m - x_2)^2 + (y_m - y_2)^2}$$

By substitution, $MS = \sqrt{(6-8)^2 + (2-(-2))^2} = \sqrt{(-2)^2 + (4)^2} = \sqrt{4+16}$

$$MS = \sqrt{20}$$

Therefore, $RM = MS$

Therefore, the coordinates of the midpoint of line segment $\overline{RS}$ are $x = 6$, $y = 2$ or $(6,2)$.

● **PROBLEM** 632

Problem: Find the coordinates of the midpoints of AC and BC, of triangle ABC, with coordinates $A(6,7)$, $B(-11,0)$, $C(1,-5)$.

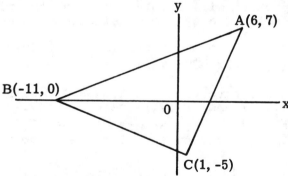

Solution: Recall the midpoint formula. For the points (x_1, y_1) and (x_2, y_2), the midpoint coordinates (x_0, y_0) are given by

$$x_0 = \frac{1}{2}(x_1 + x_2), \quad y_0 = \frac{1}{2}(y_1 + y_2).$$

For the midpoint D of $\overline{AC}$ we have

$$x_D = \frac{1}{2}(6+1) = \frac{7}{2}, \quad y_D = \frac{1}{2}(7 + (-5)) = 1.$$

For the midpoint E of $\overline{BC}$, we find

$$x_E = \frac{1}{2}(-11+1) = -5; \quad y_E = \frac{1}{2}(0-5) = \frac{-5}{2}.$$

Hence, D has coordinates $(\frac{7}{2}, 1)$, and E has $(-5, \frac{-5}{2})$.

The midpoints of the sides of a triangle are (2,5), (4,2), (1,1). Find the coordinates of its three vertices.

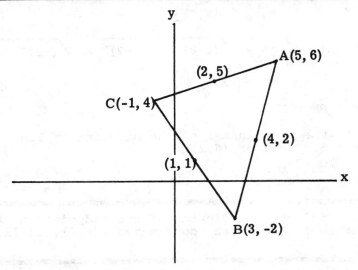

Solution: This problem will involve applying the midpoint formula and then solving a system of simultaneous equations.

For any segment, with endpoints (a,b) and (c,d), the midpoint is given by $(\frac{a+c}{2}, \frac{b+d}{2})$.

Let the triangle have vertices A,B, and C with coordinates (x_1, y_1), (x_2, y_2) and (x_3, y_3), respectively. (2,5) is the midpoint of $\overline{AC}$. (4,2) is the midpoint of $\overline{AB}$. (1,1) is the midpoint of $\overline{CB}$. Hence, by applying the midpoint formula,

$$(2,5) = \left[\frac{x_1+x_3}{2}, \frac{y_1+y_3}{2}\right] \quad (4,2) = \left[\frac{x_1+x_2}{2}, \frac{y_1+y_2}{2}\right]$$

$$(1,1) = \left[\frac{x_2+x_3}{2}, \frac{y_2+y_3}{2}\right]$$

Now, we can write

$$\frac{x_1+x_3}{2} = 2, \qquad x_1+x_3 = 4 \qquad (1)$$

$$\frac{x_1+x_2}{2} = 4, \qquad x_1+x_2 = 8 \qquad (2)$$

$$\frac{x_2+x_3}{2} = 1 , \qquad x_2+x_3 + 2 \qquad (3)$$

Sum (1) and (2) to obtain $2x_1+x_3+x_2 = 12$.

But (3) tells us $x_3+x_2 = 2$. Hence, by substitution, we have

$$2x_1+2 = 12 \text{ and } x_1 = 5.$$

From (1), $x_3 = 4-x_1 = 4-5 = -1 = x_3$
From (2), $x_2 = 8-x_1 = 8-5 = 3 = x_2$
Also,

$$\frac{y_1+y_3}{2} = 5, \quad y_1+y_3 = 10 \qquad (4)$$

$$\frac{y_1+y_2}{2} = 2, \quad y_1+y_2 = 4 \qquad (5)$$

$$\frac{y_2+y_3}{2} = 1, \quad y_2+y_3 = 2 \qquad (6)$$

Again sum (4) and (5) to obtain $2y_1+y_2+y_3 = 14$

But (6) tells us $y_3+y_2 = 2$. Hence, by substitution,
we have

$$2y_1+2 = 14 \text{ and } y_1 = 6.$$

From (4), $y_3 = 10-y_1 = 10-6 = 4 = y_3$

From (5), $y_2 = 4-y_1 = 4-6 = -2 = y_2$

The vertices of the triangle in this problem are A = (5,6),
B = (3,-2), and C = (-1,4).

● **PROBLEM** 634

Given, in the accompanying graph, the quadrilateral whose
vertices are A(-2,2), B(1,4), C(2,8), and D(-1,6).
(a) Find the coordinates of the midpoint of diagonal $\overline{AC}$.
(b) Find the coordinates of the midpoint of diagonal $\overline{BD}$.
(c) Show that ABCD is a parallelogram.

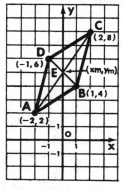

<u>Solution</u>: The x and y coordinates of the midpoint of a
line segment are given by averaging the corresponding x and
y coordinates of the endpoints of the line segment.

(a) The endpoints of $\overline{AC}$ are (-2,2) and (2,8). Therefore the coordinates of the midpoint, (x_m, y_m), will be

$$x_m = \frac{1}{2} \ (-2+2) = \frac{1}{2} \ (0) = 0$$

$$\text{and } y_m = \frac{1}{2} \ (2+8) \ = \frac{1}{2} \ (10) = 5$$

The midpoint of diagonal $\overline{AC}$ is the point (0,5).

(b) In a similar fashion, since the endpoints of diagonal $\overline{BD}$ are (1,4) and (-1,6), the coordinates of the midpoint, represented by (x'_m, y'_m), are given by

$$x'_m = \frac{1}{2} \ (1+(-1)) = \frac{1}{2} \ (0) = 0$$

$$y'_m = \frac{1}{2} \ (4+6) = \frac{1}{2} \ (10) = 5$$

The midpoint of diagonal $\overline{BD}$ is the point (0,5).

(c) If the diagonals of a quadrilateral bisect each
other, then the quadrilateral is a parallelogram. Since the
point E(0,5) is the midpoint of each diagonal and the intersection point of both diagonals, we can conclude that the
diagonals do, in fact, bisect each other.

Therefore, since ABCD is a quadrilateral whose diagonals bisect each other, it is a parallelogram.

● **PROBLEM** 635

Prove that the diagonals of a parallelogram bisect each other.

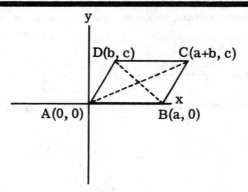

Solution: The figure above is a parallelogram with vertices
A(0,0), B(a,0), C(a+b,c), and D(b,c). For diagonal $\overline{DB}$ to bi-
sect diagonal $\overline{AC}$, the intersection of $\overline{DB}$ and $\overline{AC}$ must be the
midpoint of $\overline{AC}$. For diagonal $\overline{AC}$ to bisect diagonal $\overline{DB}$, the
intersection of $\overline{AC}$ and $\overline{DB}$ must be the midpoint of $\overline{DB}$. Since
the intersection of two noncoincident lines is a unique point,
the intersection of $\overline{AC}$ and $\overline{DB}$ must be the midpoint of both.
Therefore, if we show that the midpoint of $\overline{DB}$ is the same
point as the midpoint of $\overline{AC}$, then the diagonals indeed bisect
each other.

According to the Midpoint Formula, the midpoint of a segment determined by endpoints (x_1, y_1) and (x_2, y_2) is the point

$$\left(\frac{x_1 + x_2}{2}, \frac{y_1 + y_2}{2}\right).$$

Therefore, the midpoint of

$$\overline{AC} = \left(\frac{a+b+0}{2}, \frac{c+0}{2}\right), \qquad \text{or} \quad \left(\frac{a+b}{2}, \frac{c}{2}\right).$$

The midpoint of

$$\overline{DB} = \left(\frac{b+a}{2}, \frac{c+0}{2}\right) \qquad \text{or} \quad \left(\frac{a+b}{2}, \frac{c}{2}\right).$$

Since the two midpoints are equal, the diagonals of the parallel bisect each other.

● **PROBLEM** 636

The points A(2,-2), B(-8,4), and C(5,3) are vertices of a right triangle. Show that the midpoint of the hypotenuse is equidistant from the three vertices.

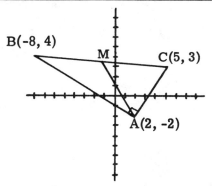

Solution: ∢BAC, as the graph shows, is the right angle of △BAC. As such, $\overline{BC}$ is the hypotenuse of the triangle. If we let M be the midpoint of $\overline{BC}$, then we wish to show MB = MC = MA.

We can find M by using the midpoint formula. The proof can be completed by calculating MB, MC, and MA with the general formula for distance between two points. The midpoint of a segment whose endpoints are (a,b) and c,d) is given by $\left(\frac{a+c}{2}, \frac{b+d}{2}\right)$. B = (-8,4) and C = (5,3), Therefore,

$$M = \left(\frac{-8+5}{2}, \frac{4+3}{2}\right) = \left(-\frac{3}{2}, \frac{7}{2}\right), \text{ by the midpoint formula.}$$

The general formula for distance between two points,

$$(x_1, y_1) \text{ and } (x_2, y_2), \text{ is } d = \sqrt{(x_1 - x_2)^2 + (y_1 - y_2)^2}$$

$$A = (2,-2), \quad B = (-8,4), \quad C = (5,3) \text{ and } M = \left(\frac{-3}{2}, \frac{7}{2}\right)$$

$$MA = \sqrt{\left(-\frac{3}{2}-2\right)^2 + \left(\frac{7}{2}+2\right)^2} = \sqrt{\left(-\frac{7}{2}\right)^2 + \left(\frac{11}{2}\right)^2} = \sqrt{\frac{170}{4}} = \frac{\sqrt{170}}{2}$$

$$MB = \sqrt{\left(-\frac{3}{2}+8\right)^2 + \left(\frac{7}{2}-4\right)^2} = \sqrt{\left(\frac{13}{2}\right)^2 + \left(-\frac{1}{2}\right)^2} = \sqrt{\frac{170}{4}} = \frac{\sqrt{170}}{2}$$

$$MC = \sqrt{\left(-\frac{3}{2}-5\right)^2 + \left(\frac{7}{2}-3\right)^2} = \sqrt{\left(-\frac{13}{2}\right)^2 + \left(\frac{1}{2}\right)^2} = \sqrt{\frac{170}{2}} = \frac{\sqrt{170}}{2}$$

Since MA, MB, and MC are all $\frac{\sqrt{170}}{2}$ units in length,

we conclude that MA = MB = MC and the midpoint of the hypotenuse is equidistant from the vertices. This will be the case in every right triangle.

CHAPTER 36

SLOPE

Basic Attacks and Strategies for Solving Problems in this Chapter. See pages 571 to 584 for step-by-step solutions to problems.

The slope of a line is a number representing the direction of a line. Pictured below are lines l_1, l_2, l_3, l_4, l_5, and l_6. The slopes of these lines are given in the table below.

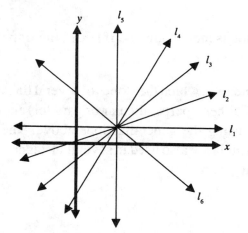

Line	Slope
l_1	0
l_2	$\dfrac{1}{2}$
l_3	1
l_4	2
l_5	undefined
l_6	-1

The slope of the line through $A(x_1, y_1)$ and $B(x_2, y_2)$ is given by the equation $m = \dfrac{y_2 - y_1}{x_2 - x_1}$ where $x_2 \neq x_1$. In calculating the slope of a line, it makes no difference which two points on the line are selected, and it makes no difference which point is classified as point A and which point is classified as point B. Also, the slope of a segment is the same as the slope of the line which contains it. More specifically, the slope of $\overline{AB}$ is the same as the slope of $\overleftrightarrow{AB}$.

In the figure below, $l_1 \parallel l_2$. If the slope of line l_1 is m_1 and the slope of line l_2 is m_2, $m_1 = m_2$.

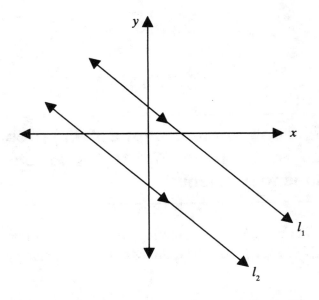

For perpendicular lines the situation is more complex. If $l_3 \perp l_4$, the slope of l_3 is m_3 and the slope of l_4 is m_4, then $m_3 \bullet m_4 = -1$.

This chapter contains several problems in which the slopes of several lines are calculated. Then establishment of either a perpendicular or parallel relationship between two lines is made using the generalizations above. The midpoint formula and the distance formula are also used in several instances. Those formulas are discussed in Chapter 35.

Step-by-Step Solutions to Problems in this Chapter, "Slope"

● **PROBLEM** 637

What is the slope of the line that passes through the origin and point (1,2)? the slope of a horizontal line? a vertical line?

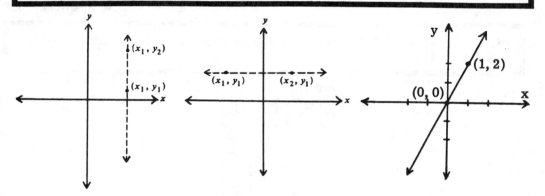

Solution: The slope of the line determined by two points, (x_1,y_1) and (x_2,y_2) such that $x_1 \neq x_2$ and is the same as m, the slope of any segment of the line. The line that passes through the origin and point (1,2) is determined by the points (0,0) and (1,2), Therefore, the slope of the line is the same as m, the slope of the line segment between (0,0) and (1,2).

$$\text{(i)} \quad m = \frac{\Delta y}{\Delta x} = \frac{y_2 - y_1}{x_2 - x_1} = \frac{2 - 0}{1 - 0} = 2 \ .$$

A horizontal line is, by definition, parallel to the x—axis. Therefore all the ordinates must be equal. Let (x_1,y_1) and (x_2,y_2) be two distinct points on the horizontal line. Then the slope of the line equals

$$\text{(ii)} \quad m = \frac{\Delta y}{\Delta x} = \frac{y_2 - y_1}{x_2 - x_1} \ .$$

Since (x_1,y_1) and (x_2,y_2) lie on the horizontal line,

$y_2 = y_1$. Therefore, the numerator $y_2 - y_1 = 0$, and $m = 0$. The slope of a horizontal line equals zero.

A vertical line is, by definition, parallel to the y-axis, and, consequently, all abscissas are equal: For all points (x_1, y_1) and (x_2, y_2) on the line, $\underline{x_1 = x_2}$.

Therefore, we cannot apply the above theorem since the theorem requires that the line be determined by two points $(_1, y_1)$ and (x_2, y_2) such that $x_1 \neq x_2$. No two such points exist on the vertical line and we are left to conclude that a vertical line has no slope.

If we did try to find the slope using the formula, then we would obtain $m = \frac{\Delta y}{\Delta x} = \frac{y_2 - y_1}{x_2 - x_1}$. For a vertical line, $x_2 = x_1$; thus $x_2 - x_1 = 0$ and $m = \frac{y_2 - y_1}{0}$. Because the denominator is zero, the value of m is not defined.

● **PROBLEM** 638

Find the slope of the line passing through the points $C(-3,1)$ and $D(-1,4)$.

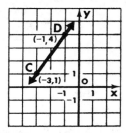

Solution: By definition, the slope of a line segment whose endpoints are (x_1, y_1) and (x_2, y_2), where $x_1 \neq x_2$, is equal to the ratio of the difference of the y-values, $y_2 - y_1$, and the difference of the x-values, $x_2 - x_1$. This is written as $m = \frac{y_2 - y_1}{x_2 - x_1}$. These points are plotted in the accompanying graph.

If we let $C = (x_1, y_1) = (-3,1)$ and $D = (x_2, y_2) = (-1,4)$, then,

$$\text{Slope of } \overleftrightarrow{CD} = \frac{y_2 - y_1}{x_2 - x_1} = \frac{4 - 1}{-1 - (-3)} = \frac{3}{2}$$

The slope of $\overleftrightarrow{CD}$ is $\frac{3}{2}$.

Prove that points A(2,3), B(4,4), and C(8,6) are collinear.

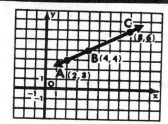

<u>Solution</u>: Three points are collinear if the slope of the line segment between the first and second point is equal to the slope of the line segment between the second and third points.

Let $A = (x_1, y_1) = (2,3)$

        $B = (x_2, y_2) = (4,4)$

        $C = (x_3, y_3) = (8,6)$

These three points have been plotted on the accompanying graph. If the slope of $\overline{AB}$ equals the slope of $\overline{BC}$, the points are collinear.

Slope of $\overline{AB} = \dfrac{y_2 - y_1}{x_2 - x_1} = \dfrac{4 - 3}{4 - 2} = \dfrac{1}{2}$

Slope of $\overline{BC} = \dfrac{y_3 - y_2}{x_3 - x_2} = \dfrac{6 - 4}{8 - 4} = \dfrac{2}{4} = \dfrac{1}{2}$

Since the slope of $\overline{AB}$ = slope of $\overline{BC}$, the points A, B, and C all lie on the same straight line, $\overleftrightarrow{AC}$, and are, therefore, collinear.

Find the slopes of the sides of triangle ABC with A(6,7), B(-11,0), and C(1,-5).

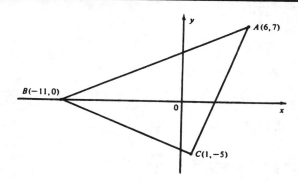

<u>Solution</u>: By definition, the slope m of the line segment determined by $P_1(x_1,y_1)$ and $P_2(x_2,y_2)$ is

$$m = \frac{y_2 - y_1}{x_2 - x_1}$$

If $x_1 \neq x_2$. If $x_1 = x_2$, the slope of the line segment is not defined. Using this formula with the given values, we find:

$$m_{AB} = \frac{7 - 0}{6 - (-11)} = \frac{7}{17}$$

$$m_{BC} = \frac{-5 - 0}{1 - (-11)} = \frac{-5}{12}$$

$$m_{AC} = \frac{7 - (-5)}{6 - 1} = \frac{12}{5}$$

are the slopes of the sides $\overline{AB}, \overline{BC}, \overline{AC}$.

● **PROBLEM** 641

Show by means of slopes, that the quadrilateral pictured in the accompanying graph, whose vertices are the points A(1,1), B(3,-2), C(4,1), and D(2,4), is a parallelogram. (See the figure).

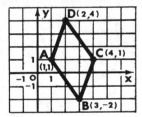

<u>Solution</u>: If we can prove that the opposite sides of ABCD are parallel, then we can conclude that ABCD is a parallelogram. To prove that the members of a pair of opposite sides are parallel, it is sufficient to show that they have the same slope. $\overline{AD}$ and $\overline{BC}$ are one pair of opposite sides, and $\overline{AB}$ and $\overline{DC}$ are the other pair. The slope formula is given, for any two points (x_1,y_1) and $x_2,y_2)$, by

$$m = \frac{y_2 - y_1}{x_2 - x_1} .$$

Let

$$A = (x_1,y_1) = (1,1)$$

$$B = (x_2,y_2) = (3,-2)$$

$$C = (x_3,y_3) = (4,1)$$

$$D = (x_4,y_4) = (2,4),$$

Then,

Slope of $\overline{AB} = \frac{y_2 - y_1}{x_2 - x_1} = \frac{-2 - 1}{3 - 1} = \frac{-3}{2}$

574

Slope of $\overline{DC} = \dfrac{y_4 - y_3}{x_4 - x_3} = \dfrac{4 - 1}{2 - 4} = -\dfrac{3}{2}$

Therefore, $\overline{AB} \parallel \overline{DC}$.

Slope of $\overline{AD} = \dfrac{y_4 - y_1}{x_4 - x_1} = \dfrac{4 - 1}{2 - 1} = \dfrac{3}{1} = 3$

Slope of $\overline{BC} = \dfrac{y_3 - y_2}{x_3 - x_2} = \dfrac{1 - (-2)}{4 - 3} = \dfrac{3}{1} = 3$

Therefore, $\overline{AD} \parallel \overline{BC}$.

Since the opposite sides of ABCD are parallel to each other, we can conclude that ABCD is a parallelogram.

● **PROBLEM** 642

For A(0,0), B(a,0), D(b,c), find the coordinates of C in the first quadrant such that quadrilateral ABCD is a parallelogram.

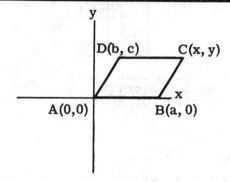

Solution: Since quadrilateral ABCD is to be a parallelogram, $\overline{AD} \parallel \overline{BC}$ and $\overline{CD} \parallel \overline{AB}$. Therefore,

(i) slope of $\overline{AD}$ = slope of $\overline{CB}$

(ii) $\dfrac{0 - c}{0 - b} = \dfrac{y - 0}{x - a}$ or $\dfrac{c}{b} = \dfrac{y}{x - a}$

Crossmultiplying, we obtain:

(iii) $c(x-a) = yb$

Similarly for sides $\overline{CD}$ and $\overline{AB}$,

(iv) $\dfrac{y - c}{x - b} = \dfrac{0 - 0}{0 - a} = 0$

Crossmultiplying, we obtain:

(v) $y - c = 0(x-b) = 0$

(vi) $y = c$

Substitution this result in (iii):

(vii) $c(x-a) = yb = (c)b$

575

Dividing by c,

        (viii) x - a = b

        (ix) x = a + b.

Then the coordinates of C are (a+b,c).

● **PROBLEM** 643

Prove, by means of slope, that the triangle plotted in the accompanying graph, whose vertices are A(0,2), B(2,3), and C(1,5), is a right triangle.

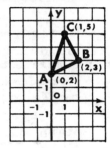

Solution: We can prove that lines $\overline{CB} \perp \overline{AB}$, making $\triangle ABC$ a right triangle, by demonstrating that the slope of $\overline{CB}$ is the reciprocal of the slope of $\overline{AB}$.

Let $A = (x_1, y_1) = (0,2)$

        $B = (x_2, y_2) = (2,3)$

        $C = (x_3, y_3) = (1,5)$,

then,

$$\text{Slope of } \overline{AB} = \frac{y_2 - y_1}{x_2 - x_1} = \frac{3 - 2}{2 - 0} = \frac{1}{2}$$

$$\text{Slope of } \overline{CB} = \frac{y_3 - y_2}{x_3 - x_2} = \frac{5 - 3}{1 - 2} = -\frac{2}{1}$$

The slope of $\overline{AB}$ is the negative reciprocal of the slope of $\overline{CB}$ because $(\frac{1}{2})(-2) = 1$. Therefore, $\overline{AB} \perp \overline{BC}$. Hence, $\triangle ABC$ is a right triangle because it contains a right angle.

● **PROBLEM** 644

Find the acute angle of the parallelogram whose vertices are A(-2,1), B(1,5), C (10,7), and D(7,3).

Solution: In order to solve this problem, we must employ a formula concerning the tangent of an angle formed by two lines intersecting on the cartesian plane. From analytic geo-

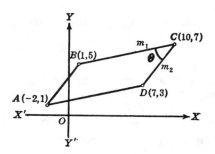

metry, we have $\tan \theta = \dfrac{m_2 - m_1}{1 + m_1 m_2}$, $m_1 m_2 \neq -1$ where m_1 is the

initial slope and m_2 is the terminal slope for the angle θ. θ is always specified by an arc in a counterclockwise direction. The base of the angle, (the line at the tail end of the arc), is the initial line which has initial slope, m_1, and the ray to which the arc points is the terminal line with terminal slope, m_2. Acute angle ∢BCD is directed from side $\overline{BC}$ to $\overline{CD}$. Hence, $\overline{BC}$ is the initial side and $\overline{CD}$ is the terminal side. By the definition of slope,

$$\text{Slope of } \overline{BC} = m_1 = \frac{7 - 5}{10 - 1} = \frac{2}{9} \text{ and}$$

$$\text{Slope of } \overline{CD} = m_2 = \frac{7 - 3}{10 - 7} = \frac{4}{3} .$$

Therefore, if we let m∢BCD = θ, by substituting into the formula given from analytic geometry, we obtain:

$$\tan \theta = \frac{\frac{4}{3} - \frac{2}{9}}{1 + \frac{4}{3} \cdot \frac{2}{9}} = \frac{\frac{12}{9} - \frac{2}{9}}{1 + \frac{8}{27}} = \frac{10}{9} \cdot \frac{27}{35} = \frac{270}{315} = \frac{6}{7} .$$

Since $\tan \theta = \frac{6}{7} = .8571$, from a standard tangent table we find that $\theta = 40°36'$. Therefore, the acute angle of parallelogram ABCD measures $40°36'$.

● **PROBLEM** 645

Find the tangent of the acute angle, θ_1, between the intersecting lines. ℓ_1: $2x + 3y - 6 = 0$ and ℓ_2: $4x - y + 3 = 0$.

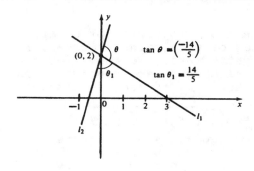

Solution: In the preceeding problem, we used the trigono-
metric formula telling us that $\tan\theta = \frac{m_2 - m_1}{1 + m_1 m_2}$. There, we
were careful to note the direction of the angle and which of
the rays of the angle had slope m_1 and which had m_2.

However, now we can be less formal and still solve the
problem. If ℓ_1 has slope m_1, then $m_1 = -\frac{2}{3}$. If ℓ_2 has slope
m_2, then $m_2 = 4$. By substitution we find,

$$\tan\theta = \frac{4 + \frac{2}{3}}{1 + 4\left(-\frac{2}{3}\right)} = \frac{12 + 2}{3 - 8} = -\frac{14}{5} .$$

We know that $\tan\theta = -\frac{14}{5}$ implies that θ is obtuse. However,
when two lines intersect, the obtuse and acute angles formed
are supplements of each other. Hence, acute angle $\theta_1 =$
$180 - \theta$.

We have the identity $\tan(180 - \theta) = -\tan\theta$. Since
$\tan\theta = -\frac{14}{5}$ and $\theta_1 = 180 - \theta$, we conclude $\tan\theta_1 = \frac{14}{5}$. This
is our final answer.

● **PROBLEM** 646

Given A(-4,-2), B(1,-3), and C(3,1) find the coordinates of D,
in the 2nd quadrant such that quadrilateral ABCD is a parallelogram.

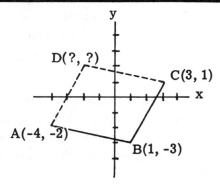

Solution: The many properties of a parallelogram allow us
several methods of proceeding; however, not all of them are
equally straightforward. We know, for example, that opposite
sides of a parallelogram are congruent and, therefore, AD = CB
and DC = AB. However, to use the distance formula

$$d = \sqrt{(x_1 - x_2)^2 + (y_2 - y_1)^2}$$

would yield two equations with square roots of unknowns.
Similarly, any attempt to show congruence of angle would bring
more complications than it would solve. The best method is to

use the property which states that opposite sides of a parallelogram are parallel. In coordinate geometry, this reduces
down to the problem of showing that the slopes of the opposite
sides are equal. Let the coordinates of D be (x,y)

Slope of $\overline{AB}$ = $\dfrac{-2 - (-3)}{-4 - 1}$ = $-\dfrac{1}{5}$

Slope of $\overline{DC}$ = $\dfrac{y - 1}{x - 3}$

Slope of $\overline{CB}$ = $\dfrac{1 - (-3)}{3 - 1}$ = $\dfrac{4}{2}$ = 2

Slope of $\overline{DA}$ = $\dfrac{y - (-2)}{x - (-4)}$ = $\dfrac{y + 2}{x + 4}$

By the properties of the parallelogram, $\overline{AB}$ || $\overline{DC}$

   (i) slope of $\overline{AB}$ = slope of $\overline{DC}$

   (ii) $-\dfrac{1}{5}$ = $\dfrac{y - 1}{x - 3}$

By crossmultiplication,

   (iii) x - 3 = -5(y-1) = -5y + 5

   (iv) x + 5y - 3 = 5

   (v)   x + 5y = 8

Since $\overline{DA}$ and $\overline{CB}$ are also opposite sides,

   (vi) slope of $\overline{DA}$ = slope of $\overline{BC}$

   (vii)   2 = $\dfrac{y + 2}{x + 4}$

   (viii) y + 2 = 2(x+4) = 2x + 8

   (ix) -2x + y + 2 = 8

   (x) -2x + y = 6

   Equations (v) and (x) form a system of two simultaneous
equations in two unknowns. We can eliminate x from the equation by multiplying equation (v) by 2 and adding the result to
equation (x):

   (xi)   2x + 10y = 16
   (x)   + -2x +   y = 6
   (xii)   11y = 22
   (xiii)   y = $\dfrac{1}{11}$ $\cdot$ 22 = 2

Substitution y = 2 into equation (v), we obtain:

   (xiv) x + 5(2) = 8

   (xv) x = 8 - 5(2) = 8 - 10 = -2.

Therefore, D = (-2,2) for ABCD to be a parallelogram.

Prove that two lines ℓ_1 and ℓ_2, with slopes m_1 and m_2, respectively, are perpendicular if and only if $m_1 m_2 = -1$, i.e. they are negative reciprocals of each other.

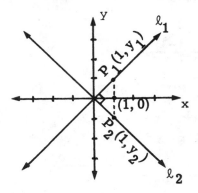

Solution: Since this is an "if and only if" proof we must prove both of the following statements: (1) If $\ell_1 \perp \ell_2$, then $m_1 m_2 = -1$ and (2) if $m_1 m_2 = -1$, then $\ell_1 \perp \ell_2$.

Part 1. Here we are given $\ell_1 \perp \ell_2$. The proof loses no generality by letting ℓ_1 intersect ℓ_2 at the origin.

We must calculate the slopes of ℓ_1 and ℓ_2 and then show their product is -1. To do this, we must select two points from each of the line segments. The point $(0,0)$ is on both lines. For another point from ℓ_1, select $P_1(1, y_2)$ By the definition of slope,

$$m_1 = \frac{y_1}{1} = y_1 \quad \text{and} \quad m_2 = \frac{y_2}{1} = y_2$$

Hence, the desired results to be proved transform to

$$y_1 y_2 = -1.$$

Since $\ell_1 \perp \ell_2$, $\triangle OP_1 P_2$ is a right triangle. We can apply the Pythagorean Theorem in an attempt to complete the proof. Accordingly, we obtain $(P_1 P_2)^2 = (OP_1)^2 + (OP_2)^2$.

The required lengths can be determined using the general formula for distance between two points.

$$P_1 P_2 = \sqrt{(1-1)^2 + (y_1 - y_2)^2} = y_1 - y_2$$

$$OP_1 = \sqrt{(1-0)^2 + (y_1 - 0)^2} = 1 + y_1^2$$

$$OP_2 = \sqrt{(1-0)^2 + (y_2 - 0)2} = 1 + y_2^2$$

By substitution,

$$(y_1 - y_2)^2 = (1 + y_1^2) + (1 + y_2^2)$$

$$y_1^2 - 2y_1y_2 + y_2^2 = 2 + y_1^2 + y_2^2$$

$$-2y_1y_2 = 2 \qquad y_1y_2 = -1$$

Since $y_1 = m_1$ and $y_2 = m_2$, we have $m_1m_2 = -1$, the desired result.

Part 2. We are given $m_1m_2 = -1$. The steps discussed in Part 1 can be traced backward and the converse of the Pythagorean theorem can be used to allow us to conclude OP_1P_2 is a right triangle and, hence, $\ell_1 \perp \ell_2$.

● **PROBLEM** 648

$\overline{CD}$ is a diameter of the circle whose center is the point P (2,1). If the coordinates of C are (0,-2), find the co-ordinates of D. (See figure).

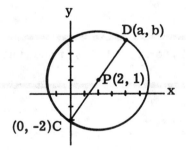

Solution: Since P is the center of a circle whose diameter is $\overline{CD}$, P must be the midpoint of $\overline{CD}$. As the midpoint, the vertical and horizontal coordinates of P must equal the average of the vertical and horizontal coordinates of the endpoints of the line segment $\overline{CD}$. Let

$$
\begin{array}{llll}
C = (x_1, y_1) & \text{where} & x_1 = 0 & y_1 = -2 \\
D = (x_2, y_2) & \text{where} & x_2 = a & y_2 = b \\
P = (x_m, y_m) & \text{where} & x_m = 2 & y_m = 1
\end{array}
$$

To solve for the unknowns a and b, we must set up the equations for the coordinates of the midpoint of the segment, and substitute the known quantities, and determine the un-knowns. At the midpoint of CD,

$$x_m = \tfrac{1}{2}(x_1 + x_2).$$

By substitution, $\qquad 2 = \tfrac{1}{2}(0 + a)$

$$4 = a$$

$$y_m = \tfrac{1}{2}(y_1 + y_2).$$

By substitution,
$$1 = \tfrac{1}{2}(-2 + b)$$

$$2 = -2 + b$$

$$4 = b.$$

To verify that the point (4,4) is D, calculate the distances CP and PD. If CP = PD, then D = (4,4).

$$CP = \sqrt{(x_1-x_m)^2 + (y_1-y_m)^2} = \sqrt{(0-2)^2 + (-2-1)^2}$$

$$CP = \sqrt{(-2)^2 + (-3)^2} \qquad = \sqrt{4 + 9} = \sqrt{13}$$

$$PD = \sqrt{(x_m-x_2)^2 + (y_m-y_2)^2} = \sqrt{(2-4)^2 + (1-4)^2} = \sqrt{(-2)^2 + (-3)^2}$$

$$= \sqrt{4 + 9} = \sqrt{13}.$$

Hence, CP = PD.

The coordinates of D are a = 4, b = 4 or (4,4).

● **PROBLEM** 649

Given: A(0,0), B(6,0), and C(3,3), find the equation for the median to side $\overline{AB}$.

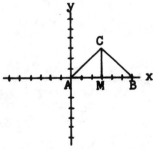

Solution: To find the equation of a line, we need two points on the line. Let $\overline{CM}$ be the median to $\overline{AB}$. Then vertex C is one point on the median. Similarly M, the midpoint of $\overline{AB}$, is also on the median. Using the Midpoint Formula,

$$M = \left[\frac{x_1+x_2}{2}, \frac{y_1+y_2}{2}\right] = \left[\frac{0+6}{2}, \frac{0+0}{2}\right] = (3,0).$$

Thus, with M(3,0), C(3,3), the equation of the median is given by the Point-slope form.

Form: $y-y_1 = m(x-x_1)$, where (x_1,y_1) is point M(3,0) and

the slope $m = \frac{\Delta y}{\Delta x} = \frac{3-0}{3-3} = \frac{3}{0}$ or undefined.

The only straight lines whose slopes are undefined are vertical lines. The equation of a vertical line is x=c, where c is the abscissa of any point on the line. Since the slope of $\overleftrightarrow{CM}$ is undefined, $\overleftrightarrow{CM}$ is a vertical line containing points (3,3) and (3,0). 3 is the common abscissa. Therefore, the equation of $\overleftrightarrow{CM}$ is x=3.

There is one last step. x=3 is an equation of a line. A median is a line segment. Therefore, we must make it clear that the endpoints are (3,3) and (3,0) - that the ordinates must be between 3 and 0. Equation of the median:

$$x=3 \text{ where } 0 \le y \le 3.$$

● **PROBLEM** 650

Given A(1,2), B(2,4), C(5,0), show that the medians are concurrent at a point and find the point.

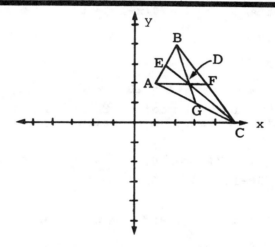

Solution: To find the point of concurrency, it is sufficient to find the intersection of any two medians. To show concurrency, we show that the point of intersection is also a point on the third median.

First, we find the equations of the medians. To find the equations of the medians, we find two points on the medians and then apply the point-slope form. The two points will be the vertex from which the median is drawn and the midpoint of the side to which it is drawn.

E is the midpoint of $\overline{AB}$ = $(\frac{1+2}{2}, \frac{2+4}{2})$ = $(\frac{3}{2}, 3)$

F is the midpoint of $\overline{BC}$ = $(\frac{2+5}{2}, \frac{4+0}{2})$ = $(\frac{7}{2}, 2)$

G is the midpoint of $\overline{AC}$ = $(\frac{1+5}{2}, \frac{2+0}{2})$ = $(3,1)$

To use the point-slope form, we also need the slopes of the medians. Apply the slope definitions:

slope of $\overline{CE} = \dfrac{\Delta y}{\Delta x} = \dfrac{0-3}{5-3/2} = \dfrac{-3}{7/2} = -\dfrac{6}{7}$

slope of $\overline{AF} = \dfrac{\Delta y}{\Delta x} = \dfrac{2-2}{1-7/2} = 0$

slope of $\overline{BG} = \dfrac{\Delta y}{\Delta x} = \dfrac{4-1}{2-3} = \dfrac{3}{-1} = -3$

The vertex point will be used in the point-slope form. Thus, the equations of the medians are:

Equation of $\overline{CE}$: $\quad y-0 = -\dfrac{6}{7}(x-5) = -\dfrac{6}{7}x + \dfrac{30}{7}$

$$\dfrac{6}{7}x + y = \dfrac{30}{7}$$

$$6x + 7y = 30$$

Equation of $\overline{AF}$: $\quad y-2 = 0(x-1) = 0$

$$y = 2$$

Equation of $\overline{BG}$: $\quad y-4 = -3(x-2) = -3x + 6$

$$3x + y = 6+4 = 10$$

With the equations of medians $\overline{CE}$ and $\overline{AF}$, we can solve for the point of intersection. We substitute $y = 2$ in the equation of $\overline{CE}$:

$$6x + 7(2) = 30 \quad \text{or} \quad 6x = 30-14 \quad \text{or}$$

$$X = \dfrac{30-14}{6} = \dfrac{8}{3}$$

Thus, $y = 2$ and $x = \dfrac{8}{3}$, and the point of concurrency is

$$\left(\dfrac{8}{3}, 2\right).$$

To verify concurrency of the three medians, we substitute the intersection point into the equation for median $\overline{BG}$. If this leads to a contradiction, then the intersection point is not on $\overline{BG}$, and the medians are not concurrent. If there is no contradiction, then the point common to medians $\overline{CE}$ and $\overline{AF}$ is also common to $\overline{BG}$ and the medians are concurrent.

Equation of $\overline{BG}$: $\quad 3x + y = 10$

By substitution,

$$3\left(\dfrac{8}{3}\right) + 2 = 10$$

$$8 + 2 = 10$$

$$10 = 10$$

There is no contradiction. $\left(\dfrac{8}{3}, 2\right)$ is the point of concurrency.

CHAPTER 37

LINEAR EQUATIONS

Basic Attacks and Strategies for Solving Problems in this Chapter. See pages 585 to 605 for step-by-step solutions to problems.

An important characteristic of a line is its slope, and knowledge about that topic is a prerequisite for the problems of this chapter. The introduction to Chapter 36 includes a brief description of slope, and the relationships between slopes of parallel and perpendicular lines.

Intercepts are mentioned in several problems in this chapter. For line l pictured below, 2 is the x-intercept and -3 is the y-intercept:

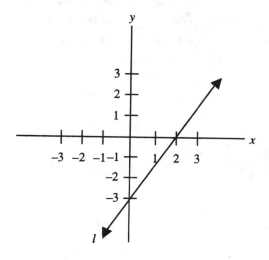

In general, if $(0, y_1)$ is a point on a line, then the y-intercept is y_1, and if $(x_1, 0)$ is a point on the line, then x_1 is the x-intercept.

A single line can be expressed by an equation, which has many equivalent forms:

(1) If a, b, and c are specific real numbers, then an equation of the form $ax + by = c$ is said to be in general form.

(2) If m is the slope of a line and if (x_1, y_1) is a point on the line, then an equation of the form $y - y_1 = m(x - x_1)$ is said to be in point-slope form. This form is discussed initially in Problem 654 and then used in several problems later in the chapter.

(3) If m is the slope of a line and b is the y-intercept of the line, then an equation of the form $y = mx + b$ is said to be in slope-intercept form. This form is established in Problem 655 and is then used in several problems later in the chapter.

When it is necessary to find the slope of a line given an equation of the line, it is easy to find the slope by first finding the slope-intercept form. On the other hand, when finding an equation of a line given the slope and a point on the line, the point-slope form of a line can be used.

The last part of the chapter includes some problems in which an equation of a line is given and the line is to be graphed. Since two points "determine" a line, this can be accomplished by finding the coordinates of two points on the line, plotting the points, and drawing the line through the two points. But it is good to graph three points. With only two points a line can always be drawn, even if one or both of the points were calculated incorrectly; three points will be noncollinear if there is a mistake.

THE EQUATION OF A LINE

● **PROBLEM** 651

Prove that the equation for the line which passes through the origin and which has slope m is y = mx.

Solution: Since the given line passes through the origin, the point (0,0) lies on the line. Let (x,y) be the coordinates of another point on the given line. By definition the slope of a line is equal to the difference of the ordinates of any 2 points lying on the line, divided by the difference of the abscissas of the same 2 points. That is, if (x_1,y_1) and (x_2,y_2) are the coordinates of 2 points on the line, then the slope of the line, m, is given by

$$m = \frac{y_2 - y_1}{x_2 - x_1}$$

In this case, $(x_1,y_1) = (0,0)$ and $(x_2,y_2) = (x,y)$. Hence

$m = \frac{y}{x}$. Multiplying both sides by x, we obtain

y = mx. This is the equation of the given line.

● **PROBLEM** 652

A boy walks at an average rate of 2 miles per hour. (a) How far has he walked after 1 hour; 2 hours; 3 hours; 4 hours? (b) Draw a graph showing how far he has walked at any time from 0 to 4 hours after he started. (Let the units on the x-axis denote the number of hours and let the units on the y-axis denote the number of miles). (c) Interpret each point of the graph and the slope. (d) Write the equation of the line.

Solution: a) The boy walks at an average rate of 2 miles per hour. Assuming he keeps up this rate in 1 hour he travels 2 miles, in 2 hours he travels 4 miles, in 3 hours he travels 6 miles, and in 4 hours he travels 8 miles.

b) Writing the answers to part (a) in tabular

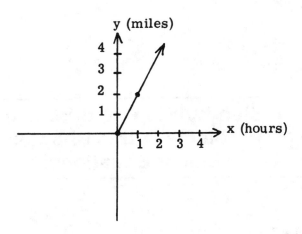

form, we see a pattern emerge regarding the relationship between the distance the boy travels and the time elapsed:

The distance travelled by the boy is seen to be, in general, the product of the rate (2 miles/hr.) and the time elapsed. Denoting the time elapsed by x, and the distance travelled by y, we obtain

$$y = (2 \text{ miles/hr}) \; x. \qquad\qquad (1)$$

Using this equation, we may draw the graph shown in the figure. (In this figure, the x-axis represents the time elapsed in hours, and the y-axis represents the distance travelled). Note that the graph is a straight line, and that x>0, since negative time has no meaning.

c) Each point (x,y) on the graph represents the distance travelled (y) after a time (x) has elapsed. The slope can be measured by using the formula

$$\text{slope} = \frac{y_2 - y_1}{x_2 - x_1}$$

where (x_1, y_1) and (x_2, y_2) are any two points on the graph. Choosing (x_1, y_1) = (0 hours, 0 miles) and (x_2, y_2) = (1 hour, 2 miles) we find that

$$\text{slope} = \frac{2 \text{ miles} - 0 \text{ miles}}{1 \text{ hour} - 0 \text{ hours}} = 2 \text{ miles/hour}.$$

This is the rate at which the boy travels.

d) From equation (1), part (b), the equation of the graph is y = (2 miles/hr) x.

● **PROBLEM** 653

a) Draw a line with slope 3 through the origin. b) Write the coordinates of the points whose x coordinates are -2, -1, 0, 1, 2, 3. c) Write the equation of the line. d) Describe the set of coordinates of the line two ways using set builder notation. e) Verify that the coordinates of part (b) satisfy the equation of part (c).

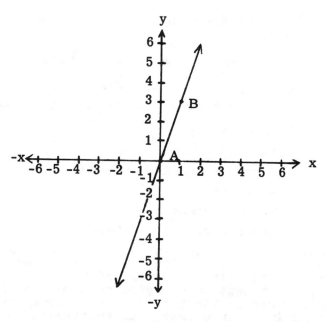

Solution: a) The fact that the line passes through the origin indicates that the point (0,0) lies on the line. This is indicated by point A in the figure. Since the slope of the line is 3, all points whose y and x distances from (0,0) are in the ratio of 3 to 1 also lie on the line. One such point is (1,3), indicated in the figure as point B. Since two points determine a line, $\overleftrightarrow{AB}$ is the line which satisfies the given c conditions, as shown in the figure.

b) Let the coordinates of a point be (x,y). Then the y distance of (x,y) from (0,0) is y - o = y. The x distance of (x,y) from (0,0) is x - 0 = x. From part (a), we realize that all points which lie on $\overleftrightarrow{AB}$ satisfy the relationship $\frac{y}{x} = \frac{3}{1}$

Multiplying both sides by x,

y = 3x   (1)

We may therefore set up the following table:
The entries in the last column are the coordinates of the points with the absissas given in the first column.

x	y = 3x	(x, y)
-2	-6	(-2, -6)
-1	-3	(-1, -3)
0	0	(0, 0)
1	3	(1, 3)
2	6	(2, 6)
3	9	(3, 9)

c) The equation of the line is given, from part (c), in equation (1).

d) An example of set builder notation is

$$S = \{(x,y) \mid x \in R \text{ and } y \in R\}.$$

The various symbols used have the meanings given below:

" = "	consists of
"{ }"	the set of
"(x,y)"	coordinate pairs, x,y
" | "	such that
" e"	is an element of
" R"	the set of real numbers

The above expression, therefore; reads, "S consists of the set of coordinate pairs x,y such that x is an element of the set of real numbers and y is an element of the set of real numbers." Now, the points (x,y) lying on $\overleftrightarrow{AB}$ may be described in two alternative ways. First, the line AB consists of all coordinate pairs (x,y) such that x is an element of the set of real numbers, and y is 3x. That is,

$$S = \{(x,y) \mid x \in R, y = 3x\},$$

where we have labelled the set of points lying on $\overleftrightarrow{AB}$ as S. Alternatively, the line $\overleftrightarrow{AB}$ consists of all coordinate pairs (x,3x) such that x is an element of the set of real numbers (Here we have used the fact that y = 3x). In set builder notation,

$$S = \{(x,3x) \mid x \in R\}.$$

e) By the table of part (b) and equation (1), all the coordinates of part (b) satisfy the equation of part (c).

● **PROBLEM** 654

Write the equation of the line which passes through the points $P_1(-1,-2)$ and $P_2(5,1)$.

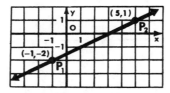

<u>Solution</u>: A line has only one slope and the slope of any segment on the line must equal the slope of the line. Let m be the slope of the line, and $A(x_1,y_1)$ be a point on the line.

Then it must be true for any other point B(x,y) on the line, that the slope of the segment $\overline{AB}$ equals the slope of the line m. Since slope is $\Delta y/\Delta x$, we obtain

(i) $\quad m = \dfrac{y - y_1}{x - x_1}$

Multiplying both sides by $(x-x_1)$, we obtain

(ii) $(y-y_1) = m(x-x_1)$.

Every point (x,y) on the line must satisfy the equation and a point that does not satisfy the equation cannot be a point on the line. Thus, (ii) is the equation of the line with slope m and point $A(x_1,y_1)$. This form of the equation of the line is known as the point slope form.

To put the line described in the problem statement in point slope form, we obtain a point on the line and a slope. We are given that $P_1(-1,-2)$ is a point on the line. Thus $x_1 =$ -1 and $y_1 = - 2$. To find the slope, remember that the slope af a line equals the slope of any segment of the line. Since P_1 and P_2 are points on the line, $\overline{P_1P_2}$ is a segment of the line. Thus,

(ii) m = slope $\overline{P_1P_2}$

Since slope $P_1P_2 = \frac{-2 - 1}{-1 - 5} = \frac{-3}{-6} = \frac{1}{2}$, m = $\frac{1}{2}$.

Substituting these values in the point slope form (equation (ii)):

(iii) $y + 2 = \frac{1}{2}(x+1)$

● **PROBLEM** 655

Write an equation of the line whose slope is $\frac{1}{3}$ and y-intercept b.

Solution: Remember that the y-intercept is the point at which the line intersects the y-axis. Furthermore, every point on the y-axis has an x-coordinate of 0. Therefore, if we are given that the y-intercept of a line is b, we know that $(0,b)$ is a point on the line. Given point $(0,b)$ and slope m, we can use the point slope theorem to solve for the equation of the line. Remember that, according to the point slope theorem,

(i) $(y-y_1) = m(x-x_1)$.

Substituting in our values, we obtain:

(ii) $y - b = m(x-0)$

(iii) $y = mx + b$.

This form of the equation of a line, involving the slope and y-intercept of the line, is called the slope intercept form.

Using this new form and given that m = $\frac{1}{3}$ and b = -2, we can write the equation of the line as

(iv) $y = \frac{1}{3} x + (-2)$

(v) $y = \frac{1}{3} x - 2$

We may stop here or continue to rework the equation into standard form ax + by = c

(vi) $x - 3y = 6$.

● **PROBLEM** 656

Find the slope and y-intercept of the line whose equation is 4x + 2y = 5.

<u>Solution</u>: By transforming the equation into the slope intercept form, y = mx + b, we will be able to read off the desired data. In the form y = mx + b, m is the slope and b is the y-intercept.

$$4x + 2y = 5$$

$$2y = -4x + 5$$

$$y = -\frac{4}{2}x + \frac{5}{2} = -2x + \frac{5}{2}.$$

Therefore, Slope = -2   y-intercept = $\frac{5}{2}$

● **PROBLEM** 657

Write an equation of the line which is parallel to 6x + 3y = 4, and whose y-intercept is -6.

<u>Solution</u>: We employ the slope intercept form for the equation to be written, since we are given the y-intercept. Our task is then to determine the slope.

We are given the equation of a line parallel to the line whose equation we wish to find. We also know that the slopes of two parallel lines are equal. Hence, by finding the slope of the given line, we will also be finding the unknown slope. To find the slope of the given equation 6x + 3y = 4, we transform the equation 6x + 3y = 4 into slope intercept form.

$$6x + 3y = 4$$

$$3y = -6x + 4$$

$$y = -\frac{6}{3} x + \frac{4}{3}$$

$$y = -2x + \frac{4}{3}.$$

Therefore, the slope of the line we are looking for is -2.

The y-intercept is -6. Applying the slope intercept form, y = mx + b, to the unknown line, we obtain,

$$y = -2x - 6$$

as the equation of the line.

● **PROBLEM** 658

Find the equation of the straight line passing through the point (4,-1) and having an angle of inclination of 135°.

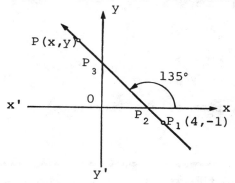

Solution: We are given one point on the line. If we had the slope of the line, we could apply the point slope theorem to determine the equation.

To determine the slope we will make use of the fact that the angle of inclination is 135°. The angle of inclination is the angle between the positive x-axis and the part of the line above the x-axis with the interior of the angle lying to the right of the line.

We shall show that the slope of the line = $-$ tan $\angle P_3P_2O$ = tan (angle of inclination).

Given points $P_2(x',0)$ and $P_3(0,y')$ on the line we can represent the slope m as

$$m = \frac{y' - 0}{0 - x'} = -\frac{y'}{x'}$$

In right triangle P_3OP_2, y' and x' are the respective lengths of sides $\overline{OP_3}$ and $\overline{OP_2}$. Hence, the tangent of $\angle P_3P_2O$ is the

$$\frac{\text{length of leg opposite } \angle P_3P_2O}{\text{length of leg adjacent } \angle P_3P_2O} \quad \text{or}$$

$$\tan \angle P_3P_2O = \frac{y'}{x'}$$

Since m = $-\frac{y'}{x'}$ and

591

$$\tan \langle P_3 P_2 O = \frac{y'}{x'} ,$$

by substitution,

$$m = - \tan \langle P_3 P_2 O$$

Exterior angle $<P_3 P_2 X$ is given as 135° and can be observed to be the supplement of $\langle P_3 P_2 O$. Accordingly, $m\langle P_3 P_2 O = 180° - 135° = 45°$. Hence, by substitution, $m = - \tan 45° = 1$. Now we know a point on the line, $(4, -1)$ and the slope $m = -1$. The point - slope theorem is $(y-y_1) = m(x-x_1)$. Making one final substitution, we obtain

$$(y+1) = -1(x-4)$$

$$y+1 = -x + 4$$

$$y = -x + 3$$

Therefore, the equation of the line is given by $y = -x + 3$ or $y + x = 3$.

● **PROBLEM** 659

Write an equation of the straight line, whose slope is 2, and which passes through point (2,3).

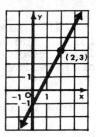

Solution: The method that will be used in this problem to determine the equation employs the use of the point slope form for the equation of a line. (It follows directly from the definition of the slope).

By definition, the slope, m, is given by $m = \frac{y_2 - y_1}{x_2 - x_1}$, where (x_1, y_1) and (x_2, y_2) are any 2 points on the line.

In this problem, we are given one point and the slope of the line. If we let (x,y) represent any other point on the line, then, by substitution of the slope formula, we can find a relation between x and y. This relation between x any y, for any arbitrary point, will be precisely the equation of the line.

Let $(x_1, y_1) = (2,3)$, (x,y) be an arbitrary point, and $m = 2$. Then, since

$$m = \frac{y-y_1}{x-x_1}$$

By substitution, $2 = \frac{y-3}{x-2}$.

$$2(x-2) = y - 3 \quad\quad\quad (1)$$
$$2x - 4 = y - 3$$
$$2x - 1 = y$$

The equation of the line is $y = 2x - 1$. Note the equation $(y-y_1) = m(x-x_1)$ is the equation commonly referred to as the point-slope form. Hence, (1) is the point-slope form for the equation of the line described.

● **PROBLEM** 660

Let a, b, h, and k be real numbers, and $h \neq 0$, $k \neq 0$.
(a). Show that if $A(a,b)$, $B(a+h, b+k)$, and $C(a+2h, b+2k)$ are three points, then they are collinear.
(b). Show that $P(a+nh, b+nk)$ is a point of the line of part (a) for any real number n.

<u>Solution</u>: A, B, and C are on the same line, (therefore are collinear) if the slopes of segments $\overline{AB}$ and $\overline{BC}$ are equal, by definition, the slope of a line segment is equal to m, where $m = \frac{y_2 - y_1}{x_2 - x_1}$ and (x_1, y_1) and (x_2, y_2) are 2 points lying on the segment. Hence,

$$m\overline{AB} = \frac{(b + k) - b}{(a + h) - a} = \frac{k}{h} \text{ and } m\overline{BC} = \frac{(b + 2k) - (b + k)}{(a + 2h) - (a + h)} = \frac{k}{h} .$$

We have used points A and B to find $m\overline{AB}$, and points B and C to calculate $m\overline{BC}$. Since $m\overline{AB} = m\overline{BC}$, A, B, and C are collinear.
(b) To show that P lies on $\overleftrightarrow{ABC}$, we must find the equation of line $\overleftrightarrow{ABC}$ and check that P satisfies the equation. Let (x,y) be one point of $\overleftrightarrow{ABC}$. Another point lying on $\overleftrightarrow{ABC}$ is $A(a,b)$. The slope of $\overleftrightarrow{ABC}$ is the same as the slope of segment $\overline{AB}$, since A, B, and C are collinear. From part (a), this is k/h. Using the definition of slope, we obtain,

$$\frac{y - b}{x - a} = \frac{k}{h} .$$

Multiplying both sides by $(x - a)$,

$$y - b = \frac{k}{h}(x-a)$$

Adding b to both sides,

$$y = \frac{k}{h}(x-a) + b \quad\quad\quad (1)$$

This is the equation of $\overleftrightarrow{ABC}$. If we substitute the coordinates of point P into equation (1) and obtain an identity, then we will know that P lies on $\overleftrightarrow{ABC}$. Doing this, we find

$$b + nk = \frac{k}{h}((a+nh) - a) + b$$

$$b + nk = \frac{k}{h}(nh) + b$$

$$nk = nk$$

$$1 = 1$$

Since this is an identity, P lies on $\overleftrightarrow{ABC}$.

● **PROBLEM** 661

Problem: Find the equation of the line through $P_1(3,-2)$ perpendicular to the line $4x + 2y - 1 = 0$.

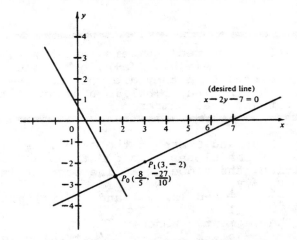

Solution: We can rewrite this equation in slope-intercept form to obtain the slope of the given line, from which we can calculate the slope of the perpendicular line. Thus, $4x + 2y - 1 = 0$ can be rewritten as $2y = -4x + 1$ or, dividing by 2, $y = -2x + \frac{1}{2}$ which is slope-intercept form ($y = mx + b$). Here, $m = -2$, from which we know that the slope m' of the $\perp$ line must be $\frac{1}{2}$ (recall that, if two lines l_1, l_2 are $\perp$, then m_1, m_2, their respective slopes, are related by the equation $m_1 m_2 = -1$, or $m_2 = \frac{-1}{m_1}$. Hence, $m' = \frac{-1}{-2} = \frac{1}{2}$.) Now, using the point-slope form $y - y_1 = m(x-x_1)$, we can obtain the desired equation. $m = \frac{1}{2}$ and the point through which the line is to pass is $(3,-2)$. Substituting these constants we obtain the solution, $y + 2 = \frac{1}{2}(x - 3)$. (In slope-intercept form, this equation reads $y = \frac{1}{2}x - \frac{7}{2}$). Finally, collecting all terms on the left and multiplying by -2, we obtain the normal form of the equation, $x - 2y - 7 = 0$.

● **PROBLEM** 662

Find the equation of the perpendicular bisector of the line segment whose endpoints are $(-2,1)$ and $(3,-5)$.

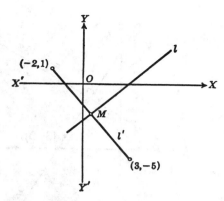

Solution: The line which is the perpendicular bisector of a segment passes through the midpoint of the segment and is perpendicular to it. For any segment given by the endpoints (a,b) and (c,d) the midpoint = $(\frac{a+c}{2}, \frac{b+d}{2})$. The slope of a line perpendicular to a given line of slope m is $-\frac{1}{m}$.

By finding the slope and a point on the line, we can determine the equation.

The slope of the given segment, m, with endpoints (-2,1) and (3,-5), is m = $\frac{-5-1}{3+2}$ = $-\frac{6}{5}$. Hence, the slope of the line perpendicular is $\frac{5}{6}$. The midpoint of the given segment is $(\frac{3-2}{2}, \frac{-5+1}{2})$ = $(\frac{1}{2},-2)$.

Hence, the equation of the line can be found using the point-slope form, $(y-y_1) = m(x-x_1)$. By substitution, this becomes $(y - (-2)) = \frac{5}{6}(x - \frac{1}{2})$

$$y + 2 = \frac{5}{6}x - \frac{5}{12}$$

$$y = \frac{5}{6}x - \frac{29}{12}$$

This equation is the answer. However, it can be simplified by multiplying through by 12 to remove the fractions. We obtain 12y = 10x - 29 as a final answer.

● **PROBLEM** 663

Find the equation of the perpendicular bisector of that portion of the straight line 5x + 3y - 15 = 0 which is intercepted by the coordinate axes.

Solution: Prior to finding the midpoint of the segment and the negative reciprocal of its slope, which are essential to determining the equation of the perpendicular bisector, we must more clearly describe the segment.

To determine the slope of the segment, let us rearrange

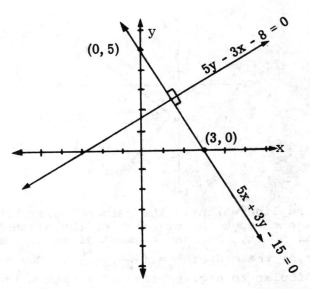

the equation into the form y = mx + b, where m is the slope and b is the y-intercept.

$$5x + 3y - 15 = 0$$
$$3y = -5x + 15$$
$$y = -\frac{5}{3}x + 5.$$

Hence, the slope, m, is $-\frac{5}{3}$. The coordinate axes provide the boundary of the segment in question. Since the y-intercept is 5, one endpoint is (0,5).

The other endpoint can be found by setting y = 0 and solving for x.

$$0 = -\frac{5}{3}x + 5$$
$$-5 = -\frac{5}{3}x$$
$$5\left(\frac{3}{5}\right) = x$$
$$3 = x.$$

Hence, the other endpoint is (3,0).

The perpendicular bisector of the segment will have slope $-\frac{1}{m}$, or $\frac{3}{5}$. Also, it will pass through the midpoint of the given segment, $\left(\frac{0+3}{2}, \frac{5+0}{2}\right)$ or $\left(\frac{3}{2}, \frac{5}{2}\right)$.

We apply the point-slope form, $(y-y_1) = m(x-x_1)$, to obtain the equation of the perpendicular bisector. Hence,

$$\left(y - \frac{5}{2}\right) = \frac{3}{5}\left(x - \frac{3}{2}\right)$$
$$y - \frac{5}{2} = \frac{3}{5}x - \frac{9}{10}$$

$$y = \frac{3}{5}x + \frac{16}{10}.$$

Multiplying through by 5 to remove the fractions, we obtain, for the equation of perpendicular bisector, $5y - 3x - 8 = 0$.

● **PROBLEM** 664

Problem: Find the point of intersection of the two lines, $4x + 2y - 1 = 0$, $x - 2y - 7 = 0$.

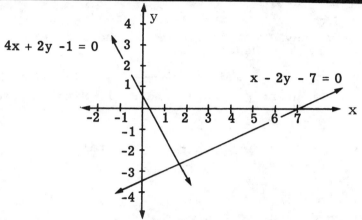

Solution: If $P_0(x_0, y_0)$ is the required point, and if we use the two slope intercept forms of these equations, we see that x_0 and y_0 must satisfy them both. The slope intercept form of $4x + 2y - 1 = 0$ is $y = -2x + \frac{1}{2}$ (just solve for y). Similarly, the slope intercept form of $x - 2y - 7 = 0$ is $y = \frac{1}{2}x - \frac{7}{2}$. Plugging in (x_0, y_0), the points to be determined, yields two expressions for y_0,

$$y_0 = -2x_0 + \frac{1}{2} \quad \text{and} \quad y_0 = \frac{1}{2}x_0 - \frac{7}{2}.$$

Substituting the second value for y_0 back into the first equation gives us an expression in x_0 alone:

$$\frac{1}{2}x_0 - \frac{7}{2} = -2x_0 + \frac{1}{2}.$$

Collecting like terms yields $\frac{5}{2}x_0 = \frac{8}{2}$, or $x_0 = \frac{8}{5}$. Substituting this value into either of the slope-intercept forms gives $y_0 = -\frac{27}{10}$. Thus, the point of intersection is $(\frac{8}{5}, -\frac{27}{10})$.

● **PROBLEM** 665

Let $A(a,c)$, $A'(a,c')$, $B(b,d)$, $B'(b,d')$ be four non-collinear points. Prove that $\overline{AB} \parallel \overline{A'B'}$ if and only if $c - c' = d - d'$.

Solution: Since this is an "if and only if" theorem, it can be proved by proving both of the following two restatements of the original theorem:

I) If $\overleftrightarrow{AB} \parallel \overleftrightarrow{A'B'}$, then c - c' = d - d'.

II) If c - c' = d - d', then $\overleftrightarrow{AB} \parallel \overleftrightarrow{A'B'}$.

We shall prove each of these statements in turn.

Statement I:

If $\overleftrightarrow{AB} \parallel \overleftrightarrow{A'B'}$, then the slope of $\overleftrightarrow{AB}$ (m) must equal the slope of A'B'(m'). That is, m = m'. But, by definition of slope, $m = \dfrac{y_2 - y_1}{x_2 - x_1}$, where (x_1, y_1) and (x_2, y_2) are the co-ordinates of any 2 points lying on the line whose slope is to be calculated. Hence,

$$m = \frac{d - c}{b - a} \text{ and } m' = \frac{d' - c'}{b - a} \ .$$

(Here, we have calculated m' by using points A' and B'.) Since m = m', we obtain:

$$\frac{d - c}{b - a} = \frac{d' - c'}{b - a}$$

Multiplying both sides by b - a, we obtain,

$$d - c = d' - c'$$

Adding c to both sides:

$$d = d' + c - c'$$

Subtracting d' from both sides:

$$d - d' = c - c' \text{ as was to be shown.}$$

Statement II:

We are now given the fact that c - c' = d - d'. From part (I), this may be written as d - c = d' - c'. We may divide both sides of this equation by b - a, obtaining

$$\frac{d - c}{b - a} = \frac{d' - c'}{b - a} \ .$$

Note that b ≠ a, for if it were, A,A', B and B' would be collinear, and this is forbidden by the statement of the problem. Hence, we need not worry about having divided the first equation by zero. From part (I), we recognize that $\dfrac{d - c}{b - a} = m$, the slope of $\overleftrightarrow{AB}$. Furthermore, $\dfrac{d' - c'}{b - a} = m'$, the slope of $\overleftrightarrow{A'B'}$. We have, therefore, shown that m = m'. In other words, $\overleftrightarrow{AB} \parallel \overleftrightarrow{A'B'}$.

● **PROBLEM** 6.6

Given a line, ℓ, with the equation ax + by = c and a point external to the line, P, with the coordinates (x_1, y_1), show that the distance, d, between the point and the line is given by the formula

$$d = \left| \frac{ax_1 + by_1 - c}{\sqrt{a^2 + b^2}} \right|$$

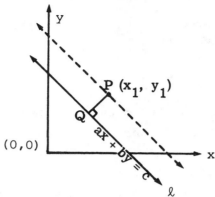

Solution: The distance between the point P and line ℓ is the length of the perpendicular $\overline{PQ}$. The procedure we will follow is to first determine the equation of $\overleftrightarrow{PQ}$. Then, we find the intersection of $\overleftrightarrow{PQ}$ and ℓ - point Q. Using the distance formula, we can then find the distance from point P to Q. The distance PQ is thus the required d.

To find the equation of $\overleftrightarrow{PQ}$, we find its slope. Remember that PQ ⊥ ℓ. Therefore, (slope $\overline{PQ}$) · (slope of ℓ) = -1, or slope $\overline{PQ}$ = $-\dfrac{1}{\text{slope } ℓ}$. Since the equation for ℓ is ax + by = c, the slope of ℓ equals $-\dfrac{a}{b}$ [To see this, solve the equation for y: ax + by = c; by = c - ax; $y = \dfrac{c-ax}{b}$; $y = -\dfrac{a}{b}x + \dfrac{c}{b}$. Thus the equation is in the form y = mx + b. The slope $m = -\dfrac{a}{b}$]. Therefore, slope $\overline{PQ}$ = $-\dfrac{1}{-(a/b)}$ = $\dfrac{b}{a}$.

We know the slope of $\overline{PQ}$, $\dfrac{b}{a}$, and a point on $\overline{PQ}$, point $P(x_1, y_1)$. We obtain the equation for $\overline{PQ}$ using the point slope formula.

(i) $(y-y_1) = \dfrac{b}{a}(x-x_1)$

(ii) $bx - ay = bx_1 - ay_1$.

The expression on the right is a constant and is rather cumbersome to carry around in calculation. Let $K = bx_1 - ay_1$. Then the equation for $\overline{PQ}$ is bx - ay = K.

For the second part of the proof, we find the coordinates of $Q(x_2, y_2)$. Because Q is on $\overleftrightarrow{QP}$, it must be true that

(iii) $bx_2 - ay_2 = K$.

Because Q is on ℓ, it must also satisfy the equation ax + by = c.

(iv) $ax_2 + by_2 = c$.

Multiplying the first equation by b and the second by a, we obtain:

(v) $b^2 x_2 - aby_2 = bK$

(vi) $a^2 x_2 + aby_2 = ac.$

Adding the two, we obtain an expression for x_2 in terms of the known a, b, K, and c.

(vii) $(b^2 + a^2)x_2 = bK + ac$

(viii) $x_2 = \dfrac{bK + ac}{b^2 + a^2}$.

Using equations (iv) and (viii), we solve for y_2.

(ix) $y_2 = \dfrac{c - ax_2}{b} = \dfrac{c - a\dfrac{bK + ac}{a^2 + b^2}}{b} = \dfrac{ca^2 + cb^2 - abK - a^2 c}{b(a^2 + b^2)}$

$= \dfrac{cb^2 - abK}{b(a^2 + b^2)} = \dfrac{cb - aK}{a^2 + b^2}$.

Thus, the coordinates of Q are $\left(\dfrac{bK + ac}{a^2 + b^2}, \dfrac{cb - aK}{a^2 + b^2}\right)$.

The desired distance d = PQ. Using the distance formula, we have

$d = PQ = \sqrt{(x_2 - x_1)^2 + (y_2 - y_1)^2}$

$= \sqrt{\left(\dfrac{bK + ac}{a^2 + b^2} - x_1\right)^2 + \left(\dfrac{cb - aK}{a^2 + b^2} - y_1\right)^2}$.

Substitute $x_1 = x_1 \dfrac{a^2 + b^2}{a^2 + b^2}$ and $K = bx_1 - ay_1$

$= \sqrt{\left(\dfrac{b^2 x_1 - aby_1 + ac - a^2 x_1 - b^2 x_1}{a^2 + b^2}\right)^2 + \left(\dfrac{cb - abx_1 + a^2 y_1 - y_1 a^2 - y_1 b^2}{a^2 + b^2}\right)^2}$

$= \dfrac{1}{a^2 + b^2} \sqrt{(-aby_1 + ac - a^2 x_1)^2 + (cb - abx_1 - y_1 b^2)^2}$

$= \dfrac{1}{a^2 + b^2} \sqrt{(a(-by_1 + c - ax_1))^2 + (b(-by_1 + c - ax_1))^2}$

$= \dfrac{1}{a^2 + b^2} \sqrt{a^2(-by_1 + c - ax_1)^2 + b^2(-by_1 + c - ax_1)^2}$

600

$$= \frac{1}{a^2+b^2} \sqrt{(a^2+b^2)(-ax_1-by_1+c)^2}$$

$$\frac{|-ax_1-by_1+c|}{\sqrt{a^2+b^2}} = \left| \frac{ax_1+by_1-c}{\sqrt{a^2+b^2}} \right|$$

GRAPHING EQUATIONS

● **PROBLEM** 667

a) Draw the graphs of S = {(x,y) | y = x} and T = {(x,y) | y = -x}.
b) Draw the graph of U = {(x,y) | y = | x |}.

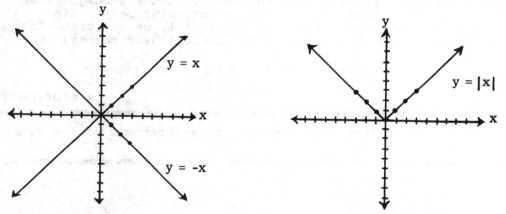

Solution: a) The graph of the set S consists of all points (x,y) such that y = x. The points (0,0), (1,1), (2,2) and (3,3) all satisfy this requirement. The graph, a straight line, is shown in figure (a).

The graph of the set T consists of all points (x,y) such that y = -x. The point (0,0), (1,-1), (2,-2) and (3,-3) satisfy this requirement. The graph, a straight line, is shown in figure (a).

b) The set U consists of all points (x,y) such that y = |x| The points (0,0), (1,1) (-1,1), (2,2) (-2,2), (3,3) and (-3,3) all satisfy this requirement. The graph of U is shown in figure (b).

● **PROBLEM** 668

Graph {(x,y): y = x +1}

Solution: Since there are no restrictions placed on the values of x and y, we assume the domain to be the set of real numbers, and therefore it is possible that the graph contains an infinite number of points. Although this may appear difficult to graph, in actuality, most of the points follow a fairly

predictable pattern. By graphing a few points, we can guess
the graph of the remaining points.

In choosing points to graph, it is usually wise to keep
the values of x small and around the origin to stay on a rea-
sonably small piece of graph paper) and also to keep the val-
lues of x in consecutive order (so that the pattern is easier
to see). Here, we choose x ={ -2, -1, 0, 1, 2}

$$\text{for} \quad x = -2, \quad y = (-2) + 1 = -1$$

$$x = -1, \quad y = (-1) + 1 = 0$$

$$x = 0, \quad y = (0) + 1 = 1$$

$$x = 1, \quad y = (1) + 1 = 2$$

$$x = 2, \quad y = (2) + 1 = 3$$

The best approximation is seen to be the straight line
$\overleftrightarrow{AB}$. To check ourselves, we choose any point on the line,
(-3,-2) for example. Since -2 = (-3) + 1, we see that the
new points added to our graph by $\overleftrightarrow{AB}$ are indeed members of

$$\{(x,y): y = x + 1\}.$$

● **PROBLEM** 669

Find (a) the x-intercept and (b) the y-intercept of the graph
of the equation 3x - 2y = 12.

x	-2	-1	0	1	2
y	-1	0	1	2	3

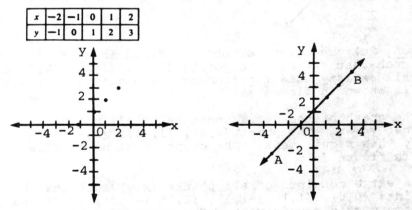

Solution: (a) The x-intercept of a graph is the value of x
that the graph takes on as it intersects the x-axis. The y-
coordinate of this point is zero (as are the y coordinates of
all points lying directly on the x-axis).

Since a straight line will only cross each axis once, by
substituting y = 0 in the equation, we can find this sole
point of intersection, i.e. the x-intercept. Doing this, we
find, 3x + 2(0) = 12, 3x = 12, x = 4
Therefore, the x-intercept is 4.

(b) Applying an argument similar to that given
above, we can conclude that the y-intercept is the graph's

sole point of intersection with the y-axis. As such, the value of x is zero. We can find this y value, the y-intercept, by substituting x = 0 in the given equation, and solving the equation for y. We obtain,

$$3(0) - 2y = 12, \quad -2y = 12, \quad y = -6$$

Therefore, the y-intercept is -6.

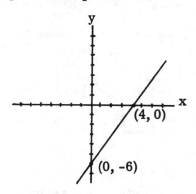

● **PROBLEM** 670

Draw the graph of x + 3y = 6.

<u>Solution</u>: The graph of an equation is the graph of the solution set of that equation; namely, the ordered pairs (x,y) which satisfy the given equation.

We will first determine several points in the solution set, by direct calculation, and plot these on the Cartesian plane. The horizontal axis, by convention, is the x-axis and the vertical axis is the y-axis. Connecting the points will give us the shape of the graph.

Note that since both variables of the equation, x and y, appear raised to the first power, the equation is linear. This means that its graph is a straight line. Because two points determine a line we need only find two points in the solution set in order to draw the graph. Choose two values of x, arbitrarily, and substitute them, one at a time, to solve for the y-value that corresponds to each of these x-values.

Let $\quad$ x = 0, then 0 + 3y = 6; y = 2 .

Let $\quad$ x = 6, then 6 + 3y = 6; y = 0.

It is convenient to put the results in a table such as the one that follows:

x	y
0	2
3	1
6	0

603

We now plot these two points, (0,2) and (6,0), on graph paper, as shown, and draw a straight line through them. We obtain a graph that represents the entire solution set of x + 3y = 6.

● **PROBLEM** 671

> Draw the graph of 3y - 2x = -6, using its slope and y-intercept.

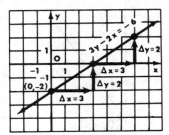

Solution: Prior to discussing the actual graphing, let us transform the equation into slope intercept form and read off the necessary information.

$$3y - 2x = -6$$

$$3y = 2x - 6$$

$$y = \frac{2}{3}x - 2$$

The equation is now in the standard form y = mx + b, where m is the slope of the line, and b is its y-intercept. Therefore, the slope is $\frac{2}{3}$ and the y-intercept is -2. (The y-intercept is the point at which the graph crosses the y axis. At the y-intercept, x = 0. Hence, in our case, (0,-2) is the y-intercept.)

To start drawing the graph, plot the point (0,-2). (See figure).

Since the slope = $\frac{\text{change in y-coordinate}}{\text{change in x-coordinate}}$, knowing the slope will allow us to determine other points on the graph. If slope = $\frac{\Delta y}{\Delta x}$ = $\frac{2}{3}$, then another point on the line will lie two units above, Δy, and 3 units to the right, Δx, of the y-intercept. This point will be (0 + 3, -2 + 2), or (3,0). This point is plotted in the accompanying graph.

Similarly, a third point would be 2 units above, and 3 units to the right, of (3,0). Hence, (3 + 3, 0 + 2), or (6,2) is another point on the graph. This has been plotted.

To conclude the graph, draw a straight line which passes through the three points plotted.

Note that three points were used but two were sufficient to determine the line.

Find, graphically, the common solution for the system of equations:

$$\begin{cases} x + y = 4 \\ y = x + 2. \end{cases}$$

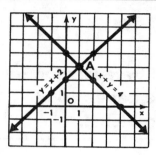

<u>Solution</u>: Determine the locus of points described by each equation, separately. Then the common solution of the system will be given by the intersection of the two loci.

The first equation, x + y = 4, is an equation of a straight line. By choosing arbitrary values of x, we can find the corresponding values of y.

If x =	0	2	4
Then y =	4	2	0

The points are plotted on the accompanying graph and we see that the locus x + y = 4 is the negatively sloped line.

The second locus, described by the equation y = x + 2, can be found in a similar manner. Select several values of x; determine the corresponding values of y, and plot the points.

If x =	- 1	0	2
Then y =	1	2	4

The resulting graph is the positively sloped line in the accompanying figure.

The graphs intersect at point A and, therefore, point A is the common solution for the system of equations.

The coordinates of A are (1, 3). Therefore, x = 1, y = 3 is the common solution of the given system of equations.

CHAPTER 38

AREA

Basic Attacks and Strategies for Solving Problems in this Chapter. See pages 606 to 614 for step-by-step solutions to problems.

The problems in this chapter involve finding the area of polygons (triangles or quadrilaterals), given either the coordinates of the vertices or information which will enable you to find the coordinates of the vertices. The four area formulas given below are used.

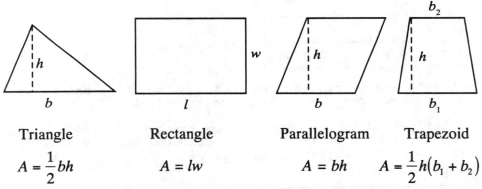

Triangle	Rectangle	Parallelogram	Trapezoid
$A = \dfrac{1}{2}bh$	$A = lw$	$A = bh$	$A = \dfrac{1}{2}h(b_1 + b_2)$

In several problems, the area of the given figure is determined by finding and using the area of related figures. For example, in the figure below, if we know that the area of rectangle $ABCD$ is 32, the area of $\triangle EAF$ is 4, the area of $\triangle FBC$ is 8 and the area of $\triangle ECD$ is also 8. Then, the area of $\triangle CEF$ is $32 - 4 - 8 - 8 = 12$.

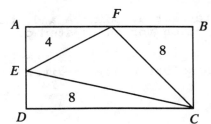

A couple of problems involve finding the area of a triangle in which one side is parallel to one of the coordinate-axes. When that is the case, that side can be classified as the base, and it is easy to find the altitude to that base. A similar approach can be taken for a parallelogram with two sides parallel to a coordinate axis, as is the case in Problem 677.

● **PROBLEM** 673

(a) On the same set of axes, draw the graph of each of the following equations (1) x = 6, (2) y = x, (3) y = 2x. (b) Find the points of intersection. (c) Consider the triangle whose vertices are the points of intersection found in (b) and find the area of this triangle.

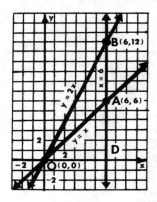

Solution: (a) The three required lines have been drawn on the accompanying graph.

x = 6 is the equation of the line ⊥ the x-axis 6 units to the right of the y-axis, $\overleftrightarrow{BD}$.

y = x is the line that includes all points in which the x and y coordinates are equal. It includes (0, 0), (1, 1), (2, 2) etc; on the graph it is line $\overleftrightarrow{OA}$.

y = 2x is the equation for the line containing the points (0, 0), (1, 2), (2, 4), (3, 6), etc. On the graph it is line $\overleftrightarrow{OB}$.

(b) The graph x = 6 will have one point of intersection with y = x and one with y = 2x. Also, the finite sloped lines will intersect once with each other.

x = 6 will intersect y = x at (6, 6)

x = 6 will intersect y = 2x at (6, 12)

and y = x will intersect y = 2x at the one point where x = 2x, namely x = 0, y = 0, or (0, 0).
Therefore, the vertices have coordinates O(0, 0), A(6, 6), and B (6, 12).

(c) We will employ the formula A = ½bh, in a triangle, to solve this part.

Area of △AOB = ½ × length of side $\overline{BA}$ × length of altitude $\overline{OD}$ to $\overleftrightarrow{BA}$

(Note that △ABO is obtuse. Thus altitude $\overline{OD}$ is external to the triangle.)

BA = the distance between (6, 12) and (6, 6), or the absolute value of the differences in the y coordinate.

$$|12 - 6| = 6 = BA$$

To find OD, note that the altitude $\overline{OD}$ is parallel to to the x-axis. Thus, its length equals the absolute value of the difference of its abscissas. Therefore,

$$OD = |6 - 0| = 6.$$

Accordingly, by substitution,
Area of △AOB = ½ (6)(6) = ½ (36) = 18.
Therefore, Area of △AOB = 18.

● **PROBLEM 674**

Plot the points A (- 2, 3), B (1, 5) and C (4, 2) and find the area of △ABC.

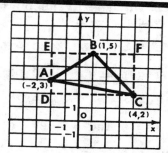

Solution: After plotting the points, as the graph shows, we find that no side is parallel to either of the axes. Therefore, when the altitude is drawn, its length is not easily determined. This makes the standard area formula for a triangle inapplicable.

To overcome this problem, we can inscribe the triangle in a figure whose sides are parallel to the axis, the lengths easier to calculate, and whose area formula can be applied. Any area within this figure but outside of the figure of main concern must be subtracted from the larger area in order to obtain the desired results.

In this problem, construct rectangle EFCD by drawing lines parallel to the x-axis through B and C and lines parallel to the y-axis through A and C.

The area of $\triangle ABC$ = area of rectangle EFCD - sum of the areas of right triangles, CDA, BEA and BFC.

Now, the endpoints of all line segments, whose lengths are needed for the calculation, will match in either the x or y coordinates. The length will be equal to the absolute value of the difference of the non-matching coordinates of the two endpoints for each segment. We will calculate the relevant lengths and then the area.

Since, A = (- 2, 3), B = (1, 5), C = (4, 2)

D = (- 2, 2), E = (- 2, 5), F = (4, 5)

Therefore,

$$ED = |5 - 2| = 3$$

$$EF = |4 - (-2)| = 6$$

$$EA = |5 - 3| = 2$$

$$EB = |1 - (-2)| = 3$$

$$BF = |4 - 1| = 3$$

$$FC = |5 - 2| = 3$$

$$DC = |4 - (-2)| = 6$$

Recall,

$$AD = |3 - 2| = 1$$

Area of $\triangle ABC$ = area of rectangle EFCD - (area of rt. $\triangle CDA$ + area of rt. $\triangle BEA$ + area of rt. $\triangle BFC$).

We must now determine the areas on the right side then substitute for the desired results.

Area of rectangle EFCD = bh = DC × ED = 6 × 3 = 18

Area of rt. $\triangle CDA$ = ½ leg × leg = ½(AD×DC)= ½(1×6) = 3

Area of rt. $\triangle BEA$ = ½ leg × leg = ½(EB×EA)= ½(2×3) = 3

Area of rt. $\triangle BFC$ = ½ leg × leg = ½(BF×FC)=½(3×3) = 4.5

Therefore, by substitution,

Area of $\triangle ABC$ = 18 - (3 + 3 + 4.5) = 18 - 10.5

Therefore, Area of $\triangle ABC$ = 7.5 sq. units.

● PROBLEM 675

Find the area of the triangle whose vertices are A (3, 2), B (7, 2) and C (6, 5).

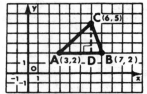

Solution: In this problem we will be applying the area of a triangle formula, i.e. A = ½ bh. The lengths of the various segments are given by the distance between the endpoints of the segments.

Since A and B have the same ordinate, $\overline{AB}$ is parallel to the x-axis and its length is given by the absolute value of the difference between the abscissas of A and B, i.e., |7 - 3| = 4. Therefore, AB = b = 4.

Vertex C is 3 units above side $\overline{AB}$, as the figure shows, therefore the length of altitude to $\overline{AB}$ must be 3.

The perpendicular distance between a point and a line parallel to the x-axis is given by the absolute value of the difference between the ordinate of the point and the ordinate of any point on the line, i.e., |5 - 2| = 3. Therefore, CD = h = 3.

Area = ½ × b × h

By substitution, A = ½ × 4 × 3 = 6.

Therefore, the area of ΔABC is 6.

● **PROBLEM** 676

Find the coordinates of the foot of the altitude to side $\overline{AC}$ of the triangle whose vertices are given by A (- 2, 1), B (4, 7), and C (6, - 3). From this, find the length of the altitude and then the area of the triangle.

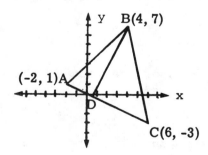

Solution: The triangle described in this problem has been drawn on the accompanying graph.

The altitude $\overline{BD}$ to side $\overline{AC}$ is, by definition, perpendicular to $\overline{AC}$. We can find the coordinates of D, by solving the equations for $\overline{AC}$ and for $\overline{BD}$ simultaneously to find their point of intersection. This intersection is point D.

The slope of $\overline{AC}$ is given by, m, the difference between the ordinates of the endpoints divided by the difference between the abscissas. A = (- 2, 1), C = (6,-3)

$$m = \frac{1 - (-3)}{-2 - 6} = \frac{4}{-8} = -\frac{1}{2}$$

Hence, by the point slope form $(y - y_1) = m(x - x_1)$, the equation of $\overline{AC}$ can be determined.

$$(y - 1) = -\tfrac{1}{2}(x + 2)$$

$$y - 1 = -\tfrac{1}{2}x - 1$$

$$y = -\tfrac{1}{2}x \quad \text{is the equation of } \overline{AC}.$$

Since $\overline{BD} \perp \overline{AC}$, slope of $\overline{BD} = -\frac{1}{m} = -\frac{1}{-\frac{1}{2}} = 2$.

Knowing B = (4, 7) and slope = 2, we can again apply the point slope form to find the equation of $\overline{BD}$:

$$(y - 7) = 2(x - 4)$$

$$y - 7 = 2x - 8$$

$$y = 2x - 1.$$

The intersection of $y = -\tfrac{1}{2}x$ and $y = 2x - 1$ can be found by substituting $y = -\tfrac{1}{2}x$ into $y = 2x - 1$ to determine x and then plugging that value of x into either equation to determine the y-coordinate.

$$-\tfrac{1}{2}x = 2x - 1 \qquad \text{By substitution,}$$

$$1 = \frac{5}{2}x \qquad\qquad y = -\tfrac{1}{2}\left(\frac{2}{5}\right)$$

$$\frac{2}{5} \cdot 1 = x \qquad\qquad y = -\frac{1}{5}$$

$$x = \frac{2}{5}$$

Hence, the coordinates of the foot of the altitude are $\left(\frac{2}{5}, -\frac{1}{5}\right)$.

The length of $\overline{BD}$ can be **found** by using the general formula for distance between two points: B = (4, 7), $D = \left(\frac{2}{5}, -\frac{1}{5}\right)$.

Hence, $BD = \sqrt{\left(4 - \frac{2}{5}\right)^2 + \left(7 + \frac{1}{5}\right)^2} = \sqrt{\left(\frac{18}{5}\right)^2 + \left(\frac{36}{5}\right)^2}$

$$BD = \sqrt{\frac{324 + 1296}{25}} = \sqrt{\frac{1620}{25}} = \frac{\sqrt{81}\ \sqrt{4}\ \sqrt{5}}{\sqrt{25}} = \frac{18\ \sqrt{5}}{5}\ .$$

The length of the altitude is, $\frac{18\ \sqrt{5}}{5}$ units.

The area of a triangle is ½(base) × (altitude).
Area of $\triangle ABC$ = ½ (AC)(BD). We know BD = $\frac{18\ \sqrt{5}}{5}$. We

must determine AC. Again we apply the distance formula.
A (− 2, 1), C = (6, − 3).

$$AC = \sqrt{(6 + 2)^2 + (-3 - 1)^2} = \sqrt{8^2 + (-4)^2}$$

$$= \sqrt{64 + 16} = \sqrt{80}$$

$$AC = \sqrt{16}\ \sqrt{5}\ = 4\ \sqrt{5}$$

Hence, area of $\triangle ABC = \frac{1}{2} \left(\frac{18\ \sqrt{5}}{5}\right)\ (4\ \sqrt{5})$

$$= \frac{1}{2} \left(\frac{18 \times 4 \times 5}{5}\right)$$

$$= ½(72).$$

Area of $\triangle ABC = 36.$

● **PROBLEM** 677

Points A (− 4, − 2), B (2, − 2), C (4, 3), and D (− 2, 3)
are the vertices of quadrilateral ABCD. (a) Plot these
points on graph paper and draw the quadrilateral. (b) What
kind of quadrilateral is ABCD? (c) Find the area of quadrilateral ABCD.

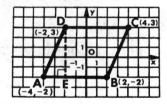

Solution: (a) Consult the accompanying graph to check the
plotting you did.

(b) Since most of the special cases of quadrilaterals
involve one or both of the pairs of opposite sides being
parallel, let us calculate the slope of each side and try
to conclude something about the parallelness of any two
sides.

If the endpoints of each side are given by P_1 =
(x_1, y_1), and $P_2 = (x_2, y_2)$, then the slope of

$$\overline{P_1P_2} = \frac{y_2 - y_1}{x_2 - x_1}$$

Therefore, since A = (- 4, - 2), B = (2, - 2), C = (4, 3) D = (- 2, 3)

Slope of $\overline{AB}$ = $\frac{-2 - (-2)}{2 - (-4)}$ = $\frac{0}{6}$ = 0

Slope of $\overline{BC}$ = $\frac{3 - (-2)}{4 - 2}$ = $\frac{5}{2}$

Slope of $\overline{CD}$ = $\frac{3 - 3}{-2 - 4}$ = $\frac{0}{-6}$ = 0

Slope of $\overline{DA}$ = $\frac{3 - (-2)}{-2 - (-4)}$ = $\frac{5}{2}$

Therefore, the slope of $\overline{AB}$ = the slope of $\overline{CD}$ and the slope of $\overline{BC}$ = slope of $\overline{DA}$. Since the slopes of opposite sides are equal, they are parallel. A quadrilateral, all of whose opposite sides are parallel, is, by definition, a parallelogram.

(c) The area of a parallelogram is given by A = bh.

Choose $\overline{AB}$ as the base of $\square$ ABCD. Since its endpoints have the same ordinate, the line is parallel to the x-axis and its length is the absolute value of the difference of the abscissas of the endpoints. Therefore,

AB = |2 - (- 4)| = 6.

Draw altitude $\overline{DE}$ to base $\overline{AB}$. Since $\overline{DE} \perp \overline{AB}$ and $\overline{AB} \parallel$ x-axis, the length of $\overline{DE}$ is calculated by the absolute value of the difference of the ordinates of D and any point on $\overline{AB}$. Therefore,

DE = |3 - (- 2)| = 5.

Then   A = 6 × 5 = 30.

Therefore, the area of parallelogram ABCD is 30.

● **PROBLEM** 678

Find the area of the polygon whose vertices are A (2, 2), B (9, 3), C (7, 6), and D (4, 5).

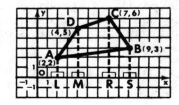

<u>Solution</u>: The polygon described in this question, and drawn on the accompanying graph, does not have any opposite sides parallel to each other, in which case it cannot be a type of parallelogram or trapezoid. Additionally, it is not regular, since no two sides are congruent. Therefore, no standard area formula for quadrilaterals can be applied.

We must therefore use a method for solution in which figures are constructed which have straightforward formulas for area and the length of whose sides and altitudes are known. This will enable us to indirectly determine the area of the given figure.

The usual approach is to extend lines from the vertices, perpendicular to one of the axis and calculate the area of the largest figure bounded by these lines and the sides of the polygon. We then subtract any area lying in this region not in the polygon, whose area we desire to find.

In this problem we draw line $\overleftrightarrow{AL}$, $\overleftrightarrow{DM}$, $\overleftrightarrow{CR}$ and $\overleftrightarrow{BS}$ all perpendicular to the x-axis. Since they are all perpendicular to the same line, they are all parallel to each other and, as such, will provide the bases of several trapezoids that will be used in this calculation. The trapezoids are LMDA, MRCD, RSBC and LSBA.

The area of polygon ABCD can be found by adding the areas of trapezoids LMDA, MRCD, RSBC and subtracting the area of trapezoid LSBA from the sum.

The formula for the area of a trapezoid with altitude h and bases b_1 and b_2 is

$$A = \tfrac{1}{2} h \ (b_1 + b_2).$$

Area of ABCD = (area of trapezoid LMDA + area of trapezoid MRCD + area of trapezoid RSBC) - area of trapezoid LSBA.

We will calculate the area of each trapezoid and substitute it into the above to find the area of **ABCD.**

The lengths of the bases of the trapezoids will be given as the distance from each vertex to the x-axis; and the altitudes as the horizontal distance between the bases along the x-axis.

Let us determine all necessary base and altitude lengths. The base lengths will equal the y-component of the vertex from which it is drawn, since that represents the vertical distance the vertex is above the x-axis.

Therefore, AL = 2, DM = 5, CR = 6, and BS = 3.

The altitude lengths will be given by the absolute value of the difference between the x-components of the two "vertex points" in each trapezoid.

Therefore,   LM = $|4 - 2|$ = 2

$$MR = |4 - 7| = 3$$
$$RS = |9 - 7| = 2$$
$$LS = |2 - 9| = 7$$

Accordingly,

Area of trapezoid LMDA = $\frac{1}{2}$(LM)(AL + DM) = $\frac{1}{2}$ (2)(2 + 5)

$$= \frac{1}{2} (2)(7) = 7.$$

Area of trapezoid MRCD = $\frac{1}{2}$(MR)(DM + CR) = $\frac{1}{2}$(3)(5 + 6)

$$= \frac{1}{2} (3)(11) = 16.5$$

Area of trapezoid RSBC = $\frac{1}{2}$(RS)(CR + BS) = $\frac{1}{2}$(2)(6 + 3)

$$= \frac{1}{2}(2)(9) = 9$$

Area of trapezoid LSBA = $\frac{1}{2}$(LS)(AL + BS) = $\frac{1}{2}$(7)(2 + 3)

$$= \frac{1}{2} (7)(5) = 17.5$$

By substitution,

Area of ABCD = (7 + 16.5 + 9) - 17. 5 = 32.5 - 17.5

Therefore, the area of ABCD = 15.

CHAPTER 39

LOCUS

Basic Attacks and Strategies for Solving Problems in this Chapter. See pages 615 to 627 for step-by-step solutions to problems.

The word **locus** means place or location. The problems in this chapter involve finding the locus of all points which meet certain conditions. Three significant generalizations about these kinds of problems are

(1) The locus of all points in a plane equidistant from two points in that plane is the line perpendicular to the segment whose endpoints are the two given points. This generalization is used in Problems 679, 680, and 682 and then proved in Problem 684.

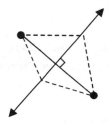

(2) The locus of all points in the interior of an angle which are equidistant from the sides of the angle is the bisector of the angle.

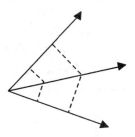

(3) The locus of all points in a plane which are a given distance from a given point in that plane is a circle with the given point as the center and the given distance as the radius.

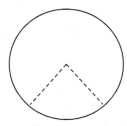

In several instances the locus is to be described in terms of an equation. When that is the case, relationships from coordinate geometry must be applied. Here are several examples.

a. The distance between $P_1(x_1, y_1)$ and $P_2(x_2, y_2)$ is given by
$$P_1P_2 = \sqrt{(x_1 - x_2)^2 + (y_1 - y_2)^2}.$$

b. An equation for the line which passes through $P_1(x_1, y_1)$ and which has slope m is $y - y_1 = m(x - x_1)$.

c. If two perpendicular lines have slopes m_1 and m_2, then $m_1 m_2 = -1$.

d. The midpoint of $\overline{PQ}$ is $R\left(\dfrac{x_1 + x_2}{2}, \dfrac{y_1 + y_2}{2}\right)$ where P is the point at (x_1, y_1) and Q is the point at (x_2, y_2).

In Problem 687, a completion of the square technique is applied. This technique is not described in an earlier portion of this book, so it will be reviewed here.

First, it is well known that for variable x and real number a
$$(x + a)^2 = x^2 + 2ax + a^2 \text{ and}$$
$$(x - a)^2 = x^2 - 2ax + a^2.$$

Note that in either case the third term can be obtained by dividing the coefficient of x by 2, and squaring the result. For example, if you want to convert $x^2 - 6x$ into a perfect square, divide -6 by 2, and square the result. Specifically, $(-6) \div 2 = -3$ and $(-3)^2 = 9$. Then, add 9 to $x^2 - 6x$ and you get $x^2 - 6x + 9 = (x - 3)^2$.

Step-by-Step Solutions to Problems in this Chapter, "Locus"

Write an equation of the locus of points equidistant from the points $P_1(2, 2)$ and $P_2(6, 2)$.

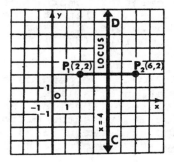

<u>Solution</u>: The given points act as the endpoints of line segment $\overline{P_1P_2}$. The locus of points equidistant from the endpoints of a line segment must be the perpendicular **bisector** of the segment.

P_1 and P_2 have the same ordinate (y-value) and, as such, $\overline{P_1P_2}$ is parallel to the x-axis. Since the x-axis $\perp$ the y-axis, a line perpendicular to $\overline{P_1P_2}$ must be parallel to the y-axis. All lines parallel to the y-axis have the same abscissa (x value).

Therefore, if we find one point that is a bisector of $\overline{P_1P_2}$, every point on the Cartesian plane, with the same abscissa as that point, will lie in the locus of points equidistant from the endpoints.

The point (4, 2) is the midpoint of segment $\overline{P_1P_2}$. Therefore, all points in the required locus will have abscissa, x = 4. (The ordinate is not important and can take on any value.)

Line $\overleftrightarrow{CD}$, in the accompanying graph, is parallel to the y-axis, because it is 4 units from it at all points.

(Parallel lines, by definition, are at all points equidistant.)

Therefore, the equation of $\overleftrightarrow{CD}$ is x = 4.

● **PROBLEM** 680

Write an equation for the locus of points equidistant from
(3, 3) and (4, 4).

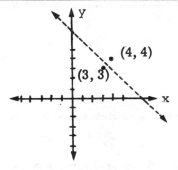

Solution: A perpendicular bisector is defined as the
locus of points that are equidistant from the endpoints
of a given segment. Therefore, the perpendicular bisector
of the segment with endpoints (3, 3) and (4, 4) is the locus
that is required.

FINDING THE EQUATION: A perpendicular bisector is a
line. To find the equation of a line, we require a point
on the line and the slope of the line. Because the
perpendicular bisector includes the midpoint of the segment,
we know that the midpoint between (3, 3) and (4, 4) is a
point on the line. By the Midpoint Formula, we obtain

$$\left(\frac{3 + 4}{2}\ ,\ \frac{3 + 4}{2}\right) = \left(\frac{7}{2}\ ,\ \frac{7}{2}\right)$$

Because the perpendicular bisector is perpendicular
to the segment, we know m, the slope of the bisector, is the
negative reciprocal of the slope of the segment.

Slope of segment = $\frac{\Delta y}{\Delta x} = \frac{3 - 4}{3 - 4} = \frac{-1}{-1}$ = + 1. Therefore, the

slope of the line must be $-\left(\frac{1}{+1}\right)$ = - 1.

Because we have the slope, - 1, and a point, $\left(\frac{7}{2}\ ,\ \frac{7}{2}\right)$,

of the line, we can find the equation of the line using the
Point-Slope Form. An equation of the line that contains
point (x_1, y_1) and has a slope m is $y - y_1 = m(x - x_1)$.
Thus, by substitution, the equation for the locus is

$$y - \frac{7}{2} = - 1 \cdot \left(x - \frac{7}{2}\right)\ .$$

Simplifying, this becomes x + y = 7.

Write an equation of the locus of points in which the ordinate of each point is 3 more than 4 times the abscissa of that point.

<u>Solution</u>: Let (x, y) represent an arbitrary point in the locus described.

We are given that the ordinate, y, of the point, is 3 more than 4 times the abscissa, x of the point. The algebraic statement of these conditions will be the equation of the locus:

$$y = 4x + 3$$

Since this equation is in the form y = ax + b, which is the general equation of a straight line, therefore the locus is a straight line given by the equation

$$y = 4x + 3$$

(a) Describe the locus of points 2 units from the y-axis and write an equation of this locus. (b) Describe the locus of points equidistant from the points $P_1(-4, 2)$ and $P_2(-4, 6)$ and write an equation for this locus. (c) Find the number of points which satisfy both conditions stated in (a) and (b) and give the coordinates of each point.

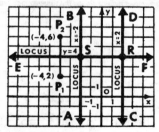

<u>Solution</u>: (a) For a locus to be a constant 2 units from the y-axis, it must be a straight line parallel to the y-axis. This derives from the definition that parallel lines are everywhere equidistant.

There are two lines that satisfy these conditions for a locus. The graph shows lines $\overleftrightarrow{AB}$ and $\overleftrightarrow{CD}$, which are both 2 units from the y-axis. One has a positive x value, the other a negative **x value**.

An equation of $\overleftrightarrow{CD}$ is x = 2. An equation of $\overleftrightarrow{AB}$ is x = -2. Both lines satisfy the locus conditions.

(b) Since P_1 and P_2 determine a line segment, all points equidistant from the endpoints will lie on the perpendicular bisector of the segment. P_1 and P_2 have the same abscissa and, thus, are parallel to the y-axis. Since the x-axis $\perp$ y-axis, a line perpendicular to $\overline{P_1P_2}$ must be parallel to the x-axis. All lines parallel to the x-axis contain points all of whom have the same ordinate, i.e., they are everywhere equidistant from the x-axis.

The midpoint of $\overline{P_1P_2}$ will tell us what this unique ordinate is. The point (- 4, 4) is the midpoint between (- 4, 2) and (- 4, 6). The perpendicular bisector has the determining equation of y = 4, i.e. all points with y = 4 lie on the locus.

y = 4 is the line $\overleftrightarrow{EF}$ in the accompanying graph.

(c) The points which satisfy conditions (a) and (b) will be the points of intersection between the loci. From the graph we see that loci (a) and loci (b) intersect at two points, R and S. Therefore, there are two points which satisfy both conditions (a) and (b).

We know that all points in loci (b) have y = 4 and in loci (a) have x = 2 or x = - 2. Therefore, since point S results from x = - 2 and y = 4 intersecting, its co-ordinates are (- 2, 4). By similar reasoning, the co-ordinates of R are (2, 4).

These results could have been found, and can be verified, by reading the coordinates on the graph.

● **PROBLEM** 683

What is the probable locus of points equidistant from the endpoints of a given line segment $\overline{AB}$? (See figure.)

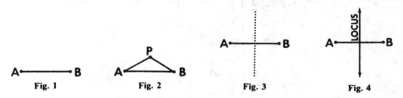

Fig. 1   Fig. 2   Fig. 3   Fig. 4

Solution: A locus is the set of all points, and only those points, that satisfy a given condition or set of conditions. The conditions that describe a locus may plot out a standard geometric shape. In this problem, we must determine, specifically, what the locus is, given the constraint that the set of locus points is the set of all points equidistant from the endpoints of a given line

segment, $\overline{AB}$.

First, construct a diagram containing the fixed lines or points which are given. In this problem, line segment $\overline{AB}$ is fixed, and drawn as shown in Fig. 1.

Second, determine those conditions which must be met, then locate a point which satisfies the given conditions. (One such point has been located in Fig. 2.) The condition to be met in the given problem is that all points P in the locus are equidistant from A and B, or PA = PB.

Third, locate several other points which meet the given condition, and which, through their proximity, will develop the shape or nature of the locus. The above process has been developed for this example. The results are re-corded in Fig. 3.

Fourth, draw a smooth line, either straight or curved, through these points. This line appears to be the locus.

A straight line appears to be the locus in this specific problem. It has been drawn in Fig. 4.

Finally, we must name the geometric figure assumed to be the locus. The probable locus in this example is apparently a straight line which is also the perpendicular bisector of the given line segment $\overline{AB}$.

● **PROBLEM** 684

Prove that the locus of points equidistant from the ends of a given line segment is the perpendicular bisector of the line segment. Prove the converse of this statement.

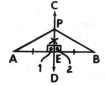
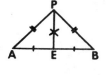

Figure (a)　　　　　Figure (b)

Solution: To prove that a given locus takes on the characteristics of a geometric figure, we must prove two statements. First, prove that if a point is on the locus, it satisfies the given condition. Then, prove either the converse or inverse of the given statement. The converse states that if the given conditions are satisfied, the point is on the locus. The inverse states that if the point is not on the locus then it does not satisfy the conditions. In this problem, we will prove the initial statement and its converse.

In order to prove that the locus is the correct one, we will (a) prove the statement "If a point is on the perpendicular bisector of a line segment, it is equidistant from the ends of the line segment," and, (b) the converse of this statement, namely, "If a point is equidistant from the ends of a line segment, it is on the perpendicular bisector of the line segment."

(a) Proof of statement, "If a point is on the perpendicular bisector of a line segment, it is equidistant from the ends of the line segment ":

Draw the perpendicular bisector $\overleftrightarrow{CD}$ of $\overline{AB}$. Select a point P on $\overleftrightarrow{CD}$ and prove PA = PB.

This can be done by showing $\overline{PA}$ and $\overline{PB}$ are corresponding parts of ΔAPE and ΔBPE, proved congruent by the S.A.S. Postulate.

STATEMENT	REASON
1. $\overleftrightarrow{CD}$ is the $\perp$ bisector of $\overline{AB}$ and P is a point on $\overleftrightarrow{CD}$.	1. Given.
2. $\overline{AE} \cong \overline{EB}$	2. A bisector divides a line segment into two congruent parts.
3. $\sphericalangle$ 1 and $\sphericalangle$ 2 are right angles	3. Perpendicular lines intersect and form right angles.
4. $\sphericalangle$ 1 $\cong$ $\sphericalangle$ 2	4. All right angles are congruent.
5. $\overline{PE} \cong \overline{PE}$	5. Reflexive property of congruence.
6. ΔAPE $\cong$ ΔBPE	6. S.A.S. $\cong$ S.A.S.
7. $\overline{PA} \cong \overline{PB}$	7. Corresponding sides of congruent triangles are congruent.
8. PA = PB	8. Congruent segments are equal in length.

(b) Proof of the converse "If a point is equidistant from the ends of a line segment, it is on the perpendicular bisector of the line segment."

Draw a point O, off $\overline{AB}$, such that PA = PB. If we let E be the midpoint of AB and can then show $\overleftrightarrow{PE}$, the bisector of $\overline{AB}$, is perpendicular to $\overline{AB}$, then the proof is done. We will show $\overleftrightarrow{PE} \perp \overline{AB}$ by proving congruence between adjacent angles $\sphericalangle$ PEA and $\sphericalangle$ PEB. $\sphericalangle$ PEA and $\sphericalangle$ PEB will be shown to be corresponding angles of ΔAPE and ΔBPE, these proved congruent by the S.S.S. $\cong$ S.S.S. Postulate.

STATEMENT	REASON
1. Let E be the midpoint of $\overline{AB}$.	1. Every line segment has one and only one midpoint.
2. Draw $\overleftrightarrow{PE}$.	2. One and only one straight line can be drawn between two points.

620

3. $\overline{AE} \stackrel{\sim}{=} \overline{EB}$.

3. A midpoint divides a line segment into two congruent parts.

4. $\overline{PA} \stackrel{\sim}{=} \overline{PB}$.

4. Given.

5. $\overline{PE} \stackrel{\sim}{=} \overline{PE}$

5. Reflexive property of congruence.

6. $\triangle APE \stackrel{\sim}{=} \triangle BPE$.

6. S.S.S. $\stackrel{\sim}{=}$ S.S.S.

7. $\sphericalangle PEA \stackrel{\sim}{=} \sphericalangle PEB$.

7. Corresponding angles of congruent triangles are congruent.

8. $\overleftrightarrow{PE} \perp \overline{AB}$.

8. If two lines intersect forming congruent adjacent angles, the lines are perpendicular (theorem).

9. $\overleftrightarrow{PE}$ is the $\perp$ bisector of $\overline{AB}$.

9. If a line is perpendicular to a line segment and bisects the line segment, then the line is the perpendicular bisector of the line segment.

● **PROBLEM** 685

What is the probable locus of points, in the interior of an angle, which is equidistant from the sides of the given angle? (See figure.)

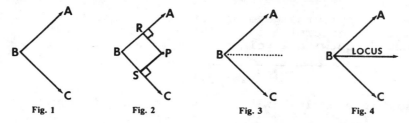

Fig. 1   Fig. 2   Fig. 3   Fig. 4

<u>Solution</u>: In order to establish the probable locus of points, we must follow the procedure outlined in the preceding problems.

Begin, as shown in Figure (1), by drawing the given angle, $\sphericalangle ABC$.

The conditions which must be satisfied by the locus are: (1) that all points, P, must be equidistant from $\overrightarrow{BA}$ and $\overrightarrow{BC}$, the sides of $\sphericalangle ABC$; and (2) that all points must lie in the interior of the angle. In Figure 2, one such point has been located and drawn such that $\overline{PR} \stackrel{\sim}{=} \overline{PS}$ when $\overline{PR} \perp \overline{BA}$ and $\overline{PS} \perp \overline{BC}$. Perpendicularity is implied here because a perpendicular segment is the shortest distance from a point outside a line to a line. Therefore, the distance between a point and a line is defined as the perpendicular distance.

Several other points, which are close enough together to develop the shape of the locus and which also satisfy the required conditions, have been shown in Figure 3.

The line drawn through the points plotted in Figure 4 appears to be a straight line. The probable locus, then, is shown to be a ray which is also the angle bisector of ∢ ABC.

● **PROBLEM** 686

(a) Write an equation of the locus of points whose distance from the origin is 5. (b) Determine whether the point $(-3,4)$ is on the locus.

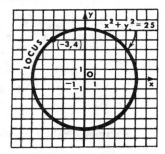

Solution: (a) Let (x, y) represent an arbitrary point in the locus. Therefore, (x, y) must be 5 units from the origin, $(0, 0)$.

We can apply the general formula for distance between two points to determine the equation of the locus. The distance, d, between 2 points (x_1, y_1) and (x_2, y_2) is

$$d = \sqrt{(x_2 - x_1)^2 + (y_2 - y_1)^2}.$$

In this example, $(x_2, y_2) = (x, y)$ and $(x_1, y_1) = (0, 0)$. Hence,

$$d = \sqrt{(x - 0)^2 + (y - 0)^2}$$

But d must equal 5.

By substitution, $5 = \sqrt{x^2 + y^2}$, which implies that

$$25 = x^2 + y^2$$

Since every point (x, y) is equidistant from the origin, we can conclude, by definition, that the locus is a circle of radius 5 that is centered at the origin. Therefore, we conclude that $x^2 + y^2 = 25$ is the equation for a circle centered at the origin, of radius 5. The circle is graphed in the figure.

(b) If the point $(-3, 4)$ is to be on the locus, $x = -3$, $y = 4$ must satisfy the equation $x^2 + y^2 = 25$.

We must substitute the values to see if this is the case:

$$(-3)^2 + (4)^2 \overset{?}{=} 25$$

$$(9) + (16) \overset{?}{=} 25$$

$$25 = 25$$

$(-3, 4)$ is on the locus.

● **PROBLEM** 687

Describe the locus of points determined by the equation

$$x^2 + y^2 + z^2 = 2x.$$

(Hint: Complete the square in x.)

<u>Solution</u>: Let us apply the hint first and then see what help this gives us in answering the question.

$$x^2 - 2x + y^2 + z^2 = 0.$$

Completing the square we obtain: $(x - 1)^2 + y^2 + z^2 - 1 = 0$, or:

(1) $\qquad (x - 1)^2 + y^2 + z^2 = 1.$

The square of the distance, d, between any two points $A(x_1, y_1, z_1)$ and $B(x_2, y_2, z_2)$ in 3-space is given by

$$d^2 = (x_1 - x_2)^2 + (y_1 - y_2)^2 + (z_1 - z_2)^2.$$

If A is allowed to be any arbitrary point (x, y, z) then we have

(2) $\quad (x - x_2)^2 + (y - y_2)^2 + (z - z_2)^2 = d^2.$

Assuming (x_2, y_2, z_2) and d are given, this is an equation for all points a distance d from (x_2, y_2, z_2). In other words, a sphere of radius d centered at (x_2, y_2, z_2).

By inspection we see that equation (2) is similar to equation (1). In fact, if we take $(x_2, y_2, z_2) = (1, 0, 0)$, and d = 1, the equations are identical. Hence, the locus described by equation (1) and the original equation is a sphere centered at $(1, 0, 0)$ with radius d.

● **PROBLEM** 688

What is the probable locus of the midpoints of the radii of a given circle?

<u>Solution</u>: We must first draw the given circle with its center at point O, as shown in figure 1.

The condition which must be satisfied in this problem

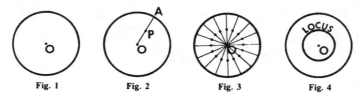

Fig. 1   Fig. 2   Fig. 3   Fig. 4

is that all points in the locus, P, must be midpoints of the radii of circle O.

One radius, $\overline{OA}$, has been drawn in figure 2. Its midpoint, P, has been located such that OP = PA.

Various other radii should now be drawn. All must have their midpoints located, as in figure 3. The radii should be drawn sufficiently close so that their midpoints convey the shape of the locus.

Now, we must construct a line, which in this case is a smooth curve, through the points which appear to be the locus. See figure 4.

It can now be said that the probable locus appears to be a circle with the same center as the given circle, the radius of which is one-half that of the given circle.

● **PROBLEM** 689

Locate the points that are a given distance, d, from a given point S, and are also equidistant from the ends

of line segment $\overline{AB}$.

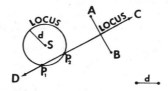

<u>Solution</u>: Conditions for two loci are given in this problem. The problem can be interpreted as one of locating the intersection points of these two loci.

The first condition, that the points of the locus be a given distance d from point S, describes a circle, the length of whose radius is d. This locus is the circle shown in the figure, with center at S.

The locus of points equidistant from the end of the line segment $\overline{AB}$ is the perpendicular bisector of $\overline{AB}$. In the figure, this is line $\overleftrightarrow{CD}$.

If d is less than the distance from S to $\overleftrightarrow{CD}$, then the circle locus and the locus given by $\overleftrightarrow{CD}$ do not intersect at any points and no points satisfy the two conditions.

If d is equal to the distance from S to $\overleftrightarrow{CD}$, then there is one point where both loci will intersect since $\overleftrightarrow{CD}$ will be the tangent of the circle centered at S with radius d. One point will satisfy both conditions.

If d is greater than the distance from S to $\overleftrightarrow{CD}$, then the locus $\overleftrightarrow{CD}$ is a secant to the circular locus and, accordingly, the two loci intersect at two points and these two points, P_1 and P_2, satisfy both conditions. This final case is exhibited in the figure.

● **PROBLEM** 690

Two concentric circles have radii whose lengths are 2 in. and 6 in. Line m is drawn, in the accompanying figure, tangent to the smaller circle. (a) Describe fully the locus of points equidistant from the two circles. (b) Describe fully the locus of points at a given distance d from line m. (c) How many points are there which satisfy the conditions given in both parts (a) and (b) if: (1) d < 2 in.? (2) d = 2 in.? (3) d = 6 in.? (4) d > 6 in.?

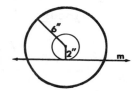

Solution: (a) The locus of points equidistant from the two concentric circles is a third concentric circle, the length of whose radius is 4 in.

The middle circle represents this locus.

(b) The locus of points at a given distance d from the given line m is a pair of lines, parallel to m, at the distance d from m. The 2 outer parallel lines shown comprise this locus.

(c) The accompanying figures, numbered 1 - 4, will be helpful in solving this problem.

(1) d < 2 in.: In this case, there are 4 points of intersection of the loci. In figure 1, they are P_1, P_2, P_3, P_4.

(2) d = 2 in.: As the distance between the parallel lines widens, the points of intersection approach each other. In case (1), both parallel lines were secants of the circular locus. However, when d = 2 in., one line becomes a tangent. Therefore, as fig. (2) shows, there are 3 points that satisfy both conditions: P_1, P_2 and P_3.

(3) d = 6 in.: The lowest of the lines in the parallel line locus, as shown in fig. (3), now lies completely outside of the circular locus and has no points of intersection with the latter. The upper line is tangent to the circle and provides the only point of intersection,

P₁, that satisfies both conditions.

    (d) d > 6 in.: The distance from line m to the outer-most edge of the circular locus is 6 in. Therefore, when the parallel lines are over 6 in. away from line m, there will be no point which satisfies both conditions.

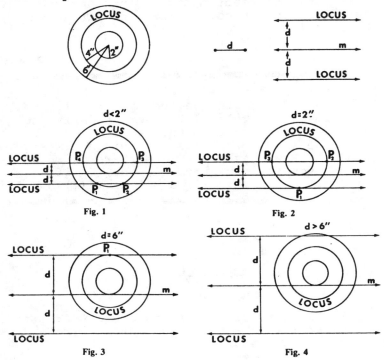

Fig. 1   Fig. 2

Fig. 3   Fig. 4

● **PROBLEM** 691

What is the locus of points traced by the center of a circle with radius R_2 that rolls around a second circle, the radius of which is R_1?

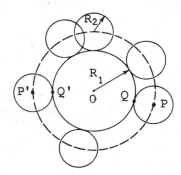

Solution: The figure shows the situation described. ⊙ P is tangent to ⊙ O at point Q. In order to find the required locus, we will use the fact that if 2 circles are tangent at the same point, the point of tangency and the centers of the two circles are collinear. Note OP = PQ + QO. PQ is a radius of ⊙ P and, therefore, equals a constant R_2. QO is the radius of ⊙ O and, therefore, equals a constant R_1. Thus, OP is a constant length $R_1 + R_2$.

As the circle P travels around circle O with new centers P' and new points of tangency Q', but P'Q' remains R_2 and OQ' is still R_1, and thus, P'O remains $R_1 + R_2$. The locus of points is, thus, the set of all points a distance $R_1 + R_2$ away from point O: a circle with center O and radius $R_1 + R_2$.

● **PROBLEM** 692

A point moves so that the sum of the squares of its distances from two given fixed points is a constant. Find the equation of its locus and show that it is a circle.

<u>Solution</u>: After determining the location of the two fixed points we will use the distance formula to prove the required result.

For simplicity, and without any restrictions, we may take the origin, O(0,0), as one of the two fixed points and the point A(a,0), $a \neq 0$, on the x-axis as the other, as in the figure shown. Let P(x,y) represent the point in motion. Then, according to the problem, we must prove P satisfies the geometric condition (1) $(PO)^2 + (PA)^2 = k$, where k is a positive constant.

By the general formula for distance between two points, $PO = \sqrt{(x-0)^2 + (y-0)^2} = \sqrt{x^2 + y^2}$ and $PA = \sqrt{(x-a)^2 + (y-0)^2} = \sqrt{(x-a)^2 + y^2}$.

Hence, the geometric condition (1) may be expressed analytically by the equation $\left[\sqrt{x^2 + y^2}\right]^2 + \left[\sqrt{(x-a)^2 = y^2}\right]^2 = x^2 + y^2 + (x-a)^2 + y^2 = k$. Expanding, we obtain

$$x^2 + y^2 + x^2 - 2ax + a^2 + y^2 - k = 0$$

$$x^2 + y^2 - ax + \frac{a^2}{2} - \frac{k}{2} = 0.$$

We will try to put this locus equation into the standard form of a circle, $(x-b)^2 + (y-c)^2 = r^2$, where (b,c) is the center and r is the length of the radius. If this can be done, then the locus is a circle.

Regroup and factor the last equation to arrive at

$$(x - \frac{a}{2})^2 + y^2 = \frac{k}{2} - \frac{a^2}{4}.$$

Hence, the locus is a circle centered at $(\frac{a}{2},0)$ with radius equal to $\sqrt{\frac{k}{2} - \frac{a^2}{4}}$, or $\frac{1}{2}\sqrt{2k - a^2}$.

Note that $2k - a^2$ must be > 0, or $k > \frac{a^2}{2}$ for the locus to be a circle. If $k = \frac{a^2}{2}$, the locus is the point $(\frac{a}{2},0)$. If $k < \frac{a^2}{2}$, no locus exists, since $\sqrt{2k - a^2}$ would not be defined.

CHAPTER 40

COORDINATE PROOFS

Basic Attacks and Strategies for Solving Problems in this Chapter. See pages 628 to 639 for step-by-step solutions to problems.

In an earlier chapter, it was mentioned that it is possible to prove standard geometry theorems using coordinate methods rather than more conventional techniques. This approach is illustrated in Problems 694, 697, 698, 699, 700, and 702. These proofs are called **analytic** proofs, and the more traditional proofs are called **synthetic** proofs.

As was the case in previous chapters, several coordinate geometry generalizations must be applied. These are listed below.

1. The distance between $P(x_1, y_1)$ and $Q(x_2, y_2)$ is

$$PQ = \sqrt{(x_1 - x_2)^2 + (y_1 - y_2)^2}.$$

2. The midpoint of $\overline{PQ}$ is $R\left(\dfrac{x_1 + x_2}{2}, \dfrac{y_1 + y_2}{2}\right)$, where P is the point at (x_1, y_1) and Q is the point at (x_2, y_2).

3. If lines l_1 and l_2 have slopes m_1 and m_2 and $m_1 = m_2$, then $l_1 \parallel l_2$.

4. If lines l_1 and l_2 have slopes m_1 and m_2 and $m_1 m_2 = -1$, then $l_1 \perp l_2$.

5. If the longest side of a triangle has length c, if the other two sides of the triangle have lengths a and b, and if $c^2 = a^2 + b^2$, then the triangle is a right triangle. This is the converse of the Pythagorean Theorem and is proved in Problem 699.

● **PROBLEM** 693

The vertices of ΔABC, when drawn on the Cartesian plane, are A (- 3, 0), B (3, 0), and C (0, 2). Prove that ΔABC is an isosceles triangle.

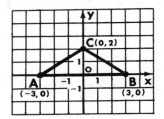

<u>Solution</u>: If we let the origin be point O, and prove ΔCOA $\cong$ ΔCOB, then, by corresponding parts, we can conclude $\overline{CA} \cong \overline{CB}$. This will be sufficient to show ΔABC is isosceles, since an isosceles triangle is defined to be one in which two sides are congruent. The SAS $\cong$ SAS method will be used.

Since, by definition of the Cartesian plane, the y-axis ⊥ x-axis; thus ∢ COA and ∢ COB are right angles and, they are congruent.

OA = 3 units and OB = 3 units and, because their lengths are equal, they are, therefore, congruent.

We now have congruence between one angle in each triangle and one corresponding adjacent side. The other adjacent side, $\overline{OC}$, is common to both triangles and, by reflexivity of congruence, is congruent to itself.

Therefore, ΔCOA $\cong$ ΔCOB by SAS $\cong$ SAS

Thus, $\overline{CA} \cong \overline{CB}$, because corresponding sides of congruent triangles are congruent.

Therefore, ΔABC is isosceles because it is a triangle which has two congruent sides.

● **PROBLEM** 694

Prove that the segment formed by joining the midpoints of two sides of a triangle is parallel to the third side and has a length equal to half that of the third side.

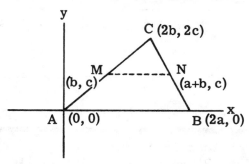

Solution: There may be a certain amount of doubt that the figure above is completely general. If we prove the theorem for the above figure, have we really proven it for all triangles? It would seem that this is a very special triangle, indeed, that has one side on the X-axis and one vertex at the origin.

We could prove the theorem for a triangle anywhere in the X-Y plane, but the notation is less complicated here. Draw a triangle anywhere in the plane and we can translate it and rotate it so that one side of it is on the X-axis and one vertex is at the origin. True we will have changed its position and direction, but the shape of the triangle will remain the same, its sides will have the same length; its angles will have the same measure. If the line between the midpoints of two sides is parallel to the third side, that too is unchanged. In short, ABC can be any triangle, and this proof is just as general as the geometric proof of the midline of a triangle.

GIVEN: ΔABC: A(0, 0); B (2a, 0); C (2b, 2c); M and N are the midpoints of $\overline{AC}$ and $\overline{CB}$ respectively.

PROVE: $\overline{MN} \parallel \overline{AB}$; and MN = ½ AB.

PROOF: If M is the midpoint of $\overline{AC}$, then by the midpoint formula, M= $\left(\dfrac{2b + 0}{2}, \dfrac{2c + 0}{2} \right)$ or M =(b, c).

Similarly, N is the midpoint of $\overline{BC}$. Therefore, N = $\left(\dfrac{2b + 2a}{2}, \dfrac{2c + 0}{2} \right)$ or N=(b + a, c).

To show $\overline{MN} \parallel \overline{AB}$, we find the slopes $\overline{MN}$ and $\overline{AB}$.

Slope of $\overline{MN} = \frac{\Delta y}{\Delta x} = \frac{c - c}{(b + a) - b} = 0$. Because $\overline{AB}$ lies on the X-axis, slope of $\overline{AB} = 0$. Since the two slopes are equal $\overline{AB} \parallel \overline{MN}$.

To show MN = ½ AB, we find the lengths of each using the distance formula.

$$AB = \sqrt{(2a - 0)^2 + (0 - 0)^2} = 2a.$$

$$MN = \sqrt{(b - (a + b))^2 + (c - c)^2} = \sqrt{(-a)^2 + 0^2}$$

$$= \sqrt{a^2} = a$$

Therefore, AB = 2 · MN or MN = ½ AB, and the proof is complete.

● **PROBLEM** 695

A triangle has vertices at A (4, - 1), B (5, 6), and C (1, 3). Plot the points, join them with line segments, and prove that the resulting triangle is an isosceles right triangle (see Figure).

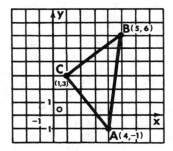

Solution: The triangle can be seen fully constructed in the accompanying graph.

We can prove that the triangle is an isosceles right triangle by proving that the lengths of two sides are the same, and that the Pythagorean Theorem holds. This theorem states that the square of the length of the longest side of a right triangle, the hypotenuse, is equal to the sum of the squares of the lengths of the other two sides.

The length of each side may be calculated by determining the distance between the two endpoints of that side.

Let AB = the distance between points A and B

BC = the distance between points B and C

CA = the distance between points C and A

Also let $A = \left(x_a, y_a\right)$ where $x_a = 4$, $y_a = -1$

$$B = \left(x_b, y_b\right) \text{ where } x_b = 5, \quad y_b = 6$$

$$C = \left(x_c, y_c\right) \text{ where } x_c = 1, \quad y_c = 3$$

For any two points (x_1, y_1) and (x_2, y_2), the distance between them is given by

$$d = \sqrt{(x_1 - x_2)^2 + (y_1 - y_2)^2}$$

Therefore, $\quad AB = \sqrt{\left(x_a - x_b\right)^2 + \left(y_a - y_b\right)^2}$

By substitution, $\quad AB = \sqrt{(4 - 5)^2 + (-1 - 6)^2}$

$$= \sqrt{(-1)^2 + (-7)^2} = \sqrt{1 + 49} = \sqrt{50}$$

$$BC = \sqrt{\left(x_b - x_c\right)^2 + \left(y_b - y_c\right)^2}$$

By substitution, $\quad BC = \sqrt{(5 - 1)^2 + (6 - 3)^2} = \sqrt{(4)^2 + (3)^2}$

$$= \sqrt{16 + 9} = \sqrt{25} = 5$$

$$CA = \sqrt{\left(x_c - x_a\right)^2 + \left(y_c - y_a\right)^2}$$

By substitution, $\quad CA = \sqrt{(1 - 4)^2 + (3 - (-1))^2}$

$$= \sqrt{(-3)^2 + (4)^2} = \sqrt{9 + 16}$$

$$CA = \sqrt{25} = 5$$

Since BC = CA, $\triangle$ABC is an isosceles triangle.

If $(AB)^2 = (BC)^2 + (CA)^2$, the Pythagorean Theorem holds, and $\triangle$ABC is also a right triangle.

By substitution, $\quad (\sqrt{50})^2 \overset{?}{=} (5)^2 + (5)^2$

$$50 \overset{?}{=} 25 + 25$$

$$50 = 50$$

Therefore, $\triangle$ABC is an isosceles right triangle.

● **PROBLEM** 696

Show that the points A (2, -2), B (- 8, 4), and C (5, 3) are the vertices of a right triangle and find its area.

Solution: We can prove that a triangle is a right triangle by showing that the square of the length of the hypotenuse is equal to the sum of the squares of the lengths of both legs, i.e. showing that the Pythagorean Theorem holds.

The general formula for distance between two points will allow us to determine the required length. The formula tells us that for any two points (x_1, y_1) and (x_2, y_2) the length of the segment between them, d, is given by

$$d = \sqrt{(x_2 - x_1)^2 + (y_2 - y_1)^2}.$$

Hence, since A = $(2, -2)$, B = $(-8, 4)$, and C = $(5, 3)$, we can apply the formula to find:

$$AB = \sqrt{(-8-2)^2 + (4-(-2))^2} = \sqrt{(-10)^2 + (6)^2} = \sqrt{136}$$

$$BC = \sqrt{(5-(-8))^2 + (3-4)^2} = \sqrt{(13)^2 + (-1)^2} = \sqrt{170}$$

$$CA = \sqrt{(5-2)^2 + (3-(-2))^2} = \sqrt{(3)^2 + (5)^2} = \sqrt{34}$$

$\overline{BC}$ is the longest side of $\triangle ABC$ and, if $\triangle ABC$ is a right triangle, $\overline{BC}$ is the hypotenuse. Therefore, if $(BC)^2 = (AB)^2 + (CA)^2$, then $\triangle ABC$ is a right triangle.

By substitution, we obtain

$$(\sqrt{170})^2 = (\sqrt{136})^2 + (\sqrt{34})^2$$

$$170 = 136 + 34 = 170.$$

Hence, $\triangle ABC$ is a right triangle.

The area of a right triangle is equal to the one-half the product of the lengths of its legs.

Ergo, area of $\triangle ABC = \frac{1}{2} (\sqrt{136})(\sqrt{34}) = \frac{1}{2} \sqrt{4624}$

$$= \frac{1}{2} \sqrt{16} \sqrt{289} = \frac{1}{2} (4)(17) = \frac{1}{2}(68).$$

Therefore, area of $\triangle ABC$ is 34 sq. units.

● **PROBLEM** 697

Prove analytically that if the diagonals of a parallelogram are equal, the figure is a rectangle.

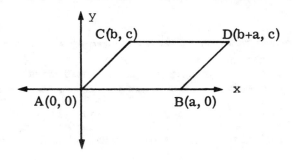

632

<u>Solution</u>: In the figure shown, ABDC is a general parallelogram positioned on the Cartesian plane in such a way that one side corresponds to the x-axis and one vertex to the origin. Here we are given AD = BC and asked to prove ABCD is a rectangle. To complete the proof, we will employ the fact that a rectangle is a parallelogram with one right angle.

The length of the two diagonals can be represented by using the distance formula.

$$AD = \sqrt{(b + a - 0)^2 + (c - 0)^2} = \sqrt{b^2 + 2ab + a^2 + c^2} \text{ and}$$

$$BC = \sqrt{(b - a)^2 + (c - 0)^2} = \sqrt{b^2 - 2ab + a^2 + c^2}$$

We are given AD = BC. Hence,

$$\sqrt{b^2 + 2ab + a^2 + c^2} = \sqrt{b^2 - 2ab + a^2 + c^2}$$

$$b^2 + 2ab + a^2 + c^2 = b^2 - 2ab + a^2 + c^2$$

$$2ab = - 2ab$$

This can only happen if a = 0 or if b = 0. a cannot equal zero or points A and B would coincide and ABDC would not exist as a quadrilateral. Hence, b = 0 is the only possible conclusion in this problem.

We know, then, that the coordinates of ABDC take the form A (0, 0), B (a, 0), C (0, c), and D (a, c). This tells us that $\overline{AC}$ is a segment of the y-axis and $\overline{AB}$ is a segment of the x-axis. Since the two axes are perpendicular, $\overline{AC} \perp \overline{AB}$. Hence ∢ CAB is a right angle and ABCD is a rectangle.

● **PROBLEM** 698

Prove analytically that any angle inscribed in a semi-circle is a right angle.

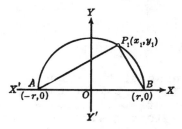

<u>Solution</u>: We will prove that the inscribed angle is a right angle by showing that it is formed by the intersection of two perpendicular line segments.

We draw a semicircle of radius r centered at the origin, as shown in the figure, the equation of the semicircle is $x^2 + y^2 = r^2$.

Let P_1 have the coordinates (x_1, y_1) and lie any-where on the semicircle. Since r is the radius of the semi-circle, the coordinates of A and B are $(-r, 0)$ and $(r, 0)$.

To prove that $\overline{P_1A} \perp \overline{P_1B}$ by analytical methods, we can show that the slopes of $\overline{P_1A}$ and $\overline{P_1B}$, m_1 and m_2 respectively, are in the relation $m_1m_2 = -1$.

By the definition of slope,

Slope of $\overline{P_1A} = m_1 = \dfrac{y_1 - 0}{x_1 - (-r)} = \dfrac{y_1}{x_1 + r}$

Slope of $\overline{P_1B} = m_2 = \dfrac{y_1 - 0}{x_1 - r} = \dfrac{y_1}{x_1 - r}$

Hence, $m_1m_2 = \left(\dfrac{y_1}{x_1 - r}\right)\left(\dfrac{y_1}{x_1 + r}\right) = \dfrac{y_1{}^2}{x_1{}^2 - r^2}$

Since P_1 lies on a circle, the coordinates (x_1, y_1) satisfy the equation of the semicircle. Hence, $x_1{}^2 + y_1{}^2 = r^2$. It follows then that $y_1{}^2 = r^2 - x_1{}^2$, and $-y_1{}^2 = x_1{}^2 - r^2$.

Therefore, by substitution, $m_1m_2 = \dfrac{y_1{}^2}{-y_1{}^2} = -1$.

Hence, $\overline{P_1A} \perp \overline{P_1B}$ and $\sphericalangle AP_1B$ is a right angle.

● **PROBLEM** 699

State and prove the converse of the Pythagorean Theorem analytically.

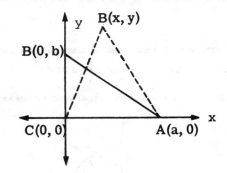

<u>Solution</u>: The converse of the Pythagorean Theorem can be stated as follows: If the sum of the squares of the length of two shorter sides of a triangle is equal to the square of the length of the longer side, then the triangle is a right triangle.

Let a = the length of one of the shorter sides

b = the length of the other short side

c = the length of the longest side.

Hence, if $a^2 + b^2 = c^2$, we are asked to prove $\triangle ABC$ is a right triangle.

We can conclude $c = \sqrt{a^2 + b^2}$. Notice that this expression is in the form of the general distance formula.

Let us begin to construct $\triangle ABC$ by placing the side of length a along the x-axis with one endpoint at the origin. The other endpoint must be $A(a, 0)$. Call the origin $C(0, 0)$.

Let the third vertex B be the arbitrary point (x, y). If B is on the y-axis, then $\triangle ABC$ will be a right triangle.

Assume AB = c and since CA = a, then BC = b, according to the given.

Hence, by the general distance formula,

$$c = \sqrt{(x - a)^2 + y^2} \quad \text{and} \quad b = \sqrt{x^2 + y^2}$$

We are given $c^2 = a^2 + b^2$. By substituting the expressions for c and b into this equation, we can solve for x.

$$(x - a)^2 + y^2 = a^2 + (x^2 + y^2)$$

$$x^2 - 2ax + a^2 + y^2 = a^2 + x^2 + y^2$$

Hence, $- 2ax = 0$.

Since $a \neq 0$, we conclude $x = 0$ must be the case.

We know $b = \sqrt{x^2 + y^2}$. By substitution, we can solve for y. $b = \sqrt{0^2 + y^2} = \sqrt{y^2}$, $y^2 = b^2$. Ergo, $y = b$.

The coordinates of point B are $(0, b)$.

The angle opposite the side of length c is formed by the intersection of two perpendicular lines, the coordinate axes. Hence, $\sphericalangle BCA$ is a right angle, and $\triangle ABC$ is a right triangle.

● **PROBLEM** 700

Prove analytically that the lines joining the midpoints of the adjacent sides of any quadrilateral form a parallelogram.

Solution: In plane geometry we can prove a quadrilateral is a parallelogram by proving both pairs of opposite sides are parallel. This condition, in analytic geometry, amounts to proving that opposite sides have the same slope, since line segments with the same slope will always be parallel.

On the accompanying graph, an arbitrary quadrilateral has been positioned with one side on the x-axis and

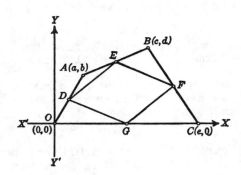

vertices A (a, b), B (c, d), C (e, 0), and O (0, 0) as
shown. The midpoints of the sides are E, F, G, and D.

We want to prove that quadrilateral DEFG is a
parallelogram. The slopes of the sides of this figure can
readily be determined once we know the coordinates of
D, E, F, and G.

Since these points are the midpoints of segments
whose endpoints are known, by the midpoint formula, their
coordinates can be determined.

D is the midpoint of $\overline{AO}$. Hence, $D = \left(\dfrac{a}{2}, \dfrac{b}{2}\right)$.

E is the midpoint of $\overline{AB}$. Hence $E = \left(\dfrac{a + c}{2}, \dfrac{b + d}{2}\right)$.

F is the midpoint of $\overline{BC}$. Hence, $F = \left(\dfrac{c + e}{2}, \dfrac{d}{2}\right)$.

G is the midpoint of $\overline{CO}$. Hence, $G = \left(\dfrac{e}{2}, 0\right)$.

Now we can proceed to calculate the slopes of the
sides of DEFG. Recall, for any segment with endpoints
(x_1, y_1) and (x_2, y_2) the slope equals $\dfrac{y_2 - y_1}{x_2 - x_1}$.

Slope of $\overline{DE} = \dfrac{\dfrac{b}{2} - \dfrac{b + d}{2}}{\dfrac{a}{2} - \dfrac{a + c}{2}} = \dfrac{\dfrac{b - b - d}{2}}{\dfrac{a - a - c}{2}} = \dfrac{-d}{-c} = \dfrac{d}{c}$

Slope of $\overline{FG} = \dfrac{0 - \dfrac{d}{2}}{\dfrac{e}{2} - \dfrac{c + e}{2}} = \dfrac{-\dfrac{d}{2}}{\dfrac{e - c - e}{2}} = \dfrac{-d}{-c} = \dfrac{d}{c}$

Hence, $\overline{DE} \parallel \overline{FG}$ since their slopes are equal.

Slope of $\overline{EF} = \dfrac{\dfrac{b + d}{2} - \dfrac{d}{2}}{\dfrac{a + c}{2} - \dfrac{c + e}{2}} = \dfrac{\dfrac{b + d - d}{2}}{\dfrac{a + c - c - e}{2}} = \dfrac{b}{a - e}$

636

Slope of $\overline{DG} = \dfrac{\dfrac{b}{2} - 0}{\dfrac{a}{2} - \dfrac{e}{2}} = \dfrac{b}{a - e}$

By similar reasoning, $\overline{EF} \parallel \overline{DG}$.

Therefore, DEFG is a parallelogram since all of its opposite sides are parallel.

● **PROBLEM** 701

As seen on the accompanying graph, ABCD is a quadrilateral with vertices at A (2, 2), B (5, - 2), C (9, 1), and D (6, 5). Prove that the geometric figure is a rhombus.

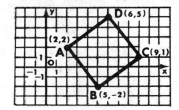

Solution: Since a rhombus is a quadrilateral in which all four sides are congruent, i.e. of equal lengths, we can complete the proof by determining the distances between any two adjacent vertices and showing that they are all equal. (The distance between any two adjacent vertices is the length of the side of the quadrilateral.)

Let AB = the distance between points A and B

    BC = the distance between points B and C

    CD = the distance between points C and D

    DA = the distance between points D and A.

Also let $A = \left(x_a, y_a\right)$ where $x_a = 2$, $y_a = 2$

    $B = \left(x_b, y_b\right)$ where $x_b = 5$, $y_b = -2$

    $C = \left(x_c, y_c\right)$ where $x_c = 9$, $y_c = 1$

    $D = \left(x_d, y_d\right)$ where $x_d = 6$, $y_d = 5$.

For any two points, (x_1, y_1) and (x_2, y_2), the general formula for distance between them is given by

$$d = \sqrt{(x_1 - x_2)^2 + (y_1 - y_2)^2}$$

Therefore,

$$AB = \sqrt{\left[x_a - x_b\right]^2 + \left[y_a - y_b\right]^2}$$

By substitution, $AB = \sqrt{(2 - 5)^2 + (2 - (-2))^2}$

$$= \sqrt{(-3)^2 + (4)^2} = \sqrt{9 + 16}$$

$$AB = \sqrt{25} = 5.$$

$$BC = \sqrt{\left[x_b - x_c\right]^2 + \left[y_b - y_c\right]^2}$$

By substitution, $BC = \sqrt{(5 - 9)^2 + (-2 - 1)^2}$

$$= \sqrt{(-4)^2 + (-3)^2} = \sqrt{16 + 9}$$

$$BC = \sqrt{25} = 5$$

$$CD = \sqrt{\left[x_c - x_d\right]^2 + \left[y_c - y_d\right]^2}$$

By substitution, $CD = \sqrt{(9 - 6)^2 + (1 - 5)^2} = \sqrt{(3)^2 + (-4)^2}$

$$= \sqrt{9 + 16}$$

$$CD = \sqrt{25} = 5.$$

$$DA = \sqrt{\left[x_d - x_a\right]^2 + \left[y_d - y_a\right]^2}$$

By substitution, $DA = \sqrt{(6 - 2)^2 + (5 - 2)^2}$

$$= \sqrt{(4)^2 + (3)^2} = \sqrt{16 + 9}$$

$$DA = \sqrt{25} = 5.$$

Therefore, $\quad AB = BC = CD = DA.$

We have shown that ABCD is an equilateral quadrilateral and have thereby proved that it is a rhombus.

● **PROBLEM** 702

Prove, in the framework of coordinate geometry, that the sum of the squares of the distances from any point in the plane to two opposite vertices of any rectangle is equal to the sum of the squares of its distances from the other two vertices.

Solution: Place a rectangle in the most convenient position on the Cartesian plane. This has been done in the figure shown. The vertices are given by the coordinates A (0, 0), B (a, 0), C (0, c), and D (a, c).

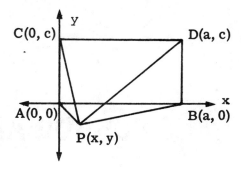

Select an arbitrary point P with coordinates (x, y). We are asked to prove

$$(PA)^2 + (PD)^2 = (PC)^2 + (PB)^2$$

We can determine the lengths PA, PD, PC and PB by applying the general formula for distance between two points. For any two points, (x_1, y_1) and (x_2, y_2), the distance between them, d, is given by

$$d = \sqrt{(x_1 - x_2)^2 + (y_1 - y_2)^2}$$

Hence,

$$PA = \sqrt{(x - 0)^2 + (y - 0)^2} = \sqrt{x^2 + y^2}$$

$$PD = \sqrt{(x - a)^2 + (y - c)^2}$$

$$PB = \sqrt{(x - a)^2 + (y - 0)^2} = \sqrt{(x - a)^2 + y^2}$$

$$PC = \sqrt{(x - 0)^2 + (y - c)^2} = \sqrt{x^2 + (y - c)^2}$$

We now substitute these expressions into the equation:

$$(PA)^2 + (PD)^2 = (PC)^2 + (PB)^2.$$

If equality holds, then the equation must be true,

(i) $(\sqrt{x^2+y^2})^2 + (\sqrt{(x-a)^2+(y-c)^2})^2 = (\sqrt{x^2+(y-c)^2})^2 + (\sqrt{(x-a)^2+y^2})^2$

(ii) $x^2+y^2+(x-a)^2+(y-c)^2 = x^2+(y-c)^2+(x-a)^2+y^2$

639

CHAPTER 41

POLAR/ANALYTIC COORDINATES

> **Basic Attacks and Strategies for Solving Problems in this Chapter. See pages 640 to 670 for step-by-step solutions to problems.**

This chapter applies some rather sophisticated aspects of trigonometry. These applications go beyond the right triangle trigonometry described in Chapter 22.

In the standard **rectangular coordinate system**, each point is associated with an ordered pair of real numbers which correspond to linear distances in different directions. In this chapter, emphasis will be placed on the **polar-coordinate system** in which one number of an ordered pair signifies linear distance, and the other signifies angular distance. Consider $\overrightarrow{OI}$ below. This is called the **polar axis**, and O is called the **pole**. Then, for any point P in the plane, its position is described by its distance from the pole and the angle $\overrightarrow{OP}$ makes with the polar axis. For the point P in the diagram below, the polar coordinates are (r, θ), where r is the distance from O to P, and θ is the radian measure (see Problem 703) of the angle formed.

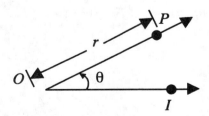

It is agreed that angles measured by a counterclockwise rotation have positive measure and those by a clockwise rotation have a negative measure.

Examine the diagram below. The radii of the concentric circles pictured are 1, 2, and 3. The polar coordinates for point P are $\left(2, \dfrac{\pi}{4}\right)$, since P is two units from O and the measure of the angle is $\dfrac{\pi}{4}$ radians.

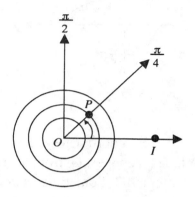

One of the unusual aspects of the polar-coordinate system is the ambiguous coordination of points. Suppose you consider that $\overrightarrow{OP}$ originated from a negative rotation. Then P could be associated with $\left(2, \dfrac{-7\pi}{4}\right)$. Also $\overrightarrow{OP}$ could result from rotations numerically greater than 2π radians. From this perspective $\left(2, \dfrac{\pi}{4}\right), \left(2, \dfrac{9\pi}{4}\right), \left(2, \dfrac{17\pi}{4}\right), \left(2, \dfrac{-7\pi}{4}\right), \left(2, \dfrac{-15\pi}{4}\right)$, and $\left(2, \dfrac{-23\pi}{4}\right)$ would all be associated with this same point.

In graphing polar-coordinate equations, it is often not very efficient to merely graph points. It is better to plot certain points and make use of the properties that a graph has. For example, for the graph below, the polar axis and the $\dfrac{\pi}{2}$-axis (sometimes called the 90°-axis) are both symmetry lines. It also has symmetry with respect to the pole. To graph this figure, if it were known that the symmetries described above existed, points could be plotted just between O and $\dfrac{\pi}{2}$, and then the rest of the graph could be obtained by applying the symmetries. Conditions for symmetry are described and justified in Problem 705.

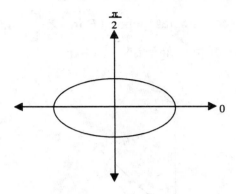

It was mentioned that in polar-coordinates the angle measure must be given in radians. The technique for converting from degree measure to radian measure is described in Problem 703. Sometimes it is desirable to convert from a rectangular-coordinate system equation to a polar-coordinate system equation, and that process is illustrated in Problems 715 and 716. The reverse conversion is illustrated in Problem 717.

GRAPHING POLAR COORDINATES

Convert the measures 30°, 45°, 60°, 90°, 180° from degrees to radians.

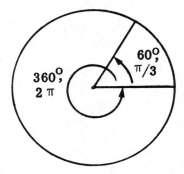

Solution: There are two systems for measuring angles and arc-length, degree-measure and radian-measure. Degree measure is most useful when working with many angles, since 360, the number of degrees in a full circle, is easily divisible into fractions without remainders (360 = 2 · 2 · 2 · 3 · 3 · 5). On the other hand, when relating arc lengths to other distance measures, radian measure is the preferred system: Arc length = (angle in radians)(radius).

In our example, the most direct means of conversion will be to express the degree measure as a fraction of 360°, and multiply by 2π, i.e., the conversion factor in the transformation. This conversion factor is actually an adjustment of units and equals 2π/360 radians per degree.

To understand how the conversion factor is derived, let D be the measure of the angle in degrees and R be the measure of the angle in radians. Note that the given angle is always a certain fraction of the circular angle regardless of the units used in measuring the angle. A circular angle has 360° or 2π radians.

For example, a vertex angle of an equilateral triangle must always be one sixth of a circular angle: $\frac{60°}{360°} = \frac{1}{6}$ and $\frac{\pi/3}{2\pi} = \frac{1}{6}$. Thus, it must be always true that:

(I) $\qquad \dfrac{D}{360} = \dfrac{R}{2\pi}$

Solving for R, we obtain

(II) $\qquad R = \dfrac{2\pi}{360} D$

Thus given any degree measure of an angle D, multiplying by $\frac{2\pi}{360}$ yields the radian measure. Therefore,

$30° \cdot \dfrac{2\pi}{360} = \dfrac{1}{12} \cdot 2\pi = \dfrac{\pi}{6}$ radians.

$45° \cdot \dfrac{2\pi}{360} = \dfrac{1}{8} \cdot 2\pi = \dfrac{\pi}{4}$ radians.

$60° \cdot \dfrac{2\pi}{360} = \dfrac{1}{6} \cdot 2\pi = \dfrac{\pi}{3}$ radians.

$90° \cdot \dfrac{2\pi}{360} = \dfrac{1}{4} \cdot 2\pi = \dfrac{\pi}{2}$ radians.

$180° \cdot \dfrac{2\pi}{360} = \dfrac{1}{2} \cdot 2\pi = \pi$ radians.

● **PROBLEM** 704

Plot each of the following points - the context is polar coordinates. (a) $(3, \pi/6)$; (b) $(3, -(\pi/6))$; (c) $(3, 5\pi/6)$; (d) $(3, 7\pi/6)$; (e) $(-3, \pi/6)$.

Solution: In Cartesian coordinate systems, we are given two lines, the x and y axes, and the position of points (x_0, y_0) is specified by the perpendicular distance of the point from the x-axis, y_0, and the perpendicular distance of the point from the y-axis, x_0. In polar coordinates, we are given a center point called the pole and a line through the pole called the polar axis. The position of point (r, θ) is specified by the length of the straight line segment connecting the point to the pole, r, and the angle formed between the straight line segment and the polar axis, θ.

In the Cartesian coordinate system, the point of constant x form a vertical line; the points of constant y form a horizontal line. In polar coordinates, the points with constant r are the set of points equidistant from the pole at a distance r and, therefore, the locus of points with constant r form a circle of radius r with the pole as the center. The points of constant θ form a ray with its end point at the pole intersecting the polar axis at an angle of θ.

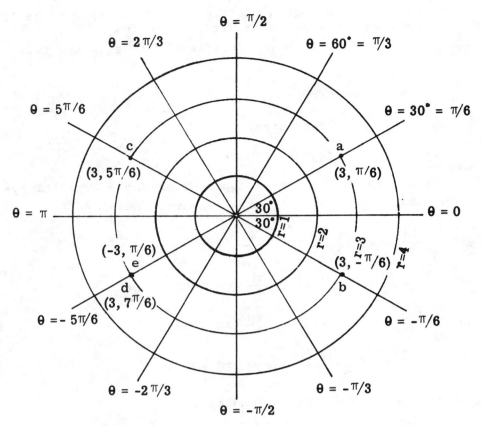

Whenever we graph point (x_0, y_0) in the Cartesian system, what we are doing (although you may never have thought of it in quite this fashion) is to consider the vertical line of constant x_0 and to consider the horizontal line of constant y_0. The point of intersection is the desired point. Similarly, when plotting (r_0, θ_0) in polar coordinates, we find the circle of radius r_0 and the ray with endpoint at the origin making an angle of θ_0 with the polar axis. The point where they intersect is the desired point.

The first point we are asked to plot is $(3, \pi/6)$. We find the circle of radius 3 and the ray of $\pi/6$. The intersection is the first point, a.

The second point is $(3, - \pi/6)$. To understand the idea of a negative angle, consider the polar axis as the base of an angle and the ray as the far side. Consider the arrow directed from the base to the far side. If the arrow is counterclockwise, the angle is positive. If the arrow is clockwise, the angle is negative. Thus, to find the ray of $\theta = - \pi/6$, draw the ray such that the angle between the polar axis and the ray is $\pi/6$ and also such that an arrow from the polar axis to the ray points in clockwise direction. The intersection point of the ray thus obtained and the circle $r = 3$ is the desired point (see the accompanying figure for point labelled b).

At this point, we may wonder if this is really a valid procedure. Is the ray that we have just drawn not also a ray we would draw for θ = 11 π/6. That is true. The confusion arises from the fact that, unlike Cartesian coordinates where every point has a distinct name, in polar coordinates a point may have several different coordinate pairs that are totally different in appearance. Among the many equivalences are

(1) Angles that differ by a multiple of 2π are the same angle. Note that 2π is the measure of a complete circle in the same way twelve hours is the measure of a complete clock. Just as from looking at a clock it is impossible to distinguish between 3:00 and 3:00 + 12 hours, it is impossible to distinguish between angles that differ by 2π. Therefore, the ray of θ = π/3 and the ray of θ = 2π + π/3 are indistinguishable.

Similarly, rays θ = π/3 and θ = 1000 (2π) + π/3 are the same ray, and (3, π/3) and (3, 2 1/3 π) are the same point.

(2) A negative angle - θ yields the same ray as the angle 2π - θ. Thus - π/6 and 11 π/6 are the same angle. This result follows immediately from what we showed in (1). The difference between 2π - θ and - θ is (2π - θ) - (- θ) = 2π - θ + θ = 2π. Since the two angles differ by 2π, the two angles are the same.

(3) Given a point with a negative radius (- k, θ), it is equivalent to (k, π + θ). Consider the polar axis as a number line. If point (3, θ = 0) is the point three units to the right on the polar axis, then (- 3, θ = 0) is the point three units to the left on the polar axis. Note that this is also the point (3, θ = π). Note that both points are points on the circle r = 3, but the point (- 3, θ = 0) is a half circle rotation from (3, θ = 0). A half rotation is the same as adding π to the given angle. Thus, the change of the radius from negative to positive can be counteracted by changing the angle from θ to θ + π.

With these properties in mind, we can plot the rest of the points. Note in (e), (- 3, π/6). From property (3), we know that (- 3, π/6) = (3, π/6 + π) = (3, 7π/6).

● **PROBLEM** 705

Find the conditions which the equation of a polar figure must satisfy for the figure to be (a) symmetric about the polar axis, (b) symmetric about the 90°-axis, (c) symmetric about the pole.

<u>Solution</u>: We know that an object is symmetric about something if for every part of that object there is a corresponding part of the object on the opposite side of that something. In more concrete terms, we let P_1 be a point on the curve C; $\overline{P_1A}$ be the perpendicular from point P_1 to the line ℓ. Now we construct perpendicular $\overline{AP_2}$ congruent to $\overline{AP_1}$ in the opposite half plane. If for every P_1 on curve C, P_2 is also a point on C, then the curve C is said to be symmetric about the line ℓ.

Symmetry about a point has a similar definition. We let P_1 be a point on the curve, and O be the given point. We locate point P_4 such that O is the midpoint of $\overline{P_1P_4}$. If for every P_1 on the curve, P_4 is also on the curve, then the curve is said to be symmetric about the point O.

(a) Suppose P_1 is any point on the given curve; $\overline{P_1A}$ the altitude to the polar axis, and P_2 the point such that $P_1A = P_2A$ and $PA_2 \perp$ polar axis. Then the given curve is symmetric about the polar axis only if P_2 is also a point on the curve. This condition just stated is sufficient for symmetry.

However, we wish to find a condition that we could use given only the equation of the polar figure. Therefore, we express the condition in terms of r and θ.

Suppose P_1 has the coordinates (r_1, θ_1), and P_2 the coordinates (r_2, θ_2). To relate the coordinates of P_1 and P_2, note that $\triangle OAP_1 \stackrel{\sim}{=} \triangle OAP_2$ by the SAS Postulate $(\overline{OA} \stackrel{\sim}{=} \overline{OA}, \overline{P_1A} \stackrel{\sim}{=} \overline{P_2A}, \sphericalangle OAP_1$ and $\sphericalangle OAP_2$ are right angles). Thus, by corresponding parts, $\overline{OP_1} \stackrel{\sim}{=} \overline{OP_2}$ and $\sphericalangle P_1OA \stackrel{\sim}{=} \sphericalangle P_2OA$. Thus, $r_2 = OP_2 = OP_1 = r_1$.

In addition, remember that a polar angle is positive if directed counterclockwise from the base (θ = 0) and negative if directed clockwise. Therefore, $\theta_2 = -m \sphericalangle P_2OA = -m \sphericalangle P_1OA = -\theta_1$. Thus, $P_2(r_1, -\theta_1)$. We now restate the condition.

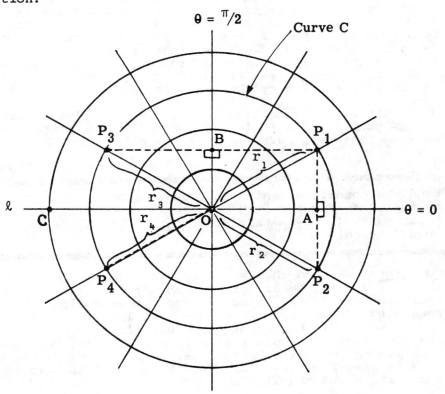

644

(c) In the figure, point P_1 is a point on the curve; $\overline{P_1O} \cong \overline{OP_4}$; and $\overrightarrow{P_1OP_4}$. Then the given curve is symmetric about the pole only if P_4 is also a point on the curve. Suppose the coordinates of P_1 and P_4 are (r_1, θ_1) and (r_4, θ_4).

Note that $\overline{P_1O} \cong \overline{OP_4}$. Therefore, $r_4 = OP_4 = OP_1 = r_1$. Furthermore, $\overrightarrow{P_1OP_4}$ and line $\theta = 0$ form opposite angles ∢ P_4OC and ∢ P_1OA. Because opposite angles are congruent, ∢ $P_1OA \cong$ ∢ P_4OC. Thus, $\theta_4 = m$ ∢ $P_4OC + m$ ∢ $COA =$ $= m$ ∢ $P_1OA + \Pi = \theta_1 + \pi$.

A graph is symmetric about the pole if, for every point (r, θ) on the graph, $(r, \theta_1 + \pi)$ is also on the graph.

A graph is symmetric about the polar axis, if every point (r, θ) on the graph, $(r, -\theta)$ is also on the graph.

(b) Suppose P_1 is any point on the given curve; $\overline{P_1B}$ the altitude to the 90°-axis; and P_3 the point such that $P_1B = P_3B$ and $\overline{P_3B}$ $\perp$ 90°-axis. The given curve is symmetric about the 90°- axis only if P_3 is also a point on the curve.

Suppose P_1 has coordinates (r_1, θ_1) and $P_3 (r_3, \theta_3)$. To relate the coordinates of the two points, note that $\triangle OP_1B \cong \triangle OP_3B$ by the SAS Postulate ($\overline{BP_3} \cong \overline{BP_1}$, $\overline{BO} \cong \overline{BO}$, and ∢ OBP_3 and ∢ OBP_1 are right **angles**). By corresponding parts, $\overline{OP_3} \cong \overline{OP_1}$ and ∢ $BOP_1 \cong$ ∢ BOP_3. Thus, $r_3 = P_3O = P_1O = r_1$, and $\theta_3 = m$ ∢ $P_3OA = m$ ∢ $P_3OB + m$ ∢ $BOA = m$ ∢ $P_3OB + \pi/2 =$ $(\pi/2 - m$ ∢ $P_1OA) + \pi/2 = \pi - m$ ∢ $P_1OA = \pi - \theta_1$. Thus, a given curve is symmetric about the 90°-axis only if, for every point (r, θ) on the graph, $(r, \pi - \theta)$ is also on the graph.

● **PROBLEM** 706

Given the figure below, is it possible to add points to the graph so that the final figure will have (a) symmetry about the polar axis, (b) symmetry about the 90°-axis, (c) symmetry about the pole? If so, add these points.

<u>Solution</u>: (a) For there to be symmetry about the polar axis, point $(r, -\theta)$ must be on the completed graph whenever (r, θ) is a point on the graph. We are given that every point in the figure of Figure 1 is a point of the final graph. Consider point A in Figure 2, with coordinates $(r', -\pi/4)$. For the final figure to be symmetric, point $(r', -(-\pi/4)$ or $(r', \pi/4)$ must also be on the **graph**. **Thus,** we add point A' $(r', \pi/4)$ to the graph. We continue until every point in the lower figure has a corresponding point in the upper half. The complete graph is symmetric about the polar axis.

Note that the upper figure has exactly the same shape as the lower figure except that it is upside down. This characteristic can be exploited in graphing other polar

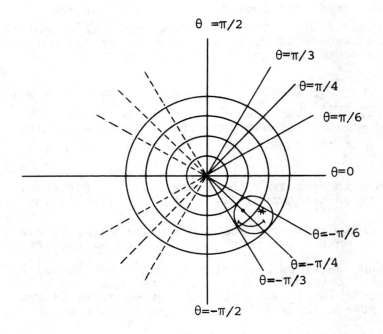

Figure 1

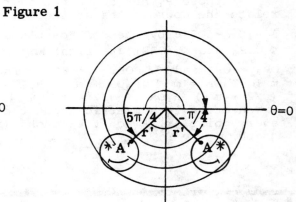

Figure 2

Figure 3

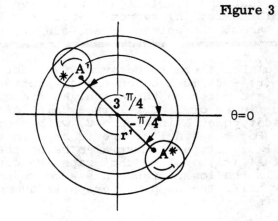

Figure 4

figures. If we know the figure is symmetric about the polar axis, we need only plot the half of the graph - from $\theta = 0$ to π (instead of $\theta = 0$ to 2π) and then sketching the other half by "flipping upside down" the half we already have.

For there to be symmetry about the 90°-axis, the point $(r, \pi - \theta)$ must be on the completed graph whenever (r, θ) is a point on the graph. Proceeding as in part (a), we add a point $A'(r', \pi - \theta)$ for every point $A(r', \theta)$ on the original figure. The result (shown in Figure 3) is a mirror image of the original figure (notice the position of eyes) - the leftmost points are now the rightmost points.

This characteristic can be exploited in graphing other polar figures. If we know that the figure is symmetric about the 90°-axis, we need only plot the half of the graph - from $\theta = - \pi/2$ to $\pi/2$. We can then sketch the rest of the figure by drawing the mirror image of the plotted half in the left side plane.

For there to be symmetry about the pole, point $(r, \pi + \theta)$ must be on the completed graph whenever (r, θ) is on the graph. Proceeding as in part (a), we add a point $A'(r', \pi + \theta)$ for every $A(r', \theta)$ on the original figure. The result is a figure that can be obtained from the original figure by (1) drawing the mirror image of the original in the left plane, and (2) "flipping upside down" the mirror image about the polar axis.

In exploiting the pole symmetry for graphing other polar figures, the two step flip may be rather hard to see. The better way is to graph any half of the graph ($- 3\pi/4 < \theta < \pi/4$, for example) and, for each point, locate the point directly opposite the points in the graph (the point such that the pole is the midpoint of the point and the graphed point).

With polar axis symmetry, we need only graph $0 \leq \theta \leq \pi$. With 90° axis symmetry, we need only graph $- \pi/2 \leq \theta \leq \pi/2$. If a figure has both 90°-axis and polar axis, note that if we plot $0 \leq \theta \leq \pi/2$, by the polar axis symmetry, we also know $- \pi/2 \leq \theta \leq 0$. Thus, we know $- \pi/2 \leq \theta \leq \pi/2$. Because of the 90°-axis symmetry, then we know $\pi/2 \leq \theta \leq 3\pi/2$. Therefore, given 90°-axis and polar axis symmetry, we need only graph $0 \leq \theta \leq \pi/2$ to plot the whole graph.

Similarly, given all three types of symmetry, it can be shown that we need only plot $0 \leq \theta \leq \pi/4$.

● **PROBLEM** 707

Graph $r^2 = 4 \cos 2\theta$.

Solution: Before blindly plotting points, let us examine the equation for simplifying factors such as symmetries.

(1) Polar-Axis Symmetry. A figure is symmetric about the polar axis, if for every point (r, θ) on the graph $(r, - \theta)$ is also a point of the graph. Suppose (r_0, θ_0) is on the

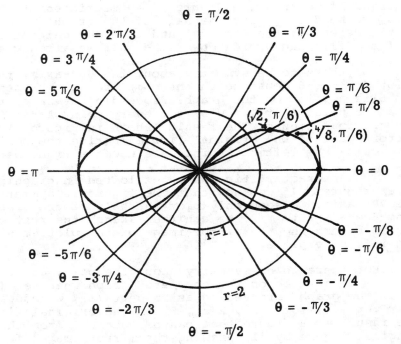

At the top of the figure, the following angle labels appear around the polar graph:

$\theta = \pi/2$, $\theta = 2\pi/3$, $\theta = \pi/3$, $\theta = 3\pi/4$, $\theta = \pi/4$, $\theta = 5\pi/6$, $\theta = \pi/6$, $\theta = \pi/8$, $(\sqrt{2}, \pi/6)$, $(\sqrt[4]{8}, \pi/6)$, $\theta = \pi$, $\theta = 0$, $r=1$, $\theta = -\pi/8$, $\theta = -5\pi/6$, $\theta = -\pi/6$, $\theta = -3\pi/4$, $\theta = -\pi/4$, $r=2$, $\theta = -2\pi/3$, $\theta = -\pi/3$, $\theta = -\pi/2$

graph. Thus, $r_0^2 = 4 \cos 2\theta_0$. Since $\cos x = \cos (-x)$, it must be true that $r_0^2 = 4 \cos (2\theta_0) = 4 \cos (-2\theta_0) = 4 \cos 2(-\theta_0)$. Thus, if (r_0, θ_0) is a point on the graph, then $(r_0, -\theta_0)$ is also on the graph. The figure $r^2 = 4 \cos 2\theta$ is symmetric about the polar axis.

    (2) Symmetry about the 90°-axis. A figure is symmetric about the 90-axis if, for every point (r, θ) on the graph, $(r, \pi - \theta)$ is also a point on the graph. Suppose (r_0, θ_0) is a point on the graph. Then we must show $r_0^2 = 4 \cos 2(\pi - \theta_0)$ is also a point on the graph. Working backwards, we use the identity $\cos 2\psi = 2 \cos^2 \psi - 1$ to show that $r_0^2 = 4 \cos 2 (\pi - \theta_0)$ is true if and only if $r_0^2 = 4(2 \cos^2 (\pi - \theta_0) - 1)$. Since $\cos (\pi - \theta) = -\cos \theta$, $r_0^2 = 4(2 \cos^2 (\pi - \theta_0) - 1)$ is true if and only if $r_0^2 = 4(2 (-\cos \theta_0)^2 - 1) = 4 (2 \cos^2 \theta_0 - 1) = 4 \cos 2\theta_0$. But, we know that, since (r_0, θ_0) is a point on the graph, it must be true that $r_0^2 = 4 \cos 2\theta_0$. Thus, $r_0^2 = 4 \cos 2(\pi - \theta_0)$. Therefore, every other equation must also be true. Thus, $(r, \pi - \theta)$ is a point on the graph if (r, θ) is on the graph. The figure is thus symmetric about the 90°-axis.

    (3) Symmetry with respect to the pole. A figure is symmetric about the pole if, for every point (r, θ) on the graph, $(r, \pi + \theta)$ is also a point on the graph. Working backward as above, we obtain $r_0^2 = 4 \cos 2(\pi + \theta_0) = 4 (2 \cos^2 (\pi + \theta_0) - 1) = 4(2 [-\cos \theta_0]^2 - 1) = 4 (2 \cos^2 \theta_0 - 1) = 4 \cos 2\theta_0$. Since (r_0, θ_0) is given to be on the graph, then $r_0^2 = 4 \cos 2\theta_0$ is a true statement; and, thus, for every point (r_0, θ_0) on the graph, the point $(r_0, \theta_0 + \pi)$ is also on the graph. The figure is symmetric about the pole.

    We now have three symmetries, and thus, we need only

consider values of θ between 0 and π/4. Two other important results are:

(1) The curve is bounded. Note cos x $\leq$ 1. Therefore, 4 cos^2 2θ $\leq$ 4. Since r^2 = 4 cos^2 2θ, it must be true that r $\leq$ 2. Since there is an upper limit to r, the curve is bounded.

(2) To find the intercepts with the polar axis, either r = 0, or θ = 0, or θ = π. If r = 0, then 0^2 = 4 cos 2θ or cos 2θ = 0. Since only cos ± π/2 has a cosine of 0, 2θ = ± π/2, or θ = ± π/4. Thus, (0, π/4), and (0, - π/4) are intercepts with the polar axis, If θ = 0, then r^2 = 4 cos 0 = 4 or r = ± 2 and (2, 0) and (- 2, 0) are points on the graph. Solving for r with θ = π yields the same points. Thus the axis intercepts are (± 2, 0), and (0, ±π/4).

Using the above information (the symmetries and the intercepts), and with the points calculated below, we obtain the accompanying graph.

For convenience of calculations, it is best to express r in terms of θ. Thus, if r^2 = 4 cos 2θ, then r = 2 $\sqrt{\cos 2\theta}$. For θ = π/8, r = 2 $\sqrt{\cos \pi/4}$ = 2 $\sqrt{\sqrt{2}/2}$ = $\sqrt{4 \cdot \sqrt{2}/2}$ = $\sqrt{2\sqrt{2}}$ = $\sqrt{\sqrt{8}}$ = $\sqrt[4]{8}$.

Thus, ($\sqrt[4]{8}$, π/8) is a point on the graph for θ = π/6, r = 2 $\sqrt{\cos \pi/3}$ = 2$\sqrt{\tfrac{1}{2}}$ = $\sqrt{2}$.

Thus, ($\sqrt{2}$, π/6) is a point on the graph.

● **PROBLEM** 708

Graph the curve r = 2 cos θ.

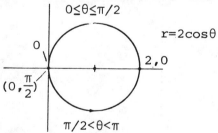

Solution: We graph the polar curve in five steps. First, we rewrite the equation in convenient form - r solved in terms of θ. Thus has already been done for us.

(1)   r = 2 cos θ.

Second, we find the bounding values for r and the symmetries of the figure. Because |cos θ| $\leq$ 1, |r| $\leq$ 2, and r is bounded. Every point of the figure must lie within or on the circle centered at the origin with radius 2.

649

Because r = 2 cos (- θ) = 2 cos θ, (r, - θ) is
implied by (r, θ) and the figure is symmetric about the
polar axis. Because r = 2 cos (π - θ) = - 2 cos θ,
(r, π - θ) does not follow from (r, θ), and the figure is
not symmetric about the 90°-axis. In the same way, r =
2 cos (π + θ) = - 2 cos θ ≠ 2 cos θ, and there is no
symmetry about the pole. Because of the symmetry about
the polar axis, we need only plot points from θ = 0 and
θ = π. The graph from θ = π to θ = 2π is just a mirror
image of the top.

Third, we locate the intercepts at the pole,
(0, π/2) and (0, 3π/2), at the polar axis, (2, 0) and
(- 2, π), and at the 90° axis, (0, π/2) and (0, 3π/2).
Note that a minus sign preceeding the r coordinate has the
effect of a rotation of π; hence, (- 2, π) = (2, 2π) =
(2, 0).

Fourth, we find the general behavior of the graphs as
θ changes. Recall, the shape of the cos θ curve in rectangular coordinates to aid in doing this.

θ	r = 2 cos θ
increasing from 0 to π/2	decreases from 2 to 0
increasing from π/2 to π	decreases from 0 to - 2.

Fifth, we plot specific points to aid in drawing.

θ	r = 2 cos θ
0	2
π/6	$\sqrt{3} \overset{\sim}{=} 1.732$
π/4	$\sqrt{2} \overset{\sim}{=} 1.4142$
π/3	1
π/2	0
2π/3	- 1
3π/4	$- \sqrt{2}$
5π/6	$- \sqrt{3}$
π	- 2

The resulting curve resembles a circle. (By using a
coordinate transformation, it can be proved to be a
circle.)

NOTE: Examining the table of values, we notice that
the r value of an angle equals the negative r value of
the supplement. Thus, for θ = π/3, r = 1; and θ = 2π/3,
r = - 1. To understand why this happens, we note that in
searching for symmetries, we found that there was no
symmetry about the 90°-axis because (r, θ) implied r =
2 cos θ and (r, π - θ) implied r = - 2 cos θ. Instead of

a symmetry, there was an anti-symmetry. This explains the values in the table. Furthermore, rewriting r = - 2 cos θ as - r = 2 cos θ, we see that (- r, π - θ) is a point on the graph whenever (r, θ) is on the graph.

But a negative sign in front of the radius has the effect of a rotation of π. Thus, (- r, π - θ) equals (r, π - θ + π) or (r, 2π - θ) = (r, - θ), or a negative symmetry about the 90°-axis implies symmetry about the polar axis.

● **PROBLEM** 709

Graph r = sin 3θ.

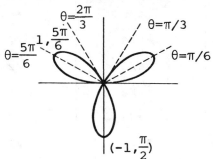

Solution: We graph the figure in five steps. First, we obtain the equation for r in term θ.

(1)   r = sin 3θ

Second, we find the boundary values of r and the symmetries. Note |sin 3θ| < 1. Thus, - 1 < r < 1. Because r = sin 3(- θ) = - sin 3θ ≠ sin 3θ, (r, - θ) being on the graph is not implied by (r, θ). Thus, there is no polar axis symmetry. Because r = sin 3(π - θ) = sin 3θ, (r, π - θ) is implied by (r, θ). Thus, there is symmetry about the 90°-axis. Because r = sin 3(π + θ) = - sin 3θ ≠ sin 3θ, (r, π + θ) is not implied by (r, θ), and there is no symmetry about the pole. Because there is symmetry about the 90°-axis, we need only consider values between - π/2 and π/2.

Third, we find the intercepts. The pole intercepts are found by setting r = 0: (0, 0), (0, ± π/3), and (0, ± 2π/3). The polar axis intercepts are found by setting θ = 0, and π: (0, 0) and (0, π). The 90°- axis intercepts are found by setting θ = ± π/2: (- 1, π/2) and (1, - π/2). [Note that these points are the same.]

Fourth, we find out the general behavior of the curve. Recall the shape of sin θ in rectangular coordinates.

θ	3θ	r = sin 3θ
Increasing from	Increasing from	Decreasing from

$-\pi/2$ to $-\pi/3$	$-3\pi/2$ to $-\pi$	1 to 0
$-\pi/3$ to $-\pi/6$	$-\pi$ to $-\pi/2$	0 to -1
$-\pi/6$ to 0	$-\pi/2$ to 0	-1 to 0
0 to $\pi/6$	0 to $\pi/2$	0 to 1
$\pi/6$ to $\pi/3$	$\pi/2$ to π	1 to 0
$\pi/3$ to $\pi/2$	π to $3\pi/2$	0 to -1

As a final step, we solve for particular points to make the sketch as precise as possible.

θ	3θ	$r = \sin 3\theta$
$\pi/18$	$\pi/6$	$\frac{1}{2} = .500$
$\pi/12$	$\pi/4$	$\sqrt{2}/2 = .707$
$\pi/9$	$\pi/3$	$\sqrt{3}/2 = .866$
$\pi/6$	$\pi/2$	$1 = 1.00$
$2\pi/9$	$2\pi/3$	$+ \sqrt{3}/2 = + .866$
$\pi/4$	$3\pi/4$	$+ \sqrt{2}/2 = .707$
$5\pi/18$	$5\pi/6$	$\frac{1}{2} = .500$
$\pi/3$	π	$0 = .000$

This curve is an example of a rose petal curve. The general equations of such a curve are

$$r = a \sin (n\theta) \qquad \text{and} \qquad r = a \cos (n\theta)$$

where n is a positive integer.

The number of leaves of the curve is equal to n if n is an odd integer. If n is even, the number of leaves is 2n. (See figure.)

● **PROBLEM** 710

Graph the curve $r = 2 + 2 \cos \theta$.

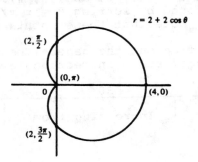

<u>Solution</u>:　In graphing any polar equation, we first rewrite the equation in convenient form - which, in most cases, consists of expressing r in terms of θ. Here the equation is already given as r in terms of θ. There is no further simplification except for factoring out a 2:

(i) $r = 2(1 + \cos θ)$.

The second step is to get a general picture about the figure - its symmetries, its bounding values, etc. A figure is bounded if there is a limit to the size of r. Note that $|\cos θ|$ is always ≤ 1. Thus, $r \leq 2 (1 + 1)$ or $r \leq 4$. The figure is bounded.

There are three types of symmetry: (1) symmetry about the polar axis; (2) about the 90°-axis; and (3) about the pole. For a figure to be symmetrical about the polar axis, $(r, - θ)$ must be a point on the figure each time $(r, θ)$ is on the figure. Suppose $(r, θ)$ is a point on the graph. Then, $r = 2(1 + \cos θ)$. Since $\cos θ = \cos (- θ)$, it is also true that $r = 2(1 + \cos (- θ))$. Thus, $(r, - θ)$ is always a point on the graph whenever $(r, θ)$ is a point on the graph. Thus, the figure is symmetric about the polar axis.

For symmetry about the 90°-axis, $(r, π - θ)$ must be on the graph whenever $(r, θ)$ is on the graph. Given $r = 2(1 + \cos θ)$, we wish to show $r = 2(1 + \cos (π - θ))$. Rewriting $\cos (π - θ)$ as $- \cos θ$, we have $r = 2(1 - \cos θ)$. Since the two expressions are not equivalent, the figure is not symmetric about the 90°-axis.

For symmetry about the pole, $(r, π + θ)$ must be on the graph if $(r, θ)$ is on the graph. Given $r = 2(1 + \cos θ)$, we would like to show $r = 2(1 + \cos (π + θ))$. Since $\cos (π + θ) = \cos π \cos θ - \sin π \sin θ = - \cos θ$, this is equivalent to proving $r = 2(1 - \cos θ)$. Recall, $\cos π = - 1$ and $\sin π = 0$. Since the two expressions are not equivalent, there is no symmetry about the pole.

Because the figure is symmetric about the polar axis, we need only plot points from $θ = 0$ to $θ = π$.

The third step is to find the intercepts. A point is an intercept of the polar axis if the θ coordinate is 0 or π. We substitute these values in equation (i) and obtain (4, 0) and (0, π). For the 90°-axis intercepts, we set θ equal to π/2 and 3π/2. Thus, (2, π/2) and (2, 3π/2) are the 90°-axis intercepts. Finally, there are the pole intercepts for which r = 0. Substituting r = 0, we obtain (0, π).

We have the intercepts and the symmetries. The next step is to get some idea of the curve's behavior as θ changes. We do this by reviewing the basic pattern of the cos θ curve in rectangular coordinates.

θ	2 cos θ	$r = 2 + 2 \cos θ$
increasing from 0 to π/2	decreases from 2 to 0	decreases from 4 to 2
increasing from π/2 to π	decreases from 0 to - 2	decreases from 2 to 0.

Finally, we plot some convenient points to make the graph more exact.

θ =	r = 2(1 + cos θ) =
π/6	2(1 + √3/2) $\overset{\sim}{\sim}$ 3.73
π/4	2(1 + √2/2) $\overset{\sim}{\sim}$ 3.414
π/3	2(1 + ½) = 3
2π/3	2(1 - ½) = 1
5π/6	2(1 - √3/2) $\overset{\sim}{\sim}$ .268

● **PROBLEM** 711

Determine the intercepts and symmetry of the polar curve r = θ, for θ ≥ 0.

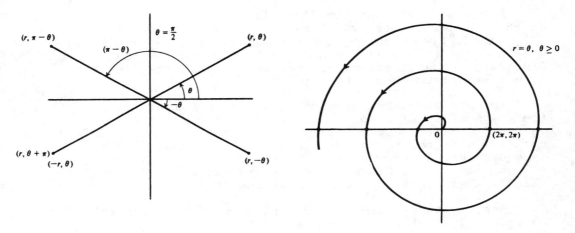

Solution: In polar coordinates points are given by specifying two numbers, r, the distance from the origin or pole, and θ, an angle of rotation. By convention, θ is measured from the horizontal axis, proceeding counter-clockwise. The horizontal axis is called the polar axis. See Figure 1.

Given an equation in terms of r and θ, we can graph the set of points whose values satisfy the equation.

An important case is that for which r can be written as an explicit function of θ, r = f(θ). In this case, we can find the intercepts of the curve with the polar axis by substituting integer multiples of π radians for θ, i.e. θ = nπ, n = 0, 1, 2, ... Similarly, we can find the intercept with the pole perpendicular to the polar axis, the π/2 axis, by setting θ = (2n + 1) π/2, (n = 0, 1, 2, ...). Negative values for θ need not be considered, because of the domain of the function in this problem. In other cases, we could consider intercepts for n = - 1, - 2, - 3, ...

In the present case, as θ increases, r increases without bound; the curve intercepts the polar axis infinitely often. Also, each value of n will determine a separate point on the π/2 axis. Thus, there are infinitely many intercepts for both axes.

On the other hand, let us examine the curve for various types of symmetry. The relevant types of symmetry are those with respect to the polar axis, with respect to the π/2 axis, and with respect to the pole.

The curve is not symmetric with respect to any of these:

A curve is only symmetric with respect to the polar axis if f(θ) = f(- θ). In this case, f(- θ) = f(θ) only for θ = π; hence this symmetry search fails.

Symmetry with respect to the π/2 axis is defined by f(π - θ) = f(θ). This is not valid for the curve r = θ, e.g. f(π - π/4) = 3π/4 ≠ f(π/4) = π/4.

Finally, symmetry with respect to the pole is defined by f(θ + π) = f(θ), which is false for all values of this function, e.g., f(π + π) = 2π ≠ f(π) = π. Hence, the graph shows no properties of symmetry whatsoever.

The curve, when sketched, is seen to be a spiral. If we had considered negative values for θ, we would have arrived at an oppositely oriented spiral. (See figure 2.)

● PROBLEM 712

The conchoid of Nicomedes has the equation r = csc θ + 2. Sketch it.

Solution: In graphing the polar equation, we first rewrite the equation in convenient form. Here, the equation is already given as r in terms of θ and no simplification is necessary:

(i) $r = \csc θ + 2 = \dfrac{1}{\sin θ} + 2.$

The second step is to determine the symmetries of the graph and the bounding values of r. A figure is bounded if there is a limit to the size of r. Note that for θ = 0 and θ = π, sin θ = 0 and 1/sin θ approaches infinity. Since $r = \dfrac{1}{\sin θ} + 2$, there is no limit on the size of r, and as θ → 0 or θ → π, r → ∞.

Testing for the three symmetries, we find:

(1) No symmetry about the polar axis. (r, θ) implies $r = \dfrac{1}{\sin θ} + 2$. (r,-θ) implies $r = \dfrac{1}{\sin (- θ)} + 2 = - \dfrac{1}{\sin θ} + 2.$

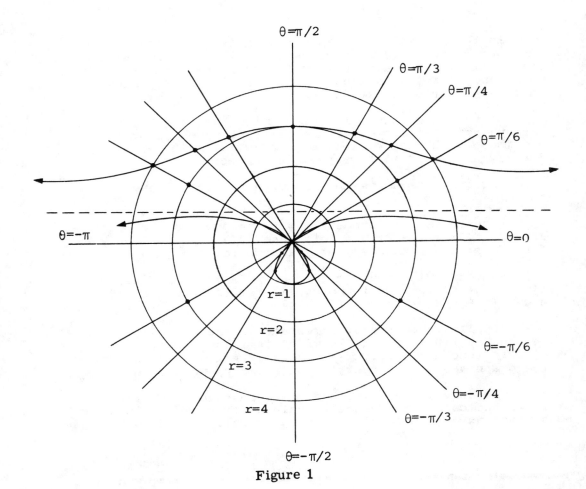

Figure 1

The second expression does not automatically follow from the first implying no polar axis symmetry is present.

    (2) Symmetry about the 90°-axis. (r, θ) implies $r = \dfrac{1}{\sin \theta} + 2$. $(r, \pi - \theta)$ implies $r = \dfrac{1}{\sin (\pi - \theta)} + 2$. Since $\sin (\pi - \theta) = \sin \pi \cos \theta - \cos \pi \sin \theta = \sin \theta$, the expression becomes $r = \dfrac{1}{\sin \theta} + 2$. Thus, $(r, \pi - \theta)$ is a point on the graph, each time (r, θ) is a point on the graph and 90°-axis symmetry exists.

    (3) No symmetry about the pole. (r, θ) implies $r = \dfrac{1}{\sin \theta} + 2$. $(r, \pi + \theta)$ implies $r = \dfrac{1}{\sin (\pi + \theta)} + 2$. Since $\sin (\pi + \theta) = \sin \pi \cos \theta + \cos \pi \sin \theta = - \sin \theta$, the expression becomes $r = - \dfrac{1}{\sin \theta} + 2$. The two expressions are not equivalent. There is no symmetry about the pole.

    Because of the 90°-axis symmetry, we need only consider $- \pi/2 < \theta < \pi/2$. (The graph of the left side is a mirror image of the right.)

656

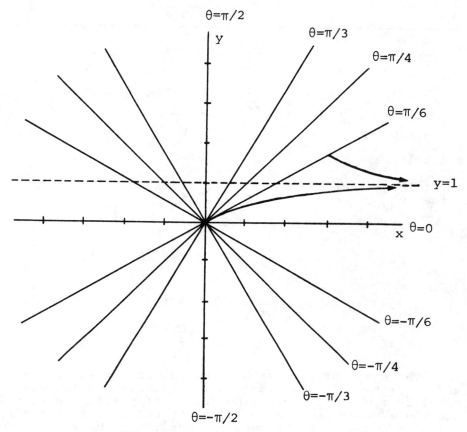

Figure 2

For the third step, we search for the polar axis, 90°-axis, and pole intercepts. To find the polar axis intercepts, we set θ = 0, and θ = π. In both cases, r is infinite. Thus, there are no polar axis intercepts.

To find the 90°-axis intercepts, we set θ = ± π/2, and obtain (3, π/2) and (1, - π/2). To find the pole intercepts, we set r = 0. If $0 = \frac{1}{\sin \theta} + 2$, then sin θ = - ½. The pole intercepts are (0, - π/6) and (0, 7π/6).

The fourth step is to examine the general behavior of the curve from the general behavior of sin θ.

θ	sin θ	$\frac{1}{\sin \theta}$	$r = \frac{1}{\sin \theta} + 2$
increasing from - π/2 to 0	increases from -1 to 0	decreases from - 1 to - ∞	decreases from 1 to - ∞
increasing from 0 to π/2	increases from 0 to 1	decreases from ∞ to 1	decreases ·from ∞ to 3

Finally, to be as precise as possible, we create a table of values.

θ	$r = \dfrac{1}{\sin \theta} + 2$
$-\pi/3$	$-\dfrac{1}{\sqrt{3}/2} + 2 = 2 - \dfrac{2}{3}\sqrt{3} \approx .84$
$-\pi/4$	$-\dfrac{1}{\sqrt{2}/2} + 2 = 2 - \sqrt{2} \approx .586$
$-\pi/6$	0
a little less than 0	$-\infty$
a little greater than 0	∞
$\pi/6$	$\dfrac{1}{\frac{1}{2}} + 2 = 4$
$\pi/4$	$\dfrac{1}{\sqrt{2}/2} + 2 = 2 + \sqrt{2} = 3.414$
$\pi/3$	$\dfrac{1}{\sqrt{3}/2} + 2 = 3.15$
$\pi/2$	3

The shape of the graph is now clear for $-\pi/2 < \theta < -\pi/6$. and $\pi/6 > \theta > \pi/2$. However, it is difficult to see what is meant by $r \to \infty$ for decreasing values of positive θ ($\theta \to 0^+$) and $r \to -\infty$ for increasing values of negative θ ($\theta \to 0^-$), i.e. $-\pi/6 < \theta < \pi/6$. We must analyze the problem more deeply to determine its ultimate shape.

Certain things are more easily expressed in one co-ordinate system than another. Lines, parabolas, and circles not centered at the origin are easier to deal with in Cartesian coordinates; limacons, cardioids, and circles centered at the origin are easier in polar coordinates. To see more clearly the shape of a given equation, we may have to switch from Cartesian to polar, polar to Cartesian, or some mixture of the two. Here, we wish to examine the behavior of $r = \dfrac{1}{\sin \theta} + 2$ for small θ. Remember that if (x, y) and (r, θ) represent the same point, then $x = r \cos \theta = \left(\dfrac{1}{\sin \theta} + 2\right) \cos \theta = \dfrac{\cos \theta}{\sin \theta} + 2 \cos \theta$ and $y = r \sin \theta = \left(\dfrac{1}{\sin \theta} + 2\right) \sin \theta = 1 + 2 \sin \theta$.

θ	$\sin \theta$	$\cos \theta$	$x = \dfrac{\cos \theta}{\sin \theta} + 2 \cos \theta$	$y = 1 + 2 \sin \theta$
θ increasing from $-\pi/6$ to 0	increasing from $-\frac{1}{2}$ to 0	increasing from $\sqrt{3}/2$ to 1	decreasing from 0 to $-\infty$	increasing from 0 to 1

θ decreasing from π/6 to 0	decreasing from ½ to 0	increasing from √3/2 to 1	increasing from 2√3 to ∞	decreasing from 2 to 1

From the above graph, we see that for very small $|\theta|$, $|x|$ will be very large; y will be close to 1. For even smaller $|\theta|$, $|x|$ will be much larger; but y will still be close to 1. But we know in Cartesian coordinates that this behavior (little change in y for large changes in x) is that of a horizontal line. Thus, as $|\theta|$ gets smaller, the graph approaches the horizontal line y = 1, and y = 1 is the equation of the asymptote.

CONVERSION OF POLAR TO CARTESIAN COORDINATES

● **PROBLEM** 713

Suppose a Cartesian coordinate system and a polar coordinate system are so superimposed that the origin and the x-axis of the Cartesian coincide with the pole and polar axis of the polar coordinate system. Show that Cartesian pair (x', y') and polar coordinate pair (r, θ) specify the same point if and only if x' = r cos θ and y' = r sin θ.

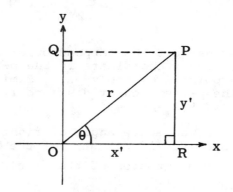

Solution: Suppose point P has the Cartesian coordinates (x', y') and polar coordinates (r, θ). If a point is specified by the Cartesian coordinate pair (x', y'), then the perpendicular distance of the point from the x axis, PR, equals y'. Similarly, the perpendicular distance of the point from the y axis, PQ, equals x'.

By inspection of the accompanying figure, we see that quadrilateral RPQO is a rectangle. Thus, ΔPOR is a right triangle with OR = x' and PR = y'. The hypotenuse PO is the distance of the point P from the origin. From the definiton of polar coordinates, we know that the distance equals r.

We can derive the required results by applying the sine and cosine ratios for the angle θ.

$$\sin \theta = \frac{y'}{r} \qquad \text{and} \qquad \cos \theta = \frac{x'}{r}$$

By cross-multiplying, we obtain

$$y' = r \sin \theta \qquad \text{and} \qquad x' = r \cos \theta.$$

● **PROBLEM** 714

In the accompanying figure, the equation of the circle is $x^2 + y^2 = 1$. Let $P_1(x_1, y_1)$ be a point on the circle. (a) show that $\cos \alpha = x_1$ and $\sin \alpha = y_1$. (b) Using the law of cosines, show that $\cos (\alpha - \beta) = \cos \alpha \cos \beta + \sin \alpha \sin \beta$.

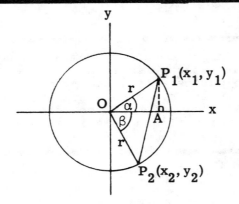

Solution: (a) Construct the perpendicular from P_1 to the x-axis. Remember the x coordinate is the perpendicular distance of the point from the y-axis. Similarly, the y coordinate is the perpendicular distance from the x-axis. Thus, $OA = x_1$ and $P_1A = y_1$.

The sine of α, an acute angle of right $\triangle OAP_1$, equals length of the opposite side/length of the hypotenuse or $\frac{AP_1}{OP_1}$. From the given equation of the circle we know OP_1 is the radius = 1. Thus,

$\sin \alpha = \frac{y_1}{1} = y_1$. Similarly,

$\cos \alpha = \dfrac{\text{length of the adjacent side}}{\text{length of the hypotenuse}} = \dfrac{OA}{OP_1} = \dfrac{x_1}{1} = x_1$.

(b) Note $\sphericalangle \alpha - \sphericalangle \beta = \sphericalangle P_1OP_2$ ($\sphericalangle \beta$ is an angle of negative measure and, by subtracting it from $\sphericalangle \alpha$, we obtain the angle equal to the sum of the absolute degree measures of $\sphericalangle \alpha$ and $\sphericalangle \beta$, $\sphericalangle P_1OP_2$,) and that $\sphericalangle P_1OP_2$ is an angle of $\triangle P_1OP_2$. After calculating the lengths of the three sides of the triangle, we can use the law of cosines to find $\cos P_1OP_2$, which is equal to $\cos (\alpha - \beta)$. Sides $\overline{P_1O}$ and $\overline{P_2O}$ are radii of the circle. Therefore, $P_1O = P_2O = 1$. To find P_1P_2, we use the distance formula.

$$P_1P_2 = \sqrt{(x_1 - x_2)^2 + (y_1 - y_2)^2}$$

We substitute these values into the Law of Cosines.

(i) $P_1P_2{}^2 = P_1O^2 + P_2O^2 - 2 \cdot P_1O \cdot P_2O \cdot \cos P_1OP_2$

(ii) $(x_1 - x_2)^2 + (y_1 - y_2)^2 = 1^2 + 1^2 - 2 \cos P_1OP_2$

(iii) $x_1{}^2 - 2x_1x_2 + x_2{}^2 + y_1{}^2 - 2y_1y_2 + y_2{}^2 = 2 - 2 \cos P_1OP_2$

(iv) $(x_1{}^2 + y_1{}^2) + (x_2{}^2 + y_2{}^2) - 2x_1x_2 - 2y_1y_2 = 2 - 2 \cos P_1OP_2$

Note that (x_1, y_1) and (x_2, y_2) are points on the circle and must satisfy the equation $x^2 + y^2 = 1$. Thus,

$$x_1{}^2 + y_1{}^2 = x_2{}^2 + y_2{}^2 = 1.$$

(v) $2 - 2x_1x_2 - 2y_1y_2 = 2 - 2 \cos P_1OP_2$

(vi) $- 2x_1x_2 - 2y_1y_2 = - 2 \cos P_1OP_2$

(vii) $\cos P_1OP_2 = x_1x_2 + y_1y_2$

But from part (a), we know that $x_1 = \cos \alpha$, $y_1 = \sin \alpha$. Similarly, $x_2 = \cos \beta$ and $y_2 = \sin \beta$.

(viii) $\cos (\alpha - \beta) = \cos \alpha \cos \beta + \sin \alpha \sin \beta$.

● **PROBLEM** 715

Given the equation $x^2 - y^2 = 1$, translate to polar coordinates and draw a sketch of the curve using the polar equation.

<u>Solution</u>: To translate a Cartesian coordinate equation to polar coordinates, we use the following relations:

(i) $x = r \cos \theta;$ $y = r \sin \theta$

Substituting into the expression $x^2 - y^2 = 1$, we have

(ii) $(r \cos \theta)^2 - (r \sin \theta)^2 = 1.$

(ii) $r^2 (\cos^2 \theta - \sin^2 \theta) = 1$

Since $\cos 2\theta = \cos^2 - \sin^2 \theta$

(iii) $r^2 = \dfrac{1}{\cos 2\theta} = \sec 2\theta$

In graphing polar figures, we first rewrite the expression in convenient form: r in terms of θ, if possible.

(i) $r = \pm \sqrt{\sec 2\theta}$

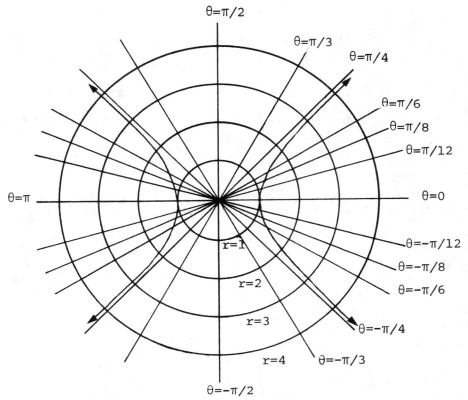

The second step is to find the upper bound on r and the symmetries. The square root function limits $\sqrt{\sec 2\theta}$ to values greater than 0. Even so, since $-\infty \leq \sec 2\theta \leq \infty$, $0 \leq \sqrt{\sec 2\theta} \leq \infty$. But, there is no maximum value for r.

Checking for symmetries, we find:

(1) Polar axis symmetry. $(r, -\theta)$ is a point on the graph only if $r = \pm \sqrt{\sec 2(-\theta)}$. Since $\sec \theta = \dfrac{1}{\cos \theta} =$

$\dfrac{1}{\cos (-\theta)} = \sec (-\theta)$, then $r = \pm \sqrt{\sec 2\theta}$. But this is always true if (r, θ) is a point on the graph. Since, $(r, -\theta)$ is implied by (r, θ), the graph is symmetrical about the polar axis.

(2) 90°-axis symmetry. $(r, \pi - \theta)$ is a point on the graph only if

$$r = \pm \sqrt{\sec 2(\pi - \theta)} = \pm \sqrt{\dfrac{1}{\cos 2(\pi - \theta)}} = \pm \sqrt{\dfrac{1}{2 \cos^2 (\pi - \theta) - 1}}$$

$$= \pm \sqrt{\dfrac{1}{2 (-\cos \theta)^2 - 1}} = \pm \sqrt{\dfrac{1}{2 \cos^2 \theta - 1}} = \pm \sqrt{\dfrac{1}{\cos 2\theta}}$$

$$= \pm \sqrt{\sec 2\theta}.$$

However, $r = \pm \sqrt{\sec 2\theta}$ is always true if (r, θ) is on the graph. Thus, $(r, \pi - \theta)$ follows from (r, θ) and there is 90°-axis symmetry.

(3) Pole symmetry $(r, \pi + \theta)$ is a pole on the graph only if

$$r = \pm \sqrt{\sec 2(\pi + \theta)} = \pm \sqrt{\frac{1}{\cos 2(\pi + \theta)}}$$

$$= \pm \sqrt{\frac{1}{2 \cos^2(\pi + \theta) - 1}} = \pm \sqrt{\frac{1}{2(-\cos\theta)^2 - 1}}$$

$$= \pm \sqrt{\frac{1}{2 \cos^2\theta - 1}} = \pm \sqrt{\frac{1}{\cos 2\theta}} = \pm \sqrt{\sec 2\theta}.$$

Since $(r, \pi + \theta)$ is implied by (r, θ), the graph is symmetric about the pole.

Since the graph has all three symmetries, we need only consider θ between 0 and $\pi/4$.

For the third step, we find the intercepts. To find the polar axis intercepts, we set $\theta = 0$. (Since there is 90°-axis symmetry, we need not set $\theta = \pi$. Every axis intercept on the left corresponds to one on the right.) **We obtain** $(+1, 0)$ and $(1, \pi)$. To find the 90°-axis intercepts, we set $\theta = \pi/2$ (since there is polar axis symmetry, we need not set $\theta = -\pi/2$) and obtain $r = \pm \sqrt{-1}$, which has no real solution. There is no 90°-axis intercept. To find the pole intercepts, we set $r = 0$, and obtain $0 = \pm \sqrt{\sec 2\theta}$. Since secant never equals 0, there is no solution and the graph never intersects the pole.

For the fourth step, we find the general behavior of the graph.

θ	$\sec 2\theta$	$r = \pm \sqrt{\sec 2\theta}$
Increases from 0 to $\pi/4$	Increases from 1 to ∞	Increases from ± 1 to $\pm \infty$.

For the final step, we obtain a table of values.

θ	$\sec 2\theta$	$r = \pm \sqrt{\sec 2\theta}$
0	1	± 1
$\pi/12$	$\frac{2}{3}\sqrt{3}$	$\pm \sqrt{\frac{2}{3}\sqrt{3}} = \pm \sqrt{\sqrt{\frac{4}{3}}}$ $= \pm \sqrt[4]{\frac{4}{3}} \overset{\sim}{=} \pm 1.07$
$\pi/8$	$\sqrt{2}$	$\pm \sqrt{\sqrt{2}} = \pm \sqrt[4]{2} \overset{\sim}{=} \pm 1.19$
$\pi/6$	2	$\pm \sqrt{2} \overset{\sim}{=} \pm 1.41$
$\pi/4$	∞	$\pm \infty$

> Given the line y = mx + b in Cartesian coordinates, find its polar equation.

Solution: To convert from Cartesian to polar coordinates, we use the following identities:

(1) $x = r \cos \theta$ $y = r \sin \theta$

By substitution, we have:

(2) $r \sin \theta = m(r \cos \theta) + b$

Solving for r, we obtain $r \sin \theta - m r \cos \theta = b$, or,

(3) $r = \dfrac{b}{\sin \theta - m \cos \theta}$

> Express the equation r = 2 cos θ in rectangular coordinates.

Solution: Recall the transformation equations for the conversion of polar to rectangular coordinates.

$r^2 = x^2 + y^2$ $x = r \cos \theta$

$\tan \theta = \dfrac{y}{x}$ $y = r \sin \theta$

Multiplying both sides of r = 2 cos θ by r, we obtain

$r^2 = 2 r \cos \theta$

But r cos θ is simply x and $r^2 = x^2 + y^2$, by the above equations. Thus, by substitition, the equation becomes

$x^2 + y^2 = 2x.$

Rearranging, we have

$x^2 - 2x + y^2 = 0$

Upon completing the square in x, we obtain:

$(x - 1)^2 + y^2 - 1 = 0$ or,

$(x - 1)^2 + y^2 = 1.$

Hence, the equation is an equation of a circle with center at (1, 0) and radius 1.

TRANSLATION & ROTATION

By a rotation of the coordinate axes, transform the equation

$$9x^2 - 24xy + 16y^2 - 40x - 30y = 0$$

into another equation lacking the cross product term.
(Plot the locus and draw both sets of axes.)

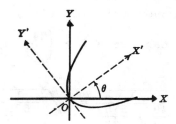

Solution: In a problem of this type where rotation of
axes are required, we must apply the formulas

$x = x' \cos \theta - y' \sin \theta$ and $y = x' \sin \theta + y' \cos \theta$

where θ is the angle through which the axes are rotated
and (x, y) and (x', y') are the coordinates of the point P
before and after rotation, respectively. (This procedure
is used principally to eliminate cross-terms in second degree
equations.)

We are not given the required angle of rotation and
must devise a procedure to transform the equation without
this fact.

If we let θ = the angle of rotation, then by substitution into the original equation, we obtain

$9(x' \cos \theta - y' \sin \theta)^2 - 24(x' \cos \theta - y' \sin \theta) \cdot$

$\qquad (x' \sin \theta + y' \cos \theta) + 16(x' \sin \theta + y' \cos \theta)^2$

$\quad - 40(x' \cos \theta - y' \sin \theta) - 30(x' \sin \theta + y' \sin \theta) = 0.$

After expansion and collection of terms, the equation
assumes the form

$*(9 \cos^2 \theta - 24 \cos \theta \sin \theta + 16 \sin^2 \theta)x'^2$

$+ (14 \sin \theta \cos \theta + 24 \sin^2 \theta - 24 \cos^2 \theta)x'y'$

$+ (9 \sin^2 \theta + 24 \sin \theta \cos \theta + 16 \cos^2 \theta)y'^2$

$- (40 \cos \theta + 30 \sin \theta)x' + (40 \sin \theta - 30 \cos \theta)y' = 0.$

We are told that this equation is not supposed to have a x'y' term. Therefore,

$$14 \sin \theta \cos \theta + 24 \sin^2 \theta - 24 \cos^2 \theta = 0$$

Two standard trigonometric identities tell us $2 \sin \theta \cos \theta = \sin 2\theta$ and $\sin^2 \theta - \cos^2 \theta = - \cos 2\theta$. Hence, by substitution,

$$14 \sin \theta \cos \theta + 24(\sin^2 \theta - \cos^2 \theta) = 7 \sin 2\theta - 24 \cos 2\theta = 0.$$

and $\dfrac{\sin 2\theta}{\cos 2\theta} = \dfrac{24}{7}$ or $\tan 2\theta = \dfrac{24}{7}$.

All angles of rotation are between 0 and 90°. Hence, $0 < 2\theta < 180°$, and tangent and cosine will agree in sign.

Again, by trigonometric identity,

$$\frac{1}{\cos^2 2\theta} = 1 + \tan^2 2\theta = 1 + \left(\frac{24}{7}\right)^2$$

$$\cos 2\theta = \sqrt{\frac{1}{\frac{49}{49} + \frac{576}{49}}} = \sqrt{\frac{1}{\frac{625}{49}}} = \frac{7}{25}$$

We can now make use of the half angle formulas of trigonometry to find the values of $\sin \theta$ and $\cos \theta$ which are needed to evaluate (*) and find the transformed equation.

$$\sin \theta = \sqrt{\frac{1 - \cos 2\theta}{2}} = \sqrt{\frac{1 - \frac{7}{25}}{2}} = \frac{3}{5}$$

$$\cos \theta = \sqrt{\frac{1 + \cos 2\theta}{2}} = \sqrt{\frac{1 + \frac{7}{25}}{2}} = \frac{4}{5}$$

If these values of $\sin \theta$ and $\cos \theta$ are substituted in (*), we have

$$\left[9 \left(\frac{4}{5}\right)^2 - 24 \left(\frac{3}{5}\right) \left(\frac{4}{5}\right) + 16 \left(\frac{3}{5}\right)^2\right] x'^2$$

$$+ \left[14 \left(\frac{4}{5}\right) \left(\frac{3}{5}\right) + 24 \left(\frac{3}{5}\right)^2 - 24 \left(\frac{4}{5}\right)^2\right] x'y'$$

$$+ \left[9 \left(\frac{3}{5}\right)^2 + 24 \left(\frac{3}{5}\right) \left(\frac{4}{5}\right) + 16 \left(\frac{4}{5}\right)^2\right] y'^2$$

$$- \left[40 \left(\frac{4}{5}\right) + 30 \left(\frac{3}{5}\right)\right] x' + \left[40 \left(\frac{3}{5}\right) - 30 \left(\frac{4}{5}\right)\right] y' = 0$$

Hence, the required equation in reduced form is

$$25y'^2 - 50x' = 0 \qquad \text{or} \qquad y'^2 - 2x' = 0$$

This equation is in the standard form for a parabola. The parabola and axes are drawn on the accompanying coordinate grid.

Transform the equation $2x^2 + \sqrt{3}\,xy + y^2 = 4$ by rotating the coordinate axes through an angle of 30°. Plot the locus and show both sets of axes.

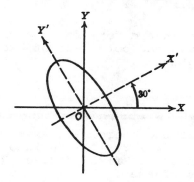

Solution: In this problem we will have to call on several formulas derived with the help of certain key trigonometric identities.

When the axes are rotated through an angle θ about the origin as a fixed point and the coordinates of any point P (x, y) are transformed into (x', y') then the equations of transformation from the old to the ' new coordinates are given by

$$x = x'\,\cos\,\theta - y'\,\sin\,\theta$$

$$y = x'\,\sin\,\theta - y'\,\cos\,\theta.$$

The expression for x and y can be substituted into the given equation to perform the transformation. In this problem, θ = 30°. Hence, since

$$x = x'\,\cos\,30° - y'\,\sin\,30° = \frac{\sqrt{3}}{2}\,x' - \frac{1}{2}\,y'$$

and $y = x'\,\sin\,30° + y'\,\cos\,30° = \frac{1}{2}\,x' + \frac{\sqrt{3}}{2}\,y'$,

by substitution, we obtain:

$$2\left(\frac{\sqrt{3}}{2}\,x' - \frac{1}{2}\,y'\right)^2 + \sqrt{3}\left(\frac{\sqrt{3}}{2}\,x' - \frac{1}{2}\,y'\right)\left(\frac{1}{2}\,x' + \frac{\sqrt{3}}{2}\,y'\right)$$

$$+ \left(\frac{1}{2}\,x' + \frac{\sqrt{3}}{2}\,y'\right)^2 = 4$$

Expanding this gives us

$$2\left(\frac{3}{4}\,x'^2 - \frac{\sqrt{3}}{2}\,x'y' + \frac{1}{4}\,y'^2\right) + \sqrt{3}\left(\frac{\sqrt{3}}{4}\,x'^2 + \frac{1}{2}\,x'y' - \frac{\sqrt{3}}{4}\,y'^2\right)$$

$$+ \left(\frac{1}{4}\,x'^2 + \frac{\sqrt{3}}{2}\,x'y' + \frac{3}{4}\,y'^2\right) = 4$$

which reduces to

$$\frac{10}{4} x'^2 + \frac{1}{2} y'^2 = 4.$$

Multiply by 2 to simplify further. Hence, we obtain

$$5x'^2 + y'^2 = 8.$$

This is the standard form for the equation of an ellipse. The ellipse and both sets of axes are shown on the accompanying graph.

● **PROBLEM** 720

> By a translation of the coordinate axes, transform the equation $x^2 - 4y^2 + 6x + 8y + 1 = 0$ into another equation to eliminate terms of the first degree. Plot the locus, and show both sets of axes.

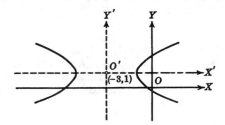

Solution: Since we are told that we must translate the coordinate axes in order to remove the first degree terms, let us review the rules for this procedure. Let us say that the translated origin is (h, k). If point P is represented before and after translation by (x, y) and (x',y'), respectively, then we can deduce that x = x' + h and y = y' + k and substitute these values to find the new equation.

In this problem we are not given the new coordinates of the translated origin. We must solve for these values given the restriction that the translated equation cannot contain a first order term. Hence, if we let (h, k) be the coordinates of the new origin, then, by substitution, we obtain the transformed equation.

$$(x' + h)^2 - 4(y' + k)^2 + 6(x' + h) + 8(y' + k) + 1 = 0$$

which, after expansion and collection of terms, assumes the form

$$(*) \quad x'^2 - 4y'^2 + (2h + 6)x' - (8k - 8)y' + h^2 - 4k^2$$
$$+ 6h + 8k + 1 = 0$$

Therefore, in order that no first order term appears in the translated equation

$$2h + 6 = 0 \qquad \text{and} \qquad 8k - 8 = 0$$

Therefore, $\quad h = -3 \quad$ and $\quad k = 1$.

Hence, the origin that will give us the required equation will be $(-3, 1)$.

We can substitute this into (*) to obtain

$$x'^2 - 4y'^2 + (2(-3) + 6)x - (8(1) - 8)y' + (-3)^2$$
$$- 4(1)^2 + 6(-3) + 8(1) + 1 = 0$$

which reduces to $\quad x'^2 - 4y'^2 - 4 = 0$.

This is the form of the equation of a hyperbola and the hyperbola, along with both sets of axes, is shown on the accompanying graph.

● **PROBLEM** 721

Transform the equation $X^3 - 3X^2 - Y^2 + 3X + 4Y - 5 = 0$ by translating the coordinate axes to a new origin at $(1, 2)$. Plot the locus and show both sets of axes.

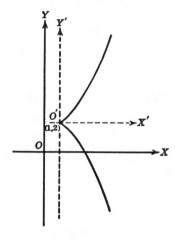

Solution: When translating the coordinate axes, each variable in the original equation becomes a function of the translation. If the axes are translated to a new origin at, say, (h, k) and any point P has coordinates of (X, Y) before the translation and (X', Y') after the translation, then we know $X = X' + h$ and $Y = Y' + k$.

These equations of transformation tell us that for every X and Y in the original equation we can substitute $X' + h$ and $Y' + k$, respectively, to find the equivalent equation on the translated axes.

If we proceed with the substitution of X and Y into the given equation, we obtain

$(X' + 1)^3 - 3(X' + 1)^2 - (Y' + 2)^2 + 3(X' + 1) +$

$$4(Y' + 2) - 5 = 0$$

$X'^3 + 3X'^2 + 3X' + 1 - (3X'^2 + 6X' + 3) - (Y'^2 + 4Y' + 4) + 3X' + 3 + 4Y' + 8 - 5 = 0$

$$X'^3 - Y'^2 = 0$$

This is an easier equation to plot. Select several points in the (X',Y') system that satisfy the equation so that we can observe the shape of the locus.

X'	0	1	1	2	2
Y'	0	1	-1	$\sqrt{8}$	$-\sqrt{8}$

This locus is shown on the accompanying graph. Notice that it is symmetric around the X'-axis.

This is also the graph of the original equation relative to the X and Y axes.

CHAPTER 42

CONIC SECTIONS: CIRCLES AND ELLIPSES

Basic Attacks and Strategies for Solving Problems in this Chapter. See pages 671 to 704 for step-by-step solutions to problems.

A **circle** is defined to be the locus of all points in a plane at a given distance from a given point. In Problem 726, it is established that a circle with center a (a, b) and radius r, has as an equation, $(x-a)^2 + (y-b)^2 = r^2$, and an equation of that type is an equation of a circle with center at (a, b) with radius r. This information is particularly helpful in graphing an equation. For example, suppose you were to graph the equation $2x^2 + 2y^2 + 4x - 7y + 2 = 0$. This seems difficult, but in Problem 728 it is established that this is an equation for a circle with center at $\left(-1, \dfrac{7}{4}\right)$ and radius $\dfrac{7}{4}$. With this more concise information, graphing the circle is easy.

An **ellipse** is defined as the locus of all points, the sum of whose distances from two fixed points (called foci) is a constant. If $(c, 0)$ and $(-c, 0)$ are the foci and the sum of the distances is $2a$, then a picture of the resulting ellipse follows:

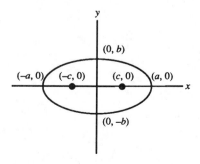

In this case, the vertices of the major axis are $(a, 0)$ and $(-a, 0)$, and the length of the semimajor axis is a. Similarly, the vertices of the minor axis are $(0, b)$ and $(0, -b)$, and the length of the semiminor axis is b. In this case the center of the ellipse is at the origin. It can be shown that $c^2 = \sqrt{a^2 - b^2}$. The eccentricity e is defined as $e = \dfrac{c}{a}$, and this number is an indication of the shape of the ellipse. The standard form for the equation of the ellipse described above is

$$\frac{x^2}{a^2} + \frac{y^2}{b^2} = 1$$

For an ellipse with foci at $(0, c)$ and $(0, -c)$, then the standard form for its equation is

$$\frac{x^2}{b^2} + \frac{y^2}{a^2} = 1$$

where, again, a is the length of the semimajor axis and b is the length of the semiminor axis.

As indicated in the solution to Problem 740, a nice way to visualize an ellipse is to think about it as a squashed circle, and the eccentricity of the ellipse is an indication of the amount of squashing. In fact, a circle is an ellipse with an eccentricity of 0. The two equations given above are for ellipses with center at the origin. Equations for ellipses not centered at the origin are given at the bottom of page 689.

A skill that is needed for problems involving circles or ellipses is completion of the square. That procedure is described in the introduction to Chapter 39, and a review of that topic might be desirable.

Step-by-Step Solutions to Problems in this Chapter, "Conic Sections: Circles and Ellipses"

CIRCLES

● PROBLEM 722

Write the equation of the circle with center C at the origin and with radius 7.

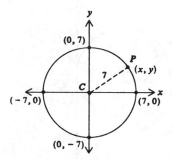

Solution: A circle is defined as the collection of all points that are a given distance (i.e. the radius) from the center. Therefore, if P is a point on ⊙C, then PC = 7. To convert this expression into an equation in terms of x's and y's, we use the distance formula. Let the coordinates of P, any point on the circle C, be (x,y). Since the center is at the origin, C = (0,0). Therefore, $PC = \sqrt{(x-0)^2 + (y-0)^2} = \sqrt{x^2 + y^2}$. But the radius PC must equal 7. Thus, $\sqrt{x^2 + y^2} = 7$. This alone would be adequate to describe the circle; however, it is good practice never to leave unknowns in radical form. Squaring both sides, the finished equation is $x^2 + y^2 = 49$.

● PROBLEM 723

Write the equation of the circle with its center at the origin and with radius of length 3.

Solution: The points lying on a circle, are, by definition,

all equidistant from the center of the circle. The distance from a point on the circle to the center of the circle is the radius of the circle, in our case, 3.

Let P(x,y) be a typical point lying on the circle. Using the distance formula, the distance from P to the center of the given circle (the origin) is $\sqrt{x^2 + y^2}$. But, this distance must be equal to the radius of the circle, 3. Hence,

$$3 = \sqrt{x^2 + y^2}.$$

Squaring both sides of this equation, we obtain the equation of the circle

$$x^2 + y^2 = 9.$$

● **PROBLEM** 724

Write an equation for the circle with center A(-3,-2) and radius 3.

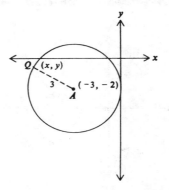

Solution: Let Q be an arbitrary point on the circle. Then QA is a radius and QA = 3. To convert this given information into a general equation involving x's and y's, we use the distance formula. Let the coordinates of Q be (x,y). The coordinates of the center A are given to be (-3,-2). Thus, if we let (x,y) be the coordinates of point Q, then

$$QA = \sqrt{(x-(-3))^2 + (y-(-2))^2} = \sqrt{(x+3)^2 + (y+2)^2}$$

But QA = 3; therefore, $\sqrt{(x+3)^2 + (y+2)^2} = 3$. Squaring both we obtain obtain $(x+3)^2 + (y+2)^2 = 9$. The equation is generally left in this form. Multiplying out $(x+3)^2$ and $(y+2)^2$ does not lead to any significant simplification, and often just contributes to the confusion.

● **PROBLEM** 725

Problem: Find the equation of a circle C with center at (2,1) and radius r = 3.

672

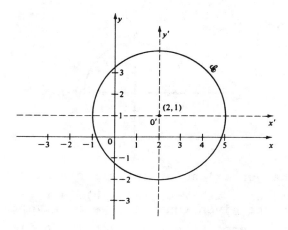

Solution: The length of the line segment determined by the center and a point (x,y) on the circle is given by $\sqrt{(x-2)^2 + (y-1)^2} = 3$. (This follows from the definition of length and the Pythagorean formula.) Squaring both sides gives $(x-2)^2 + (y-1)^2 = 9$. If the origin were translated to (2,1), with the new axes remaining parallel, respectively, to the original, so that $x' = x - 2$, $y' = y - 1$, then the equation of the circle would be $x'^2 + y'^2 = 9$.

● **PROBLEM** 726

Write the equation of a circle with center at C(a,b) and radius r.

Solution: By definition, each point lying on a circle is an equal distance from the center of the circle. The distance from the center of the circle to a point on the circle is called the radius of the circle. Using the distance formula, we may obtain the equation of the circle described in the statement of the problem. The radius, r, must equal the distance from C(a,b) to a typical point P(x,y) of the circle. Hence,

$$r = \sqrt{(x-a)^2 + (y-b)^2}.$$

Squaring both sides of (1), we obtain

$$(x-a)^2 + (y-b)^2 = r^2$$

for the equation of a circle with radius r and center located at C(a,b).

● **PROBLEM** 727

Draw the graph of the curve defined by $x^2 + y^2 = 25$.

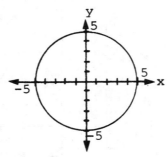

Solution: The general equation of a circle with center at C(a,b) and radius r is $(x-a)^2 + (y-b)^2 = r^2$. Comparing this equation with the given equation, we observe that a = 0, b = 0, $r^2 = 25$. Hence, the radius, being a distance, is always positive is 5, and the center of the circle is at C(0,0). With this information, we can, without further calculation, sketch the figure shown in the diagram.

● **PROBLEM** 728

Problem: Does the equation $2x^2 + 2y^2 + 4x - 7y + 2 = 0$ determine a circle?

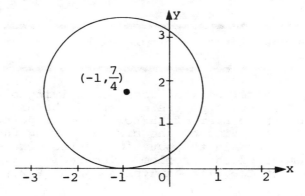

Solution: The method to be used in solving this problem is called "completing the square," and involves regrouping terms of x and y so as to obtain the form of the equation of a circle, $(x-h)^2 + (y-k)^2 = a^2$. First, dividing by 2 and regrouping, we have

$$(x^2 + 2x) + (y^2 - \tfrac{7}{2}y) = -1.$$

Now, we add a constant to both sides that, when distributed into the x and y terms, will complete each square (i.e., will result in a perfect square.) Notice that if 1 is added to $x^2 + 2x$, we will have a perfect square; also, if $\frac{49}{16}\left(=(\frac{-7}{4})^2\right)$ is added to $y^2 - \frac{7}{2}y$ we will have another perfect square.

Then, adding, $1 + \frac{49}{16} = 4\frac{1}{16}$ to each side, gives

$$(x^2 + 2x + 1) + (y^2 - \frac{7}{2}y + \frac{49}{16}) = -1 + 1 + \frac{49}{16} \quad \text{or}$$

$$(x + 1)^2 + (y - \frac{7}{4})^2 = (\frac{7}{4})^2, \text{ the equation of a circle with}$$

center at $(-1, \frac{7}{4})$ and radius $\frac{7}{4}$.

● **PROBLEM 729**

Problem: Is the equation $2x^2 + 2y^2 + 4x - 7y + 10 = 0$ the equation of a circle?

Solution: We will answer this question by completing the square and seeing whether the equation has any real solutions. Dividing by two and regrouping,

$$(x^2 + 2x) + (y^2 - \frac{7}{2}y) = -5.$$

Completing the squares,

$$(x^2 + 2x + 1) + (y^2 - \frac{7}{2}y + \frac{49}{16}) = -5 + 1 + \frac{49}{16}.$$

Simplifying,

$$(x + 1)^2 + (y - \frac{7}{4})^2 = \frac{-15}{16}.$$

The left hand side is positive for all real values of x,y. So this equation has no real solution.

● **PROBLEM 730**

Write the equation of the family of all concentric circles whose common center is the point (-3,5). Draw three members of the family, specifying the value of the parameter in each case.

Solution: The standard form of the equation of all circles centered at (-3,5) is given by $(x+3)^2 + (y-5)^2 = r^2$ where r is the parameter equalling the radius length of the circle.

To draw three members of this family, let r take on three separate values and construct each circle. The accompanying graph contains the three circles.

When r = 1, the circle is given by $(x+3)^2 + (y-5)^2 = 1$. When r = 2, the circle is given by $(x+3)^2 + (y-5)^2 = 4$. When r = 3, the circle is given by $(x+3)^2 + (y-5)^2 = 9$.

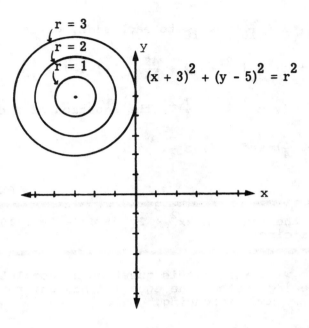

$(x + 3)^2 + (y - 5)^2 = r^2$

Find the equation of the circle that goes through the points
(1,3), (-8,0) and (0,6).

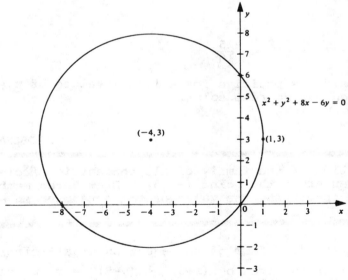

$x^2 + y^2 + 8x - 6y = 0$

(−4,3)

(1,3)

<u>Solution</u>: We are asked to find the equation of a circle.
There are two forms: (1) The standard form $(x-h)^2 + (y-k)^2$
$= a^2$; (2) the general form $x^2 + y^2 + Dx + Ey + F = 0$.
Both are equivalent; however, the standard form is more convenient when we are dealing with the coordinates of the
center point (h,k) or the length of the radius a. As we can
see, all three numbers h, k, and a, can be readily obtained

given the standard form. The drawback of the standard form is its factored format which, in problems which do not ask for the center coordinates or the radius, is just extra work to calculate. In these cases, the multiplied out form, the general form, is more convenient.

Here we are asked to find the equation given three points. We are given three pairs of (x,y) and asked to find the three equation unknowns. If we were to use standard form, the unknowns would be h, k, and a. The unknowns would be squared. Moreover, to solve the equations we would have to multiply out the $(x-h)^2$ and $(y-h)^2$ terms. It would be long and difficult.

If we use the second method, though, the unknowns would be the unsquared terms D, E, and F. Furthermore, there are no multiplications (other than x^2 and y^2) necessary. Thus, even though we may wish to express the equation of the circle in standard form later on, it is easier to solve for the general form equation first.

Suppose the equation of the circle is $x^2 + y^2 + Dx + Ey + F = 0$. Because point $(1,3)$ is on the circle it must satisfy the equation. Then,

(i) $1^2 + 3^2 + D(1) + E(3) + F = 0$

or D + 3F + F = -10.

Similarly since $(-8,0)$ and $(0,6)$ are also points on the circle, we have:

(ii) $(-8)^2 + 0^2 + D(-8) + E(0) + F = 0$ or

-8D + F = -64.

(iii) $0^2 + 6^2 + D(0) + E(6) + F = 0$ or 6E + F = -36.

We thus have three equations and three unknowns.

(iv) D + 3E + F = -10

(v) -8D   + F = -64.

(vi)   6E + F = -36.

We make this system a system of two equations and two unknowns by eliminating D. Multiply (iv) by 8 and add it to (v). We have:

(vii) 24E + 9F = -144

(viii) 6E + F = - 36   (see (vi))

Now eliminate E to form one equation and one unknown. Multiply (viii) by -4 and add the result to (vii) to obtain

(ix) 5F = 0   or   F = 0.

Substituting this result in (viii), we find E. Since $6E + F = -36$, then $6E + 0 = -36$ or $E = \frac{-36}{6} = -6$. Substituting the value for F in (v), we find $-8D + F = -64$. Thus, $-8D = -64$ or $D = \frac{-64}{-8} = 8$. The general form of the equation is

(x) $\quad x^2 + y^2 + 8x - 6y = 0.$

Usually, it is standard practice to give the equation in standard form.

(xi) $\quad (x^2 + 8x) + (y^2 - 6y) = 0$

(xii) $\quad (x^2 + 8x + 16) + (y^2 - 6y + 9) - 16 - 9 = 0$

(xiii) $\quad (x+4)^2 + (y-3)^2 = 25.$

● **PROBLEM** 732

Euclid defined a circle as the locus of points equidistant from a given point. Apollonius, on the other hand had an alternate definition: Given two points, A and B, and a constant $k \neq 1$, the set of all points P such that $PA = k \cdot PB$ is a circle. Consider points $A(0,0)$ and $B(b,0)$, and the constant k. Show that the Apollonian definition does indeed lead to the equation of the circle. Also, find the coordinates of the center and the radius of the circle.

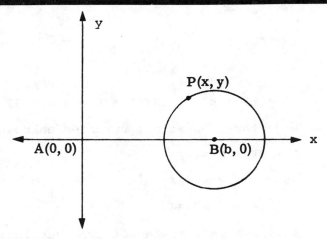

<u>Solution</u>: Let $P(x,y)$ be any point such that $AP = k \cdot PB$. Then, by using the distance formula, we know that

(i) $\quad \sqrt{(x-0)^2 + (y-0)^2} = k\sqrt{(x-b)^2 + (y-0)^2}$

(ii) $\quad x^2 + y^2 = k^2[(x-b)^2 + y^2]$

(iii) $\quad x^2 + y^2 = k^2(x^2 + y^2) - 2k^2xb + k^2b^2$

(iv) $(k^2-1)x^2 - 2k^2bx + k^2b^2 + (k^2-1)y^2 = 0$

(v) $x^2 - \dfrac{2k^2b}{k^2-1}x + \dfrac{k^2}{k^2-1}b^2 + y^2 = 0.$

From (v), we see that the coefficients of x^2 and y^2 are equal. Thus, the figure will be a circle as long as the radius does not turn out to be negative or zero. To put (v) into standard form, we first complete the square in x.

(vi) $\left[x^2 - \dfrac{2k^2b}{k^2-1}x + \left(\dfrac{k^2b}{k^2-1}\right)^2\right] + y^2 + \dfrac{k^2b^2}{k^2-1} - \left(\dfrac{k^2b}{k^2-1}\right)^2 = 0$

(vii) $\left[x - \dfrac{k^2b}{k^2-1}\right]^2 + y^2 = \dfrac{k^2b^2}{k^2-1}\left[\dfrac{k^2}{k^2-1} - 1\right] = \left[\dfrac{kb}{k^2-1}\right]^2.$

(viii) $\left[x - \dfrac{k^2b}{k^2-1}\right]^2 + y^2 = \left(\dfrac{kb}{k^2-1}\right)^2.$

Thus, the locus of points P is a circle with center $\left(\dfrac{k^2b}{k^2-1}, 0\right)$ and radius $\left(\dfrac{kb}{k^2-1}\right)$

● **PROBLEM** 733

Find the equation of the circle that goes through the points (1,2) and (3,4) and has radius a = 2.

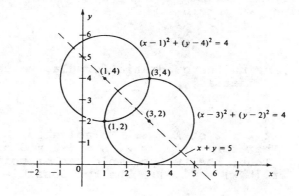

<u>Solution</u>: The two forms of the equation of a circle are the standard and general forms. Because we are given the radius, we will use the standard form. Suppose the equation of the desired circle is $(x-h)^2 + (y-k)^2 = a^2$, where (h,k) is the center and a is the radius. We are given that the radius a = 2. Furthermore, because points (1,2) and (3,4) are points on the circle, they must satisfy the circle equation. This fact will allow us to find (h,k) and fully determine the circle. Hence,

(i) $(1-h)^2 + (2-k)^2 = 2^2$

(ii) $(3-h)^2 + (4-k)^2 = 2^2.$

Multiplying out, we obtain:

(iii) $1 - 2h + h^2 + 4 - 4k + k^2 = 4$

(iv) $9 - 6h + h^2 + 16 - 8k + k^2 = 4.$

Simplifying, we obtain

(v) $-2h + h^2 - 4k + k^2 = -1$

(vi) $-6h + h^2 - 8k + k^2 = -21.$

To deal with two unknowns with squared terms is too difficult. If we could find a relation between h and k, then we could substitute in equation (v) and obtain a quadratic equation in <u>one</u> variable. We can deal with this by using the quadratic equation.

To find a relation between h and k, note that if you subtract (vi) from (v), the squared terms subtract out. We thus obtain,

(vii) $+4h + 4k = 20.$

Thus, solving for h in terms of k, we have $h = \dfrac{20-4k}{4} =$ 5 - k. Substituting this result in equation (v), we obtain:

(viii) $-2(5-k) + (5-k)^2 - 4k + k^2 = -1.$

Simplifying, we obtain

(ix) $k^2 - 6k + 8 = 0.$

Instead of using the quadratic formula, we see this can be factored. That is,

(x) $(k-4)(k-2) = 0.$

Thus, either k = 2 or k = 4. To solve for h, remember that h = 5-k. Therefore, if k = 2, h = 5-2 = 3 and if k = 4, h = 5 - 4 = 1. Thus, the two possible center points are (h,k) = (3,2) or (h,k) = (1,4). The two possible equations, therefore, are

(xi) $(x-1)^2 + (y-4)^2 = 4$

(xii) $(x-3)^2 + (y-2)^2 = 4.$

Note: It may seem odd that there are two possible answers. but from a geometric viewpoint, it is logical. Given two points and a radius length, there are two possible circles that can be constructed, the centers of the circles lying on

opposite sides of the line determined by the two given
points.

● **PROBLEM** 734

Problem: Find the points of intersection (if any) of the
circles C_1 and C_2 where

$$C_1: \quad x^2 + y^2 - 4x - 2y + 1 = 0$$

$$C_2: \quad x^2 + y^2 - 6x + 4y + 4 = 0.$$

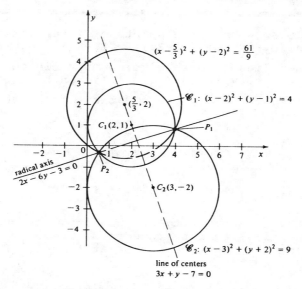

Solution: The analytic method reduces this problem to a
routine, although possibly messy, problem in algebra. We
seek the points of intersection of the circle; thus, an obvious first step is to subtract the second equation from
the first, giving:

$$2x - 6y - 3 = 0.$$

This is the equation of a straight line. The geometric
significance of $2x - 6y - 3 = 0$ is that any points common
to C_1 and C_2 must also lie on this line. It is a straightforward process to find points of intersection of this line
with either C_1 or C_2. We can solve for x in terms of y,

$$x = 3y + \frac{3}{2}$$

substitute back into the equation C_1, find the two real
values of y, then use $x = 3y + \frac{3}{2}$ again to find the corresponding values of x. Using this or an equivalent procedure,
we obtain as points of intersection

$$P_1\left(\frac{9}{4} + \frac{3\sqrt{135}}{20}, \frac{1}{4} + \frac{\sqrt{135}}{20}\right) \approx (3.99, 0.83) \qquad \text{and}$$

$$P_2\left(\frac{9}{4} - \frac{3\sqrt{135}}{20}, \frac{1}{4} - \frac{\sqrt{135}}{20}\right) \approx (0.51, -0.33).$$

● **PROBLEM** 735

Given that two circles $x^2 + y^2 + D_1x + E_1y + F_1 = 0$ and $x^2 + y^2 + D_2x + E_2y + F_2 = 0$ intersect at two points, show that the equation for the line determined by the points of intersection is $(D_1-D_2)x + (E_1-E_2) + (F_1-F_2) = 0$.

<u>Solution</u>: To show that the points of intersection lie on the equation $(D_1-D_2)x + (E_1-E_2) + (F_1-F_2) = 0$, we show that any point that lies on both circles must lie on the line. Suppose points (x_1,y_1) and (x_2,y_2) lie on both circles. Then (x_1,y_1) must satisfy the equations for both circles. That is,

$$\text{(i)} \quad x_1^2 + y_1^2 + D_1x_1 + E_1y_1 + F_1 = 0$$

$$\text{(ii)} \quad x_1^2 + y_1^2 + D_2x_1 + E_2y_1 + F_2 = 0.$$

By subtraction property of equality, if equation (i) and equation (ii) are true statements, then equation (i) minus equation (ii) must be a true statement. Subtracting (ii) from (i), we have:

$$\text{(iii)} \quad (D_1-D_2)x_1 + (E_1-E_2)y_1 + (F_1-F_2) = 0.$$

Thus, point (x_1,y_1) satisfies the equation. $(D_1-D_2)x + (E_1-E_2)y + (F_1-F_2) = 0$. Similarly, we can show (x_2,y_2) satisfies the given equation. Thus, both points of intersection lie on the line with the given equation. Since two distinct points determine a line, the line

$$(D_1-D_2)x + (E_1-E_2)y + (F_1-F_2) = 0$$

is the line determined by the points of intersection.

● **PROBLEM** 736

Find the equation of the line tangent to the circle $x^2 + y^2 - 10x + 2y + 18 = 0$ and having a slope equal to 1.

<u>Solution</u>: Since we are given the slope of the tangent line, to apply the slope-intercept form of a straight line

682

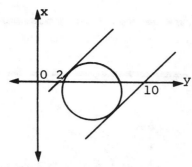

in finding the equation of the tangent line, we need to calculate the y-intercept.

The slope-intercept form is y = mx + b where m is the
slope and b is the y-intercept. Hence, we can substitute
m = 1, the given, to find that all lines tangent with slope
equal to 1 will have the form y = x + b. Our task, now, is
to find b.

We can solve for b by substituting y = x + b into the
equation for the circle and finding an equation for b by
considering the tangency characteristics. By substitution,
$x^2 + (x+b)^2 - 10x + 2(x+b) + 18 = 0$ and
$x^2 + (x^2+2xb+b^2) - 10x + 2x + 2b + 18 = 0$ which reduces to
$2x^2 + (2b-8)x + (b^2+2b+18) = 0$. To derive an equation for
b, proceed to solve the quadratic equation for x. For the

general quadratic form $ax^2 + bx + c = 0$, $x = \dfrac{-b \pm \sqrt{b^2-4ac}}{2a}$.

There can be only one point of intersection between the
circle and the line, since the line is a tangent. Hence, x
can only have one value. Therefore, the discriminant,
$b^2 - 4ac$ must equal zero. In our problem,

$$b^2 - 4ac = (2b-8)^2 - 4(2)(b^2+2b+18) = 0$$

$$4b^2 - 32b + 64 - 8b^2 - 16b - 144 = 0$$

$$-4b^2 - 48b - 80 = 0$$

$$-4(b^2 + 12b + 20) = 0$$

To find b such that y = x + b is a tangent line, it is
sufficient to solve,

$$b^2 + 12b + 20 = 0$$

$$(b + 10)(b + 2) = 0.$$

Hence, b = -10 or b = -2. Therefore, there are two possible
lines tangent to the given circle with slope equal to one.
Their equations are y = x - 10 and y = x - 2. The accompanying figure represents this circumstance.

Find the equation of the tangent to the circle
$x^2 + y^2 - 8x - 6y + 20 = 0$ at the point $(3,5)$.

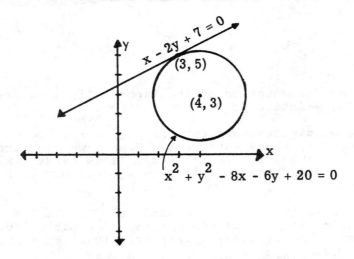

Solution: We are given one point on the tangent line, but can find the equation of the line only after we have determined the slope.

If we let m = the slope of the tangent, then, according to the point-slope form, the equation is $(y-5) = m(x-3)$, or $y = mx - 3m + 5$.

To find the slope, m, we proceed as if we were finding the intersection of the line and the circle. We substitute $y = mx - 3m + 5$ into the equation of the circle. This will enable us to extract an equation that will allow us to solve for the slope. After substitution, we have

$$x^2 + (mx - 3m + 5)^2 - 8x - 6(mx - 3m + 5) + 20 = 0$$

$$x^2 + (m^2x^2 - 6m^2x + 9m^2 + 10mx - 30m + 25) - 8x - 6mx + 18m$$
$$- 30 + 20 = 0$$

which reduces to

$$(m^2+1)x^2 - (6m^2 - 4m + 8)x + (9m^2 - 12m + 15) = 0.$$

Anytime we have an equation of the form $ax^2 + bx + c$ = 0, its solutions are given by $x = \dfrac{-b \pm \sqrt{b^2-4ac}}{2a}$. If there is to be only one solution for x, as we want to find a unique tangency point, then $b^2 - 4ac = 0$. This fact will allow us to determine the slope of the tangent line.

In our problem, $b^2 - 4ac = 0$ is the same as

$$(6m^2 - 4m + 8)^2 - 4(m^2 + 1)(9m^2 - 12m + 15) = 0$$

$$(36m^4 - 48m^3 + 112m^2 - 64m + 64) -$$

$$(36m^4 - 48m^3 + 96m^2 - 48m + 60) = 0 \text{ which reduces to}$$

$16m^2 - 16m + 4 = 0.$ By factoring, $4(4m^2 - 4m + 1) = 0.$
It is sufficient to solve $4m^2 - 4m + 1 = 0$ to find the value
of m so that the line is tangent to the circle.

$$4m^2 - 4m + 1 = (2m - 1)^2 = 0.$$

As such, $2m - 1 = 0$ and $m = \frac{1}{2}$.

Now that we know the slope of the line passing through
(3,5) that intersects the circle at only one point, we can
write the equation of the tangent line. It is
$(y-5) = \frac{1}{2}(x-3)$. Multiply by 2 to obtain $2y - 10 = x - 3$.

Hence, the equation of the tangent is $x - 2y + 7 = 0$.
(See the accompanying figure.)

● **PROBLEM** 738

Find the equation of the line drawn from the point (8,6)
tangent to the circle $x^2 + y^2 + 2x + 2y - 24 = 0$.

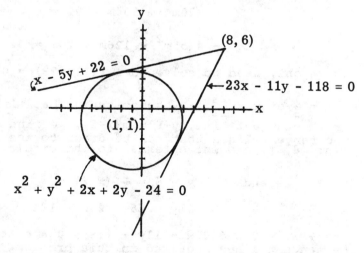

<u>Solution</u>: Given a point through which the line passes, we
can fully describe the line only after we have found its
slope. According to the point-slope form of a linear equation, the equations of the lines passing through (8,6) is
given by $y - 6 = m(x - 8)$, where m is the slope. When m is
the slope of the tangent, the equation, $y - 6 = m(x - 8)$,
is the equation of the tangent. The equation can be rewrit-
tem as $y = mx - 8m + 6$.

To find the unique intersection point of the line and the circle, the tangency point, we can substitute $y = mx - 8m + 6$ into the equation for the circle and solve for x. Since the line is a tangent line, there can only be one value for x. This last fact will assist us in determining the slope of the line. Hence,

$$x^2 + (mx - 8m + 6)^2 + 2x + 2(mx - 8m + 6) - 24 = 0$$

$$x^2 + (m^2x^2 - 16m^2x + 12mx + 64m^2 - 96m + 36) + 2x + (2mx$$
$$- 16m + 12) - 24 = 0$$

which reduces to,

$$(m^2 + 1)x^2 - (16m^2 - 14m - 2)x + (64m^2 - 112m + 24) = 0.$$

By the quadratic formula,

$$x = \frac{-[-(16m^2 - 14m - 2)] \pm \sqrt{[-(16m^2 - 14m - 2)]^2 - 4(m^2 + 1)(64m^2 - 112m + 24)}}{2(m^2 + 1)}$$

Since x is a coordinate of a tangency point, it can only take on one value. Hence, the discriminant must equal zero. By setting the discriminant equal to zero we can solve for m and, thereby, determine fully the equation of the tangent line.

$$[-(16m^2 - 14m - 2)]^2 - 4(m^2 + 1)(64m^2 - 112m + 24) = 0$$

$$(256m^4 - 448m^3 + 132m^2 + 56m + 4) - (256m^4 - 448m^3 + 352m^2 - 448m + 96) = 0$$

$$- 220m^2 + 504m - 92 = 0$$

$$- 4(55m^2 - 126m + 23) = 0.$$

By factoring we only need to solve $(5m-1)(11m-23) = 0$.

Hence, $m = \frac{1}{5}$ or $m = \frac{23}{11}$. By substituting back into our equation for the tangent line, $y - 6 = m(x-8)$, we find that there are two tangent lines that can be drawn to the circle from the point (8,6), a point external to the circle.

$$y - 6 = \frac{1}{5}(x-8) \qquad \text{and} \qquad y - 6 = \frac{23}{11}(x-8)$$

$$5y - 30 = x - 8 \qquad\qquad\qquad 11y - 66 = 23x - 184.$$

Hence, $x - 5y + 22 = 0$ and $23x - 11y - 118 = 0$ are the equations of the tangent lines required in this problem. The figure shows the circle and its tangents.

● **PROBLEM** 739

The equations of two circles, C_1 and C_2, are $x^2 + y^2 + 7x -$

$10y + 31 = 0$ and $x^2 + y^2 - x - 6y + 3 = 0$ respectively. Find the equation of the circle, C_3, which passes through the intersection of the circles C_1 and C_2 and has its center on the line ℓ with equation $x-y-2=0$.

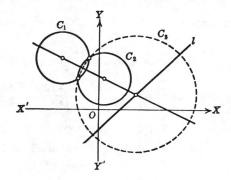

Solution: Let us assume that C_1 and C_2 intersect at the points (x_1,y_1) and (x_2,y_2). Since the coordinates cause both equations to equal zero, we know they will solve the equation $x^2 + y^2 + 7x - 10y + 31 + k(x^2 + y^2 - x - 6y + 3) = 0$, where k is a parameter.

Since the points of intersection satisfy this equation, the desired circle must be one of the family of circles specified by the equations.

The center of the circle, when found, will fully determine the circle.

Combine the coefficients of the equation to obtain

$$(k+1)x^2 + (k+1)y^2 + (7-k)x - (10+6k)y + (31+3k) = 0.$$

The value of k will be determined by the condition that the center of C_3 is to lie on the line with equation $x - y - 2 = 0$.

If we put the above equation of a circle into the standard form $(x-a)^2 + (y-b)^2 = r^2$ where (a,b) is the center, then we will be able to find the value of k using the fact that the center is on the given line. We have

$$x^2 + y^2 + \left(\frac{7-k}{k+1}\right)x - \left(\frac{10+6k}{k+1}\right)y + \left(\frac{31+3k}{k+1}\right) = 0 \qquad (*)$$

$$x^2 + \left(\frac{7-k}{k+1}\right)x + y^2 - \left(\frac{10+6k}{k+1}\right)y = -\left(\frac{31+3k}{k+1}\right).$$

By completing the squares we can put this equation in standard form.

Add $\left(\frac{7-k}{2(k+1)}\right)^2 + \left(\frac{10+6k}{2(k+1)}\right)^2$ to both sides of the equation.

$$\left[x^2 + \frac{7-k}{k+1}x + \left(\frac{7-k}{2(k+1)}\right)^2\right] + \left[y^2 - \frac{10+6k}{k+1}y + \left(\frac{10+6k}{2(k+1)}\right)^2\right]$$

$$= -\frac{31+3k}{k+1} + \left(\frac{7-k}{2(k+1)}\right)^2 + \left(\frac{10+6k}{2(k+1)}\right)^2.$$

Let the right hand side = r^2

$$\left[x - \left(-\frac{7-k}{2(k+1)}\right)\right]^2 + \left(y - \frac{10+6k}{2(k+1)}\right)^2 = r^2.$$

Hence, the circle has center at $\left(-\frac{7-k}{2(k+1)}, \frac{10+6k}{2(k+1)}\right)$ which

equals $\left(\frac{k-7}{2(k+1)}, \frac{3k+5}{k+1}\right)$. Since this point must satisfy the equation $x - y - 2 = 0$, we can substitute and solve for x.

$$\frac{k-7}{2(k+1)} - \frac{3k+5}{k+1} - 2 = 0$$

$$\frac{k-7 - 2(3k+5) - 2(2(k+1))}{2(k+1)} = \frac{k-7 - 6k - 10 - 4k - 4}{2(k+1)} = \frac{-9k - 21}{2(k+1)} = 0.$$

Hence, $-9k - 21 = 0$ or $k = -\frac{7}{3}$. Now we can substitute $k = -\frac{7}{3}$ into any of the forms of our circle equation to obtain our final result. We have chosen to substitute into (*). Hence,

$$C_3: \quad x^2+y^2 + \left[\frac{7+\frac{7}{3}}{-\frac{7}{3}+1}\right]x - \left[\frac{10-6(\frac{7}{3})}{-\frac{7}{3}+1}\right]y + \left[\frac{31-3(\frac{7}{3})}{-\frac{7}{3}+1}\right] = 0$$

$$x^2+y^2 + \left[\frac{\frac{28}{3}}{-\frac{4}{3}}\right]x - \left[\frac{-4}{-\frac{4}{3}}\right]y + \left[\frac{24}{-\frac{4}{3}}\right]$$

$C_3: \quad x^2 + y^2 - 7x - 3y - 18 = 0$ with center at $\left(\frac{7}{2}, \frac{3}{2}\right)$. All three circles have been drawn in the accompanying figure.

ELLIPSES

● PROBLEM 740

For the ellipse $36x^2 + 64y^2 = 2304$, sketch the ellipse and find the value of e and the coordinates of the vertices and foci.

Solution: One way to visualize the ellipse is to think of it as a squashed circle (Figure 1) where certain diameters $\overline{AB}$ become smaller and certain diameters $\overline{CD}$ become

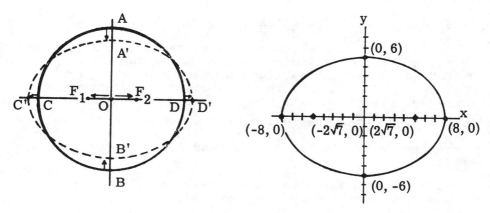

Figure 1

Figure 2

longer. The diameter that becomes shortest is called the
minor axis (A'B' in Figure 1). The diameter that becomes
the longest is called the major axis (C'D'). The length
of half the minor axis (the shortened radius) is always
denoted by b; the length of half the major axis is a.
For ellipses, it is always true that the major and minor
axes are perpendicular to each other. The squashing caused
us to define two axes analogous to the diameter. The
flattening also causes us to define two points called
foci analogous to the center of the circle. In a circle,
a point is on a circle if the distance between the point
and circle center is a certain amount. In an ellipse, a
point is on the ellipse if the distance between the point
and one focus plus the distance between the point and the
other focus is a certain amount.

To help measure the squashing, the idea of eccentricity
is introduced. The eccentricity, e, of an ellipse is defined as the distance the foci are from the center divided
by the length of the major axis. For the case of only a
slightly squased ellipse, the foci will be very near the
center and the eccentricity will be close to zero. For
very flattened ellipses, the foci will be very spread out
and close to the endpoints of the major axis. For this
case, the eccentricity will be close to one.

By the introduction of squashing, many ellipses and
many equations can be found. We limit ourselves in this
book to the discussion of two types of squashing: (1)
Squashing so that the major axis is parallel to the x-axis;
and (2) Squashing so that the major axis is parallel to
the y-axis. This limitation allows us to remember only
two possible equations for the ellipse.

The general equation of an ellipse is either

(1) $\dfrac{(x - h)^2}{a^2} + \dfrac{(y - k)^2}{b^2} = 1$

(if the major axis is parallel to the x-axis) or

(2) $\dfrac{(x - h)^2}{b^2} + \dfrac{(y - k)^2}{a^2} = 1$

(if the major axis is parallel to the y-axis).

The center of the ellipse lies at point (h, k). The vertices are the endpoints of the major and minor axes. For case (1), the vertices of the major axis are (h + a, k) and (h - a, k). The vertices of the minor axis are (h, k + b) and (h, k - b). The constant c is defined as $\sqrt{a^2 - b^2}$. The foci of the ellipse are located at (h + c, k) and (h - c, k). For an ellipse, the eccentricity, e is defined as c/a, and is always between 0 and 1.

For the given equation, $36x^2 + 64y^2 = 2304$, place it in standard form by dividing by 2304.

(i) $\dfrac{x^2}{64} + \dfrac{y^2}{36} = 1$.

From inspection, we see that h = k = 0. Also, we see that the denominator of the x^2 term > the denominator of the y^2 term. Thus, the equation is of form case (I) and the major axis is coincident with the x-axis. Furthermore, a = $\sqrt{64}$ = 8; b = $\sqrt{36}$ = 6; and c = $\sqrt{a^2 - b^2}$ = $\sqrt{28}$ = 2 $\sqrt{7}$.

Since h = k = 0, the center is the point (0, 0). The vertices are (h + a, k) = (8, 0); (h - a, k) = (- 8, 0); (h, k + b) = (0, 6); and (h, k - b) = (0, - 6). The foci are located at (h + c, k) = (2 $\sqrt{7}$, 0) and (h - c, k) = (- 2$\sqrt{7}$, 0). The eccentricity,

$$e = \frac{c}{a} = \frac{2 \sqrt{7}}{8} = \frac{\sqrt{7}}{4} \ .$$

To sketch the ellipse, locate the vertex points. Connect the points with oval curves. Keep in mind that the ellipse is symmetric about both axes.

● **PROBLEM** 741

In the equation of an ellipse,

$$4x^2 + 9y^2 - 16x + 18y - 11 = 0,$$

determine the standard form of the equation, and find the values of a, b, c, and e.

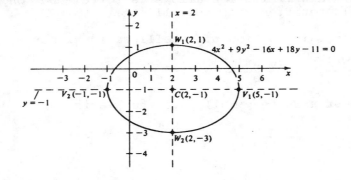

Solution: By completing the squares, we can arrive at the standard form of the equation, from which the values of the parameters can be determined. Thus,

$$4(x^2 - 4x + 4) + 9(y^2 + 2y + 1) = 36.$$

or $4(x - 2)^2 + 9(y + 1)^2 = 36$. Dividing by 36,

$$\frac{(x - 2)^2}{9} + \frac{(y + 1)^2}{4} = 1.$$

Thus, the center of the ellipse is at $(2, -1)$. Comparing this equation with the general form,

$$\frac{x^2}{a^2} + \frac{y^2}{b^2} = 1, \text{ where } a > b, \text{ we see that } a = 3, b = 2.$$

$$c = \sqrt{a^2 - b^2} = \sqrt{5}.$$

Finally, $e = \frac{c}{a} = \frac{\sqrt{5}}{3} \approx 0.745.$

● **PROBLEM** 742

Find the equation of the ellipse which has vertices V_1 $(-2, 6)$, V_2 $(-2, -4)$, and foci F_1 $(-2, 4)$, F_2 $(-2, -2)$. (See figure.)

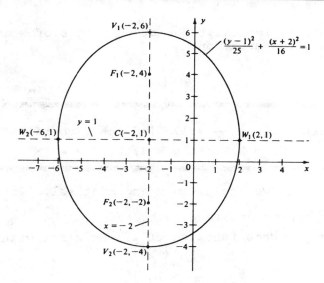

Solution: The major axis is on the line $x = -2$, and the center is at $(-2, 1)$. Hence, a, the length of the semimajor axis, equals the difference between the y-coordinates of V_1 (say) and the center, i.e., a = 5. From the coordinates of the foci, c = 3. Since $b^2 = a^2 - c^2$, $b = \sqrt{25 - 9} = 4$, and the ends of the minor axis, on y = 1, are at W_1 (2, 1) and W_2 (-6, 1). The equation can now be written, in the form

$$\frac{(y - k)^2}{a^2} + \frac{(x - h)^2}{b^2} = 1$$

$$\frac{(y - 1)^2}{25} + \frac{(x + 2)^2}{16} = 1.$$

● **PROBLEM** 743

Show that $x = 5 \cos \theta$ and $y = 3 \sin \theta$ satisfies

$$\frac{x^2}{25} + \frac{y^2}{9} = 1.$$

Solution: If $x = 5 \cos \theta$, $y = 3 \sin \theta$ satisfies the equation

$$\frac{x^2}{25} + \frac{y^2}{9} = 1,$$

then we will obtain an identity when we substitute these values of x and y into the equation. Doing this, we find

$$\frac{(5 \cos \theta)^2}{25} + \frac{(3 \sin \theta)^2}{9} = \frac{25 \cos^2 \theta}{25} + \frac{9 \sin^2 \theta}{9}$$

$$\cos^2 \theta + \sin^2 \theta = 1$$

$$1 = 1$$

Since this is an identity, the given values of x and y satisfy the given equation.

● **PROBLEM** 744

For the equation $\frac{x^2}{25} + \frac{y^2}{9} = 1$, find the y coordinates when
a) $x = 2$; b) $x = 3$; c) $x = 4$; d) $x = 5$; e) $x = 6$.

Solution: Since we are given x and must solve for y in each case, it is convenient to put the equation $\frac{x^2}{25} + \frac{y^2}{9} = 1$ in the form y = some function of x. We start with

$$\frac{x^2}{25} + \frac{y^2}{9} = 1 \qquad (1)$$

Subtracting $\frac{x^2}{25}$ from both sides of this equation,

$$\frac{y^2}{9} = 1 - \frac{x^2}{25} \qquad (2)$$

Taking the square root of both sides of equation (2)

692

$$\frac{y}{3} = \sqrt{1 - \frac{x^2}{25}} \tag{3}$$

Now, multiplying both sides of equation (3) by 3, we get

$$y = 3\sqrt{1 - \frac{x^2}{25}}.$$

We may now proceed with the calculations:

a) $x = 2$, $y = 3\sqrt{1 - \frac{4}{25}} = \frac{3}{5}\sqrt{21}$

b) $x = 3$, $y = 3\sqrt{1 - \frac{9}{25}} = \frac{12}{5}$

c) $x = 4$, $y = 3\sqrt{1 - \frac{16}{25}} = \frac{9}{5}$

d) $x = 5$, $y = 3\sqrt{1 - \frac{25}{25}} = 0$

e) $x = 6$, $y = 3\sqrt{1 - \frac{36}{25}} = \left(\frac{3\sqrt{11}}{5}\right)i.$

Note that in part e, we obtain an imaginary value of y for x = 6. Hence, there is no real pair of coordinates (x,y) when x = 6. Realizing that the given equation describes an ellipse, this means that no point on the ellipse has x co-ordinate equal to 6, for all coordinate pairs corresponding to points on any geometric figure must be real.

● **PROBLEM** 745

a) For the equation $\frac{(x-1)^2}{25} + \frac{y^2}{9} = 1$, find the y-coordinates when x = -4, -3, -2, -1, 0, 1, 2, 3, 4, 5 and 6.

b) Plot the points (x,y) obtained in (a), and sketch the results.

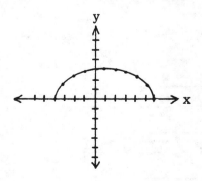

Solution: Since we are trying to find the y coordinates for given x coordinates, it is convenient to solve the equation

$\frac{(x-1)^2}{25} + \frac{y^2}{9} = 1$ for y as a function of x.

We start with

$$\frac{(x-1)^2}{25} + \frac{y^2}{9} = 1.$$

Subtracting $\frac{(x-1)^2}{25}$ from both sides of this equation yields

$$\frac{y^2}{9} = 1 - \frac{(x-1)^2}{25}.$$

Taking the square root of both sides

$$\frac{y}{3} = \sqrt{1 - \frac{(x-1)^2}{25}}. \tag{1}$$

Multiplying both sides by 3 gives us the desired equation:

$$y = 3\sqrt{1 - \frac{(x-1)^2}{25}}. \tag{2}$$

We may find the y coordinates corresponding to the given x coordinates by substituting the latter into equation (1). The results are listed below:

$$x = -4, \quad y = 3\sqrt{1 - \frac{(-5)^2}{25}} = 0.$$

$$x = -3, \quad y = 3\sqrt{1 - \frac{(-4)^2}{25}} = \frac{9}{5}$$

$$x = -2, \quad y = 3\sqrt{1 - \frac{(-3)^2}{25}} = \frac{12}{5}$$

$$x = -1, \quad y = 3\sqrt{1 - \frac{(-2)^2}{25}} = \frac{3\sqrt{21}}{5}$$

$$x = 0, \quad y = 3\sqrt{1 - \frac{(-1)^2}{25}} = \frac{3\sqrt{24}}{5}$$

$$x = 1, \quad y = 3\sqrt{1 - 0} = 3$$

$$x = 2, \quad y = 3\sqrt{1 - \frac{(1)^2}{25}} = \frac{3\sqrt{24}}{5}$$

$$x = 3, \quad y = 3\sqrt{1 - \frac{(2)^2}{25}} = \frac{3\sqrt{21}}{5}$$

$$x = 4, \quad y = 3\sqrt{1 - \frac{(3)^2}{25}} = \frac{12}{25}$$

$$x = 5, \quad y = 3 \sqrt{1 - \frac{(4)^2}{25}} = \frac{9}{5}$$

$$x = 6, \quad y = 3 \sqrt{1 - \frac{(5)^2}{25}} = 0.$$

b). A plot of the points (x,y) found in part (a) is indicated in the figure. Note that we have drawn the upper portion of an ellipse. Since the points plotted satisfy the equation $\frac{(x-1)^2}{25} + \frac{y^2}{9} = 1$, we may advance the educated guess that this is the equation of an ellipse, as, indeed, it is.

● **PROBLEM** 746

The equation of an ellipse is $\frac{x^2}{a^2} + \frac{y^2}{b^2} = 1$. Discuss what happens if $a = b = r$.

<u>Solution</u>: If $a = b = r$, the given equation becomes

$$\frac{x^2}{a^2} + \frac{y^2}{b^2} = \frac{x^2}{r^2} + \frac{y^2}{r^2} = 1. \tag{1}$$

Multiplying the last branch of the equality in equation (1) by r^2 yields

$$x^2 + y^2 = r^2. \tag{2}$$

This is the equation of a circle with center at $C(0,0)$ and radius r. Hence, we see that a circle is a special case of an ellipse.

● **PROBLEM** 747

By definition, if an ellipse has a foci F_1 $(-c,0)$ and F_2 $(c,0)$, and P (x,y) is a point on the ellipse, then $PF_1 + PF_2 = k$, where k is a constant such that $k > F_1F_2 = 2c$. Assuming that the above holds, and defining a constant b such that $b^2 = a^2 - c^2$, and a constant as such that $a = \frac{k}{2}$, prove that the equation of the ellipse is $\frac{x^2}{a^2} + \frac{y^2}{b^2} = 1$.

<u>Solution</u>: Using the definition of an ellipse,

$$PF_1 + PF_2 = k = 2a \tag{1}$$

The distance formula yields

$$PF_1 = \sqrt{(x+c)^2 + y^2} \tag{2}$$

$$PF_2 = \sqrt{(x-c)^2 + y^2} \tag{3}$$

Substituting (2) and (3) in (1)

$$\sqrt{(x+c)^2 + y^2} + \sqrt{(x-c)^2 + y^2} = 2a.$$

Subtracting $\sqrt{(x-c)^2 + y^2}$ from both sides of this equation

$$\sqrt{(x+c)^2 + y^2} = 2a - \sqrt{(x-c)^2 + y^2}$$

Squaring both sides, and expanding the terms under the radicals

$$x^2 + 2cx + c^2 + y^2 =$$

$$4a^2 - 4a\sqrt{x^2-2cx + c^2 + y^2} + x^2 - 2cx + c^2 + y^2.$$

Cancelling x^2, c^2, and y^2 from both sides

$$2cx = 4a^2 - 4a\sqrt{x^2 - 2cx + c^2 + y^2} - 2cx.$$

Subtracting $-4a\sqrt{x^2 - 2cx + c^2 + y^2}$ and $2cx$ from both sides

$$4a\sqrt{x^2 - 2cx + c^2 + y^2} = 4a^2 - 4cx.$$

Dividing both sides by 4a

$$\sqrt{x^2 - 2cx + c^2 + y^2} = a - \frac{c}{a}x.$$

Squaring both sides

$$x^2 - 2cx + c^2 + y^2 = a^2 - 2cx + \frac{c^2}{a^2}x^2.$$

Transposing the left side of this equation to the right side, and regrouping we obtain

$$x^2\left(1 - \frac{c^2}{a^2}\right) + (c^2 - a^2) + y^2 = 0. \tag{4}$$

But

$$b^2 = a^2 - c^2.$$

Multiplying both sides by -1

$$c^2 - a^2 = -b^2. \tag{5}$$

Dividing both sides by a^2

$$\frac{c^2}{a^2} - 1 = \frac{-b^2}{a^2}.$$

Adding 1 to both sides

$$\frac{c^2}{a^2} = 1 - \frac{b^2}{a^2} \qquad (6)$$

Using equations (5) and (6) in (4), we obtain

$$x^2\left[1 - \left(1 - \frac{b^2}{a^2}\right)\right] + (-b^2) + y^2 = 0$$

or $\quad x^2 \frac{b^2}{a^2} - b^2 + y^2 = 0.$

Adding b^2 to both sides

$$x^2 \frac{b^2}{a^2} + y^2 = b^2.$$

Multiplying both sides by $\frac{a^2}{b^2}$

$$x^2 + \frac{a^2 y^2}{b^2} = a^2.$$

Dividing both sides by a^2 yields the required equation

$$\frac{x^2}{a^2} + \frac{y^2}{b^2} = 1.$$

● **PROBLEM** 748

A single-lane highway must pass under a series of bridges. It is proposed that the bridges be shaped as semi-ellipses with the height equal to the width. The builder feels he must allow room for a 6 foot wide, 12 foot high truck to pass under it. What is the lowest bridge that can be built to serve this purpose.

Solution: The accompanying figure has been drawn to visually depict the problem statement. Since the bridge must be an ellipse, we have drawn the figure on a co-ordinate grid. Let semi-ellipse MPN be a representative bridge and the rectangle shown be a truck of the size given driving exactly in the center of the lane. We are told MN = OP.

We want to find OP. Notice that this length is half the length of the major axis of the total ellipse

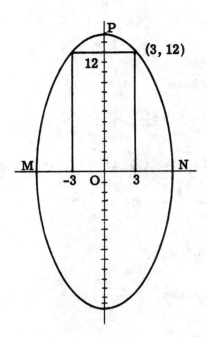

from which the bridge design comes.

If we let a = OM, semi-minor axis

then   2a = OP, semi-major axis.

Since we wish to see whether or not the upper vertex of the truck will touch the ellipse, we need to find a general formula for any point on the ellipse and then determine the value of (a) that will allow the truck to just miss touching the bridge.

The general formula for any point (x, y) on the ellipse is

$$\frac{x^2}{a^2} + \frac{y^2}{4\,a^2} = 1$$

The upper vertex of the truck is (3, 12). If (a) is such that (3, 12) is on the ellipse the truck will touch the bridge. Hence, any value of a just slightly greater than this value will be a solution to the problem.

Substitute (3, 12) in the equation for the ellipse and solve for a.

$$\frac{3^2}{a^2} + \frac{12^2}{4\,a^2} = 1$$

$$\frac{4\,(9) + 144}{4\,a^2} = 1$$

$$\frac{180}{4\,a^2} = 1$$

$$a^2 = \frac{180}{4} = 45$$

$$a = 3\sqrt{5}, \qquad 2a = 6\sqrt{5}.$$

Therefore, the truck will be able to pass under an arch bridge with height just greater than $6\sqrt{5}$ feet, or 13 ft. 5 in.

● **PROBLEM** 749

The latus rectum of an ellipse is the chord through either focus perpendicular to the major axis. Show that the length of the latus recta of ellipse

$$\frac{x^2}{a^2} + \frac{y^2}{b^2} = 1$$

is given by the formula $(2b^2)/a$.

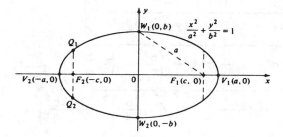

Solution: In the accompanying figure, ellipse

$$\frac{x^2}{a^2} + \frac{y^2}{b^2} = 1$$

has foci $(-c, 0)$ and $(c, 0)$ and the latus rectum $\overline{Q_1Q_2}$. We are asked to show $Q_1Q_2 = 2b^2/a$. To find the length of Q_1Q_2, we first find the coordinates of Q_1 and Q_2. To find $Q_1(x_1, y_1)$, note that both foci lie on the x-axis. Since the major axis is the line determined by the foci, the major axis is the x-axis. Further, since the latus recta are perpendicular to the major axis, this implies that the latus rectum Q_1Q_2 is a vertical line. Since every point on a vertical line has the same x-coordinate, then $Q_1(x_1,y_1)$ and $Q_2(x_2,y_2)$ must have the same x as $F_2(-c, 0)$. Thus, $x_1 = x_2 = -c$. To complete locating Q_1 and Q_2, we now find their y-coordinates. Note that Q_1 is a point on the ellipse and thus (x_1,y_1) must satisfy the equation

$$\frac{x^2}{a^2} + \frac{y^2}{b^2} = 1.$$

(i) $\qquad \frac{x_1^2}{a^2} + \frac{y_1^2}{b^2} = 1$

(ii) $\qquad \frac{(-c)^2}{a^2} + \frac{y_1^2}{b^2} = 1$

Solving for y_1, we obtain

699

$$\frac{y_1{}^2}{b^2} = 1 - \frac{c^2}{a^2} \quad \text{or,}$$

$$y_1 = b \sqrt{1 - \frac{c^2}{a^2}} = b \sqrt{\frac{a^2 - c^2}{a^2}} = \frac{b}{a} \sqrt{a^2 - c^2}.$$

In earlier problems, it was shown that a, b, and c must satisfy the equation $a^2 + b^2 = c^2$. Thus, $a^2 - c^2 = b^2$, and $\sqrt{a^2 - c^2} = b$. Therefore,

$$y_1 = \frac{b}{a} \sqrt{a^2 - c^2} = \frac{b^2}{a}$$

and the coordinates of Q_1 are $(-c, b^2/a)$.

By a similar method, we can show that the coordinates of Q_2 are $(-c, -b^2/a)$. Having the coordinates of Q_1 and Q_2, we can now solve for the length of the latus rectum Q_1Q_2.

$$Q_1Q_2 = \sqrt{(x_1 - x_2)^2 + (y_1 - y_2)^2}$$

$$= \sqrt{(-c - (-c))^2 + \left[\frac{b^2}{a} - \left(-\frac{b^2}{a}\right)\right]^2}$$

$$= \sqrt{0^2 + \left(\frac{2b^2}{a}\right)^2} = \sqrt{\left(\frac{2b^2}{a}\right)^2} = \left(\frac{2b^2}{a}\right)$$

Thus, the length of the latus rectum is $\frac{2b^2}{a}$.

● **PROBLEM** 750

Consider a point $p_1(x_1, y_1)$ on the ellipse $b^2x^2 + a^2y^2 = a^2b^2$. A tangent to the ellipse at p_1 is a line through p_1 with no other point on the ellipse. Prove that if $y_1 \neq 0$, there is a tangent at p_1, its slope is m = $(-b^2x_1)/(a^2y_1)$ and its equation can be put in the form

$$\frac{x_1 x}{a^2} + \frac{y_1 y}{b^2} = 1.$$

Solution: This problem has two parts: (A) show that the tangent to the ellipse has slope m = $(-b^2x_1)/(a^2y_1)$, and (B) show the equation of the tangent is

$$\frac{x_1 x}{a^2} + \frac{y_1 y}{b^2} = 1.$$

(A) With the given information, we can determine the equation, y = mx + z, of the tangent line. (We use z instead of b for the y intercept to avoid confusion). To determine the equation, we must find the values of m and z. To solve for the two unknowns, we need a system of two equations in two unknowns.

The first equation derives from the given information

that the tangent line ℓ contains point $p_1(x_1,y_1)$. Thus, (x_1, y_1) must satisfy the equation of ℓ or:

(i) $y_1 = mx_1 + z$

To find the second relationship, we use the fact that the tangent and the ellipse have only one common point.

Consider any line $y = mx + z$ intersecting the ellipse. We find the general formula for the intersection points in terms of m and z. Then we find the special relationship that must exist between m and z for there to be only one point of intersection. Combining this relation with the equation (i), we can then solve for our particular m and z.

Suppose (X, Y) are points of intersection of the line and the ellipse. Then X and Y must satisfy the equations:

(i) $Y = MX + Z$

(ii) $\dfrac{X^2}{a^2} + \dfrac{Y^2}{b^2} = 1.$

We have two simultaneous equations in two unknowns. Substituting the value Y in equation (i) into equation (ii), we obtain an equation totally in terms of X.

(iii) $\dfrac{X^2}{a^2} + \dfrac{(MX + Z)^2}{b^2} = 1$

Multiplying out and simplifying,

(iv) $\dfrac{X^2}{a^2} + \dfrac{M^2 X^2}{b^2} + \dfrac{2MXZ}{b^2} + \dfrac{Z^2}{b^2} = 1$

(v) $X^2 \left[\dfrac{1}{a^2} + \dfrac{M^2}{b^2}\right] + X \left(\dfrac{2MZ}{b^2}\right) + \left[\dfrac{Z^2}{b^2} - 1\right] = 0.$

Using the quadratic formula $x = \dfrac{-b \pm \sqrt{b^2 - 4ac}}{2a}$, we would normally obtain two values for X,

$x_1 = \dfrac{-b + \sqrt{b^2 - 4ac}}{2a}$ and $x_2 = \dfrac{-b - \sqrt{b^2 - 4ac}}{2a}$.

This would imply two different points of intersection (x_1, y_1) and (x_2, y_2). For there to be only one inter-section point, and thus a tangent line, it must be true that $x_1 = x_2$ or $\dfrac{-b + \sqrt{b^2 - 4ac}}{2a} = \dfrac{-b - \sqrt{b^2 - 4ac}}{2a}$.

This is true only if $\sqrt{b^2 - 4ac} = 0$. Thus, we wish to find the relationship between M and Z that must always exist for $\sqrt{b^2 - 4ac} = 0$. In equation (v), $a = \left(\dfrac{1}{a^2} + \dfrac{M^2}{b^2}\right)$, $b = \dfrac{2MZ}{b^2}$, and $c = \dfrac{Z^2}{b^2} - 1$. Thus,

(vi) $0 = \sqrt{b^2 - 4ac}$

$$= \sqrt{\left(\frac{2MZ}{b^2}\right)^2 - 4\left(\frac{Z^2}{b^2} - 1\right)\left(\frac{1}{a^2} + \frac{M^2}{b^2}\right)}$$

(vii) $\quad 0^2 = \frac{4\,M^2Z^2}{b^4} - \frac{4\,Z^2}{a^2b^2} - \frac{4\,Z^2M^2}{b^4} + \frac{4}{a^2} + \frac{4\,M^2}{b^2}$

(viii) $\quad \frac{Z^2}{a^2b^2} = \frac{1}{a^2} + \frac{M^2}{b^2}$

(ix) $\quad Z^2 = b^2 + a^2M^2.$

Thus, if a line $y = Mx + z$ is to be tangent to the ellipse $\frac{x^2}{a^2} + \frac{y^2}{b^2} = 1$, then M and Z must satisfy the above equation. Since we wish the given ℓ to satisfy this equation, the slope m and intercept z must also satisfy equation (i). Thus, we have our two equations involving m and z.

(x) $\quad y_1 = mx_1 + z$

(xi) $\quad z^2 = b^2 + a^2m^2$

From equation (x), we have $z = y_1 - mx_1$. Substituting this result in (xi), we can solve for (xi) for **m.**

(xii) $\quad (y_1 - mx_1)^2 = b^2 + a^2m^2$

(xiii) $\quad y_1^2 - 2mx_1y_1 + m^2x_1^2 = b^2 + a^2m^2$

(xiv) $\quad m^2(a^2 - x_1^2) + m(2x_1y_1) + (b^2 - y_1^2) = 0.$

Using the quadratic equation to solve for m, we obtain

(xv) $\quad m = \dfrac{-(2x_1y_1) \pm \sqrt{4x_1^2y_1^2 - 4(b^2 - y_1^2)(a^2 - x_1^2)}}{2(a^2 - x_1^2)}$

(xvi) $\quad m = \dfrac{-(2x_1y_1) \pm 2\sqrt{x_1^2y_1^2 - b^2a^2 + b^2x_1^2 + a^2y_1^2 - y_1^2x_1^2}}{2(a^2 - x_1^2)}$

(xvii) $\quad m = \dfrac{-x_1y_1 \pm \sqrt{b^2x_1^2 + a^2y_1^2 - b^2a^2}}{a^2 - x_1^2}$

Note that x_1 and y_1 are points on the ellipse. Therefore, $\frac{x_1^2}{a^2} + \frac{y_1^2}{b^2} = 1$ or $b^2x_1^2 + a^2y_1^2 - a^2b^2 = 0$. Thus the expression under the radical is 0, and (xvii) becomes

(xviii) $\quad m = \dfrac{-x_1y_1}{a^2 - x_1^2}$

To get this into the desired form of the problem statement, note that since $\frac{x_1^2}{a^2} + \frac{y_1^2}{b^2} = 1$, $x_1^2 = a^2 - \frac{a^2}{b^2}y_1^2$

The denominator $a^2 - x_1^2$ thus becomes $\frac{a^2}{b^2} y_1^2$ and thus,

(xix) $m = \dfrac{- x_1 y_1}{\frac{a^2}{b^2} y_1^2} = - \dfrac{b^2}{a^2} \dfrac{x_1}{y_1}$.

(B) The equation of the tangent is given by the formula $y = mx + z$ where $m = - \dfrac{b^2}{a^2} \dfrac{x_1}{y_1}$. To find z, remember from part (A) that $z^2 = b^2 + a^2 m^2$. Thus,

$$z = \sqrt{b^2 + a^2 m^2} = \sqrt{b^2 + a^2 \left(+ \frac{b^4}{a^4} \frac{x_1^2}{y_1^2} \right)}$$

$$= b \sqrt{1 + \frac{b^2}{a^2} \frac{x_1^2}{y_1^2}}$$

The equation of the tangent is

$$y = - \frac{b^2}{a^2} \frac{x_1}{y_1} x + b \sqrt{1 + \frac{b^2}{a^2} \frac{x_1^2}{y_1^2}}$$

To rework this into a more simplified form, note the

$$b \sqrt{1 + \frac{b^2}{a^2} \frac{x_1^2}{y_1^2}} \quad \text{equals} \quad \frac{b}{ay_1} \sqrt{a^2 y_1^2 + b^2 x_1^2}$$

$$= \frac{b}{ay_1} (ab) \sqrt{\frac{y_1^2}{b^2} + \frac{x_1^2}{a^2}} = \frac{b^2}{y_1} \sqrt{1} = \frac{b^2}{y_1} . \quad \text{Thus,}$$

(xx) $y = - \dfrac{b^2}{a^2} \dfrac{x_1}{y_1} x + \dfrac{b^2}{y_1}$

Multiplying both sides by $\frac{y_1}{b^2}$ and bringing the x term to the left we obtain:

(xxi) $\dfrac{x_1 x}{a^2} + \dfrac{y \, y_1}{b^2} = 1.$

● PROBLEM 751

Find the area of the ellipses

(a) $\dfrac{x^2}{9} + \dfrac{y^2}{25} = 1$ (b) $\dfrac{x^2}{144} + \dfrac{y^2}{256} = 1$ (c) $\dfrac{x^2}{64} + \dfrac{y^2}{49} = 1$

(d) $\dfrac{x^2}{81} + \dfrac{y^2}{16} = 1.$

<u>Solution</u>:   The area of an ellipse whose equation is

$\dfrac{x^2}{a^2} + \dfrac{y^2}{b^2} = 1$ is $\pi ab.$

Applying this fact to the given equations, we find

(a) $\dfrac{x^2}{9} + \dfrac{y^2}{25} = 1$,　　$a^2 = 9$,　　$a = 3$

　　　　　　　　　　　$b^2 = 25$,　$b = 5$

　　　　　　　Area　$= \pi ab = 15\pi$

(b) $\dfrac{x^2}{144} + \dfrac{y^2}{256} = 1$,　$a^2 = 144$,　$a = 12$

　　　　　　　　　　　$b^2 = 256$,　$b = 16$

　　　　　　　Area　$= \pi ab = 192\pi$

(c) $\dfrac{x^2}{64} + \dfrac{y^2}{49} = 1$,　　$a^2 = 64$,　　$a = 8$

　　　　　　　　　　　$b^2 = 49$,　　$b = 7$

　　　　　　　Area　$= \pi ab = 56\pi$

(d) $\dfrac{x^2}{81} + \dfrac{y^2}{16} = 1$,　　$a^2 = 81$,　　$a = 9$

　　　　　　　　　　　$b^2 = 16$,　　$b = 4$

　　　　　　　Area　$= \pi ab = 36\pi$

CHAPTER 43

CONIC SECTIONS: PARABOLAS AND HYPERBOLAS

> **Basic Attacks and Strategies for Solving Problems in this Chapter. See pages 705 to 731 for step-by-step solutions to problems.**

A **parabola** is defined as the locus of all points in a plane whose distances from a fixed point and a fixed line are equal. The point is called the **focus** and the line is called the **directrix**. Suppose the fixed point is $(p, 0)$ and the fixed line is $x = -p$. Here is a picture of such a parabola

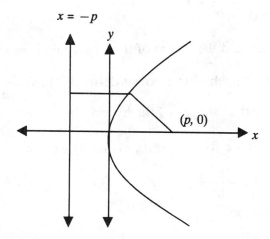

The standard equation of such a parabola is $y^2 = 4px$. In this case, the x-axis is an axis of symmetry. On the other hand, if the focus is at $(0, p)$ and if $y = -p$ is the directrix, then the standard equation is $x^2 = 4py$ and in this case, the y-axis is an axis of symmetry. In both cases, the standard equations can be derived just by using the distance formula, and in both cases the vertex at parabola is at the origin. As indicated in Problem 752, a parabola with vertex at (h, k) and with axis of symmetry parallel to the x-axis has $(y - k)^2 = 4p(x - h)$ as an equation, and if the axis of symmetry is parallel to the y-axis then the parabola has $(x - h)^2 = 4p(y - k)$ as an equation.

A **hyperbola** is defined as the locus of all points, the difference of whose distances from two fixed points is a constant. The fixed points are called foci. Here is a picture of a hyperbola with foci at $(c, 0)$ and $(-c, 0)$ and with fixed difference of $2a$.

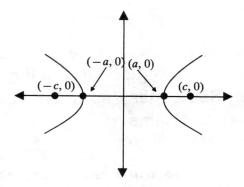

Using only the distance formula and simple algebraic techniques, it is easy to derive the following equation:

$$\frac{x^2}{a^2} - \frac{y^2}{b^2} = 1$$

In this case, $b^2 = c^2 - a^2$, and the center of the hyperbola is said to be at the origin. The eccentricity of such a hyperbola is defined to be $e = \frac{c}{a}$ and it gives a general description of the "shape" of the hyperbola.

On the other hand, if the foci are at $(0, c)$ and $(0, -c)$, then

$$\frac{y^2}{a^2} - \frac{x^2}{b^2} = 1$$

is the standard equation.

When the center of the parabola is at (h, k), then the two equations above become

$$\frac{(x - h)^2}{a^2} - \frac{(y - k)^2}{b^2} = 1 \text{ and}$$

$$\frac{(y - k)^2}{a^2} - \frac{(x - h)^2}{b^2} = 1$$

Step-by-Step Solutions to Problems in this Chapter, "Conic Sections: Parabolas and Hyperbolas"

PARABOLAS

● PROBLEM 752

Problem: Find the equation of a parabola that has vertex at $(-1,2)$, axis of symmetry parallel to the x-axis, and goes through the point $P_1(-3,-4)$.

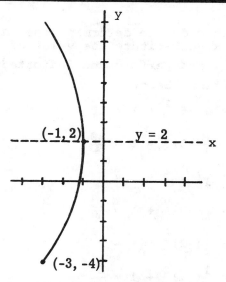

Solution: Recalling the equation of a parabola with axis of symmetry parallel to the x-axis, $(y-k)^2 = 4p(x-h)$, we can substitute the given value for the vertex $(-1,2)$ for (h,k) yielding $(y-2)^2 = 4p(x+1)$. To find the value of p_1, we must use the only other piece of information that we are given: namely, that the curve passes through the point $(-3,-4)$. Substituting in $(y-2)^2 = 4p(x+1)$, we solve for p, $36 = 4p(-2)$, $4p = -18$. Thus, the desired equation is

$$(y-2)^2 = -18(x+1).$$

a) For x = -6, -5, -4, -3, -2, -1, 0, 1, 2, 3, 4, 5, 6 determine the corresponding values of y from the equation y = $\frac{1}{4}$x^2 + 1. b) Plot the corresponding points and sketch the curve.

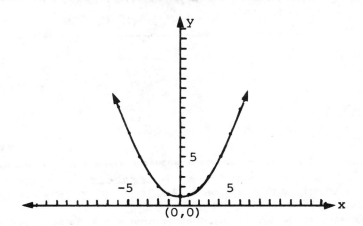

Solution: a) In order to determine the values of y for the given values of x, substitute the value of x into the expression y = $\frac{1}{4}$x^2 + 1, and perform the indicated operations. We list the calculations below:

x = -6 , y = $\frac{1}{4}$(-6)^2 + 1 = 10

x = -5 , y = $\frac{1}{4}$(-5)^2 + 1 = $\frac{29}{4}$

x = -4 , y = $\frac{1}{4}$(-4)^2 + 1 = 5

x = -3 , y = $\frac{1}{4}$(-3)^2 + 1 = $\frac{13}{4}$

x = -2 , y = $\frac{1}{4}$(-2)^2 + 1 = 2

x = -1 , y = $\frac{1}{4}$(-1)^2 + 1 = $\frac{5}{4}$

x = 0 , y = $\frac{1}{4}$(0)^2 + 1 = 1

Now, instead of blindly solving for y for the positive values of x, note that in each calculation above, the value of x used is squared. Hence, whenever x equals a number or its opposite, we must get the same value of y. (For instance, for x = 6, y = 10, etc.)

In order to plot the points (x,y) found above, we first set up a set of labelled Cartesian coordinate axes, as shown in the figure. To plot the point (-6,10), we go along the x

706

axis until we get to -6, and then we stay at this position and more vertically until we locate 10 on the y axis. Our final position, marked by a dot, locates the point (-6,10).

Plotting the rest of the points found in part (a), and connecting them with smooth curves, we find that the figure sketched is a parabola. Since the points used to draw this figure satisfy the equation $y = \frac{1}{4}x^2 + 1$, this must be the equation of the parabola sketched.

● **PROBLEM** 754

Plot points of the curve corresponding to $y = \frac{1}{4}(x-2)^2$ for $x = 4, 3, 2, 1, 0$ and sketch the curve.

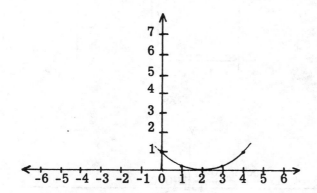

Solution: The equation of the curve is given to us as $y = \frac{1}{4}(x-2)^2$. In order to calculate the y coordinates corresponding to the given x coordinates, we substitute each x coordinate into the equation of the curve, and calculate y. This is done below:

$$x = 4 \quad , \quad y = \frac{1}{4}(4-2)^2 = 1$$

$$x = 3 \quad , \quad y = \frac{1}{4}(3-2)^2 = \frac{1}{4}$$

$$x = 2 \quad , \quad y = \frac{1}{4}(2-2)^2 = 0$$

$$x = 1 \quad , \quad y = \frac{1}{4}(1-2)^2 = \frac{1}{4}$$

$$x = 0 \quad , \quad y = \frac{1}{4}(0-2)^2 = 1.$$

We plot each point found above, connecting them by a smooth curve. (See figure). This gives us the parabola shown. Since the plotted points satisfy the given equation, the latter must be the equation of the parabola sketched.

In the accompanying figure, a bridge is built with its sup-
porting structure in the shape of a parabola. If the focus
is located 200 feet below the center of the bridge and 100
feet from either side, how thick must the supports be at
either side?

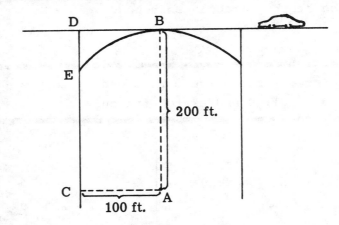

Figure 1

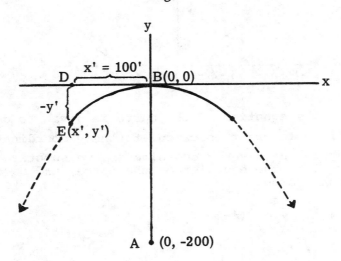

Figure 2

Solution: The problem can be reduced to that of Fig. 2. We
wish to find the support thickness y'. We know the general
equation of a parabola with the y-axis as the axis of symmetry, (0,p) as the focus, and the origin as the vertex is

(i) $x^2 = 4py$.

In this case, p = -200. Thus, the equation of the parabola is

(ii) $x^2 = -800y$.

To find the support thickness, y', remember that (x',y') is a point on the parabola. x' is the distance from the vertex from the left side; thus, x' = 100. Substituting this result in (ii), we obtain

(ii) $(100)^2 = -800y'$

(iii) $y' = \dfrac{100^2}{-800} = -12.5$.

Thus, at its thickest, the bridge support is 12.5 ft. below the top of the bridge.

● **PROBLEM** 756

Write the equation of the parabola whose focus has coordinates (0,2) and whose directrix has equation y = -2.(See figure)

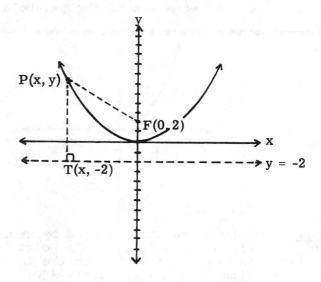

Solution: Since, by definition, each point lying on a parabola is equidistant from both the focus and directrix of the parabola, the origin must lie on the specific parabola described in the statement of the problem (see figure).

To find the equation of the parabola, choose a point P(x,y) lying on the parabola (see figure). By definition, then, the distance PT must equal the distance PF, where T lies on the directrix, directly below P. Since T also lies on y = -2, it has coordinates (x,-2). Using the distance formula, we find PF = PT

$$\sqrt{x^2 + (y-2)^2} = \sqrt{(y+2)^2}.$$

Squaring both sides of this equation, we obtain

$$x^2 + (y-2)^2 = (y+2)^2.$$

Expanding this, we get

$$x^2 + y^2 - 4y + 4 = y^2 + 4y + 4.$$

Subtracting y^2, $-4y$, and 4 from each side of this equation yields

$$x^2 = 8y.$$

Dividing both sides of this equation by 8 gives the equation of the parabola,

$$y = \frac{1}{8}x^2.$$

● **PROBLEM** 757

Prove that if $F(0,1)$ is the focus, and the line $y = -1$ is the directrix, then the equation of the parabola is $y = \frac{1}{4}x^2$.

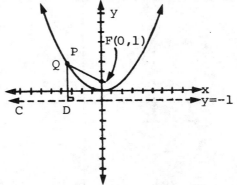

Solution: We may solve this problem by recalling the definition of a parabola. Let $P(x,y)$ be a point lying on the parabola. Then, the distance from P to the focus (PF) must equal the distance from P to the directrix (PD). (See figure). However, in order to calculate PD, we must know the coordinates of point D. Looking at the figure, we observe that D lies directly below P. Hence, it must have x-coordinate x. Furthermore, D lies on line $y = -1$, and, therefore, must have y-coordinate -1. The coordinates of D are the $D(x,-1)$.

Using the fact that PF = PD along with the distance formula, we find PF = PD

$$\sqrt{x^2 + (y-1)^2} = \sqrt{(y+1)^2}.$$

Squaring both sides of this equation, and expanding the quantities under the radicals yields

$$x^2 + y^2 - 2y + 1 = y^2 + 2y + 1.$$

Subtracting $y^2 - 2y + 1$ from each side of the last equation, we obtain

710

$$x^2 = 4y.$$

Dividing both sides by 4 gives us the equation of the parabola

$$y = \frac{1}{4}x^2.$$

● **PROBLEM** 758

Problem: Consider the equation of a parabola
$x^2 - 4x - 4y + 8 = 0$. Find the focus, vertex, axis of symmetry, and the directrix.

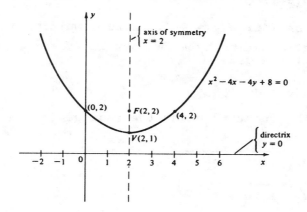

Solution: Completing the square in x gives
$(x^2 - 4x + 4) - 4y + 8 = 4$ or $(x - 2)^2 = 4(y - 1)$. from which we can read off the required information. The vertex is at (2,1), the minimum value for the equation. The axis of symmetry is, therefore, the line x = 2. The focus is at (2,2), and the directrix is y = 0 (see figure.) Upon translation of coordinates, the parabola can be written as $x'^2 = 4y'$, where x' = x - 2, y' = y - 1. Comparing with the standard equation $x^2 = 4py$, we see that p = 1.

● **PROBLEM** 759

Show that if the pair of numbers (x,y) satisfies $y = \frac{1}{4}x^2$, then the distance FP from F(0,1) to P(x,y) is equal to the distance PQ from P(x,y) to Q(x,-1).

Solution: By the distance formula, the distance FP is

$$FP = \sqrt{x^2 + (y-1)^2}. \tag{1}$$

The distance PQ is

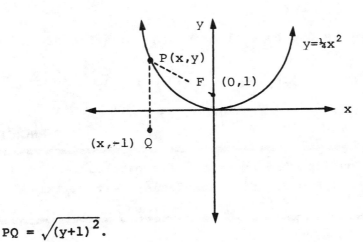

$$PQ = \sqrt{(y+1)^2}.\tag{2}$$

But $P(x,y)$ satisfies $y = \frac{1}{4}x^2$. Hence, multiplying both sides of this last equation by 4, we obtain $4y = x^2$. Using this in equation (1), we get

$$FP = \sqrt{4y + (y-1)^2}.\tag{3}$$

Expanding the quantity under the radical in equation (3), we find

$$FP = \sqrt{4y + y^2 - 2y + 1}$$

or $FP = \sqrt{y^2 + 2y + 1}$

Factoring the quantity under the radical in the last equation, we obtain

$$FP = \sqrt{(y+1)^2}.\tag{4}$$

Comparison of equations (2) and (4) shows that $FP = PQ$.

● **PROBLEM** 760

Show that if the pair of numbers (x,y) satisfies $y = \frac{1}{4d}x^2$, then the distance FP from $F(0,d)$ to $P(x,y)$ is equal to the distance PQ from $P(x,y)$ to $Q(x,-d)$.

Solution: By the distance formula, the distances FP and PQ are

$$FP = \sqrt{x^2 + (y-d)^2}\tag{1}$$

$$PQ = \sqrt{(y+d)^2}.\tag{2}$$

But $P(x,y)$ satisfies $y = \frac{1}{4d}x^2$. Multiplying both sides of this equation by 4d, we find $x^2 = 4dy$. Substituting this in equation (1) yields

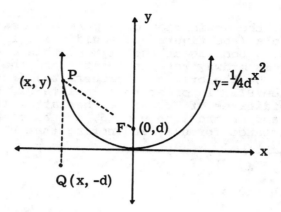

$$FP = \sqrt{4dy + (y-d)^2}.$$

Expanding the quantity under the radical gives us

$$FP = \sqrt{4dy + y^2 + d^2 - 2dy}$$

or $\quad FP = \sqrt{y^2 + d^2 + 2dy}$

Factoring the argument of the square root yields

$$FP = \sqrt{(y+d)^2}. \tag{3}$$

Comparing equations (2) and (3), we find that FP = PQ.

● **PROBLEM** 761

Prove that if $F(0,d)$ is the focus and the line $y = -d$ is the directrix, then the equation of the parabola is $y = \frac{1}{4d}x^2$. (See figure).

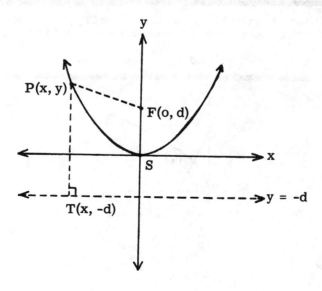

Solution: By the definition of a parabola, each point P(x,y) of the parabola (see figure) is equidistant from the directrix and the focus. Looking at the figure, this means specifically that point S is midway between point F and the line y = -d. Hence, S must be the origin. Furthermore, distance PF must equal distance PT. In order to calculate PT, however, we must know the coordinates of T. Since the latter lies directly below point P and on the line y = -d, it has coordinates (x,-d). Using the distance formula, and noting that PF = PT, we find

$$\sqrt{x^2 + (y-d)^2} = \sqrt{(y + d)^2}.$$

Squaring both sides

$$x^2 + (y-d)^2 = (y+d)^2.$$

Expanding

$$x^2 + y^2 + d^2 - 2dy = y^2 + 2dy + d^2.$$

Subtracting y^2, d^2, and 2dy from each side,

$$x^2 - 4dy = 0.$$

Adding 4dy to both sides of this equation,

$$x^2 = 4dy.$$

Dividing both sides by 4d, we obtain the equation of the parabola

$$y = \frac{1}{4d}x^2.$$

● **PROBLEM** 762

Find, both analytically and graphically, the points of intersection of the two curves whose equations are

$$2x + y - 4 = 0 \quad \text{and} \quad y^2 - 4x = 0.$$

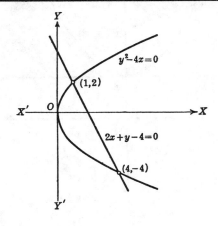

<u>Solution</u>: By using one equation to express y in terms of x, then substituting this into the second equation, we can form an equation in one variable, x. When we solve this equation, the values of x will be the abscissas of the points of intersection. The ordinates, y-values, can be found by substitution back into either original equation.

Since $2x + y - 4 = 0$, $\quad y = 4 - 2x$.

Hence, by substitution $\quad (4 - 2x)^2 - 4x = 0$

$$16 - 16x + 4x^2 - 4x = 0$$

which reduces to $\qquad x^2 - 5x + 4 = 0$.

Factoring, we obtain $\qquad (x - 4)(x - 1) = 0$.

This tells us $x = 4$ or $x = 1$.

Substituting these values into $2x + y - 4 = 0$ gives us

$$y = 4 - 2(4) \qquad \text{and} \qquad y = 4 - 2(1)$$

$$y = -4 \qquad\qquad\qquad y = 2$$

Hence, analytically we have found the points of intersection are $(1, 2)$ and $(4, -4)$.

Graphically, the points of intersection can be obtained by tracing the two given curves and reading the coordinates of their intersection points. The graphs are shown in the accompanying figure.

● **PROBLEM** 763

Consider the parabola $y^2 = 4px$. A tangent to the parabola at point P_1 (x_1, y_1) is defined as the line that intersects the parabola at point P_1 and nowhere else.

(a) Show that the slope of the tangent line is $\frac{2p}{y_1}$.

[Hint: Let the slope be m. Find the equation of the line passing through P_1 with slope m. What are the points of intersection of the tangent line and the parabola? For what values of m, would there be only one intersection point?]

(b) Find the equation of the tangent line.

(c) Prove that the intercepts of the tangent line are $(-x_1, 0)$ and $(0, \frac{1}{2}y_1)$.

<u>Solution</u>: (a) We know three things about the tangent line ℓ. First, ℓ is a line, therefore, its equation is of the form

715

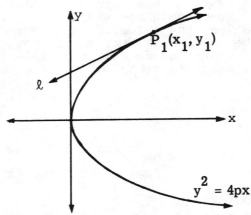

$$y^2 = 4px$$

$y = mx + b$. Second, P_1 is on ℓ. Therefore, (x_1, y_1) must satisfy the equation for ℓ. Therefore, $y_1 = mx_1 + b$. Third, ℓ is tangent to the parabola at point P_1. This implies two things: (1) P_1 is a point on the parabola. Thus, $y_1^2 = 4px_1$; and (2) P_1 is the only point common to both ℓ and the parabola.

To determine the line we must find m and b.

We first wish to find m. We know that ℓ and $y^2 = 4px$ intersect at only one point. Therefore, we solve for the intersection points of ℓ and $y^2 = 4px$ in terms of m. The value of m that permits only one point of intersection is the correct one.

Any point (X, Y) that is a point of intersection of the line and the parabola must satisfy the equations

(i) $Y = mX + b$

(ii) $Y^2 = 4pX$.

We have two equations in two unknowns X and Y. The first expresses Y in terms of X. Substituting this expression in (ii) does not change the validity of the second equation. In addition, it makes the second equation a single equation with a single unknown.

(iii) $(mX + b)^2 = 4pX$.

Multiplying out and simplifying, we obtain:

(iv) $m^2x^2 + (2mb - 4p)X + b^2 = 0$.

Then, by the quadratic formula, we can obtain a value for X in terms of m, p, and b.

(v) $X = \dfrac{-(2mb-4p) \pm \sqrt{(2mb-4p)^2 - 4m^2b^2}}{2m^2}$

Because ℓ is a tangent, there should exist only one possible intersection point and, therefore, only one value of X. For equation (V) to have a single value, $\sqrt{(2mb - 4p)^2 - 4m^2b^2}$ must equal 0. In other words, for ℓ to be a tangent line,

(vi)　　$(2mb-4p)^2 - 4m^2b^2 = 0$

(vii)　　$4m^2b^2 - 16mbp + 16p^2 - 4m^2b^2 = 0$

(viii)　　$16p^2 = 16mbp$　or　$p = bm$.

Thus, we know $p = mb$. p is known; m and b are what we are solving for. To find values for m and b, we find another relationship between m and b.

Note that point $P_1(x_1, y_1)$ is a point on ℓ. Therefore, $y_1 = mx_1 + b$.

With this second set of simultaneous equations, $p = mb$ and $y_1 = mx_1 + b$, we can solve for m. Since $p = mb$, $b = \frac{p}{m}$. Substituting this in $y_1 = mx_1 + b$, we obtain $y_1 = mx_1 + \frac{p}{m}$. Multiplying by m, we obtain the expression $m^2x_1 - my_1 + p = 0$. By the quadratic formula, we have $m = \dfrac{y_1 \pm \sqrt{y_1^2 - 4x_1 p}}{2x_1}$. Note that (x_1, y_1) are points on the parabola; and therefore, $y_1^2 - 4x_1p = 0$. The radical $\sqrt{y_1^2 - 4x_1p} = 0$, and thus $m = \dfrac{y_1}{2x_1}$.

We are asked to show, though, that $m = \frac{2p}{y_1}$. Although not immediately obvious, this is the same as our answer:

$$m = \frac{y_1}{2x_1} = \frac{y_1}{2x_1} \cdot \frac{\frac{y_1}{4x_1}}{\frac{y_1}{4x_1}} = \frac{y_1^2/4x_1}{y_1/2} \left(\text{since } y_1^2 = 4px_1, \ p = \frac{y_1^2}{4x_1} \right)$$

$$m = \frac{p}{y_1/2} = \frac{2p}{y_1}$$

Thus the slope m of the tangent of the parabola $y^2 = 4px$ at point $P_1(x_1, y_1)$ is $\frac{2p}{y_1}$.

(b)　The equation of the tangent line ℓ is $y = mx + b$. From part (a), we know $m = \frac{2p}{y_1}$. Furthermore, we know from part (a) that $p = mb$. Thus, $b = \frac{p}{m} = \frac{p}{(2p/y_1)} = \frac{y_1}{2}$, and the equation of

ℓ becomes $Y = \left(\dfrac{2p}{y_1}\right)x + \left(\dfrac{y_1}{2}\right)$. Another acceptable form of the equation

is obtained by letting $p = \dfrac{y_1^2}{4x_1}$: $Y = \left[\dfrac{y_1}{2x_1}\right]x + \dfrac{y_1}{2}$.

(c) Given the equation in part (b), we know that the y-intercept is given by b. Since $b = \dfrac{y_1}{2}$, the y-intercept is $\left(0, \dfrac{y_1}{2}\right)$.

$\left[\text{An alternate method is to realize that the y-intercept is obtained by setting } x = 0. \text{ Thus, } Y = \left(\dfrac{y_1}{2x_1}\right) \cdot 0 + \left(\dfrac{y_1}{2}\right) = \dfrac{y_1}{2}\right]$.

To find the x-intercept, we set y = 0. Thus,

$$0 = \left(\dfrac{y_1}{2x_1}\right)x + \left(\dfrac{y_1}{2}\right), \quad \text{or} \quad \left(\dfrac{y_1}{2x_1}\right)x = -\dfrac{y_1}{2}.$$

Dividing by $\dfrac{y_1}{2}$ and multiplying by x_1, we obtain $X = -x_1$. Thus, $(-x_1, 0)$ is the x-intercept.

HYPERBOLAS

• **PROBLEM** 764

Draw the graph of the curve whose equation is xy = 4.

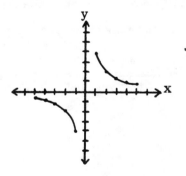

<u>Solution</u>: The easiest way to do this is to set up a chart. We shall plot points with x coordinates lying between x = - 5 and x = 5. Substituting these values of x into the e-quation xy = 4, we find the corresponding values of y. Actually, it is easier if we solve the given equation for y in terms of x.

Dividing both sides of xy = 4 by x, we obtain the required equation

$$y = \dfrac{4}{x} .$$

The chart of coordinates follows:

x	y = 4/x
5	4/5
4	1
3	4/3
2	2
1	4
- 1	- 4
- 2	- 2
- 3	- 4/3
- 4	- 1
- 5	- 4/5

The figure shows that the curve, whose equation is $xy = 4$, is an hyperbola.

● **PROBLEM** 765

Graph the hyperbola $y^2 - x^2 = 4$. What are the equations of the asymptotes? Draw the asymptotes.

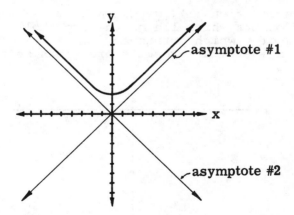

<u>Solution</u>: We will first find the equations of the asymptotes of the given hyperbola, as these can be used in graphing the required figure. In order to do this, we must find the limiting form of $y^2 - x^2 = 4$ as the hyperbola approaches its asymptotic form.

In order to see how the hyperbola approaches its asymptotic form, we notice that as we trace a hyperbola to the right (or to the left), it becomes arbitrarily close to a set of intersecting straight lines. These straight lines are the asymptotes of the hyperbola. We approach the asymptotes as the x coordinate of a point on the hyperbola gets very large.

Using the equation $y^2 - x^2 = 4$, we can express this analytically. First, for convenience, we solve for y in terms of x. Adding x^2 to both sides of the above equation, we find

$$y^2 = x^2 + 4$$

Taking the square root of both sides

$$y = \sqrt{x^2 + 4} \tag{1}$$

As was said above, as the hyperbola approaches its limiting form, x gets very large. In this limit, 4 is negligible in comparison to x^2. Using this fact in equation (1), we obtain

$$y = \sqrt{x^2} \quad \text{or} \quad y = \pm x. \tag{2}$$

This equation is obeyed by a point on the hyperbola as the latter appraoches its limiting form. Hence, (2) gives the equations of the asymptotes of $y^2 - x^2 = 4$.

Since the asymptotes are straight lines, and 2 points determine a straight line, we need to find 2 points satisfying y = x to draw the first asymptote, and 2 points satisfying y = - x to draw the second asymptote. Two points satisfying y = x are (0, 0) and (1, 1). Two points satisfying y = - x are (0, 0) and (1, - 1). The graphs of these equations are shown in the figure.

Using equation (1), we can now actually plot the values of points which lie on the hyperbola. We calculate the values of y which correspond to values of x between x = 4 and x = - 4. The results are shown below, in chart form:

x	$y = \sqrt{x^2 + 4}$
4	$\sqrt{20}$
3	$\sqrt{13}$
2	$\sqrt{8}$
1	$\sqrt{5}$
0	2
- 1	$\sqrt{5}$
- 2	$\sqrt{8}$
- 3	$\sqrt{13}$
- 4	$\sqrt{20}$

These points are plotted in the figure. By definition of the asymptotes, we may extend the branches of the hyperbola as shown in the figure.

Determine the intercepts, find the asymptotes, and locate the foci of the following hyperbolas:

(a) $x^2 - \dfrac{y^2}{4} = 1.$ (b) $\dfrac{y^2}{16} - \dfrac{x^2}{4} = 1.$

<u>Solution</u>: Before jumping into complicated calculation, let's be sure we know what we are looking for. The intercepts are the points at which the curve crosses the x or y axis. Tracing a point on the curve as the curve crosses the x-axis, we note that the y-coordinate of the curve is 0. Hence, in order to determine the "x intercept" of a curve, set $y = 0$ in the equation of the curve and solve for x. The x-intercept is then $(x, 0)$. Similarly, in order to find the "y intercept" of a curve, set $x = 0$ in the equation of the curve, and solve for y. The y intercept is then $(0, y)$.

If the equation of the hyperbola is $\dfrac{x^2}{a^2} - \dfrac{y^2}{b^2} = 1$, the asymptotes are given by $y = \pm \dfrac{b}{a} x$.

If the foci are labelled $F_1(-c, 0)$, and $F_2(c, 0)$ the value of c is given by $c^2 = a^2 + b^2$.

If the equation of the hyperbola is: $\dfrac{y^2}{b^2} - \dfrac{x^2}{a^2} = 1$, the asymptotes are also $y = \pm \dfrac{b}{a} x$.

If the foci are labelled $F_1 (0, -c)$ and $F_2(0, c)$, c is given by $c^2 = a^2 + b^2$.

(a) $x^2 - \dfrac{y^2}{4} = 1$ (1)

To find the x-intercepts, set $y = 0$ in (1), and solve for x. This yields

$$x^2 + 0 = 1.$$

Hence, $x = \pm 1$ and the x-intercepts are $(\pm 1, 0)$. The y-intercepts are found by setting $x = 0$ in (1). This gives us

$$-\frac{y^2}{4} = 1$$

Multiplying both sides by -4

$$y^2 = -4.$$

Since the square root of a negative number is imaginary,

there are no y-intercepts.

Equation (1) is of the form $\frac{x^2}{a^2} - \frac{y^2}{b^2} = 1$, with asymptotes given by $y = \pm \frac{b}{a}$ x. In this case, $a^2 = 1$ and $b^2 = 4$. Hence, $a = \pm 1$, $b = \pm 2$. The asymptotes are then $y = \pm 2x$.

Finally, the foci are $F_1(-c,0)$ and $F_2(c,0)$ with $c^2 = a^2 + b^2$. In our case, $c^2 = 4 + 1 = 5$, and $c = \pm\sqrt{5}$. Hence, the foci are $(-\sqrt{5}, 0)$ and $(\sqrt{5}, 0)$.

(b) $\frac{y^2}{16} - \frac{x^2}{4} = 1$ (2)

To find the y-intercepts, set x = 0. This yields

$$\frac{y^2}{16} = 1$$

Multiplying both sides by 16

$y^2 = 16$ or $y = \pm 4$.

The y-intercepts are therefore (0, ± 4). To find the x-intercepts, set y = 0 in (2) This gives

$$- \frac{x^2}{4} = 1.$$

Multiplying both sides by - 4, we obtain

$$x^2 = - 4.$$

Since the square root of a negative number is imginary, this hyperbola has no x-intercepts.

The given equation is of the form $\frac{y^2}{b^2} - \frac{x^2}{a^2} = 1$, with asymptotes $y = \pm \frac{b}{a}$ x. In our specific case, $a^2 = 4$, $b^2 = 16$. Hence, $a = \pm 2$, $b = \pm 4$, and $y = \pm 2x$ are the **equations** of the asymptotes.

The foci are $F_1(0, - c)$ and $F_2(0, c)$, with $c^2 = a^2 + b^2$. For our problem, $c^2 = 20$ or $c = \pm \sqrt{20}$. Hence, the foci are $(0, - \sqrt{20})$ and $(0, \sqrt{20})$.

● **PROBLEM** 767

Show informally that $y = \pm(b/a)x$ are the equations of the asymptotes of the hyperbola whose equation is

$$\frac{x^2}{a^2} - \frac{y^2}{b^2} = 1.$$

<u>Solution</u>: As we trace an hyperbola far to the right (or to the left), we notice that the hyperbola become arbitrarily

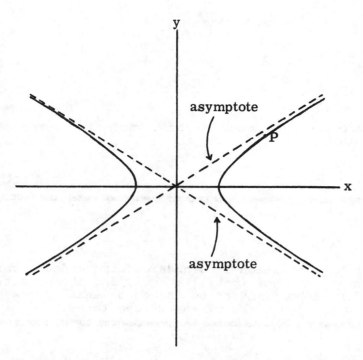

close to a set of intersecting straight lines (see figure).
These lines are the asymptotes of the hyperbola.

Note from the figure that as P moves out along the
hyperbola, it approaches the asymptotes. We may use this
fact, along with the equation of the hyperbola, to find
the equations of the asymptotes. First, we solve

$\frac{x^2}{a^2} - \frac{y^2}{b^2} = 1$ for y in terms of x. Adding $- \frac{x^2}{a^2}$

to both sides of this equation, we obtain

$- \frac{y^2}{b^2} = - \frac{x^2}{a^2} + 1$

Multiplying both sides by $- b^2$ yields:

$$y^2 = \frac{b^2}{a^2} x^2 - b^2 \qquad\qquad (1)$$

Factoring b^2 from the right side of this equation, and
taking the square root of both sides, we obtain

$$y = \pm b \sqrt{\frac{x^2}{a^2} - 1} \qquad\qquad (2)$$

The (-) sign appears in (2) because this is also a
solution of equation (1). Now, as P moves out along the
hyperbola, its x coordinate gets very large. Since a is a
fixed constant; sooner or later x > a. This means that
$x^2 \gg a^2$; or $\frac{x^2}{a^2} \gg 1$.

Hence, we may neglect 1 in comparison with $\frac{x^2}{a^2}$ in equation (2). We then find

$$y = \pm b \sqrt{\frac{x^2}{a^2}}$$

$$y = \pm \frac{b}{a} x$$

These are the equations of the asymptotes of the given hyperbola.

● **PROBLEM** 768

Consider the equation

$$x^2 - 4y^2 + 4x + 8y + 4 = 0.$$

Express this equation in standard form, and determine the center, the vertices, the foci, and the eccentricity of this hyperbola. Describe the fundamental rectangle and find the equations of the 2 asymptotes.

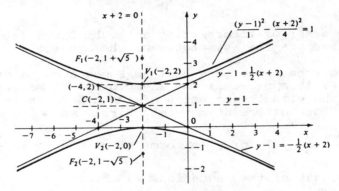

<u>Solution</u>: Rewrite the equation by completing the squares, i.e.,

$$(x^2 + 4x + 4) - 4(y^2 - 2y + 1) = -4$$

or $$(x + 2)^2 - 4(y - 1)^2 = -4$$

or, dividing, rearranging terms,

$$\frac{(y - 1)^2}{1} - \frac{(x + 2)^2}{4} = 1.$$

The center, located at (h, k) in the equation

$$\frac{(y - k)^2}{a^2} - \frac{(x - h)^2}{b^2} = 1 \quad \text{is, therefore, at}$$

(- 2, 1). Furthermore, a = 1, b = 2. Thus,

$$c = \sqrt{1^2 + 2^2} = \sqrt{5}, \text{ and } e = \sqrt{5}.$$

The vertices are displaced ± a from the center

while the foci are displaced ± c (along the transverse axis). Therefore, the vertices are (- 2, 1 ± 1) and the foci are (- 2, 1 ± $\sqrt{5}$).

By definition, the fundamental rectangle is the rectangle whose vertices are at (h ± b, k ± a). Hence, in this example, the coordinates of the vertices of the rectangle are (0, 2), (- 4, 2), (- 4, 0), and (0, 0). The equations of the two asymptotes are determined by finding the slopes of the lines passing through the center of the hyperbola and two of the vertices of its fundamental rectangle (see figure). Then,

$$m = \frac{\Delta y}{\Delta x} = \pm \tfrac{1}{2}$$

gives the two slopes and the point-slope form, choosing the point (- 2, 1) which is common to both asymptotes, gives

$$y - 1 = \pm \tfrac{1}{2}(x + 2).$$

● **PROBLEM** 769

Find the equation of the hyperbola with vertices V_1 (8, 0), V_2 (2, 0) and eccentricity e = 2.

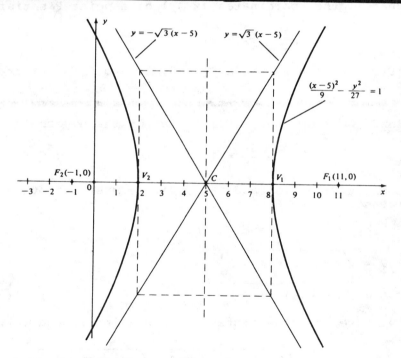

Solution: There are two basic forms of the equation of an hyperbola that is not rotated with respect to the coordinate axes:

$$\frac{(x - h)^2}{a^2} - \frac{(y - k)^2}{b^2} = 1 \quad \text{and} \quad \frac{(y - k)^2}{a^2} - \frac{(x - h)^2}{b^2} = 1.$$

Which form is appropriate depends upon whether the transverse axis is parallel to the x-axis or to the y-axis, respectively. To determine the equation of an hyperbola, then, it is necessary to first discover which equation applies, then to solve for the constants h, k, a, c. The information about the vertices implies that the transverse axis is the x- axis.

Thus, the first form of the equation for an hyperbola applies. In this case, the center, which is the average of the vertices is at (5, 0). The distance between the vertices is 2a = 6; therefore a = 3. In order to determine the value of b, we use the relation between eccentricity, e, c, and a: e = c/a. Thus, c = e · a = 2 · 3 = 6.

But b = $\sqrt{c^2 - a^2}$ = $\sqrt{6^2 - 3^2}$ = $\sqrt{27}$ ≈ 5.2
Substituting for h, k, a, b, we have

$$\frac{(x - 5)^2}{9} - \frac{y^2}{27} = 1,$$

or $3x^2 - y^2 - 30x + 48 = 0.$

● **PROBLEM** 770

Show that if the coordinates (x, y) of a point P satisfy

$$\frac{x^2}{9} - \frac{y^2}{16} = 1,$$

then $\left|F_1P - F_2P\right|$ = 6, where $F_1(-5, 0)$ and $F_2(5, 0)$ are the foci.

Solution: We know that P(x, y) lies on the hyperbola

$$\frac{x^2}{9} - \frac{y^2}{16} = 1,$$

with foci at $F_1(-5,0)$ and $F_2(5, 0)$. We must show that the distance $\left|F_1P - F_2P\right|$ = 6.

We may write F_1P and F_2P as

$$F_1P = \sqrt{(x + 5)^2 + y^2} \tag{1}$$

$$F_2P = \sqrt{(x - 5)^2 + y^2} \tag{2}$$

where we have used the distance formula in (1) and (2). Hence,

$$\left|F_1P - F_2P\right| = \sqrt{(x + 5)^2 + y^2} - \sqrt{(x - 5)^2 + y^2} \tag{3}$$

We cannot show that $\left|F_1P - F_2P\right|$ = 6 unless we can relate x and y to some known value. We can do this because we know that the coordinates x and y satisfy

$$\frac{x^2}{9} - \frac{y^2}{16} = 1.$$

We must solve this equation for y^2 (or x^2), and substitute the result into equation (3). We start with

$$\frac{x^2}{9} - \frac{y^2}{16} = 1$$

Multiplying both sides by -16,

$$y^2 - \frac{16}{9} x^2 = -16$$

Adding $\frac{16}{9} x^2$ to both sides,

$$y^2 = \frac{16}{9} x^2 - 16 \qquad (4)$$

Substituting (4) in (3)

$$|F_1P - F_2P| = \sqrt{(x+5)^2 + \frac{16}{9} x^2 - 16} - \sqrt{(x-5)^2 + \frac{16}{9}x^2 - 16}$$

Expanding under the radicals in this last equation,

$$|F_1P - F_2P| = \sqrt{x^2 + 10x + 25 + \frac{16}{9} x^2 - 16} - \sqrt{x^2 - 10x + 25 + \frac{16}{9}x^2 - 16}$$

Simplifying,

$$|F_1P - F_2P| = \sqrt{\frac{25}{9} x^2 + 10x + 9} - \sqrt{\frac{25}{9} x^2 - 10x + 9}$$

Factoring $\frac{25}{9}$ from each term under each radical,

$$|F_1P - F_2P| = \sqrt{\frac{25}{9}\left(x^2 + \frac{90}{25} x + \frac{81}{25}\right)} - \sqrt{\frac{25}{9}\left(x^2 - \frac{90}{25} x + \frac{81}{25}\right)}$$

Now, $\left(x^2 + \frac{90}{25}x + \frac{81}{25}\right) = \left(x + \frac{9}{5}\right)^2$

$\left(x^2 - \frac{90}{25}x + \frac{81}{25}\right) = \left(x - \frac{9}{5}\right)^2$

Using the last 2 expressions in the equation for $|F_1P - F_2P|$ yields

$$|F_1P - F_2P| = \sqrt{\frac{25}{9}\left(x + \frac{9}{5}\right)^2} - \sqrt{\frac{25}{9}\left(x - \frac{9}{5}\right)^2}$$

Taking the indicated square roots, we obtain

$$|F_1P - F_2P| = \frac{5}{3}\left(x + \frac{9}{5}\right) - \frac{5}{3}\left(x - \frac{9}{5}\right)$$

$$|F_1P - F_2P| = \frac{90}{15} = 6$$

Notice that the term in x disappears. This is very convenient; if our final expression were to involve x, we would have to leave it in that form. There is no other equation given in the problem which relates x with known quantities.

By definition, if an hyperbola **has** foci $F_1(-c, 0)$ and $F_2(c, 0)$, and $P(x, y)$ is a point on the hyperbola, then $|PF_1 - PF_2| = k$, where k is a constant such that $k < F_1F_2 = 2c$.

Assuming that the above holds, and defining a constant such that $a = K/2$. and a constant b such that $b^2 = c^2 - a^2$, prove that the equation of the hyperbola is

$$\frac{x^2}{a^2} - \frac{y^2}{b^2} = 1.$$

<u>Solution</u>: From the given facts,

$$|PF_1 - PF_2| = k = 2a. \tag{1}$$

By the distance formula,

$$PF_1 = \sqrt{(x + c)^2 + y^2} \tag{2}$$

$$PF_2 = \sqrt{(x - c)^2 + y^2} \tag{3}$$

Using (3) and (2) in (1), we obtain

$$\left|\sqrt{(x + c)^2 + y^2} - \sqrt{(x - c)^2 + y^2}\right| = 2a \tag{4}$$

Since any real number (positive or negative) times itself must be positive, we may write, using equation (4),

$$\left(\left|\sqrt{(x+c)^2+y^2} - \sqrt{(x-c)^2+y^2}\right|\right)^2 = \left(\sqrt{(x+c)^2+y^2} - \sqrt{(x-c)^2+y^2}\right)^2 = 4a^2.$$

Expanding the last branch of the last equation,

$$(x+c)^2+y^2+(x-c)^2+y^2-2\sqrt{((x+c)^2+y^2)((x-c)^2+y^2)} = 4a^2.$$

Expanding the last equation again, and regrouping terms

$$2x^2 + 2y^2+2c^2-2\sqrt{((x+c)^2+y^2)((x-c)^2+y^2)} = 4a^2.$$

Subtracting $-2\sqrt{((x+c)^2+y^2)((x-c)^2+y^2)}$ and $4a^2$ from both sides of the last equation yields

$$2x^2+2y^2+2c^2-4a^2 = 2\sqrt{((x+c)^2+y^2)((x-c)^2+y^2)}$$

Dividing both sides by 2

$$x^2 + y^2 + c^2 - 2a^2 \quad = \sqrt{((x + c)^2 + y^2)((x - c)^2 + y^2)}$$

Squaring both sides

$$(x^2 + y^2 + c^2 - 2a^2)^2 = \left((x + c)^2 + y^2\right)\left((x - c)^2 + y^2\right)$$

Expanding the right side

$$(x^2 + y^2 + c^2 - 2a^2)^2 = (x^2 + y^2 + c^2 + 2cx)(x^2 + y^2 + c^2 - 2cx)$$

Expanding both sides using the distributive law,

$$(x^2 + y^2 + c^2)^2 + 4a^4 - 4a^2(x^2 + y^2 + c^2) = (x^2 + y^2 + c^2)^2 - 4c^2x^2$$

Cancelling like terms on both sides and expanding the remaining terms on the left side of the last equation, we find

$$4a^4 - 4a^2x^2 - 4a^2y^2 - 4a^2c^2 = -4c^2x^2 \tag{5}$$

But, we know that

$$c^2 - a^2 = b^2$$

Adding a^2 to both sides of this equation,

$$c^2 = a^2 + b^2 \tag{6}$$

Using equation (6) in equation (5)

$$4a^4 - 4a^2x^2 - 4a^2y^2 - 4a^4 - 4a^2b^2 = -4a^2x^2 - 4b^2x^2$$

Subtracting $-4a^2x^2$ and $-4b^2x^2$ from both sides of the last equation

$$-4a^2y^2 - 4a^2b^2 + 4b^2x^2 = 0$$

Dividing through by $4a^2b^2$ gives

$$-\frac{y^2}{b^2} - 1 + \frac{x^2}{a^2} = 0$$

Adding 1 to both sides of the last equation yields the desired result

$$\frac{x^2}{a^2} - \frac{y^2}{b^2} = 1.$$

SOLID GEOMETRY

CHAPTER 44

LOCUS

Basic Attacks and Strategies for Solving Problems
in this Chapter. See pages 732 to 751 for step-by-
step solutions to problems.

As indicated in the introduction to Chapter 39, **locus** refers to a place or location. In Chapter 39, the locus problems concerned two-dimensional figures; the problems in this chapter refer to three-dimensional figures.

Several problems in this chapter refer to spheres. A **sphere** is the locus of all points in space and equally distant from a fixed point. The fixed point is called the **center** of the sphere. A segment having one endpoint on the sphere and having, as the other endpoint the center of the sphere is called a **radius** of the sphere. A segment which contains the center of the sphere and whose endpoints are on the sphere is called a **diameter**. Just as with circles, for a given sphere, all radii are congruent and all diameters are congruent. The symbol r is used to denote either a radius, or the length of a radius. Context will indicate whether a segment or the length of a segment is intended. A **section** of a sphere is the intersection of the sphere and a plane. A helpful way to visualize this is to imagine "sawing through" a spherical object. In Problem 777, it is established that the intersection of a plane and a sphere is a circle. When the intersection of the plane and the sphere is a circle such that the radius of the circle is identical to the radius of the sphere, the circle is said to be a **great circle**. In Problem 778, it is established that, if the plane intersecting the sphere contains the center of the sphere, the resulting circle is a great circle.

In the case of three-dimensional geometry, sometimes it is possible to "translate" a problem into a two-dimensional mode and to apply two-dimensional geometry theorems. For example, Problems 780 and 781 both involve spheres, but in both instances the Pythagorean Theorem is used. Also, many two-dimensional geometry theorems have three-dimensional geometry counterparts, and often the proofs of such pairs of theorems are almost identical. For example, there is a theorem which states that chords which are equally distant from the center of a circle are congruent, and Problem 782 is a three-dimensional version of that theorem. There is a theorem which states that the locus of all points equidistant from the sides of an angle is the bisector of that angle, and Problem 787 states that the locus of all points equidistant from the faces of a dihedral angle is the plane bisecting the dihedral angle. Thus, as an angle is formed in two dimensions by intersecting lines, a dihedral angle is formed in three dimensions by intersecting planes.

Step-by-Step Solutions to Problems in this Chapter, "Locus"

● **PROBLEM** 772

Graph {(x, y) : y ≥ |x|} where x and y are members of the set {- 3, - 2, - 1, 0, 1, 2, 3} .

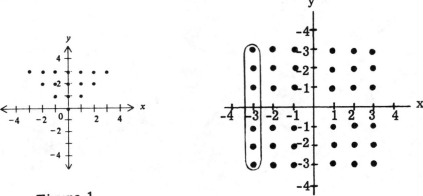

Figure 1

Figure 2

Solution: In this example, x and y are limited to a finite set of values. Therefore, there are a finite number of points. Thus for this graph, we need only to list all ordered pairs in the set, and plot each point, to complete the graph.

Each point on the graph must satisfy the equation

$$y \geq |x|.$$

To list all the points, we first choose a possible value of x, - 3 for example, and find all the possible values of y such that y ≥ |- 3|, or y ≥ 3. y is limited to the set {- 3, - 2, - 1, 0, 1, 2, 3}. Therefore, the only possible value of y is 3. Point (- 3, 3) is, consequently, a point on the graph. Next, choose another value of x, - 2 for example. For y = 3, it is true that y ≥ |- 2|. Also for y = 2, it is true that y ≥ |- 2|. Therefore, points (- 2, 2) and (- 2, 3) are on the graph. We continue

choosing possible values of x and finding the corresponding
y's such that y $\geq$ |x|, until all x's have been chosen. When
all x's have been tested, then our list of points is complete. The final list of points is (- 3, 3), (- 2, 3),
(- 2, 2), (- 1, 3), (- 1, 2), (- 1, 1), (0, 3), (0, 2),
(0, 1), (0, 0), (1, 3), (1, 2), (1, 1), (2, 3), (2, 2),
(3, 3). The graph is shown in Fig. 1.

WHY THIS METHOD WORKS: If x and y must be members of
the set {- 3, - 2, - 1, 0, 1, 2, 3},then the graph of all
possible points in the set would be as in Fig. 2. By
selecting x = - 3, we were actually testing all the
circled points in Fig. 2, and listing only those that
satisfy the relation y $\geq$ |x|. By repeating the procedure
for every possible x, we see that all possible points are
tested.

● **PROBLEM** 773

Draw the graph of XY if X = {1, 2} and Y = {3, 4}.

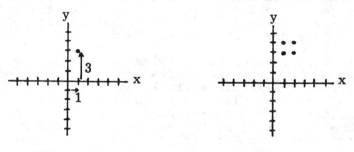

Figure 1   Figure 2

Solution: The Cartesian product of X and Y is the set of
all ordered pairs (x, y) where x belongs to X and y belongs
to Y. To list all possible ordered pairs, select an element
from X, for example 1. Now list all possible combinations
of 1 with the elements of Y: (1, 3), (1, 4). Take another
element from X, 2, and list all possible combinations with
Y: (2, 3), (2, 4). Since there are no more elements in X,
our list is complete: (1, 3), (1, 4), (2, 3), (2, 4). A
quick way to ensure that we have not omitted any elements
is to multiply the number of elements in X, 2, with the
number of elements in Y, 2. 2 × 2 = 4. Since we have four
elements already listed, we are assured that the list is
complete.

To graph the Cartesian product, we plot each of the
points individually. Given the point (x, y), we **count** x
spaces to the right on the x axis if the number is positive
(if x is negative, count |x| spaces to the left) and then
count y spaces up if the number is positive (if y is
negative, count |y| spaces down) and draw a small dot
to indicate the ordered pair. To graph (1, 3), for example, from the origin (0, 0) move right one space and
then directly up three spaces. The result is shown in
Fig. 1.

The completed graph is shown in Fig. 2.

What is the set theory definition of:

    a)   a circle with center P in a plane M, which has radius length r?

    b)   a sphere with center P and radius length r?

    c)   the interior of a sphere with center P and radius length r?

    d)   the exterior of a sphere with center P and radius length r?

Solution: To give a set theory definition of any locus of points, we must enumerate all the defining characteristics of the points which comprise the locus in question.

a) Here, the locus is a circle of radius r, with center at P, lying in plane M. Since the circle lies in M, all the points, X, of the circle lie in M. (i.e. $X \in M$). Furthermore, all the points X are a fixed distance, r, from P (XP = r). Hence, denoting the given locus by C, we may write $C = \{X \mid XP = r \text{ and } X \in M\}$. (C is the set of all points X such that X is a distance r from P, and X lies in M.).

b) In this part of the problem, the given locus is a sphere with center at P and radius length r. The defining characteristic of all points, X, lying on the sphere, is that they are a fixed distance r from the center (P). That is, XP = r, for all X. Hence, calling the given locus S, $S = \{X \mid XP = r\}$ (i.e. S is the set of all points X such that the distance from X to P is r).

c) The given locus is the interior of a sphere with center at P and radius length r. All points, X, comprising this locus have the distinguishing characteristic that their distance from the center of the sphere, XP, is less than r. That is, XP < r. (Note that XP ≤ r gives the interior plus the surface of the sphere.). Calling the given set I, we obtain $I = \{X \mid XP < r\}$ (I is the set of points X such that the distance of X from P is less than r).

d) Here, the locus is the exterior of a sphere with center at P and radius length r. The points, X, of this locus are characterized by the fact that their distance from P is greater than r (XP > r). Letting the given locus be E, we may write $E = \{X \mid XP > r\}$ (E is the set of points X such that the distance of X from the center of the sphere is greater than r).

Describe the locus which is the set of points in a plane lying a distance of 3 units from a line in the plane. What is the locus if we remove the condition that the given points and line lie on the same plane?

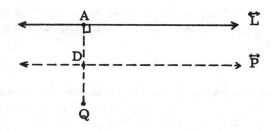

Solution: All points 3 units from the line on one side and all points 3 units from the line on the other side satisfy the given condition. The locus is thus a pair of lines lying on the opposite sides of the given line, each parallel to the given line.

To prove that this is indeed the locus, we must prove:

    (1)   Any point of the locus satisfies certain given conditions;

    (2)   Any point satisfying the given conditions is in the locus.

Since parallel lines are everywhere equidistant, we are assured that all the points are 3 units from the line. To show that all points fulfilling the given conditions are included in the parallel lines, consider a point Q, 3 units away from the given line $\overleftrightarrow{L}$, that does not lie on the parallel line $\overleftrightarrow{P}$. (See figure)

Point A is chosen such that $\overline{QA} \perp \overleftrightarrow{L}$.
Point D is the intersection of $\overline{QA}$ and $\overleftrightarrow{P}$.

Therefore, QA and DA are the distances of points Q and A from line $\overleftrightarrow{L}$.

QA = 3 from our assumptions.
DA = 3 because all points on $\overleftrightarrow{P}$ are 3 units from $\overleftrightarrow{L}$.

But if D is between points Q and A, then QA = DA = 3 can only be true if point Q is point D.

However, this contradicts our assumption that point Q does not lie on parallel line $\overleftrightarrow{P}$.

Thus, there is no point 3 units from line $\overleftrightarrow{L}$ that does not lie on the two parallel lines and the two parallel lines are an exact description of the locus.

If the given points and line are not required to lie on the same plane, then the locus is a cylindrical surface around the line. That all points on the cylindrical surface are three units from the line follows from the definition of cylindrical surface.

A cylindrical surface is the set of all lines parallel to a given line (at a fixed distance from that line) lying in the same curved surface that does not contain the given line. To show that the cylindrical surface includes all points satisfying the condition, we consider a point not on the surface, yet satisfying the condition. By considering the plane determined by line L and this omitted point Ω, we can reduce the three dimensional case to the two dimensional case shown before.

Describe the locus of points equidistant from two concentric spheres, the length of whose radii are 2 and 6.

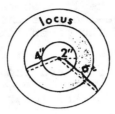

Solution: The locus of points that satisfy the given conditions must lie between the surfaces of the two spheres.

Since the original two spheres are concentric, a radius of the larger sphere will be concurrent with a radius of the smaller sphere. The distance between the two surfaces is the length of the larger radius minus the length of the smaller radius, i.e. 6 - 2 = 4. One point in the locus must be at the midpoint of this distance, or 2 units from the inner spherical surface (or 4 units from the center). Repeating this process for every radius will result in our plotting a sphere of radius 4. Therefore, the locus described is a sphere of radius 4, with the same center as the given spheres.

Show if a plane intersects a sphere, the intersection is a circle.

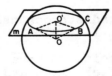

Solution: The condition given in this problem states that the points described lie on the intersection of a sphere and a plane. To show the locus of points is a circle, as suggested by the problem, we must show

(1) Every point in the intersection of the plane and the sphere is a point on a circle;

(2) Every point on the circle determined by at least three points of the intersection belongs to the intersection.

To show the first part, consider the accompanying figure. Plane m intersects sphere O in section ABC. From the center of O draw $\overline{OO'} \perp m$. Consider $\triangle AO'O$ and $\triangle BO'O$. Since $\overline{OO'} \perp m$, $\sphericalangle AO'O$ and $\sphericalangle BO'O$ are right angles. $\overline{AO}$ and $\overline{BO}$ are radii of the sphere and thus $\overline{AO} \cong \overline{BO}$. By reflexivity, $\overline{OO'} \cong \overline{OO'}$. By the Hypotenuse-Leg Theorem, $\triangle AO'O \cong \triangle BO'O$. By corresponding parts $\overline{AO'} \cong \overline{O'B}$. Since A and B were any points on the intersection, this means point O' is equidistant from all points of the intersection. Thus, every point of the intersection is contained in the circle with center O' and radius $\overline{AO'}$.

In the first part, we showed every point of the intersection lies on the circle with center O' and radius $\overline{AO'}$. In the second part, we show all points on the circle are also points of the intersection. Note that the circle is coplanar with m. Thus, every point Q on the circle is on m. Since $\odot O'$ is a circle, $\overline{QO'} = \overline{AO'}$. Furthermore, since $\overline{OO'} \perp m$, and $\overline{QO'}$ is on plane m, $\overline{QO'} \perp \overline{OO'}$. By the Pythagorean Theorem, in rt. $\triangle QOO'$, $QO^2 = QO'^2 + OO'^2$. Since $QO'^2 = AO'^2$, $QO^2 = AO'^2 + OO'^2$. But in rt. $\triangle AO'O$, $AO^2 = AO'^2 + OO'^2$. Thus, $QO^2 = AO^2$ and points Q and A are equidistant from O. By definition of the sphere, since A is a point on the sphere, point Q is also on the sphere. Thus, every point Q is on the intersection of the plane and the sphere.

Therefore, if a plane intersects a sphere, the intersection is a circle.

● **PROBLEM** 778

In the figure, if plane M passes through the center, P, of the sphere S, prove that the intersection set of S and M is a great circle of S.

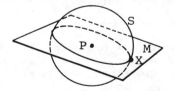

(c)

Solution: If S is a sphere of radius r, then any circle lying on S, with radius r, is a great circle of S. We will first show that the intersection set of S and M lies on sphere S. Then we show that the intersection set is a circle of radius r, and therefore is a great circle see figure.)

Since all the points of S lie on the sphere of radius r, the intersection set of S and M lies on the sphere. Hence, these points must also be members of S.

To show that the set of points is a great circle of P, note that (1) because the set of points are members of M, the set is collinear; (2) because the set of points are members of S, for every point X of the set of points, XP = radius of sphere; (3) P is on plane M.

For a set of coplanar points, if there exists a point P

coplanar with the set, such that the point is equidistant from every point of the set, then the set is a circle of center P and radius equal to the distance between each point and P. Therefore, the intersection set is a circle of radius XP = radius of sphere.

The intersection set is thus a circle on sphere S with radius equal to the radius of the sphere - in short, a great circle of S.

● PROBLEM 779

Planes R and S, shown in the figure, are parallel planes, a distance of 8 in. apart. Point P is 2 in. from plane R and 6 in. from plane S.

   (a)   Describe the locus of points equidistant from plane R and plane S.

   (b)   Describe the locus of points 3 in. from point P.

   (c)   Describe the locus of points which are equidistant from planes R and S and 3 in. from point P.

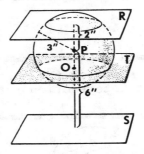

Solution: (a)   Since planes R and S are parallel and, therefore, everywhere equidistant, all points equidistant from both planes must lie midway between R and S at all places. There will definitely be three noncollinear points exactly midway between R and S. Therefore, the locus of points described is a plane 4 in. from both planes R and S and, since the locus is everywhere equidistant from R and S, it is parallel to the given planes.

   (b)   By definition, the locus of points at a constant distance away from a fixed point is a sphere (spherical surface). In this problem, the sphere is centered at point P and has a radius of length 3 in.

   (c)   As is standard in locus problems, a locus of points containing both the characteristics of being equidistant from planes R and S, and 3 in. from point P will be the intersection of each of the loci as described separately in (a) and (b).

   Since plane T is 4 in. from plane R, and the outermost point on the spherical surface is 5 in. from R, we are guaranteed some points of intersection. Anytime a plane intersects a spherical surface, the locus of points in the intersection can be described as a circle (or a point).

In this problem, the circle of intersection is a small circle of sphere P, centered at O, as the figure shows.

● **PROBLEM** 780

In a sphere whose radius is 13 units, find the length of a radius of a small circle of the sphere if the plane of the small circle is 5 units from the plane passing through the center of the sphere.

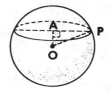

Solution: As the figure shows, the length of the small radius is AP. This is the quantity we wish to determine.

As it turns out, if O is the center of the sphere and $\overline{OA}$ is the perpendicular to the plane of the small circle of the sphere, then $\overline{OA}$ = 5 and A is the center of the small circle. Since $\overline{OA}$ is perpendicular to the plane of the small circle, and $\overline{AP}$ lies in this plane, $\overline{OA} \perp \overline{AP}$ and $\triangle OAP$ is a right triangle. The hypotenuse of this triangle is equal to the radius of the sphere.

The length of one leg and the hypotenuse are known and, using the Pythagorean Theorem, we can find the length of the second leg, $\overline{AP}$. This leg will be the radius of the small circle of the sphere.

In rt. $\triangle OAP$, $(\overline{AP})^2 + (\overline{OA})^2 = (\overline{OP})^2$.

Since $\overline{OA}$ = 5 and $\overline{OP}$ = 13, by substitution,

$$(\overline{AP})^2 + (5)^2 = (13)^2$$

$$(AP)^2 = (13)^2 - (5)^2 = 169 - 25 = 144$$

$$\overline{AP} = 12.$$

Therefore, the length of the radius of the small circle is 12.

● **PROBLEM** 781

What is the radius of the circle of intersection of a plane with a sphere of radius 25 if the plane is 24 units from the center of the sphere?

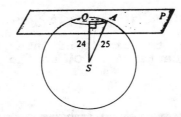

Solution: If A is a point on the circle Q, then the unknown is QA, the radius of the circle Q. Forming the triangle △SQA, we note that AS is a radius of the sphere and, therefore, AS = 25. In addition, $\overline{QS}$ must be perpendicular to $\overline{QA}$ because $\overline{QS}$ is perpendicular to the entire plane P. In general, a line that connects the center of a sphere and the center of the circle of intersection of the sphere with a plane, is perpendicular to the intersecting plane. The length of this line is the distance from the center of the sphere to the plane of intersection. Hence, $\overline{QS}$ = 24.

Thus, we have a right triangle with the length of the hypotenuse and one of the legs known. To find the length of the other leg, we use the Pythagorean Theorem.

(i) $c^2 = a^2 + b^2$

(ii) $AS^2 = SQ^2 + QA^2$

Substituting in AS = 25, SQ = 24,

(iii) $25^2 = 24^2 + QA^2$

(iv) $QA^2 = 25^2 - 24^2 = 625 - 576 = 49$

(v) $QA = \sqrt{49} = 7$.

● **PROBLEM** 782

Given: Planes P and Q intersect sphere S in ⊙A and ⊙B, repectively; M is a point of ⊙A; N is a point of ⊙B: $\overline{AS} \cong \overline{BS}$.

Prove: ⊙A ≅ ⊙B.

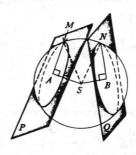

Solution: To prove two circles congruent, we must prove that their radii are congruent, in this case, $\overline{AM}$ and $\overline{BN}$. Whenever two line segments are to be equated, it is wise to look for congruent triangles. The congruent triangles for which we search must contain $\overline{AM}$ and $\overline{BN}$. In addition, since we are given that $\overline{AS} \cong \overline{BS}$, the triangles should also contain these segments. ΔAMS and ΔBNS satisfy these requirements. In proving ΔAMS ≅ ΔBNS, we shall make use of three facts: (1) $\overline{AS} \cong \overline{BS}$ (given); (2) ∡ MAS and ∡ NBS are right angles (the line connecting the center of the sphere and the center of the intercepted circle is perpendicular to the intercepting plane); and (3) $\overline{MS} = \overline{NS}$ (all radii are equal).

Statements	Reasons
1. Planes P and Q intersect sphere S in ⊙A and ⊙B, respectively; M is a point of ⊙A; N is a point of ⊙B; $\overline{AS} \cong \overline{BS}$.	1. Given.
2. $\overline{MS} \cong \overline{NS}$.	2. All radii of a sphere are congruent. (Definition of a sphere.)
3. $\overline{AS} \perp$ plane P. $\overline{BS} \perp$ plane Q.	3. If a line contains the center of a sphere, and the center of a circle of intersection of the sphere with a plane not through the center of the sphere, then the line is perpendicular to the intersecting plane.
4. $\overline{AS} \perp \overline{AM}$. $\overline{BS} \perp \overline{BN}$.	4. A line perpendicular to a plane is perpendicular to every line in the plane that intersects the line.
5. ∡A and ∡B are right angles.	5. Definition of perpendicular lines.
6. ΔMAS and ΔNBS are right triangles.	6. Definition of right triangle.
7. ΔMAS ≅ ΔNBS.	7. If the hypotenuse and a leg of one right triangle are congruent to the hypotenuse and a leg of another right triangle, then the two triangles are congruent.
8. $\overline{MA} \cong \overline{NB}$.	8. Correspoinding sides of congruent triangles are congruent.
9. ⊙A ≅ ⊙B.	9. Circles are congruent if and only if their radii are congruent.

Show that the locus of points equidistant from two given
points is the plane perpendicular to the line segment joining them at their midpoint.

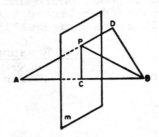

Solution: To prove that a certain set of points is a locus,
we must show (1) all points in the set satisfy the conditions,
and (2) all points that satisfy the conditions are contained
in the set.

We satisfy the first condition by showing that any point
P on m is equidistant from A and B (by showing $\triangle APC \cong \triangle BPC$).
We satisfy the second condition by selecting a point outside
the plane D on the B-side of the half-space and showing that
AD > DB. Thus, any point not on the plane cannot be equidis-
tant from both points.

Given: Points A and B, with plane m perpendicular to
$\overline{AB}$ and bisecting $\overline{AB}$ at C.
To Prove: The locus of points equidistant from A and B
is plane m $\perp \overline{AB}$ at its midpoint C.

Statements	Reasons
1. From P, any point in plane m, draw $\overline{PA}$, $\overline{PB}$ and $\overline{PC}$.	1. Two points determine a line.
2. $\overline{AB} \perp \overline{PC}$.	2. If a line is $\perp$ to a plane, then it is $\perp$ to any line in the plane passing through the point of intersection.
3. $\overline{PC} \cong \overline{PC}$.	3. Reflexive Property.
4. $\overline{AC} \cong \overline{BC}$.	4. Definition of a midpoint.
5. $\angle PCA \cong \angle PCB$.	5. All right angles are congruent.
6. $\triangle PAC \cong \triangle PBC$.	6. The SAS Postulate.
7. PA = PB.	7. Corresponding parts of congruent triangles are equal in length.

Thus, every point on plane m is equidistant from A and B.

8. From D, any point outside m in the B-halfspace, draw $\overline{DA}$ and $\overline{DB}$.	8. Two points determine a line.
9. DP + PB > DB.	9. The sum of two sides of a triangle ($\triangle$PDB) exceeds the third.
10. DP + PA > DB or DA > DB.	10. Substitution Postulate.
11. Only points on m will be in the locus.	11. Postulate of Elimination.

Any point outside plane m is not equidistant from A and B.

● **PROBLEM** 784

Show that if a plane intersects two parallel planes, then it intersects them in two parallel lines.

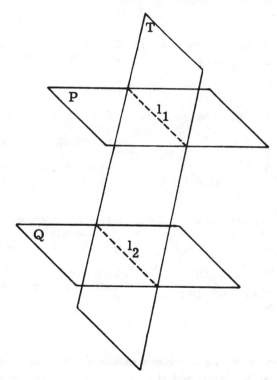

Solution: Two lines are parallel if they are (1) coplanar and (2) they never intersect. Two noncoincident nonparallel planes intersect in a line. Therefore, the intersection of a plane T with two parallel planes are two lines that (1) are both on plane T and (2) never intersect because the planes in which the lines lie are parallel. As such, the two lines of intersection will be parallel.

Given: Plane T intersects parallel planes P and Q.

Prove: The intersection consists of two parallel lines.

743

Statements	Reasons
1. Plane T intersects parallel planes P and Q.	1. Given.
2. The intersection of T and P is line ℓ_1. The intersection of T and Q is line ℓ_2.	2. The intersection of two non-parallel, noncoincident planes is a line coplanar to both planes. (Plane T cannot be parallel or coincident to P or Q because it would contradict our assumption that T intersects P and Q).
3. ℓ_1 and ℓ_2 are coplanar.	3. From 2, we know that both ℓ_1 and ℓ_2 lie on plane T.
4. Either (1) ℓ_1 intersects ℓ_2 (2), or ℓ_1 does not intersect ℓ_2.	4. Either a statement or its negation is true.
5. Case (1) ℓ_1 intersects ℓ_2. Let r be a point on ℓ_1 and ℓ_2.	5. Definition of intersection of two lines.
6. r lies on P. r lies on Q.	6. If a line lies on a plane, then all points of the line are common to the plane.
7. There is no point r such that r lies on P and r lies on Q.	7. Parallel planes have no point in common.
8. ℓ_1 does not intersect ℓ_2.	8. Since statement (1) leads to a contradiction, the negation must be true.
9. $\ell_1 \parallel \ell_2$ or the intersection consists of two parallel lines.	9. Two coplanar lines that do not intersect are parallel.

● **PROBLEM** 785

We are given plane R, and points A and B which are not in plane R, as the figure shows. (Line segment $\overline{AB}$ is not perpendicular to plane R.) (a) Describe the locus of points equidistant from A and B. (b) Describe the locus of points in plane R equidistant from A and B.

<u>Solution</u>: (a) In plane geometry, by definition, the locus of points equidistant from the endpoints of a segment is the perpendicular bisector of the segment. In solid geometry, the locus will be the plane, which contains the perpendicular bisectors of the segment.

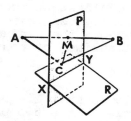

The locus described is the plane P perpendicular to $\overline{AB}$ at the midpoint of $\overline{AB}$, M.

(b)   The locus of points in plane R, equidistant from A and B, are those points that are in both plane R and the plane found in part (a). This is the intersection of planes R and P. If two planes intersect, they intersect in a line. Therefore, the locus of points equidistant from A and B in plane R is the straight line $\overline{XY}$.

● **PROBLEM** 786

Given a plane V and a line k perpendicular to V at point Q, a point P on k is at a distance d from Q. M is the midpoint of $\overline{PQ}$. Describe fully the locus of points:

(a)   at a distance $\frac{1}{2}$d from k

(b)   at a distance $\frac{1}{2}$d from V

(c)   at a distance $\frac{1}{2}$d from M

(d)   that satisfy conditions a and b
(e)   that satisfy conditions b and c.

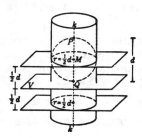

Solution:

(a)   The locus of points equidistant from a line in space is a cylindrical surface with its axis being the line and the radius the given distance. Thus, the locus is a cylindrical surface with axis k and radius $\frac{1}{2}$d.

(b)   Remember that parallel planes are everywhere equidistant. The set of points a distance $\frac{1}{2}$d from plane V is two parallel planes, one in each half-space, each a distance $\frac{1}{2}$d from V.

(c)   The locus of points equidistant from a given point
      forms a sphere with the given point as the center
      and the given distance as the radius. Thus, the
      locus is a spherical surface of center M and radius
      $\frac{1}{2}$d.

(d)   The points in this locus must lie on the cylindri-
      cal surface and on one of the two parallel planes.
      We are given that k ⊥ V. Since the two planes are
      ∥ to V, k is also ⊥ to the two planes. If a cylin-
      der intersects a plane perpendicular to its axis,
      then the intersection is a circle. (If the plane
      were not ⊥ to the axis, then the intersection might
      be another conic section.) Since there are two
      planes, the locus is two circles each having center
      on line k and a radius of $\frac{1}{2}$d.

(e)   The points in this locus must lie on the sphere with
      center M and on one of the two parallel planes.
      This can be rephrased as the intersection of the
      sphere with one plane plus the intersection of the
      sphere and the other plane. Consider the plane in
      the same halfspace (created by V) as M. Every
      point in the plane is $\frac{1}{2}$d from V. Furthermore, since
      $\overline{MQ}$ is on k and thus $\overline{MQ}$ ⊥ V, then MQ is the distance
      of M from V. Since M is a midpoint of $\overline{PQ}$,
      MQ = $\frac{1}{2}$PQ = $\frac{1}{2}$d. Thus, the center of the sphere is
      also $\frac{1}{2}$d from plane V and must lie on the parallel
      plane. The intersection of a sphere and a plane
      passing through its center is a great circle. The
      intersection of the sphere and one parallel plane
      is a circle of center M and radius $\frac{1}{2}$d.

          The intersection of the sphere and the parallel
      plane on the opposite side of V is the empty set.
      Note that the distance between sphere center M and
      any point on the plane is d. Since the radius of
      the sphere is only $\frac{1}{2}$d, then every point on the plane
      is exterior to the sphere and there is no intersec-
      tion.
          Thus, the total locus is a circle of center M
and radius $\frac{1}{2}$d.

● **PROBLEM** 787

Show that the locus of points within a dihedral angle and
equidistant from its faces is the plane bisecting the dihedral angle.

Solution: To prove a set of points is a locus, we must prove
(1) all points of the set satisfy the conditions; and (2) all

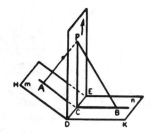

points that satisfy the conditions are contained in the locus.

Part I: We show that every point in the plane bisecting the dihedral angle is equidistant from the faces of the dihedral angle. In the figure, dihedral angle m-DE-n is bisected by plane r. Point P is any point on r. PA and PB are the distances from P to the faces m and n, respectively. We must show that P is equidistant from m and n or PA = PB. We can do this by proving congruent triangles.

Given: Plane r bisects dihedral angle m-DE-n; P is any point on r.

To prove: P is equidistant from m and n.

Statements	Reasons
1. Plane r bisects dihedral angle m - DE - n.	1. Given.
2. Through any point P in r, draw $\overleftrightarrow{PA} \perp$ m and $\overleftrightarrow{PB} \perp$ n.	2. Through an external point, one and only one perpendicular to a given plane can be constructed.
3. Pass a plane through $\overleftrightarrow{PA}$ and $\overleftrightarrow{PB}$ perpendicular to planes m and n, intersecting m in $\overleftrightarrow{AC}$, r in $\overleftrightarrow{PC}$, and n in $\overleftrightarrow{BC}$.	3. Two intersecting lines determine a plane; two planes intersect in a straight line If a line ($\overleftrightarrow{PA}$, $\overleftrightarrow{PB}$) is $\perp$ plane (m,n), then every plane (APCB) containing the line is $\perp$ given plane.
4. $\overleftrightarrow{DC} \perp$ plane APB.	4. If two intersecting planes are $\perp$ to a third plane, then their line of intersection is $\perp$ to that plane.
5. $\overleftrightarrow{DC} \perp \overleftrightarrow{AC}$, $\overleftrightarrow{DC} \perp \overleftrightarrow{PC}$, $\overleftrightarrow{DC} \perp \overleftrightarrow{BC}$.	5. If a line is perpendicular to a plane (APCB), then it is $\perp$ to every line in the plane passing through its foot.
6. ∢PCA and ∢PCB are plane angles of dihedral angles m-DE-n and n-DE-r.	6. The plane angle of a dihedral angle is formed by two lines, one in each face, drawn $\perp$ to the edge of the dihedral angle at the same point.

7. $\angle PCA \cong \angle PCB$.	7. Two plane angles are equal if their dihedral angles are equal. (Recall plane r bisects dihedral angle m-DE-n).
8. $\overleftrightarrow{PA} \perp \overleftrightarrow{AC}$, $\overleftrightarrow{PB} \perp \overleftrightarrow{BC}$ or $\angle PAC$ and $\angle PBC$ are right angles.	8. If a line is $\perp$ to a plane, then it is $\perp$ to every line in the plane passing through its foot.
9. $\overline{PC} \cong \overline{PC}$.	9. Reflexive Property.
10. $\triangle PAC \cong \triangle PBC$.	10. The AAS Postulate.
11. PA = PB and P is equidistant from A and B.	11. Corresponding parts of congruent triangles are congruent.

Part 2. We will now show that any point equidistant from the faces of a dihedral angle lies in a plane bisecting the angle. We do this in a proof very similar to Part 1. Note that we are given that PA = PB this time, instead of $\angle PCA \cong \angle PCB$. Thus, instead of proving $\triangle PAC \cong \triangle PBC$, using $\angle PCA \cong \angle PCB$, to show PA = PB; we will prove $\triangle PAC \cong \triangle PBC$, using PA = PB, to show $\angle PCA \cong \angle PCB$. Thus, dihedral angles can be shown to be $\angle m - DE - r$ and $\angle n - DE - r$ are congruent and the plane determined by $\overleftrightarrow{DE}$ and P bisects $\angle m - DE - n$.

Given: Any point P equidistant from m and n in the interior of m - DE - n.

Prove: Plane r, determined by P and $\overleftrightarrow{DE}$ bisects dihedral angle m - DE - n.

Statements	Reasons
1. Through P, any point equidistant from m and n, draw $\overline{PA} \perp$ m and $\overline{PB} \perp$ n.	1. From an external point, exactly one perpendicular to a given plane can be drawn.
2. $\overleftrightarrow{PA}$ and $\overleftrightarrow{PB}$ determine a plane $\perp$ m and $\perp$ n, intersecting m in $\overleftrightarrow{AC}$ and n in $\overleftrightarrow{BC}$.	2. Two intersecting lines determine a plane. If a line is $\perp$ to a plane, then every plane containing the line is $\perp$ to this given plane. Two planes intersect in a straight line.
3. Draw $\overline{PC}$.	3. Two points determine a line.
4. $\overleftrightarrow{DC} \perp$ plane PAB.	4. If two intersecting planes are $\perp$ to a third plane, then their intersection is $\perp$ to that plane.
5. $\overleftrightarrow{DC} \perp \overleftrightarrow{AC}$, $\overleftrightarrow{DC} \perp \overleftrightarrow{PC}$, $\overleftrightarrow{PC} \perp \overleftrightarrow{BC}$.	5. If a line is $\perp$ to a plane, then it is $\perp$ to every line in the plane passing through the point of intersection.

6.	⊀PCA and ⊀PCB are the plane angles of dihedral angles m - DE - r and n - DE - r.	6.	The plane angle of a dihedral angle is formed by two lines, one in each face, ⊥ to the edge of the dihedral angle at the same point.
7.	PA = PB.	7.	Given.
8.	$\overleftrightarrow{PA} \perp \overleftrightarrow{AC}$ and $\overleftrightarrow{PB} \perp \overleftrightarrow{BC}$ or ⊀PAC and ⊀PBC are right angles.	8.	Step 1. Also, if a line is ⊥ to a plane, then it is ⊥ to every line in the plane that passes through the point of intersection.
9.	$\overline{PC} \cong \overline{PC}$.	9.	Reflexive Property.
10.	ΔPAC ≅ ΔPBC.	10.	Hypotenuse-Leg Theorem.
11.	⊀PCA ≅ ⊀PCB.	11.	Corresponding parts of congruent triangles are congruent.
12.	⊀m - DE - r ≅ ⊀n - DE - r.	12.	Two dihedral angles are congruent if their plane angles are congruent.
13.	Plane r bisects ⊀m - DE - n.	13.	The bisector of an angle divides that angle into two congruent angles.

● **PROBLEM** 788

A chord through the end of a diameter of a circle is extended the length of the diameter. Find and sketch the locus of the ends of all such line segments.

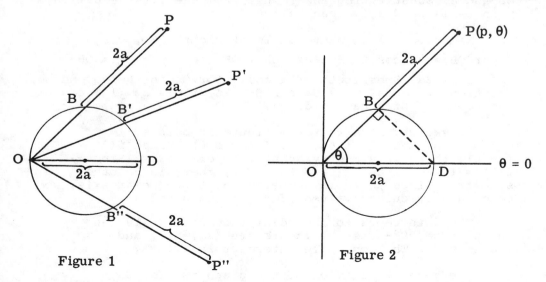

Figure 1

Figure 2

<u>Solution</u>: Given a condition to be satisfied, the locus of the condition is the set of points that satisfy the condi-

tion. In Fig. 1, $\overline{OD}$ is a **diameter** and point P is a point such that the part of the segment $\overline{OP}$ external to the cirle, $\overline{BP}$, has length OD. Thus, P satisfies the description in the problem statement and must be a point on the locus. But points P' and P" also satisfy the description and must also be points of the locus. To specify the locus completely, we must derive an equation such that every point in the locus must satisfy the equation and every point not in the locus must not satisfy the equation.

Whenever a locus is described as the end of a line of varying length revolving around a fixed point, the equation of the locus is more easily derived in polar rather than Cartesian coordinates. The pole should be chosen at a con-venient place, usually the fixed point around which the line revolves. Then a point (p, θ) satisfying the condition is chosen and relations between p, θ, and the given values are obtained from the geometric and trigonometric properties of the situation.

Here, choose one endpoint of the given diameter as the pole O and the diameter $\overline{OD}$ as the polar axis. (See Figure 2.) Let OD = 2a, $\overline{OB}$ be an arbitrary chord, and $\overline{OP}$ be an extension of $\overline{OB}$ such that BP = 2a. Therefore, point P (p, θ) is a point of the locus.

To solve for a relation involving unknowns p and θ and the given radius a, note that since $\overline{OD}$ is a diameter, $\overset{\frown}{OBD}$ is a semicircle. Since ΔOBD is inscribed in semicircle $\overset{\frown}{OBD}$, ΔOBD is a right triangle, and θ = ∢ BOD is an acute angle of right ΔOBD.

To solve for an equation involving p, θ, and a, first solve for OB in terms of θ and a. Next, find p in terms of OB and a. Substituting the result from the first equation, p in terms of θ and a can be found.

Because ϑ is an acute angle of right ΔOBD, cos θ = $\frac{OB}{2a}$ or OB = 2a cos θ. **Because** OP = OB + BP, p = OP = OB + BP = OB + 2a. Substituting 2a cos θ for OB in the second equation, we obtain p = 2a cos θ + 2a or p = 2a (1 + cos θ), the equation of a cardioid.

To sketch, note that the figure is bounded: 0 ≤ p ≤ 4a. Let r = f(θ) = 2a(1 + cos θ). Since f(θ) = f(- θ), there is symmetry about the polar axis. Since f(θ) ≠ f(π - θ) and f(θ) ≠ f(π + θ), there is no other symmetry. Since there is symmetry about the polar axis, we need only calculate values of θ between 0 and π.

The intercepts to the polar axis are (4a, 0) and (0, π). The 90°-axis intercepts are (2a, π/2) and (2a, - π/2). The lone pole intercept is (0, π).

The general behavior of the graph is: for θ increasing from 0 to π, r decreases from 4a to 0. Some specific points are calculated below. The graph is as

750

shown in Figure 2.

θ	$\cos \theta$	$r = 2a(1 + \cos \theta)$
$\pi/6$	$\sqrt{3}/2$	$(2 + \sqrt{3})a \cong 3.73\,a$
$\pi/4$	$\sqrt{2}/2$	$(2 + \sqrt{2})a \cong 3.41a$
$\pi/3$	$1/2$	$3a$
$2\pi/3$	$-1/2$	a
$3\pi/4$	$-\sqrt{2}/2$	$(2 - \sqrt{2})a \cong .586a$
$5\pi/6$	$-\sqrt{3}/2$	$(2 - \sqrt{3})a \cong .268a$

CHAPTER 45

INTERSECTIONS OF LINES AND PLANES

Basic Attacks and Strategies for Solving Problems in this Chapter. See pages 752 to 773 for step-by-step solutions to problems.

In a formal geometry course, a plane is left undefined, but a common axiom is the following:

Three noncollinear points determine a plane.

Problem 789 states that a line and a point not on the line determine a plane. The proof given relies on the axiom given above.

It is a common practice to use a parallelogram to represent a plane. Examine the figure below. It illustrates two parallel planes, and a third plane intersecting them. This picture illustrates that when two planes intersect the intersection is a line.

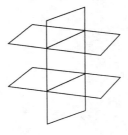

There are many everyday models of planes. For example, in most rooms the floor and any wall represent intersecting planes. Their intersection is the line along the floor. The floor and ceiling represent parallel planes, and each wall is perpendicular to them. This relationship nicely models that two parallel planes are equidistant everywhere (Problem 792).

The last three problems in this chapter concern polyhedral angles. In analyzing the relationship among polyhedral angles, it sometimes helps to imagine that a polyhedron has been cut and opened up. For example, consider the pyramid pictured below. Suppose we cut out $\triangle ACD$, then cut along BC, and then open up the resulting three faces.

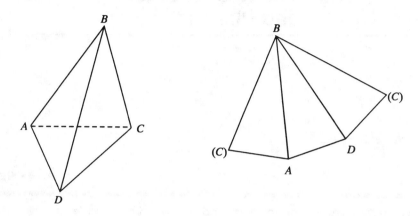

This illustrates that $m \not\angle CBA + m \not\angle ABD + m \not\angle DBC < 360°$. If this were not the case, there would not be enough "room" to fold these faces in to form the pyramid (Problems 803 and 805).

As was the case in the previous chapter, most of the proofs given here relate back to corresponding proofs from plane geometry.

Step-by-Step Solutions to
Problems in this Chapter,
"Intersections of Lines
and Planes"

LINES & PLANES

● PROBLEM 789

Prove that a plane is determined by a line and a point not on
that line. Also show that the plane contains the given line.

Solution: To say that a plane is determined in this way is
to say that there is one and only one such figure containing
a line, ℓ, and a point C, not on line ℓ.

We will prove there are two points A and B on the line
which, together with C, determine a plane. A most basic property of a plane is that a plane is determined by three points
not on the same straight line.

We will also be able to conclude that line ℓ is contained
in the plane, because a line passing through two points in a
plane, lies entirely in the plane.

Statement	Reason
1. Line ℓ is a straight line.	1. Given.
2. A and B are two points on line ℓ.	2. Every straight line contains at least two points.
3. Point C is not on line ℓ.	3. Given.
4. There is one and only one plane P that contains points A,B and C.	4. A plane is determined by three points which are not on the same line.
5. Plane P contains line ℓ.	5. If a straight line passes through two points in a plane, then the line lies entirely in the plane.

752

Prove that there exists a plane M which is perpendicular to a line L at a point P on L.

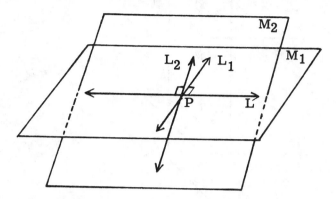

Solution: The idea behind this solution is to find a plane, M, perpendicular to a line L at a point P on L.

Given line L, there are many planes containing L. Choose 2 of these planes, and label them M_1 and M_2 (See figure). Within plane M_1, there is only one line perpendicular to L and passing through point P (L_1). Similarly, within plane M_2, there is only one line perpendicular to L and passing through point P (L_2). Now, $L \perp L_1$ at P and $L \perp L_2$ at P. But, if a line is perpendicular to 2 intersecting lines at their point of intersection, it is perpendicular to the plane containing the 2 lines. Hence, L is perpendicular to the plane containing L_1 and L_2 at the point P. Therefore, there does exist a plane (call it M) which is perpendicular to a line L at a point P on L.

Show that if a line is perpendicular to one of two parallel planes, then it is perpendicular to the other.

Solution: For a line to be perpendicular to a plane, we show that the line is perpendicular to every line in the plane that passes through the point of intersection. Suppose $\overleftrightarrow{AB} \perp$ plane P at point A, plane P ∥ plane Q, and ℓ_1 intersects plane Q at point B. Then for every line $\overleftrightarrow{BD}$ in plane Q, there is a line $\overleftrightarrow{AC}$ in plane P such that $\overleftrightarrow{AC} \parallel \overleftrightarrow{BD}$. Then, interior angles on the same side of the transversal must be supplementary. Since for every $\overleftrightarrow{AC}$, $\overleftrightarrow{AC} \perp \overleftrightarrow{AB}$, then $\overleftrightarrow{AB} \perp \overleftrightarrow{BD}$ for every $\overleftrightarrow{BD}$, and $\overleftrightarrow{AB}$ must be perpendicular to plane Q.

Given: Plane P ∥ plane Q; $\overleftrightarrow{AB} \perp P$ at A. $\overleftrightarrow{AB}$ intersects Q at point B.

Prove: $\overleftrightarrow{AB} \perp Q$.

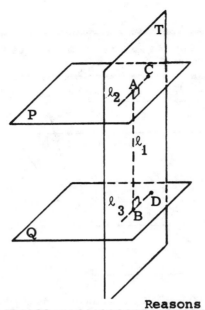

Statements	Reasons
1. Plane P ∥ plane Q; $\overleftrightarrow{AB} \perp P$ at A, $\overleftrightarrow{AB}$ inter- sects Q at point B.	1. Given.
2. Point D is any point in plane Q.	2. A plane contains an infi- nite number of points.
3. Plane T is the plane determined by $\overleftrightarrow{AB}$ and $\overleftrightarrow{BD}$.	3. Two intersecting lines determine a plane.
4. Plane T intersects P at line $\overleftrightarrow{AC}$.	4. The intersection of two nonparallel, noncoincident planes is a line common to both planes.
5. $\overleftrightarrow{AC} \parallel \overleftrightarrow{BD}$.	5. If a plane intersects two parallel planes, then it intersects them in two parallel lines.
6. $\overline{BA} \perp \overline{AC}$.	6. A line is perpendicular to a plane if, and only if, it is perpendicular to every line in the plane that passes through the point of intersection.
7. ∢BAC is a right angle.	7. Perpendicular lines inter- sect to form right angles.
8. m∢BAC = 90°.	8. All right angles measure 90°.
9. m∢BAC + m∢ABD = 180 .	9. Interior angles of parallel lines on the same side of

10. m∢ABD = 90°.

10. Subtraction Property of Equality.

11. $\overleftrightarrow{AB}$ ⊥ plane Q.

11. Since D could be any point in the plane, $\overleftrightarrow{AB}$ is perpendicular to every line $\overleftrightarrow{BD}$ in plane Q. By definition, $\overleftrightarrow{AB}$ is perpendicular to Q.

● **PROBLEM** 792

Show that parallel planes are everywhere equidistant.

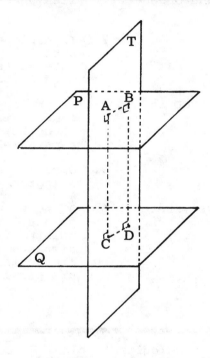

Solution: Let A and B be points on plane P, where P and Q are parallel planes. Let $\overleftrightarrow{AC}$ and $\overleftrightarrow{BD}$ be lines perpendicular to Q where C and D are the points of intersection of these lines with plane Q. Then AC and BD are the distances from plane P to Q. We show that quadrilateral ABCD is a parallelogram and therefore that AC = BD and the planes are everywhere equidistant.

Given: Plane P ∥ plane Q.

Show: P and Q are everywhere equidistant.

Statements	Reasons
1. Plane P ∥ plane Q.	1. Given.

2. Let A and B be points on plane P.

2. A plane contains an infinite number of points.

3. Draw $\overleftrightarrow{AC}$, $\overleftrightarrow{BD} \perp Q$ such that C and D are points on plane Q.

3. From an external point to a given plane, one and only one line may be drawn perpendicular to the plane.

4. $\overleftrightarrow{AC} \parallel \overleftrightarrow{BD}$.

4. Two lines perpendicular to the same plane are parallel.

5. $\overleftrightarrow{AC}$ and $\overleftrightarrow{BD}$ lie on plane T.

5. Two parallel lines determine a plane.

6. Plane T contains $\overleftrightarrow{AB}$, and $\overleftrightarrow{CD}$. Plane P contains $\overleftrightarrow{AB}$. Plane Q contains $\overleftrightarrow{CD}$.

6. If two points of a line lie on the plane, then the line lies on the plane.

7. $\overleftrightarrow{AB} \parallel \overleftrightarrow{CD}$.

7. If a plane intersects two parallel planes, then it intersects them in two parallel lines.

8. Quadrilateral ABCD is parallelogram.

8. If the opposite sides of a quadrilateral are parallel, then the quadrilateral is a parallelogram.

9. $\overline{AC} \cong \overline{BD}$.

9. Opposite sides of a parallelogram are congruent.

10. P and Q are everywhere equidistant

10. A and B were arbitrary points. If for any two points of one plane, the distance to a given plane is constant, then planes are parallel.

● **PROBLEM** 793

Given that $\overleftrightarrow{AB}$ is perpendicular to plane P, $\overleftrightarrow{BC}$ and $\overleftrightarrow{BD}$ lie in plane P, and $\overline{BC} \cong \overline{BD}$, prove that $\overline{AC} \cong \overline{AD}$.

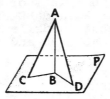

Solution: To prove that $\overline{AC} \cong \overline{AD}$, we must prove that the triangles in which these lines are corresponding sides, ΔABC and ΔABD, are congruent. This will be done using the S.A.S. Postulate.

In order to prove that the included angles are congruent,

756

(i.e. ∢ABC ≅ ∢ABD), we will have to apply the definition of a line perpendicular to a plane. This definition tells us that a line perpendicular to a plane at a point is perpendicular to every line, in the plane, which passes through that point.

Statement	Reason
1. $\overleftrightarrow{AB} \perp$ plane P.	1. Given.
2. $\overleftrightarrow{BC}$, $\overleftrightarrow{BD}$ lie in plane P.	2. Given.
3. $\overleftrightarrow{AB} \perp \overleftrightarrow{BC}$, $\overleftrightarrow{AB} \perp \overleftrightarrow{BD}$.	3. Definition of a line perpendicular to a plane.
4. ∢ABD and ∢ABC are right angles.	4. Perpendicular lines intersect forming right angles.
5. ∢ABC ≅ ∢ABD.	5. All right angles are congruent.
6. $\overline{BC} \cong \overline{BD}$.	6. Given.
7. $\overline{AB} \cong \overline{AB}$.	7. Reflexive Property of Congruence.
8. △ABC ≅ △ABD.	8. S.A.S. Postulate.
9. $\overline{AC} \cong \overline{AD}$.	9. Corresponding sides of congruent triangles are congruent.

● **PROBLEM** 794

In the accompanying diagram plane p is perpendicular to plane s, intersecting s in line $\overline{QR}$. Line $\overline{EF}$, in plane s, is perpendicular to $\overline{QR}$ at E. Line $\overline{AB}$ is drawn parallel to $\overline{EF}$ through any point A in p not on $\overline{QR}$. Line $\overline{AE}$ is drawn. Prove angle BAE is a right angle.

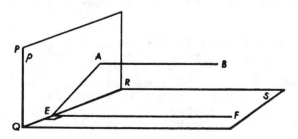

Solution: We can prove that $\overline{EF} \perp \overline{AE}$ and then, since $\overline{EF} \parallel \overline{AB}$, conclude that $\overline{AB}$ must be $\perp \overline{EA}$. Thus, proving that ∢BAE is a right angle.

Statement	Reason
1. Plane p $\perp$ plane s $\overline{FE} \perp \overline{QR}$.	1. Given.
2. $\overline{FE} \perp$ plane p.	2. If two planes are $\perp$ to each

757

other, then a line drawn in one of them ⊥ to their intersection is ⊥ to the other.

3. $\overline{FE} \perp \overline{EA}$.

3. A line ⊥ to a plane is ⊥ to every line in the plane passing through its foot.

4. $\overline{AB} \parallel \overline{EF}$ and lies in the same plane as $\overleftrightarrow{EF}$.

4. Given, and a plane is determined by two parallel lines.

5. $\overline{AB} \perp \overline{EA}$.

5. In a plane, a line ⊥ to one of two parallel lines is ⊥ to the other.

6. ∢BAE is a right angle.

6. Perpendicular lines intersect to form right angles.

● **PROBLEM** 795

If two planes are perpendicular to each other, prove that a line drawn in one of them perpendicular to their intersection is perpendicular to the other plane.

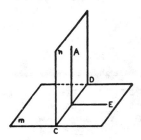

<u>Solution</u>: In the accompanying figure, plane m ⊥ plane n and m intersects n in $\overleftrightarrow{CD}$. Line $\overleftrightarrow{AB}$ is in n and $\overleftrightarrow{AB} \perp \overleftrightarrow{CD}$. We must show $\overleftrightarrow{AB} \perp$ m.

We can show a line is perpendicular to a plane by showing it is perpendicular to two lines in the plane. Here, $\overleftrightarrow{CD}$, the intersection of m and n, is one line in m that is perpendicular to $\overleftrightarrow{AB}$. To find the second line, remember that if two planes are perpendicular, their dihedral angle is a right angle. Draw $\overleftrightarrow{BE}$ in plane m perpendicular to $\overleftrightarrow{CD}$. Then, ∢ABE is a dihedral angle and m∢ABE = 90°. Thus, $\overleftrightarrow{AB} \perp \overleftrightarrow{BE}$, a second line in plane m. Thus, $\overleftrightarrow{AB} \perp$ m.

Given: Plane m ⊥ plane n: $\overleftrightarrow{CD}$ is their line of intersection; $\overline{AB}$ in n ⊥ CD.

To Prove: $\overleftrightarrow{AB} \perp$ m.

Statements	Reasons
1. Plane m ⊥ plane n; $\overleftrightarrow{CD}$ is their line of intersection: $\overline{AB}$ in n ⊥CD.	1. Given.

2. Draw $\overline{BE}$ in m $\perp \overleftrightarrow{CD}$ at B.

2. In a plane, a perpendicular can be drawn to a point in a line.

3. ∢ABE is the plane angle of dihedral angle A-CD-E.

3. The plane angle of a dihedral angle is formed by two lines, one in each face, drawn $\perp$ to the edge of the dihedral angle at the same point.

4. n $\perp$ m.

4. Given.

5. ∢ABE is a right angle; $\overline{AB} \perp \overline{BE}$.

5. The measure of a dihedral angle is given by its plane angle. A right angle is formed by the intersection of two perpendicular lines.

6. $\overline{AB} \perp \overline{CD}$.

6. Given.

7. $\overline{AB} \perp$ m.

7. If a line is $\perp$ to two intersecting lines at their point of intersection, then it is $\perp$ to the plane of the lines.

● **PROBLEM** 796

ABCD is a rectangle. $\overline{PA}$ is perpendicular to the plane of the rectangle and the line $\overleftrightarrow{PB}$ is drawn. Prove that ∢PBC is a right angle.

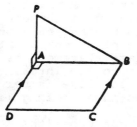

Solution: We will prove ∢PBC is a right angle by showing $\overline{BC} \perp$ plane PAB. Since a line perpendicular to a plane is perpendicular to every line in the plane passing through its foot, we can conclude $\overline{BC} \perp \overline{PB}$ making ∢PBC a right angle.

Statement	Reason
1. $\overline{PA} \perp$ plane ABCD.	1. Given.
2. $\overline{PA} \perp \overline{AD}$.	2. If a line is $\perp$ to a plane, then it is $\perp$ to every line in the plane passing through its foot.
3. ABCD is a rectangle	3. Given.
4. $\overline{AD} \perp \overline{AB}$.	4. Definition of rectangle.

5. AD ⊥ plane PAB.	5. If a line is ⊥ to two inter-secting lines at their point of intersection, then it is ⊥ to the plane determined by the lines.
6. $\overline{BC}$ ∥ $\overline{AD}$.	6. Definition of rectangle.
7. $\overline{BC}$ ⊥ plane PAB.	7. If one of two parallel lines is ⊥ to a plane, then the other is also ⊥ to the plane.
8. $\overline{BC}$ ⊥ $\overline{PB}$.	8. Reason 2.
9. ∢PBC is a right angle.	9. Perpendicular lines form right angles.

● **PROBLEM** 797

If ABCD and AFED are both squares, and, as the figure shows, $\overleftrightarrow{AF}$ and $\overleftrightarrow{AB}$ lie in plane P, then prove $\overleftrightarrow{DA}$ ⊥ plane P.

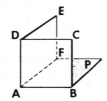

Solution: Two intersecting lines determine a plane. A line perpendicular to a plane at a point is perpendicular to all lines in that plane which pass through the point. Therefore, a line perpendicular to a plane at a point is perpendicular to any 2 lines in the plane which intersect at that point. The converse of this last statement is also true. Namely, if a line is perpendicular to two intersecting lines at their point of intersection, the line is perpendicular to the plane determined by these two intersecting lines.

Based on this, we can prove that $\overleftrightarrow{DA}$ ⊥ $\overleftrightarrow{AB}$ and $\overleftrightarrow{DA}$ ⊥ $\overleftrightarrow{AF}$. $\overleftrightarrow{AB}$ and $\overleftrightarrow{AF}$ intersect and lie in plane P. By the above discussion we can conclude that $\overleftrightarrow{DA}$ ⊥ plane P.

Statement	Reason
1. ABCD is a square.	1. Given.
2. $\overleftrightarrow{DA}$ ⊥ $\overleftrightarrow{AB}$.	2. Two consecutive sides of a square are perpendicular to each other.
3. AFED is a square.	3. Given.
4. $\overleftrightarrow{DA}$ ⊥ $\overleftrightarrow{AF}$.	4. Same as reason 2.
5. $\overleftrightarrow{AF}$ and $\overleftrightarrow{AB}$ intersect and lie in plane P.	5. Given

6. $\overleftrightarrow{DA} \perp$ plane P.

6. If a line is perpendicular to each of two intersecting lines at the point of intersection, then the line is perpendicular to the plane determined by these two intersecting lines.

● **PROBLEM** 798

Line k is parallel to plane P. Prove that a plane perpendicular to line k is also perpendicular to P.

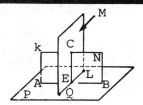

Solution: The best way to present this proof will be in two column format. We will show that the plane perpendicular to line k contains a line perpendicular to two intersecting lines in the parallel plane and is, therefore, perpendicular to that plane.

Statement	Reason
1. Line k ∥ plane P plane M ⊥ line k.	1. Given
2. C is the intersection of line k with plane M.	2. The intersection of a line and a plane is a point.
3. $\overline{QL}$ is the intersection of planes M and P.	3. The intersection of two planes is a straight line.
4. Drop $\overline{CE} \perp \overline{QL}$ from point C.	4. A perpendicular may be drawn in a plane, from a point to a line.
5. Pass a plane through lines k and $\overleftrightarrow{CE}$ intersecting plane P in line $\overleftrightarrow{AB}$.	5. Two intersecting lines determine a plane.
6. Line k ∥ $\overleftrightarrow{AB}$.	6. If a line is parallel to a plane, then it is parallel to the intersection of that plane with any plane containing the line.
7. Line k ⊥ $\overline{CE}$.	7. A line perpendicular to a plane is perpendicular to every line in the plane passing through its foot.

761

8. $\overline{AB} \perp \overline{CE}$.

8. In a plane, if a line is perpendicular to one of two parallel lines, then it is perpendicular to the other.

9. $\overleftrightarrow{CE} \perp \overline{QL}$.

9. Construction.

10. $\overleftrightarrow{CE} \perp$ plane P.

10. If a line is perpendicular to two intersecting lines at their point of intersection, then it is perpendicular to the plane determined by the lines.

11. Plane M $\perp$ plane P.

11. If a line is perpendicular to a given plane, then every plane which contains this line is perependicular to the given plane.

● **PROBLEM** 799

Show that two lines perpendicular to the same plane are parallel.

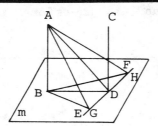

Solution: In the accompanying figure, $\overleftrightarrow{AB}$ and $\overleftrightarrow{CD}$ are perpendicular to plane m. We must prove $\overleftrightarrow{AB} \parallel \overleftrightarrow{CD}$. This can be done in two steps: first prove $\overleftrightarrow{AB}$ and $\overleftrightarrow{CD}$ are coplanar and then show $\overleftrightarrow{AB} \parallel \overleftrightarrow{CD}$. To show $\overleftrightarrow{AB}$ and $\overleftrightarrow{CD}$ are coplanar, we relate $\overleftrightarrow{AB}$ and $\overleftrightarrow{CD}$ to two other lines, $\overleftrightarrow{AD}$ and $\overleftrightarrow{BD}$.

We use the theorem stating that all lines perpendicular to a line at a point are coplanar. We find a line such that $\overleftrightarrow{BD}$, $\overleftrightarrow{AD}$, and $\overleftrightarrow{CD}$ are all perpendicular to it at point D. Then $\overleftrightarrow{CD}$ is coplanar to the plane of $\overleftrightarrow{BD}$ and $\overleftrightarrow{AD}$. Since points A and B must lie in the plane of $\overleftrightarrow{BD}$ and $\overleftrightarrow{AD}$, so must the line $\overleftrightarrow{AB}$. Since $\overleftrightarrow{AB}$ and $\overleftrightarrow{CD}$ both lie in the plane determined by $\overleftrightarrow{BD}$ and $\overleftrightarrow{AD}$, $\overleftrightarrow{AB}$, and $\overleftrightarrow{CD}$ are coplanar. To show that $\overleftrightarrow{AB}$ and $\overleftrightarrow{CD}$ are parallel, we show $\overleftrightarrow{AB} \perp \overleftrightarrow{BD}$ and $\overleftrightarrow{CD} \perp \overleftrightarrow{BD}$. Therefore, $\overleftrightarrow{AB} \parallel \overleftrightarrow{CD}$.

Statements	Reasons
1. Draw $\overline{BD}$ and $\overline{AD}$.	1. Two points determine a line.
2. In plane m, construct $\overline{EF} \perp \overline{BD}$.	2. At any point on a line, a $\perp$ may be drawn.
3. On $\overline{EF}$, construct $\overline{HD} \cong$	3. Congruent line segments

$\overline{DG}$ so that $\overline{BD}$ is $\perp$ bisector of $\overline{HG}$.

may be constructed on a line.

4. Draw $\overline{AH}$, $\overline{GA}$, $\overline{BH}$, and $\overline{BG}$.

4. Two points determine a line.

5. $\overline{BH} \cong \overline{BG}$.

5. A point on the $\perp$ bisector of a line segment is equidistant from the ends of the line segment.

6. $\overline{AH} \cong \overline{AG}$.

6. If line segments drawn from a point in the $\perp$ to a plane meet the plane at equal distances from the foot of the $\perp$, then the line segments are equal. (This can be shown by proving $\triangle ABG \cong \triangle ABH$ by SAS.)

7. $\overline{AD} \perp \overline{HG}$.

7. The line determined by two points equidistant from the endpoints of the segments is the perpendicular bisector of the segment.

8. $\overline{CD} \perp \overline{EF}$.

8. If a line is $\perp$ to a plane then it is $\perp$ to any line in the plane passing through its foot.

9. $\overline{BD}$, $\overline{AD}$ and $\overline{CD}$ lie in the same plane, plane CDB.

9. All the $\perp$'s to a line at a point in the line lie in a plane which is $\perp$ to the line at the given point.

10. $\overline{AB}$ lies in the plane of $\overline{BD}$ and $\overline{AD}$.

10. A line which joins two points in a plane lies wholly within the plane.

11. $\overline{AB} \perp \overline{BD}$ and $\overline{CD} \perp \overline{BD}$.

11. If a line is $\perp$ to a plane, then it is $\perp$ to every line in the plane passing through its foot.

12. $\overline{AB} \parallel \overline{CD}$.

12. Two coplanar lines perpendicular to a third line are parallel to each other.

● **PROBLEM** 800

Show that if a line is perpendicular to each of two intersecting lines at their point of intersection, then the line is perpendicular to the plane determined by them.

Solution: There are many ways of defining a plane. Consider two distinct points in space, Q and R. The focus of all points equidistant from the two points would be one line, the perpendicular bisector, if we are limited to a plane.

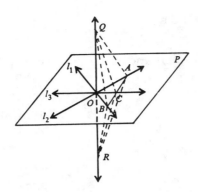

However, for three dimensional space, there are different
perpendicular bisectors. The set of all perpendicular bisectors form a plane themselves. An alternate definition of
a plane would therefore, be the set of all perpendicular bisectors of a segment in space. Here, we are given a line and
a plane determined by two intersecting lines. We choose endpoints on the perpendicular line, such that the two intersecting lines are perpendicular bisectors of the chosen segment.
We show that any line coplanar with the intersecting lines and
intersecting the perpendicular line must also be a perpendicular bisector of the segment and, thus, every line in the plane
is perpendicular to the perpendicular line.

Given: Lines ℓ_1, and ℓ_2 determine plane P and intersect
at O, $\overleftrightarrow{QR}$ is not in plane P; $\overleftrightarrow{QR} \perp \ell_1$ and $\overleftrightarrow{QR} \perp \ell_2$.

Prove: $\overleftrightarrow{QR}$ is perpendicular to any other line in plane
P that passes through O; or, $\overleftrightarrow{QR} \perp$ plane P.

Statements	Reasons
1. ℓ_1 and ℓ_2 determine plane P and intersect at O; $\overleftrightarrow{QR} \perp \ell_1$, $\overleftrightarrow{QR} \perp \ell_2$.	1. Given.
2. Let Q be a point on $\overleftrightarrow{QR}$ in one half-space and R the point on $\overleftrightarrow{QR}$ in the opposite half-space such that OQ = OR.	2. A plane divides space into two half spaces. Also, the Point Uniqueness Postulate (From a given point A on a given line and in a given direction, there is only one point B, such that AB equals a given length).
3. ℓ_1 and ℓ_2 are both perpendicular bisectors of QR.	3. Definition of perpendicular bisectors.
4. Let ℓ_3 be any line in plane P, containing O, but distinct from ℓ_1 and ℓ_2. (If $\overline{AB}$ meets ℓ_3, label the intersec-	4. Through any point in a plane an infinite number of lines can be drawn.

tion C, where A lies
in ℓ_2 and B lies in
ℓ_1).

5.	$QA = RA$; $QB = RB$.	5.	A line is the perpendicular bisector of a segment if and only if it is the set of all points equidistant from the endpoints of the segment.
6.	$\overline{AB} \cong \overline{AB}$.	6.	Every segment is congruent to itself.
7.	$\triangle QBA \cong \triangle RBA$.	7.	The SSS Postulate.
8.	$\angle QBA \cong \angle RBA$.	8.	Corresponding angles of congruent triangles are congruent.
9.	$\overline{BC} \cong \overline{BC}$.	9.	Every segment is congruent to itself.
10.	$\triangle QBC \cong \triangle RBC$.	10.	The SAS Postulate.
11.	$\overline{CQ} \cong \overline{CR}$.	11.	Corresponding parts of congruent triangles are congruent.
12.	Point C is on the perpendicular bisector of $\overline{QR}$.	12.	The perpendicular bisector of a segment contains all points equidistant from the endpoints of the segment.
13.	ℓ_3 is $\overleftrightarrow{CO}$ and ℓ_3 is a perpendicular bisector of $\overline{QR}$.	13.	If a line contains two points equidistant from the endpoints of a segment, then the line is the perpendicular bisector of the segment.
14.	$\ell_3 \perp \overleftrightarrow{QR}$.	14.	Definition of perpendicular bisector.
15.	$\overleftrightarrow{QR} \perp$ plane P.	15.	A line is perpendicular to a plane if it is perpendicular to every line in the plane that passes through the point of intersection.

● **PROBLEM** 801

Show that if each of two intersecting planes is perpendicular to a third plane, then their intersection is perpendicular to the plane.

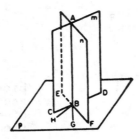

Solution: In the accompanying figure, planes n and m ⊥ p.
n and m intersect at $\overleftrightarrow{AB}$; n and p intersect at $\overleftrightarrow{EF}$; and m and p
intersect at $\overleftrightarrow{CD}$. We must show $\overleftrightarrow{AB}$ ⊥ p, and we prove this by
finding two lines in p that are perpendicular to $\overleftrightarrow{AB}$.

From an earlier theorem, we know that if two planes are
perpendicular to each other, a line drawn in one plane perpendicular to their intersection is perpendicular to the other
plane. Consider line $\overline{BH}$ in p such that $\overline{BH}$ ⊥ $\overleftrightarrow{EF}$. Since $\overleftrightarrow{EF}$ is
the intersection of planes n and p, $\overline{BH}$ ⊥ n. A line perpendi-
cular to a plane is perpendicular to every line in the plane
that intersects the perpendicular. Therefore, $\overline{BH}$ ⊥ $\overleftrightarrow{AB}$.
Similarly, a line in p perpendicular to $\overleftrightarrow{CD}$, $\overline{BG}$, is perpendicular
to m and thus, $\overline{BG}$ ⊥ $\overleftrightarrow{AB}$. Remember that a line perpendicular to two lines in a plane is perpendicular to the plane.
Since $\overleftrightarrow{AB}$ ⊥ $\overline{BH}$ and $\overleftrightarrow{AB}$ ⊥ $\overline{BG}$, and $\overline{BH}$ and $\overline{BG}$ are lines of plane p,
$\overleftrightarrow{AB}$ ⊥ p.

● **PROBLEM** 802

In the figure, a triangular pyramid V-ABC has $\overline{VC}$ perpendicular
to base ABC. The midpoints of $\overline{VA}$, $\overline{AC}$, and $\overline{CB}$ are R, S, and T,
respectively. The plane through R, S, and T intersects $\overline{VB}$ in
W. Prove RSTW is a rectangle.

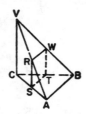

Solution: This solution can best be developed in the form of
a two column proof. We will prove RSTW is a rectangle by
proving that the quadrilateral has a pair of opposite sides
parallel and congruent and that one vertex angle is a right
angle.

Statement	Reason
1. $\overline{VC}$ ⊥ plane ABC.	1. Given.
2. $\overline{VC}$ ⊥ $\overline{CA}$.	2. A line perpendicular to a plane is perpendicular to every line in the plane passing through its foot.

766

3. R is the midpoint of
   $\overline{VA}$. S is the midpoint
   of $\overline{CA}$.

3. Given.

4. $\overline{RS} \parallel \overline{VC}$ and RS = $\frac{1}{2}$VC.

4. If a line joins the midpoints of two sides of a
   triangle, then it is parallel to the third side and
   equal to one half the length
   of that side.

5. $\overline{RS} \perp$ plane ABC.

5. If one of two parallel lines
   is perpendicular to a plane,
   then the other line is also
   perpendicular to the plane.

6. $\overline{RS} \perp \overline{ST}$.

6. Reason 2.

7. $\overline{VC} \parallel$ plane RSTW.

7. If two lines are parallel, then
   every plane containing one
   and only one of the lines is
   parallel to the other line.

8. $\overline{VC} \parallel \overline{WT}$.

8. If a line is parallel to a
   plane, then it is parallel
   to the line of intersection
   of that plane with any plane
   containing the line.

9. W is the midpoint
   of $\overline{VB}$.

9. If a line bisects one side
   of a triangle and is parallel to a second side, then
   it bisects the third side.

10. T is the midpoint of
    $\overline{CB}$.

10. Given.

11. WT = $\frac{1}{2}$CV.

11. The line joining the mid-
    points of two sides of a
    triangle is equal to $\frac{1}{2}$ the
    third side.

12. RS = WT.

12. Things equal to the same
    quantity are equal to each
    other.

13. $\overline{RS} \parallel \overline{WT}$.

13. If two lines are parallel
    to a third line then they
    are parallel to each other.

14. $\angle$RST is a right angle.

14. Perpendicular lines inter-
    sect to form right angles
    (See Step 5.).

15. RSTW is a rectangle.

15. A quadrilateral in which
    a pair of opposite sides
    are parallel and equal and
    which contains a right
    angle is a rectangle.

POLYHEDRAL ANGLES

> Prove that the sum of the measures of any two faces of a
> trihedral angle is greater than the measure of the third angle
> face.

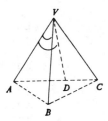

Solution: A polyhedral angle is the figure formed by three or
more planes that intersect in one point. The point of intersection of all the planes is called the vertex of the polyhedral angle. The planes themselves are called the faces; the
lines that form the intersection of the faces are called edges;
and the plane angles formed by the intersection of the two
edges of a face are face angles. The trihedral angle has three
face angles. The three face angles are related somewhat like
the sides of a triangle. To see this, have a plane that is not
a face intersect the tetrahedral angle. The intersection is a
triangle, the sides of which are partially determined by the
size of the face angles that subtend them. Therefore, the restrictions upon the sides of a triangle should have some analogy to the restrictions upon the face angles.

The triangle inequality for plane triangles requires that
the sum of the measures of any two sides of a triangle be
greater than the measure of the third side. Here, we are asked
to prove the comparable theorem for trihedral angles. It is
not surprising that we use the Triangle Inequality Theorem.

> Given: Trihedral ∢V-ABC; ∢AVC is the face angle of great-
> est measure.
>
> Prove: m∢AVB + m∢BVC > m∢AVC.
>
> Proof Outline: We find point D on the plane VAC such
> that ∢AVC = ∢AVD + ∢DVC and ∢AVD ≅ ∢AVB. Then,
> to show m∢AVB + m∢BVC > m∢AVC, we need only show
> m∢BVC > m∢DVC. By Triangle Inequality Theorem,
> we show BC > DC and that ∢BVC of ΔBVC > ∢DVC of
> ΔDVC.

Statements	Reasons
1. Trihedral ∢V-ABC; ∢AVC is the face angle of greater measure.	1. Given.
2. Draw $\overline{AC}$.	2. Two distinct points deter-

768

3. Construct, in face VAC, ∢AVD ≅ ∢AVB, such that ADC.

3. If an angle A is larger than a given angle, then a ray can be constructed such that it forms, with one side of ∢A, an angle congruent to the given angle and such that every point of the ray is in the interior of ∢A.

4. VD ≅ VB.

4. On a given line, a segment can be determined such that it is congruent to a given segment. (There is no restriction on point B, other than it is on edge VB of the trihedral angle. Therefore, we arbitrarily choose point B such that VD = VB.)

5. VA ≅ VA.

5. Every segment is congruent to itself.

6. △VAB ≅ △VAD.

6. SAS Postulate.

7. AB ≅ AD

7. Corresponding sides of congruent triangles are congruent.

8. AB + BC > AC.

8. Triangle Inequality Theorem.

9. AC = AD + DC.

9. Definition of Betweenness.

10. AB + BC > AD + DC or BC > DC.

10. Substitution Postulate and Subtraction Postulate.

11. VC ≅ VC.

11. Every segment is congruent to itself.

12. m∢BVC > m∢DVC.

12. If two sides of a triangle (△VDC) are congruent to two sides of a second triangle (△BVC), then the included angle of the first triangle is greater than the included angle of the second triangle if the third side of the first triangle is greater than the third side of the second.

13. m∢AVC = m∢AVD + m∢DVC.

13. If D is a point on the interior of ∢ABC, then m∢ABC = m∢ABD + m∢DBC (Angle Sum Postulate.)

14. m∢BVC + m∢AVB > m∢AVD + m∢DVC.

14. Substitution from Steps 12 and 3. Also, equals added to both sides of an in-

equality does not change
the inequality.

15. m∢BVC + m∢AVB > m∢AVC.   15. Substitution Postulate.

● **PROBLEM** 804

From a point within the dihedral angle formed by the intersection of two planes, perpendicular lines are drawn to each
plane. (a) Prove that the plane determined by these perpendiculars is perpendicular to the edge of the dihedral angle.
(b) If the number of degrees in the dihedral angle is represented by n, express in terms of n the number of degrees in
the angle formed by the two perpendicular lines.

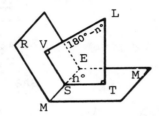

Solution: In the accompanying figure, point L is in the in-
terior of dihedral angle R-$\overline{ME}$-N. $\overline{LT}$ ⊥ plane N and $\overline{LV}$ ⊥ plane
R have been drawn.

We wish to prove plane LVT is ⊥ $\overline{ME}$, the edge of the dihedral angle. To do this remember that if a plane is perpendicular to two planes, then it is also perpendicular to their
line of intersection. We will show that the plane LTV is perpendicular to planes N and R and thus is perpendicular to $\overline{ME}$.

Statement	Reason
1. $\overline{LT}$ ⊥ plane N. $\overline{LV}$ ⊥ plane R.	1. Given.
2. $\overline{LV}$ and $\overline{LT}$ determine plane LVT.	2. Two intersecting lines de-termine a line.
3. Plane LTV ⊥ plane N. Plane LTV ⊥ plane R.	3. A plane containing a line perpendicular to a given plane is perpendicular to the given plane.
4. Plane LTV ⊥ $\overline{ME}$.	4. If a plane is ⊥ to two planes, then it is ⊥ to their line of intersection.

(b) We wish to find the measure of ∢VLT. This angle is part
of a quadrilateral LVTS. The sum of the measures of the angles
of a quadrilateral is 360°.

Since angles V and T each measure 90°, they sum to 180°
and the remaining angles of LVTS, ∢VST and ∢VLT, sum to 360° -

770

$180° = 180°$. We are given m∢VST = n, therefore, we can conclude
m∢VLT = 180° - n.

● **PROBLEM** 805

Show that the sum of the measures of the face angles of any
convex polyhedral angle is less than 360°. An informal argument will suffice.

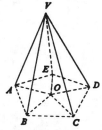

Solution: The only theorem about face angles we have learned
thus far is that, for trihedral angles, the sum of any two
face angles must be greater than the measure of the third.
Here, we use the theorem to relate the supplements of the face
angles to the interior angles of a convex polygon. Let a
plane cut the edges of the polyhedral angle at points A, B, C,
D, E. Let O be any point in the polygon ABCDE. Draw the seg-
ments $\overline{OA}$, $\overline{OB}$, $\overline{OC}$, $\overline{OD}$, and $\overline{OE}$.

Call the n triangles with common vertex O the O-triangles.
Call the n triangles with common vertex V (the triangles formed
by the faces of the polyhedral angle) the V-triangles. Consider the base angles of the V triangles, ∢ABV, ∢BAV, ∢VBC,
∢VCB, etc., and the base angles of the O-triangles, ∢OAB, ∢OBA,
∢OBC, ∢OCB, etc. We show that (1) the sum of the base angles
of the V triangles > the sum of the base angles of the O triangles; and, therefore, (2) the sum of the vertex angles (face
angles) of V-triangles is less than the sum of the vertex
angles (∢AOB, ∢BOC, ∢COD, etc.) of O-triangles. Since the vertex angles of O triangles sum to 360°,because a circle centered
at O can be drawn, the sum of face angles must be less than 360°.

    Given: Polyhedral angle ∢V-ABCDE...with n faces; the sum
            of the measures of the face angles equals S.
            Plane P intersects ∢V-ABCDE...at points A, B, C,
            D, E... .

    Prove: S < 360.

Statements	Reasons
1. Polyhedral angle ∢V-ABCDE...with n faces; the sum of the measures of the face angles equals S. Plane P intersects ∢V-ABCDE...at point A, B, C, D, E...	1. Given.
2. Select any point O in	2. In the interior of the poly-

the interior of convex polygon ABCDE...

gon, there are an infinite number of points.

3. Draw segments from O to each vertex of polygon.

3. Two points determine a line segment. (From this point on, the triangles formed by the faces of the polyhedral angle will be referred to as V-triangles. Those formed by the n triangles in polygon ABCDE, will be called ...the O-triangles. The vertex angles of the V triangles are the face angles. The vertex of the O-triangles have point O as the vertex.)

4. In trihedral ∢B - AVC, m∢ABV + m∢VBC > m∢ABC.

4. The sum of the measures of any two face angles of a trihedral angle is greater than the measure of the third.

5. m∢ABC = m∢ABO + m∢OBC.

5. Angle Sum Postulate.

6. m∢ABV + m∢VBC > m∢ABO + m∢OBC.

6. Substitution Postulate. (Result: Two V-triangle base angles > two O-triangle base angles.)

7. m∢BCV + m∢DCV > m∢BCO + m∢OCD and m∢CDV + m∢EDV > m∢ODC + m∢EDO, etc.

7. Repeat steps 4-6 for trihedral angles ∢C-BVD, ∢D-CVE, etc.

8. m∢BAV + m∢ABV + m∢CBV + m∢BCV... > m∢ABO + m∢OBC + m∢OCB + m∢OCD...

8. If a > b and c ≥ d, then a + c > b + d. [Result: from Steps 6 and 7, we have that (sum of all V-triangle base angles) > (sum of all O-triangle base angles).]

9. m∢AOB + m∢OAB + m∢OBA = = 180. m∢BOC + m∢OBC + m∢OCB = 180. m∢COD + m∢OCD + m∢ODC = 180.
   .
   .
   .
   (rest of the O-triangles)

9. The sum of the measures of the angles of a triangle is 180°. (Note, the equations are in the form of (vertex angle) + (base angles) = 180°.)

10. (m∢AOB + m∢BOC + m∢COD+...) + (m∢OAB + m∢OBC + m∢OCD + m∢OBA + m∢OCB+...) = 180 · n.

10. Addition Property of Equality. (Note there were n triangles, and thus the total number of degrees is 180 · n. Note the expression in the first parenthesis consists o. all O-triangle vertex angles

The second expression consists of all O-triangle base
angles.)

11. $m\angle AVB + m\angle VAB + m\angle VBA$
    $= 180$. $m\angle BVC + m\angle VBC$
    $+ m\angle VCB = 180$.

    .
    .
    .

    (rest of V-triangles)

11. The measures of the angles
    of a triangle sum to 180°.

12. $(m\angle AVB + m\angle BVC + m\angle CVD + ...) +$
    $(m\angle VAB + m\angle VBC + \angle VCD + m\angle VBA + m$
    $VCB + ...) = 180 \cdot n$.

12. Addition Property of Equal-
    ity. (Note the vertex
    angles are in the first
    parenthesis.)

13. $(m\angle AVB + m\angle BVC +$
    $m\angle CVD...) + (m\angle VAB +$
    $m\angle VBC + ...) = (m\angle AOB$
    $+ m\angle BOC + m\angle COD...) +$
    $(m\angle OAB + m\angle OBC...)$.

13. Transitivity Property.
    Both expressions in Step
    10 and 12 equal $180 \cdot n$.

14. $m\angle AVB + m\angle BVC +$
    $m\angle CVD... > m\angle AOB +$
    $m\angle BOC + m\angle COD...$.

14. If $a > b$ and $c = d$, then
    $c - a < d - b$. In this
    case a and b correspond to
    Step 8.

15. $m\angle AOB + m\angle BOC... =$
    $360°$.

15. The measure of the sum of
    angles that form a circle
    is 360°.

16. $m\angle AVB + m\angle BVC +$
    $m\angle CVD = S$.

16. Given.

17. $S < 360°$.

17. Substitution of Steps 15
    and 16 in 14.

CHAPTER 46

RECTANGULAR SOLIDS, PRISMS, AND PYRAMIDS

> **Basic Attacks and Strategies for Solving Problems in this Chapter. See pages 774 to 791 for step-by-step solutions to problems.**

The figures below are all prisms. A **prism** is a polyhedron in which two congruent polygonal faces are in parallel planes, and the other faces are parallelograms. The two parallel faces are called the **bases**. The bases of the prism on the left are pentagons, so it is called a **pentagonal prism**, while the bases of the figures on the right are triangles, so they are called **triangular prisms**. In the prism on the left and also the one on the right, all the faces except the bases are rectangles. Prisms of this kind are called **right prisms**. A rectangular right prism is usually called a **rectangular solid**. A picture of such a prism is shown in Problem 806.

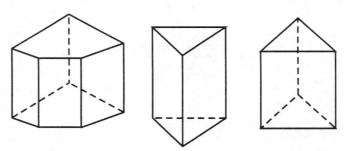

A **pyramid** is a polyhedron with one face called a base, with all other faces being triangles which share a common vertex. If the base of a pyramid is a regular polygon and if all the triangular faces are congruent, then the pyramid is said to be a **regular pyramid**. As was the case with prisms, pyramids are classified in terms of their base. The pyramid pictured on the next page is a square pyramid and is regular.

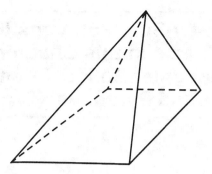

Sometimes it helps to visualize polyhedra by having the faces cut and opened up. Suppose that for the square pyramid pictured above, you cut so as to separate the triangular faces, and then opened up the figure. Here is what you would see.

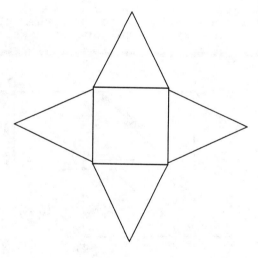

This is called a net for the pyramid. In the solution given for Problem 806, Figure 2 is a net for the rectangular solid pictured in Figure 1.

RECTANGULAR SOLIDS & PRISMS

● **PROBLEM** 806

A spider on the ceiling in one corner of a 18' × 15' × 8' room sees a fly on the floor at the opposite corner (See Figure 1.). Find the shortest path that the spider can take.

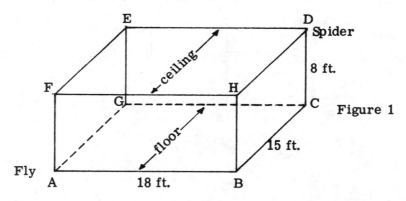

Figure 1

Solution: The elimination of confusing and irrelevant factors is the main difficulty in distance problems. For example, a person may drive in as straight a line as possible from New York to California, but because of mountains, valleys, and potholes in the road, the distance travelled will be greater than the straight line distance between New York and California. To determine the straight line distance, one must be able to eliminate the mountains and valleys.

In this problem, we are not concerned with the number of direction changes or transitions from wall to ceiling.

Therefore, we re-draw the picture, eliminate the wall-ceiling angles and wall-wall angles - this problem's equivalent of mountains and valleys. (See Figure 2.)

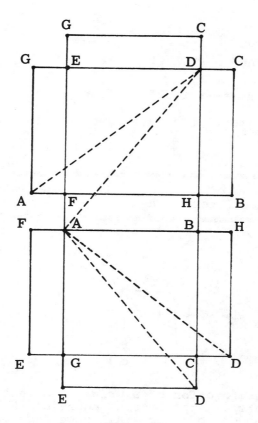

Figure 2

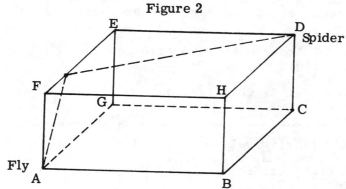

Figure 3

[Note that in flattening out the room, for completeness, three of the walls are represented twice. Thus, there are two locations D for the spider and two positions A for the fly.]

The problem reduces to finding the shortest distance between two points in a plane. Any one of the four dotted lines in Figure 2 is correct. Using the Pythagorean Theorem, we obtain:

(i) $AD^2 = AH^2 + DH^2$

(ii) $AD^2 = (AF + FH)^2 + (15 \text{ ft.})^2$

(iii) $AD^2 = (8 \text{ ft.} + 18 \text{ ft.})^2 + (15 \text{ ft.})^2$

(iv) $AD^2 = 676 \text{ ft.}^2 + 225 \text{ ft.}^2$

(v) $AD = \sqrt{901 \text{ ft.}^2} \simeq 30 \text{ ft.}$

The shortest path is illustrated in Figure 3.

● **PROBLEM** 807

Show that the plane of a right section of a prism is perpendicular to all its lateral edges.

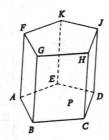

<u>Solution</u>: A right section of a prism is a section formed by a plane which cuts all the lateral edges of the prism and is perpendicular to one of them. All edges are parallel to each other. Therefore, if one edge is perpendicular, all of them are.

Given: Prism P with lateral edges $\overline{AF}$, $\overline{BG}$, $\overline{CH}$, $\overline{DJ}$, $\overline{EK}$...; right section R; edge $\overline{AF} \perp$ R.

Prove: $\overline{BG} \perp$ R, $\overline{CH} \perp$ R, ...

Statements	Reason
1. Prism P with lateral edges $\overline{AF}$, $\overline{BG}$, $\overline{CH}$, $\overline{DJ}$, $\overline{EK}$...; right section R; edge $\overline{AF} \perp$ R.	1. Given.
2. AF ‖ BG ‖ CH ‖ JD...	2. The lateral edges of a prism are congruent and parallel.
3. $\overline{BG} \perp$ R, $\overline{CH} \perp$ R, $\overline{JD} \perp$ R...	3. If a plane is perpendicular to one of a set of parallel lines, then it is perpendicular to every one of the parallel lines.

● **PROBLEM** 808

The lateral edges of a prism are congruent and parallel. Prove this.

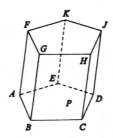

<u>Solution</u>: A prism is a polyhedron whose faces consist of two
parallel and congruent polygons, called bases, and the parallelograms, called lateral faces, formed by connecting pairs
of corresponding vertices of the parallel polygons. Since
the lateral edges are opposite sides of parallelograms, by
transitivity, we can show they are all congruent.

   Given: Prism P with lateral edges $\overline{AF}$, $\overline{BG}$, $\overline{CH}$, $\overline{DJ}$, $\overline{EK}$.

   Prove: $\overline{AF} \cong \overline{BG} \cong \overline{CH} \cong \overline{DJ} \cong \overline{EK}$

           $\overline{AF} \parallel \overline{BG} \parallel \overline{CH} \parallel \overline{DJ} \parallel \overline{EK}$

Statements	Reasons
1. Prism P with lateral edges $\overline{AF}$, $\overline{BG}$, $\overline{CH}$, $\overline{DJ}$, $\overline{EK}$...	1. Given.
2. Quadrilateral FGBA, GHCB, HCDJ... are parallelograms.	2. The lateral faces of a prism are parallelograms.
3. $\overline{AF} \parallel \overline{BG}$, $\overline{BG} \parallel \overline{CH}$,...	3. Opposite sides of a parallelogram are parallel.
4. $\overline{AF} \parallel \overline{BG} \parallel \overline{CH}$...	4. If a line is parallel to a given line, then it is parallel to all lines parallel to the given line.
5. AF $\cong$ BG, BG $\cong$ CH,...	5. Opposite sides of a parallelogram are congruent.
6. $\overline{AF} \cong \overline{BG} \cong \overline{CH}$...	6. Segments congruent to a given segment are congruent (Transitive Property).

● **PROBLEM** 809

Show that every section of a prism made by a plane parallel
to the bases is congruent to the bases.

<u>Solution</u>: By inspection, we see that the plane cuts the prism
into two prisms and, therefore, the section cut by the plane
is a base of both prisms and congruent to both. To prove
this we must show the polygon formed by the plane and the

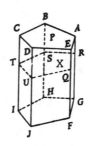

prism must have sides and angles congruent to the corresponding sides and angles of the base. We prove the sides congruent by showing that the faces of the "new prism" are parallelograms and, since the opposite sides of a parallelogram are congruent, corresponding sides of the polygons are congruent. To show the angles are congruent, we prove that all triangles in the new polygon formed by three consecutive vertices are congruent to the triangles formed by the corresponding three vertices in the base. Thus, the interior angles are congruent by definition of triangle congruency.

Given: Prism P intersects plane X in polygon QRSTU...;
        Plane X ∥ bases of P.

Prove: Polygon QRSTU... ≅ base of P.

Statements	Reasons
1. Prism P intersects plane X in polygon QRSTU...; plane X ∥ bases of P.	1. Given.
2. $\overline{UJ}$ ∥ $\overline{QF}$ ∥ $\overline{RG}$...	2. The lateral edges of a prism are parallel.
3. $\overline{UQ}$ ∥ $\overline{JF}$, $\overline{QR}$ ∥ $\overline{FG}$...	3. If a plane intersects two parallel planes, then they intersect in two parallel lines.
4. Quadrilaterals UQFJ, QRGF, TUJI... are parallelograms.	4. If both pairs of opposite sides of a quadrilateral are parallel, then the quadrilateral is a parallelogram.
5. $\overline{UQ}$ ≅ $\overline{JF}$, $\overline{QR}$ ≅ $\overline{FG}$, $\overline{TU}$ ≅ $\overline{IJ}$...	5. Opposite sides of a parallelogram are congruent.
6. $\overline{TQ}$ ∥ $\overline{IF}$, $\overline{UR}$ ∥ $\overline{JG}$...	6. If a plane intersects two parallel planes, then they intersect in two parallel lines.
7. Quadrilaterals TIFQ, UJGR,... are parallel.	7. If both pairs of opposite sides of a quadrilateral are parallel, then the quadrilateral is a parallelogram.

8. $\overline{TQ} \cong \overline{IF}$, $\overline{UR} \cong \overline{JG}$.

8. Opposite sides of a paral-lelogram are congruent.

9. $\triangle TUQ \cong \triangle IJF$, $\triangle UQR \cong \triangle JFG...$

9. The SSS Postulate.

10. $\sphericalangle TUQ \cong \sphericalangle IJF$, $\sphericalangle UQR \cong \sphericalangle JFG...$

10. Corresponding angles of con-gruent triangles are con-gruent.

11. Polygon QRSTU... $\cong$ base of P.

11. If the corresponding angles and the corresponding sides of two polygons are congru-ent, then the two polygons are congruent.

PYRAMIDS

● **PROBLEM** 810

In a regular square pyramid, the length of each side of the square base is 12 in., and the length of the altitude is 8 in.

(a) Find the length of the slant height of the pyramid.

(b) Find, in radical form, the length of the lateral edge of the pyramid.

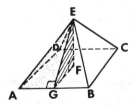

<u>Solution</u>: Both parts of this example will involve an appli-cation of the Pythagorean Theorem in solid geometry. If h is the length of the hypotenuse, and a and b the lengths of the other legs of a right triangle, then, $h^2 = a^2 + b^2$.

(a) The slant height is the perpendicular from the ver-tex of the pyramid to any side of the base. Since each face of a regular pyramid is an isosceles triangle, the slant height bisects the base. To find the slant height, note that slant height $\overline{EG}$ is the hypotenuse of $\triangle EFG$. The altitude of the pyramid, $\overline{EF}$, as shown in the figure, is perpendicular to the plane of the base ABCD, by definition. As such, $\overline{EF} \perp \overline{FG}$, because $\overline{FG}$ lies in the plane of the base and intersects $\overline{EF}$ at the latter's point of intersection with the base. The al-titude must be drawn to the center of the base. Therefore, FG equals one-half the length of a side of the base.

$$FG = \frac{1}{2}(12) \text{ in.} = 6 \text{ in.}$$

Since △EFG contains a right angle, it is a right triangle. $\overline{EG}$ is the hypotenuse, or slant height. Therefore,

$$(EG)^2 = (EF)^2 + (FG)^2.$$

By substitution, $(EG)^2 = (8 \text{ in.})^2 + (6 \text{ in.})^2 = (64 + 36) \text{ in.}^2$

$$= 100 \text{ in.}^2$$

$$EG = \sqrt{100} \text{ in.} = 10 \text{ in.}$$

Therefore, the slant height is 10 in.

(b)　The lateral edge of a pyramid is the intersection of any two adjacent lateral faces. It is given, in this problem, by the hypotenuse of the right triangle formed when the slant height is drawn, namely hypotenuse $\overline{EA}$ of right triangle EAG. Since the slant height $\overline{EG}$ is the altitude to the base of isosceles triangle AEB, G is the midpoint of $\overline{AB}$. Therefore, AG equals one-half the length of the side of the base, i.e. AG = $\frac{1}{2}$(12) in. = 6 in.

In rt. △EGA, $(EA)^2 = (EG)^2 + (AG)^2.$

By substitution, $(EA)^2 = (10 \text{ in.})^2 + (6 \text{ in.})^2 = (100 + 36) \text{ in.}^2$

$$= 136 \text{ in.}^2$$

$$EA = \sqrt{136} \text{ in.} = \sqrt{4} \cdot \sqrt{34} \text{ in.} = 2\sqrt{34} \text{ in.}$$

Therefore, the length of the lateral edge is $2\sqrt{34}$ in.

● **PROBLEM** 811

Show that the lateral edges of a regular pyramid are congruent.

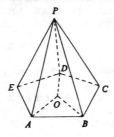

Solution:　A regular pyramid is a pyramid the base of which forms a regular polygon whose center coincides with the projection of the vertex onto the base. Therefore, (see figure) $\overline{PO}$ is perpendicular to the plane of the base. Where we want to show congruence, it is wise to look for congruent triangles. Here, the common factors are (1) $\overline{PO}$ is centrally located and, therefore, available to be a congruent side; (2) O is the center of the base polygon. Therefore, AO = OB = OC..., (3)

$\overline{PO}$ is given to be perpendicular to the base plane; therefore, ⊀POB, ⊀POA, ⊀POC... are all right angles. Therefore, ΔPAO ≅ ΔPBO ≅ ΔPCO... by the SAS Postulate, and the congruence of lateral edges follows from corresponding parts.

Given: Regular pyramid P with base ABCDE... and the center of base O.

Prove: $\overline{PA} \cong \overline{PB} \cong \overline{PC}$.

Statements	Reasons
1. Regular pyramid P with base ABCDE...and the center of base O.	1. Given.
2. $\overline{AO} \cong \overline{OB} \cong \overline{OC} \cong$ ...	2. The center of a regular polygon is equidistant from the vertices.
3. $\overline{PO} \cong \overline{PO}$.	3. Every segment is congruent to itself.
4. $\overline{PO} \perp$ plane EABCD.	4. The projection of the vertex of a regular pyramid coincides with the center of the base.
5. ⊀POA ≅ ⊀POB ≅ ⊀POC...	5. All right angles are congruent.
6. ΔPOA ≅ ΔPOB ≅ ΔPOC...	6. The SAS Postulate.
7. $\overline{PA} \cong \overline{PB} \cong \overline{PC}$...	7. Corresponding sides of congruent triangles are congruent.

● **PROBLEM** 812

Prove: If a pyramid is cut by a plane parallel to its base, as shown in the diagram, then:

(1) $\dfrac{VA'}{VA} = \dfrac{VB'}{VB} = \dfrac{VC'}{VC} = \dfrac{VD'}{VD} = \dfrac{VE'}{VE} = \dfrac{VK'}{VK}$

(2) $\dfrac{A'B'}{AB} = \dfrac{B'C'}{BC} = \dfrac{C'D'}{CD} = \dfrac{D'E'}{DE} = \dfrac{E'A'}{EA} = \dfrac{VA'}{VA} = \cdots = \dfrac{VK'}{VK}$

(3)   Polygon A'B'C'D'E' ∿ Polygon ABCDE.

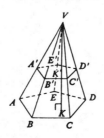

<u>Solution</u>: There are two parts to the proof. First we show that the plane divides each face of the pyramid into similar triangles and, therefore, the lateral edges and the altitude are divided proportionally. Second to show the polygon created by the intersection of the plane and the pyramid is similar to the base polygon by showing that all corresponding angles are congruent.

Given: Pyramid V-ABCDE... cut by a plane parallel to the base, intersecting the lateral edges <u>in</u> A', B', C', D', E', ... and the altitude $\overline{VK}$ in K'.

Prove: (1) $\dfrac{VA'}{VA} = \dfrac{VB'}{VB} = \ldots \dfrac{VK'}{VK}$

(2) $\dfrac{A'B'}{AB} = \dfrac{VA'}{VA} \ldots = \dfrac{VK'}{VK}$.

(3) Polygon A'B'C'D'E' $\sim$ Polygon ABCDE.

Statements	Reasons
1. Pyramid V-ABCDE... cut by a plane parallel to the base, intersecting the lateral edges in A',B',C', D',E' and the altitude $\overline{VK}$ in K'.	1. Given.
2. $\overleftrightarrow{A'K'} \parallel \overleftrightarrow{AK}$.	2. If a plane (in this case, the plane determined by the intersecting lines $\overleftrightarrow{AV}$ and $\overleftrightarrow{AK}$) intersects a pair of parallel planes, then the intersection is a pair of parallel lines.
3. $\dfrac{VA'}{VA} = \dfrac{VK'}{VK}$.	3. A line ($\overleftrightarrow{A'K'}$) that intersects two sides of a triangle ($\triangle VAK$) and is parallel to the third side, divides the two sides proportionally.
4. $\dfrac{VB'}{VB} = \dfrac{VK'}{VK}$, $\dfrac{VC'}{VC} = \dfrac{VK'}{VK}$,...	4. Repeat the procedure of Steps 3 and 4 for triangle $\triangle VBK$, $\triangle VCK$...
5. $\dfrac{VA'}{VA} = \dfrac{VB'}{VB} = \dfrac{VC'}{VC} \ldots = \dfrac{VK'}{VK}$.	5. Transitivity Postulate.
6. $\sphericalangle AVB = \sphericalangle A'VB'$.	6. Reflexive Property.
7. $\triangle A'VB' \sim \triangle AVB$.	7. If two sides of a triangle are similar to two sides of a second triangle, and the

	included angle of the first triangle is congruent to the included angle of the second triangle, then the two triangles are similar.
8. $\dfrac{A'B'}{AB} = \dfrac{VA'}{VA}$.	8. Corresponding sides of similar triangles are similar.
9. $\dfrac{B'C'}{BC} = \dfrac{VB'}{VB}$ $\dfrac{C'D'}{CD} = \dfrac{VC'}{VC}$, ...	9. Repeat the procedure of Steps 6, 7, and 8 for $\triangle BVC$, $\triangle CVD$, ...
10. $\dfrac{A'B'}{AB} \quad \dfrac{B'C'}{BC} \quad \dfrac{C'D'}{CD}$ $... = \dfrac{VA'}{VA} = ... =$ $\dfrac{VK'}{VK}$.	10. Transitivity Postulate and Substitution from Steps 8, 9, and 5.
11. $\overset{\longleftrightarrow}{AC} \parallel \overset{\longleftrightarrow}{A'C'}$.	11. If a plane (determined in this case by intersecting lines $\overset{\longleftrightarrow}{AV}$ and $\overset{\longleftrightarrow}{CV}$) intersects a pair of parallel planes, the intersection is a pair of parallel lines.
12. $\triangle A'VC' \sim \triangle AVC$.	12. If a line intersects two sides of a triangle ($\triangle AVC$) and is parallel to the third side, then the line divides the two sides proportionally and the triangle formed with the line as the third side (A'VC') is similar to the original triangle.
13. $\dfrac{A'C'}{AC} = \dfrac{VA'}{VA}$.	13. The sides of similar triangles are proportional.
14. $\dfrac{A'C'}{AC} = \dfrac{VA'}{VA} = \dfrac{VB'}{VB}$	14. Transitivity Postulate. (from steps 13 and 10)
15. $\triangle A'B'C' \sim \triangle ABC$.	15. The SSS-Similarity Theorem (If three sides of a triangle are proportional to three sides of a second triangle, then the two triangles are similar.)
16. $\sphericalangle A'B'C' \cong \sphericalangle ABC$.	16. Corresponding angles of similar triangles are congruent.
17. $\sphericalangle B'C'D' \cong \sphericalangle BCD$, $\sphericalangle C'D'E' \cong \sphericalangle CDE$...	17. Repeat the procedure of Steps 12 - 16.
18. Polygon A'B'C'D'E' $\sim$ Polygon ABCDE.	18. Two polygons are similar if all corresponding angles

are congruent and all corresponding sides are pro-
portional.

● **PROBLEM** 813

If two pyramids have congruent altitudes and bases with
equal areas, show that sections parallel to the bases at
equal distances from the vertices have equal area.

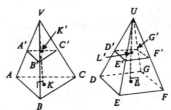

Solution: Planes parallel to the base intersect a pyramid in
a polygon similar to the base, with the ratio of similitude
dependent on the height of the pyramid and the distance between the polygon and the base. From an earlier theorem, we
know that the ratio of the areas of similar polygons is proportional to the square of the ratio of similitude.

Given two sections parallel to the bases at equal distances and two pyramids with congruent altitudes, the ratio
of similitudes is equal if the ratio of areas is equal. If
we are further told that the bases of the pyramids have equal
areas, then the area of the sections must also be equal.

Given: Pyramids V-ABC and U-DEFG, where $\overline{VK}$ is an alti-
tude of V-ABC and $\overline{UL}$ an altitude of U-DEFG;
$\overline{VK} \cong \overline{UL}$. Area of $\triangle ABC$ = Area of DEFG; plane
A'B'C' ‖ plane ABC and meets $\overline{VK}$ at K'; plane
D'E'F'G' ‖ plane DEFG and meets $\overline{UL}$ in L';
$\overline{VK'} \cong \overline{UL'}$.

Prove: Area of $\triangle A'B'C'$ = Area of quadrilateral D'E'F'G'.

Statements	Reasons
1. (See above).	1. Given.
2. $\triangle A'B'C' \sim \triangle ABC$ quadrilateral DEFG $\sim$ D'E'F'G'.	2. If a pyramid is cut by a plane parallel to its base, then the section is a poly-gon similar to the base.
3. The ratio of simili-tude of $\triangle A'B'C'$ and $\triangle ABC = \frac{A'B'}{AB}$. The ratio of similitude of quad D'E'F'G' and DEFG $= \frac{D'E'}{DE}$.	3. The ratio of similitude of two similar polygons is the proportion of the lengths of corresponding sides.

4. The ratio of similitude of $\triangle A'B'C'$ and $\triangle ABC = \dfrac{VK'}{VK}$. The ratio of similitude of quad $D'E'F'G'$ and $DEFG = \dfrac{UG'}{UG}$.

4. If a pyramid is cut by a plane parallel to its base, then the ratio of the lengths of the altitudes is equal the ratio of similitude of the base and the section.

5. $\dfrac{\text{Area of } A'B'C'}{\text{Area of } ABC} = \left(\dfrac{VK'}{VK}\right)^2$.
$\dfrac{\text{Area of quad } D'E'F'G'}{\text{Area of quad } DEFG} = \left(\dfrac{UG'}{UG}\right)^2$.

5. The ratio of the area of two similar polygons is the square of the ratio of similitude.

6. $\dfrac{VK'}{VK} = \dfrac{UG'}{UG}$

6. Division Property of Equality.

7. $\dfrac{\text{Area of } \triangle A'B'C'}{\text{Area of } \triangle ABC} = \dfrac{\text{Area of quad } D'E'F'G'}{\text{Area of quad } DEFG}$.

7. Transitivity Postulate.

8. Area of $\triangle ABC$ = Area of quad $DEFG$.

8. Given.

9. Area of $\triangle A'B'C'$ = Area of quad $D'E'F'G'$.

9. Multiplication Property of Equality.

● **PROBLEM** 814

Show that the perpendicular from the vertex to the base of a regular pyramid contains only points that are equidistant from the faces.

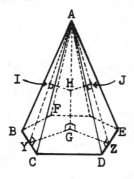

Solution: In the figure, regular pyramid A-BCDEF... has regular polygon BCDEF... as a base. $\overline{AG}$ is the perpendicular from vertex A to the base polygon. $\overline{AY}$ and $\overline{AZ}$ are slant heights. What we must show is that any point on $\overline{AG}$, point H for example, is equidistant from each face of the pyramid. To do this, we prove that the distance from the point H to any two arbitrary sides is equal. Since H is equidistant from any two sides, it is equidistant from every side.

Let the two arbitrary faces be △ABC and △ADE, and let the perpendicular from H to △ABC be $\overline{HI}$, and the perpendicular from H to △ADE be $\overline{HJ}$. What we must show is HI = HJ.

Note that HI and JH are corresponding parts of △HIA and △HJA. In △HIA and △HJA, there are two known correspondences: Since $\overline{HI}$ and $\overline{HJ}$ are perpendiculars to the faces, ∢HIA and ∢HJA are right angles. Thus ∢HIA = ∢HJA. Furthermore, $\overline{HA} \cong \overline{HA}$. To prove congruence of triangles, though, we need another correspondence.

The other correspondence is ∢IAH ≅ ∢JAH and this follows from (a) proving △YAG ≅ △ZAG and therefore, ∢YAG ≅ ∢ZAG; and (b) ∢YAG and ∢IAH are the same angle and ∢ZAG and ∢JAH are the same angle.

Thus our method of attack is:

(1) Prove △YAG ≅ △ZAG and thus ∢YAG ≅ ∢ZAG.

(2) Show that $\overline{YIA}$ and $\overline{ZJA}$. Since $\overline{AHG}$, this means ∢YAG and ∢IAH have the same sides and therefore are the same angle. Similarly for ∢ZAG and ∢JAH.

(3) With three correspondences available to us, we show △HAI ≅ △HAJ.

(4) By corresponding parts, HI = HJ.

Therefore, we will have shown that any point on the perpendicular from the vertex to the base of a regular pyramid is equidistant from any two faces (and thus equidistant from every face) of the pyramid.

Given: Regular pyramid A-BCDEF...; AG ⊥ plane BCDEF...; $\overline{AHG}$; Point I is on the face ABC and point J is on face ADE such that $\overline{IH}$ ⊥ plane ABC and $\overline{HJ}$ ⊥ plane ADE.

Prove: IH = HJ.

Statements	Reasons
1. (See above).	1. Given
2. Point G is the center of regular polygon BCDE...	2. The projection of the vertex of a regular pyramid onto the base is the center of the regular polygon that forms the base.
(1) We show △YAG ≅ △ZAG by the SAS Postulate.	
3. Locate Y such that $\overline{GY}$ ⊥ $\overline{BC}$ and Z such that $\overline{GZ}$ ⊥ $\overline{DE}$.	3. From an external point, only one perpendicular can be drawn to a given line.

4. $\overline{GY} \cong \overline{GZ}$.	4. The center of a regular polygon is equidistant from the sides.
5. ∢AGY and ∢AGZ are right angles.	5. If a line is perpendicular to a plane, then it forms a right angle with every line in the plane that intersects it.
6. ∢AGY ≅ ∢AGZ.	6. All right angles are congruent.
7. $\overline{AG} \cong \overline{AG}$.	7. A segment is congruent to itself.
8. △YAG ≅ △ZAG.	8. The SAS Postulate.
9. ∢YAG ≅ ∢ZAG.	9. Corresponding parts of congruent triangles are congruent.

(2) To show $\overline{YIA}$ or that points Y, I, and A are collinear, we show that the perpendicular $\overleftrightarrow{HI}$ must lie in the plane YAHG. Since I is also on plane ABC, I is on the intersection of planes YAHG and ABC. Since the intersection of two planes is a line, point I must lie on $\overleftrightarrow{YA}$, the intersection of YAHG and ABC. Thus $\overline{YIA}$. A similar procedure is used to show $\overline{AJZ}$.

10. Plane YAHG ⊥ plane BCDEF...	10. If a plane (YAHG) contains a line ($\overleftrightarrow{AG}$) perpendicular to a given plane (BCDEF), then the two planes are perpendicular.
11. $\overleftrightarrow{BC}$ ⊥ plane YAHG.	11. If a line ($\overleftrightarrow{BC}$) in one of two perpendicular planes is perpendicular to the line of intersection ($\overleftrightarrow{GY}$), then it is perpendicular to the other plane.
12. Plane ABC ⊥ plane YAHG.	12. If a plane (ABC) contains a line ($\overleftrightarrow{BC}$) perpendicular to a given plane (YAHG), then the two planes are perpendicular.
13. I is on $\overleftrightarrow{YA}$ or $\overleftrightarrow{YIA}$.	13. Given two perpendicular planes, the perpendicular to one plane from a point on the second plane must intersect the second plane at the line of intersection of the two planes.
14. J is on $\overleftrightarrow{AZ}$ or $\overleftrightarrow{AJZ}$.	14. Repeat the procedure 10 to 13 for face ADE, and length HJ.
15. m∢YAG = m∢IAH and m∢ZAG = m∢JAH.	15. Two angles with the same two rays for sides are

16. ∢IAH ≅ ∢JAH.	16. Restatement of Step 9.
17. $\overline{AH}$ ≅ $\overline{AH}$.	17. A segment is congruent to itself.
18. ∢HIA and ∢HJA are right angles.	18. The perpendicular from an external point to a plane creates right angles with every line in the plane that intersects it.
19. ∢HIA ≅ ∢HJA.	19. All right angles are congruent.
20. ΔHIA ≅ ΔHJA.	20. The AAS Postulate.
21. $\overline{HI}$ = $\overline{HJ}$.	21. Corresponding parts of congruent triangles are congruent.

Note: In the course of this proof, a subsidiary theorem was proved which should not be overlooked:

(i) In a regular pyramid, the plane formed by the perpendicular from the vertex and any slant height is perpendicular to the face plane.

and a corollary to this,

(ii) The perpendicular to a face of a regular pyramid, drawn from any point on the pyramid altitude, intersects the face on the slant height.

● **PROBLEM** 815

Given that the perpendicular from the vertex to the base of a regular pyramid contains only points equidistant from the faces, show the converse. That is, show that if a point in the interior of a regular pyramid is equidistant from the faces, then it lies on the perpendicular drawn from the vertex.

Solution: In the accompanying Figure 1, regular pyramid A-BCD has altitude $\overline{AG}$. We wish to show that every point H equidistant from the faces of A-BCD must be on the perpendicular $\overline{AG}$. The best method here is to use indirect proof; that is, assume that there is a point equidistant from the faces not on the perpendicular $\overline{AG}$ and then disprove this supposition.

Suppose point H is equidistant from the faces of A-BCD and does not lie on perpendicular $\overline{AG}$. Through point H, construct the plane B'C'D' parallel to the base ΔBCD. Let the intersection of the perpendicular $\overline{AG}$ and plane B'C'D' be the point H'.

Consider the properties of the tetrahedron thus formed, A-B'C'D'. (Remember that a pyramid with the same vertex as the given pyramid and with a base formed by a plane parallel

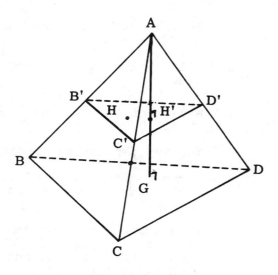

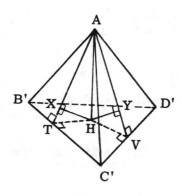

Figure 1   Figure 2

to the base of the given pyramid, is a similar pyramid.)
Since A-BCD is a regular pyramid, A-B'C'D' is also a regular
pyramid. Furthermore, since $\overline{AH'G}$ is perpendicular to one of
two parallel planes, BCD, it must be perpendicular to the
other parallel plane. Thus, $\overline{AH'} \perp$ plane B'C'D'. Therefore,
AH' is the altitude of regular pyramid A-B'C'D' and H' must
be the center of the base polygon B'C'D'.

We will now show that point H is also the center of the
base polygon B'C'D'. Thus, regular polygon B'C'D' has two
centers H and H'. However, a regular polygon can only have
one center. Thus our supposition that there existed any
point H must be wrong. Therefore, no point outside the perpendicular $\overline{AG}$ can be equidistant from the faces; which proves
that if a point in the interior of the regular pyramid is
equidistant from the faces, then it lies on the perpendicular
drawn from the vertex.

To show that H, a point assumed to be equidistant from
the faces, must be the center of the base polygon, we show
that H is equidistant from every side of the polygon. We do
this by showing H equidistant from two arbitrary faces AB'C'
and AC'D'. We show (1) HT = HV, and (2) $\overline{HT} \perp \overline{B'C'}$ and $\overline{HV} \perp \overline{D'C'}$,
(showing that HT and HV are the distances from H to
the faces).

Given: Regular pyramid A-B'C'D', with point H on base
        B'C'D' such that HX = HY where HX $\perp$ plane AB'C'
        and HY $\perp$ plane AC'D'.

Prove: H is equidistant from sides $\overline{B'C'}$ and $\overline{C'D'}$.

Statements	Reasons
1. (See above).	1. Given.
2. ∢AXH and ∢AYH are right angles.	2. If a line is perpendicular to a plane, then it forms right angles with every line

789

in the plane that intersects it.

3. ⊄AXH ≅ ⊄AYH.

3. All right angles are congruent.

4. $\overline{AH} ≅ \overline{AH}$.

4. A segment is congruent to itself.

5. ΔAXH ≅ ΔAYH.

5. The Hypotenuse-Leg Theorem.

6. ⊄TAH = ⊄VAH.

6. Corresponding parts of congruent triangles are congruent.

7. ⊄AHT and ⊄AHV are right angles.

7. If a line ($\overline{AH}$) is perpendicular to a plane (B'C'D'), then it forms right angles with every line in the plane that intersects H.

8. ΔHAT ≅ ΔHAV.

8. The ASA Postulate.

9. $\overline{HT} ≅ \overline{HV}$.

9. Corresponding parts of congruent triangles are congruent.

Now we show that HT and HV are distances.

10. Plane ATH ⊥ plane AB'C'.

10. If a plane (A H) contains a line ($\overline{XH}$) ⊥ to a plane (AB'C'), then the two planes are ⊥.

11. Plane ATH ⊥ plane B'C'D'.

11. If a plane (A H) contains a line ($\overline{AH}$) ⊥ to a plane (B'C'D'), then the two planes are ⊥.

12. Plane AB'C' intersects plane B'C'D' in line $\overleftrightarrow{B'C'}$.

12. If two planes intersect, they intersect in a line.

13. $\overleftrightarrow{B'C'}$ ⊥ plane ATH.

13. If two planes are ⊥ to a given plane, then their intersection is ⊥ to the given plane.

14. $\overleftrightarrow{B'C'}$ ⊥ $\overline{HT}$.

14. If a line is ⊥ to a plane, then it is ⊥ to every line in the plane that intersects it.

15. HT is the distance from H to $\overline{BC}$.

15. The distance between a point and a line is the length of the perpendicular segment from the point to the line.

16. HV is the distance from H to $\overline{CD'}$.

16. Repeat steps 10-15 substituting planes AHV, AC'D',

17. H is equidistant from   17. Steps 15, 16, and 9.
    sides $\overline{B'C'}$ and $\overline{C'D'}$.

   The same procedure can be used to show H is equidistant
from any two sides of base polygon B'C'D'. Therefore, point
H is equidistant from every side of the polygon and is thus
the center of the polygon.

   Earlier we showed that H' was the center. Since a polygon can have only one center point, our assumption (that point
H exists equidistant from the faces of A-BCD but not on $\overline{AG}$)
is false.

   Therefore, every point in the interior of the regular
pyramid equidistant from the faces must lie on the perpendicular from the vertex.

CHAPTER 47

SPHERES

> **Basic Attacks and Strategies for Solving Problems in this Chapter. See pages 792 to 805 for step-by-step solutions to problems.**

The introduction to Chapter 44 includes a discussion of spheres. It might be desirable to review those pages prior to continuing with this chapter.

The problems in this chapter concern the geometry of the sphere, which is very different from the geometry of a plane. Examine the picture below.

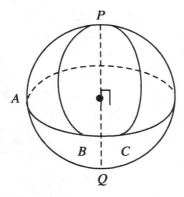

The arc that contains A, B, and C is part of a great circle (intersection of the sphere and a plane which contains its center). To every great circle is associated a pair of **poles**, which are the endpoints of the diameter of the sphere perpendicular to the plane containing the great circle. For the earth, the north and south poles are the geometric poles of the great circle commonly called the equator. In the diagram, P and Q are poles of the great circle containing A and B. Consider $\overset{\frown}{AB}$. There is only one great circle which contains both A and B. Although it would seem that there are many curved paths on the sphere which join A and B, $\overset{\frown}{AB}$ on a sphere is defined as the arc $\overset{\frown}{AB}$ of the great circle which contains A and B. An arc between two points on a sphere is intuitively equivalent to a line segment between two points in a plane: the shortest path between them. Like normal arcs, spherical arcs have both angular and linear measure.

Just as, in a plane, two segments with a common endpoint form an angle, we will say that a **spherical angle** is formed by two arcs on a sphere which share an endpoint. For instance, in the picture on the previous page, $\angle BPC$ is a spherical angle formed by $\overset{\frown}{PB}$ and $\overset{\frown}{PC}$. Each arc of a spherical angle is part of a great circle, and there is a tangent to each great circle at the point of their intersection. The spherical angle formed by thearcs is defined to have the same measure as the angle formed by thetwo tangent lines. Since P is a pole of the great circle containing B and C, $m \angle BPC = m \overset{\frown}{BC}$. This is proven in Problem 817.

Consider now points A, B, and C on a sphere, not all on the same great circle. The **spherical triangle** ABC is the union of the three arcs $\overset{\frown}{AB}$, $\overset{\frown}{BC}$, and $\overset{\frown}{AC}$.

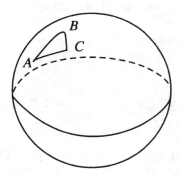

While planar triangles have angle sums of 180°, spherical triangles can have angle sums anywhere from 180° to 540° (Problem 818). Also, since the sides of a spherical triangle are arcs, its perimeter can be measured both in degrees and linear units (see Problem 819).

Each side of a spherical triangle lies on a great circle. A spherical triangle is **polar** to a second triangle if the vertices of the first triangle are poles of the great circles containing the sides of the second triangle. For example, let A', B', and C' be poles of the great circles of $\overset{\frown}{BC}$, $\overset{\frown}{AC}$, and $\overset{\frown}{AB}$, respectively. Then, spherical triangle $A'B'C'$ is polar to spherical triangle ABC. Problem 823 proves that, also, spherical triangle ABC is polar to spherical triangle $A'B'C'$.

Step-by-Step Solutions to Problems in this Chapter, "Spheres"

The photographer N. Larger was giving his geometer friend, E. Lips, a tour of the darkroom. "What's this?" asked Lips, pointing to two tomato cans in the wall. "O, that," said Larger proudly. "I had trouble with termites and they ate two holes in the wall, one square shaped and one round (see Figure 1a). I found that I could plug up both holes with exactly the same shape. By putting the tomato can in standing up, I could completely block out light from the square shaped hole. By putting the tomato can in on its side I could completely plug up the circular hole. One kind of shape fills two kinds of holes. Pretty good, huh?"

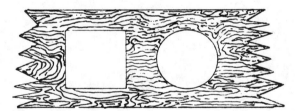

Figure 1A

At that moment, though, the termites ate another hole through the wall in the shape of a triangle. (see Figure 1b) Larger was crestfallen. "There's no shape in the world that will plug up all <u>three</u> holes."

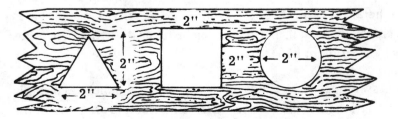

Figure 1B

The geometer smiled. "Well, it's a pretty odd looking figure, but I can think of one."

What is the figure?

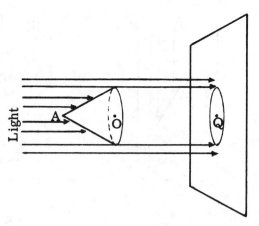

Figure 2

Solution: When dealing with one object fitting through other objects, we consider projection, the outer boundaries of the object. To see what is meant by this, consider the shadows created by the cone A-⊙O. (see Figure 2). When deciding whether the cone will fit through a circular hole of radius 2, we decide whether the shadow ⊙O will fit through the circular hole. It is immaterial to us in Figure 2 that point A juts out and that A-⊙O is not a planar surface. As long as ⊙O is the outer boundary, we need only worry about flat surface ⊙Q. In more technical terms, we are concerned with the projection of the cone to the plane parallel to the base.

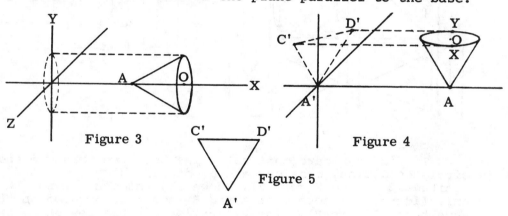

Figure 3

Figure 5

Figure 4

The projection of a point onto a plane is the intersection of the perpendicular from the point to the plane and the plane. The projection of a solid is the set of the projections of every point in the solid. Note in Figure 3, by drawing the perpendiculars from points of the cone to the Y-Z plane, we outline the shadow, a circle. Another important fact to remember is that a figure may have more than one projection - just as an object may have another shadow. Suppose the cone was so placed that the base was parallel to the XZ-plane (Figure 4). Then the projection is Figure 5.

A third important fact about projections is that the lengths of the projection are not necessarily equal to the lengths of the original object. A person's shadow is hardly ever the same height as the person. In the same way, segments $\overline{A'C'}$ and $\overline{A'D'}$ are not congruent to slant heights $\overline{AX}$ and $\overline{AY}$.

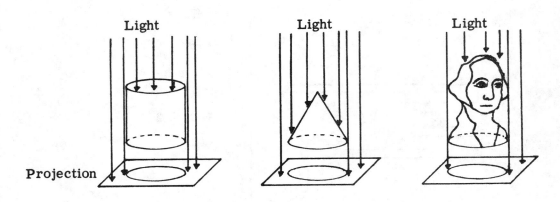

Figure 6

In this particular problem, we must find a figure which has three projections corresponding to the holes cut in the wood. To satisfy the criteria that the object fit exactly through the circular hole, any figure with a circular base (and that does not have "protruding" parts) will suffice (see Figure 6).

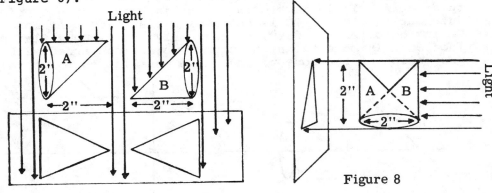

Figure 7

Figure 8

To find a figure that satisfies both the triangle and circle condition, consider a cone of base diameter 2 and altitude 2 (see Figure 7). However, the cone is not the only figure. Suppose cone A and B were superimposed on the same base (Figure 8). The projection of this new shape in the horizontal direction is also a triangle. Inserting a thin triangular wedge (Figure 9) does not change the left-right projection. In addition, the projection of Figure 9 onto the plane of this page is a square. Thus, the solid of Figure 9 is the desired solid.

Figure 9

● **PROBLEM** 817

Show that a spherical angle is measured by the arc of the great circle which has the vertex of the angle as a pole and is included between the sides, extended if necessary, of the angle.

<u>Solution</u>: In the figure, sphere O contains spherical angle
APB. PD and PE are tangents to arcs PA and PB at point P.
By definition, the measure of the spherical angle is equal
to the measure of the angle formed by the lines that are
tangent to the arcs at the point of intersection. Thus, by
formal definition, the measure of spherical angle ∢APB equals
m∢DPE. For purposes of proofs and calculations, this is an
awkward definition.

The proof we will present relates the measure of the
spherical angle to an arc of the great circle, a much more ac-
cessible measure. In the figure, let the great circle whose
pole is P intersect arcs AP and BP at points A and B. (Notice
that this does not restrict the proof in any way. The exact
location of A does not matter as long as it remains on the
same arc.) We wish to show that m⌢AB = m∢AOB = m∢APB = m∢DPE.
We do this by showing that the sides of ∢AOB are parallel to
the sides of ∢DPE and thus ∢AOB ≅ ∢DPE. Since, by definition,
m∢APB = m∢DPE, we then have m∢AOB = m∢APB.

Given: Spherical angle APB on sphere O, ⌢AB of the great
circle whose pole is P and sides PA and PB which
include the spherical angle.

Prove: m∢APB = m⌢AB.

Statements	Reasons
1. Draw diameter POP' and radii OA And OB of sphere O.	1. A diameter of a sphere is a line segment passing through the center of the sphere with endpoints on the sphere. Also, two points determine a line.
2. Draw PD tangent to ⌢PA, a point P and PE tangent to ⌢PB at point P.	2. A tangent can be constructed to a given circle at a given point on it.
3. ⌢PA and ⌢PB are quadrants.	3. The polar distance of a great circle is a quadrant.
4. ∢POA and ∢POB are right angles.	4. A central angle is measured by its intercepted arcs.
5. AO ⊥ PO, BO ⊥ PO.	5. Lines are perpendicular if they form right angles.
6. DP ⊥ PO, EP ⊥ PO.	6. A tangent and radius are ⊥ at the point of contact.
7. AO ∥ DP, BO ∥ EP.	7. In a plane, if two lines are ⊥ to a third line, then they are parallel to each other.

8.	∢AOB ≅ ∢DPE or m∢AOB = m∢DPE		8.	If the sides of two nonco-planar angles are parallel and extend in the same direction, then the angles are congruent.
9.	m∢AOB = mAB.		9.	A central angle is measured by its intercepted arc.
10.	m∢DPE = m∢APB.		10.	A spherical angle is equal in degree measure to the angle formed by the lines which are tangent to the arcs at their point of intersection.
11.	m∢APB = m$\overset{\frown}{AB}$.		11.	Substitution and Transitivity.

● **PROBLEM** 818

Show that the angle sum of a spherical triangle is greater than 180° and less than 540°.

Solution: In the accompanying figure, spherical triangles △ABC and △A'B'C' are polar to each other. From an earlier theorem, we know that, in polar triangles, each angle of one has the same measure as the supplement of the side lying opposite it in the other. Thus, the sum of the three vertex angles of one triangle and three sides of the triangle polar to it is the sum of three supplementary pairs or 3 · (180°) = 540°. Thus, the angle sum of the first triangle is dependent on the sum of the degree measures (degree sum) of the sides of the second triangle.

Remember that each side of a spherical triangle is an arc of a great circle. Thus the degree measure of a side is the degree measure of the arc or, equivalently, the degree measure of the central angle that subtends it. For a very small second triangle, the arc measure of the sides can be very small. In this way the degree sum of the second triangle can be made arbitrarily close to zero. Since (three vertices of △1) + (degree sum of △2) = 540°, this means the angle sum of △1 can be made arbitrarily close to 540°. Thus, the upper limit for the angle sum of a spherical triangle is 540°.

Notice that as the degree sum of △2 increases, the angle sum of △1 decreases. Therefore, to find the minimum value of the angle sum of △1, we find the maximum degree sum of △2. But the sum of the sides of any spherical polygon is always less than 360°; therefore, the maximum for the degree sum of △2 = 360 , and the minimum angle sum of △1 = 540° - 360° = 180°. Thus, 180° < angle sum of spherical triangle < 540°.

Note: The key to showing the lower limit is 180° is the theorem that states: the sum of the sides of a spherical polygon is less than 360°. To justify this important theorem, draw the segments connecting each vertex of a spherical polygon to the center of the sphere to form a polyhedral angle. Since each side is an arc of a great circle, the measure of each side is the measure of a face angle of the polyhedral angle. By a theorem proved earlier, we know that the sum of the face angles is always less than 360°. Therefore, the sum of the sides of a spherical polygon must also always be less than 360°.

Arguing from a more intuitive approach, we might say that all arcs that compose the sides of a polygon laid end to end should not form an entire great circle. Since the meausre of a great circle is 360°, the degree sum of the sides of a polygon at most should be less than 360°.

● **PROBLEM** 819

On a sphere of radius 9 inches, the perimeter of a spherical triangle is 12π inches. The sides of the triangle are in the ratio 3:4:5.

   a)   Find the sides of the triangle in degrees.
   b)   Find the angles of its polar triangle.
   c)   Find the area of the polar triangle in square inches. (answer may be left in terms of π.)
   d)   A zone on this sphere is equal in area to the polar triangle. Find the number of inches in the altitude of the zone.

Solution:

   (a)   The perimeter is given in inches; we wish to find the lengths of the sides of the triangle in degrees. Therefore, we first convert the perimeter into degrees, p°, and then use the ratio 3:4:5 to find the number of degrees in each side.

   If the radius is 9, then the circumference of a great circle is 2πr = 2π(9) = 18π. Thus, a length of 18π would span 360°. Since the length is proportional to the number of degrees spanned, we have the ratio:

$$\frac{p°}{360} = \frac{\text{length}}{\text{circumference}} = \frac{12\pi}{18\pi}$$

$$p° = \frac{12\pi}{18\pi} \ 360° = \frac{2}{3} \cdot 360° = 240°.$$

   Suppose the three sides are a, b, and c. Because they are in a 3:4:5 ratio, then there is some number x such that a = 3x, b = 4x, and c = 5x. The perimeter equals the sum of

797

the sides. Therefore, p° = a + b + c. 240° = 3x + 4x + 5x = 12x or x = $\frac{240°}{12}$ = 20°. Then, a = 3(20°) = 60°; b = 4(20°) = 80°; and c = 5(20°) = 100°. The measure of the sides are 60°, 80°, and 100°.

(b) By an earlier theorem, we know that the angles of the polar triangle are supplementary to the sides of the spherical triangles opposite them. Therefore, if the sides of the triangle are 60°, 80°, and 100°, the angles of the polar triangles are (180 -60°), (180 -80°), and (180°-100°) or 120°, 100°, and 80°.

(c) The area of a spherical triangle of angles a, b, and c is is given by the formula $\frac{a+b+c-180}{720}$ s. s, the surface area of the sphere is $4\pi r^2 = 4\pi(9)^2$. Thus, the area of the triangle is $\frac{120+80+100-180}{720}(4\pi(9)^2) = \frac{120}{720}4\pi(9)^2 = \frac{2}{3}\pi(9)^2 = 54\pi$.

(d) The zone of a sphere is the region formed when two parallel planes intersect a sphere. That part of the sphere between the planes is called a zone and the distance between the planes is the altitude of the zone. The surface area of the zone is always equal to the lateral surface area of a cylinder of height equal to the altitude of the zone and base congruent to a great circle of the intercepted sphere. We are given that the surface area of the zone equals the area of the triangle. Thus,

$$(2\pi r)h = 54\pi.$$

Since r = 9, h = $\frac{54\pi}{2\pi(9)}$ = 3 inches.

● PROBLEM 820

The number of degrees in the angles of a spherical triangle are in the ratio of 3:4:5 . The area of the triangle is equal to the area of a zone of altitude 3 on the same sphere. If the radius of the sphere is 15, find each angle of the triangle.

Solution: We wish to find the angles A, B, and C, of the triangle. We are given that the angles are in the ratio of 3:4:5. Thus, there is some x such that m∢A = 3x; m∢B = 4x; and m∢C = 5x. We can then calculate the area of the triangle in terms of x. We are told that the area of the triangle is equal to the area of a zone. Since the radius and altitude are given, the area of a zone is a known constant. Thus, the area of the triangle is a constant. Hence, we will have an expression involving x that is equal to a constant. We can solve for x and, from this, obtain m∢A, m∢B, and m∢C.

First we will relate x to the area of the triangle. The area, A_T, of the spherical triangle is given by the formula

$$A_T = \frac{\text{angle excess of } \Delta}{\text{total number of degrees in a sphere}} \cdot \left(\text{total surface area}\right)$$

$$A_T = \frac{m\angle A + m\angle B + m\angle C - 180}{720°} \cdot (4\pi r^2)$$

Let m∢A = 3x, m∢B = 4x, m∢C = 5x, and r = 15. Then,

$$A_T = \frac{3x+4x+5x-180}{720°} \cdot 4\pi(15)^2 = \frac{12x-180}{720°} \; 4\pi \cdot 225 = 15\pi(x-15).$$

Secondly, we find the area of the zone. A zone is the section of a sphere cut by two parallel planes. Its lateral area equals the lateral area of a cylinder (with base equal to a great circle of the sphere) cut by the same parallel planes. Thus, a zone of altitude h in a sphere of radius r would have the lateral area of a cylinder of base radius r and height h, or 2πrh.

$$A_Z = 2\pi rh = 2\pi(15)(3) = 90\pi$$

We are given that the area of the zone equals the area of the triangle. Equating the two expressions, we have a method to solve for x.

$$A_T = A_Z$$

$$15\pi(x-15) = 90\pi$$

$$x - 15 = 6 \quad \text{or} \quad x = 21.$$

Thus, m∢A = 3(21) = 63; m∢B = 4(21) = 84; and m∢C = 5(21) = 105. The three angle measures are 63°, 84°, and 105°.

● **PROBLEM** 821

Show that if a point on a sphere is at a distance of a quadrant from each of two other points on the sphere, not the extremities of a diameter, then the point is a pole of the great circle passing through these points.

Solution: Let's begin by defining our terms. In a sphere of radius r, the great circle is a circle, all of whose points lie on the sphere, and which has a radius equal to the radius of the sphere. All great circles have a common center and that center is the center of the sphere. Most planar and linear measurements of the sphere involve the great circles in some way. The circumference of a sphere, for example, is the circumference of any great circle. In addition, the shortest distance between any two points of a sphere is found by drawing the great circle determined by the two points. The part of the circumference between the two points is the geodesic - or path of shortest distance. Examples of great circles are the equator of the earth and the prime meridian.

A quadrant of a sphere is defined to be one-fourth the great circle. It is a sector of a circle of radius r, and the measure of its central angle is $\frac{360}{4} = 90°$. The distance of a quadrant is the arc of the quadrant. The North Pole and any point on the equator are a distance of a quadrant away from each other.

For every great circle, there are two points on the sphere, one on each side of the great circle, that are a quadrant away from the circumference of the great circle. These are the poles of the great circle. The segment with the two poles as endpoints is called the axis and is a diameter of the sphere. Thus, the axis intersects the great circle at the center of the sphere. In addition, this diameter is perpendicular to the plane of the great sphere.

In the accompanying figure, sphere O has great circle ABC as a great circle. $\overset{\frown}{PA}$ and $\overset{\frown}{PB}$ are quadrants.

We must show that P is a pole of ⊙ABC. To show that a point of a sphere is a pole a great circle, it is sufficient to show that the axis of the point $\overline{POP'}$ is perpendicular to the plane of the great circle. We show PO ⊥ AO and PO ⊥ OB. Since $\overline{PO}$ is perpendicular to two lines in plane AOB, $\overline{PO}$ ⊥ plane of the great circle.

  Given: Points P, A, and B on the sphere O; $\overset{\frown}{AB}$ of a great
          circle ABC; and the quadrants $\overset{\frown}{PA}$ and $\overset{\frown}{PB}$.

  To Prove: P is a pole of ⊙ABC.

Statements	Reasons
1. Draw $\overline{AO}$, $\overline{BO}$ and diameter $\overline{POP'}$.	1. Two points determine a line; a diameter is a line segment passing through the center and having its endpoints on the sphere.
2. $\overset{\frown}{PA}$ and $\overset{\frown}{PB}$ are quadrants.	2. Given.
3. ∢POA and ∢POB are right angles; or $\overline{PO}$ ⊥ $\overline{OA}$, $\overline{PO}$ ⊥ $\overline{OB}$.	3. The central angle of a quadrant is a right angle.
4. $\overline{PO}$ ⊥ plane ABC (or $\overline{PP'}$ ⊥ ABC).	4. If a line is perpendicular to each of two intersecting lines at their point of intersection, then it is per-

pendicular to the plane of
the lines.

5. PP' is the axis of
⊙ABC.

5. The axis of a circle of a
sphere is the diameter of
the sphere which is perpendicular to the plane of the
circle.

6. P is a pole of ⊙ABC.

6. The extremities of the axis
are the poles of the circle.

● **PROBLEM** 822

Given two polar triangles, show that each angle of one polar
triangle has the same measure as the supplement of the side
lying opposite it in the other.

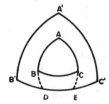

<u>Solution</u>: In the accompanying figure, △ABC and A'B'C' are
polar triangles. We must show that ∢A and B'C' are supplementary. (The proof for the other vertices follows from an
analogous argument.)

To relate ∢A and $\overarc{B'C'}$, extend arcs $\overarc{AB}$ and $\overarc{AC}$ so that
they intersect B'C' at points D and E. We now solve for $\overarc{B'C'}$
and ∢A in terms of DE and use the results to show that $\overarc{B'C'}$
and ∢A are supplementary.

We can solve for $\overarc{B'C'}$ in terms of $\overarc{DE}$ by first expressing
$\overarc{B'E}$ and $\overarc{C'D}$ in terms of $\overarc{DE}$ and then using this to relate $\overarc{B'C'}$
and $\overarc{DE}$. It may seem extraneous to have intermediate unknown
lengths $\overarc{B'E}$ and $\overarc{C'D}$, but this is the crucial point. $\overarc{B'E}$ and
$\overarc{C'D}$ are not unknown; they can be determined. Since △ABC and
△A'B'C' are polar triangles, B' is the pole of $\overarc{ACE}$ and C' is
the pole of $\overarc{ABD}$. Since the distance between the pole and
any point on the great circle is a quadrant, m$\overarc{B'E}$ = 90° and
m$\overarc{C'D}$ = 90°.

Note that m$\overarc{B'C'}$ = m$\overarc{B'D}$ + m$\overarc{DC'}$ = m$\overarc{B'D}$ + 90°. We wish to
find m$\overarc{B'C'}$ in terms of m$\overarc{DE}$. Therefore, we eliminate $\overarc{B'D}$.
Since m$\overarc{B'D}$ = m$\overarc{B'E}$ - m$\overarc{DE}$ = 90° - m$\overarc{DE}$, m$\overarc{B'C'}$ = (90° - m$\overarc{DE}$) +
90° = 180° - m$\overarc{DE}$.

To find m∢A in terms of m$\overarc{DE}$, note that, since △ABC and
A'B'C' are polar, A is the pole of $\overarc{B'DEC'}$. The measure of a
spherical angle, ∢A, equals the measure of the arc, $\overarc{DE}$, intercepted by the sides of the angle, $\overarc{AB}$ and $\overarc{AC}$, on the great
circle whose pole is the vertex of the angle, ⊙B'DEC'. Thus,
m∢A = m$\overarc{DE}$.

Therfore, m∢A + m$\overarc{B'C'}$ = [m$\overarc{DE}$] + [180° - m$\overarc{DE}$] = 180°.
Thus, ∢A and $\overarc{B'C'}$ are supplementary.

Show that if one spherical triangle is the polar triangle of another, then the second is the polar triangle of the first.

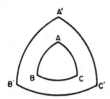

Solution: By definition, spherical triangle ABC is said to be the polar triangle of spherical triangle A'B'C' if the vertices of A'B'C' are the poles of the sides of ABC. That is, A' is the pole of great circle ⊙BC; B', the pole of ⊙AC; and C', the pole of ⊙AB. Here, we shall show that the relation is symmetric: if the above holds, then A must be a pole of ⊙B'C', B a pole of ⊙A'C', and C a pole of ⊙A'B'.

To show this, we use the fact that, if a point on a sphere is at a distance of a quadrant from each of two other points not the extremities of a diameter, then the point is a pole of the great circle passing through these points.

Given: Spherical ∆ABC and its polar ∆A'B'C'.

To Prove: ∆ABC is the polar triangle of ∆A'B'C'.

Statements	Reasons
1. B is the pole of A'C' and C is the pole of A'B' (BA', BC', CA', and CB' are quadrants).	1. If one spherical triangle is the polar triangle of another, then the vertices of the second are the poles of the sides of the first. A pole is a quadrant away from any point on a great circle.
2. A' is a quadrant's distance from B and C (along A'B and A'C).	2. The polar distance of a great circle (along ⊙A'C' and ⊙A'B') is a quadrant.
3. A' is a pole of $\overarc{BC}$.	3. If a point on a sphere is at a distance of a quadrant from each of two other points not the extremities of a diameter, then the point is the pole of a great circle passing through these points.
4. Similarly B' is the pole of $\overarc{AC}$ and C' the pole of $\overarc{AC}$.	4. Repeat Steps 1-3.
	5. One spherical triangle is the polar triangle of another if the vertices of the second are the poles of the sides of the first.
5. ∆ABC is the polar triangle of A'B'C'.	

The most remote spot of ocean, the point at 48°30' S and 125°30' W in the Pacific, is 1660 miles from the nearest land Pitcairn Island. Approximate the area of this oceanic expanse as (1) a circle; (2) a zone of a sphere. Compare the two. (The radius of the earth is 3960 mi.)

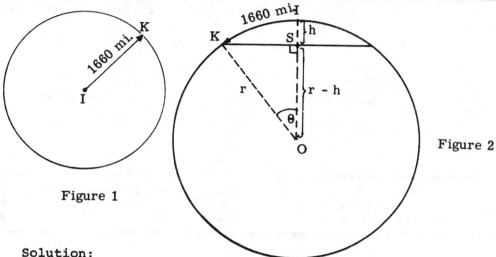

Figure 2

Figure 1

Solution:

(1)   The area of a circle with radius r equals πr^2. Therefore, the area of the watery expanse is $\pi(1660 \text{ mi.})^2 = 2.76 \times 10^6 \pi \text{ mi.}^2$

(2)   The area of a zone is given by the formula $2\pi rh$. To find h, we first find θ. With θ, we can find $\overline{OS}$. Since h is the radius minus $\overline{OS}$, we can find h if we know θ.

To find θ, remember that the arc length and central angle measure are proportional

$$\text{(i)} \quad \frac{\theta}{360} = \frac{\widehat{mIK}}{\text{circum. of earth}}.$$

The circumference of the earth (of radius 3960 miles) is $2\pi r = 2\pi(3,960 \text{ mi.})$. IK is given to be 1660 mi.

$$\text{(ii)} \quad \frac{\theta}{360} = \frac{1660 \text{ mi.}}{2\pi(3960 \text{ mi.})}$$

$$\text{(iii)} \quad \theta = 360 \frac{1660}{2\pi \cdot 3,960} = \frac{830}{11\pi} = 24°.$$

To find $\overline{OS}$, note that in right $\triangle KSO$:

$$\text{(iv)} \quad \cos \theta = \frac{OS}{KO} = \frac{r-h}{r} = \frac{3,960 \text{ mi} - h}{3960}.$$

Consulting the trigonometric tables, we find cos 24° =

.914. Substituting this value in and multiplying both sides by 3,960, we obtain:

$$(v) \quad (.914)(3,960) = 3,960 - h$$

$$(vi) \quad h = 3,960 - (.914)(3,960) = 3,960(1-.914)$$

$$= 3,960(.086) = 341 \text{ mi.}$$

With a value for h, we can now solve for the area of the zone:

$$(vii) \text{ Area} = 2\pi rh = 2\pi(3960 \text{ mi.})(341 \text{ mi.}) =$$

$$2.7 \times 10^6 \text{ mi.}^2$$

The zone approximation is smaller than the planar approximation by 2%. In either case, the area covered by the water is greater than any single nation.

● **PROBLEM** 825

The slant height ℓ of a certain right circular cone is equal to the diameter of the base. A sphere is inscribed in the cone. Express in terms of ℓ, the radius of this sphere.

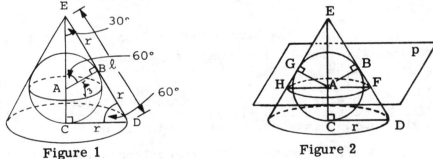

Figure 1   Figure 2

Solution: In Figure 1, sphere A is inscribed in the cone with vertex E and base circle C. We will first show that the perpendicular to the base of the cone passes through the center of the sphere. Second, we will show that the points of intersection with the side of the cone, B, bisect the slant height of the cone. Then, $\overline{EB}$ will be known, ⊀AEB can be found, and, using the proportions of right $\triangle ABE$, we can find the radius $\overline{AB}$.

To show that the center of the sphere is on the perpendicular, $\overline{EC}$ is involved and usually can be assumed. For the sake of completeness, though, we will outline the proof here. In Figure 2, let point A be the center of the sphere; radii $\overline{AG}$ and $\overline{AB} \perp$ to the side of cone; the plane P passing through A parallel to base ⊙C intersects the cone in a circle. H and F are points on the circle; H, A, G, and E are coplanar; B, A, F, and E are coplanar. $\overline{GA} \cong \overline{AB}$, ⊀AGE and ⊀ABE are right angles; $\overline{AE} \cong \overline{AE}$. Thus, $\triangle AGE \cong \triangle ABE$, and $GE \cong BE$. Because the little cone with vertex E and ⊙A as base is a right circular

804

cone, $\overline{HE} \cong \overline{FE}$. Thus, HE - GE = FE - BE or $\overline{HG} \cong \overline{FB}$. By SAS, $\triangle HAG \cong \triangle FAB$. Thus, $\overline{HA} = \overline{AF}$. Since H and F are any points on ⊙A, A is the center of the circle. Since the perpendicular $\overline{EC}$ of the cone intersects every circle (⊙A) formed by the intersection of a right circular cone and a plane parallel to the base, point A is on $\overline{EC}$.

To complete the second part, note that in Figure 1, $\overline{BD}$ and $\overline{CD}$ are tangents to sphere A from external point D. Thus, $\overline{BD} \cong \overline{CD}$. Since $\overline{CD}$ is a radii of the cone base, then $\overline{CD} = r$, and $\overline{BD} = r$. Note that slant height $\overline{ED}$ equals the base diameter 2r. Since $\overline{ED} = \overline{EB} + \overline{BD}$, then $\overline{EB} = \overline{ED} - \overline{BD} = 2r - r = r$.

To find ∢AEB, note in right $\triangle CED$, the side opposite ∢AEB, CD, equals one-half the hypotenuse $\overline{DE}$. Thus, m∢AEB = 30°.

In right $\triangle AEB$, $\overline{AB}$ is a radius of the sphere, EB = r and m∢AEB = 30°. To relate these three, note that

$\tan 30 = \dfrac{AB}{EB} = \dfrac{\text{radius of the sphere}}{r}$. Since $\tan 30° = \dfrac{\sqrt{3}}{3}$,

we obtain $\dfrac{\sqrt{3}}{3} = \dfrac{\text{radius of sphere}}{r}$ or radius of the sphere $= \dfrac{\sqrt{3}}{3}r$.

Since r, the radius of the cone base, equals $\frac{1}{2}$ the slant

height, ℓ, then the radius of the inscribed sphere $= \dfrac{\sqrt{3}}{3} \dfrac{\ell}{2} = \dfrac{\sqrt{3}}{6}\ell$.

CHAPTER 48

SURFACE AREAS

Basic Attacks and Strategies for Solving Problems
in this Chapter. See pages 806 to 828 for step-by-
step solutions to problems.

The introduction to Chapter 46 includes a discussion of rectangular solids, prisms, and pyramids, and it might be wise to review those pages before continuing with this chapter.

Surface area obviously refers to the area of the surface. For a prism, the faces other than the bases are called **lateral faces**, and the surface area of a prism is the sum of surface areas of the bases and the lateral faces. Consider the right pentagonal prism pictured below. The figures on the right show this prism cut into three parts.

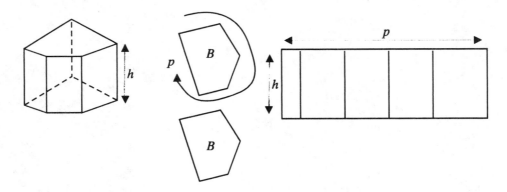

The surface area of the prism is $2B + ph$, where B is the area of the base, p is the perimeter of a base, and h is the length of a lateral edge. This illustrates the case for a pentagon. This procedure can be generalized to a right prism with any polygon for a base.

The surface area of a pyramid is the sum of the areas of the base and the lateral faces. In the case of a regular pyramid, Problem 833 establishes that the lateral surface area is $\frac{1}{2}hp$, where p is the perimeter of the base and h is the slant height of any of the triangular faces. Thus the surface area of a regular pyramid is $B + \frac{1}{2}hp$, where B is the area of the base.

The formula for the surface area of a sphere ($s = 4\pi r^2$) is given as a postulate in this chapter. An argument illustrating why this formula is reasonablei is given in Problem 837.

The formula for the surface area of a right circular cylinder is $2\pi r^2 + \pi rh$. This formula is justified in Problem 843. The formula for the surface area of a cone is $\pi r^2 + \pi rl$, and this formula is established in Problem 845.

RECTANGULAR SOLIDS & CUBES

● PROBLEM 826

Find the surface area of a cube when each edge is of length
a) 1; b) 2.

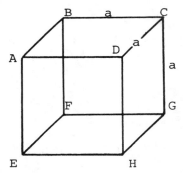

Solution: The figure shows a cube of edge length a. The surface area of the cube, A(C), is given by the sum of the areas of each face of the cube.

Notice that the cube has 6 faces (ABCD, EFGH, ADHE, BFGC, BAEF, and DCGH), and each face is a square of edge length a. Hence,

$$A(C) = 6a^2$$

where a^2 is the area of one face.

a). If a = 1, A(C) = 6

b). If a = 2, A(C) = 24.

If the total surface area of the cube is 150 sq. in., find the length of an edge of the cube.

Solution: We have previously established that for a rectangular solid the total area, or T, is given by T = S + 2B, where S is the lateral area, and B is the base area. A cube is a specific case of a rectangular solid. Let e = the length of an edge. e is the same for all edges of a cube. If h is the cube's altitude, and p is the perimeter of the base, S = hp, For a cube, h = e and p = 4e. Therefore,

$$S = e(4e) = 4e^2 \qquad B = e^2.$$

As such, $T = 4e^2 + 2(e^2) = 6e^2$.

In this case, $T = 150 \text{ in.}^2$. By substitution,

$$T = 150 \text{ in.}^2 = 6e^2$$
$$25 \text{ in.}^2 = e^2$$
$$5 \text{ in.} = e.$$

Therefore: the length of an edge at the cube is 5 in.

Find the lateral area and the total surface area of a rectangular solid in which the dimensions of the base are 5 in. and 4 in. and the length of the altitude is 3 in.

Solution: The lateral area of a rectangular solid is the sum of the areas of its lateral faces, i.e. those perpendicular to the base. The total surface area is the lateral area plus the sum of the areas of the two bases.

Let the dimensions of the base be, in general, ℓ and w, and let the length of the altitude be h. Since opposite faces of a rectangular solid are of equal area, and all lateral faces have a height h, two opposing faces will have dimensions ℓ by h while two other opposing faces will have dimensions w by h. Based on the concept that the area of a rectangle is equal to the product of the lengths of its sides, it is correct to determine the lateral surface area by the equation

$$\ell h + wh + \ell h + wh = \text{Lateral Area.}$$

Factoring, we obtain lateral area = $h(\ell+w+\ell+w)$. Since $\ell + w + \ell + w = p$ (the perimeter of the base), then S (the lateral area of a rectangular solid) = hp. The total surface area includes the sum of the areas of the two bases plus the lateral area.

Total area = $hp + 2(\ell w)$ or S + 2B, where B = ℓw.

In this problem, the perimeter of the base = (5+4+5+4) in. = 18 in. The altitude of the solid is h = 3 in. Lateral area of the solid = S = hp = (3×18) in.^{2} = 54 in.^{2}. Area of the base = (5×4) in.^{2} = (20) in.^{2} = B. Total surface area = S + 2(B) = $\left(54 + 2(20)\right)$ in.^{2} = 94 in.^{2}. Therefore, Lateral area = 54 sq. in.; Total area = 94 sq. in.

● **PROBLEM** 829

In the rectangular solid shown, $\overline{DA}$ = 4 in., $\overline{DC}$= 3, and $\overline{GC}$ = 12.

(a) Find the length of $\overline{CA}$, a diagonal of the base ABCD.

(b) Using the result found in part (a), find the length of $\overline{GA}$, a diagonal of the solid.

(c) If $\overline{DA}$ = ℓ, $\overline{DC}$ = w, and $\overline{GC}$ = h, represent the length of $\overline{GA}$ in terms of ℓ, w, and h.

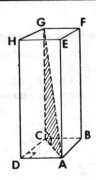

Solution: All three parts of this problem will involve applying the Pythagorean Theorem in a solid geometry setting. Recall that the Pythagorean Theorem tells us that in a right triangle, the square of the length of the longest side, the hypotenuse, is equal to the sum of the squares of the lengths of the legs.

ABCD is a rectangle because the base of a rectangular solid is a rectangle. The diagonal of a rectangle partitions it into two right triangles. Knowing the length of two adjacent sides, $\overline{DC}$ and $\overline{DA}$, the length of the diagonal can be determined, as mentioned, by the Pythagorean Theorem.

In right $\triangle CDA$, $(CA)^2 = (DA)^2 + (DC)^2$. By substitution, $(CA)^2 = (4)^2 + (3)^2 = 16 + 9 = 25$

CA $= \sqrt{25} = 5$. Therefore, CA = 5.

(b) Since the edge of a rectangular solid is perpendicular to the base of the solid, the edge is perpendicular to any line passing through its foot. (This follows from the definition of a line being perpendicular to a plane.) Therefore, $\overline{GC} \perp \overline{CA}$, and $\triangle GCA$ is a right triangle in which $\angle GCA$ is a right angle.

We want to find the length of $\overline{GA}$, the hypotenuse of $\triangle GCA$, knowing the lengths of the other two legs. Once again, the Pythagorean Theorem is suggested.

In right $\triangle GCA$, $(GA)^2 = (CA)^2 + (GC)^2$.

By substitution, $(GA)^2 = (5)^2 + (12)^2 = 25 + 144 = 169$,

$$GA = \sqrt{169} = 13.$$

Therefore, $\overline{GA}$, the diagonal of the solid, measures 13.

(c) In this part we must follow the same logic and reasoning as above. However, instead of substituting in numerical measurements, we are asked to be more general and substitute letters as variables for actual lengths.

To find GA, given DA = ℓ, DC = w, and GC = h, we must first calculate CA.

As in part (a), $(CA)^2 = (DA)^2 + (DC)^2$.

Again, by substitution, $(CA)^2 = \ell^2 + w^2$. $CA = \sqrt{\ell^2 + w^2}$.

Now that we have represented $\overline{CA}$, we can proceed as in (b). Therefore, $(GA)^2 = (CA)^2 + (GC)^2$. Substitute to obtain

$$(GA)^2 = \left[\sqrt{\ell^2 + w^2}\right]^2 + h^2$$
$$(GA)^2 = \ell^2 + w^2 + h^2$$
$$GA = \sqrt{\ell^2 + w^2 + h^2}$$

Therefore, GA $= \sqrt{\ell^2 + w^2 + h^2}$.

Note: Given the length, ℓ, and width, w, of the base and the height, h, of the rectangular solid, the length of the diagonal of the solid, d, is given by $d = \sqrt{\ell^2 + w^2 + h^2}$.

TETRAHEDRONS, PYRAMIDS & PRISMS

Find the surface area of a regular tetrahedron when each edge
is of length a) 1; b) 2.

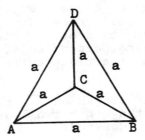

<u>Solution</u>: The illustration shows a regular tetrahedron. A
tetrahedron is a closed space figure with 4 triangular faces,
and a regular tetrahedron is one in which the 4 triangular
faces are congruent.

The surface area, S, of this polyhedron is equal to the
sum of the areas of its 4 faces. That is,

S = Area($\triangle$ADB) + Area($\triangle$BCD) + Area($\triangle$ACD) + Area($\triangle$ACB).   (1)

But,

$$\triangle ADB \cong \triangle BCD \cong \triangle ACD \cong \triangle ACB.$$

Therefore, we may rewrite (1) as

$$S = 4 \text{ Area}(\triangle ADB).\qquad\qquad(2)$$

Note that $\triangle$ADB is an equilateral triangle of edge length a.
The area of an equilateral triangle is equal to $\sqrt{3}/4$ times
the square of its edge length. Hence,

$$\text{Area}(\triangle ADB) = \sqrt{3}/4\left(a^2\right)$$

and, using this in (2),

$$S = a^2\sqrt{3}.$$

a) If a = 1, s = $\sqrt{3}$.

b) If a = 2, s = $4\sqrt{3}$.

In the figure, let $\angle$A'C'B', $\angle$ACB, $\angle$OCA, and $\angle$OCB be right
angles. Furthermore, OC' = k, OC = h, and the plane contain-
ing region A'C'B' is parallel to the base (ABC).

a) Show that A'C'/AC = k/h

b) Show that B'C'/BC = k/h

c) What is the area of triangular region A'B'C'?

d) What is the area of triangular region ABC?

e) Combine the results of parts (a) to (d) to obtain the formula $k^2/h^2 = A$ (region A'B'C')/A (region ABC).

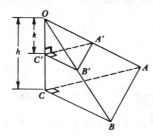

Solution:

a) To obtain the indicated equation, we prove that $\triangle A'OC' \sim \triangle AOC$ and set up a proportion between the sides of these 2 triangles involving $\overline{A'C'}$, $\overline{AC}$, $\overline{OC'}$ and $\overline{OC}$.

First, the plane containing A'B'C' is parallel to the plane containing ABC. Since $\overline{A'C'}$ and $\overline{AC}$ lie in the plane containing points O, C, and A, they are not skew and, therefore, $\overline{A'C'} \parallel \overline{AC}$. Therefore, corresponding angles are congruent. Hence, $\angle OC'A' \cong \angle OCA$ and $\angle OA'C' \cong \angle OAC$. This means that $\triangle OC'A' \sim \triangle OCA$, by the A.A. (angle angle) Similarity Theorem. By definition of similarity,

$$\frac{A'C'}{AC} = \frac{OC'}{OC} = \frac{k}{h}.$$

b) We use a procedure, similar to that used in part (a), in relation to $\triangle OB'C'$ and $\triangle OBC$. Again, the planes which contain A'C'B' and ACB are parallel. $\overline{C'B'}$ and $\overline{CB}$ lie in the plane containing OCB. Hence, $\overline{C'B'}$ and $\overline{CB}$ are not skew, and $\overline{C'B'} \parallel \overline{CB}$. This means that corresponding angles are congruent, and $\angle OC'B' \cong \angle OCB$ and $\angle OB'C' \cong \angle OBC$. By the A.A. Similarity Theorem, $\triangle OB'C' \sim \triangle OBC$, implying that

$$\frac{B'C'}{BC} = \frac{OC'}{OC} = \frac{k}{h}.$$

c) The area of $\triangle A'B'C'$ is equal to ½ the product of its base and altitude. Since $\angle A'C'B'$ is a right angle (see figure) we let $\overline{B'C'}$ be the altitude and $\overline{A'C'}$ be the base. Hence,

$$\text{Area } \triangle A'B'C' = \frac{1}{2}(A'C')(B'C').$$

From parts (a) and (b)

$$A'C' = \frac{k}{h}(AC)$$

$$B'C' = \frac{k}{h}(BC)$$

whence

$$\text{Area } \triangle A'B'C' = \frac{1}{2} k^2/h^2 (AC)(BC).$$

$$= k^2/h^2 \text{ (area } \triangle ABC).$$

d) Using arguments similar to those detailed in part (c),

$$\text{Area } \triangle ABC = \frac{1}{2}(AC)(BC)$$

$$= h^2/k^2 \text{ (area } \triangle A'B'C')$$

since $\overline{AC} \perp \overline{BC}$, as shown in the figure.

e) Combining the results of parts (c) and (d),

$$\frac{A(\text{region } A'B'C')}{A(\text{region } ABC)} = \frac{\text{Area } \triangle A'B'C'}{\text{Area } \triangle ABC} = \frac{\frac{1}{2}k^2/h^2 (AC)(BC)}{\frac{1}{2}(AC)(BC)}$$

$$\frac{A(\text{region } A'B'C')}{A(\text{region } ABC)} = \frac{k^2}{h^2}$$

● **PROBLEM** 832

Find the lateral area and the total surface area of a regular triangular pyramid if each edge of the base measures 6 in. and each lateral edge of the pyramid measures 5 in. (Answer may be left in radical form. See figure.)

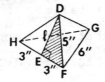

Solution: The lateral area is the sum of the areas of the lateral faces of the pyramid. The total surface area is the lateral area plus the area of the base.

Since the pyramid is regular, all the faces are congruent isosceles triangles, each of whose area is given by $\frac{1}{2}bh$, where b is the length of the base of the $\triangle$, and h is the height. There are three such faces. We shall examine $\triangle DHF$, shown in the figure.

The altitude, $\overline{DE}$, of $\triangle DHF$, is perpendicular to $\overline{HF}$. Also, $\overline{DE}$ bisects $\overline{HF}$ because DHF is isosceles. Since HF = 6, EF = HE = 3. We need $\overline{DE}$ to find the area. By the Pythagorean Theorem, since $\triangle DEF$ is a right triangle,

$$(FD)^2 = (DE)^2 + (EF)^2$$

By substitution, $(5 \text{ in.})^2 = (DE)^2 + (3 \text{ in.})^2$

$$(DE)^2 = 25 \text{ in.}^2 - 9 \text{ in.}^2 = 16 \text{ in.}^2$$

$$DE = 4 \text{ in.}$$

Area of $\triangle DEF = \frac{1}{2}(DE)(HF)$

By substitution, Area of $\triangle DEF = \frac{1}{2}(4)(6)\text{in.}^2 = \frac{1}{2}(24) \text{ in.}^2$

$$= 12 \text{ in.}^2$$

Lateral Area $= 3(\text{Area of } \triangle DEF) = 3(12)\text{in.}^2 = 36 \text{ in.}^2$

The base is an equilateral triangle, since each edge of the base measures 6 in. Therefore,

$$\text{Base Area} = \frac{s^2\sqrt{3}}{4}$$

(the area of an equilateral triangle), where s is the length of an edge of the triangle. By substitution,

$$\text{Base Area} = \frac{(6)^2\sqrt{3} \text{ in.}^2}{4} = 9\sqrt{3} \text{ in.}^2$$

Total Surface Area = Lateral Area + Base Area = $(36 + 9\sqrt{3})\text{in.}^2$

Therefore, Lateral Area = 36 sq. in.; Total Surface Area = $(36 + 9\sqrt{3})$ sq. in.

● **PROBLEM** 833

Prove: The lateral area of a regular pyramid is equal to one-half the product of its slant height and the perimeter of its base.

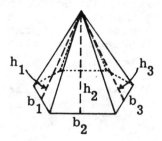

<u>Solution</u>: The lateral area of any polyhedral figure is the sum of the area of the faces. The area of the face of base b_1 and height h_1 is $\frac{1}{2}b_1h_1$. The lateral area is thus $\frac{1}{2}b_1h_1 + \frac{1}{2}b_2h_2 + \cdots$. Since all the slant heights of a regular pyramid are equal, $h = h_1 = h_2 = \cdots$, the expression can be factored as $\frac{1}{2}h(b_1 + b_2 + \cdots)$. Since $b_1 + b_2 + \cdots$ is

the perimeter of the base, the lateral area equals $\frac{1}{2}h \cdot p$ where h is the slant height and p is the perimeter of the base.

● **PROBLEM** 834

The base of a right prism is a regular hexagon with area $24\sqrt{3}$. If the lateral faces of the prism are squares, what is the lateral area?

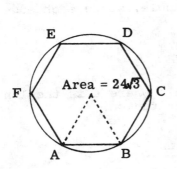
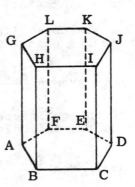

<u>Solution</u>: The area of a prism = e·p where e is the length of the edge and p is the perimeter of a right section. Since this is a right prism, the bases themselves are right sections. Thus p = perimeter of hexagon = 6s (where s is the length of the side of the hexagon). Also, because each lateral face is a square, the edge e must equal the length of the side of the base s. Thus, area = e·p = s·(6s) = $6s^2$.

We use the given area of the hexagon to solve for s. The area of a regular polygon is $\frac{1}{2}a \cdot p$ where a is the length of the apothem and p is perimeter. To solve for a and p would be lengthy and unnecessary. We note, instead, that the area of a regular hexagon is the sum of the areas of six equilateral triangles, each of side s. (In fact, this division of a regular polygon into congruent triangles is how the formula $\frac{1}{2}ap$ is usually proved.) Therefore,

Area of base = $24\sqrt{3}$ = 6 · (Area of equilateral triangle of

$$\text{side s)} = 6 \cdot \left(\frac{s^2\sqrt{3}}{4}\right) = \frac{3}{2}\sqrt{3}\ s^2$$

$$24\sqrt{3} = \frac{3}{2}\sqrt{3}\ s^2$$

$$s^2 = \frac{24\sqrt{3}}{3/2\sqrt{3}} = \frac{2 \cdot 24}{3} = 16$$

$$s = \sqrt{16} = 4.$$

From above, we showed that the lateral area of the prism = e·p = s·(6s) = $6s^2$ = $6(4)^2$ = 96.

Show that the lateral area of a prism is equal to the product of the perimeter of a right section and the length of a lateral edge.

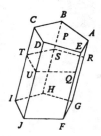

Solution: The lateral area of a prism equals the sum of the areas of the lateral faces, which, by definition, are parallelograms. The area of the nth parallelogram is $a_n b_n$ where a_n is the altitude and b_n is the lateral edge that forms the base. The lateral area of the entire prism equals $a_1 b_1 + a_2 b_2 + \ldots + a_n b_n$. The lateral edges of a prism are congruent, $b_1 = b_2 = \ldots b_n$. Therefore, the expression for lateral area can be expressed as $b_1(a_1 + a_2 + \ldots a_n)$. We show that the altitudes form the sides of a right section of the prism.

Therefore, $a_1 + a_2 + \ldots + a_n$ is the perimeter of the right section and the proof is complete.

Given: Prism P with right section QRSTU...; e the length of the lateral edge; Z the lateral area; w the perimeter of the right section.

Prove: $Z = e \cdot w$.

Statements	Reasons
1. Prism P with right section QRSTU...; e the length of the lateral edge; Z the lateral area; w the perimeter of the right section.	1. Given
2. $\overline{DJ}$, $\overline{EF}$, $\overline{AG}$... $\perp$ QRSTU.	2. The right section of a prism is perpendicular to each lateral edge.
3. $\overline{DJ} \perp \overline{UQ}$, $\overline{EF} \perp \overline{QR}$...	3. A line that is perpendicular to a plane is perpendicular to each line in the plane that passes through the point of intersection.

4.	$\overline{UQ}$, $\overline{QR}$, $\overline{RS}$...are altitudes of parallelograms DJFE, EFGA, BAGH, etc.	4.	An altitude of a parallelogram is a line segment whose endpoints lie on opposite sides of the parallelogram and which is perpendicular to those sides.
5.	Area of DJFE = UQ · EF. Area of EFGA = QR · AG.	5.	The area of a parallelogram equals the length of the base times the length of the altitude drawn to the base.
6.	Lateral Area of P = UQ · EF + QR · AG + RS · BH...	6.	The lateral area of a prism is the sum of the areas of the lateral faces.
7.	EF = AG = BH...	7.	The lateral edges of a prism are congruent.
8.	Lateral Area of P = EF · (UQ + QR + RS...).	8.	Substitution Postulate and Distributive Property.
9.	e = EF.	9.	Given.
10.	w = UQ + QR + RS...	10.	The perimeter of a polygon is the sum of the lengths of its sides.
11.	Lateral Area of P = Z = e · w.	11.	Substitution Postulate.

● **PROBLEM** 836

Find the surface area of a regular icosahedron when each edge is of length a) 3; b) 5.

Solution: A regular icosahedron is a closed space figure which is composed of the union of 20 faces. Each of the 20 faces is an equilateral triangle, and each is congruent to the remaining 19.

The surface area, S, of a regular icosahedron is equal to the sum of the areas of its 20 faces. Since all the faces are congruent to each other,

$$S = 20(\text{area of 1 face}).\qquad(1)$$

But each face is an equilateral triangle. Hence,

$$\text{area of 1 face} = a^2\sqrt{3}/4.\qquad(2)$$

where a is the edge length of one side of the equilateral triangle. Using (2) in (1),

$$S = a^2 5\sqrt{3}.$$

a) If a = 3, $S = 45\sqrt{3}$.

b) If a = 5, $S = 125\sqrt{3}$.

SPHERES

Find the surface area of a sphere of radius 1.

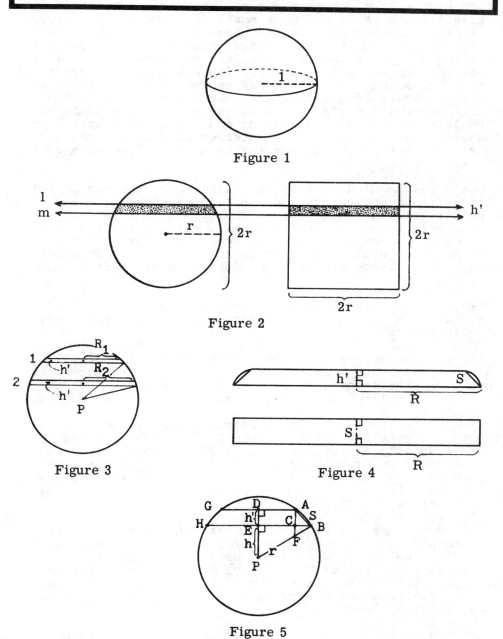

Figure 1

Figure 2

Figure 3

Figure 4

Figure 5

<u>Solution</u>: The surface area of a sphere is given by the for-
mula $4\pi r^2$. This can be remembered as the area of four great
circles of the sphere, but this gives no clue to its deriva-
tion. Consider, instead, a right cylinder with the base con-

gruent to the great circle of the sphere and with the height of the cylinder equal to the diameter. Then, the lateral surface area of the cylinder equals (circumference of base) times (height) or $(2\pi r) \cdot (2r) = 4\pi r^2$. The surface area of a sphere equals the lateral surface area of the smallest right cylinder into which the sphere can be inscribed.

In Figure 2, a sphere and a cylinder are intersected by two parallel planes, ℓ and m, a small distance h' apart such that ℓ and m are parallel to the bases of the cylinder. We will argue that the lateral surface areas of the regions cut by the two planes are equal regardless of where the plane intersects the sphere and cylinder. Since both sphere and plane are of equal height, 2r, for every region of the cylinder cut, there is a region of equal lateral surface area cut on the sphere. Since the surface area of the sphere equals the sum of the lateral area of the regions and the lateral area of the cylinder equals the sum of its intersected regions, the surface area of the sphere equals the lateral surface area of the cylinder.

This may be hard to believe. After all, the region cut on the cylinder will always be a cylinder of radius r and height h', but the region cut on the sphere is constantly changing. In Figure 3, region 2 has a much larger radius than 1. It doesn't seem possible that the surface area of both are equal to the lateral area of congruent cylinders. Remember, though, that surface area is dependent on slant height, not the altitude h'.

Regions, such as 1, lose surface area because of shrinking R, but such regions gain surface area because the slant height increases.

Suppose we approximate the region 1 as a cylinder. (See Figure 4.) For very small h', this is a very good approximation. Then, the lateral surface area equals $2\pi Rs$. We now find R and s in terms of h' and r, the radius of the sphere.

In Figure 5, ⊙P is a great circle of the sphere that intersects the parallel planes in chords $\overline{GA}$ and $\overline{HB}$. D and E are the midpoints of parallel chords $\overline{AG}$ and $\overline{HB}$. Therefore, the segment drawn from the center P, $\overline{PD}$, is collinear with E and perpendicular to both $\overline{AG}$ and $\overline{HB}$. DE is therefore a distance between $\overline{AG}$ and $\overline{BH}$. Since the distance between the parallel planes is h', DE = h'.

$\overline{AB}$ is the segment whose endpoints are the intersection of the parallel lines and the great circle. Therefore, AB is the slant height, and AB = s. EB is the radius of the cylindrical region; thus EB = R. $\overline{PB}$ is a radius of the great circle and, therefore, the radius of the sphere; consequently, PB = r. Point F is so chosen on $\overline{PB}$, such that $\overline{AF} \perp$ HB. EP is the arbitrary length h.

Having explained the figure, we proceed. By Pythagorean Theorem, $R^2 = r^2 - h^2$ or $R = \sqrt{r^2 - h^2}$. To find s, note that for very small h', point A approaches point B; and $\overset{\leftrightarrow}{AB}$, to a very good approximation, approaches the tangent line. Thus, for very small h', ABF is close to a right angle. Henceforth, we shall use the notation "$\sim$" to show approximate equality.

∢ABF ≃ ∢PEB because both are right angles. Since $\overline{DP}$ ⊥ $\overline{HB}$ and $\overline{AF}$ ⊥ $\overline{HB}$, then $\overline{DP}$ ∥ $\overline{AF}$. By corresponding angles, ∢DPF ≅ ∢AFB. By the A-A Similarity Theorem ΔPEB ∿ ΔFBA.

Since the sides of similar triangles are proportional, $\frac{s}{r} = \frac{h'}{R}$. Thus $s = \frac{r}{R}h'$.

We substitute our values for s and R in the expression for the surface area of the region, $2\pi Rs$. Therefore, area $= 2\pi R(\frac{r}{R}h') = 2\pi rh'$.

The lateral area of the region in the cylinder equals (perimeter of the base) · (height) or $(2\pi r) \cdot h = 2\pi rh$. The surfaces have equal lateral areas.

This is not a proof, of course. Estimating the region cut in the sphere as a cylinder and estimating ∢FBA as a right angle immediately invalidated the argument as a proof. However, the reasoning behind the approximations were correct, and for infinitely small h', the approximations are exact. (This argument is the basis for the proof in limit theory.) At any rate, the argument above offers an insight into the expression.

Getting back to the original problem, the surface area of a sphere of radius 1 equals $4\pi r^2 = 4\pi(1)^2 = 4\pi$.

● **PROBLEM** 838

Find the surface area of a sphere whose radius measures 14 in. [Use $\pi = \frac{22}{7}$] (See figure.)

Solution: The surface area of a sphere is, by postulate, equal to four times the area of one of its great circles.

Therefore, $S = 4\pi r^2$, where r is the radius of a great circle of the sphere. Since the radius of a great circle of a sphere is equal to the radius of the sphere itself, r = 14 in. By substitution,

$$S = 4(\frac{22}{7})(14)^2 \text{ in.}^2$$

$$S = 2464 \text{ in.}^2$$

Therefore, area of the sphere = 2464 sq. in.

The ratio of the altitude of a zone to the diameter of the sphere on which it is drawn is 1:5. The area of the zone is 80π.

    a)   Find the area of the sphere.
    b)   If one of the bases of the zone is a great circle, find the area of the other base.

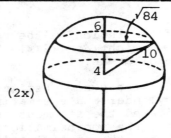

(2x)

Solution:

    (a) To find the area of the sphere we use the formula $S = 4\pi r^2$. To use the formula, though, we must solve for r. We solve for r using the given zone area. We know that the zone area and r are related by the formula,

A_{zone} = (circumference of the great circle) • (zone height)

   80π = 2πr • h.

    This equation can be solved for r if we can eliminate the second variable h. From the given, though, we know that h:d = 1:5 or $h = \frac{1}{5}d$. Since the diameter d is twice the radius, we have $h = \frac{1}{5}(2r) = \frac{2}{5}r$. Substituting this result in the zone area formula, we have

$$80\pi = 2\pi r \cdot (\tfrac{2}{5}r) = \tfrac{4}{5}\pi r^2$$

$$r^2 = 80\pi \cdot (\tfrac{5}{4\pi}) = 100$$

$$r = 10.$$

    Thus the radius of the sphere is 10 and its area equals $4\pi r^2 = 4\pi(10)^2 = 400\pi$.

    (b) To find the area of the smaller base, we use the formula πr_1^2, where r_1 is the radius of the smaller base, To find r_1, consider the line drawn from the center of the sphere perpendicular to the plane of the circle. By an earlier theorem, we know that this line intersects the circle at its center. Thus, a right triangle is formed with hypotenuse equal to the radius of the sphere, one leg equal to the altitude of the zone and one leg equal to the radius of the smaller base. The radius of the sphere equals 10 (see Part a). The altitude of the zone can be found by the relationship used in Part a, namely $h = \frac{2}{5}r$, where r is the radius of the sphere. So $h = \frac{2}{5}(10) = 4$. Then, by the Pythagorean Theorem, we have

(radius of circle)^2 = (radius of sphere)^2 - (altitude of zone)^2

$$= 10^2 \qquad\qquad - 4^2$$

$$= 84.$$

Since the area of the circle = π(radius of circle)^2. we have $\pi(84) = 84\pi$.

● **PROBLEM** 840

A sphere has radius 7. What is the area enclosed by a spherical triangle whose angles have measures a = 100°, b = 120°, c = 140°? (Take $\pi = \frac{22}{7}$.)

Solution: If a, b, and c are the measures of the angles of a spherical triangle, and S is the surface area of a sphere, then the area, A, enclosed by the sphereical triangle is

$$A = \frac{(a+b+c-180°)S}{720°}.$$

For our case,

$$A = \frac{(100°+120°+140°-180°)S}{720°}$$

$$A = \frac{S}{4}. \tag{1}$$

But $S = 4\pi r^2$, where r is the radius of the sphere. Hence, (1) becomes

$$A = \frac{4\pi r^2}{4} = \pi r^2.$$

Noting that r = 7 yields

$$A = \pi(7)^2 = \left[\frac{22}{7}\right](49)$$

$$A = (22)(7) = 154.$$

● **PROBLEM** 841

A sphere whose diameter is 12 feet is illuminated by a point source of light 18 feet from the center of the sphere. Find the area of the portion of the sphere which is illuminated. [Answer may be left in terms of π.]

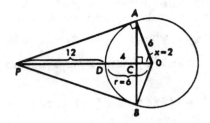

Solution: The region illuminated is a portion of a sphere whose boundary is a circle. Thus, the area is a zone and the area of the surface illuminated is the area of a zone. The area of a zone equals (height of zone) · (circumference of great circle).

To find the circumference of the great circle, we use the formula $C = 2\pi r$. Since the diameter equals 12 ft., the radius $= \frac{1}{2}(12) = 6$ ft. Thus the circumference is $2\pi(6$ ft.$) = 12\pi$ ft.

To find the altitude $\overline{DC}$ of the zone, we use the fact that $\overline{DC} = \overline{DO} - \overline{CO}$. Note that $\overline{DO}$ is a radii; thus, $\overline{DO} = 6$ ft. To find $\overline{CO}$, note that $\overline{AC} \perp \overline{PO}$. Therefore, $\overline{AC}$ is an altitude of right triangle $\triangle PAO$ and thus side AO is the mean proportional of adjacent segment CO and hypotenuse PO.

$$\frac{CO}{AO} = \frac{AO}{PO}$$

Since PO = 18 ft., and AO = 6 ft., we have

$$\overline{CO} = \frac{AO^2}{PO} = \frac{36}{18} = 2 \text{ ft.}$$

Then, $\overline{DC} = \overline{DO} - \overline{CO} = 6 - 2 = 4$ ft. With the altitude DC of the zone and the great sphere circumference, we can solve for the zone area.

$$A_{zone} = (12\pi \text{ ft.})(4 \text{ ft.}) = 48\pi \text{ sq. ft.}$$

The total area illuminated is 48π sq. ft.

CYLINDERS & CONES

● PROBLEM 842

A right circular cylinder has a base whose diameter is 7 and height is 10. What is the surface area of the cylinder, not including the bases?

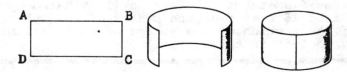

822

Solution: From the figure, we see that calculating the required surface area is equivalent to calculating the area of the rectangle shown. The area of the rectangle, K, is

$$K = (AB)(BC).$$

But BC is the height of the cylinder, and BC = 10. AB is equivalent to the circumference of the base of the cylinder. If the base has diameter d, the circumference of the base is πd, which equals 7π in our case. Hence, AB = 7π. Therefore,

$$K = (7\pi)(10) = 70\pi$$

is the required surface area.

● **PROBLEM** 843

Find, in terms of π, the lateral area and the total surface area of a right circular cylinder if the radius of its base measures 5 in. and its altitude measures 8 in. (See figure).

Solution: The lateral face of a right circular cylinder is actually a rectangle, whose base length (b) is the circumference of the base of the cylinder, and whose height (h) is the height of the cylinder. Therefore, the lateral area is equal to bh. In the case of a right circular cylinder, b = 2πr, the circumference of the base, where r is the base radius.

Lateral Area = 2πrh.

By substitution, Lateral Area = $2\pi(5 \text{ in.})(8 \text{ in.}) = 80\pi \text{ in.}^2$ The total surface area equals the lateral area plus the sum of the area of the two bases.

Area of one Base = Area of a circle = πr^2

$$= \pi(5)^2 \text{ in.}^2 = 25\pi \text{ in.}^2$$

Area of the two Bases = $2(25\pi \text{ in.}^2) = 50\pi \text{ in.}^2$

Total Surface Area = $(80\pi + 50\pi)\text{in.}^2 = 130\pi \text{ in.}^2$

Therefore, Lateral area = 80π sq. in.

Total Surface area = 130π sq. in.

In the figure shown, B, C, and D are right angles. AB = ED = a. $\overset{\frown}{AE}$ is a quadrant of a circle whose radius is 2a. Find in terms of a the total area of the solid formed by rotating the figure through 360° about $\overline{AB}$ as an axis. (Leave answer in terms of π.)

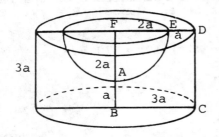 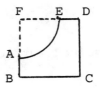

Solution: We can find the total surface area of the solid that is generated, by summing up the areas of the surfaces comprising the solid. The solid generated is a cylinder with a hemisphere of radius 2a scooped out of the top.

From the planar figure, since AB = ED = a and AF = EF = 2a, we can conclude BC = DC = 3a.

T = total area = lateral area of a cylinder + the area of the base (a circle) + the surface area of a hemisphere + the area of a ring.

Lateral Area of a cylinder = $2\pi rh$

Base area = πr^2

Surface area of a hemisphere = $2\pi r^2$

Area of a ring = $\pi(r^2_{cyl.} - r^2_{sphere})$

The radius of the base of the cylinder = 3a and the height also = 3a. Therefore,

Lateral Area of cylinder = $2\pi(3a)(3a) = 18\pi a^2$ and

Base area = $\pi(3a)^2 = 9\pi a^2$.

The radius of the hemisphere is 2a. Therefore, surface area of the hemisphere = $2\pi(2a)^2 = 8\pi a^2$ and Area of the ring = $\pi(9a^2 - 4a^2) = 5\pi a^2$. It follows, by substitution, that

$T = 18\pi a^2 + 9\pi a^2 + 8\pi a^2 + 5\pi a^2 = 40\pi a^2$.

Therefore, the total surface area of the solid generated is $40\pi a^2$.

Find the lateral area and the total area of a right circular cone in which the radius measures 14 in. and the slant height measures 20 in. [Use $\pi = \frac{22}{7}$].

Solution: The lateral surface of a right circular cone, with base radius r, when laid flat, appears to be a sector of a large circle. As such, its area is equal to the area of that sector.

The radius of the sector is equal to the slant height, ℓ, of the cone. The arc length intercepted is equal to the circumference of the base of the cone, $2\pi r$.

Area of the Sector

= (Area of entire circle)$\left(\dfrac{\text{Central Angle measure}}{\text{Measure of full circle}}\right)$.

Let θ = Central angle measure.

Area of the Sector = $\pi\ell^2(\dfrac{\theta}{2\pi}) = \ell^2\dfrac{\theta}{2}$.

All arcs and angles are measured in radians. In radians,

$\theta = \dfrac{\text{arc length}}{\text{radius}} = \dfrac{2\pi r}{\ell}$.

Therefore, area of the sector = $\ell^2(\dfrac{1}{2})(\dfrac{2\pi r}{\ell})$.

Area of the sector = $\pi r\ell$ = Lateral area of the cone.

Hence, by substitution,

Lateral Area of the cone = $\pi(14)(20)$ in.^{2} = $(\dfrac{22}{7})(14)(20)$ in.^{2}

$= 880$ in.^{2}

Total Area = Lateral area + Base area

$= \pi r\ell + \pi r^2$

$= 880$ in.$^2 + \dfrac{22}{7}(14)^2$ in.^{2}

$= (880 + 616)$ in.$^2 = 1496$ in.^{2}

Therefore, Lateral Area = 880 sq. in.

Total Area = 1496 sq. in.

a) Find the lateral surface area of a right circular cone whose base has radius 3 yds. and whose slant height is 7 yds.
b) Find its base area.

Solution:

a) The surface area of a right circular cone with slant height s and base radius r is

$$A = \pi rs.$$

For our problem

$$A = (\pi)(3 \text{ yds.})(7 \text{ yds.})$$

$$A = 65.97 \text{ yd.}^2$$

b) The area of the base, of radius r, is

$$A = \pi r^2.$$

or

$$A = \pi(3 \text{ yd.})^2 = 28.26 \text{ yd.}^2$$

The flat pattern in the accompanying diagram is used to make a lampshade. It is the lateral surface of the frustum of a right circular cone.

The two concentric circles have radii of 12 and 4 inches and m∢AOB = 60°. The minor sector AOB and the remainder of the interior of the circle are removed and discarded. $\overline{AC}$ is fastened to $\overline{BD}$ with no space or overlap.

(a) Find the radius of each of the two bases of the lampshade.

(b) Find the area of the outer surface of the lampshade. [Leave answer in terms of π.]

(c) Find the altitude of the lampshade. [Leave in radical form.]

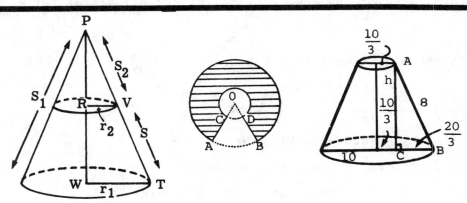

<u>Solution:</u>

(a) The smaller base of the lampshade will be formed out of the portion of the smaller circle remaining after the 60° sector has been removed. Since 60° is $\frac{1}{6}$ of the total measure of the circle, the smaller base of the lampshade will be a circle whose circumference will be $\frac{5}{6}$ of the original small circle.

Since the circumference is a linear function of the radius, the ratio of the circumferences of any two circles will be equal to the ratio of the radii. Thus, the radius $\overline{OC}$ of the smaller circle will be $\frac{5}{6}$ of the original smaller radius of 4. Therefore, the radius of the smaller base of the lampshade is $\frac{5}{6}(4)$ or $\frac{10}{3}$ inches.

By similar logic, the radius of the larger base is $\frac{5}{6}$ of the original large radius of 12. The larger base has a radius of $\frac{5}{6}(12)$ or 10 inches.

Therefore, the radii of the bases of the lampshade are $\frac{10}{3}$ in. and 10 in.

(b) The area of the outer surface of the lampshade has the form of the surface area of a frustum of a cone. Therefore, its area can be found by first looking at a frustum in general.

Assume that a frustum of a cone is formed with base radii r_1 and r_2 and slant height s. The entire cone has slant height s_1 and the upper cone is left with slant height s_2.

The lateral area of the entire cone is, in general, π(base radius)(slant height). In this case, $L_1 = \pi r_1 s_1$.

The lateral area of the upper cone is $L_2 = \pi r_2 s_2$. Hence, the lateral area of the frustum is $L = L_1 - L_2 = \pi(r_1 s_1 - r_2 s_2)$. Since we are only given data about the frustum, we don't know s_1 and s_2 and, as such, must try to eliminate them.

We note that $\triangle PWT$ is a right triangle and since $r_2 \parallel r_1$ (this is the case in all frustums of right circular cones). $\triangle PRV \sim \triangle PWT$, by the AA Similarity Theorem. Therefore,

$$\frac{s_2}{s_1} = \frac{r_2}{r_1} \qquad \text{and} \qquad s_2 = \frac{s_1 r_2}{r_1}.$$

Now, by substituting into the equation for L, we obtain

$$L = \pi\left(r_1 s_1 - \frac{s_1 r_2^2}{r_1}\right) = \frac{\pi}{r_1}(r_1^2 s_1 - s_1 r_2^2) = \frac{\pi s_1}{r_1}(r_1^2 - r_2^2)$$

$$= \frac{\pi s_1}{r_1}(r_1^2 - r_2^2) = \frac{\pi s_1}{r_1}(r_1 - r_2)(r_1 + r_2)$$

$$= \left(\pi s_1 - \frac{\pi s_1 r_2}{r_1} \right)(r_1 + r_2).$$

Substituting $s_2 = \frac{s_1 r_2}{r_1}$ into the last equation,

$$L = \pi(s_1 - s_2)(r_1 + r_2).$$

We see that $s = s_1 - s_2$. Therefore,

$$L = \pi s(r_1 + r_2).$$

In our lampshade problem we know $r_1 = 10$, $r_2 = \frac{10}{3}$ and $s = CA$ = DB = the difference between the original two radii, 12 - 4, or 8.

By substitution,

$$L = \pi(8)(10 + \frac{10}{3}) = \pi(8)(\frac{40}{3}) = \frac{320}{3}\pi.$$

Therefore, the area of the outer surface is $\frac{320\pi}{3}$ sq. in.

(c) We can find the altitude, AC, by recongizing that $\triangle ABC$ is a right triangle and then applying the Pythagorean Theorem. The hypotenuse of $\triangle ABC$ is 8 in., since it corresponds to the slant height of the frustum.

The top radius is $\frac{10}{3}$ in and the bottom radius is 10 in. By projecting the top radius onto the bottom we see that $CB = 10 - \frac{10}{3} = \frac{20}{3}$.

Let h = the length of the altitude. Then,

$$h^2 = 8^2 - (\frac{20}{3})^2 = 64 - \frac{400}{9} = \frac{576 - 400}{9} = \frac{176}{9}$$

$$h = \sqrt{\frac{176}{9}} = \frac{\sqrt{16}\sqrt{11}}{3} = \frac{4}{3}\sqrt{11}.$$

Therefore, the altitude of the frustum is $\frac{4}{3}\sqrt{11}$ in.

CHAPTER 49

VOLUMES I

Basic Attacks and Strategies for Solving Problems in this Chapter. See pages 829 to 851 for step-by-step solutions to problems.

The introduction to Chapter 46 includes a discussion of rectangular solids, prisms, and pyramids. It might be wise to review those pages prior to continuing with this chapter.

Volume refers to the number of unit-sized cubes it wouldtake to fill up a solid. Examine the rectangular solid below. Notice that in the bottom layer there are 8 rows with 4 cubes in each row for a total of 32 cubes in the first layer. Since there are five layers and $5 \times 32 = 160$, the volume of the rectangular solid is 160 cubic units.

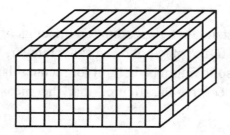

Note that if the dimensions of the base of this solid are ℓ and w_ℓ then ℓ would refer to the number of cubes in each row and w would refer to the number of rows, so there would be ℓw cubes in the first layer. If the third dimension of this rectangular solid were h, then h would be the number of layers; thus, the standard volume formula for a rectangular solid is

$$V = \ell wh.$$

Since ℓw is also the area of the base of the rectangular solid, if we let $B = \ell w,$

another volume formula is
$$V = Bh.$$

Consider the right prism pictured below.

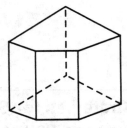

Suppose that the area of the base is B and the altitude is h. This means that B indicates the number of cubes in the first layer, and h is the number of layers. Thus, the volume formula for any right prism is
$$V = Bh.$$

This formula can be extended to include all prisms, not just right prisms, as a result of Cavalieri's Principle. The proof of this fact is established in Problem 856.

The volume formula for a pyramid is $V = \dfrac{1}{3}Bh$, where B again represents the area of the base, and h is the altitude. This formula is developed in Problem 857.

Be sure to understand that, depending on the units of measurement, the number of cubes in a figure may not be an exact integer. While the intuitive picture of cubes in a solid might be helpful for rectangular solids, remember when dealing with less regular figures that the measurements are real numbers.

Step-by-Step Solutions to
Problems in this Chapter,
"Volumes I"

RECTANGULAR SOLIDS & CUBES

● **PROBLEM** 848

How many cubic feet are contained in a packing case which is
a rectangular solid 4 ft. long, 3 ft. wide and $3\frac{1}{2}$ ft. high?

Solution: The volume of a rectangular prism having a base
area B and an altitude h is V = Bh. Since the base of the
rectangular solid is a rectangle, the area of the base (of
length ℓ and width w) is B = ℓw. By substitution, B = 3 × 4
= 12. Since h = $3\frac{1}{2}$:

Volume of the solid = Bh = 12 × $3\frac{1}{2}$ = 42.

Therefore, the packing case contains 42 cu. ft.

● **PROBLEM** 849

A 20 × 40-ft. swimming pool is 4 ft. deep and is filled to the
brim. a) How many cubic feet of water are in the pool? b)
If one gallon is 0.13 cubic feet, how many gallons of water
are in the pool? c) If water weighs approximately 62.4
lb/cu. ft., what is the approximate weight of water in the
pool?

<u>Solution</u>:

a) The pool is a rectangular parallelepiped filled with water. Hence, the volume of water = the volume of the pool. The volume of the pool is equal to the product of its edge lengths, since the pool is a rectangular parallelepiped. In our case,

Volume of water = volume of pool

Volume of water = (20 ft.)(40 ft.)(4 ft.)

Volume of water = 3,200 ft.^{3} = 3200 cu. ft.

b) Suppose 3,200 cu. ft. equals x gallons. We know that 0.13 cu. ft. equals 1 gallon. Hence, setting up a proportion

$$\frac{x}{3200 \text{ cu. ft.}} = \frac{1 \text{ gallon}}{0.13 \text{ cu. ft.}}$$

or
$$x = \frac{3200 \text{ gallons}}{0.13} = 24615.4.$$

c) The statement of the problem tells us that 1 cu. ft. of water weighs 62.4 lbs. Let 3200 cu. ft. of water weigh x lbs. Setting up a proportion

$$\frac{x}{3200 \text{ cu. ft.}} = \frac{62.4 \text{ lbs.}}{1 \text{ cu. ft.}}$$

$$x = (3200)(62.4 \text{ lbs.})$$

$$= 199,680 \text{ lbs.}$$

● **PROBLEM** 850

A room is 12 ft. wide, 15 ft. long, and 8 ft. high. If an air conditioner changes the air once every five minutes, how many cubic feet of air does it change per hour?

<u>Solution</u>: The entire volume of air in the room is processed in 5 minutes. The rate of processing is, therefore,

$$R = V/5 \text{ min}$$

where V is the volume of the room. Since the room is a rectangular parallelepiped, its volume is equal to the product of the lengths of its edges, or (12 ft.)(15 ft.)(8 ft.) = 1440 cu. ft. in our case. Hence, the rate of processing is

$$R = \frac{1440 \text{ cu. ft.}}{5 \text{ min.}}.$$

Since we want the rate in cubic feet per hour, we change 5 minutes to hours. Because, 60 minutes is 1 hour, 5 minutes is 1/12 of an hour. Hence,

$$R = \frac{1440 \text{ cu. ft.}}{1/12 \text{ hr.}} = 17,280 \text{ cu. ft./hr.}$$

What are the dimensions of a solid cube whose surface area is numerically equal to its volume?

Solution: The surface area of a cube of edge length a is equal to the sum of the areas of its 6 faces. Since a cube is a regular polygon, all 6 faces are congruent. Each face of a cube is a square of edge length a. Hence, the surface area of a cube of edge length a is

$$S = 6a^2.$$

The volume of a cube of edge length a is

$$V = a^3.$$

We require that A = V, or that

$$6a^2 = a^3$$

or

$$a = 6.$$

Hence, if a cube has edge length 6, its surface area will be numerically equal to its volume.

Find the volume of a cube whose total area is 54 sq. in.

Solution: The volume, V, of a cube, whose edge is length e, is given by the product of the area of the base, e^2, and the height of the cube, e. Therefore, $V = e^3$.

The area of any face is e^2. Therefore, the total area of the 6 faces is $6e^2$. This fact will enable us to calculate the length of the edge, given the total area

$$54 \text{ in.}^2 = 6e^2$$
$$9 \text{ in.}^2 = e^2$$
$$3 \text{ in.} = e.$$

Since $V = e^3$, by substitution,

$$V = (3 \text{ in.})^3 = 27 \text{ in.}^3$$

Therefore, the Volume of the cube is 27 in.^{3}

● **PROBLEM** 853

If a cube has edge length a, what happens to

a) the cube's surface area as a is increased by a factor x?
b) the cube's volume as a is increased by a factor x?

Solution:

a) The surface area of a cube is $6a^2$, where a is the edge length of the cube. If a is changed to xa, the new surface area is $6x^2a^2$. Hence, if we change the edge length of a cube by a factor of x, we change the cube's surface area by a factor of x^2.

b) The volume of a cube of edge length a is a^3. If we change a by a factor x, the cube's volume is changed to x^3a^3 (i.e. by a factor of x^3).

PRISMS

● **PROBLEM** 854

The base of a right prism, as shown in the figure, is an equilateral triangle, each of whose sides measures 4 units. The altitude of the prism measures 5 units. Find the volume of the prism. [Leave answer in radical form.]

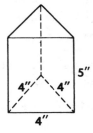

Solution: We imagine the prism as a stack of equilateral triangles, congruent to the base of the prism. Let each of these triangles be one unit of measure thick. We can then calculate the area of the base, B, and multiply it by the number of bases needed to complete the height of the prism, h, to obtain the volume of the prism. Therefore, V = Bh. All prism volumes can be thought of in this way.

In this particular problem, the base is an equilateral triangle. Therefore $B = \frac{s^2\sqrt{3}}{4}$, where s is the length of a side of the base. By substitution, $B = \frac{(4)^2\sqrt{3}}{4} = 4\sqrt{3}$. Since the prism is 5 units high,

$$V = Bh = (4\sqrt{3})5 = 20\sqrt{3}.$$

Therefore, the volume of the prism is $20\sqrt{3}$ cu. units.

● **PROBLEM** 855

A wooden form for a small dam is in the shape of a right prism whose base is an isosceles trapezoid as shown in the adjacent figure. AB is 3 ft., CD is 5 ft., h is 8 ft., and the dam is 44 ft. long. The form has been filled with concrete to a depth of 2 ft. Find, to the nearest cubic yard, the additional number of cubic yards of concrete needed to fill the form.

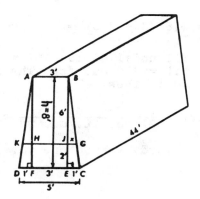

Solution: In the figure, ABCD is the base of the wooden form. $\overline{KG}$ is the line parallel to base $\overline{DC}$, two feet from $\overline{DC}$, marking the surface level of the concrete. We are asked to find the _unfilled_ volume of the wooden form; that is, the volume of the prism with base ABGK. The volume equals $B\ell$, where B = area of the base and ℓ = altitude. ℓ, the altitude is known.

Since the extra concrete affects only the dimensions of the base, the length of the empty part of the prism equals the length of the original prism and ℓ = 44 ft.

The base area B, (ABGK), though, is definitely affected by the loss of two feet. The area is given by the formula $\frac{1}{2}h(b_1 + b_2)$, where b_1 and b_2 are lengths of the base sides and h the altitude. With $\overline{KG}$ as the base instead of $\overline{CD}$, both h and b_2 change. Instead of h = 8 ft., the loss of two feet makes h = 8 - 2 = 6 ft.

To calculate the change in the length of b_2, draw altitude $\overline{AF}$ and $\overline{BE}$. Then the new b_2, KG, = KH + HJ + JG. HJ is

833

known; it is 3 (since quadrilateral ABJH is a parallelogram). To find JG, we relate $\overline{JG}$ to $\overline{EC}$ by proving $\triangle JBG \sim \triangle EBC$. Note $\sphericalangle JBG \cong \sphericalangle EBC$. Furthermore, since $\overline{KG} \parallel \overline{DC}$, $\sphericalangle BJG = \sphericalangle BEC$. By the A-A Similarity Theorem, $\triangle JBG \sim \triangle EBC$ and

$$\frac{BJ}{JG} = \frac{BE}{EC}.$$

BE, as an altitude of ABCD, = 8 ft. BJ = 8 - 2 = 6 ft. Because ABCD is an isosceles trapezoid, DF = EC = $\frac{1}{2}$(DC - FE). Since ABEF is a parallelogram, FE = AB = 3 and thus EC = $\frac{1}{2}$(5-3) = 1 ft. Substituting in our values for BJ, BE, and EC, we obtain

$$\frac{6}{JG} = \frac{8}{1} \qquad \text{or} \qquad JG = \frac{3}{4} \text{ ft.}$$

By a similar procedure, we can show HK = .75 ft. Thus, the new b_2, KG, = KH + HJ + JG = .75 + 3 + .75 = 4.5 ft. With b_1, b_2 and h known, we can find the area of ABGK.

$$B = \frac{1}{2}(6)(3 + 4.5) = 22.5 \text{ sq. ft.}$$

Thus, the volume still unfilled is = Bℓ = (22.5 sq. ft.)(44 ft.) = 990 cu. ft. The answer, however, must be in cubic yards. Thus,

$$V = 990 \text{ cu. ft.} = \frac{990 \text{ cu. ft.}}{27 \text{ cu. ft./cu. yd.}} = 36\frac{1}{3} \text{ cu. yd.}$$

To the nearest cubic yard, V = 36 cu. yd.

● **PROBLEM** 856

Show that two prisms have equal volumes if their bases have equal areas and their altitudes are equal.

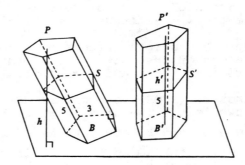

Solution: In defining the volumes of polyhedral regions, three postulates are needed: (1) to each polyhedral region, there corresponds a unique positive number - in other words, every polyhedral region has one and only one volume; (2) the volume of a rectangular solid equals the product of its length, width, and height - this defines a standard from which other volume formulas can be derived; and (3) (Cavalieri's Principle) if two solid regions have equal altitudes, and if the sections, made by planes parallel to the base of each solid and at the same distance from each base, are always equal in area, then the volumes of the solid regions are equal - this provides a means of relating the volumes of all figures from the second postulate.

For this problem, we show that all sections cut by planes parallel to the bases of a prism are congruent to the bases, and thus, for prisms, Cavalieri's Postulate can be reduced to two requirements. For two prisms to have equal volumes, they must have (1) equal base areas, and (2) equal altitudes.

Given: Prisms P and P' with bases B and B' such that area of B equals area of B'; and altitudes h and h' such that h = h'.

Prove: Volume of P = Volume of P'.

Statements	Reasons
1. Prisms P and P' with bases B and B' such that area of B = area of B'; and altitudes h and h' such that h = h'.	1. Given.
2. Construct two planes parallel to the bases of P and P' making sections S and S' at equal distances d and d' from the bases B and B'.	2. In a given space, only one plane can be drawn a given distance from the given plane.
3. S ≅ B. S' ≅ B'.	3. Every section of a prism made by a plane parallel to the bases is congruent to the bases.
4. Area of S = Area of B Area of S' = Area of B'.	4. If two figures are congruent, then their areas are equal.
5. Area of S = Area of S'.	5. Transitivity Postulate.
6. Volume of P = Volume of P'.	6. If two solid regions have equal altitudes and if the sections made by planes parallel to the base of each solid and at the same distance from each base are always equal in area, then the volumes are equal. (Cavalieri's Principle.)

PYRAMIDS & TETRAHEDRONS

● **PROBLEM** 857

> Show that the volume of any pyramid is equal to one-third the product of the area of the base and altitude. Assume that the volume of a triangular pyramid equals one-third the product of the base area and the altitude.

<u>Solution</u>: From any vertex of the n-gon that forms the base, draw the n-3 diagonals to the other vertices. This separates the n-gon into n-2 triangles, each with areas $b_1, b_2 \ldots b_{n-2}$.

The n-2 triangular pyramids formed by the triangles as bases and the original vertex as vertex, comprise the entire original pyramid. Therefore, the volume V of the original pyramid equals the sum of the volumes of the n-2 triangular pyramids. Since the volume of the triangular pyramid equals $\frac{1}{3}a \cdot b$, we obtain:

$$V = \frac{1}{3}a_1 \cdot b_1 + \frac{1}{3}a_2 \cdot b_2 \ldots + \frac{1}{3}a_{n-2} \cdot b_{n-2}.$$

$a_1, a_2 \ldots a_{n-2}$ are the altitudes drawn from the vertex to the planes of the base triangles. Since all the base triangles are coplanar, all altitudes must be congruent, or $a_1 = a_2 \ldots = a_{n-2} = a$. Factoring out the a in each volume term, we obtain:

$$V = \frac{1}{3}ab_1 + \frac{1}{3}ab_2 + \frac{1}{3}ab_3 + \ldots + \frac{1}{3}ab_{n-2} = \frac{1}{3}a \, (b_1 + b_2 \ldots + b_{n-2}).$$

Since the n-2 triangles comprise the base, the sum of the areas of the triangles must equal the base area, b.

$$V = \frac{1}{3}a \cdot (b_1 + b_2 \ldots + b_{n-2}) = \frac{1}{3}a \cdot b.$$

● **PROBLEM** 858

> Given: Triangular pyramid O-ABC with base area b and altitude length a.
>
> Prove: Volume O-ABC = $\frac{1}{3}b \cdot a$.

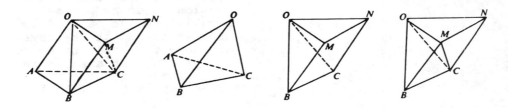

Solution: We know that the volume of a prism of base area b and altitude length a equals b · a. Therefore, if we show that three triangular pyramids congruent to the given one can form a prism of base area b and altitude length a, then the volume of each of these pyramids is equal to $\frac{1}{3}$ that of the prism, or

$V = \frac{1}{3}ba$.

Statements	Reasons
1. Triangular pyramid O-ABC with volume V, base area b, and altitude length a.	1. Given.
2. Locate point N on the opposite side of plane OBC as point A such that quadrilateral NOAC is a parallelogram. Locate point M on the opposite side of plane OBC as point A such that quadrilateral MOAB is a parallelogram.	2. Given three distinct, non-collinear points in space, there is at least one point in space such that the four points form a parallelogram.
3. $\overline{MB} \parallel \overline{AO}$. $\overline{NC} \parallel \overline{AO}$.	3. Opposite sides of a parallelogram are parallel.
4. $\overline{MB} \cong \overline{AO}$. $\overline{NC} \cong \overline{AO}$.	4. Opposite sides of a parallelogram are congruent.
5. $\overline{MB} \parallel \overline{NC}$.	5. Lines parallel to the same line are parallel.
6. $\overline{MB} \cong \overline{NC}$.	6. Transitivity Postulate.
7. Quadrilateral MNCB is a parallelogram.	7. If a pair of opposite sides of a quadrilateral are parallel and congruent, then the quadrilateral is a parallelogram.
8. $\overline{MN} \parallel \overline{BC}$. $\overline{OM} \parallel \overline{AB}$. $\overline{ON} \parallel \overline{AC}$.	8. Opposite sides of a parallelogram are parallel.
9. $\overline{MN} \cong \overline{BC}$. $\overline{OM} \cong \overline{AB}$. $\overline{ON} \cong \overline{AC}$.	9. Opposite sides of a parallelogram are congruent.
10. $\triangle OMN \cong \triangle ABC$.	10. The SSS Postulate.
11. $\triangle OMN \parallel \triangle ABC$.	11. Two polygons are parallel if their corresponding sides are parallel.

12.	Figure OMN - ABC is a prism.	12. The polyhedral region formed by joining the corresponding vertices of parallel and congruent polygons is a prism whose bases are the polygons.

We now divide O-MNCB into two congruent pyramidal regions O-MNC and O-MCB, and show O-MNC or C-OMN has the same volume as O-ABC. Since volume O-ABC = volume O-MNC = volume O-MCB, volume O-ABC = $\frac{1}{3}$ volume of the prism.

13.	$\overline{MB} \cong \overline{NC}$. $\overline{MN} \cong \overline{BC}$.	13. Opposite sides of a parallelogram are congruent.
14.	$\overline{MC} \cong \overline{MC}$.	14. All line segments are congruent to themselves.
15.	$\triangle MBC \cong \triangle MNC$.	15. The SSS Postulate.
16.	Area of $\triangle MBC$ = Area of $\triangle MNC$.	16. Congruent triangles have equal areas.
17.	Altitude of O-MBC = altitude of O-MNC.	17. From a given external point (O) to a given plane (NMBC), only one line perpendicular to the plane (the altitude) can be drawn. Since the bases $\triangle MBC$ and $\triangle MNC$ are coplanar, the altitudes drawn from O must be the same.
18.	Volume O-MBC = Volume O-MNC.	18. If two pyramids have bases of equal areas and altitudes of equal lengths, then the volumes are equal.
19.	Altitude of O-ABC = altitude of C-MNO	19. Parallel planes (OMN and ABC) are everywhere equidistant.
20.	Area of $\triangle OMN$ = Area of $\triangle ABC$.	20. If two triangles are congruent, then their areas are equal. (See Steps 9 and 10 for proof of congruence.)
21.	Volume of O-ABC = Volume of C-MNO.	21. Two pyramids, whose bases have the same area and whose altitudes are of equal length, have equal volume.
22.	Volume of C-MNO = Volume of O-MNC.	22. The volume of a polyhedral region equals itself.
23.	Volume of OMN-ABC = Volume C-MNO + Volume O-MBC + Volume O-ABC.	23. The volume of a polyhedral figure divided into several regions equals the sum of the volumes of the regions.

24. Volume of OMN—ABC = 3 · Volume O-ABC.

24. Substitution Postulate (Steps 18, 21).

25. Volume of OMN - ABC = a · b.

25. The volume of a prism equals the base area, b, times the altitude length, a.

26. a · b = 3(Volume O-ABC).

26. Substitution Postulate.

27. Volume O-ABC = $\frac{1}{3}$a · b.

27. Given real numbers, a, b, c with c ≠ 0; if a = b, then $\frac{a}{c} = \frac{b}{c}$.

● **PROBLEM** 859

Let $\overline{ABCD}$ be a regular tetrahedron with each edge of length 2. Let $\overline{AE}$ be perpendicular to the base, and assume that CE = $\frac{2}{3}$CF. (See figure). In △BCD, $\overline{CF} \perp \overline{BD}$. a) What is the length of CF? b) What is the length of CE? c) What is the length of AE? d) What is the base area of △BCD? e) What is the volume of the tetrahedron?

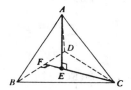

Solution:

a) We can prove that △CFB ≅ △CFD by the A.A.S. (angle angle side) Postulate in order to show DF = BF. Since we know DB, we can find DF. By using the Pythagorean Theorem we can determine CF. ∢CFB ≅ ∢CFD because $\overline{CF} \perp \overline{BD}$ and all right angles are congruent. Since all the edges of ABCD are of length 2, △DBC is equilateral. By definition, then, ∢FBC ≅ ∢FDC, and $\overline{DC} \cong \overline{BC}$. By the A.A.S. Postulate, △CFB ≅ △CFD, hence DF = BF. But DF + BF = DB. Thus,

$$DF = \frac{1}{2}DB.$$

Since DB = 2,

$$DF = 1.$$

Applying the Pythagorean Theorem to △DFC gives

$$(DF)^2 + (FC)^2 = (DC)^2$$

or

$$FC = \sqrt{(DC)^2 - (DF)^2}.$$

Noting that DC = 2 and DF = 1

$$FC = \sqrt{4 - 1} = \sqrt{3}$$

b) By the statement of the problem,

$$CE = \frac{2}{3}CF.$$

From part (a), CF = $\sqrt{3}$. Hence,

$$CE = \frac{2}{3}\sqrt{3} = \frac{2\sqrt{3}}{3}.$$

c) Using the Pythagorean Theorem, for $\triangle AEC$,

$$(AE)^2 + (EC)^2 = (AC)^2$$

or
$$AE = \sqrt{(AC)^2 - (EC)^2}.$$

By realizing that AC is an edge of ABCD, we note that AC = 2. From part (b), EC = $\frac{2}{\sqrt{3}}$. Therefore,

$$AE = \sqrt{4 - 4/3}$$

$$AE = \sqrt{8/3}.$$

d) The area of $\triangle BCD$ is $\frac{1}{2}$ the product of its altitude (FC) and its base (DB). Hence,

$$Area = \frac{1}{2}(FC)(DB).$$

Noting that FC = $\sqrt{3}$ from part (a), and that DB = 2,

$$Area = (\frac{1}{2})(\sqrt{3})(2) = \sqrt{3}.$$

e) The volume of a tetrahedron of base area A and altitude h is

$$V = \frac{1}{3}Ah.$$

In our case h = AE = $\sqrt{8/3}$ from part (c). Hence,

$$V = (\frac{1}{3})(\sqrt{3})\left[\frac{\sqrt{8}}{\sqrt{3}}\right]$$

$$V = \frac{2\sqrt{2}}{3}.$$

● **PROBLEM** 860

The slant height of a regular pyramid A-BCD is 2. If the altitude is of length 1, find the volume.

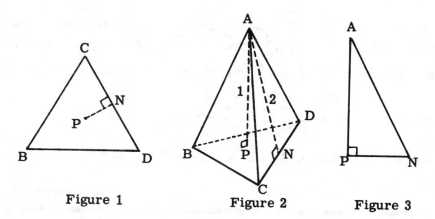

| Figure 1 | Figure 2 | Figure 3 |

<u>Solution</u>: The volume of pyramid A-BCD equals one third the base area times the altitude. The altitude equals 1. Since this is a regular pyramid and only three points of its base are specified, the base must be a regular or equilateral triangle. The area of the base, therefore, equals $\frac{S^2\sqrt{3}}{4}$, where S is the length of a side. To find S, we show:

(1) $NC = \frac{1}{2}CD = \frac{1}{2}S$;   (2) $BN = 3 \cdot PN$;   (3) $S = BC = \sqrt{BN^2 + NC^2}$

(1) Since A-BCD is a regular pyramid, its lateral edges are all congruent; therefore, face ADC is an isosceles triangle and, since the altitude drawn from the vertex angle of an isosceles triangle bisects the base, $CN = \frac{1}{2}CD = \frac{1}{2}S$.

(2) Since A-BCD is a regular pyramid, the altitude $\overline{AP}$ must intersect the base at the center of the base, or the point equidistant from all sides. But, by an earlier theorem, this point is also the point of concurrency of all angle bisectors of the triangle. Therefore, BP bisects ∢B. Consider equilateral triangle △BCD as an isosceles triangle of vertex B. Then, the angle bisector of the vertex angle bisects the base and therefore, angle bisector $\overline{BP}$ is also the median drawn to side $\overline{CD}$. By similar reasoning, considering △BCD as an isosceles triangle with C and D as vertex angles, we show that the angle bisectors of an equilateral triangle are the same as the medians of the triangle. Therefore, P, the point of concurrency of the angle bisectors, is the point of concurrency of the medians as well. Since the distance of the point of concurrency of medians from a given side is one third the length of the median, $PN = \frac{1}{3} \cdot BN$.

To find PN, note that PN is a leg in right triangle APN. By Pythagorean Theorem, $PN^2 = AN^2 - AP^2 = 2^2 - 1^2 = 3$. $PN = \sqrt{3}$.

(3) The median drawn to the base of an isosceles triangle is the perpendicular bisector of the base. Therefore, ∢BNC is a right angle and △BNC <u>is</u> a right triangle. We wish to find the length of a side $\overline{BC}$. By Pythagorean Theorem, $BC^2 = BN^2 + NC^2$. BC = S. From (1), we have $NC = \frac{1}{2}S$. From

(2), we have BN = 3 · PN = $3\sqrt{3}$. Substituting in values, we obtain $s^2 = (3\sqrt{3})^2 + (\frac{1}{2}s)^2$ or $s^2 = 27 + \frac{1}{4}s^2$, or $\frac{3}{4}s^2 = 27$.

Multiplying both sides by $\frac{4}{3}$, we obtain $s^2 = 36$ or $S = 6$.

$$\text{Area of base} = \frac{s^2\sqrt{3}}{4} = \frac{6^2\sqrt{3}}{4} = 9\sqrt{3}.$$

$$\text{Volume of A–BCD} = \frac{1}{3}a \cdot b = \frac{1}{3}(1)(9\sqrt{3}) = 3\sqrt{3}.$$

● **PROBLEM** 861

A pyramid has a square base measuring 5 in. on each side and a height of 10 in. What is its volume?

Solution: The volume of a pyramid with base area b and height h is

$$V = \frac{1}{3}bh.$$

Since the base of the given pyramid is a square of edge length 5 in., its area is $(5 \text{ in.})^2 = 25 \text{ in.}^2$. The height of the pyramid is 10 in. Then,

$$V = (\frac{1}{3})(25 \text{ in.}^2)(10 \text{ in.})$$

$$V = \frac{250}{3} \text{ in.}^2 = 83.33 \text{ in.}^2.$$

● **PROBLEM** 862

Find the volume of a pyramid whose base is an equilateral triangle with side 10 and whose altitude is 20.

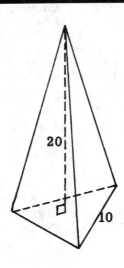

<u>Solution</u>: The volume of a pyramid = $\frac{1}{3}$a · b. a, the altitude, equals 20. b, the area of the base, equals the area of the equilateral triangle of side 10. $b = \frac{s^2\sqrt{3}}{4} = \frac{(10)^2\sqrt{3}}{4} = \frac{100\sqrt{3}}{4} = 25\sqrt{3}$.

The volume of the pyramid = $\frac{1}{3}(20)(25\sqrt{3}) = \frac{500\sqrt{3}}{3}$.

● **PROBLEM** 863

Find the volume of a regular square pyramid if each edge of the base measures 10 in., and the slant height of the pyramid measures 13 in. (see figure).

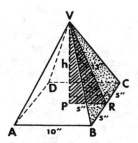

<u>Solution</u>: The volume of a regular square pyramid is equal to $\frac{1}{3}$ × (Area of the base) × (measure of the altitude). The area A, of the base, which is a square of side S, is given by $A = S^2$. In this problem S = 10 in. By substitution, $A = (10 \text{ in.})^2$ = 100 sq. in. = Area of the base.

The length of the altitude is unknown, but can be calculated by the Pythagorean Theorem. When we draw altitude $\overline{VP}$, right triangle VPR is formed. Then,

$$(VR)^2 = (VP)^2 + (PR)^2.$$

VR is the slant height, and is given as 13 in. PR = 5 in., since the altitude intersects the base at its midpoint, and we are told the base is 10 in. wide.

Therefore, by substitution,

$$(13 \text{ in.})^2 = (VP) + (5 \text{ in.})^2$$

$$169 \text{ in.}^2 = (VP)^2 + 25 \text{ in.}^2$$

$$144 \text{ in.}^2 = (VP)^2 \Rightarrow VP = 12 \text{ in.} = \text{altitude.}$$

Therefore, since $V = \frac{1}{3}$(base Area)(altitude), by substitution, $V = \frac{1}{3}(100)(12)\text{in.}^3 = 400 \text{ in.}^3$.

Therefore, the volume of pyramid = 400 cu. in.

Find the volume of a frustum of a pyramid if the area of the bases are b and b' and the altitude is h.

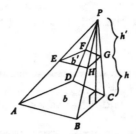

<u>Solution</u>: The frustum of a pyramid is the figure formed by the base of the pyramid, a section of the pyramid parallel to the base and the surface of the pyramid between the base and the section. Another way of viewing this is to consider the section of the pyramid parallel to the base. The part "above" the section forms a little pyramid with the section as its base. The part of the pyramid below the section forms a frustum. Therefore, the volume of the frustum equals the volume of the larger pyramid minus the smaller pyramid. (See figure. Note: b is the area of quadrilateral ABCD, b' is the area of EFGH, h is the altitude of the frustum, and h' is the altitude of the smaller pyramid.)

Volume of P-EFGH = $\frac{1}{3}$(base area)(height) = $\frac{1}{3}$b'h'

Volume of P-ABCD = $\frac{1}{3}$(base area)(height) = $\frac{1}{3}$b(h + h')

Volume of frustum = $\frac{1}{3}$b(h + h') - $\frac{1}{3}$b'h'

$$= \frac{1}{3}[bh + h'(b-b')].$$

We are not finished yet, because our formula contains an h'. The formula that is the answer must only be in terms of b, b' – the area of the bases – and h, the altitude. We solve for h' in terms of these three terms and then eliminate h' from the equation.

To solve for h', we take advantage of the similarities between the little pyramid P-EFGH and the larger one P-ABCD. From an earlier theorem, we know that when a plane parallel to the base cuts a pyramid, the polygon formed is similar to the base with its ratio of similitude equal to the ratio of the altitudes. Therefore,

$$\frac{\text{altitude of P-EFGH}}{\text{altitude of P-ABCD}} = \frac{\text{side of EFGH}}{\text{corresponding side of ABCD}} \quad \text{or}$$

$$\frac{h'}{h + h'} = \frac{EH}{AB} = \text{ratio of similitude.}$$

We wish to find h' in terms of h, b, and b' and therefore we must get rid of $\frac{EH}{AB}$, the ratio of similitude. By earlier theorem, the ratio of the areas of similar polygons equals the square of the ratio of similitude. Therefore, $\frac{\text{area of EFGH}}{\text{area of ABCD}} = \left(\frac{EH}{AB}\right)^2$. But area of EFGH = b' and area of ABCD = b. Therefore, $\frac{b'}{b} = \left(\frac{EH}{AB}\right)^2$ or $\frac{EH}{AB} = \frac{\sqrt{b'}}{\sqrt{b}}$. We can substitute this result in the equation above to obtain a relation solely in terms of h', h, b', and b.

$$\frac{h'}{h + h'} = \frac{\sqrt{b'}}{\sqrt{b}} \ .$$

By crossmultiplication, $h'\sqrt{b} = (h + h')\sqrt{b'} = h\sqrt{b'} + h'\sqrt{b'}$

$$h'\sqrt{b} - h'\sqrt{b'} = h\sqrt{b'}$$

$$h'(\sqrt{b} - \sqrt{b'}) = h\sqrt{b'}$$

$$h' = \frac{h\sqrt{b'}}{\sqrt{b} - \sqrt{b'}} \ .$$

By substituting in the volume expression, we obtain:

Volume = $\frac{1}{3}[bh + h'(b - b')]$

$$= \frac{1}{3}\left[bh + \frac{\sqrt{b'}}{\sqrt{b} - \sqrt{b'}} \cdot h \cdot (b - b')\right].$$

Factoring out an h and rationalizing the denominator, we have the final result:

Volume = $\frac{1}{3}h\left[b + \frac{\sqrt{b'}}{\sqrt{b} - \sqrt{b'}}(b - b')\right]$

$$= \frac{1}{3}h\left[b + \frac{\sqrt{b'}}{\sqrt{b} - \sqrt{b'}} \cdot \frac{(\sqrt{b})^2 - (\sqrt{b'})^2}{1}\right]$$

$$= \frac{1}{3}h\left[b + \frac{\sqrt{b'}}{\sqrt{b} - \sqrt{b'}} \frac{(\sqrt{b} + \sqrt{b'})(\sqrt{b} - \sqrt{b'})}{1}\right]$$

$$= \frac{1}{3}h[b + \sqrt{b'}(\sqrt{b} + \sqrt{b'})]$$

$$= \frac{1}{3}h[b + b' + \sqrt{bb'}].$$

● **PROBLEM** 865

Show that the volume of a regular tetrahedron of side s is given by the formula $\frac{\sqrt{2}}{12}s^3$.

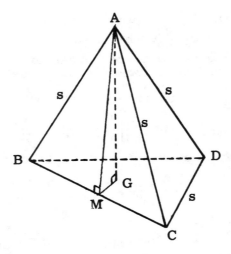

Solution: In the accompanying figure, regular tetrahedron A-BCD has slant height $\overline{AM}$ and perpendicular to the base, $\overline{AG}$. We wish to find the volume. The volume of any pyramid equals $\frac{1}{3}Bh$ where B = the base area and h = altitude. Thus, volume

$$V = \frac{1}{3}(\text{area of } \triangle BCD) \cdot (AG).$$

Since every face of a regular tetrahedron is an equilateral triangle, the area of $\triangle BCD = \frac{\sqrt{3}}{4}s^2$.

To find AG, consider right $\triangle AMG$. We find AM and MG in terms of s. By using the Pythagorean Theorem, we can then find AG. To find AM, remember that, by definition, a slant height is always perpendicular to the side of the base that it intersects. Thus, $\triangle ABM$ is a right triangle. Furthermore, since all edges of a regular tetrahedron are congruent, $\overline{AB} \cong \overline{BC}$ and $\triangle ABC$ is isosceles. (In an isosceles $\triangle$, the altitude to the base is the $\perp$ bisector of the base.) $BM = \frac{1}{2}BC = \frac{1}{2}s$. Using the Pythagorean Theorem in $\triangle ABM$, we have $AM^2 = AB^2 - BM^2 = s^2 - \frac{1}{4}s^2 = \frac{3}{4}s^2$ or $AM = \frac{\sqrt{3}}{2}s$.

To find MG, note that the altitude of a regular pyramid intersects the base polygon at its center. Thus, point G is the center of equilateral $\triangle BCD$, and the point of concurrency of the medians. Since the medians are trisected by the point of concurrency, $MG = \frac{1}{3}MD$. Using a procedure similar to the one above to find AM, we find $MD = \frac{\sqrt{3}}{2}s$. Thus,

$$MG = \frac{1}{3}MD = \frac{1}{3}\left(\frac{\sqrt{3}}{2}s\right) = \frac{\sqrt{3}}{6}s.$$

Thus, hypotenuse $\overline{AM}$ and leg $\overline{MG}$, of right $\triangle AMG$, are known. By the Pythagorean Theorem,

$$AG^2 = AM^2 - MG^2 = \left(\frac{\sqrt{3}}{2}s\right)^2 - \left(\frac{\sqrt{3}}{6}s\right)^2 = \left(\frac{3}{4} - \frac{1}{12}\right)s^2 = \frac{2}{3}s^2.$$

Thus, AG $= \sqrt{\frac{2}{3}}s$ and the volume formula becomes:

$$V = \frac{1}{3}\left[\frac{\sqrt{3}}{4}s^2\right]\left(\sqrt{\frac{2}{3}}s\right) = \frac{\sqrt{2}}{12}s^3.$$

• **PROBLEM** 866

Derive a formula for the volume of a regular octahedron in terms of the edge e.

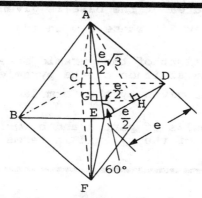

Solution: The octahedron A-BCDE-F can be considered as two congruent square pyramids. Thus, its volume $V = 2 \cdot (\frac{1}{3}Bh)$ where B is the base area and h is the altitude of each pyramid.

The base is a square of edge e. Therefore, $B = e^2$. To find the altitude h, note that h = AG, a leg of right $\triangle$AGH. If we find AH and GH, we can use the Pythagorean Theorem to find AG.

Note, $\overline{AH}$ is leg of right $\triangle$AEH. Since each face of the octahedron is an equilateral triangle, m$\angle$AEH = 60°. In a 30-60-90 right triangle, the side opposite the 60° angle is $\frac{\sqrt{3}}{2}$ times the hypotenuse. Since AE = e, then AH $= \frac{\sqrt{3}}{2}e$.

To find GH, note that A-BCDE is a regular pyramid. Thus, the perpendicular from vertex A to base BCDE intersects the base square BCDE at its center. Since the center of a square is at a distance of half the length of a side from the square's side, GH $= \frac{e}{2}$.

Knowing GH, and AH, we can find the altitude AG, using the Pythagorean Theorem.

$$\ddot{A}G^2 = AH^2 - GH^2 = \left(\frac{\sqrt{3}}{2}e\right)^2 - \left(\frac{e}{2}\right)^2 = \frac{1}{2}e^2$$

$$AG = \frac{\sqrt{2}}{2}e.$$

Thus, $h = \frac{\sqrt{2}}{2}e$, $B = e^2$, and the volume of the octahedron

is $2 \cdot \left(\frac{1}{3}e^2 \cdot \frac{\sqrt{2}}{2}e\right) = \frac{\sqrt{2}}{3}e^3$.

● **PROBLEM** 867

A regular pyramid has a pentagon for its base.

(a) Show that the area of the base is given by the formula $\beta = \frac{5}{4}e^2 \tan 54°$ where e = an edge of the base.

(b) If the slant height of the pyramid makes an angle of 54° with the altitude of the pyramid, show that the altitude is given by the formula $h = \frac{e}{2}$.

(c) Using the formulas of part a and b, write a formula for the volume of the pyramid in terms of the base edge e.

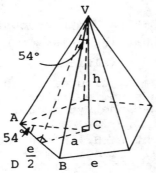

Solution:

(a) The area of a regular polygon A, whose perimeter is p and whose apothem is a, is given by the formula $A = \frac{1}{2}ap$. The perimeter of a regular n-gon equals n times the length of any one side. For this pentagon, p = 5e.

To find the length of the apothem a, draw the perpendicular from the center of the pentagon, C, to any side $\overline{AB}$, intersecting $\overline{AB}$ at point D. D is the midpoint of $\overline{AB}$. (To see this, remember that the center of a regular polygon is also the center of the circumscribing circle. Thus, $\overline{AC}$ and $\overline{BC}$ are radii of that circle. $\overline{AC} = \overline{BC}$ and $\triangle ABC$ is isosceles. $\overline{CD}$ is then the altitude drawn from the vertex angle of an isosceles triangle and therefore bisects the base.) Thus, $AD = \frac{1}{2}AB = \frac{e}{2}$. Furthermore, m∢CAD = 54°. (To see this, note that all five central angles of the pentagon are congruent. Therefore, $m∢ACB = \frac{360°}{5} = 72°$. Since $\triangle ABC$ is isosceles, ∢CAD ≅ ∢CBA. By using the fact that the angle sum of a triangle is 180°, we

848

have m∢CAD + m∢CBA + m∢ACB = 180°, or 2m∢CAD = 180° – 72° or m∢CAD = 54°.)

In right △ACD, then, a is the length of the leg opposite known ∢CAD. Leg AD is also known. The relationship between these three quantities is tan ∢CAD = $\dfrac{\text{length of opposite leg}}{\text{length of adjacent leg}}$ = $\dfrac{CD}{AD}$ or tan 54° = $\dfrac{a}{e/2}$. Solving for a, we obtain a = $\dfrac{e}{2}$ tan 54°.

We have a and p. Thus, the base area equals $\dfrac{1}{2}$ap = $\dfrac{1}{2}(\dfrac{e}{2}$ tan 54°)(5e) = $\dfrac{5}{4}e^2$ tan 54°.

(b) To find h, we construct another triangle, △VDC. To show that △VDC is indeed a right triangle, we remember that the pyramid is regular. The perpendicular to the base of a regular pyramid passes through the center of the base. Since $\overline{VC}$ is the segment connecting the vertex to the center of the base, it is perpendicular to the base and ∢VCD is a right angle. In △VDC, we are given leg $\overline{DC}$ = a = $\dfrac{e}{2}$ tan 54°. Furthermore, ∢DVC is given to be 54°. Relating height $\overline{VC}$ to these quantities, we have

tan 54° = $\dfrac{\text{length of opposite leg}}{\text{length of adjacent leg}}$ = $\dfrac{DC}{VC}$ = $\dfrac{\frac{e}{2} \tan 54°}{\overline{VC}}$. Thus,

$$h = VC = \dfrac{\frac{e}{2} \tan 54°}{\tan 54°} = \dfrac{e}{2}.$$

(c) The volume of any pyramid is given by the formula $\dfrac{1}{3}$bh where b = base area and h = height of the pyramid. From part (a), we know the base area to be $\dfrac{5}{4}e^2$ tan 54°. From part (b), the height is $\dfrac{e}{2}$. Thus,

$$V = \dfrac{1}{3}bh = \dfrac{1}{3}(\dfrac{5}{4}e^2 \tan 54°)(\dfrac{e}{2}) = \dfrac{5}{24}e^3 \tan 54°.$$

● **PROBLEM** 868

The lateral edge of a regular square pyramid makes an angle of θ degrees with its projection on the base. The length of the side of the base is S and the height of the pyramid is h. A plane is passed between the base and the vertex p units from the base and parallel to it so as to cut off a smaller pyramid.

(a) Express S in terms of h and θ.

(b) Derive an expression for the volume of the smaller pyramid in terms of h, θ, and p.

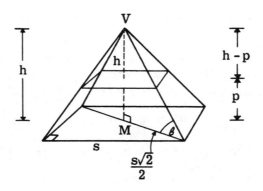

Solution:

(a) The lateral edge of a square pyramid projects onto the base along the diagonal of the square. $\triangle VMA$, formed by the lateral edge, its projection on the base, MA, and the altitude, VM, of the pyramid is a right triangle with θ as one of the acute angles.

If we can express MA in terms of S, then, by applying the tangent ratio, we can gain our desired results.

Let d = length of the whole diagonal, then $d^2 = 2S^2$ by the Pythagorean Theorem. Therefore, $d = S\sqrt{2}$. Since the altitude bisects the diagonal of the base in a regular square pyramid, $MA = \frac{1}{2}d = \frac{S\sqrt{2}}{2}$.

$$\text{Tan } \theta = \frac{\text{length of leg opposite } \theta}{\text{length of leg adjacent } \theta} = \frac{h}{\frac{S\sqrt{2}}{2}} = \frac{2h}{S\sqrt{2}}$$

$$S\sqrt{2} \text{ Tan } \theta = 2h$$

$$S = \frac{2h}{\sqrt{2} \text{ Tan } \theta} \cdot \frac{\sqrt{2}}{\sqrt{2}}$$

Therefore, $S = \frac{h\sqrt{2}}{\text{Tan } \theta}$.

(b) The volume of a pyramid is given by the formula $V = \frac{1}{3}Bh$ where B = the area of the base and h = the altitude. The large pyramid has $B = S^2$ and altitude h. Therefore, $V_{large} = \frac{1}{3}S^2h$.

Since the smaller pyramid is formed by a plane cutting the larger pyramid parallel to the base, we know by a theorem that the smaller pyramid will be similar to the larger one, and the ratio of each corresponding length equals the ratio

of the altitudes. Thus, $\dfrac{V_{small}}{V_{large}} = \dfrac{\frac{1}{3}B'h'}{\frac{1}{3}Bh} = \dfrac{\frac{1}{3}S'^2h'}{\frac{1}{3}S^2h} = \left(\dfrac{S'}{S}\right)^2\left(\dfrac{h'}{h}\right)$.

But $\dfrac{S'}{S} = \dfrac{h'}{h}$. Thus $\dfrac{V_{small}}{V_{large}} = \left(\dfrac{h'}{h}\right)^3$. Therefore, the ratio of the volumes of two similar pyramids will be equal to the cube of the ratio of any corresponding linear measures.

Let V' = the Volume of the smaller pyramid, and V = the Volume of the larger pyramid. We will now proceed to derive the expression for V' as directed in the question.

Since the plane passes p units above the base, the altitude of the smaller pyramid is (h-p) units.

Hence, $\dfrac{V'}{V} = \dfrac{(h-p)^3}{h^3}$ and $V' = \dfrac{(h-p)^3}{h^3} V$. Substitute $V = \frac{1}{3}s^2 h$.

$$V' = \frac{(h-p)^3 \left(\frac{1}{3}s^2 h\right)}{h^3} = \frac{1}{3} \frac{(h-p)^3 s^2}{h^2}$$

but $S = \dfrac{2h}{\sqrt{2}\ \text{Tan}\ \theta}$ from part (a). Therefore, by substitution,

$$V' = \frac{1}{3} \frac{\left[\dfrac{2h}{\sqrt{2}\ \text{Tan}\ \theta}\right]^2 (h-p)^3}{h^2} = \frac{\left(\frac{1}{3}\right)\left[\dfrac{4h^2}{2\ \text{Tan}^2\ \theta}\right](h-p)^3}{h^2}$$

Therefore, $V' = \dfrac{2(h-p)^3}{3\ \text{Tan}^2\ \theta}$.

CHAPTER 50

VOLUMES II

Basic Attacks and Strategies for Solving Problems in this Chapter. See pages 852 to 876 for step-by-step solutions to problems.

The problems in this chapter include finding the volume of spheres, right-circular cylinders, and right-circular cones.

The formula for the volume of a sphere, $V = \dfrac{4}{3}\pi r^3$, is justified in Problem 869. A right-circular cylinder is pictured below:

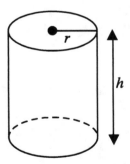

The formula for the volume of this type solid is very similar to the volume formula for a right prism. In Chapter 49, it was argued that the volume formula for a right prism is $V = Bh$, where B is the area of the base and h is the altitude. It was agreed that B refers to the number of cubes in the "first layer," and h indicates the number of layers. The same argument holds for a cylinder, and so a volume formula for the cylinder is $V = Bh$. However, this time the base is a circle, therefore $B = \pi r^2$. The standard volume formula for a right-circular cylinder is $V = \pi r^2 h$.

The volume formulas for a prism and a pyramid are $V = Bh$ and $V = \frac{1}{3}Bh$, respectively. This same relationship holds between a right-circular cylinder and a right-circular cone. More specifically, the volume formulas for a cylinder and cone are $V = \pi r^2 h$ and $V = \frac{1}{3}\pi r^2 h$, respectively. No formal justification of the last formula is given in this book.

However, consider the figures below:

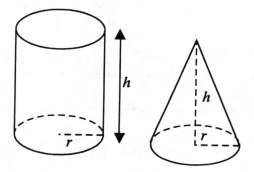

The cylinder and cone both have r as the radius and h as the height. It is at least somewhat reasonable that the cone would only "hold" about $\frac{1}{3}$ as much as the cylinder.

Step-by-Step Solutions to
Problems in this Chapter,
"Volumes II"

SPHERES

● **PROBLEM** 869

Find the volume of sphere of radius 2.

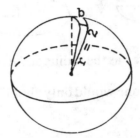

Solution: The volume of a sphere equals $\frac{4}{3}\pi r^3$. To see this, consider any polygon on the surface of the sphere. Draw the radii of the sphere to the vertices. The resulting solid resembles a pyramid with altitude roughly equal to the radius and a base that is not a plane surface but slightly spherical. By choosing polygons of smaller and smaller areas, the base becomes less curved and more like a plane figure; and, thus, the volume of the pyramidlike figure approaches $\frac{1}{3}r \cdot b$ where b is the area of the base polygon and r, the radius of the circle, is approximately the altitude. Suppose we divide the sphere into n pyramidlike figures. Then, if $b_1, b_2, \ldots, b_n$ are the areas of each base, the volume of the sphere equals the sum of the volumes of the n pyramids:

$$V = \frac{1}{3}rb_1 + \frac{1}{3}rb_2 + \ldots + \frac{1}{3}rb_n.$$

Factoring out $\frac{1}{3}r$, we obtain

$$V = \frac{1}{3}r(b_1 + b_2 + \ldots b_n).$$

Since the n pyramids comprise the entire volume, the n bases must sum to the total surface area; that is, $(b_1 + b_2 \ldots + b_n) = 4\pi r^2$. Thus,

$$V = \frac{1}{3}r(4\pi r^2) = \frac{4}{3}\pi r^3.$$

Getting back to the problem, the sphere of radius 2 has volume $\frac{4}{3}\pi(2)^3$ or $\frac{32}{3}\pi$.

● **PROBLEM** 870

A metal sphere is melted and recast into a hollow spherical shell whose outer radius is 277 cm. The radius of the hollow interior of the shell is equal to the radius of the original sphere. Find, to the nearest centimeter, the radius of the original sphere.

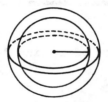

radius=x

Inner radius=x

Outer radius=.277 cm.

Solution: Since the same amount of metal will be used in both the sphere and the shell, the volume of the sphere must equal the volume of the shell. In general, the formula for the volume of a sphere is $V = \frac{4}{3}\pi r^3$, where r = the radius of the sphere.

Let x = the radius of the hollow interior of the shell and the original sphere.

The spherical shell is equivalent to a sphere with a 277 cm. radius that has had a sphere of radius x carved out of its center.

$$V_{\text{spherical shell}} = \frac{4}{3}\pi(277)^3 - \frac{4}{3}\pi x^3$$

$$V_{\text{original sphere}} = \frac{4}{3}\pi x^3.$$

Recall that $V_{\text{spherical shell}} = V_{\text{original sphere}}$

$$\frac{4}{3}\pi(277)^3 - \frac{4}{3}\pi x^3 = \frac{4}{3}\pi x^3$$

$$(277)^3 - x^3 = x^3$$

$$(277)^3 = 2x^3$$

$$x^3 = \frac{(277)^3}{2}$$

$$x = \frac{277}{\sqrt[3]{2}} = \frac{277}{1.26} = 219.8.$$

Therefore, the radius of the original sphere to the nearest centimeter is 220 cm.

Let S be a solid sphere whose radius is of length 7 in. (a) Find V(S), the volume of the sphere. (b) Find the circumference of a great circle of S. (c) Find the area of the region enclosed by a great circle of S. (d) Find the surface area of S.

Solution:

a) The volume of a sphere of radius length r is

$$V(S) = \frac{4}{3}\pi r^3.$$

In our problem, r = 7 in., and

$$V(S) = (\frac{4}{3})(\pi)(7 \text{ in.})^3 = 1436.76 \text{ in.}^3$$

b) A great circle of S is a circle which lies on S and has the same radius length as S. For our problem, r = 7 in. The circumference of the great circle, C, is

$$C = 2\pi r = (2)(\pi)(7 \text{ in.})$$

$$C = 43.98 \text{ in.}$$

c) The area, A, of the region enclosed by the great circle is

$$A = \pi r^2$$

since r is the radius of the great circle. Hence,

$$A = \pi(7 \text{ in.})^2 = 153.94 \text{ in.}^2.$$

d) The surface area of S, A(S), is equal to 4π times the radius length squared. That is,

$$A(S) = 4\pi r^2$$

$$A(S) = (4)(\pi)(7 \text{ in.})^2 = 615.75 \text{ in.}^2$$

Find, to the nearest cubic inch, the volume of a sphere whose radius measures 7 in. Use $\pi = \frac{22}{7}$ (see figure).

Solution: In geometry, it is postulated that the volume of a sphere is equal to $\frac{1}{3}$ the product of its surface area and the length of its radius. The surface area, in turn, is 4 times the area of any great circle of the sphere. Since a great circle of a sphere has the same radius length as the sphere, the surface area (s) of a sphere of radius r is

$$S = (4)(\pi r^2) = 4\pi r^2.$$

Therefore, the volume (V) of the sphere is

$$V = \frac{1}{3}rS = \frac{1}{3}r \cdot 4\pi r^2$$

$$V = \frac{4}{3}\pi r^3.$$

In our case, r = 7 in., $\pi = \frac{22}{7}$, and

$$V = (\frac{4}{3})(\frac{22}{7})(7 \text{ in.})^3$$

$$V = (\frac{4}{3})(22)(49 \text{ in.}^3)$$

$$V = 1437\frac{1}{3} \text{ in.}^3.$$

Therefore, to the nearest cubic inch,

$$V = 1437 \text{ in.}^3.$$

● **PROBLEM** 873

Given a sphere of radius r, find the volume of the regular tetrahedron that circumscribes the sphere.

Solution: In Figure 1, sphere P of radius r is circumscribed by regular tetrahedron ABCD. Sphere P is tangent with face ABC at point E; face ABD at F; face ACD at G; and face BCD at H. s is the length of the edge of tetrahedron ABCD. We wish

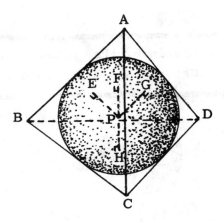

Figure 1

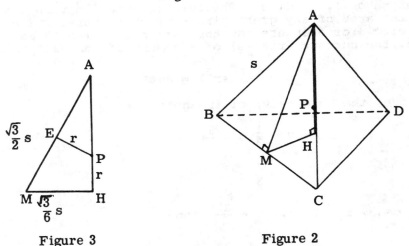

Figure 3

Figure 2

to find the volume V of ABCD in terms of the radius r. From
an earlier proof, we know that the volume of a tetrahedron is
$\frac{\sqrt{2}}{12}s^3$ where s is the length of the edge. Therefore, we find s
in terms of r; substitute in the tetrahedron volume formula;
and the problem is solved.

We will find the relationship of the edge AB = s and the
sphere radius PE = r. We do this by considering triangle AMH
in Figure 2, formed by the slant height $\overline{AM}$ and the perpendicular from the vertex A to the base $\overline{AH}$. We can find AM and MH.
Furthermore, from an earlier proof, we know that (1) since P,
the sphere center, is equidistant from the faces of the regular pyramid (the distances to the faces from an inscribed
sphere are radii. Since all radii are congruent, the center
is equidistant), and since all points equidistant from the
faces of a regular pyramid lie on the perpendicular from the
vertex, then P must lie on the perpendicular $\overline{AH}$, or $\overleftrightarrow{APH}$; and
(2) the perpendicular ($\overline{PE}$) to the face of a regular pyramid
(A-BCD...) from a point (P) on the vertex perpendicular ($\overline{AH}$)
intersects the face (ABC) on its slant height ($\overleftrightarrow{AM}$). Therefore, $\overleftrightarrow{AEM}$.

Combining all these facts, we construct Figure 3. Note that $\angle AEP$ and $\angle AHM$ are right angles (thus, $\angle AEP \cong \angle AHM$) and $\angle A \cong \angle A$. Therefore, $\triangle AEP \sim \triangle AHM$ by the A-A Similarity Theorem. Because the sides of similar triangles are proportional, we have

(i) $\dfrac{EP}{AP} = \dfrac{HM}{AM}$.

Because $\overline{EP}$ is a radius, $EP = r$. To find AM, note that $\overline{AM}$ is a leg of right $\triangle ABM$. ($\triangle ABM$ is a right triangle because a slant height of a face is always $\perp$ to the base.) We know that $AB = s$. Furthermore, because face $\triangle ABC$ is an equilateral triangle (every face of a regular tetrahedron is an equilateral triangle) perpendicular $\overline{AM}$ bisects base $\overline{BC}$, and $BM = \frac{1}{2}BC = \frac{1}{2}s$. By the Pythagorean Theorem, $AM^2 = BA^2 - BM^2 = s^2 - \left(\frac{1}{2}s\right)^2 = \frac{3}{4}s^2$, or $AM = \frac{\sqrt{3}}{2}s$.

To find MH, remember that the perpendicular from the vertex of a regular pyramid intersects the base polygon at its center. Therefore, H is the center of the base equilateral $\triangle BCD$ — that is H is the point of concurrency of the altitudes, angle bisectors, and medians. (Note in an equilateral triangle, these are the same three lines.) Because H is the concurrency point of the medians, it divides the median DM such that $HD = \frac{2}{3}DM$ or $HM = \frac{1}{3}DM$. By considering right $\triangle BDM$, we show by the Pythagorean Theorem that $DM^2 = BD^2 - BM^2 = s^2 - \left(\frac{1}{2}s\right)^2 = \frac{3}{4}s^2$ or $DM = \frac{\sqrt{3}}{2}s$. Since $HM = \frac{1}{3}DM$, then $HM = \frac{1}{3}\left(\frac{\sqrt{3}}{2}s\right) = \frac{\sqrt{3}}{6}s$.

Returning to the equation (i) and substituting in our values, we obtain:

(ii) $\dfrac{r}{AP} = \dfrac{\frac{\sqrt{3}}{6}s}{\frac{\sqrt{3}}{2}s}$ or $r = \dfrac{\frac{\sqrt{3}}{6}s}{\frac{\sqrt{3}}{2}s}AP = \frac{1}{3}AP$.

Refer back to Figure 3. We have discovered that $r = \frac{1}{3}AP$ or $AP = 3r$. Since $\overline{PH}$ is a radius of the sphere, $PH = r$, and thus $AH = AP + PH = 3r + r = 4r$. In right $\triangle AMH$, we have hypotenuse AM expressible in terms of s, leg MH expressible in terms of s, and leg AH expressible in terms of r. By using Pythagorean Theorem, we can relate s to r.

(iii) $AM^2 = MH^2 + AH^2$

(iv) $\left(\dfrac{\sqrt{3}}{2}s\right)^2 = \left(\dfrac{\sqrt{3}}{6}s\right)^2 + (4r)^2$

(v) $\dfrac{3}{4}s^2 = \dfrac{1}{12}s^2 + 16r^2$.

(vi) $16r^2 = \dfrac{3}{4}s^2 - \dfrac{1}{12}s^2 = \left(\dfrac{9}{12} - \dfrac{1}{12}\right)s^2 = \dfrac{2}{3}s^2$

857

(vii) $\frac{3}{2} \cdot 16r^2 = s^2$

(viii) $24r^2 = s^2$

(ix) $s = \sqrt{24}r = 2\sqrt{6}r$.

Now, we can use the volume formula.

(x) $V = \frac{\sqrt{2}}{12}s^3 = \frac{\sqrt{2}}{12}(2\sqrt{6}r)^3 = 8\sqrt{3}r^3$.

CYLINDERS

● **PROBLEM** 874

Find, in terms of π, the volume of a right circular cylinder if the radius of its base measures 4 in. and its altitude measures 5 in.

Solution: If we picture the base of the cylinder as having a depth of one unit of measure, we can then calculate the volume by determining the area of the base and multiplying it by the height of the cylinder. In effect, this multiplication amounts to stacking the bases up to the height of the cylinder.

Area of the base = πr^2, where r is the radius of the circular base. Therefore, the volume is given by

$$V = \pi r^2 h.$$

By substitution, $V = \pi (4)^2 5$ in.^{3} = 80π in.^{3}.

Therefore, volume of the cylinder = 80π in.^{3}.

● **PROBLEM** 875

Find the volume of a right circular cylinder of height 6 if it has the same lateral surface area as a cube of edge 3.

Solution: The volume of a cylinder equals $\pi r^2 h$. The height, h, equals 6. The radius can be found by relating the lateral surface area of the cube and the cylinder. The lateral surface area of the cylinder equals (base perimeter) · (height) = $2\pi rh$. The lateral surface area of the cube which has no

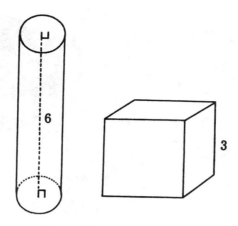

bases equals $4s^2$. By equating these two expressions, we obtain

$$2\pi rh = 4s^2.$$

By substitution, $2\pi r(6) = 4(3)^2$

$$r = \frac{4(3)^2}{2\pi(6)} = \frac{36}{2\pi(6)} = \frac{3}{\pi}.$$

Therefore,

Volume of cylinder = (Area of base) · (height)

$$= \pi r^2 \qquad\qquad \cdot h$$

$$= \pi\left(\frac{3}{\pi}\right)^2 \qquad\quad \cdot 6$$

$$= \frac{54}{\pi}.$$

CONES

● PROBLEM 876

a) Find the volume of a solid right circular cone whose height is 4 ft. and whose base has radius 3 ft.

b) What is the slant height of this cone?

c) Find the surface area of this cone.

Solution: Figure (A) shows the cone. s is the measure of the slant height, h is the measure of the height, and r is the measure of the radius of the base.

a). The volume of a solid right circular cone whose height is h, and whose base has radius r, is

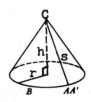

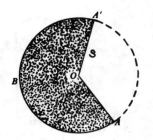

Figure A   Figure B

$$V = \frac{1}{3}\pi r^2 h.$$

In our case, r = 3 ft., h = 4 ft., and

$$V = (\frac{1}{3})(\pi)(9 \text{ ft.}^2)(4 \text{ ft.})$$

$$V = 37.70 \text{ ft.}^3.$$

b). The slant height of a right circular cone can be calculated by the Pythagorean Theorem. Look at the right triangle with sides h, r, and s, in figure (A). Applying the Pythagorean Theorem to this triangle,

$$s^2 = h^2 + r^2$$

or

$$s = \sqrt{h^2 + r^2}$$

$$s = \sqrt{16 \text{ ft.}^2 + 9 \text{ ft.}^2} = 5 \text{ ft.}$$

c). Figure (B) shows the region obtained by cutting the cone of figure (A) along $\overline{AC}$. This region has the same area as the surface area of the cone. The area of the region can be calculated from the following proportion

$$\frac{\text{Area(region)}}{\text{Area(circle O)}} = \frac{\text{length } \overset{\frown}{ABA'}}{\text{length of circle O}}$$

$$\text{Area(region)} = \frac{(\pi s^2)(2\pi r)}{(2\pi s)}.$$

Here we have used the fact that the length of $\overset{\frown}{ABA'}$ is $2\pi r$, since $\overset{\frown}{ABA'}$ is the boundary of the circular base of the cone. Hence,

$$\text{Area(region)} = \pi rs.$$

In our case, s = 5 ft. and r = 3 ft. Then

$$\text{Area(region)} = 15\pi \text{ ft.}^2 = 47.12 \text{ ft.}^2.$$

The surface area of the cone is therefore 47.12 ft.^{2}.

860

A cone is generated by rotating a right triangle with sides 3, 4, and 5 about the leg whose measure is 4. Find the total area and volume of the cone.

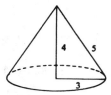

Solution: The total area of the cone equals the (1) lateral surface area plus (2) the area of the base.

1) The lateral surface area equals one half the product of its (a) slant height and (b) the circumference of its base. This was derived earlier from the fact that the lateral surface can be viewed as a sector of a circle.

a) The slant height of the cone is the length of any element (line segments on the surface of the coin joining the vertex to the base). By the manner in which the cone is constructed, we know slant height equals 5.

b) The circumference of the base, a circle of radius 3, equals $2\pi r$ or 6π. Thus, lateral surface area equals $\frac{1}{2}(5)(6\pi)$ = 15π.

2) The area of the base, a circle of radius 3, equals $\pi r^2 = \pi(3)^2 = 9\pi$.

The total surface area equals $15\pi + 9\pi = 24\pi$.

The volume of a cone is given by the formula $\frac{1}{3}\pi r^2 h$. (Note the similarity between this formula and the formula for the volume of a pyramid, $\frac{1}{3}a \cdot b \cdot h$.) Since r = 3 and h = 4, the volume equals $\frac{1}{3}\pi(3)^2(4) = 12\pi$.

● **PROBLEM** 878

Find, in terms of π, the volume of the right circular cone whose base measures 6 in. and whose altitude measures 8 in. (See figure).

Solution: The volume of a right circular cone is equal to
$\frac{1}{3}$ the product of the area of its base and the measure of its
altitude. Since a right circular cone has a circular base,
the area of the base is πr^2, where r is the base radius. If
h is the height of the cone, and V is its volume,

$$V = \frac{1}{3}\pi r^2 h.$$

In this example, r = 6 in. and h = 8 in. Therefore,

$$V = \frac{1}{3}\pi (6 \text{ in.})^2 8 \text{ in.} = \frac{1}{3}\pi (36)(8) \text{ in.}^3$$

$$V = 96\pi \text{ in.}^3.$$

Therefore, volume of the cone is 96π cu. in.

● **PROBLEM** 879

A frustum of a solid cone is the portion of solid cone remaining when the solid cone above a given section is removed.
(See figure.) (a). Using the properties of similar triangles, what can we say about x/3 and (x + 4)/5? (b). Find
the height, x + 4, of the large cone. (c) Find the volume
of the small solid cone. (d) Find the volume of the large
solid cone. (e) Find the volume of the frustum. (f)
Taking $r_1 = 3$, $r_2 = 5$, and h = 4, compute

$V = \frac{1}{3}\pi h(r_1^2 + r_1 r_2 + r_2^2)$. Compare this with part (e).

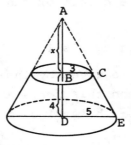

Solution:

(a) We will prove that $\triangle ABC \sim \triangle ADE$, by the A.A. (angle-
angle) Similarity Theorem. We will then be able to set up a
proportion between $\overline{AB}$, $\overline{AD}$, $\overline{BC}$, and $\overline{DE}$, thereby finding a relationship between x/3 and (x + 4)/5.

First, ∢A is common to both $\triangle ABC$ and $\triangle ADE$. Therefore,
∢BAC $\cong$ ∢DAE. Furthermore, the plane containing $\overline{BC}$ is parallel to the base of the cone. Hence, ∢ACB $\cong$ ∢AED, since they
are corresponding angles of the parallel segments $\overline{BC}$ and $\overline{DE}$.
In conclusion, $\triangle ABC \sim \triangle ADE$, and we may write

$$\frac{AB}{AD} = \frac{BC}{DE}.$$

Since AB = x, AD = x + 4, BC = 3 and DE = 5, the last equation may be rewritten as

$$\frac{x}{x+4} = \frac{3}{5}$$

or

$$5x = 3x + 12$$

or

$$x = 6.$$

Hence, $\frac{x+4}{5} = 2$ and $\frac{x}{3} = 2$ (i.e., $\frac{x+4}{5} = \frac{x}{3}$).

(b) The height of the large cone, x + 4, is, from part (a), 10.

(c) The volume of the small cone is equal to $\frac{\pi}{3}$ times the product of its height (x) and the square of the length of the radius of its base (3^2). Hence,

$$V_S = (\frac{\pi}{3})(x)(3^2)$$

$$V_S = (\frac{\pi}{3})(6)(9) = 56.52.$$

(d) The volume of the large solid cone is equal to $\frac{\pi}{3}$ times the product of its height (x + 4) and the square of the length of the radius of its base (5^2). Hence,

$$V_L = (\frac{\pi}{3})(x+4)(5^2)$$

$$V_L = (\frac{\pi}{3})(10)(25) = 261.67.$$

(e) The volume of the large cone, minus the volume of the small cone, equals the volume of the frustum, (V_F). Therefore,

$$V_F = V_L - V_S = 261.67 - 56.52$$

$$V_F = 205.15.$$

(f) If $r_1 = 3$, $r_2 = 5$, and h = 4,

$$V = \frac{1}{3}\pi h(r_1^2 + r_1 r_2 + r_2^2)$$

$$V = (\frac{\pi}{3})(4)((3)^2 + (3)(5) + (5)^2)$$

$$V = (\frac{\pi}{3})(4)(9 + 15 + 25)$$

$$V = (\frac{4\pi}{3})(49)$$

V = 205.15.

Note that this conforms to the volume of the frustum as calculated in part (e).

● **PROBLEM** 880

The largest volcanic eruption occurred when Mt. Tambor in
Indonesia expelled 36 cubic miles of lava into the atmosphere
on April 7, 1915, losing 4100 ft. of its original altitude.
Assume the original volcano to be a circular cone and the
final volcano to be a frustum. If the material lost from the
top of the mountain contained 1% of the erupted material, find
the angle of inclination of the volcano.

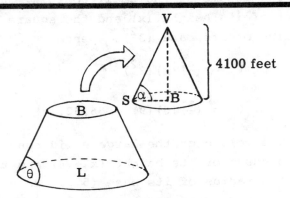

Solution: In the figure, cone V - ⊙B is exploded off the top
of the volcano with other debris. We wish to find m∢θ, the
angle of inclination. Note that V - ⊙B is a cone similar to
the original mountain. Thus, m∢θ = m∢α.

To find m∢α, note that in right ΔVSB:

$$(i) \quad \tan \alpha = \frac{VB}{SB}.$$

VB is given as 4100 ft. To find SB, note that SB is a
radius of the base circle of cone V - ⊙B. From the volume
equation, we have

$$(ii) \quad \text{Vol. of } V-\odot B = \frac{1}{3}\pi r^2 h = \frac{1}{3}\pi SB^2(4100 \text{ ft.}).$$

Remember that the volume of the cone is 1% of the expelled volume: 36 cu. miles. Thus, vol. of V - ⊙B =
$\frac{1}{100} \times$ 36 cu. mi. = .36 cu. mi.

$$(iii) \quad .36 \text{ mi.}^3 = \frac{1}{3}\pi(4100 \text{ ft.})SB^2.$$

To obtain compatible measurements, note 1 mi.^{3} =
(5280 ft.)^3. Thus, substituting and dividing by $\frac{1}{3}\pi$(4100 ft.),
we obtain

864

(iv) $\dfrac{(.36)(5280 \text{ ft.})^3}{\frac{1}{3}\pi(4100 \text{ ft.})} = SB^2$

(v) $SB = \sqrt{\dfrac{3(.36)(5280 \text{ ft.})^3}{\pi(4100 \text{ ft.})}} = 3513 \text{ ft.}$

Substituting this result in eq. (i),

(vi) $\tan \alpha = \dfrac{4100}{3513} = 1.167.$

Consulting trigonometric tables, we obtain for α, the angle of inclination, the value of 49.5°.

● **PROBLEM** 881

A manufacturer wishes to change the container in which his product is marketed. The new container is to have the same volume as the old, which is a right circular cylinder with altitude 6 and base radius 3. The new container is to be composed of a frustum of a right circular cone surmounted by a frustum of another right circular cone so that the smaller bases of the two frustums coincide. The radii of the bases of the lower frustum are $3\frac{1}{2}$ and 2, respectively, and its altitude is 4. The radius of the upper base of the upper frustum is 3. Find, to the nearest tenth, the altitude of the upper frustum.

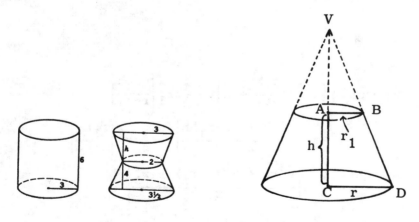

Figure 1   Figure 2

Solution: We are given enough information to find the volume of the cylinder by the formula $V = \pi r^2 h$. We know that the volume of the frustums will depend, in some way, on the radii of the bases and the altitude. Since all four radii are known, along with the cylindrical volume, we can solve for the altitude by setting the volume of the cylinder equal to the total volume of both frustums.

865

We will first derive an expression for the volume of a frustum of a right circular cone of base radii r and r_1 and altitude h.

Extend the lateral surfaces of the frustum to form a right circular cone at vertex V. We have two cones, one with base of radius r and the other with base of radius r_1. The volume of the frustum will then be the volume of the large cone minus the area of the smaller cone. The volume of a right circular cone is $V = \frac{1}{3}\pi r^2 h$, where h = the altitude and r = the base radius.

If we let x = the height of the small cone then $x + h$ = the height of the larger cone.

$$\text{Volume of frustum} = \frac{1}{3}\pi r^2 (x+h) - \frac{1}{3}\pi r_1^2 x.$$

We now must express x in terms of the given information about the frustum.

$\triangle VAB \sim \triangle VCD$ by the A.A. Similarity Theorem. (They both contain a right angle and both contain $\sphericalangle AVB$.) Therefore, the ratio of similitude between corresponding sides is equal for all sides. As such, $\frac{r}{r_1} = \frac{x+h}{x}$. We can solve for x and eliminate the unknown

$$rx = r_1 x + r_1 h$$

$$(r - r_1)x = r_1 h$$

$$x = \frac{r_1 h}{r - r_1}.$$

Hence, by substitution,

$$\text{Volume of frustum} = \frac{1}{3}\pi r^2 \left[\frac{r_1 h}{r-r_1} + h \right] - \frac{1}{3}\pi r_1^2 \left[\frac{r_1 h}{r-r_1} \right]$$

$$= \frac{1}{3}\pi r^2 \left[\frac{r_1 h + (r-r_1)h}{r - r_1} \right] - \frac{1}{3}\pi r_1^2 \left[\frac{r_1 h}{r-r_1} \right]$$

$$= \frac{1}{3}\pi h \left[\frac{r^2 r_1 + r^2(r-r_1) - r_1^3}{r - r_1} \right]$$

$$= \frac{1}{3}\pi h \left[\frac{r_1(r^2-r_1^2) + r^2(r-r_1)}{r - r_1} \right]$$

$$= \frac{1}{3}\pi h \left[\frac{r_1(r+r_1)(r-r_1) + r^2(r-r_1)}{r - r_1} \right]$$

$$\text{Volume of frustum} = \frac{\pi h}{3}[r^2 + r_1^2 + r_1 r].$$

The original cylindrical container has a volume equal to

866

$\pi r^2 h$. By substituting $r = 3$ and $h = 6$ we obtain $V_{cyl.} = \pi(3)^2 6 = 54\pi$.

The lower frustum has radii 2 and $3\frac{1}{2}$ and altitude 4. If we substitute this in the formula just derived, we find the

$$V_{lower\ frustum} = \frac{\pi(4)}{3}\left[(2)^2 + \left[\frac{7}{2}\right]^2 + (2)\left[\frac{7}{2}\right]\right]$$

$$= \frac{4\pi}{3}\left[4 + \frac{49}{4} + 7\right]$$

$$= \frac{4\pi}{3}\left[\frac{16}{4} + \frac{49}{4} + \frac{28}{4}\right] = \frac{4\pi}{3}\left(\frac{93}{4}\right) = 31\pi.$$

Let h = the unknown altitude of the upper frustum.

We are given that the upper base has radius 3. The lower base coincides with the upper base of the lower frustum and therefore has radius equal to 2.

$$V_{upper\ frustum} = \frac{\pi(h)}{3}\left((2)^2 + (3)^2 + (2)(3)\right)$$

$$= \frac{\pi(h)}{3}(4 + 9 + 6) = \frac{\pi h}{3}(19) = \frac{19}{3}\pi h.$$

We can solve for h by substituting the three volumes we found into

$$V_{cyl.} = V_{lower\ frustum} + V_{upper\ frustum}$$

$$54\pi = 31\pi + \frac{19}{3}\pi h$$

$$54 - 31 = \frac{19}{3}h$$

$$23 = \frac{19}{3}h$$

$$h = \frac{23 \cdot 3}{19} = \frac{69}{19}$$

$$h = 3.63.$$

Therefore, the altitude of the upper frustum of the new container is 3.6.

● PROBLEM 882

In the accompanying figure, right triangle ABC is inscribed in semicircle BCA. $\overline{CD} \perp \overline{BA}$. CD = p and BA = 2q. Express the total volume generated by the shaded region if the semicircle is rotated 360° about the diameter $\overline{AB}$.

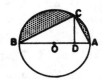

<u>Solution</u>: A circle or a semicircle rotated about its diameter results in a sphere. A triangle rotated about a side results in one or two right circular cones. The volume generated by the shaded region is that volume of the sphere not contained in the cones. Thus, Volume of generated solid = Volume of sphere - Volume of cone generated by right $\triangle$BCD - Volume of cone generated by right $\triangle$ACD.

The volume of a sphere equals $\frac{4}{3}\pi r^3$, where r is the radius. Since diameter $\overline{AB}$ has length 2q, the radius of the sphere = $\frac{1}{2}(2q) = q$. Volume of sphere = $\frac{4}{3}\pi q^3$.

The volume of a right circular cone is $\frac{1}{3}\pi r^2 h$, where r is the base radius and h the altitude. In the cone generated by right $\triangle$BDC, (since $\overline{BA}$ is the generating axis) $\overline{CD}$ is the base radius. Thus, r = CD = p. BD is unknown and cannot be deter-mined with the given information.

$$\text{Volume} = \frac{1}{3}\pi p^2 \cdot BD.$$

Similarly, the volume of the right cone generated by right $\triangle$ACD = $\frac{1}{3}\pi p^2 \cdot DA$.

The volume of the shade can now be determined.
$$V = \frac{4}{3}\pi q^3 - \frac{1}{3}\pi p^2 \cdot BD - \frac{1}{3}\pi p^2 \cdot DA$$

$$= \frac{4}{3}\pi q^3 - \frac{1}{3}\pi p^2 (BD + DA).$$
Note that BD + DA = BA = 2q. Thus,

$$V = \frac{4}{3}\pi q^3 - \frac{1}{3}\pi p^2 \cdot 2q = \frac{2\pi}{3} q(2q^2 - p^2).$$

● **PROBLEM** 883

A frustum of a cone of revolution is inscribed in a frustum of a regular pyramid whose bases are squares. Express in simplest form, in terms of π, the ratio of

    (a)   the volume of the frustum of the cone to the volume of the frustum of the pyramid
$$V_{fr} = \frac{1}{3}h(B_1 + B_2 + \sqrt{B_1 B_2}) .$$

    (b)   the lateral area of the frustum of the cone to the lateral area of the frustum of the pyramid.

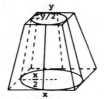

<u>Solution</u>:

(a) In the accompanying figure, the lengths of the base edges are x and y. We wish to find $\dfrac{\text{volume of cone frustum}}{\text{volume of pyramid frustum}}$.

Volume of a frustum $= \frac{1}{3}h(B_1 + B_2 + \sqrt{B_1 B_2})$

where h = altitude or distance between the bases.

B_1 and B_2 = areas of the bases.

For the volume of the frustum of the cone, we have

$$h = h; \quad B_1 = \pi r^2 = \pi \left(\frac{x}{2}\right)^2 = \frac{\pi x^2}{4}; \quad B_2 = \pi r^2 = \pi \left(\frac{y}{2}\right)^2 = \frac{\pi y^2}{4}.$$

Thus, $V_{fr.cone} = \frac{1}{3}h(B_1 + B_2 + \sqrt{B_1 B_2})$

$$= \frac{1}{3}h\left[\frac{\pi x^2}{4} + \frac{\pi y^2}{4} + \sqrt{\frac{\pi x^2}{4} \cdot \frac{\pi y^2}{4}}\right]$$

$$= \frac{1}{3}h\left[\frac{\pi x^2}{4} + \frac{\pi y^2}{4} + \frac{\pi xy}{4}\right] = \frac{\pi}{3 \cdot 4}h(x^2 + y^2 + xy).$$

For the volume of the frustum of the pyramid, we have

$$h = h; \quad B_1 = s^2 = x^2; \quad B_2 = s^2 = y^2.$$

Thus, $V_{fr.pyramid} = \frac{1}{3}h(B_1 + B_2 + \sqrt{B_1 B_2}) = \frac{1}{3}h\left[x^2 + y^2 + \sqrt{x^2 y^2}\right]$

$$= \frac{1}{3}h(x^2 + y^2 + xy).$$

The desired ratio is thus,

$$\frac{V_{fr.cone}}{V_{fr.pyramid}} = \frac{\frac{\pi}{3 \cdot 4}h(x^2 + y^2 + xy)}{\frac{1}{3}h(x^2 + y^2 + xy)} = \frac{\pi}{4}.$$

(b) We must find $\dfrac{\text{lateral area of cone frustum}}{\text{lateral area of pyramid frustum}}$. From an earlier theorem, we know that the lateral area of the frustum of a right circular cone is $S_c = \pi \ell(r + r')$ where ℓ is the slant height and r and r' the respective radii.

Thus, $S_c = \pi \ell \left(\frac{x}{2} + \frac{y}{2}\right) = \frac{\pi \ell}{2}(x + y).$

For a pyramid frustum, we know the lateral area equals one-half the product of its slant height and the sum of the perimeters of the bases. Thus, $S_p = \frac{1}{2}\ell(p + p')$. Since the perimeter of a square is four times the side, we have

$S_p = \frac{1}{2}\ell(4x + 4y) = 2\ell(x + y).$

The desired ratio is $\dfrac{S_c}{S_p} = \dfrac{\frac{\pi \ell}{2}(x + y)}{2\ell(x + y)} = \dfrac{\pi}{4}$.

In the accompanying figure, a right circular cone is constructed on the base of a hemisphere. The surface of the hemisphere is equal to the lateral surface of the cone. Show that the volume V of the solid formed can be found by the formula $V = \frac{1}{3}\pi r^3 (2 + \sqrt{3})$.

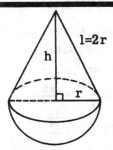

Solution: The volume of the solid formed equals the volume of the hemisphere plus the volume of the right circular cone. The volume of a hemisphere is half the volume of a sphere. Thus, $V_{hemi.} = \frac{1}{2}(\frac{4}{3}\pi r^3) = \frac{2}{3}\pi r^3$. The volume of a cone of base radius r and height h equals $\frac{1}{3}\pi r^2 h$. (Note in this case that the base of the cone is the great circle of the hemisphere. Thus, r, the base radius of the cone, equals r, the radius of the hemisphere.) The desired volume, V, of the solid is, therefore, $V = \frac{2}{3}\pi r^3 + \frac{1}{3}\pi r^2 h = \frac{\pi}{3}r^2 (2r + h)$.

The problem, though, asks us to express the volume totally in terms of r, not r and h.

Consider the right triangle formed by altitude h, base radius r and slant height ℓ. By the Pythagorean Theorem, $h^2 = \ell^2 - r^2$. r is known; and ℓ and r are related by the surface area formulas. Thus, ℓ can be expressed in terms of r and the results substituted in the Pythagorean equation above to express h in terms of r.

Surface Area of hemisphere = $2\pi r^2$.

Lateral Area of cone = $\pi r \ell$.

By the given, the two surface areas are equal. Then, $2\pi r^2 = \pi r \ell$ or $\ell = 2r$.

Using the Pythagorean equation, we have $h^2 = (2r)^2 - r^2 = 4r^2 - r^2 = 3r^2$. Thus, $h = \sqrt{3}r$.

We substitute this in our expression for V and obtain $V = \frac{\pi}{3}r^2 (2r + \sqrt{3}r) = \frac{\pi}{3}r^3 (2 + \sqrt{3})$.

An edge of a regular tetrahedron is s. Show that the volume of the inscribed cone is $\frac{\pi s^3 \sqrt{6}}{108}$.

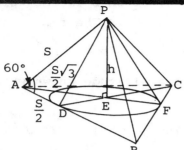

Solution: In the accompanying figure, P-ABC is the regular tetrahedron. The inscribed cone has vertex P and base circle ⊙E. We wish to find the volume of the inscribed cone. The volume of a cone is given by the formula $V = \frac{1}{3}Bh$ where B is the area of base circle ⊙E and h is the altitude PE.

To find the altitude h, note that the cone and the tetrahedron have the same vertex P. Also, the bases of the cone and the tetrahedron are coplanar. Since from external point P only one perpendicular can be drawn perpendicular to the base plane, the altitude of the cone must equal the height of the tetrahedron. To find the altitude PE, we use the Pythagorean Theorem. In △APE, $PE^2 = AP^2 - AE^2$. AP is known. Unfortunately AE is not. To find AE we take advantage of the unique properties of point E.

Remember that P-ABC is a regular tetrahedron, and thus all faces are equilateral triangles. Thus, △ABC is equilateral. Remember further that the altitude of a regular pyramid intersects the base at its center, and thus if $\overline{PE} \perp$ plane ABC, E must be the center of △ABC. As a center point of the regular polygon, E is the point of concurrency of the angle bisectors and the perpendicular bisectors. Since △ABC is equilateral (thus the angle bisectors are the medians), E is also the point of concurrency of the medians. Since the medians concur at their trisection point, $EF = \frac{1}{3}AF$ or $AE = \frac{2}{3}AF$.

AF, though unknown, can quickly be found by applying the Pythagorean Theorem to △AFB (note that median $\overline{AF}$ of equilateral △ABC is the perpendicular bisector of side $\overline{BC}$). AB = s, $BF = \frac{1}{2}s$. Then, $AF^2 = AB^2 - BF^2 = s^2 - \left(\frac{1}{2}s\right)^2$ and $AF = \frac{\sqrt{3}}{2}s$. Since $AE = \frac{2}{3}AF$, then $AE = \frac{\sqrt{3}}{3}s$.

Substituting back in the original equation of △APE, we have $PE^2 = AP^2 - AE^2 = s^2 - \left(\frac{\sqrt{3}}{3}s\right)^2 = \frac{2}{3}s^2$ or altitude $PE = \sqrt{\frac{2}{3}}s$.

For the second piece required to apply the area formula, we must find the area of the circular base. Remember that the center of the inscribed circle of $\triangle ABC$ is the concurrency point of the perpendicular bisectors. For equilateral triangles, the perpendicular bisectors, medians, angle bisectors, and altitudes are identical. Thus, point E is the center of the circle and EF is the radius. But, since E is the trisection point of median, $EF = \frac{1}{3}AF = \frac{1}{3}\left(\frac{\sqrt{3}}{2}s\right) = \frac{\sqrt{3}}{6}s$. The radius of the base is thus $\frac{\sqrt{3}}{6}s$, and the area of the base $B = \pi r^2 = \pi\left(\frac{\sqrt{3}}{6}s\right)^2 = \frac{\pi}{12}s^2$.

With the base area and altitude known we can now find the volume in terms of s.

$$V = \frac{1}{3}Bh = \frac{1}{3}\left(\frac{\pi}{12}s^2\right)\left(\sqrt{\frac{2}{3}}s\right) = \frac{\pi}{36}s^3\sqrt{\frac{2}{3}} = \frac{\pi}{108}s^3\sqrt{6}.$$

● **PROBLEM** 886

An isosceles triangle, each of whose base angles is θ and whose legs are a, is rotated through 180°, using as an axis its altitude to the base.

(a)   Find the volume V of the resulting solid, in terms of θ and a.

(b)   Using the formula found in part a, find V, to the nearest integer if a = 5.2 and θ = 27°. (Use π = 3.14)

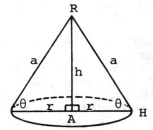

<u>Solution</u>:

(a)   The resulting figure is a right circular cone. Its volume, V, is given by $V = \frac{1}{3}\pi r^2 h$ where h is the altitude of the cone and r is the radius of the base. First, we express r and h in terms of the given angle, θ, and the given length, a.

In the figure, we see that when the altitude to the base of an isosceles triangle is drawn, a right triangle, $\triangle ARH$, is formed. Since we know one acute angle and the length of the hypotenuse, we can express the length of the other legs by applying several trigonometric ratios.

We can calculate the length of the leg opposite θ, the altitude, with the help of the sine ratio.
$\sin \theta = \dfrac{\text{opposite}}{\text{hypotenuse}}$. By substitution, $\sin \theta = \dfrac{h}{a}$.

Therefore, h = a sin θ.

The length of the leg adjacent angle θ, the radius, can be found using the cosine ratio. $\cos \theta = \dfrac{\text{adjacent}}{\text{hypotenuse}}$. By substitution, $\cos \theta = \dfrac{r}{a}$. Therefore, r = a cos θ.

We have now expressed r and h in terms of θ and a. By substituting these expressions in the volume formula, $V = \frac{1}{3}\pi r^2 h$, we can find the volume in terms of a and θ.

$$V = \frac{1}{3}\pi r^2 h.$$

By substitution, $V = \frac{1}{3}\pi (a \cos \theta)^2 (a \sin \theta)$

$$= \frac{1}{3}\pi a^2 \cos^2 \theta \, a \sin \theta.$$

Therefore, $V = \frac{1}{3}\pi a^3 \cos^2 \theta \sin \theta.$

(b) In this part we substitute a = 5.2 and θ = 27 into the formula found in part (a) and proceed with the calculation. We use π = 3.14.

$$V = \frac{1}{3}(3.14)(5.2)^3 (\cos 27°)^2 (\sin 27°).$$

According to a standard table of trigonometric functions cos 27° = .8910 and sin 27° = .4540. Hence, by substitution,

$$V = \frac{1}{3}(3.14)(5.2)^3 (.8910)^2 (.4540)$$

$$V = 53.04.$$

Therefore, V = 53 to the nearest integer.

● **PROBLEM** 887

In rectangle ABCD with diagonal AC, let x represent the length of side $\overline{AB}$ and y the length of side $\overline{BC}$. As the figure shows, the rectangle is revolved through 360° and about $\overline{AB}$ as an axis.

(a) Find in terms of x and y the volume of the solid generated by triangle ABC.

(b) Find in terms of x and y the solid generated by triangle ADC.

(c) If there are 150 cubic inches in the volume generated by triangle ABC, how many cubic inches are in the volume generated by triangle ADC?

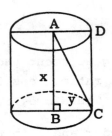

Solution:

(a) Triangle ABC is a right triangle and when revolved about one of its legs generates a right circular cone. The volume of a cone is given by the formula $V_{cone} = \frac{1}{3}\pi r^2 h$, where r = the base radius and h = the length of the altitude. In this case x, the length of $\overline{AB}$, is the altitude and y, the length of $\overline{BC}$, is the radius in question.

Hence, by substitution,

$$V_{cone} = \frac{1}{3}\pi y^2 x.$$

(b) The volume of the solid generated by $\triangle ADC$ is equal to the volume of the cylinder generated by the rectangle minus the volume of the cone from part (a). $V_{cyl} = \pi r^2 h$, where r = radius of the base and h = the altitude. Therefore,

$$r = y \quad , \quad h = x$$
$$V_{cyl} = \pi y^2 x.$$

Volume of solid generated by $\triangle ADC$, $V' = V_{cyl} - V_{cone}$.

$$V' = \pi y^2 x - \frac{1}{3}\pi y^2 x$$
$$V' = \frac{2}{3}\pi y^2 x.$$

(c) The volume of the solid generated by $\triangle ADC$, $\frac{2}{3}\pi y^2 x$, is twice that of the solid generated by $\triangle ABC$, $\frac{1}{3}\pi y^2 x$.

Therefore, if the solid generated by $\triangle ABC$ is 150 cu. units, then the volume of the solid generated by $\triangle ADC$ is 2(150), or 300 cu. units.

● **PROBLEM** 888

Each leg of an isosceles triangle is 3m units in length and the base is 2m units. A line s is drawn through the vertex of the triangle parallel to the base. The triangle is revolved through 360° about line s with the base of the triangle always remaining parallel to s. Find, in terms of m, the volume of the resulting solid.

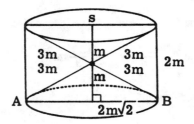

Solution: The solid generated in this problem is a right circular cylinder with two right circular cones carved out of the top and bottom. It is impractical to calculate directly the volume of this type of solid figure. However, since we do have a formula for the volume of a cylinder and a cone, the volume of the solid can be given as $V_{solid} = V_{cyl.} - V_{cones}$. $V_{cyl.} = r^2 h$ and $V_{cone} = \frac{1}{3}\pi r^2 h$, where $r = $ length of the base radius and $h = $ the altitude in both cases.

The length of the altitude of the cylinder = length of the base of the isosceles triangle = 2m.

The figure shows us that the length of the altitude of the cone = $\frac{1}{2}$ the length of the base of the triangle = $\frac{1}{2}(2m) = m$.

The base radius for both solids is the same and must be calculated as it is not given. One side of the triangle, the altitude of the cone and the radius form a right triangle and, as such, the length of the radius can be found with the aid of the Pythagorean Theorem.

Let $r = $ length of the radius.
$h = $ altitude length of cone, m.
$i = $ length of side of isosceles triangle, 3m.

Then, $r^2 = (i)^2 - (h)^2$. By substitution, $r^2 = 9m^2 - m^2$.

$$r^2 = 8m^2$$
$$r = \sqrt{8}\sqrt{m^2} = m\sqrt{4}\sqrt{2} = 2m\sqrt{2}.$$

Hence, by substitution,

$$V_{cyl.} = \pi(2m\sqrt{2})^2(2m) = 16\pi m^3 = 16\pi m^3$$

$$V_{cone} = \frac{1}{3}\pi(2m\sqrt{2})^2(m) = \frac{1}{3}\pi(8m^2)m = \frac{8}{3}\pi m^3.$$

It follows, then, that

$$V_{solid} = V_{cyl} - 2(V_{cone}) \quad \text{and, by substitution,}$$

$$V_{solid} = 16\pi m^3 - \frac{16}{3}\pi m^3$$

$$= \frac{48}{3}\pi m^3 - \frac{16}{3}\pi m^3 = \frac{32}{3}\pi m^3.$$

Therefore, $V_{solid} = \frac{32}{3}\pi m^3$.

● **PROBLEM** 889

In the accompanying figure, ABCD is a trapezoid. m∡A = m∡D = 90°, AD = DC = m, and AB = 2m. The line t, which is in the plane of ABCD, is parallel to $\overline{AD}$ and is m units from AD.

If the region ABCD is revolved through 360° about line t, a solid is generated. Express in terms of m, the volume of this solid.

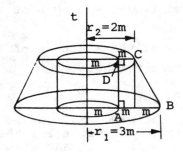
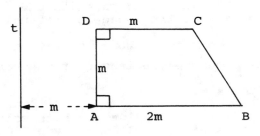

Solution: The solid formed is the frustum of a cone of re-volution less a right cylinder cut out of the center. Thus, volume of solid = (cone frustum volume) - (right cylinder volume).

To find the volume of the cone frustum, remember that $V_{fr} = \frac{1}{3}\pi h(r_1^2 + r_2^2 + r_1 r_2)$, where r_1 and r_2 are the radii of the bases. Here $r_1 = m + AB = m + 2m = 3m$, $r_2 = m + DC = m + m = 2m$, and (because $\overleftrightarrow{DA} \parallel t$) h = DA = m. Thus, $V = \frac{1}{3}\pi m((3m)^2 + (2m)^2 + (2m)(3m)) = \frac{19}{3}\pi m^3$.

To find the volume of the cylinder, remember that $V_{cylinder} = \pi r^2 h$ where h = m and r = m. Thus, $V = \pi m^3$.

The volume of the solid $= \frac{19}{3}\pi m^3 - \pi m^3 = \frac{16\pi m^3}{3}$.

CHAPTER 51

THREE-DIMENSIONAL COORDINATE/ ANALYTIC GEOMETRY

Basic Attacks and Strategies for Solving Problems in this Chapter. See pages 877 to 888 for step-by-step solutions to problems.

The problems in this chapter are similar to the problems in Chapter 35. In Chapter 35, the emphasis was on points in a plane, while in this chapter, the emphasis is on points in space. The position of points in a plane relative to a pair of perpendicular axes can be represented by an ordered pair of real numbers, while in space, there are three perpendicular axes and the position of a point is represented by an ordered triple. In three-dimensional coordinate geometry, the axes are called the x-axis, the y-axis, and the z-axis, and the three components of the ordered triple represent the distances parallel to those axes, respectively.

Many of the relationships in two-dimensional geometry carry over to three-dimensional geometry. Examine the table below. Notice the close relationships between corresponding formulas:

Points	Distance between points	Midpoint of $\overline{AB}$
$A(x_1, y_1)$ and $B(x_2, y_2)$	$d = \sqrt{(x_2 - x_1)^2 + (y_2 - y_1)^2}$	$\left(\dfrac{x_1 + x_2}{2}, \dfrac{y_1 + y_2}{2}\right)$
$A(x_1, y_1, z_1)$ and $B(x_2, y_2, z_2)$	$d = \sqrt{(x_2 - x_1)^2 + (y_2 - y_1)^2 + (z_2 - z_1)^2}$	$\left(\dfrac{x_1 + x_2}{2}, \dfrac{y_1 + y_2}{2}, \dfrac{z_1 + z_2}{2}\right)$

The three-dimensional relationships described here are used in the problems in this chapter. Also, the completion of a square technique is applied in Problem 899. That process is described in the introduction to Chapter 39.

The three-dimensional analogs of lines and circles are planes and spheres. Note the similarities between the equations of the corresponding two- and three-dimensional objects:

Two dimensions	Circle: $(x - h)^2 + (y - k)^2 = r^2$	Line: $ax + by + c = 0$
Three dimensions	Sphere: $(x - h)^2 + (y - k)^2 + (z - \ell)^2 = r^2$	Plane: $ax + by + cz + d = 0$

Step-by-Step Solutions to Problems in this Chapter, "Three-Dimensional Coordinate/ Analytic Geometry"

LINES & PLANES

● **PROBLEM** 890

Show that the point $P_1(2, 2, 3)$ is equidistant from the points $P_2(1, 4, -2)$ and $P_3(3, 7, 5)$.

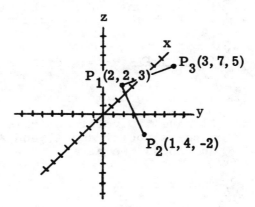

Solution: The distance, d, between any two points (x_1, y_1, z_1) and (x_2, y_2, z_2) is given by the formula

$$d = \sqrt{(x_2 - x_1)^2 + (y_2 - y_1)^2 + (z_2 - z_1)^2}$$

As a visual aid, let us plot the three points and draw the segments whose lengths we wish to show equal.

We are asked to show $P_1P_2 = P_1P_3$. By substituting into the formula given,

$$P_1P_2 = \sqrt{(2 - 1)^2 + (2 - 4)^2 + (3 - (-2))^2}$$

$$= \sqrt{(1)^2 + (-2)^2 + (5)^2} = \sqrt{30}$$

$$P_1P_3 = \sqrt{(2-3)^2 + (2-7)^2 + (3-5)^2}$$

$$= \sqrt{(-1)^2 + (-5)^2 + (-2)^2} = \sqrt{30}$$

Hence, $P_1P_2 = P_1P_3$.

● **PROBLEM** 891

Find the coordinates of the points of trisection and the midpoint of the line segment whose endpoints are $P_1(1, -3, 5)$ and $P_2(-3, 3, -4)$.

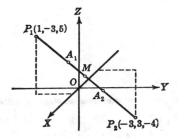

Solution: If $P_1(x_1, y_1, z_1)$ and $P_2(x_2, y_2, z_2)$ are the given endpoints of line segment $\overline{P_1P_2}$, then the coordinates (x, y, z) of a point P which divides this segment in the ratio $r = P_1P : P_2P$ are

$$x = \frac{x_1 + rx_2}{1 + r} , \quad y = \frac{y_1 + ry_2}{1 + r} , \quad z = \frac{z_1 + rz_2}{1 + r} , \quad (r \neq -1).$$

The results are analogous to the two dimensional case.

A segment will have two points of trisection and one midpoint. Let M be the midpoint and A_1 and A_2 be the points of trisection. Then, for A_1 we have

$$r = \frac{P_1A_1}{A_1P_2} = \frac{1}{2} \text{ and for } A_2 \text{ we have } r = \frac{P_1A_2}{P_2A_2} = \frac{2}{1} = 2.$$

Hence, the coordinates of A_1 are

$$x = \frac{x_1 + rx_2}{1 + r} = \frac{1 + \frac{1}{2}(-3)}{1 + \frac{1}{2}} = -\frac{1}{3}$$

$$y = \frac{y_1 + ry_2}{1 + r} = \frac{-3 + \frac{1}{2}(3)}{1 + \frac{1}{2}} = -1$$

$$z = \frac{z_1 + rz_2}{1 + r} = \frac{5 + \frac{1}{2}(-4)}{1 + \frac{1}{2}} = 2.$$

Similarly, for point A_2, we have

$$x = \frac{1 + 2(-3)}{1 + 2} = -\frac{5}{3} \; ; \; y = \frac{-3 + 2(3)}{1 + 2} = 1;$$

$$z = \frac{5 + 2(-4)}{1 + 2} = -1.$$

Therefore, the points of trisection are

$$A_1 \left(-\frac{1}{3}, -1, 2\right) \quad \text{and} \quad A_2 \left(-\frac{5}{3}, 1, -1\right).$$

For the midpoint, M, $r = \dfrac{P_1 M}{M P_2} = \dfrac{1}{1} = 1$.

Hence the coordinates of M are

$$x = \frac{1 + (-3)}{2} = -1; \quad y = \frac{-3 + 3}{2} = 0; \quad z = \frac{5 - 4}{2} = \frac{1}{2}.$$

The midpoint is M $(-1, 0, \frac{1}{2})$.

● **PROBLEM 892**

In general, three points determine a plane. Find the equation of the plane determined by D(1, 2, 1), E(2, 0,3), and F(1, - 2, 0).

<u>Solution</u>: The equation of a plane in 3-space is of the form:

$$ax + by + cz = d$$

where a, b, and c cannot all equal zero.

We know that any three noncollinear points must determine a plane. Hence, by substituting the coordinates of D, E, and F into the general form of the plane, we can solve for a, b, and c which will fully determine the plane.

After substituting, we obtain:

(1) $a + 2b + c = d$

(2) $2a \quad\quad + 3c = d$

(3) $a - 2b \quad\quad = d$

This is a linear system of equations of 4 unknowns in 3 equations which implies that there is no unique solution. This is not to say that we cannot find an e-quation of the plane. This means there is more than one equation of the plane. Suppose

$$2x + 3y + 4z = 1$$

is an equation of a plane. Multiplying the equation by 2 does not change the points that satisfy the equation. Thus,

$$4x + 6y + 8z = 2$$

is also an equation of the plane; also

$$6x + 9y + 12z = 3 \text{ and } 20x + 30y + 40z = 10.$$

More precisely, any equation of the plane ax + by+ cz + d = 0 can be rewritten (if $d \neq 0$) as a'x + b'y + c'z + 1 = 0 where a' = a/d, b' = b/d, and c' = c/d. Written in this manner, it is more apparent that the plane is an equation in 3 unknowns - not 4.

To solve the system $ax + by + cz + d = 0$, we treat d as a constant. We then solve for a, b, and c, in terms of d. Thus, $a = a'd$, $b = b'd$, and $c = c'd$, where a', b', c' are constants. Substituting our results in the equation of the plane, we obtain:

$$a'dx + b'dy + c'dz + d = 0$$

Dividing through by unknown d, we obtain

$$a'x + b'y + c'z + 1 = 0.$$

We now solve the system for a, b, and c in terms of d. Adding (1) + (2) + (3), we obtain

(4) $$4a + 4c = 3d$$

From (2), we have (5) $a = \dfrac{d - 3c}{2}$.

Substituting (5) in (4) and solving for c,

$$4 \left(\dfrac{d - 3c}{2} \right) + 4c = 3d$$

$$2d - 6c + 4c = 3d$$

$$- 2c = d$$

(6) $$c = -\dfrac{d}{2} .$$

By substitution of (6) in (5),

$$a = \dfrac{d - 3\left(-\dfrac{d}{2}\right)}{2} = \dfrac{d + \dfrac{3d}{2}}{2} = \dfrac{5d}{2} \cdot \dfrac{1}{2} = \dfrac{5d}{4}$$

Plugging this result into (3), we find

$$\dfrac{5d}{4} - 2b = d, \quad - 2b = -\dfrac{d}{4} \quad \text{or} \quad b = \dfrac{d}{8}$$

We can determine the equation of the plane by substituting the expressions for a, b, and c and then eliminating d, the as yet unspecified variable.

Hence, $\dfrac{5d}{4} x + \dfrac{d}{8} y - \dfrac{d}{2} z - d = 0.$

Divide by $d \neq 0$

$$\dfrac{5}{4} x + \dfrac{1}{8} y - \dfrac{1}{2} z - 1 = 0$$

Multiply by 8, to simplify.

Therefore, the desired equation is
$$10x + y - 4z = 8.$$

Find the equation of the plane passing through the point
(4, - 1, 1) and parallel to the plane 4x - 2y + 3z - 5 = 0.

Solution: The general form of the equation of a plane
is

$$ax + by + cz + d = 0$$

Hence, to answer this question we must determine a, b,
c, and d.

To find a, b, and c we must draw on an analogy
from the two-dimensional case.

The equation of a line is given by

$$ex + fy + k = 0.$$

By the definition of slope, any line parallel to this
line is of the form

$$ex + fy + \ell = 0.$$

Any two parallel lines have identical coefficients
preceeding the variables in their equations.

While the slope concept is inapplicable to planes,
the above rule of thumb can be extended to planes. Hence,

ax + by + cz + d = 0 is parallel to ax + by + cz + e = 0.

In this problem the plane parallel to 4x-2y+3z-5=0
will have a=4, b=-2, and c=3.

We know (4, - 1, 1) is a point on the plane. This
will allow us to find d, by substituting these coordinates
into the following equation.

$$4x - 2y + 3z + d = 0.$$

It follows that, 4(4) - 2(- 1) + 3(1) + d = 0

$$16 + \quad 2 \quad + \quad 3 \quad + d = 0$$

Hence,   d = - 21.

Therefore, the equation of the required plane is

$$4x - 2y + 3z - 21 = 0.$$

Show that the points A(- 1, - 3, 7), B(- 2, - 2, 9),
and C(1, 3, 5) are the vertices of a right triangle.

Solution: By determining the length of each side of ΔABC
and then showing the Pythagorean Theorem holds, we can prove

ΔABC is a right triangle.

We will use the distance formula to determine the lengths of the sides. For any two points $S(x_1, y_1, z_1)$ and $T(x_2, y_2, z_2)$, the length of the segment between them is given by

$$ST = \sqrt{(x_1 - x_2)^2 + (y_1 - y_2)^2 + (z_1 - z_2)^2}$$

Therefore, the lengths of the sides of ΔABC are

$$AB = \sqrt{(-1 - (-2))^2 + (-3 - (-2))^2 + (7 - 9)^2}$$

$$= \sqrt{(1)^2 + (-1)^2 + (-2)^2} = \sqrt{6}$$

$$AC = \sqrt{(-1 - 1)^2 + (-3 - 3)^2 (7 - 5)^2}$$

$$= \sqrt{(-2)^2 + (-6)^2 + (2)^2} = \sqrt{44}$$

$$BC = \sqrt{(-2 - 1)^2 + (-2 - 3)^2 + (9 - 5)^2}$$

$$= \sqrt{(-3)^2 + (-5)^2 + (4)^2} = \sqrt{50}$$

Since $\overline{BC}$ is the longest side, under Pythagoras' theorem, for ΔABC to be a right triangle

$$(BC)^2 = (AB)^2 + (AC)^2 \qquad \text{must be true.}$$

By substitution, $(\sqrt{50})^2 = (\sqrt{6})^2 + (\sqrt{44})^2$ or

$$50 = 50.$$

Hence, ΔABC is a right triangle.

● **PROBLEM** 895

Given points $P_1(x_1, y_1, z_1)$, $P_2(x_2, y_2, z_2)$ and the origin $0 (0, 0, 0)$

(a) show that $\cos \sphericalangle P_1OP_2 = \dfrac{x_1x_2 + y_1y_2 + z_1z_2}{d_1d_2}$

where $d_1 = OP_1$ and $d_2 = OP_2$. (b) Find a condition on $x_1, x_2, y_1, y_2, z_1,$ and z_2 such that $\overline{P_1O} \perp \overline{P_2O}$.

Solution: (a) $\sphericalangle P_1OP_2$ is an angle of ΔP_1OP_2. We can find the lengths of the sides of the triangle using the distance formula. Once we have the lengths of the three sides, we can use the law of cosines to solve for $\cos \sphericalangle P_1OP_2$.

$$d_1 = P_1O = \sqrt{(x_1 - 0)^2 + (y_1 - 0)^2 + (z_1 - 0)^2}$$

$$= \sqrt{x_1^2 + y_1^2 + z_1^2}$$

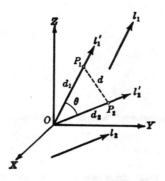

$$d_2 = P_2O = \sqrt{(x_2 - 0)^2 + (y_2 - 0)^2 + (z_2 - 0)^2}$$

$$= \sqrt{x_2{}^2 + y_2{}^2 + z_2{}^2}$$

$$d = P_1P_2 = \sqrt{(x_1 - x_2)^2 + (y_1 - y_2)^2 + (z_1 - z_2)^2}$$

By the law of cosines, we have

$$\cos \sphericalangle P_1OP_2 = \frac{d_1{}^2 + d_2{}^2 - d^2}{2d_1d_2}$$

$$= \frac{(x_1{}^2+y_1{}^2+z_1{}^2)+(x_2{}^2+y_2{}^2+z_2{}^2) - [(x_1-x_2)^2+(y_1-y_2)^2+(z_1- z_2)^2]}{2d_1d_2}$$

$$= \frac{x_1^2+y_1^2+z_1^2+x_2^2+y_2^2+z_2^2-x_1^2+2x_1x_2-x_2^2-y_1^2+2y_1y_2-y_2^2-z_1^2+2z_1z_2- z_2^2}{2d_1d_2}$$

$$= \frac{x_1x_2 + y_1y_2 + z_1z_2}{d_1d_2} \quad .$$

(b) If line segments $\overline{P_1O}$ and $\overline{P_2O}$ are perpendicular to each other, then the angle between them is 90° or 270°. But the cosine of 90° and 270° equals 0 and these are the only values (below 360°) for which cosine equals 0. Thus, a sufficient condition for two lines to be perpendicular is for the cosine of the angle formed between them to be 0. From part (a), we know that

$$\cos P_1OP_2 = \frac{x_1x_2 + y_1y_2 + z_1z_2}{d_1d_2}$$

which is equal to zero only when the numerator $(x_1x_2 + y_1y_2 + z_1z_2)$ equals zero.

Therefore, given two segments with common endpoint at the origin O and the other endpoints being $P_1(x_1, y_1, z_1)$ and $P_2(x_2, y_2, z_2)$, a sufficient condition for line segments $\overline{P_1O}$ and $\overline{P_2O}$ to be perpendicular is:

$$x_1x_2 + y_1y_2 + z_1z_2 = 0.$$

SOLIDS

Determine the intercepts of 4x + y + 2z = 8 and sketch the
portion of the graph in the octant of 3-space in which
x > 0, y > 0, and z > 0.

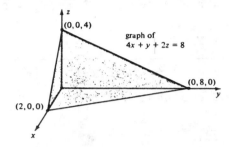

Solution: The given equation represents a plane in a
3-dimensional rectangular coordinate system.

The intercepts of the plane are given by the
points (x', 0, 0), (0, y', 0), and (0, 0, z'), where the
non-zero coordinate signifies the axis with which the
plane intersects.

Thus, if y = z = 0 then 4x = 8 and x = 2

if x = z = 0 then y = 8

if y = x = 0 then 2z = 8 and z = 4.

The three intercepts are (2, 0, 0), (0, 8, 0), and
(0, 0, 4). The points have been plotted on the accompanying
graph and the shaded region is the portion of the plane
required by the question.

The point P(2, 3, 3) is one vertex of the rectangular
parallelepiped formed by the coordinate planes and the
planes passing through P and parallel to the coordinate
planes. Find the coordinates of the other seven vertices.

Solution: In the figure shown, let PECFADBO be the
parallelepiped in question. Since the coordinates of point
P are (2, 3, 3), the values of x, y, and z from the figure
are x = 2, y = 3, and z = 3. All points in coordinate space
are given by an ordered triad of real numbers (x, y, z).

Note the following aids to naming points in space.

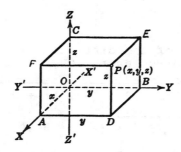

All points in xy plane have z = 0. All points in the yz plane have x = 0. All points in the xz plane have y = 0. Once in any of these three planes, the rules for naming the other two coordinates are the same as for a standard Cartesian plane.

For a point not in any of these three planes, the coordinates (x, y, z) are given in the following manner:

x = perpendicular distance, in units, to the yz plane

y = perpendicular distance to the xz plane

z = perpendicular distance to the xy plane.

Hence, according to these rules, the seven vertices other than P are

A = (2, 0, 0),   B = (0, 3, 0),   D = (2, 3, 0)

O = (0, 0, 0),   F = (2, 0, 3),   C = (0, 0, 3)

   and   E = (0, 3, 3).

● **PROBLEM** 898

Find an equation of the sphere which has the segment joining $P_1(2, -2, 4)$ and $P_2(4, 8, -6)$ for a diameter.

<u>Solution</u>: The equation of a sphere with center at C(h, k, ℓ) and radius r is the graph of

$$(x - h)^2 + (y - k)^2 + (z - \ell)^2 = r^2.$$

(**Note** that this resembles the equation of a circle in two-space.)

The center of a sphere is located at the midpoint of a diameter and the length of the radius is the distance from the center of the sphere to the endpoint of the diameter.

Given the coordinates of the endpoints of the diameter, we can use the midpoint formula to obtain the coordinates of the center C.

Since, $P_1 = (2, -2, 4)$ and $P_2 = (4, 8, -6)$,

$$C = \left(\frac{4 + 2}{2} , \frac{8 - 2}{2} , \frac{-6 + 4}{2} \right) , \text{ or } C = (3, 3, -1).$$

An application of the distance formula will allow us to obtain the radius length, CP_1. Hence

$$CP_1 = \sqrt{(3 - 2)^2 + (3 - (-2))^2 + ((-1) - 4)^2}$$

$$= \sqrt{(1)^2 + (5)^2 + (-5)^2} = \sqrt{51}.$$

By substituting $C = (3, 3, -1)$ and $CP_1 = \sqrt{51}$ back into the standard form of the equation of a sphere, we obtain the desired equation. Hence,

$$(x - 3)^2 + (y - 3)^2 + (z + 1)^2 = 51$$

is an equation of the sphere.

● **PROBLEM** 899

Find the center, radius, and volume of a sphere whose equation is

$$x^2 + y^2 + z^2 - 8x + 6y - 12z + 12 = 0.$$

Solution: The standard form of the equation of a sphere centered at (h, k, ℓ) with radius r is

$$(x - h)^2 + (y - k)^2 + (z - \ell)^2 = r^2.$$

Hence, by putting the given equation in this form we can read off the coordinates of the center and the length of the radius. We do this by completing the squares in x, y, and z.

Regrouping, we obtain

$$x^2 - 8x + y^2 + 6y + z^2 - 12z + 12 = 0$$

$$(x - 4)^2 + (y + 3)^2 + (z - 6)^2 = -12 + 61$$

$$= 49 = (7)^2$$

Hence, the center is $(4, -3, 6)$ and the radius is 7 units.

The volume, V, of a sphere is given by

$$V = \frac{4}{3} \pi r^3.$$

Consequently, by substitution, the volume of the sphere given by our value for the radius is

$$\frac{4}{3} \pi (7)^3 = (457.33) \pi.$$

Using $\pi = 3.14$, we obtain the volume 1436 cu. units.

Describe the geometric shape determined by the graph of
the equation $x^2 + y^2 = 1$ when plotted on a 3-dimensional
system of rectangular coordinates.

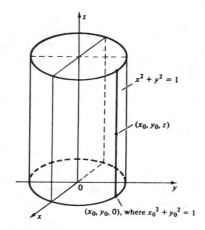

Solution: All points in a 3-dimensional system are
given by the triple (x, y, z). In this problem the determining equation contains only two variables. Consequently,
z can take on an infinite number of values corresponding to
the x and y values that satisfy the equation.

In the xy plane (z = 0), the equation traces out
a circle centered at the origin, (0, 0) with a radius of
1 unit.

For any value of z, say z = n, $-\infty < n < \infty$, the
given equation will always result in a circular graph
centered at (0, 0, n) with radius 1.

There are an infinite number of circles described
in this manner whose centers form a straight line, the
z-axis, and whose points determine a surface at a constant
1 unit from the z-axis. Hence, the geometric shape described is a right circular cylinder. (See the accompanying
figure for a graphical representation.)

● **PROBLEM** 901

Show that the equation $x^2 + y^2 + 2z^2 + 2xz - 2yz = 1$
represents a cylinder and find the equation of its
directrix.

Solution: A cylinder is a surface generated by a straight
line which moves so that it is always parallel to a given
fixed line and always passes through a given fixed curve
called the directrix. Since several curves would determine
the same cylinder, the directrix is taken to be the curve

formed by the intersection of the cylinder and a co-ordinate plane.

To prove that a given figure is a cylinder, we pass several parallel planes through the cylinder. If the curves formed by the intersection are congruent, then the figure is a cylinder. Furthermore, the curve formed by the inter-section of the figure and the coordinate plane is the directrix.

To simplify calculations, we choose as our planes the planes parallel to the XY plane. Each such plane has the general form $z = k$. The intersections of the planes and the figure are the curves

(i) $x^2 + y^2 + 2k^2 + 2kx - 2ky = 1,$ $z = k.$

Rewriting, we obtain

(ii) $(x + k)^2 + (y - k)^2 = 1,$ $z = k.$

The equations of intersection (ii) are all circles of radius 1. Since parallel planes cut congruent curves with the figure, the figure is a cylinder.

To find the directrix, we find the intersection with the XY plane, $z = 0$.

(iii) $x^2 + y^2 = 1,$ $z = 0.$

CHAPTER 52

GEOMETRY OF THE EARTH

> **Basic Attacks and Strategies for Solving Problems in this Chapter. See pages 889 to 898 for step-by-step solutions to problems.**

The earth is roughly the shape of a sphere, with a diameter of about 8,000 miles. The problems in this chapter are solved with the simplifying assumptions that the earth is a sphere and the diameter of the earth is exactly 8,000 miles.

Below is a sketch of the earth. The equator is a great circle. The points N and S are poles for the equator. This means that $\overline{NS}$ is a segment which passes through the center of this great circle, and thus, the center of the sphere. For obvious reasons, N is called the north pole and S is called the south pole.

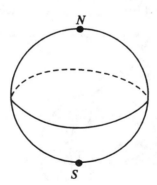

Several problems in this chapter mention longitude and latitude. Latitude is measured in degrees north or south from the equator. In the picture below, P and Q are on a great circle with the poles and O is at the center of the earth. Since $m \angle POQ = 30°$ and since P is north of the equator, the latitude for point P is 30° north

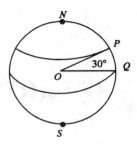

Imagine a plane cutting through the earth in such a way that the plane is parallel to the plane of the equator, and the plane passes through P. Any point on the resulting circle (called a **small circle**) would also have a latitude of 30° north. A point 30° from the equator, but in the southern hemisphere, would have a latitude of 30° south. Thus, the latitude of a point on the earth can range from 90° north to 90° south.

Longitude is measured differently. The **prime meridian** is the great semicircle which passes through the north pole, the south pole, and Greenwich, England. In the figure below, $\overset{\frown}{NQS}$ represents this prime meridian. Now point P is on another great semicircle which contains N and S. The measure of spherical $\measuredangle QNP$ is 10°. Since P is east of the prime meridian, the longitude of P is 10° east. All points of $\overset{\frown}{NPS}$ (other than N and S) have longitude 10° east. Just as the equator divides the earth into northern and southern hemispheres, the prime meridian divides the earth into east and west hemispheres. If R were 10° from $\overset{\frown}{NQS}$, and in the western hemisphere, we would say R is 10° west, not 350° east. Thus, the longitude of a point on the earth can range from 180° west to 180° east.

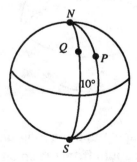

Two problems in this chapter refer to measuring angles in terms of radians. Consider the circle on the next page. If the linear measure of an arc is equal to the radius of the circle, we say that the radian measure of its subtended angle is 1 radian. In general, the radian measure of an angle is the "number of r's" in the intercepted arc. Since the entire circumference of a circle is $2\pi r$, the radian measure corresponding to an angle of 360° is 2π, which means

$$2\pi \text{ radians} = 360°,$$

$$1 \text{ radian} = \frac{360°}{2\pi}, \text{ or}$$

$$1° = \frac{\pi}{180} \text{ rad.}$$

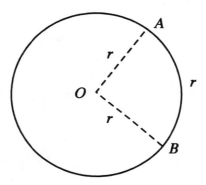

Since radians are the most common units of angle measure (and because they can usually be recognized because they contain a π), writing "radians" or "rad" is usually omitted. The formula for arc length is $S = r\theta$, where θ is the radian measure of the central angle. This formula comes directly from the definition of radian measure.

Step-by-Step Solutions to Problems in this Chapter, "Geometry of the Earth"

New York City has longitude 74°W and Portland, Oregon has longitude 122°W. (a) What is the difference in longitude between these 2 cities? (b) What fraction of 360° is this? (c) What is the actual time difference between New York City and Portland?

<u>Solution</u>: (a) The difference in longitude between the 2 cities is 122°W - 74°W = 48°W.

(b) This is 48° out of 360° or 48/360 = 2/15 of 360°.

(c) In one day, the earth rotates once. In other words, in 24 hours, a fixed meridian of the earth (say the Prime Meridian) passes through 360°. Hence, we can set up the following proportion:

$$\frac{24 \text{ hours}}{360°} = \frac{1 \text{ hour}}{x}$$

or $$x = \frac{360°}{24} = 15°.$$

Hence, for every 15° difference in the longitude of 2 locations, there is one hour time difference. From part (a), we know there is a 48° difference in the longitude of the locations of our 2 cities. This corresponds to a 48/15 = 3.20 hour time difference between New York and Portland.

● **PROBLEM** 903

Find the circumference of the 41° parallel. (In the figure, $\overline{OP}$ is the radius of the 41° parallel. Cos 41° = .755.)

<u>Solution</u>: We must find the circumference of circle O with radius $\overline{OP}$. By definition, the circumference of the circle is

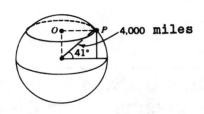

4,000 miles

$$C = 2\pi \ (OP) \tag{1}$$

Since we do not know $\overline{OP}$, we must try to find OP in terms of r, the radius of the earth, and some angle.

By definition, the cosine of an angle of a right triangle is equal to the ratio of the length of the side adjacent to the angle, so long as it is not the hypotenuse, and the hypotenuse length. In the figure,

$$\cos 41° = \frac{OP}{4,000 \text{ miles}}$$

or

$$OP = (4,000 \text{ miles}) \cos 41° \tag{2}$$

Substituting (2) into (1),

$$C = 2\pi \ (4,000 \text{ miles}) \cos 41°$$

or

$$C = (8,000 \text{ miles})(\pi)(.755)$$

$$C = 18975.2 \text{ miles.}$$

● **PROBLEM** 904

Show that a circle of a sphere is equidistant from its antipodes. That is, in the figure, show that for any 2 points C and D, of the circle, CA = DA and CB = DB. (The center of the circle is point O.)

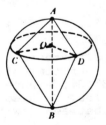

Solution: A circle of a sphere (circle O in figure) is a circle lying on the surface of the sphere. The diameter of the sphere, which is perpendicular to the plane of the circle and passes through its center is the axis of the circle ($\overline{AB}$ in figure). The endpoints of this axis are the antipodes of the circle (A and B in figure).

We shall prove that CA = DA and CB = DB by proving that $\triangle AOC \overset{\sim}{=} \triangle AOD$ and $\triangle BOC \overset{\sim}{=} \triangle BOD$.
To do this, look at triangles AOC and AOD and note

that $\overline{AB}$ which is the axis of circle O, is perpendicular to the plane of O. Hence, $\sphericalangle\,AOD \cong \sphericalangle\,AOC$, since both are right angles. Furthermore, $\overline{AO}$ is common to both triangles, and $\overline{AO} \cong \overline{AO}$. Lastly, $\overline{OC} \cong \overline{OD}$, since both are radii of circle O. By the S.A.S. Postulate, $\triangle AOC \cong \triangle AOD$, and, as a result, $\overline{CA} \cong \overline{DA}$. This implies that CA = DA.

Now, examine $\triangle BOC$ and $\triangle BOD$. Again, $\sphericalangle\,BOC \cong \sphericalangle\,BOD$; they are both right angles. From above, $\overline{OC} \cong \overline{OD}$. Lastly, $\overline{OB} \cong \overline{OB}$, since this segment is common to both triangles. By the S.A.S. Postulate, $\triangle BOC \cong \triangle BOD$. This implies that CB = DB.

● **PROBLEM** 905

What is the length of the shortest path from the North Pole to the South Pole? (Take the diameter of the earth to be 8,000 miles.)

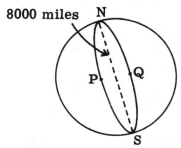

North Pole

8000 miles

South Pole

<u>Solution</u>: The figure shows a picture of the earth. We desire to go from point N to point S by the shortest possible route.

The path of shortest distance on a given surface between 2 points of the surface is called a geodesic. For a plane surface, the geodesics are lines. For a spherical surface, the geodesics are arcs of great circles. (A great circle of a sphere is any circle lying on the surface of the sphere with radius length equal to the radius length of the sphere.)

Hence, the shortest path from N to S is $\overset{\frown}{NPS}$, which is an arc of great circle NPSQ. Note that $\overset{\frown}{NPS}$ is a semi-circle. Its circumference is ½ the circumference of great circle NPSQ. That is,

distance from N to S = length $\overset{\frown}{NPS}$

distance from N to S = (½)(2πr),   (1)
where r is the radius of great circle NPSQ. But, since NPSQ

is a great circle of the earth, it has radius equal to the radius of the earth (4,000 miles). Hence, using this fact in (1) yields

distance from N to S = ($\frac{1}{2}$)(2π (4,000 miles))

= 12,566 miles.

● **PROBLEM** 906

If a jet plane flies at an altitude of 5 miles above the equator and circles the globe once, how much farther than the circumference of the equator does it fly?

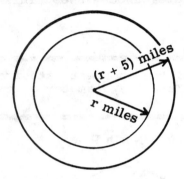

Solution: The figure shows the earth, which is a circle of radius r miles, and the path of the plane, which is a circle of radius r + 5 miles.

The plane travels a distance equal to the circumference of the large circle (C_L). Therefore, we may write

distance travelled by plane = C_L = 2π (r + 5) miles

The circumference of the equator is

C_E = (2πr) miles

Hence, the extra distance travelled over the circumference of the equator is:

C_L - C_E = (2π (r + 5) - 2πr) miles

or C_L - C_E = (2π (5)) miles

C_L - C_E = 31.4 miles.

● **PROBLEM** 907

How far apart are 2 points on the equator if their longitudes differ by 1°?

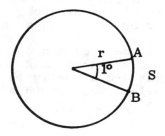

Solution: The figure shows an exaggerated view of the great circle coincident with the equator, as seen from the North Pole. Points A and B are the points of intersection of 2 meridians, 1° apart, with the great circle through the equator. (This great circle is also perpendicular to the line connecting the North and South poles.)

We must find the distance S, which is the length of arc $\overset{\frown}{AB}$. By the definition of the arc length of an arc of radius r subtended by an angle θ, we may write

$$S = r\theta, \tag{1}$$

where θ is given in radians. We know that r = 4,000 miles (the radius of the earth), and θ = 1°. But θ must be in radians. Now,

2π radians = 360°. Therefore,

$\frac{\pi}{180}$ radians = 1°.

Using these facts in (1) yields

S = (4,000 miles)(π/180 radians)

S = 69.8 miles.

(Note, from (1), θ = S/r. Since S and r are both in units of length, θ is actually dimensionless. Hence, the radian is a "dimensionless" unit. This is why our final answer comes out in terms of miles, and not miles times radians.)

In conclusion, 2 points on the equator, separated by 2 meridians 1° apart, are 69.8 miles from each other on the surface of the earth.

● **PROBLEM** 908

Bellingham, Washington, is at 49°N latitude and 122°W longitude. San Francisco, California is at 38°N latitude and 122°W longitude. What is the direct air distance from Bellingham to San Francisco?

Solution: Both cities are located at 122°W longitude. This means that both are 122° West of the meridian through Greenwich Observatory in London, England.

Bellingham is 49° North of the equator while San Francisco is 38° North of the equator. Hence, the positions of the 2 cities differ by 11° along a meridian.

Now, a meridian is part of a great circle. Furthermore, a great circle of the earth has the same radius as the earth. The 2 cities are actually 11° apart on a circle of radius r = 4,000 miles. The length of arc, S, between the 2 cities, is given by

$$S = r\theta, \tag{1}$$

where θ is in radians. Since 1° = $\pi/180$ radians, 11° = $(11/180)\,\pi$ radians. Thus, (1) becomes

$$S = (4,000 \text{ miles})(11/180 \ \pi \text{ radians})$$

$$S = 767.95 \text{ miles.}$$

● **PROBLEM** 909

If an observer could be suspended in the same spot in space over the equator and watch the world turn, how many miles of the equator would he observe to pass by in 1 hour?

Solution: The earth goes through 1 revolution (2π radians or 360°) in 24 hours. Hence, to find out how many radians it goes through in 1 hour, we set up the following proportion:

$$\frac{2\pi \text{ radians}}{24 \text{ hours}} = \frac{x}{1 \text{ hour}} \quad \text{or} \quad x = \frac{2\pi}{24} \text{ radians} = \frac{\pi}{12} \text{ radians}$$

The arclength, S, corresponding to $\pi/12$ radians is

$$S = r \ (\pi/12 \text{ radians}),$$

where r is the earth's radius at the equator. Since r = 4,000 miles,

$$S = (4,000 \text{ miles}) \left(\frac{\pi \text{ radians}}{12}\right) = 1047.2 \text{ miles.}$$

● **PROBLEM** 910

Sea captains have an old rule, "The distance, d, in miles at which an object can be seen from the surface of the sea is $\sqrt{\dfrac{3\,h}{2}}$, where h is the height of the object in feet above the sea level." Prove that this rule is a very good approximation by showing that $d = \sqrt{\dfrac{3\,h}{2} + \left[\dfrac{h}{5280}\right]^2}$.

Consider the earth to be spherical and to have a radius of 3960 miles. [1 mile = 5280 ft.]

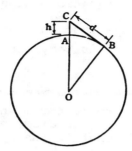

Solution: Imagine slicing through the sphere along the great circle passing through the ship and the object being sighted. Let B stand for the ship's position and C be the top of the object. Since, the ship is exactly on the surface of the earth, B is a point of tangency of the line of sight from the ship to the object.

$\overline{OB}$ is a radius to the point of tangency and, as such $\overline{OB} \perp \overline{BC}$.

Therefore, $\triangle OBC$ is a right triangle. We can apply the Pythagorean Theorem to obtain an expression for d, the length of side $\overline{CB}$ (the line of sight).

We are given OA = OB = 3960 miles.

If we let h = the height of the object in feet,

then $\dfrac{h}{5280}$ = the height in miles

and $OC = OA + \dfrac{h}{5280}$

$(OC)^2 = (OB)^2 + (CB)^2$ by the Pythagorean Theorem.

By substitution,

$$\left(3960 + \frac{h}{5280}\right)^2 = (3960)^2 + d^2$$

$$d^2 = \left(3960 + \frac{h}{5280}\right)^2 - (3960)^2$$

$$d^2 = (3960)^2 + 2\left(\frac{3960\ h}{5280}\right) + \left(\frac{h}{5280}\right)^2 - (3960)^2$$

$$d^2 = 2\left(\frac{3}{4}\right)h + \left(\frac{h}{5280}\right)^2$$

$$d = \sqrt{\frac{3\ h}{2} + \left(\frac{h}{5280}\right)^2}$$

But h is likely to be very small in comparison to 5280, which causes the square of the ratio, $\left(\dfrac{h}{5280}\right)^2$, to be even less significant. Therefore, the term $\left(\dfrac{h}{5280}\right)^2$

can be disregarded in our equation for d.

Hence, we can conclude that the sea captain's approximation of $d = \sqrt{\frac{3}{2} h}$ is very good.

● **PROBLEM** 911

The surface area of a sphere is $4\pi r^2$, where r is the radius of the sphere. Assuming that the diameter of the earth is 8,000 miles and $\pi = 22/7$, what is the surface area of the earth?

Solution: The surface area, S, of a sphere of radius r is $4\pi r^2$. In terms of the diameter, d, this becomes

$$S = 4\pi \left(\frac{d}{2}\right)^2 = \pi d^2, \tag{1}$$

where we have used the fact that d = 2r.

Assuming the earth to be a sphere, and using (1), we find that its surface area is

$$S = \pi d^2 = \left(\frac{22}{7}\right)(8,000 \text{ miles})^2$$

$$S = \left(\frac{22}{7}\right)(64,000,000 \text{ miles}^2)$$

$$S = 201,142,850 \text{ miles}^2.$$

● **PROBLEM** 912

Let points A and B be on the equator and let N be the North Pole. If spherical triangle ABN encloses an area of 33,500,000 sq. miles, what is the measure of spherical angle ANB?

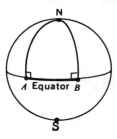

Solution: Due to the fact that the circle through A and B passes through the equator, N, the North Pole is an antipode of the circle. In addition, S, the South Pole, is an antipode of the circle. Note that $\overline{NA}$ and $\overline{NB}$ are arcs of great circles through antipodes N and S. Since a circle of a sphere and any great circle containing its antipodes meet at right angles, m ∢ NAB = m ∢ NBA = 90°, as indicated. Now, if a, b and c are the measures of the angles of a spherical triangle, and S is the surface area of the sphere, then the area, A, enclosed by the spherical triangle is

$$A = \frac{(a + b + c - 180°)}{720°} S$$

In our case, let

$$a = m \sphericalangle NAB$$

$$b = m \sphericalangle NBA$$

$$c = m \sphericalangle ANB.$$

Then, $A = \left[\dfrac{m \sphericalangle NAB + m \sphericalangle NBA + m \sphericalangle ANB - 180°)}{720°}\right] S$

$$A = \left(\frac{m \sphericalangle ANB}{720°}\right) S \tag{1}$$

But we know from the statement of the problem that A = 33,500,000 sq. miles.

The surface area of the earth is S = 201,142,850 sq. miles. Using these facts in (1) yields

$$33{,}500{,}000 \text{ sq. miles} = (m \sphericalangle ANB) \left(\frac{201{,}142{,}850 \text{ sq. miles}}{720°}\right)$$

or, $m \sphericalangle ANB = \dfrac{(33{,}500{,}000)(720°)}{201{,}142{,}850)} \sim 120°.$

● **PROBLEM** 913

Let points A and B be on the equator and let N be the North Pole. If m $\sphericalangle$ ANB = 45°, what is the surface area bounded by spherical triangle ABN?

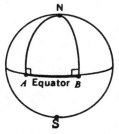

A Equator B

Solution: If a, b, and c are the measures of the angles of a spherical triangle, and S is the surface area of a sphere, then the area, A, enclosed by the spherical triangle is given by

$$A = \frac{(a + b + c - 180°)}{720°} S$$

In the figure, let m $\sphericalangle$ ANB = a, m $\sphericalangle$ NAB = b, and m ∢ NBA = c. Then,

$$A = \frac{(m \sphericalangle ANB + m \sphericalangle NAB + m \sphericalangle NBA - 180°)}{720°} \ S \qquad (1)$$

But, as indicated in the figure, m ∢ NAB = 90° and m ∢ NBA = 90°. Therefore, (1) becomes

$$A = \left(\frac{m \sphericalangle ANB}{720°}\right) S$$

The statement of the problem tells us that m ∢ ANB = 45°. Furthermore, $S = 201{,}142{,}850 \ miles^2$. Therefore,

$$A = \left(\frac{45°}{720°}\right) (201{,}142{,}850 \ miles^2)$$

$$A = \frac{201{,}142{,}850 \ miles^2}{16}$$

$$A = 12{,}571{,}428 \ miles^2.$$

SUMMARY OF ESSENTIAL GEOMETRIC THEOREMS & PROPERTIES

AXIOMS

A quantity is equal to itself (reflexive law).

If two quantities are equal to the same quantity, they are equal to each other.

If a & b are any quantities, and a = b, then b = a (symmetric law).

The whole is equal to the sum of its parts.

If equal quantities are added to equal quantities, the sums are equal quantities.

If equal quantities are subtracted from equal quantities, the differences are equal quantities.

If equal quantities are multiplied by equal quantities, the products are equal quantities.

If equal quantities are divided by equal quantities (not 0), the quotients are equal quantities.

LINES

Two points determine one and only one line.

Let A & B be two points on a line. The set of points on the line between A & B and including A & B is called the line segment AB.

A line which divides a line segment into two segments with equal measure is called a bisector of the line segment. The intersection of the two is called the midpoint of the line segment.

A line segment has only one midpoint.

If three or more points lie on the same line, they are said to be collinear.

Two lines are perpendicular if, and only if, they meet and form equal adjacent angles.

The perpendicular to a line through a point not on the line is unique.

The distance from a point to a line is the measure of the perpendicular segment drawn from the point to the line.

A perpendicular bisector of a line segment is a line that bisects and is perpendicular to the given line segment.

ANGLES

An angle is the union of two rays with a common endpoint. The common endpoint is called the vertex of the angle.

An angle bisector is a ray that divides an angle into two angles that have equal measure.

A right angle is an angle whose measure is 90°.

All right angles are equal.

A straight angle is an angle whose sides form a line. The measure of a straight angle is 180°.

All straight angles are equal.

An acute angle is an angle whose measure is larger than 0° but less than 90°.

An obtuse angle is an angle whose measure is greater than 90° but less than 180°.

Two angles whose sum is 180° are called supplementary angles. Each angle is called the supplement of the other.

Supplements of the same or equal angles are themselves equal.

Two angles whose sum is 90° are called complementary angles. Each angle is called the complement of the other.

Complements of the same or equal angles are themselves equal.

Two angles are called adjacent angles if, and only if, they have a common vertex and a common side lying between them.

A pair of nonadjacent angles formed by two intersecting lines is called a pair of vertical angles.

Vertical angles are equal.

If two lines are cut by a transversal, nonadjacent angles on opposite sides of the transversal but on the interior of the two lines are called alternate interior angles.

If two lines are cut by a transversal, nonadjacent angles on opposite sides of the transversal and on the exterior of the two lines are called alternate exterior angles.

If two lines are cut by a transversal, angles on the same side of the transversal and in corresponding positions with respect to the lines are called corresponding angles.

TRIANGLES

A triangle is a closed three-sides figure.

The points of intersection of the sides of a triangle are called the vertices of the triangle.

A right triangle is a triangle with one right angle.

The side opposite the right angle in a right triangle is called the hypotenuse of the right triangle. The other two sides are called arms or legs of the right triangle.

An equilateral triangle is a triangle all of whose sides are of equal measure.

An isosceles triangle is a triangle that has at least two sides of equal measure. The third side is called the base of the triangle.

A scalene triangle is a triangle that has no pair of sides with equal measure.

The perimeter of a triangle is the sum of the measures of the sides of a triangle.

If one of the sides of a triangle is extended, the angle formed which is adjacent to an angle of the triangle is called an exterior angle.

An exterior angle of a triangle is equal to the sum of the nonadjacent interior angles.

Two triangles are congruent if, and only if, all of their corresponding parts are equal. ($\cong$)

If the hypotenuse and an acute angle of one right triangle are equal, respectively, to the hypotenuse and an acute angle of a second right triangle, the triangles are congruent.

If three sides of one triangle are equal, respectively, to three sides of a second triangle, the triangles are congruent. (SSS = SSS)

If two sides and the included angle of one triangle are equal, respectively, to two sides and the included angle of a second triangle, the triangles are congruent. (SAS = SAS)

If two angles and the included side of one triangle are equal, respectively, to two angles and the included side of a second triangle, the triangles are congruent. (ASA = ASA)

If two angles and any side of one triangle are equal, respectively, to two angles and any side of a second triangle, the triangles are congruent. (AAS = AAS)

The sum of the measures of the interior angles of a triangle is 180°.

If two angles of one triangle are equal respectively to two angles of a second triangle, their third angles are equal.

The acute angles of a right triangle are complementary.

A triangle can have at most one right or obtuse angle.

An acute triangle is a triangle with all acute angles.

An obtuse triangle is a triangle with one obtuse angle.

If the hypotenuse and an arm of one right triangle are equal, respectively, to the hypotenuse and an arm of a second right triangle, the right triangles are congruent.

The perpendicular bisector of the base of an isosceles triangle passes through the vertex.

The bisector of the vertex angle of an isosceles triangle is the perpendicular bisector of the base of the triangle.

If a triangle has two equal angles, then the sides opposite those angles are equal.

If two sides of a triangle are equal, then the angles opposite those sides are equal.

In a right triangle, the square of the hypotenuse is equal to the sum of the squares of the other two sides.

If a triangle has sides of length a, b, and c and $c^2 = a^2 + b^2$, then the triangle is a right triangle.

In a 30°-60° right triangle, the hypotenuse is twice the length of the side opposite the 30° angle. The side opposite the 60° angle is equal to the length of the side opposite the 30° angle multiplied by $\sqrt{3}$.

In an isosceles 45° right triangle, the hypotenuse is equal to the length of one of its arms multiplied by $\sqrt{2}$.

A line segment that connects the midpoints of two sides of a triangle is parallel to the third side and half as long.

A median of a triangle is a line segment drawn from one vertex of a triangle to the midpoint of the opposite side.

The medians of a triangle meet in a point.

An altitude of a triangle is a line segment drawn from one vertex of the triangle perpendicular to the opposite side, or if necessary, to an extension of the opposite side.

The altitudes of a triangle meet in a point.

The perpendicular bisectors of the sides of a triangle meet at a point which is equally distant from the vertices of the triangle.

The angle bisectors of a triangle meet in a point which is equidistant from the sides of the triangle.

The point of intersection of the medians of a triangle is 2/3 of the way from the vertex to the midpoint of the opposite side.

PARALLELISM

PARALLEL LINES:
Two lines are called parallel lines if, and only if, they are in the same plane and do not intersect.

Parallel lines are always the same distance apart.

Two lines are parallel if they are both perpendicular to a third line.

Given a line ℓ and a point P not on line ℓ, there is only one line through point P that is parallel to line ℓ.

If two lines are cut by a transversal so that alternate interior angles are equal, the lines are parallel.

If two parallel lines are cut by a transversal, all pairs of alternate interior angles are equal.

If two lines are cut by a transversal so that corresponding angles are equal, the lines are parallel.

If two parallel lines are cut by a transversal, each pair of corresponding angles is equal.

If two lines are cut by a transversal so that two interior angles on the same side of the transversal are supplementary, the lines are parallel.

If two parallel lines are cut by a transversal, pairs of interior angles on the same side of the transversal are supplementary.

If a line is perpendicular to one of two parallel lines, it is perpendicular to the other also.

If two lines are cut by a transversal so that alternate interior angles are not equal, the lines are not parallel.

If two lines are cut by a transversal so that corresponding angles are not equal, the lines are not parallel.

If two lines are cut by a transversal so that two interior angles on the same side of the transversal are not supplementary, the lines are not parallel.

If two nonparallel lines are cut by a transversal, the pairs of alternate interior angles are not equal.

PARALLELOGRAM:

A parallelogram is a quadrilateral whose opposite sides are parallel. ($\square$)

Opposite sides of a parallelogram are equal.

Nonconsecutive angles of a parallelogram are equal.

Consecutive angles of a parallelogram are supplementary.

A diagonal of a parallelogram divides the parallelogram into two congruent triangles.

The diagonals of a parallelogram bisect each other.

If both pairs of opposite sides of a quadrilateral are equal, then the quadrilateral is a parallelogram.

If two opposite sides of a quadrilateral are both parallel and equal, the quadrilateral is a parallelogram.

If the diagonals of a quadrilateral bisect each other, then the quadrilateral is a parallelogram.

RHOMBUS:

A rhombus is a parallelogram with two adjacent sides equal.

All sides of a rhombus are equal.

The diagonals of a rhombus are perpendicular to each other.

If the diagonals of a parallelogram are perpendicular, the parallelogram is a rhombus.

The diagonals of a rhombus bisect the angles of the rhombus.

A square is a rhombus with a right angle.

RECTANGLE:

A rectangle is a parallelogram with one right angle.

All angles of a rectangle are right angles.

The diagonals of a rectangle are equal.

If the diagonals of a parallelogram are equal, the parallelogram is a rectangle.

TRAPEZOID:

A trapezoid is a quadrilateral with two and only two sides parallel. The parallel sides are called bases.

An isosceles trapezoid is a trapezoid whose nonparallel sides are equal. A pair of angles including one of the parallel sides is called a pair of base angles.

The base angles of an isosceles trapezoid are equal.

If a line joins the midpoints of two sides of a triangle, that line is parallel to and equal to one-half of the third side.

The median of a trapezoid is parallel to the bases and equal to one-half their sum.

The median of a trapezoid is the line joining the midpoints of the non-parallel sides.

If three or more parallel lines cut off equal segments on one transversal, they cut off equal segments on all transversals.

AREAS

The area of a rectangle is given by the formula $A = bh$ where b is the length of the base and h is the height of the rectangle.

The area of a triangle is given by the formula $A = 1/2\ bh$ where b is the length of a base and h is the corresponding height of the triangle.

The area of a parallelogram is given by the formula A = bh where b is the length of a base and h is the corresponding height of the parallelogram. The side to which an altitude is drawn is called a base. The length of an altitude is called a height.

The area of a rhombus is equal to one-half the product of its diagonals.

The area of a trapezoid is given by the formula A = 1/2 h(b + b') where h is the height and b and b' are the lengths of the bases of the trapezoid.

The area of a circle is given by the formula A = πr^2 where π is approximately 3.14 and r is the radius of the circle.

Any two congruent figures have the same area.

RATIOS, PROPORTIONS, SIMILARITY

A ratio is the comparison of two numbers by their indicated quotient. A ratio is a fraction, with denominators not equal to zero.

A proportion is a statement that two ratios are equal.

In the proportion a/b = c/d the numbers a & d are called the extremes of the proportion, and the numbers b & c are called the means of the proportion. The single term, d, is called the fourth proportional.

In a proportion, the product of the means is equal to the product of the extremes. (If a/b = c/d, then ad = bc.)

A proportion may be written by inversion. (If a/b = c/d, then b/a = d/c.)

The means may be interchanged in any proportion. (If a/b = c/d, then a/c = b/d.)

The extremes may be interchanged in any proportion. (If a/b = c/d, then d/b = c/a.)

A proportion may be written by addition. (If $\frac{a}{b} = \frac{c}{d}$, then $\frac{a+b}{b} = \frac{c+d}{d}$.)

A proportion may be written by subtraction. (If $\frac{a}{b} = \frac{c}{d}$, then $\frac{a-b}{b} = \frac{c-d}{d}$.)

If three terms of one proportion are equal, respectively, to three terms of a second proportion, the fourth terms are equal. Thus, if $\frac{a}{b} = \frac{c}{d} = \frac{e}{f}$, then $\frac{a+c+e}{b+d+f} = \frac{a}{b}$.

A line parallel to one side of a triangle divides the other two sides proportionally.

If a line divides two sides of a triangle proportionally, it is parallel to the third side.

The bisector of one angle of a triangle divides the opposite side in the same ratio as the other two sides.

All congruent triangles are similar.

Two triangles similar to the same triangle are similar to each other.

If two triangles have one angle of one equal to one angle of the other and the respective sides including these angles are in proportion, the triangles are similar.

If three sides of one triangle are in proportion to the three corresponding sides of a second triangle, the triangles are similar.

Two triangles are similar if, and only if, two angles of one triangle are equal to two angles of the other triangle.

The altitude on the hypotenuse of a right triangle is the mean proportional between the segments of the hypotenuse.

Two polygons are similar if, and only if, all pairs of corresponding angles are equal and all pairs of corresponding sides are in proportion.

CIRCLES

A circle is a set of points in the same plane equidistant from a fixed point called its center. A line segment drawn from the center of the circle to one of the points on the circle is called a radius of the circle.

The length of a diameter is twice the length of a radius.

A portion of a circle is called an arc of the circle.

In a circle, parallel lines intercept equal arcs.

The circumference of a circle is the distance about a circle. The circumference of a circle is given by the formula, $C = 2\pi r$ where r is the radius of the circle.

Two circles are congruent if, and only if, their radii or diameters are equal.

A semicircle is an arc of a circle whose endpoints lie on the extremities of a diameter of the circle.

An arc greater than a semicircle is called a major arc.

An arc less than a semicircle is called a minor arc.

An angle whose vertex is at the center of a circle and whose sides are radii is called a central angle.

The number of degrees in the arc intercepted by a central angle is equal to the number of degrees in the central angle. This number is called the measure of the arc.

A sector of a circle is the set of points between two radii and their intercepted arc.

The line passing through the centers of two circles is called the line of centers.

A line that intersects a circle in two points is called a secant.

A line segment joining two points on a circle is called a chord of the circle.

A chord that passes through the center of the circle is called a diameter of the circle.

An angle whose vertex is on the circle and whose sides are chords of the circle is called an inscribed angle.

The measure of an inscribed angle is equal to one-half the measure of its intercepted arc.

An angle formed by a tangent and a chord is equal to one-half the measure of the intercepted arc.

If two chords intersect within a circle, each angle formed is equal to one-half the sum of its intercepted arc and the intercepted arc of its vertical angle.

An angle formed by the intersection of two secants outside a circle is equal to one-half the difference of its intercepted arcs.

An angle formed by the intersection of a tangent and a secant outside a circle is equal to one-half the difference of the intercepted arcs.

A line drawn from the center of a circle perpendicular to a chord bisects the chord and its arc.

A line drawn from the center of a circle to the midpoint of a chord is perpendicular to the chord.

The perpendicular bisector of a chord passes through the center of the circle.

In the same circle or congruent circles equal chords have equal arcs.

In the same circle or congruent circles equal arcs have equal chords.

In the same circle or congruent circles, equal chords are equidistant from the center.

In the same circle or congruent circles, chords equidistant from the center are equal.

If two chords intersect within a circle, the product of the segments of one chord is equal to the product of the segments of the other chord.

If two circles intersect in two points, their line of centers is the perpendicular bisector of their common chord.

A line that has one and only one point of intersection with a circle is called a tangent to the circle. Their common point is called a point of tangency.

If two tangents are drawn to a circle from a point outside the circle, the tangents are of equal measure.

A line drawn from a center of a circle to a point of tangency is perpendicular to the tangent passing through the point of tangency. Also, if a line is perpendicular to a radius at its intersection with its circle, the line is tangent to the circle.

If two circles are tangent either internally or externally, the point of tangency and the centers of the circles are collinear.

If a secant and a tangent are drawn to a circle, the measure of the tangent is the mean proportional between the secant and its external segment.

If two secants are drawn to a circle from a point outside the circle, the products of the secants and their external segments are equal.

The opposite angles of an inscribed quadrilateral are supplementary.

If a parallelogram is inscribed within a circle, it is a rectangle.

POLYGONS

A polygon is a convex figure with the same number of sides as angles.

A quadrilateral is a polygon with four sides.

An equilateral polygon is a polygon all of whose sides are of equal measure.

An equiangular polygon is a polygon all of whose angles are of equal measure.

A regular polygon is a polygon that is both equilateral and equiangular.

A circle can be inscribed in any regular polygon.

A circle can be circumscribed about any regular polygon.

The center of a circle that is circumscribed about a regular polygon is the center of the circle that is inscribed within the regular polygon.

If a circle is divided into n equal arcs, the chords of the arcs will form a regular polygon.   (n>2)

If a circle is divides into n equal arcs and tangents are drawn to the circle at the endpoints of these arcs, the figure formed by the tangents will be a regular polygon.   (n>2)

An apothem of a regular polygon bisects its respective side.

The radius of a regular polygon is a line segment drawn from the center of the polygon to one of its vertices.

A central angle of a regular polygon is the angle formed by radii drawn to two consecutive vertices.

A central angle of a regular polygon is given by the formula, $x = \dfrac{2}{n} \cdot 180°$, where n is the number of sides of the polygon.

The area of a regular polygon is equal to the product of one-half its apothem and its perimeter.

The sum of the interior angles of a polygon is given by the formula $S = (n - 2) \cdot 180°$, where n is the number of sides of the polygon.

The measure of an interior angle, a, of a regular polygon is given by the formula: $a = \dfrac{(n - 2) \cdot 180°}{n}$ where n is the number of sides of of the polygon.

The sum of the exterior angles of a polygon is always 360°.

The area of a sector of a circle divided by the area of the circle is equal to the measure of its central angle divided by 360°.

INEQUALITIES

If the same quantity is added to both sides of an inequality, the sums are unequal and in the same order.   (If $a<b$, then $a + c < b + c$.)

If equal quantities are added to unequal quantities, the sums are unequal and in the same order.   (If $a< b$ and $c = d$, then $a + c < b + d$.)

If unequal quantities are subtracted from equal quantities, the differences are unequal and in the opposite order. (If $a < b$ and $c = d$, then $c - a > d - b$.)

If both sides of an inequality are multiplied by a positive number, the products are unequal and in the same order. (If $a < b$ and c is a positive number, then $ac < bc$.)

If both sides of an inequality are multiplied by a negative number, the products are unequal in the opposite order. (If $a < b$ and c is a negative number, then $ac > bc$.)

The relation "$<$" is transitive. (If $a < b$ and $b < c$, then $a < c$.)

If unequal quantities are added to unequal quantities of the same order, the sums are unequal quantities and in the same order. (If $a < b$ and $c < d$, then $a + c < b + d$.)

In a triangle an exterior angle is greater than either nonadjacent interior angle.

If two sides of a triangle are unequal, the angles opposite these sides are unequal in the same order.

If two angles of a triangle are unequal, the sides opposite these angles are unequal in the same order.

If two sides of one triangle are equal to two sides of a second triangle and the included angle of the first is greater than the included angle of the second, then the third side of the first triangle is greater than the third side of the second.

If two sides of one triangle are equal to two sides of a second triangle and the third side of the first is greater than the third side of the second, then the angle opposite the third side of the first triangle is greater than the angle opposite the third side of the second.

In the same circle or in equal circles, the greater of two central angles will intercept the greater arc.

In the same circle or in equal circles, the greater of two arcs will be intercepted by the greater of the central angles.

In the same circle or equal circles, the greater of two chords intercepts the greater minor arc.

In the same circle or equal circles, the greater of two minor arcs has the greater chord.

If two unequal chords form an inscribed angle within a circle, the shorter chord is the farther from the center of the circle.

LOCUS

A circle is the locus of points at a given distance from a fixed point, called its center, and all in the same plane.

The locus of points whose coordinates satisfy a given linear equation is a straight line.

The locus of the vertex of the right angle of a right triangle with a fixed hypotenuse is a circle with the hypotenuse as diameter.

The locus of all points equidistant from the sides of an angle is the angle bisector.

The locus of all points equidistant from two points is the perpendicular bisector of the segment joining the two points.

The locus of a point at a given distance from a given line is a pair of lines, one on each side of the given line, parallel to the given line and at the given distance from it.

COORDINATE GEOMETRY

Directed distance of a horizontal line segment from one point to a second is the x-coordinate of the second minus the x-coordinate of the first.

Directed distance of a vertical line segment from one point to a second is the y-coordinate of the second minus the y-coordinate of the first.

The slope, m, of a nonvertical line segment determined by points $P(x_1, y_1)$ and $Q(x_2, y_2)$ is given by $m = \dfrac{y_2 - y_1}{x_2 - x_1}$.

On a nonvertical line, all segments of the line have the same slope.

Two nonvertical lines are perpendicular if, and only if, their slopes are negative reciprocals.

Two nonvertical lines are parallel if, and only if, they have the same slope. Vertical lines are parallel.

A linear equation is any equation which can be written in the form of $Ax + By + C = 0$.

INDEX

Numbers on this page refer to **PROBLEM NUMBERS**, not page numbers

AA similarity theorem, 243, 249, 262, 268, 335, 349, 355, 365, 366, 525, 527, 528
AAA postulate, 243, 248, 376
AAS postulate, 110-112, 160, 194, 210, 219, 225, 234, 307
ASA postulate, 101-109, 161, 177, 180, 193, 198, 220, 235, 535
Abscissa, 653, 673, 674, 676, 681
Absolute values (for distance), 673, 675, 677, 678
Acute angle, 34, 171, 201
Acute trianlges, 595
Addition postulate, 15, 50, 68, 80, 86, 90, 131, 159, 306, 310, 350, 503, 609
Addition property:
  congruence, 221, 291
  equality, 16, 40, 50, 334
  inequality, 38, 173, 175, 177, 188, 190
Adjacent angles, 150, 162, 215, 342, 348
Adjacent interior angles, 181
Adjacent sides, of parallelograms, 209, 230
Alternate angles of hexagons, 517
Alternate exterior angles, 133
Alternate interior angles, 133, 140, 141, 143
Altitudes, 422, 426, 432, 433, 437

parallelograms, 835
pyramids, 861, 862
quadrilaterals, 225, 237, 238, 546
regular pyramids, 814, 815
rhombuses, 475
right prisms, 854
trapezoids, 480
triangles, 112, 165, 234, 269-274, 549, 673-676
Analytical proofs, 697-700
Angles, 25-29
  acute, 34
  adjacent, 150, 162, 215, 342, 348
  adjacent interior, 181
  alternate exterior, 133
  alternate interior, 133, 140, 141, 143
  bisectors, 49-51
  complementary, 24, 26
  construction, 573-591
  convex polyhedral, 805
  in degrees, 278, 282
  of depression, 428
  dihedral, 787, 795, 804
  of elevation, 427, 430
  formed by intersecting chords, 522
  formed by secant to circle, 364-377
  formed by tangents to circle, 355-363
  of incidence, 100, 235
  of inclination, 658

inscribed, 294, 295
launching, 293
measure of, 12, 13, 572
negative, 704
noncongruent, 4
obtuse, 38, 171, 201, 487
opposite, 310
polyhedral, 803
projection in pyramid, 868
in radians, 572
of reflection, 100, 235
related, 298, 300, 344
remote exterior, 371
remote interior, 308
right, 17, 42, 62
spherical, 817, 820, 912
straight, 149, 213, 215,
    284, 342, 518, 561
supplementary, 27, 28, 31
trihedral, 803, 805
vertical, 4, 63, 84, 88,
    103, 108
Angle-addition postulate, 16,
    174, 175, 377, 518, 527
Angle-Angle-Side postulate,
    110-112, 160, 194, 210,
    219, 225, 235, 307
Angle-Angle similarity theorem,
    243, 249, 262, 268, 335,
    349, 355, 365, 366
Angle-Angle-Angle postulate,
    243, 248, 376
Angle bisector, 49-51
Angle bisector theorem for
    concurrence, 606
Angle construction, 580, 581,
    583
Angle inequality, 177, 329
Angle inscription, 698
Angle-Side-Angle postulate,
    101-112, 161, 177, 180,
    193, 198, 220, 235, 535
Angle subtraction theorem, 235
Angle-sum of polygon, 492,
    493, 499
Angle-sum of quadrilaterals,
    431, 493
Angle-sum of triangle theorem,
    37, 38, 43, 44
Angle of take-off, 426

Angle-formation, near circle,
    374
Angular speed, revolutions per
    minute, 292
Antipode, 904, 912
Apollonian definition, 732
Apothem, 514, 523, 551,
    554-558
Arc:
    bisector, 602
    of circle, 385, 393
    of intersecting circles, 350
    intercepted by chords, 522
    intercepted by rays, 295,
        296
    length, 278, 282-285
    major, 311, 336, 357, 371,
        374, 375
    measure, in degrees, 330,
        336, 358, 363, 374, 375,
        433, 436
    minor, 299, 302, 311, 315,
        316, 318, 336, 357, 371,
        374, 375, 433
Area:
    circle, 378-382, 394, 396,
        548-550, 552, 553, 561,
        569
    circle circumscribed about
    a square, 553
    circle inscribed in a square,
    553
    in coordinate geometry,
    673-675, 677, 678
    ellipse, 751
    equivalent, 439, 455
    hexagon, 551, 557, 560,
    571
    minor segment, 385, 386
    obtuse triangle, 454
    octagon, 567
    parallelogram, 470-485,
    608, 677
    pentagon, 558, 564
    polygon, 678
    polygon inscribed in a cir-
    cle, 559, 560, 567
    quadrilateral, 677, 678
    ratio of polygons circum-
    scribed about a given cir-

cle, 565
ratio of similar polygons, 562, 563, 566
rational, 391, 392
rectangle, 409, 438, 440, 444-446, 546, 567
regular polygon, 551, 554-559, 566, 568-571
rhombus, 473, 476, 477
right triangle, 674, 696
sector of a circle, 383, 385, 387, 548, 845
similar triangles, 461-469
square, 440-443, 445, 545, 547, 548, 552, 561
surface, of tetrahedron, 830
trapezoid, 467, 678
triangle, 446-460, 466-468, 537, 550, 565, 567, 568, 596, 673-676
Area-addition postulate, 552
Area ratios, 461, 462
of circles, 388-390, 394
circumscribed about a given circle, 565
of similar polygons, 562, 563, 566
Arithmetic progression formula, 499
Assumptions, 4, 19, 38, 99, 162, 163, 205, 207
Asymptotes, 765
of a hyperbola, 766, 767
Axes, coordinate, 673, 674
translation of coordinate, 720, 721
Axis of symmetry of a parabola, 752, 758
Axioms, 4

Base, of isosceles triangle, 6, 98, 115, 602
Base angles, of isosceles triangles, 39, 49, 64, 67
Base angles, of isosceles right

triangles, 44
Bases of trapezoids, 237, 238, 260
Betweenness, 23, 177, 180, 277, 326
Bisectors:
of an angle, 49-51
of an arc, 602
intersection of, 51
of a line, 63, 67, 96
perpendicular, 10, 30, 32, 33
of a vertex angle, 49
Boundary of a triangle, 163

Cardioids, 712, 788
Cartesian:
coordinate axes, 753
coordinate system, 704
plane, 629, 630, 670, 693, 897
Cavalieri's postulate, 856
Center of a circle, 128
Central angles, 330, 336, 361
measure of, 278, 282, 284, 285
Central-angle theorem, 558
Chords of a circle, 288, 289, 291
Circles:
angles formed by secants, 364-377
angles formed by tangents, 355-363
arc of, 278-293, 385, 393
area of, 378-382, 394, 396, 548-550, 552, 553, 561, 569
center of, 128, 282
central angles and arcs, 278-293
chords, 288, 289, 291, 309-332
circumscribed, 507, 526-528, 536
circumscribed about a

pentagon, 522
circumscribed about a
square, area of, 553
construction of, 600, 606
equations, 722-739
externally-tangent, 333,
342, 346, 347, 360
inequality, 302, 318, 326,
331
inscribed in a square, area
of, 553
family of, 337, 348, 353,
354, 513
inequality in, 364
inscribed, 509-513, 533-
535, 537, 545, 549, 550,
552, 553, 565, 571, 606
inscribed angles and arcs,
294-308
internally tangent, 343,
345, 358
intersecting, 348-350, 352-
354
major arc, 311, 336, 357,
371, 374, 375
minor arc, 299, 302, 311,
315, 316, 318, 336, 357,
371, 374, 375, 433
radius, 128
set-theoretic definition
of, 774
tangent lines to, 333-341
tangent to other circles,
342-347
Circumference of circle, 278-
283, 285
Circumscribed circle, 507,
526-528, 536
of a pentagon, 522
Circumscribed polygon, 526
Circumscribed square, 545
Circumscribed triangle, 506,
605
construction of, 605
Circumscribing circle, 434
Circumscription, 609
Coincident lines, 163, 545
Collinearity, 21, 22, 65
Common chords, 354
Common internal tangent, 333,

360
Common solution of a system
of equations, 672
Common tangents, 334, 335,
345, 358
Complementary angles, 24, 26
Completion of the square, 728,
729, 741
Complex constructions, 616,
620, 622
Concentric circles, 292, 323,
341, 361, 690, 730
family of, 730
Conchoid of Nicomedes, 712
Conclusion, 4-11
Concurrency, 51, 133, 164,
165
of tangents, 347
Concyclic points, 351, 501-
503, 505-507
Conditional statement, 5, 7,
10, 62, 63
Cones:
angle of inclination, 880,
886
frustum, 879-881, 883,
889
slant height, 876
surface area, 876, 877
volume of, 876-889
Congruence, addition property
for, 221, 291
Congruence in coordinate geometry, 693, 702
Congruence, definition, 55, 56,
64
Congruence, transitive property of, 46, 63, 140, 203,
234, 236, 362
Congruent angles, 55-67
Congruent arcs, 284, 290, 291,
298
Congruent central angles, 290,
291
Congruent chords, 301, 306,
307, 310
Congruent circles, 328, 329,
353
Congruent line segments, 68-72
Congruent triangles, 73

Conic section curves, 722-771
Consecutive angles (of a rhom-
     bus), 431
Consecutive interior angles,
     133, 147, 191, 192
Consecutive sides, 212, 216
Construction, 614
     of an angle, 580, 581, 583
     of angles near a circle,
     374
     of a circle, 600-606
     of a circumscribed trian-
     gle, 605
     complex, 612-622
     lines and angles, 573-591
     of a tangent line, 603, 604
     of a perpendicular line,
     618
     of polygons, 607-611
     of rectangles, 608
     of triangles, 592-594, 596,
     598, 599, 612, 615
Contrapositive, 35, 36, 524
Converse statement, 7, 8, 68
Conversion factor, 703
Convex angle of a triangle, 163
Convex polygon, 491, 492
Convex polyhedral angle, 805
Coordinate axes, translation
     of, 720, 721
Coordinate geometry, 623-702
     areas in, 673-678
     distance in, 623-636
     locus, 679-692
Coordinate graph paper, 623
Coordinate proofs, 693-702
Coordinates:
     Cartesian, 704
     of a point, 770
     polar, 704, 711, 712, 715
     rectangular, 717
     3-space, 897, 900
Coplanarity, 21, 22, 133, 155
Corresponding angles and parts,
     83
Cosine ratio, 425, 426, 432,
     436, 437
Cube, 851-853
     edge of, 826
     face of, 826

     lateral surface area, 875
     surface areas, 826-829
     total surface area, 827
     volume of, 851-853
Curve sketching, 753
Cyclic polygon, 503
Cyclic quadrilateral, 502-505,
     508, 515, 524, 531, 532
     opposite angles of, 515
Cylinders:
     directrix of, 901
     lateral surface area, 837,
     875
     right, 837
     volume of, 874, 875, 888
Cylindrical surface, 775

Deduction, 1, 4
     logical, 7
Deductive reasoning, 1-3, 11
Degrees, measure, 278, 282,
     293, 375
Dependent clause, 5
Depression angle, 428
Determining loci, 684
Diagonals, 61
     of pentagons, 522
     of rectangles, 413
     of regular polygons, 526
     of rhombuses, 214, 217,
     407, 549
Diameter of a circle, 128
Diameter of an inscribed cir-
     cle in a regular polygon,
     514
Dihedral angle, 787, 795, 804,
     plane bisector of, 787
Direct proof, 11-16, 95
Directrix:
     of a cylinder, 901
     of a parabola, 756-758,
     761
Discriminant, 738
Distance, 14, 673-676, 678
     between points, 653
     between a point and a line,
     666

in coordinate geometry,
623-636
definition of, 293
determination, 625-630
formula, 648, 666, 676,
686, 695, 696, 699, 701,
702
horizontal, 425, 426, 678
perpendicular, 323, 331,
340, 675
time formula, 293
vertical, 678
Distributive property, 369, 371,
537
Division postulate, 71, 72, 95,
139
for equality, 217, 336, 337
for inequality, 170, 330
Division of segment, 587-589
Domain, 668
Duplication of segment, 573

Earth geometry, 902-913
Eccentricity:
of an ellipse, 740
of a hyperbola, 768-769
Edge:
of a cube, 826
of a rectangular solid, 828
Elements of a set, 55
Elementary proofs, 163
Elevation angle, 427, 430
Elimination, 3, 163
Ellipses:
area, 751
eccentricity of, 740
equations, 740-751
foci of, 740, 747
latus rectum of, 749
major axis, 740, 742
semi-, 748
semi-major axis, 742
vertices of, 740
Equations:
circle, 722-739
ellipse, 740-751

graphing, 667-672
hyperbola, 764-771
linear, 651-666, 672
parabola, 752-763
perpendicular bisector,
663
plane, 892, 893
slope, 666
sphere, 898
system of, 672
tangent to a circle, 736-
738, 750
transformation, 717, 719,
721
Equiangular triangles, 376
Equidistance, 115, 164
Equidistant locus of points,
783, 785
Equilateral triangles, 36, 42-
48
interior angles of, 287
permeter of, 452, 544

Equivalent areas, 439, 455
Equivalent polygons, 607
Exterior-angle theorem, 52,
54, 181-183
Exterior of an angle, 295, 296
Exterior angles, 52-54
of parallel lines, 133, 527
of a polygon, 486-489, 496,
500
remote, 371
of a triangle, 52-54, 181,
182, 184, 187, 296, 308

Exterior to a quadrilateral,
222, 231
Exterior to a triangle, 163, 173,
182
External segments, 367-370
External tangent lines, 333,
337, 339, 342, 357, 369

Externally tangent circles, 333,
342, 346, 347, 360

Extremes, product of, 242,
261, 266, 270-272, 349
Extremes (of a proportion),
464

Face:
    angles, 803, 805
    of a cube, 826
    of a pyramid, 832
Factoring, 340, 459
Factoring postulate, 217
Families of circles, 337, 348,
    353, 354, 513
Finite sloped lines, 673
Focus (foci):
    of an ellipse, 740, 747
    of a hyperbola, 766, 771
    of a parabola, 756-758,
    761
Formal proof, 63
Formulas:
    distance, 648, 666, 676,
    686, 695, 696, 699, 701
    midpoint, 632, 634-636,
    680, 694, 700
    quadratic, 30, 531
Fourth proportional, 242
Frustum:
    of a cone, 879-881, 883,
    889
    of a pyramid, 864
    of a right circular cone,
    847
    volume of, 864
Functions, trigonometric, 539,
    540, 542, 559, 560, 568-
    572
Fundamental rectangle, 768

General distance formula be-
    tween two points, 676
General equation of a parabola,
    755, 756
General form of a circle, 731
General statement, 1, 2
Geodesics, 821, 905
Geometric means, 270
Geometric proportions and
    similarities, 239-259
Geometry, earth, 902-913
Graph of a curve, 727, 764

Graphing equations, 667-672
Graphing polar coordinates,
    703-712
Graphs, 673, 674, 667-671
Great circle:
    circumference, 871
    of the earth, 905, 907,
    908, 910
    pole of, 821
    of a sphere, 778, 838

Hemisphere, volume, 884
Heron's formula, 458
Hexagons, 494, 496, 500, 517,
    518, 551, 557, 571, 834
    alternate angles of, 517
    area of, 551, 557, 560,
    571
    interior angles of, 494
    regular, 518, 519, 538
Horizontal distance, 425, 426,
    678
Hyperbola, 764, 765
    asymptotes, 766, 767
    eccentricity of, 769
    equations, 764-771
    foci of, 766, 771
    transverse axis of, 769
Hypotenuse, 43, 45
Hypotenuse-leg theorem, 126-
    132, 218, 290, 322, 327,
    341, 377
Hypothesis, 4-8, 10, 17

Icosahedron, surface area of,
    836
Identity, trigonometric, 718,
    719
If-and-only-if theorem, 172
If-then statement, 5, 55, 59
Image, virtual, 100
Incidence angle, 100, 235

Included angles of triangles, 223, 261

Indirect method of proof, 17, 38, 99, 162, 205, 206, 506

Induction, mathematical, 18-20

Inductive reasoning, 18-20

Inequalities, 52, 166-190
  in circles, 364
  in triangles, 173, 178-180, 182, 184, 186, 188-190

Inscribed angles, in circles, 294, 295

Inscribed angles, in semicircle, 527, 530

Inscribed circles, 509-513, 533-535, 537, 545, 549, 550, 552, 553, 565, 571, 606

Inscribed circle in a regular polygon, 514

Inscribed pentagon, 522

Inscribed polygons, 287, 541, 554, 559, 560, 570

Inscribed quadrilaterals, 502-506, 508

Inscribed rectangle, 546

Inscribed right triangle, 434

Inscribed sphere, 825

Inscribed square, 547, 561, 567

Inscribed triangle, 363, 530, 604, 674

Inscription, 538

Intercept:
  of a line, 655, 656
  of a polar curve, 708
  x-, 669
  y-, 669, 671

Intercepted arc:
  by chords, 522
  measure of, 515
  by rays, 295, 296

Interior angles:
  of equilateral triangle, 287
  nonadjacent, 181
  of nonconvex polygon, 492

of parallel lines, 133, 136, 142, 143, 150, 162, 163, 203, 205, 336, 351, 504

of parallelogram, 200

of polygon, 486, 487, 489, 491, 493-497, 499, 500

of regular hexagon, 494

remote, 308

sum of, for nonconvex polygons, 492

of triangle, 16, 53, 182, 308, 371

Interior of an angle, 163, 179, 180, 295, 296

Interior of a circle, 163, 326

Interior of a triangle, 163, 188

Internal segments, 368

Internal tangent lines, 333, 343

Internally tangent circles, 343, 345, 358

Intersecting chords of a circle, 522
  angle formed by, 522

Intersecting circles, 348-350, 352-354

Intersecting lines, 115, 125
  opposite angles of, 335, 350

Intersection:
  of bisectors, 51
  of loci, 604, 672, 682, 690
  of perpendicular lines, 97, 99
  point of, 163, 164
  of two circles, 762

Inverse statement, 10, 11

Isosceles right triangles, 44, 216, 223, 695, 696
  base angles of, 44

Isosceles trapezoid, 236, 237, 260, 855

Isosceles triangles, 35, 39-41
  base of, 6, 98, 115, 602
  base angles of, 39, 49, 64, 67
  median of, 6, 75, 602

Lateral area, 847
  cylinder, 837
  frustum, 847
  polyhedral figure, 833
  prism, 835
  pyramid, 832, 833
  rectangular solid, 828
  right circular cone, 845,
  846
Lateral edge of pyramid, 810
Lateral face of right circular
  cylinder, 843
Latitude, 908
Latus rectum of ellipse, 749
Launch angle, 293
Law of sines, 434
Length of apothem, 523
Length of arc, 278, 282-285
Length of line segment, 725
Limacons, 712
Line of center, 333, 345
Line perpendicular to plane,
  791-793, 797, 798, 800,
  801
Line segment, 6, 15, 67-69
  midpoint, 72, 116, 117,
  127
  construction, 574, 582,
  584, 585, 586, 590, 613
Line tangent construction, 603,
  604
Linear equations, 651-672
  graphing of, 668, 670
Linear pair, 52, 53, 59, 109,
  116, 134, 135, 172, 182,
  308, 353, 377, 486, 489
Lines, 21-24
  bisector, 63, 67, 71, 96
  construction, 573-591
  intercept, 655, 656
  intersecting, 115, 125
  noncoincident, 51, 190,
  277, 309
  nonparallel, 51, 163, 164,
  190, 277, 309
  point-slope form, 654, 658,
  659, 661, 663, 666, 676
  postulate, 97
  skew, 163
  slope, 651-653, 660, 662,

665
  slope-intercept form, 656,
  657, 661, 664, 671
Lines and Angles, 21-29, 166-
  180
Locus, in coordinate geometry,
  679-692
Locus of points, 164, 604, 672
Locus (loci), 624, 626, 679-
  691
  intersection of, 604, 672,
  682, 690
Logical deduction, 7
Logic, 4-11
Longitude, 902, 907, 908

Major arcs of a circle, 311,
  336, 357, 371, 374, 375
Major axis of an ellipse, 740,
  742
Major premise, 1, 2
Major sectors, 548
Mathematical induction, 18-20
Mean proportional, 241, 271-
  274, 349, 368, 369, 549
Mean ratio, 527
Means, geometric, 270
Means, product of, 242, 261,
  266, 270-272, 349, 365,
  366, 368, 369
Means of a proportion, 530
Measure of an angle, 12, 13
Measure, in degrees, 278, 282,
  293, 375
Measure of intercepted arc,
  515
Medians:
  to hypotenuse of right triangle, 45
  of isosceles triangle, 6,
  75, 602
  proportions in similar triangles, 276-277
  of trapezoid, 238, 359
  of triangle, 229, 238, 276,
  277

Meridian, 902, 908
Methods of proof, 1-10
Midline of a triangle, 165, 202, 277
Midpoint distance, 631-636
Midpoint formula, 632, 634-636, 680, 694, 700, 898
Midpoint of a line segment, 72, 116, 117, 127
Minimum length, 235
Minor arc of circle, 299, 302, 311, 315, 316, 318, 336, 357, 371, 374, 375, 433
Minor premise, 1, 2
Minor segment, area of, 385, 386
Mirror length, 235
Multiplication postulate, 14, 61, 277, 327
Multiplication property for inequality, 176, 178, 330

Necessary condition, 7
Negation of statement, 9, 10, 99, 164, 205
Negative angle, 704
Negative slope, 672
Nicomedes, conchoid of, 712
Nonadjacent interior angles, 181
Noncoincident lines, 51, 190, 277, 309
Noncollinearity, 146, 163, 165, 501, 506, 507
Noncongruent angles, 4
Nonconvex polygon, sum of interior angles, 492
Non-obvious rational areas, 391, 392
Nonparallel lines, 51, 163, 164, 190, 277, 309
Nonscalene triangle, 36, 48
Number line, real, 625

Obtuse angle, 38, 171, 201
Obtuse triangles, 595, 673
    area of, 454
Octagon, 495, 567
    area of, 567
Octahedron, volume of a regular, 866
Octant of 3-dimensional coordinate space, 896
Opposite angles, 310
    of intersecting lines, 335, 350
    of parallel lines, 136, 503, 505
    of quadrilaterals, 224, 225, 230, 231, 277
Ordered pairs, 670
Ordinate, y-axis, 651, 674, 676, 681
Origin, translation of, 725
Overlapping triangles, 78, 79, 86, 90, 91, 93, 104, 105, 110, 118, 121, 122, 187

Parabola, 752, 757, 763
    axis of symmetry, 752, 758
    directrix, 756-758, 761
    equations, 752-763
    focus, 756-758, 761
Parallel chords, 332
Parallel lines:
    exterior angles, 133, 527
    interior angles, 133, 136, 142, 143
    opposite angles, 136, 503, 505
Parallel planes, 792, 794
Parallelepiped, rectangular, 897
Parallelism, 133-165
Parallelograms, 61, 191-205
    area, 470-485, 608, 677
    definition, 677
    interior angles, 200
Pentagon, 11, 490, 498, 558, 559

area, 558, 564
Perimeter, 259, 269
  base of rectangular solid, 828
  equilateral triangle, 452, 544
  polygon, 436, 551, 554-558
  regular polygon, 539-542
  rhombus, 549
  squares, 442, 533, 534
  triangle, 338
Perpendicular bisector, 10, 30, 32, 33
  equation, 663
Perpendicular construction, 618
Perpendicular distance, 323, 331, 340, 675
Perpendicular lines, 6, 10, 31, 49
  intersection, 97, 99
Perpendicularity, 30-34
Planar, 163
Plane, 146-148, 163-165
  bisector of dihedral angle, 787
  Cartesian, 629, 630, 670, 693
  determined by a line and a point, 789
  determined by three points, 892
  equation, 892, 893
  parallel, 784
  perpendicular to a plane, 794, 801
Plotting points, 623, 624, 631-633, 639, 642, 644, 646, 672, 677
Points:
  betweenness postulate, 68, 118, 159, 180, 184, 190
  collinear, 21, 22, 65
  coplanar, 21, 22, 133
  of intersection, 163, 164
  locus, 164, 604, 672
  noncollinear, 146, 163, 165, 501, 506, 507
  set, 55, 667

of tangency, 334-337
  uniqueness postulate, 34, 184, 232
Polar coordinates, 703-721
  axes, 704, 706-708, 711
  conversion to Cartesian coordinates, 713-717
  graphing, 703-712
Polar curves:
  intercepts, 708
  radii, 708
Polar triangles, 822, 833
Pole, 704, 711
  of great circle, 821
Polygons, 7, 11, 514, 562, 568, 570
  area, 545-572, 678
  area of inscribed, 559, 560, 567
  area ratio of similar, 562, 563, 566
  circles circumscribing, 514-532
  construction, 607-611
  cyclic quadrilaterals, 501-508
  equivalent, 607
  exterior angles, 486-489
  four-sided, 7, 61
  inscribed, 541, 554, 559, 560, 570
  inscribed circles, 509-513
  interior angles, 490-500
  nonconvex, 492
  perimeter, 436, 533-544, 551, 554-558
  similar, 245, 257-259, 566
  sum of interior angles, 492, 493, 499
  sum of interior angles of nonconvex, 492
Polyhedral angle, 803
Polyhedral figure, lateral area, 833
Polyhedrons, 808
Positive polar angle, 705
Positive slope, 672

Numbers on this page refer to **PROBLEM NUMBERS**, not page numbers

Postulates:
addition, 15, 50, 68, 80, 86, 90, 131, 159
addition for congruence, 221, 291
addition for equality, 16, 40, 50, 334
addition for inequality, 38, 173, 175, 177, 188, 190
angle-addition, 16, 174, 175, 377, 518, 527
angle-angle-angle, 243, 248, 376
angle-angle-side, 110-112, 160, 194, 210, 219, 225, 235, 307
angle-side-angle, 101-112, 161, 177, 180, 193, 198, 220, 235, 535
area-addition, 552
Cavalieri's, 856
of contradiction, 207
division, 71, 72, 95, 139
division for equality, 217, 336, 337
division for inequality, 170, 330
of elimination, 163, 207
factoring, 217
line, 97
multiplication, 14, 61, 277, 327
multiplication of inequality, 176, 178, 330
point-betweenness, 68, 118, 159, 180, 184, 190
point uniqueness, 34, 184, 232
side-angle-angle, 324
side-angle-side, 73
side-side-side, 110, 113
substitution, 13
substitution of inequality, 167
subtraction, 13, 60, 70, 87
subtraction of equality, 16, 41, 60
subtraction of inequality, 168, 169, 171, 174

transitivity, 16, 51
Premise:
major, 1, 2
minor, 1, 2
Prisms:
definition, 808
lateral area, 835
lateral edges, 808
rectangular, 848
right, 834, 854, 855
section, 807, 809
volumes, 854-856
Projection, 816
angle, in pyramid, 868
of vertex of regular pyramid, 814, 815
Product of the extremes, 242, 261, 266, 270-272, 349
Product of the means, 242, 261, 266, 270-272, 349, 365, 366, 368, 369
Proofs, 8, 12-14, 35, 36
analytical, 697-700
by contradiction, 35, 36
coordinate, 693-702
direct, 11, 95
elementary, 163
formal, 63
indirect, 17, 38, 99, 162, 205, 206, 506
Properties of circles, 304
Properties of parallel lines, 145-165
Proportional, fourth, 242
Proportional, mean, 241, 271-274, 349, 368, 369, 549
Proportions, 239, 240
extremes, 464
means, 530
Proving lines parallel, 133-144
Proving triangles similar, 243-259
Ptolemy's theorem, 525, 531, 532
Pyramids, 810-815
altitude, 861, 862
angle projection in, 868
faces, 832
frustum, 864, 883
lateral area, 832, 833

lateral edges, 810, 811
parallel sections of, 813
projection of vertex, 814, 815
regular, 810, 811, 860, 867
slant height, 810, 833, 860
square, 810
triangular, 802, 857, 858
volume, 857, 858, 860-863, 883
volume of frustum, 864
volume of regular, 867
volume of square, 868
Pythagorean theorem, 128, 266
Pythagorean theorem applications, 397-421

Quadratic formula, 30, 531, 736, 750, 763
Quadrilaterals, 7, 25, 86, 120, 191-238
angle-sum of, 493
area, 677
cyclic, 502-505, 508, 515, 524, 531, 532
inscribed, 502-506, 508
opposite angles, 224, 225, 230, 231, 277
vertices, 677, 678

Radian measure, 703
Radius,
of circle, 128
of earth, 903, 906
of circle inscribed in regular polygon, 514
of polar curve, 708
Rational areas, 391, 392
Ratios and proportions, 239-242

of areas of polygons circumscribed about a given circle, 565
of areas of similar polygons, 562, 563, 566
of circumferences, 281
mean ratio, 527
of similar triangles, 464, 465
of similitude, 245, 269, 365, 564, 813
trigonometric, 436
Real number line, 625
Real numbers, 668
transitive property of, 56
Rectangles, 7, 71, 208, 224-235
areas of, 409, 438, 440, 444 - 446, 546, 567
construction, 608
diagonals, 413
fundamental, 768
inscribed, 546
Rectangular coordinates, 717
Rectangular parallelepiped, 849, 850, 897
Rectangular prism, 848
Rectangular solids:
edges of, 828
lateral area, 828
perimeter of base, 828
surface area, 826
volume, 848-850
Revolution, volume of, 882, 886-889
Reflexivity:
of congruence, 73, 75, 77
definition, 5, 7
of equality, 15, 50, 301
Regular hexagon, 494, 518, 519, 538
interior angles of, 494
Regular octahedron, volume of, 866
Regular pentagon, 522, 527, 541
area, 558, 564
circumscribed circle of, 522

diagonals, 522
Regular polygons, 514, 526,
527, 555
area, 551, 554-559, 566,
568-571
diagonals, 526
diameter of inscribed circle in, 514
inscribed, 287
inscribed circle in, 514
perimeter, 539-542
radius of inscribed circle
in, 514
Regular pyramid, volume of
867
Regular tetrahedron, 830, 859
Related angles, 298, 300, 344
Remote exterior angles, 371
Remote interior angles of a
triangle, 52, 181-183, 187,
308, 503, 506
Revolutions per minute, 292
Rhombus (rhombi), 206-215,
217, 431, 549, 701
altitude, 475
area, 473, 476, 477
diagonals, 210, 214, 217,
407, 549
perimeter, 549
Right angles, 17, 42, 62
Right circular cones:
lateral surface area, 845,
846
slant height, 846
Right circular cylinder, 842,
843, 900
base, 842
height, 842
lateral surface area, 842,
843
Right cylinder, 837
Right prism, 834, 854, 855
altitude, 854
Right section of a prism, 807
Right triangle, 42, 43, 46-48
area, 674, 696
inscribed, 434
isosceles, 44, 216, 223,
695, 696
median to hypotenuse,

45
in similar triangles, 268-
275
Rotation of coordinate axes, 718

SAA postulate, 324
SAS postulate, 73
SAS similarity theorem, 261,
286
SSS postulate, definition, 110,
113
SSS similarity theorem, 349,
351
Sagitta, 411
Scalene triangle, 35-38, 48, 99,
596
Secant, 365-371, 374, 375
to outside circle(s), 349
ray, 358
Sector(s), 383, 384, 386
area of, 383, 385, 387, 548
major, 548
Segment(s):
division of, 587-589
duplication, 573
external, 367-370
internal, 368
multiplication, 590
trisection points of, 891
Semicircles(s), 128, 303, 345, 348
351, 376, 434, 436, 507, 546
Semi-ellipses, 748
Semimajor axis of ellipse, 742
Semiperimeter of triangle, 460
Set-theoretic definition of circle,
774
Set(s):
elements of, 55
notation, 653
of points, 55, 667
of real numbers, 668
solution, 670
Shortest path, 806
Side-angle-side postulate, 73-
100
Side-angle-side similarity theor-

em, 261, 286, 349, 354, 505
564

Side-side-side postulate, 110,
113-125

Side-side-side similarity theorem, 349, 351

Similar polygons, 245, 257-
259, 566

Similar triangles, 8, 244, 246,
260-277, 832
areas of, 461-469
ratios of, 464, 465

Similarity theorem, 243, 248
260-262, 268, 277, 286
angle-angle, 243, 249, 262,
268, 335, 349, 355, 365, 366
525, 527, 528
side-angle-side, 261, 286,
349, 354, 505, 564
side-side-side, 349, 351

Similitude, ratio of, 245, 269,
365, 564

Sine:
law of, 434
ratio, 422-424, 432-434

Skew lines, 163

Slant height, 860, 863
of cone, 876
of pyramid, 810, 833, 860
of right circular cone, 846
of square pyramid, 863

Slope, 637-650
equation of, 666
of a line, 651-653, 660, 662
665
of a line perpendicular to
a given line, 662
negative, 672
point-slope form, 654,
658, 659, 661, 663, 666,
676
positive, 672

Slope-intercept form of a line,
656, 657, 661, 664, 671,
736

Small circle of sphere, 779

Solid coordinate geometry,
890-901

Solid geometry, locus in, 772-
788

Solid, volume of rectangular,
848-850

Solution set, 670

Speed:
angular (rpm), 292
linear (mph), 293

Sphere, 780, 781, 816-825
equation of, 898
exterior of, 774
great circle of, 778
inscribed in a cone, 825
interior of, 774
small circle of, 779
surface area of, 837-841,
871-872
volume of, 869-873, 899
zone of, 824

Spherical angle, 817, 820, 912

Spherical surface, 779, 905

Spherical triangle, 818-820,
823, 840, 912, 913

Square, 7, 216-223
area of, 440-443, 445, 545
547, 548, 552, 561
completion of the, 728, 729
741
inscribed, 547, 561, 567
perimeter of, 442, 533, 534

Square pyramid(s):
regular, 863, 868
volume of, 868

Standard form of circle, 731

Steiner-Lehmus Theorem, 524

Straight angle, 149, 213, 215,
284, 342, 518, 561

Straight line, 20, 63

Subsegment, 334

Subset, 341

Substitution postulate, 13
for inequality, 167

Subtend, 25

Subtraction postulate, 13, 60,
70, 87
for equality, 16, 41, 60
for inequality, 168, 169, 171,
174

Sufficient conditions, 7

Supplementary angles, 27, 28, 31

Surface area, 826-847
of a cone, 827-829, 876,

877
of a cube, 826
of an icosahedron, 836
lateral, of a cube, 875
lateral, of a cylinder, 875
lateral, of a right circular cone, 845, 846
of a rectangular solid, 826
of a solid, 844
of a sphere, 837-841, 871,
872, 911, 913
of a tetrahedron, 830
Syllogism, 1, 2
Symmetry, 57
about 90´-axis, 705
about polar axis, 705
about pole, 705
of polar curve, 705, 707
System of equations:
common solution, 672
linear system, 892

Tangency, point of, 334-337
Tangent lines to circle, 333,
334
external, 333, 337, 339, 342,
357, 369
internal, 333, 343
Tangent ratio, 427-431, 435,
436, 558, 569, 570, 572
Tangent ray(s), 358
Tangent circles, external, 333,
342, 346, 347, 360
Tangent circles, internal, 343,
345, 358
Tangent segment, 509
Tetrahedron, 830, 885
regular, 830, 859
surface area of, 830
volume of, 859, 865, 873
Theorem, 4
addition, 15, 50, 68, 80,
86, 90, 131, 159, 306, 310,
350, 503, 609
angle-addition, 16, 174,
175, 377, 518, 527

angle-angle-angle, 243,
248, 376
angle-angle-side, 110-
112, 160, 194, 210, 219,
225, 235, 307
angle-side-angle, 101-109,
161, 177, 180, 193, 198,
220, 235, 535
angle-subtraction, 235
angle-sum of polygon, 492,
493, 499
angle-sum of quadrilateral,
431
angle-sum of triangle, 37,
38, 43, 44
area-addition, 552
division, 71, 72, 95, 139
hypotenuse-leg, 126-132,
218, 290, 322, 327, 341,
377
multiplication, 14, 61, 277,
327
Ptolemy's, 525, 531, 532
Pythagorean, 128, 266
Steiner-Lehmus, 524
triangle-inequality, 184-187,
188-190, 325, 326
vertical-angle, 179, 180
Transformation equations, 717,
719, 721
Transitivity postulate, 16, 51
for congruence, 46, 63, 140,
203, 234, 236, 362
for equality, 12, 61, 301,
319, 333, 362, 365, 437
for inequality, 166, 187,
330, 364
for real numbers, 56
Translation, 718-721
of coordinate axes, 720,
721
of the origin, 725
Transversals, 133, 134
Transverse axis of a hyperbola,
769
Trapezoids, 7, 236-238, 260,
359
altitude, 480
area of, 467, 678
bases of, 237, 238, 260

isosceles, 236, 237, 260
median of, 238, 359
Triangle inequality theorem,
184-190
Triangle, 35-54
(30-60-90), 415-417
acute, 595
angle-sum of, 37, 38, 43,
44
area of, 446-460
area of similar, 461-469
construction of, 592-594,
596, 598, 599, 612, 615
equiangular, 376
equilateral, 36, 46-48
exterior angle of, 52-54
included angles, 223, 261
inscribed in circles, 363,
530, 604, 674
inscribed in semicircles,
528, 532
interior angles, 16, 53,
182, 308, 371
isosceles, 35
isosceles right, 44, 216,
223, 695, 696
median of, 6, 75, 229, 238
midline of, 165, 202, 277
nonscalene, 36, 48
obtuse, 673
overlapping, 78, 79, 86,
90, 91, 93, 104, 105, 110,
118, 121, 122, 187
perimeter of, 163, 338
polar, 822, 833
right, 42, 43
scalene, 35, 36, 48, 99,
596
semiperimeter of, 460
similar, 244, 246
similarity, 8
spherical, 818-820, 823,
840
Triangle similarity, 8
Triangular pyramid, 802, 857,
858
Trigonometric functions, 539,
540, 542, 559, 560, 568-
572
Trigonometric identity, 718,
719

Trigonometric ratios, 422-437
436
Trihedral angle, 803, 805
Trisection points of a segment,
891

Unequal chords, 302, 318, 331
Unequal minor arcs, 302, 318

Variables, 670
Vertex, 115
Vertex angle, 39, 98, 114, 222,
223, 230, 232, 336, 433,
498
Vertical angle, 4, 63, 84, 88,
103, 108
theorem, 179, 180
Vertical distance, 678
Vertices of an ellipse, 740
Vertices of a quadrilateral,
677, 678
Virtual image, 100
Volume, 848-889
of cones, 876-889
of a cube, 851-853
of a cylinder, 874, 875,
888
of a frustum of a pyramid,
864
of a hemisphere, 884
of a prism, 854-856
of a rectangular solid,
848-850
of a regular octahedron,
866
of a regular pyramid, 867
of revolution, 882, 886-889
of a sphere, 869-873, 899
of a square pyramid, 868
of a tetrahedron, 859, 865,
873

X-intercept, 669

Y-intercept, 669, 671

Zone of a sphere, 824

REFERENCE LIST OF FORMULAS, DEFINITIONS, AND THEOREMS

A. GEOMETRY

Formulas 1–5 refer to plane figures where:

a, b, c = sides of a triangle.
$\quad s$ = semi-perimeter = $\frac{1}{2}(a + b + c)$.
$\quad b$ = base.
b_1, b_2 = the bases of a trapezoid.
$\qquad\qquad s$ = arc of circle.

C = circumference of circle.
h = altitude.
K = area.
r = radius of circle.

1. **Triangle.** $\quad K = \frac{1}{2}bh;\ K = \sqrt{s(s - a)(s - b)(s - c)}$.
2. **Parallelogram.** $\quad K = bh$.
3. **Trapezoid.** $\quad K = \frac{1}{2}(b_1 + b_2)h$.
4. **Circle.** $\quad C = 2\pi r;\ K = \pi r^2$.
5. **Sector of a circle.** $\quad K = \frac{1}{2}sr$.

Formulas 6–10 refer to solid figures where:

B = area of base.
h = altitude.
r = radius.
s = slant height.

S = area of lateral surface.
$\quad$ = area of surface of sphere.
T = area of total surface.
V = volume.

6. **Prism.** $\quad V = Bh$.
7. **Pyramid.** $\quad V = \frac{1}{3}Bh$.
8. **Right circular cylinder.** $\quad S = 2\pi rh;\ T = 2\pi r(h + r);\ V = \pi r^2 h$.
9. **Right circular cone.** $\quad S = \pi rs;\ T = \pi r(s + r);\ V = \frac{1}{3}\pi r^2 h$.
10. **Sphere.** $\quad S = 4\pi r^2;\ V = \frac{4}{3}\pi r^3$.

B. ALGEBRA

1. Division by zero is an excluded operation.
2. If the product of two or more quantities is equal to zero, at least one of the factors must be equal to zero.
3. **The quadratic equation.** The quadratic equation

$$ax^2 + bx + c = 0, \quad a \neq 0,$$

has the two roots

$$x = \frac{-b \pm \sqrt{b^2 - 4ac}}{2a},$$

where $D = b^2 - 4ac$ is called the *discriminant*. If a, b, c are all *real* numbers, then these roots are real and equal if $D = 0$; real and unequal if $D > 0$; conjugate complex if $D < 0$.

The sum of the roots = $-\dfrac{b}{a}$; the product of the roots = $\dfrac{c}{a}$.

4. **Logarithms.** *Definition.* If N, x, and b are three quantities connected by the relation

$$N = b^x, \quad b > 0, \quad b \neq 1,$$

then the exponent x is called the *logarithm of N to the base b*, and we write the equivalent relation

$$x = \log_b N.$$

The logarithm of a negative number does not exist in the *real* number system; the logarithm of zero is undefined.

If M and N are two positive numbers, then the following three relations hold:

$$\log_b(MN) = \log_b M + \log_b N, \qquad \log_b\left(\frac{M}{N}\right) = \log_b M - \log_b N,$$

$$\log_b (M)^n = n \log_b M, \ n \text{ a real number.}$$

The following relations are to be noted:

$$\log_b 1 = 0; \quad \log_b b = 1; \quad \log_b \frac{1}{N} = -\log_b N.$$

The logarithm of a number to any base may be obtained by the relation

$$\log_a N = \frac{\log_b N}{\log_b a},$$

where $a > 0$, $a \neq 1$; $b > 0$, $b \neq 1$.

5. Determinants. *A determinant of order n* is a quantity represented by a square array of n^2 quantities, called elements, arranged in n rows and n columns.

The evaluation of determinants is given in textbooks on algebra. We may, however, note the following important properties of determinants:

Property 1. Any property of a determinant which is valid for its rows is also valid for its columns.

Property 2. The value of a determinant is unchanged if its corresponding rows and columns are interchanged.

Property 3. The sign of a determinant is changed if any two of its rows are interchanged.

Property 4. If a determinant has two identical rows, its value is zero.

Property 5. If each element of any row of a determinant is multiplied by any one number k, the value of the determinant is multiplied by k.

Property 6. The value of a determinant is unchanged if each element of any row is multiplied by any one number k and added to the corresponding element of any other row.

6. Systems of linear equations. For brevity, the theorems given here will be illustrated by systems of three linear equations; they hold, however, for systems of any number of equations.

Consider, then, the system of three *non-homogeneous linear equations* in three unknowns:

(1)
$$\begin{cases} a_1x + b_1y + c_1z = k_1, \\ a_2x + b_2y + c_2z = k_2, \\ a_3x + b_3y + c_3z = k_3, \end{cases}$$

where the k's are constants, not all zero. The determinant formed by the coefficients is called the *determinant of the system* and is generally designated by Δ, that is,

$$\Delta = \begin{vmatrix} a_1 & b_1 & c_1 \\ a_2 & b_2 & c_2 \\ a_3 & b_3 & c_3 \end{vmatrix}$$

Let Δ_j be the determinant formed from Δ by replacing the elements of the jth column by the k's. Then we have

Cramer's Rule. If $\Delta \neq 0$, *the system* (1) *has a unique solution given by*

$$x = \frac{\Delta_1}{\Delta}, \quad y = \frac{\Delta_2}{\Delta}, \quad z = \frac{\Delta_3}{\Delta}.$$

If $\Delta = 0$ and $\Delta_j \neq 0$ for at least one j, the system (1) has no solution and is said to be *inconsistent*.

If $\Delta = 0$ and $\Delta_j = 0$ for all j, the system (1) has either infinitely many solutions or no solution.

Consider next the system of three *homogeneous linear equations in three unknowns:*

$$(2) \qquad \begin{cases} a_1x + b_1y + c_1z = 0, \\ a_2x + b_2y + c_2z = 0, \\ a_3x + b_3y + c_3z = 0. \end{cases}$$

If the determinant Δ of this system is different from zero, there is only one solution by Cramer's Rule: $x = 0, y = 0, z = 0$; this is called the *trivial solution.*

THEOREM. *A system of n homogeneous linear equations in n unknowns has solutions other than the trivial solution of zeros if and only if the determinant of the system is equal to zero.*

C. TRIGONOMETRY

1. Definitions of the trigonometric functions. Let θ be an angle whose range of values is given by $-360° \le \theta \le 360°$. For the purposes of defining such an angle and its trigonometric functions it is convenient to use the rectangular coordinate system. The statements which follow apply to *each* of the four positions shown in Fig. 200.

If a line coincident with the X-axis is rotated in the XY-coordinate plane about the origin O into a new position OA, there is said to be *generated* an angle $XOA = \theta$ having OX as its *initial side* and OA as its *terminal side*. If the rotation is counterclockwise, the angle generated is said to be *positive;* for clockwise

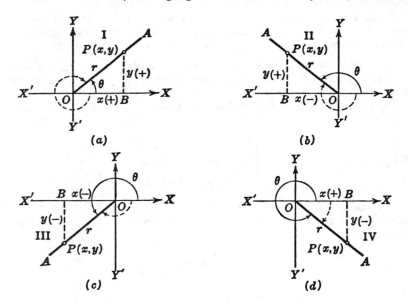

(a)   (b)

(c)   (d)

rotation (shown dotted in the figures), the angle is said to be *negative*. The angle is said to lie in the same quadrant as its terminal side.

On the terminal side OA take any point P distinct from O and having coordinates (x,y). From P drop a perpendicular PB to the X-axis. The line segment OP is called the *radius vector*, is designated by r, and is always taken as *positive*. In the triangle OPB, $OB = x$ and $PB = y$ have the signs of the coordinates of the point P, as indicated for the four quadrants. Then, irrespective of the quadrant in which θ lies, the six trigonometric functions of θ are *defined* both as to magnitude and sign by the following ratios:

$$\text{sine of } \theta: \ \sin \theta = \frac{y}{r}, \qquad\qquad \text{cosine of } \theta: \ \cos \theta = \frac{x}{r},$$

$$\text{tangent of } \theta: \ \tan \theta = \frac{y}{x}, \qquad\qquad \text{cotangent of } \theta: \ \cot \theta = \frac{x}{y},$$

$$\text{secant of } \theta: \ \sec \theta = \frac{r}{x}, \qquad\qquad \text{cosecant of } \theta: \ \csc \theta = \frac{r}{y}.$$

The definitions hold without change for positive and negative angles greater than 360° in numerical value.

2. Fundamental trigonometric identities.

$$\csc \theta = \frac{1}{\sin \theta}, \quad \sec \theta = \frac{1}{\cos \theta}, \quad \cot \theta = \frac{1}{\tan \theta}, \quad \tan \theta = \frac{\sin \theta}{\cos \theta},$$

$$\sin^2 \theta + \cos^2 \theta = 1, \quad 1 + \tan^2 \theta = \sec^2 \theta, \quad 1 + \cot^2 \theta = \csc^2 \theta.$$

3. Reduction formulas.

$$\sin (90° \pm \theta) = \cos \theta, \qquad \cos (90° \pm \theta) = \mp \sin \theta, \qquad \tan (90° \pm \theta) = \mp \cot \theta,$$

$$\sin (180° \pm \theta) = \mp \sin \theta, \quad \cos (180° \pm \theta) = - \cos \theta, \quad \tan (180° \pm \theta) = \pm \tan \theta,$$

$$\sin (270° \pm \theta) = - \cos \theta, \quad \cos (270° \pm \theta) = \pm \sin \theta, \quad \tan (270° \pm \theta) = \mp \cot \theta,$$

$$\sin (360° \pm \theta) = \pm \sin \theta, \quad \cos (360° \pm \theta) = \cos \theta, \qquad \tan (360° \pm \theta) = \pm \tan \theta.$$

4. Radian measure of angles. Let θ be a central angle intercepting an arc of length s on a circle of radius r. Then the measure of the angle θ in *radians* is *defined* by $\theta = \frac{s}{r}$. Note that, since s and r are lengths, this ratio is a *pure number*. From this definition of radian measure we have at once the conversion relation:

$$\pi \text{ radians} = 180°$$

whence

$$1 \text{ radian} = \frac{180°}{\pi} = 57.2958° \text{ (approx.)} = 57° \ 17' \ 45'' \text{ (approx.)},$$

$$1° = \frac{\pi}{180} \text{ radian} = 0.017453 \text{ radian (approx.)}.$$

5. Trigonometric functions of special angles.

Angle θ in		$\sin \theta$	$\cos \theta$	$\tan \theta$
Radians	Degrees			
0	0°	0	1	0
$\frac{\pi}{6}$	30°	$\frac{1}{2}$	$\frac{1}{2}\sqrt{3}$	$\frac{1}{3}\sqrt{3}$
$\frac{\pi}{4}$	45°	$\frac{1}{2}\sqrt{2}$	$\frac{1}{2}\sqrt{2}$	1
$\frac{\pi}{3}$	60°	$\frac{1}{2}\sqrt{3}$	$\frac{1}{2}$	$\sqrt{3}$
$\frac{\pi}{2}$	90°	1	0	

HANDBOOK of
**MATHEMATICAL,
SCIENTIFIC, &
ENGINEERING**
FORMULAS, FUNCTIONS,
GRAPHS, TABLES,
& TRANSFORMS

RESEARCH & EDUCATION ASSOCIATION

A particularly useful reference for those in math, science, engineering and other technical fields. Includes the most-often used formulas, tables, transforms, functions, and graphs which are needed as tools in solving problems. The entire field of special functions is also covered. A large amount of scientific data which is often of interest to scientists and engineers has been included.

Available at your local bookstore or order directly from us by sending in coupon below.

"The ESSENTIALS"
of Math & Science

Each book in the ESSENTIALS series offers all essential information of the field it covers. It summarizes what every textbook in the particular field must include, and is designed to help students in preparing for exams and doing homework. The ESSENTIALS are excellent supplements to any class text.

The ESSENTIALS are complete and concise with quick access to needed information. They serve as a handy reference source at all times. The ESSENTIALS are prepared with REA's customary concern for high professional quality and student needs.

Available in the following titles:

Advanced Calculus I & II
Algebra & Trigonometry I & II
Anatomy & Physiology
Anthropology
Astronomy
Automatic Control Systems /
    Robotics I & II
Biology I & II
Boolean Algebra
Calculus I, II, & III
Chemistry
Complex Variables I & II
Computer Science I & II
Data Structures I & II
Differential Equations I & II
Electric Circuits I & II
Electromagnetics I & II

Electronics I & II
Electronic Communications I & II
Fluid Mechanics /
    Dynamics I & II
Fourier Analysis
Geometry I & II
Group Theory I & II
Heat Transfer I & II
LaPlace Transforms
Linear Algebra
Math for Computer Applications
Math for Engineers I & II
Math Made Nice-n-Easy Series
Mechanics I, II, & III
Microbiology
Modern Algebra
Molecular Structures of Life

Numerical Analysis I & II
Organic Chemistry I & II
Physical Chemistry I & II
Physics I & II
Pre-Calculus
Probability
Psychology I & II
Real Variables
Set Theory
Sociology
Statistics I & II
Strength of Materials &
    Mechanics of Solids I & II
Thermodynamics I & II
Topology
Transport Phenomena I & II
Vector Analysis

If you would like more information about any of these books,
complete the coupon below and return it to us or visit your local bookstore.

RESEARCH & EDUCATION ASSOCIATION
61 Ethel Road W. • Piscataway, New Jersey 08854
Phone: (732) 819-8880   **website: www.rea.com**

Please send me more information about your Math & Science Essentials books

Name _____

Address _____

City _____ State _____ Zip _____

REA's **Problem Solvers**

The "PROBLEM SOLVERS" are comprehensive supplemental text-books designed to save time in finding solutions to problems. Each "PROBLEM SOLVER" is the first of its kind ever produced in its field. It is the product of a massive effort to illustrate almost any imaginable problem in exceptional depth, detail, and clarity. Each problem is worked out in detail with a step-by-step solution, and the problems are arranged in order of complexity from elementary to advanced. Each book is fully indexed for locating problems rapidly.

ACCOUNTING
ADVANCED CALCULUS
ALGEBRA & TRIGONOMETRY
AUTOMATIC CONTROL
    SYSTEMS/ROBOTICS
BIOLOGY
BUSINESS, ACCOUNTING, & FINANCE
CALCULUS
CHEMISTRY
COMPLEX VARIABLES
DIFFERENTIAL EQUATIONS
ECONOMICS
ELECTRICAL MACHINES
ELECTRIC CIRCUITS
ELECTROMAGNETICS
ELECTRONIC COMMUNICATIONS
ELECTRONICS
FINITE & DISCRETE MATH
FLUID MECHANICS/DYNAMICS
GENETICS
GEOMETRY
HEAT TRANSFER

LINEAR ALGEBRA
MACHINE DESIGN
MATHEMATICS for ENGINEERS
MECHANICS
NUMERICAL ANALYSIS
OPERATIONS RESEARCH
OPTICS
ORGANIC CHEMISTRY
PHYSICAL CHEMISTRY
PHYSICS
PRE-CALCULUS
PROBABILITY
PSYCHOLOGY
STATISTICS
STRENGTH OF MATERIALS &
    MECHANICS OF SOLIDS
TECHNICAL DESIGN GRAPHICS
THERMODYNAMICS
TOPOLOGY
TRANSPORT PHENOMENA
VECTOR ANALYSIS

If you would like more information about any of these books,
complete the coupon below and return it to us or visit your local bookstore.

RESEARCH & EDUCATION ASSOCIATION
61 Ethel Road W. • Piscataway, New Jersey 08854
Phone: (732) 819-8880   **website: www.rea.com**

Please send me more information about your Problem Solver books

Name _____

Address _____

City _____ State _____ Zip _____